International Congress on Polymers in Concrete
(ICPIC 2018)

Mahmoud M. Reda Taha
Editor

# International Congress on Polymers in Concrete (ICPIC 2018)

Polymers for Resilient and Sustainable Concrete Infrastructure

With contributions from

**Girum Urgessa**
Department of Civil, Environmental, and Infrastructure Engineering,
George Mason University
Fairfax, VA, USA

**Moneeb Genedy**
Department of Civil Engineering, University of New Mexico
Albuquerque, NM, USA

 Springer

*Editor*
Mahmoud M. Reda Taha
Department of Civil Engineering
University of New Mexico
Albuquerque, NM, USA

ISBN 978-3-319-78174-7      ISBN 978-3-319-78175-4  (eBook)
https://doi.org/10.1007/978-3-319-78175-4

Library of Congress Control Number: 2018936519

© Springer International Publishing AG, part of Springer Nature 2018
This work is subject to copyright. All rights are reserved by the Publisher, whether the whole or part of the material is concerned, specifically the rights of translation, reprinting, reuse of illustrations, recitation, broadcasting, reproduction on microfilms or in any other physical way, and transmission or information storage and retrieval, electronic adaptation, computer software, or by similar or dissimilar methodology now known or hereafter developed.
The use of general descriptive names, registered names, trademarks, service marks, etc. in this publication does not imply, even in the absence of a specific statement, that such names are exempt from the relevant protective laws and regulations and therefore free for general use.
The publisher, the authors and the editors are safe to assume that the advice and information in this book are believed to be true and accurate at the date of publication. Neither the publisher nor the authors or the editors give a warranty, express or implied, with respect to the material contained herein or for any errors or omissions that may have been made. The publisher remains neutral with regard to jurisdictional claims in published maps and institutional affiliations.

Printed on acid-free paper

This Springer imprint is published by the registered company Springer International Publishing AG part of Springer Nature.
The registered company address is: Gewerbestrasse 11, 6330 Cham, Switzerland

## Diamond Sponsors

  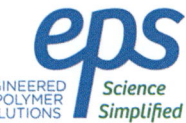

## Platinum Sponsors

## Gold Sponsors

# Contents

**Part I Keynote Papers**

1. **Concrete-Polymer Materials: How Far Have We Come, and Where Do We Need to Go?** ............................ 3
   David W. Fowler

2. **Polymer Concrete for Bridge Preservation** .................... 15
   Michael M. Sprinkel

3. **Feasibility Study of the Use of Polymer-Modified Cement Composites as 3D Concrete Printing Material** ................ 27
   Kyu-Seok Yeon, Kwan-Kyu Kim, and Jaeheum Yeon

4. **Experimental Analysis and Micromechanics-Based Prediction of the Elastic and Creep Properties of Polymer-Modified Concrete at Early Ages** .................................. 37
   Luise Göbel, Bernhard Pichler, and Andrea Osburg

5. **Durability and Long-Term Performance of Fiber-Reinforced Polymer as a New Civil Engineering Material** ................ 49
   Brahim Benmokrane and Ahmed H. Ali

6. **Nano-modified Polymer Concrete: A New Material for Smart and Resilient Structures** ........................ 61
   Mahmoud M. Reda Taha

**Part II Polymer Materials**

7. **Bio-Based Superplasticizers for Cement-Based Materials** ........ 77
   Stephan Partschefeld and Andrea Osburg

## Contents

**8  Screening Encapsulated Polymeric Healing Agents
for Carbonation-Exposed Self-Healing Concrete,
Service Life Extension, and Environmental Benefit**............  83
Philip Van den Heede, Bjorn Van Belleghem, Maria Adelaide Araújo,
João Feiteira, and Nele De Belie

**9  Synthesis and Characterization of Superabsorbent Polymer
Hydrogels Used as Internal Curing Agents: Impact of Particle
Shape on Mortar Compressive Strength**......................  91
Stacey L. Kelly, Matthew J. Krafcik, and Kendra A. Erk

**10  Evaluation of Microencapsulated Corrosion Inhibitors
in Reinforced Concrete**...................................  99
Reece Goldsberry, Jose Milla, Melvin McElwee,
Marwa M. Hassan, and Homero Castaneda

**11  The Use of Polymer Additions to Enhance the Performance
of Industrial and Residential Decorative Concrete Flooring**......  107
Michelle Sykes and Deon Kruger

**12  The Effect of Glucose on the Properties of Cement Paste**........  113
Samantha Mirante and Ali Ghahremaninezhad

**13  Microstructured Polymers and Their Influences
on the Mechanical Properties of PCC**........................  121
Alexander Flohr, Luise Göbel, and Andrea Osburg

**14  Stability of Latex in Cement Paste: Experimental Study
and Theoretical Analysis**..................................  129
Dongdong Han, Weideng Chen, and Shiyun Zhong

**15  Experimental Verification of Use of Secondary Raw Materials
as Fillers in Epoxy Polymer Concrete**.......................  135
Rostislav Drochytka and Jakub Hodul

**16  Effects of Anionic Asphalt Emulsion on Early-Age Cement
Hydration**..............................................  143
Jinxiang Hong, Wei Li, and Kejin Wang

**17  Polymer Solutions for Protection of Concrete Exposed
to Strong Alkaline or Acid Effluent on Industrial Installations**....  151
Nicolas Roche and Hervé Davaux

**18  Lightweight Filled Epoxy Resins for Timber Restoration**.........  157
Torben Wiegand and Andrea Osburg

**19  Effect of Methyl Methacrylate Monomer on Properties
of Unsaturated Polyester Resin-Based Polymer Concrete**.........  165
Kyu Seok Yeon, Nan Ji Jin, and Jung Heum Yeon

20  Analysis of Mechanical Behavior and Durability of Coatings for Use as Flooring in the Petroleum Industry .................. 173
Jane Proszek Gorninski and Jéssica Maiara de Freitas

21  Evaluation of the Performance of Engineered Cementitious Composites (ECC) Produced from Local Materials .............. 181
Gabriel Arce, Hassan Noorvand, Marwa Hassan, and Tyson Rupnow

22  Application of Phase Change Material (PCM) in Concrete for Thermal Energy Storage ............................... 187
Nengfu Tao and Hai Huang

23  Mortars with Phase Change Materials (PCM) and Stone Waste to Improve Energy Efficiency in Buildings .............. 195
Mariaenrica Frigione, Mariateresa Lettieri, Antonella Sarcinella, and José Barroso de Aguiar

24  Physical and Mechanical Properties of Cement Mortars with Direct Incorporation of Phase Change Material ........... 203
Sandra Cunha, José Aguiar, Victor Ferreira, and António Tadeu

25  Mechanical Performance of Fly Ash Geopolymeric Mortars Containing Phase Change Materials ....................... 211
M. Kheradmand, Z. Abdollahnejad, and F. Pacheco Torgal

### Part III  Polymer Concrete

26  Are Polymers Still Driving Forces in Concrete Technology? ...... 219
Lech Czarnecki, Mahmoud Reda Taha, and Ru Wang

27  Mechanical Properties of Polymer Cement-Fiber-Reinforced Concrete (PC-FRC): Comparison Based on Experimental Studies ................................................. 227
Tomasz Piotrowski, Piotr Prochoń, and Alice Capuana

28  Environmental Temperature and Humidity Adaptability of Polymer-Modified Cement Mortar ...................... 235
Ru Wang, Shaokang Zhang, and Peiming Wang

29  Innovative Polymer-Modified Pervious Concrete ............... 243
Aly M. Said and Oscar Quiroz

30  Combined Methods to Investigate the Crack-Bridging Ability of Waterproofing Membranes ............................. 249
Marius Waldvogel, Roger Zurbriggen, Alfons Berger, and Marco Herwegh

| | | |
|---|---|---|
| 31 | **Polymer Concrete for a Modular Construction System: Investigation of Mechanical Properties and Bond Behaviour by Means of X-Ray CT**................................... Franziska Vogt, Alexander Gypser, Florian Kleiner, and Andrea Osburg | 255 |
| 32 | **Bending and Crack Characteristics of Polymer Lattice-Reinforced Mortar**................................ Brian Salazar, Ian Williams, Parham Aghdasi, Claudia Ostertag, and Hayden Taylor | 261 |
| 33 | **The Influence of Specimen Shape and Size on the PCC Compressive Strength Values**............................. Joanna J. Sokołowska, Tomasz Piotrowski, and Iga Gajda | 267 |
| 34 | **Long-Term Investigation on the Compressive Strength of Polymer Concrete with Fly Ash**......................... Joanna J. Sokołowska | 275 |
| 35 | **Overlays: A Great Use for Polymer Concrete**................. David W. Fowler and David P. Whitney | 283 |
| 36 | **PC with Superior Ductility Using Mixture of Pristine and Functionalized Carbon Nanotubes**..................... AlaEddin Douba and Mahmoud Reda Taha | 291 |
| 37 | **Contribution of Concrete-Polymer Composites and Ancient Mortar Technology to Sustainable Construction**............... Dionys Van Gemert, Lech Czarnecki, Ru Wang, and Özlem Cizer | 299 |
| 38 | **Smart Monitoring of Movement and Internal Temperature Changes Within Polymer Modified Concrete Repair Patches**...... Johannes Bester, Jacques G. Engelbrecht, and Michael Grobler | 307 |
| 39 | **PIC: Does It Have Potential?**............................... David W. Fowler | 313 |
| 40 | **A Perspective on 40 Years of Polymers in Concrete History**....... Albert O. Kaeding | 321 |
| 41 | **Development Length of Steel Reinforcement in Polymer Concrete for Bridge Deck Closure**......................... Moneeb Genedy, Rahulreddy Chennareddy, Michael Stenko, and Mahmoud Reda Taha | 329 |
| 42 | **Development of Ultrarapid-Hardening Epoxy Mortar for Railway Sleepers**................................... Sunhee Hong, Jaehoon Lee, Duhyouk Kim, Junwoo Kim, and Yong Jeong | 337 |

43  Contribution of C-PC to Resilience of Concrete Structures
    in Seismic Country Japan........................... 345
    Makoto Kawakami, Mikio Wakasugi, and Fujio Omata

44  Precast Polymer Concrete Panels for Use on Bridges
    and Tunnels........................................ 353
    Michael S. Stenko

## Part IV  Polymer Fiber Concrete

45  The Effect of Combinations of Treated Polypropylene Fibers
    on the Energy Absorption of Fiber-Reinforced Shotcrete........ 363
    Johannes J. Bester, Kulani D. Mapimele, and George Fanourakis

46  Bond Performance of Steel-Reinforced Polymer (SRP) Subjected
    to Environmental Conditioning and Sustained Stress............ 369
    Wei Wang and John J. Myers

47  High-Strength, Strain-Hardening Cement-Based Composites
    (HS-SHCC) Made with Different High-Performance
    Polymer Fibers..................................... 375
    Marco Liebscher, Iurie Curosu, Viktor Mechtcherine,
    Astrid Drechsler, and Stefan Michel

48  Uniaxial Tensile Creep Behavior of Two Types of Polypropylene
    Fiber Reinforced Concrete........................... 383
    Rutger Vrijdaghs, Marco di Prisco, and Lucie Vandewalle

49  Dynamic Behavior of Textile Reinforced Polymer Concrete
    Using Split Hopkinson Pressure Bar..................... 389
    Mahmoud Abdel-Emam, Eslam Soliman, Amr Nassr,
    Wael Khair-Eldeen, and Aly Abd-Elshafy

50  Steel-Fiber Self-Consolidating Rubberized Concrete
    Subjected to Impact Loading......................... 397
    Mohamed K. Ismail, Assem A. A. Hassan, Katherine E. Ridgley,
    and Bruce Colbourne

51  Effect of Fiber Combinations on the Engineering Properties
    of High-Performance Fiber-Reinforced Cementitious
    Composites........................................ 405
    Dongyeop Han, Min-Cheol Han, Jong-Tae Lee,
    and Cheon-Goo Han

52  Application of Fibre-Reinforced Polymer-Reinforced Concrete
    for Low-Level Radioactive Waste Disposal................. 413
    Ricardo Lopes and Deon Kruger

53  Efficiency of Polymer Fibers in Lightweight Plaster............ 421
    Jakob Sustersic, Andrej Zajc, and Gregor Narobe

## Part V  Polymer Concrete with Recycled Waste

**54  Properties of Ceramic Waste Powder-Based Geopolymer Concrete** .......... 429
Sama T. Aly, Dima M. Kanaan, Amr S. El-Dieb, and Samir I. Abu-Eishah

**55  Use of Recycled Polymers in Asphalt Concrete for Infrastructural Applications** .......... 437
Sook F. Wong

**56  Influence of Method of Preparation of PC Mortar with Waste Perlite Powder on Its Rheological Properties** .......... 443
Grzegorz Adamczewski, Piotr Woyciechowski, Paweł Łukowski, Joanna Sokołowska, and Beata Jaworska

**57  Design and Manufacture of a Sustainable Lightweight Prefabricated Material Based on Gypsum Mortar with Semirigid Polyurethane Foam Waste** .......... 449
Sara Gutiérrez González, Carlos Junco, Veronica Calderon, Ángel Rodríguez Saiz, and Jesús Gadea

**58  Cement Mortars Lightened with Rigid Polyurethane Foam Waste Applied On-Site: Suitability and Durability** .......... 457
Carlos Junco, Sara Gutiérrez, Jesús Gadea, Veronica Calderón, and Ángel Rodríguez

**59  Latex-Modified Concrete Overlays Using Recycled Waste Paint** .......... 465
Aly M. Said and Oscar Quiroz

**60  Effect of Using Kaolin and Ground-Granulated Blast-Furnace Slag on Green Concrete Properties** .......... 471
Kamal G. Sharobim, Hassan A. Mohamadien, Omar M. Omar, and Mostafa M. Geriesh

**61  Lightweight Structural Recycled Mortars Fabricated with Polyurethane and Surfactants** .......... 479
Verónica Calderón, Raquel Arroyo, Matthieu Horgnies, Ángel Rodríguez, and Pablo Luis Campos

**62  Hydration in Mortars Manufactured with Ladle Furnace Slag (LFS) and the Latest Generation of Polymeric Emulsion Admixtures** .......... 485
Ángel Rodríguez, Sara Gutiérrez-González, Isabel Santamaría-Vicario, Veronica Calderón, Carlos Junco, and Jesús Gadea

**63  Chemical Resistance of Vinyl-Ester Concrete with Waste Mineral Dust Remaining After Preparation of Aggregate for Asphalt Mixture** .......... 491
Joanna J. Sokołowska and Piotr P. Woyciechowski

## Part VI  Geopolymers

**64  Performance Studies on Self-Compacting Geopolymer Concrete at Ambient Curing Condition** .................... 501
Krishneswar Ramineni, Narendra Kumar Boppana, and Manikanteswar Ramineni

**65  Effect of 3D Printing on Mechanical Properties of Fly Ash-Based Inorganic Geopolymer** ........................ 509
Biranchi Panda, Nisar Ahamed Noor Mohamed, and Ming Jen Tan

**66  Optimization of Fly Ash-Based Geopolymer Using a Dynamic Approach of the Taguchi Method** ........................ 517
Takeomi Iwamoto, Kozo Onoue, Yasutaka Sagawa, and Ryosuke Tsutsumi

**67  Microstructural and Strength Investigation of Geopolymer Concrete with Natural Pozzolan and Micro Silica** .............. 525
Muhammed Kalimur Rahman, Mohammed Ibrahim, and Luai M. Al-Hems

**68  Performance of Steel Fiber-Reinforced High-Performance One-Part Geopolymer Concrete** ........................... 533
Zahra Abdollahnejad, Tero Luukkonen, Paivo Kinnunen, and Mirja Illikainen

**69  Effect of Different Class C Fly Ash Compositions on the Properties of the Alkali-Activated Concrete** .............. 541
Eslam Gomaa, Simon Sargon, Cedric Kashosi, Ahmed Gheni, and Mohamed ElGawady

**70  Effect of Curing Temperatures on Zero-Cement Alkali-Activated Mortars** ............................... 549
Simon P. Sargon, Eslam Y. Gomaa, Cedric Kashosi, Ahmed A. Gheni, and Mohamed A. ElGawady

**71  Properties of PVA Fiber Reinforced Geopolymer Mortar** ........ 557
Wei Li and Hongjian Du

**72  Thermal Performance of Fly Ash Geopolymeric Mortars Containing Phase Change Materials** ..................... 565
M. Kheradmand, F. Pacheco Torgal, and M. Azenha

**73  Development of Fiber-Reinforced Slag-Based Geopolymer Concrete Containing Lightweight Aggregates Produced by Granulation of Petrit-T** ............................... 571
Mohammad Mastali, Katri Piekkari, Paivo Kinnunen, and Mirja Illikainen

## 74 Applications of Geopolymers in Concrete for Low-Level Radioactive Waste Containers ......................... 577
Kyle D. Poolman and Deon Kruger

## Part VII Fiber Reinforced Polymers (FRP)

## 75 Microstructure and Mechanical Property Behavior of FRP Reinforcement Autopsied from Bridge Structures Subjected to In Situ Exposure ....................................... 585
Wei Wang, Omid Gooranorimi, John J. Myers, and Antonio Nanni

## 76 The Influences of Mechanical Load on Concrete-Filled FRP Tube Cylinders Subjected to Environmental Corrosion ......... 593
Song Wang and Mohamed A. ElGawady

## 77 Finite Element Analysis of RC Beams Strengthened in Shear with NSM FRP Rods ...................................... 601
Akram R. Jawdhari and Ali Hadi Adheem

## 78 Effect of Sustained Load Level on Long-Term Deflections in GFRP and Steel-Reinforced Concrete Beams ............... 609
Stephanie L. Walkup, Eric S. Musselman, and Shawn P. Gross

## 79 Flexural Behavior and Cracks in Concrete Beams Reinforced with GFRP Bars ....................................... 617
Naser Kabashi, Cene Krasniqi, Jakob Sustersic, Arton Dautaj, Enes Krasniqi, and Hysni Morina

## 80 Flexural Rigidity Evaluation of Seismic Performance of Hollow-Core Composite Bridge Columns .................. 627
Mohanad M. Abdulazeez and Mohamed A. ElGawady

## 81 Three-Dimensional Numerical Analysis of Hollow-Core Composite Building Columns ............................. 635
Mohanad M. Abdulazeez and Mohamed A. ElGawady

## 82 Pultruded GFRP Reinforcing Bars with Carbon Nanotubes ...... 645
Rahulreddy Chennareddy, Amr Riad, and Mahmoud M. Reda Taha

## 83 On Mechanical Characteristics of HFRP Bars with Various Types of Hybridization .................................... 653
Andrzej Garbacz, Elzbieta Szmigiera, Kostiantyn Protchenko, and Marek Urbanski

## 84 Fatigue Behavior Characterization of Superelastic Shape Memory Alloy Fiber-Reinforced Polymer Composites .......... 659
Sherif M. Daghash and Osman E. Ozbulut

| 85 | Strength Performance of Concrete Beams Reinforced with BFRP Bars .......................................... 667 |

Elzbieta Szmigiera, Marek Urbanski, and Kostiantyn Protchenko

**Part VIII  Polymer Concrete with Nanomaterial**

| 86 | A Comparative Study on Colloidal Nanosilica Incorporation in Polymer-Modified Cement Mortars ....................... 675 |

Niloufar Zabihi and M. Hulusi Özkul

| 87 | Parametric Study on the Performance of UHPC and Nano-modified Polymer Concrete (NMPC) Composite Wall Panels for Protective Structures ............... 683 |

Olaniyi Arowojolu, Ahmed Ibrahim, and Mahmoud Reda Taha

| 88 | Mechanical Characterization of Polymer Nanocomposites Reinforced with Graphene Nanoplatelets .................... 689 |

Ugur Kilic, Sherif M. Daghash, and Osman E. Ozbulut

| 89 | Oil Well Cement Modified with Bacterial Nanocellulose ......... 697 |

Christian M. Martín, Ignacio Zapata Ferrero, Patricia Cerrutti, Analía Vázquez, Diego Manzanal, and Teresa M. Pique

| 90 | Effect of Incorporating Nano-silica on the Strength of Natural Pozzolan-Based Alkali-Activated Concrete ........... 703 |

Mohammed Ibrahim, Muhammed K. Rahman, Megat Azmi M. Johari, and Mohammed Maslehuddin

**Part IX  Strengthening and Restoration Using Polymers**

| 91 | Review of Polymer Coatings Used for Blast Strengthening of Reinforced Concrete and Masonry Structures ............... 713 |

Girum S. Urgessa and Mohammadjavad Esfandiari

| 92 | Evaluation of Polymer-Modified Restoration Mortars for Maintenance of Deteriorated Sewage Treatment Structures ... 721 |

Wanki Kim and Sunhee Hong

| 93 | Finite Element Modeling of CFRP-Strengthened Low-Strength Concrete Short Columns ................................... 729 |

Khaled A. Alawi Al-Sodani, Muhammed K. Rahman, Mohammed A. Al-Osta, and Ali A. H. Al-Gadhib

| 94 | Improvement Works to Existing Column Stumps by Fiber-Reinforced Polymer Strengthening System ............. 735 |

Jin Ping Lu and Sook Fun Wong

95 **Silicone Resin Enclosing Method Applied for the Maintenance of Steel Bearings**........................................ 743
Makoto Kawakami, Fujio Omata, Atsushi Toyoda, and Shingo Kato

96 **Bio-Based Polyurethane Elastomer for Strengthening Application of Concrete Structures Under Dynamic Loadings**.............. 751
Sudharshan N. Raman, H. M. Chandima C. Somarathna, Azrul A. Mutalib, Khairiah H. Badri, and Mohd. Raihan Taha

# Part I
# Keynote Papers

# Chapter 1
# Concrete-Polymer Materials: How Far Have We Come, and Where Do We Need to Go?

David W. Fowler

Concrete-polymer materials (CPMs) include polymer-impregnated concrete (PIC), polymer concrete (PC), and polymer-modified concrete (PMC) along with sulfur concrete and crack-filling resins and monomers. These materials began to be developed in the 1950s, but it was not until the early 1970s that they began to be well known in the industry. PIC is no longer being used, but PC and PMC are widely used as are the crack-filling monomers. The materials have come a long way in the nearly 50 years after they became known. Many conferences and workshops have been held, hundreds of papers have been published, and the American Concrete Institute Committee 548 has produced many special publications, guides, and specifications for the use of the materials. The future is bright. New polymers are coming on stream that offer many potential benefits for CPM. Nanotechnology will likely play an important role. Many innovative applications and processes to use CPC including 3D printing and new and exciting products using precast PC are on the horizon.

## 1 Introduction

The paper will briefly discuss the CPMs that have been developed and their current status. A review of how far have we come and a projected look ahead to see where we need to go will be presented.

D. W. Fowler (✉)
University of Texas, Austin, TX, USA
e-mail: dwf@mail.utexas.edu

© Springer International Publishing AG, part of Springer Nature 2018
M. M. Reda Taha (ed.), *International Congress on Polymers in Concrete (ICPIC 2018)*, https://doi.org/10.1007/978-3-319-78175-4_1

## 2 How Far Have We Come?

Concrete-polymer materials are one of the newest construction materials. There were a few uses in the 1950s, but it was not until the early 1970s that they became known in the industry. Polymer-impregnated concrete (PIC) was first investigated in the late 1960s, and the publication of that research provided an impetus for many researchers to begin studying the materials. American Concrete Institute Committee 548 Polymers in Concrete was the first committee to have the mission of developing and reporting information on polymers in concrete.

PIC had marvelous strength and durability properties, but full impregnation required complex facilities to dry concrete, apply a vacuum, impregnate with monomer under pressure, and cure. Partial-depth impregnation for improving durability of bridge decks appeared promising, but the process was time-consuming.

Polymer concrete (PC) received considerable interest due to its relative simplicity in production compared to PIC. A thin PC overlay could be applied to a bridge deck in much less time than was required for partial-depth impregnation. It was also found to be a very promising material for making fast and durable repairs in concrete bridge decks, pavements, and other structures. It was also successfully developed for precast elements such as building cladding, utility drains, underground utility boxes, high-voltage insulators, railroad crossing panels, and other utility products. Nanotechnology is just beginning to be explored for improving performance of PC.

Polymer-modified concrete (PMC), particularly latex-modified concrete (LMC), has been arguably the most widely used CPC. Since it serves as a modifier for portland cement concrete, it is less expensive than PC and is easier to use, although the strength and durability is not as high. It is widely used for overlays on bridge decks and parking garages and for prepackaged repair materials.

Epoxy and high molecular weight methacrylate (HMWM) for gravity-filled crack repair and epoxy injection represent a very successful use of polymers. Gravity filling in particular is much simpler than epoxy injection and has proven to be a very popular method for sealing cracks.

Sulfur concrete (SC), made of aggregates and sulfur binder, received attention due to sulfur being a waste material and due to the high strength and corrosion resistance of SC. The high temperature required to get sulfur into a molten state for mixing was the biggest impediment. The use was primarily limited to overlays in facilities requiring acid resistance. Precast sulfur components appeared promising but never received much attention in the USA.

This brings us to the question: How far have we come? If we consider that CPMs were virtually unknown before the 1970s, the industry has come a long way. PC for repairs, overlays, and precast has been quite successful. LMC is a staple for bridge and garage overlays and for prepackaged repair. Gravity-filling monomers and resins have been very popular. The information dissemination has been very good through the International Congress of Polymers in Concrete (ICPIC) that holds conferences every 3 years and publishes proceedings; through ACI 548 in its sessions at ACI conventions, ACI Special Publications, committee documents including specifications on various types of overlays and papers published in ACI journals; and through several RILEM committee publications, several Asian symposia on polymers in concrete, several conferences on polymers in concrete hosted by the University of Johannesburg (formerly Rand Afrikaans University), and other workshops and symposia held around the world.

However, some of the lofty projections made for the growth of CPM in the construction market have not been achieved. Higher costs and lack of familiarity by specifiers and contractors are reasons most often cited as the cause. There is little research being reported, and few new materials have come on the market.

## 3 Where Do We Need to Go?

The CPM industry, including researchers, resin and monomer manufacturers, equipment manufacturers, material suppliers, and contractors, must continue to improve existing materials and application techniques and to develop new and more efficient ones. Development of specifications and user guides must be accelerated; regardless of the advancements, if those who specify and use the materials are not kept informed, growth of the industry will lag. Continued dissemination of information to specifiers and users is essential.

**Materials**
In the 1970s, the potentially huge market for CPM repair materials attracted the interest of many chemical companies. Initially, methyl methacrylate (MMA) was one of the most widely used monomers in PIC and PC due to its very low viscosity, rapid curing, relatively low cost, and ability to produce PIC and PC with very good mechanical and durability properties. Its odor and flammability prior to curing were major disadvantages. HMWM was developed as a potential replacement, but its much higher cost resulted in it being limited primarily as a primer for overlays and for gravity filling of cracks. The first polyester and epoxies were not sufficiently ductile for producing good overlays; manufacturers were able to provide high elongation, low modulus resins that served very effectively. Finding new/lower-cost monomers and resins should be an objective going forward. Resins produced from recycled resins were researched and found to produce good-quality polymer concrete. However, the demand for these resins in other applications kept the cost relatively high, and they have never had a significant use in CPMs. The use of recycled resins should remain a priority for future research.

There are some exciting new polymers being developed that may have potential for use in CPC. A new hybrid polymer has been developed at Northwestern University that "has internal compartments, one like a rigid like a skeleton, another like a gel." The gel-like structure has fast response when stimulated and can be removed and regenerated. It is said to have potential in "materials with self-repair capability" [1]. Such a polymer could have potential in polymer concrete or LMC overlays, repair materials, and crack-filling materials.

IBM Research announced in 2014 two new polymers, code named Titan and Hydro, both of which come from the same reaction. Similar to the polymers developed at Northwestern, Titan is rigid, while the Hydro is a gel. Titan has about one-third the tensile strength of steel, and when combined with 2–5% carbon nanotubes, it is three times stronger than polyamides used in the manufacture of some aircraft. Hydro is flexible and can self-heal. Two pieces placed next to each other will combine into one piece without being forced together. Although very different in makeup, both polymers are recyclable. Hydro is water-soluble, and Titan can be broken down with a light acid. The polymers are reversible thermosets, long thought to be irreversible, and the ability to recycle the polymers could be a major advantage. Although IBM indicates it could be months or years before these polymers are available commercially, the CPC industry needs to be aware of the potential for future improvement in CPC materials [2].

In 2017 MIT reported a new method to measure certain types of defects in polymers called loops and a simple way to reduce the number of loops that result in a stronger polymer. A loop occurs when a chain in the polymer network binds to itself instead of another chain. This development may provide a method for increasing the strength of polymers used in precast PC and other applications [3].

The incorporation of carbon nanotubes has considerable potential for improving fatigue resistance of polymer concrete [4]. The PC mixes incorporated epoxy and 0.5%, 1.0%, and 1.5% multiwalled carbon nanotubes (MWCNTs) by weight of epoxy resin. The three levels of MWCNTs increased the fatigue strength by 61%, 100%, and 520%, respectively.

Researchers in the USA and Japan have developed a polymer that can repair itself over and over when irradiated with UV light. It is said that this is the first material in which covalent bonds repeatedly reattach, allowing completely separated pieces to be fused together. "Other materials ... can repair themselves repeatedly but lack the covalent bonds that increase materials' strength and stability" [5].

Smart and reactive polymers can adjust their shape in response to external stimuli including change in temperature or acidity/alkalinity, light, ultrasound, and chemical agents. It is anticipated that self-healing technology can be incorporated. Potential applications in CPC materials are coatings that can clean surfaces of PC precast panels, provide a warning of stresses that can result in cracking in structural PC, send an alert when CPC repair materials have failed, and in some cases repair cracks that have formed [6].

Sulfur, an inorganic polymer, was originally part of the CPC materials included in ACI Committee 548 documents, but due to lack of interest and use, sulfur is no longer included. It was used primarily for corrosion-resistant floor overlays and sumps and for precast applications. It is the author's opinion that sulfur has considerable potential due to the very large quantities that are produced and the relative low cost compared to many organic polymers used in CPC. A process for modifying sulfur by reacting it with olefinic hydrocarbon polymers was developed in 1973. Later, a process for modifying sulfur by reaction with cyclopentadiene oligomer and dicyclopentadiene was developed. Modified sulfur has much better durability when exposed to freezing and thawing and immersion in water. The production of sulfur concrete requires mixing heated graded aggregate (177–204 C) with modified sulfur cement and fine filler; the sulfur concrete is maintained at 132–141 C until placed. Special mixers are required due to the high temperatures. Sulfur concrete has a very high resistance to acid and has been used for floors, sumps, and drains that are exposed to acids. It can be placed in forms and reinforced with conventional steel reinforcing or glass fibers. Strength is developed as the sulfur concrete is cooled. Typical properties after 1-day cooling are compressive strength, 27 MPa; flexural strength, 5 MPa; absorption, 0.10%; coefficient of thermal expansion, $15 \times 10^{-6}/°C$; and modulus of elasticity, 3 to $4 \times 10^6$ GPa.

Numerous precast applications using sulfur concrete have been developed. Figure 1.1 shows sulfur concrete sewer pipe. Figure 1.2 shows a precast sump for acid waste, while Fig. 1.3 illustrates the use of sulfur concrete for precast highway median barriers. Sulfur concrete has considerable potential for pavement applications (Fig. 1.4).

The use of sulfur has considerable potential in view of the move toward minimizing the use of cement to lower $CO_2$ emissions and to utilize waste materials. The cost of sulfur should make sulfur concrete competitive with conventional concrete in countries that produced significant amounts of sulfur.

To continue to advance the use of CPC, it is essential that researchers, material suppliers, and contractors must be aware of the innovations in polymer materials and incorporate them into the family of CPC materials that can be useful to the industry.

**Fig. 1.1** Precast sewer pipe

**Fig. 1.2** Precast sump for acid waste

**Fig. 1.3** Precast median barriers

ICPIC should be the forum for reporting new advances in materials and applications of CPC.

**Innovative Applications**

Due to space limitations, only PC applications will be considered. Precast PC has been one of the most promising innovations, but it has not reached its potential. PC building cladding has been shown to be a very good use of PC starting in 1958 [7]. Figure 1.5 shows a medical building in Massachusetts that is clade with precast PC panels. The panels are thin PC with a portion of the panels being thin brick embedded in PC; the panels are attached to a 150-mm steel stud frame that serves as

**Fig. 1.4** Paving with sulfur concrete

**Fig. 1.5** PC cladding on medical building

the erection frame and the permanent wall framing (Fig. 1.6). The light weight, ability to form intricate shapes and textures, excellent durability, and ability to provide color have made them popular with architects. An intricate PC cross for a church is shown in Fig. 1.7. Figure 1.8 illustrates the use of thin precast PC column covers and curved panels. PC manholes (Fig. 1.9), sewer pipe, and lift stations are excellent applications. PC median barriers for highways should be explored to take advantage of the reduced weight and impact resistance.

**Fig. 1.6** Panel for building in Fig. 1.5

**Fig. 1.7** Intricate PC cross

**Fig. 1.8** Column covers and curved ceiling panels

**Fig. 1.9** PC sewer manholes

One of the most intriguing possibilities is the use of 3D printing for PC applications. It has been used with very good results using cementitious-based mortars and concretes. The use of PC or even LMC would provide a finished material that would be more impermeable, stronger, more durable, and, in the case of PC, cure much more quickly. Figure 1.10 shows a printer placing a wall of a concrete structure [8].

**Fig. 1.10** 3D printing of a wall of a house [10]

**Fig. 1.11** Building placed with 3D printing [9]

Figure 1.11 illustrates a small office hotel that was the first 3D printed building in Europe that met European building codes. The building had curved walls and a ripple effect to "illustrate the demand freedom that 3D printing allows [9].

A San Francisco startup has used its 3D printer to construct concrete walls on a test site in Russia to produce a small 40-m² livable house. After the walls are printed, contractors install insulation, windows, appliances, and roof for a cost of about $10,000 [10].

MIT researchers have developed a process to use 3D printing to produce concrete that has variable density to produce lighter weight materials that retain adequate load-carrying ability (Fig. 1.12). This process could be used to produce PC or PMC with greater thermal insulation qualities [11].

**Fig. 1.12** Variable density concrete using 3-D printer [11]

## 4 Conclusions

Concrete polymer materials have had a rapid introduction into the industry and have found many useful applications. The future is bright with many potential applications that include 3-D printing, precast PC including innovative cladding, utility products, and architectural materials. It is essential that the industry continue to stay abreast of new developments in materials and application methods.

## References

1. https://news.northwestern.edu/stories/2016/01/new-polymer
2. http://mashable.com/2014/05/15/ibm-new-self-healing-polymer/#epPGfqOmAaqx
3. http://news.mit.edu/2017/stronger-polymers-reduce-loops-0424
4. Daghash, S. M., Tarefder, R., & Reda Taha, M. M. (2015). In K. Sobolev & S. Shah (Eds.), *A new class of carbon nanotube: Polymer concrete with improved fatigue strength, nanotechnology in construction* (pp. 285–290). Switzerland: Springer International Publishing.
5. https://phys.org/news/2011-01-polymer-amazing-self-healing-properties.html
6. https://www.weforum.org/agenda/2015/02/5-synthetic-materials-that-will-shape-the-future/
7. Prusinski, R. C. (1978). Study of commercial development in precast polymer concrete. *ACI Special Publication, 58*, 75–102.
8. https://3dprint.com/195020/3d-printhuset-building-on-demand/
9. https://hackaday.com/2016/04/15/3d-printing-houses-from-concrete/
10. https://qz.com/924909/apis-cor-can-3d-print-and-entire-house-in-just-one-day/
11. http://www.rapidreadytech.com/2012/01/mits-natural-approach-to-3d-printing/

# Chapter 2
# Polymer Concrete for Bridge Preservation

Michael M. Sprinkel

Polymer concrete has been used for bridge preservation for more than 50 years. Latex-modified concrete overlays, developed in the 1960s, are used to extend the service life of bridge decks by providing improved adhesion and reduced permeability. In the 1980s, polymer concrete overlays made with polyester styrene, epoxy, and methacrylate binders became accepted deck protection systems for bridges and parking structures. The rapid cure allowed DOTs to install the skid-resistant deck protection overlays with short lane closure times providing minimal delays to the traveling public. Polymer concrete overlays are placed on pavement surfaces prone to accidents that can be reduced by the application of a high friction polymer surface treatment. Gravity fill polymer crack fillers were developed as a low-cost crack-sealing alternative that can help extend the life of cracked concrete decks. Accelerated bridge reconstruction in which deteriorated decks are replaced at night or other short lane closure periods is possible using prefabricated deck elements that are attached or joined using polymer concrete or rapid hardening polymer-modified concrete. Both polymer and polymer-modified concrete have and continue to be cost-effective fundamental materials for extending the life of our infrastructure, and use continues to increase.

## 1 Introduction

Concrete overlays are placed on bridge decks to reduce infiltration of water and chloride ions, thereby extending the service life, and to improve skid resistance and ride quality. Latex-modified concrete (LMC), developed in the 1960s, is used as an

---

M. M. Sprinkel (✉)
Virginia Transportation Research Council, Virginia Department of Transportation, Charlottesville, VA, USA
e-mail: Michael.Sprinkel@vdot.virginia.gov

overlay to extend the life of bridge decks by providing improved adhesion and reduced permeability [1]. In the 1980s, polymer concrete (PC) overlays made with polyester styrene, epoxy, and methacrylate binders became accepted deck protection systems for bridges and parking structures. The rapid cure allowed DOTs to install the skid-resistant deck protection overlays with short lane closure times providing minimal delays to the traveling public. LMC very-early (LMC-VE) overlays, prepared with rapid hardening cement (ASTM C928), have been placed on bridge decks by the Virginia Department of Transportation (VDOT) since 1997 [1–3]. PC overlays have been placed on pavement surfaces prone to accidents which can be reduced by the application of a high friction polymer surface treatment (HFST) [4]. The overlay is typically one layer of polymer binder and calcined bauxite which provides a bald tire skid number of 70 or higher [5]. Plastic shrinkage and drying shrinkage cracking is a major problem with hydraulic cement concrete decks. The cracks often allow chlorides and water direct access to the reinforcement, accelerating corrosion and reducing the service life of the deck. Gravity fill polymer crack fillers were developed in the 1980s as a low-cost crack-sealing alternative that can help extend the life of cracked concrete decks. Accelerated bridge reconstruction in which deteriorated decks are replaced at night or other short lane closure periods is possible using prefabricated deck elements that are attached or joined using PC, LMC-VE concrete, or rapid hardening hydraulic cement concrete. Both PC and PMC have and continue to be cost-effective fundamental materials for extending the life of our infrastructure, and use continues to increase.

This paper presents VDOT's experience with LMC, LMC-VE, and PC overlays, gravity fill crack sealing, and deck replacements with PC connections. The properties and performance of the PC and PMC applications are presented. Information is taken from selected research reports and publications. Information include construction, mixture proportions, permeability to chloride ion, shrinkage, compressive strength, bond strength, and costs.

## 2 Latex-Modified Concrete Overlays

Latex-modified concrete (LMC) overlays placed on the decks provide an economical way to extend the service life. These overlays have been used for more than 50 years. They have performed well when properly constructed, providing a service life of 30–40 years. With the increase in traffic, particularly on the interstate system and major primary roads, lanes cannot be closed for the extended periods required to repair decks and place overlays with conventional proven materials. To minimize traffic delays, contractors work at night or weekends and open repairs and overlays to traffic with as little as 3 h of curing.

VDOT has used LMC-VE concrete for deck spall repairs done at night on Interstate 81 since 1992. Based on the positive experience with repairs, VDOT worked with contractors to overlay two decks at night using LMC-VE. The first, constructed in 1997, was the Lord Delaware Bridge located in King and Queen

**Fig. 2.1** LMC-VE overlay placed at night. Fogging to increase the humidity

County. The second, constructed in 1998, was the Route 620 west-bound lane structure over I-495 in Fairfax County, shown in Fig. 2.1. [3, 6].

LMC-VE overlays can be placed at night using the following sequence. A lane is closed from 9 PM to 5 AM. The concrete surface is removed by milling, and areas to be repaired are removed with pneumatic hammers. LMC-VE repairs and/or sections of overlay are placed up to 2 PM and cured for 3 h. This sequence is repeated until all lanes are repaired and overlaid. Deck surfaces should be sound, clean, and dry to obtain high overlay bond strengths. Typical surface preparation includes milling followed by shot blasting followed by wetting the surface to achieve a saturated surface dry condition at the time the overlay is placed. Hydrodemolition is often used to prepare the surfaces on older bridges because the high-pressure water removes deteriorated concrete adjacent to and below the top mat of reinforcement and provides a heavy texture and saturated surface that provides for good bond of the overlay. In 2006, a 5000 square yard LMC-VE overlay was placed on the deck on I-64 over the Rivanna River. Another notable application in 2006 was the overlay placement on the Theodore Roosevelt Bridge (I-66 into the District of Columbia) [3]. In 2014, LMC-VE concrete was used for a deck closure pour on I-81 at Fairfield, Virginia. The pour was 120-ft long by 10-ft wide by 9-in thick. The concrete was placed in December, and LMC-VE concrete was selected for use because of low shrinkage and ambient temperatures near 50 F. While curing time for LMC-VE concrete increases with a decrease in concrete temperature, significant increases in curing time only occur at temperatures below 50 F [2]. LMC-VE is not placed at temperatures below 50 F. Furthermore, LMC-VE and conventional LMC overlays have the same mixture proportions (Table 2.1). Table 2.2 shows typical compressive strength data. LMC-VE achieves a compressive strength of 2500 psi in as little as

**Table 2.1** Mixture proportions

| Mixture | LMC-VE | LMC |
|---|---|---|
| Cement type | VE hardening | I/II |
| Cement, lb/yd$^3$ (kg/m$^3$) | 658 (390) | 658 (390) |
| Fine aggregate, lb/yd$^3$ (kg/m$^3$) | 1600 (949) | 1571 (932) |
| Coarse aggregate, lb/yd$^3$ (kg/m$^3$) | 1168 (693) | 1234 (732) |
| Latex, lb/yd$^3$ (kg/m$^3$) | 205 (122) | 205 (122) |
| Water ($w/c \leq 0.40$), lb/yd$^3$ (kg/m$^3$) | 137 (81) | 137 (81) |
| Air, percent | 3–7 | 3–7 |
| Slump, in (mm) | 4–6 (102–152) | 4–6 (102–152) |

**Table 2.2** Compressive strength, psi (MPa)

| Age | LMC-VE | LMC |
|---|---|---|
| 3 hours | 3500 (24) | – |
| 4 hours | 3800 (26) | – |
| 5 hours | 4100 (28) | – |
| 24 hours | 5500 (38) | 2500 (17) |
| 7 days | 6300 (43) | 4800 (33) |
| 28 days | 6700 (46) | 5600 (39) |

**Table 2.3** Chloride ion permeability, coulombs

| Age | LMC-VE | LMC |
|---|---|---|
| 28 days | 300–1400 | 1500–2560 |
| 1 year | 0–10 | 210–2060 |
| 3 years | – | 300–710 |
| 5 years | – | 450–500 |
| 9 years | 0–60 | 100–400 |

3 hours. The very-early strength is achieved using an ASTM C 928 cement that contains approximately 1/3 calcium sulfoaluminate and 2/3 dicalcium silicate [3, 6]. A retarder is typically added to provide an increase in working time for properly placing the concrete. The addition of too much retarder for the temperature can cause an increase in setting time and a long curing time.

Very rapid hardening concretes are often considered to be less durable than conventional hydraulic cement concretes. LMC-VE overlays can be more durable than overlays constructed with type I/II Portland cement. First, LMC-VE concrete is less prone to cracking because the drying shrinkage is less. At an age of 170 days, the length change (ASTM C157) of LMC-VE specimens was approximately 0.02%. The length change of LMC specimens was 0.06% [3, 6]. LMC-VE specimens had significantly lower permeability to chloride ions (AASHTO T277) than LMC [7]. The permeability at 28 days of age of LMC-VE overlays is low to very low, and the permeability of conventional LMC overlays is low (Table 2.3). The permeability after 1 year of LMC-VE was very low, and LMC overlays was low [6]. Like all hydraulic cement concrete overlays, some LMC and LMC-VE overlays have cracks. The most common type of crack, plastic shrinkage, is caused by placing the

**Table 2.4** Bond strength, psi (MPa)

| Age | LMC-VE | LMC |
|---|---|---|
| 1–6 months | 153–276 (1.1–1.9) | 116–260 (0.8–1.8) |
| 3–5 years | – | 200–310 (1.4–2.1) |
| 9–10 years | 176–301 (1.2–2.1) | 246–296 (1.7–2.0) |

**Table 2.5** Construction cost of overlays in 2010, $/yd² ($/m²)

| Mixture | LMC-VE | LMC |
|---|---|---|
| Overlay | 90 (108) | 83 (99) |
| Misc. | 32 (38) | 32 (38) |
| Traffic | 28 (33) | 44 (53) |
| Total | 150 (179) | 159 (190) |

overlay at evaporation rates greater than 0.05 lb/ft2/hr., failure to apply the wet burlap curing material before the concrete surface dries and a failure to keep the burlap wet during the curing period. Like any hydraulic cement concrete overlay, LMC and LMC-VE overlays have high bond strengths when constructed using good surface preparation equipment and procedures, mixture proportions, and placement and curing procedures [6]. Tensile bond strengths (ASTM C1583) between 176 and 301 psi were measured for 9-year-old overlays (Table 2.4) [6].

LMC and LMC-VE overlays bond well to decks that are milled (with milling heads that do not damage the surface), shot blasted, and wetted to achieve a saturated surface dry condition. Often, the bonding mortar brushed into the surface ahead of an LMC placement is omitted for LMC-VE because of the rapid drying of the mortar. LMC and LMC-VE overlays bond well to surfaces prepared by hydrodemolition.

Finally, LMC-VE overlays need not cost more than conventional overlays (Table 2.5) [6]. The cost of the rapid hardening cement is approximately four times that of the Type I and II cements used in LMC overlays. The reduced cost of traffic control for LMC-VE overlays more than offsets the higher cement cost. Including the cost savings associated with reduced delays to drivers can make LMC-VE cost less than LMC and other conventional overlays [6].

## 3 Polymer Concrete Overlays

PC overlays that have exhibited good performance include multiple-layer (ML) epoxy, ML epoxy urethane, methacrylate slurry, and premixed (PM) polyester styrene. The overlays are constructed in accordance with the AASHTO TF 34 guide specifications [8] and ACI standards [9, 10]. Evaluations have indicated that these overlays can have a life of 25 years, providing a cost-effective option for extending the life of concrete decks [8, 11–13]. ML overlays consist of two or more layers of unfilled polymer binder and broadcasted gap-graded, clean, dry, angular-grained aggregate, providing a thickness of about 0.25 in (6.3 mm). Slurry overlays are typically a polymer aggregate slurry struck off

**Table 2.6** Properties of binders

| Property | Epoxy | Polyester | Methacrylate | Test method |
|---|---|---|---|---|
| Viscosity, p | 7–25 | 1–5 | 11–13 | ASTM D 2393 |
| Gel time, min | 15–45 | 10–25 | 15–45 | AASHTO T 237 |
| Tensile strength, at 7 days, Ksi (MPa) | 2.1–5.2 (13.8–34.4) | 2.1–5.2 (13.8–34.4) | 0.5–1.2 (3.4–8.3) | ASTM D 638 |
| Tensile elongation, at 7 days, % | 30–80 | 30–80 | 100–200 | ASTM D 638 |

**Table 2.7** Typical aggregate gradations (percent passing sieve)

| Sieve size | ML | Slurry | PM |
|---|---|---|---|
| 0.13 mm | – | – | 100 |
| 0.10 mm | – | – | 83–100 |
| No. 4 | 100 | – | 62–82 |
| No. 8 | 30–75 | – | 45–64 |
| No.16 | 0–5 | 100 | 27–50 |
| No. 20 | – | 90–100 | – |
| No. 30 | 0–1 | 60–80 | 12–35 |
| No. 40 | – | 5–15 | – |
| No. 50 | – | 0–5 | 6–20 |
| No. 100 | – | – | 0–7 |
| No. 140 | – | 100 | – |
| No. 200 | – | 98–100 | 0–3 |
| No. 270 | – | 96–100 | – |
| No.325 | – | 93–99 | – |

with gage rakes and covered with broadcasted aggregate providing a thickness of about 0.4 in (10 mm). PM overlays are a polymer concrete mixture consolidated and struck off with a vibratory screed, providing a thickness of 0.75 in (19 mm) or more.

## 3.1 Materials

PC overlay binders are typically epoxy, epoxy urethane, polyester styrene, and methacrylate. The binders are prepared by mixing a resin with a curing agent or initiator as they are used. Binders are usually classified on the basis of the properties of the uncured and cured binder (Table 2.6) [8–10, 13].

Epoxy urethane binders have properties similar to those for epoxy with the exception that the viscosity is typically higher (35–70 poises). The aggregates are usually silica and basalt. Uniformly graded aggregates are typically used with slurry and premixed overlays, and gap-graded aggregates are used with ML overlays and are broadcast onto the top of slurry and some PM overlays (Table 2.7) [8–10,

13]. Aggregates must be dry (≤0.2% moisture). Aggregates have historically been silica sand or basalt with an angular grain, but in recent years, other aggregates have been used. Aggregates must be clean. PC is typically made by combining the binders in Table 2.6 with the aggregates in Table 2.7. Gap-graded aggregates used for ML overlays have been used to construct epoxy slurry overlays. The polymer for ML overlays is typically applied at a minimum rate of 2.5 gal (9.5 L) per 100 ft$^2$ (0.09 Km$^2$) for layer 1 and 5.0 gal (19.0 L) per 100 ft$^2$ (0.09 Km$^2$) for layer 2. For slurry and premixed overlays, the binder and aggregate are mixed together prior to placement. Typically, PC overlays are cured for 3 hours, but longer times may be required at lower temperatures. The curing time for some systems can be changed by changing the amount of curing agent or initiator.

## 3.2 Construction, Surface Preparation, and Methods of Application

PC overlays should be constructed in accordance with specifications like AASHTO TF 34 guide specifications [8] and ACI standards [9, 10]. High bond strengths can be obtained when surfaces are properly prepared. Concrete with a tensile rupture strength of less than 150 psi (1.0 MPa) should be removed and replaced. Also, concrete with a chloride ion content greater than 1.3 lb/yd$^3$ (0.77 kg/m$^3$) at the reinforcing steel should be considered for removal and replacement prior to placing the overlay. Reinforcement corrosion can cause spalling of the cover concrete and PC overlay. Cracks greater than 0.04 in (1 mm) in width should be filled, and spalls should be repaired prior to placing the PC overlay. The surface should be prepared by shot blasting and other approved cleaning practices to texture the surface and to remove detrimental materials that may interfere with the bonding or curing of the overlay. Areas that cannot be cleaned by shot blasting can be grit blasted. Test patches should be placed and tested for adhesion strength using ASTM C1583. Passing test results provide an indication that the surface preparation is adequate and materials were properly mixed, placed, and cured. The same cleaning practice, materials, equipment, and staff should be used to construct that the test patches should be used to place the PC overlay [8–10].

The overlay should be placed the same day the surface is shot blasted. Areas that are not overlaid should be shot blasted again just before the overlay is placed. ML overlays are typically constructed as shown in Fig. 2.2. The binder is spread over the shot-blasted surface using notched squeegees, and the gap-graded aggregate is broadcast to excess over the binder. Following a brief curing period, the un-bonded aggregate is removed, and a second layer is applied [8, 9]. Slurry overlays are typically constructed as shown in Fig. 2.2. A prime coat is placed ahead of the slurry placement, and gap-graded aggregate is broadcast to excess over the slurry. A seal coat is applied to the broadcasted aggregate [8, 10]. PM overlays are constructed as shown in Fig. 2.3. A prime coat is typically placed ahead of the overlay

**Fig. 2.2** Notched squeegees are used to place ML epoxy overlay (left). Slurry mixed with mortar mixers is struck off with gage rakes (right). Basalt aggregate is broadcast to excess onto both overlays

**Fig. 2.3** PM polyester concrete placement: vibrating screed is used to consolidate and strike off PC (left); continuous batching and paving are used (right)

placement. A vibratory screed is used to strike off and consolidate the concrete. Continuous batching and paving equipment has been used to place some overlays. Skid resistance is improved by placing grooves in the freshly placed concrete or by broadcasting aggregate onto the surface [8].

## 3.3 Performance of PC Overlays, Service Life, and Cost

Decks have been evaluated at various ages to provide an indication of performance [8, 11–13]. The bond strength does not change much with time. The protection provided by the overlay as indicated by AASHTO T 277 remains low to very low over the life of the overlay. The protection provided over the life of the PC overlay is similar to that provided by LMC and LMC-VE overlays. The bald tire (ASTM E 524) skid number is typically of 50–60 for new PC overlays and gradually decreases to 30s or 40s after 15–20 years. Acceptable numbers are typically being maintained throughout the life of the overlays. PC overlays have shown good resistance to wear under traffic except in states in which studded tires and chains are used. PC overlays constructed in accordance with AASHTO guide specifications and ACI standards should provide a good surface for 25 years [13]. The cost of materials, surface preparation, labor, equipment, and traffic control determine the cost of the PC overlay. Costs can be expected to vary on the basis of the local labor rates and competitive bidding. ML epoxy overlays are typically constructed in Virginia at a cost that is about 50 percent of the cost of an LMC overlay (see Table 2.5). Other PC overlays typically cost more than multiple-layer epoxy overlays and may cost about the same as an LMC or LMC-VE overlay. In summary, PC overlays constructed in accordance with AASHTO specifications and ACI standards can provide a skid-resistant wearing and protective surface for concrete bridge decks for 25 years and are a cost-effective alternative to hydraulic cement concrete overlays.

## 4  High Friction Surface Treatments

Highway agencies are being directed by FHWA toward setting up, or improving, pavement friction management programs to ensure pavement surfaces provide adequate and durable friction properties that reduce friction-related crashes [14, 15]. High friction surface treatments (HFST) are an application of epoxy and calcined bauxite aggregate to roadway surfaces to provide very high skid resistance. They may be effective when applied to surfaces in which the demand for tire-pavement friction is especially high. Rural asphalt roads with a high risk and/or record of crashes and low skid resistance are good candidates for HFST. Since 2012, the VDOT has placed HFST at locations that met this description. The epoxy is the same as that used for MLE overlays, and the single layer application rate is about 3.5 gallons (13.2 L) per 100 ft$^2$ (0.09 Km$^2$) which is approximately 50 mils which is adequate for typical asphalt surfaces [4]. The rate should be higher for surfaces with a higher macrotexture as indicated in Table 2.8. The bond strength typically exceeds the tensile strength of the asphalt surface it is placed on. The asphalt quality must be good, and the HFST should not be placed on asphalt that needs to be resurfaced [16]. Automated or semiautomated equipment is typically used.

## 5  Gravity Fill Polymer Crack Fillers

Plastic shrinkage and drying shrinkage cracking is a major problem with hydraulic cement concrete decks (Fig. 2.4). The cracks often allow chlorides and water direct access to the reinforcement, accelerating corrosion and reducing the service life of

**Table 2.8** HFST epoxy application rates as a function of asphalt texture

| Macrotexture depth, in (mm) (ASTM E 965) | Minimum epoxy application rate, mils |
|---|---|
| <0.04 (1.0) | 50 |
| 0.04–0.06 (1.0–1.5) | 75 |
| >0.06 (1.5) | 2 applications @ 50 |

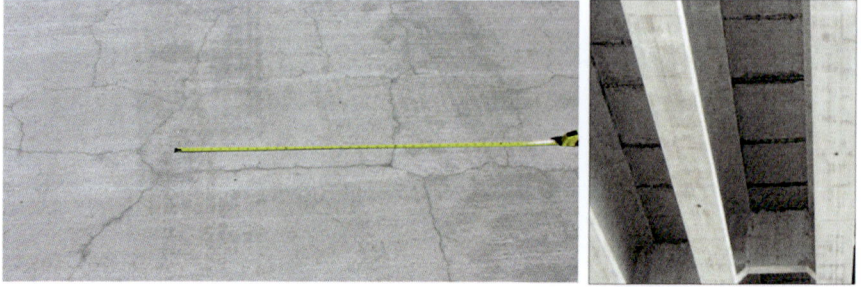

**Fig. 2.4** Plastic shrinkage cracking (left); drying shrinkage cracking (right)

**Fig. 2.5** Crack fillers can be applied to individual linear cracks or spread over the deck surface to fill plastic shrinkage cracks. Polymer fills crack in core

the deck. Gravity fill polymer crack fillers were developed in the 1980s as a low-cost crack-sealing alternative that can help extend the life of cracked concrete decks. Crack fillers can be applied to individual linear cracks or spread over the deck surface to fill plastic shrinkage cracks (Fig. 2.5).

## 6 PC and PMC Closure Pours for Precast Deck Elements

Accelerated bridge reconstruction in which deteriorated decks are replaced at night or other short lane closure periods is possible using prefabricated deck elements that are attached or joined using PC, LMC-VE concrete, or rapid hardening hydraulic cement concrete (Fig. 2.6).

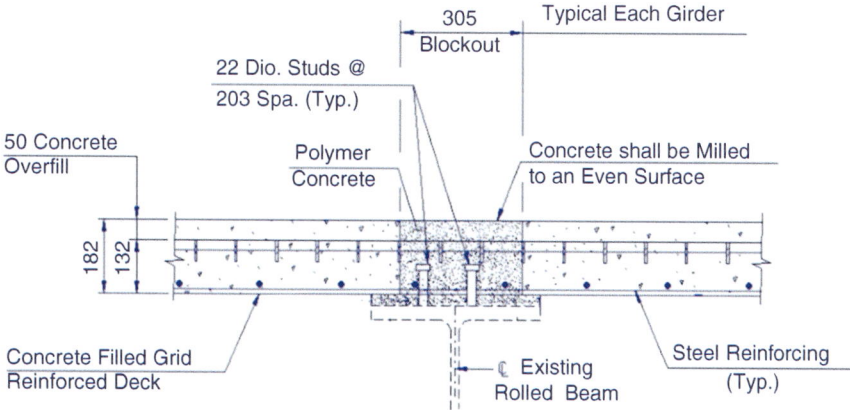

**Fig. 2.6** The deck on I-95 in Richmond, Virginia, was replaced in 2000 with night lane closures using transverse deck slabs connected by PC

## 7 Standard Practice

The use of LMC-VE and multiple-layer epoxy concrete overlays has continued to increase with time, and the record of success continues to grow. The VDOT 2016 Road and Bridge Specifications include requirements for the use of LMC, LMC-VE, and multiple-layer epoxy overlays and gravity fill crack sealers [17].

## 8 Conclusions and Recommendations

Both PC and PMC have and continue to be cost-effective fundamental materials for extending the life of our infrastructure, and use continues to increase. Departments of transportation (DOTs) should use LMC, LMC-VE, and PC overlays to extend the service life of provided bridge decks. DOTs should apply HFST to surfaces prone to frequent accidents caused by low surface friction. DOTs should fill and seal cracks with gravity fill polymer materials. DOTs should use PC to connect and anchor prefabricated bridge elements.

## References

1. Sprinkel, M. M. (1999). *Very-early-strength latex-modified concrete overlay* (pp. 18–23). Transportation Research Record 1668. Washington, DC: Transportation Research Board.
2. Sprinkel, M. M. (1998). Very-early-strength latex-modified concrete overlays, VTRC TAR 99-TAR3, Charlottesville.

3. Sprinkel, M. M. (2006). *Very-early-strength latex-modified concrete bridge overlays.* Washington, DC: TR News 247 Research Pays Off, Transportation Research Board.
4. VDOT. (2011). *Special provision for high friction epoxy aggregate surface treatment.* Richmond: Virginia Department of Transportation.
5. ASTM. (1994). *Specification for standard smooth tire for pavement skid-resistance tests,* ASTM E524-82. Philadelphia: American Society for Testing and Materials.
6. Sprinkel, M. M. (2011). *Rapid concrete bridge deck overlays,* ACI-SP-P10. Frontiers of Polymers in Concrete. Farmington Hills: American Concrete Institute.
7. Sprinkel, M. M. (2009). *Sampling and testing latex-modified concrete for permeability to chloride ion.* Transportation Research Record 2113. Washington, DC: Transportation Research Board.
8. AASHTO-AGC-ARTBA. (1995). *Guide specifications for polymer concrete bridge deck overlays.* Washington, D.C: AASHTO-AGC-ARTBA TASK FORCE 34.
9. ACI. (2007). *548.8-07: Specification for type EM (Epoxy multi-layer) polymer overlay for bridge and parking garage decks.* Farmington Hills: American Concrete Institute.
10. ACI. (2008). *548.9-08: Specification for type ES (Epoxy Slurry) polymer overlay for bridge and parking garage decks.* Farmington Hills: American Concrete Institute.
11. Sprinkel, M. M., Sellars, A., & Weyers, R. (1993). *Rapid concrete bridge deck protection, repair and rehabilitation.* Washington, DC: SHRP-S-344, Strategic Highway Research Program.
12. Sprinkel, M. M. (1997). *Nineteen year performance of polymer concrete bridge overlays.* Farmington Hills: American Concrete Institute, SP-169.
13. Sprinkel, M. M. (2003). *Twenty-five year experience with polymer concrete bridge deck overlays, SP-214-5 polymer in concrete: The first thirty years* (pp. 51–61). Farmington Hills: ACI.
14. FHWA. (2010). *Pavement friction management,* Technical Advisory T 5040.38. Washington, DC: U.S. Department of Transportation, Federal Highway Administration.
15. FHWA. (2013). *Pavement friction "key points".* Roadway Departure Safety, FHWA Safety Program website. http://safety.fhwa.dot.gov/roadway_dept/pavement/pavement_friction
16. Sprinkel, M. M., 2016, Kevin K. McGhee and Edgar David de León Izeppi, *Virginia's experience with high friction surface treatments,* TRR Transportation Research Board, Washington, DC.
17. DOT. (2016). *Road and bridge specifications.* Richmond: Virginia Department of Transportation.

# Chapter 3
# Feasibility Study of the Use of Polymer-Modified Cement Composites as 3D Concrete Printing Material

Kyu-Seok Yeon, Kwan-Kyu Kim, and Jaeheum Yeon

Recently, a 3D printing process, also known as additive manufacturing, has been introduced in the construction industry as a means of increasing productivity. To apply this technology to construction, concrete, one of the most representative construction materials, must be used as the 3D printing material. Concrete is one of the key elements associated with 3D printing in construction. However, the research on concrete that considers the unique characteristics of 3D printing technology is currently insufficient. There are few studies that address the use of polymer-modified cement composite as a 3D concrete printing material. Thus, this research used preliminary tests to evaluate the feasibility of using a polymer-modified cement composite as a 3D concrete printing material. The polymer-modified cement composite was prepared using not only Portland cement, fly ash, blast furnace (BF) slag, and silica fume, but also a water-soluble polymer. Four fresh properties – flowability, extrudability, open time, and buildability – were evaluated, and it was determined that the addition of the polymer effectively improved the performance of 3D concrete printing materials.

---

K. -S. Yeon
Department of Regional Infrastructure Engineering, Kangwon National University, Chuncheon, South Korea

K. -K. Kim
Jungbu Business Division, Korea Conformity Laboratories, Pocheon, South Korea

J. Yeon (✉)
Texas A&M University-Commerce, Department of Engineering and Technology, Commerce, TX, USA
e-mail: jaeheum.yeon@tamuc.edu

# 1 Introduction

The concepts of a three-dimensional (3D) concrete printing process originated in 1988 with the development of contour crafting by Behrokh Khoshnevis, at the University of Southern California in the USA [1]. After its development, the construction industry began considering ways of applying this technology. Hence, existing construction practices such as the labor-intensive cast-in-place method are changing to the automated construction of concrete structures [1]. Through this additive manufacturing technology, the possibility has emerged of a new construction paradigm that encompasses design flexibility, increased productivity, and the instant creation of new 3D printing materials [3]. This 3D concrete printing technology may one day enable the rapid construction of mega-scale concrete structures [3].

This 3D concrete printing technology can be broken down into three categories: 3D modeling software, concrete printer, and printing material (concrete). Among these three categories, there are currently three representative 3D printing types, based on the output modes. For example, contour crafting, discussed above, is based on extruding a cement-based paste [2]. A different 3D concrete printing technology developed by a research team at Loughborough University, in the UK, known as freeform construction, is based on extrusion of cement mortar [3]. Many researchers have focused on the 3D printer itself. Consequently, there is a significant body of research that addresses 3D printing machines. However, few studies associated with 3D printing materials currently exist, and to date there has been no study investigating the applicability of polymer-modified concrete with an additive manufacturing (extrusion) system. A polymer-modified cement composite was developed to examine the feasibility of using polymer-modified concrete as a 3D concrete printing material. Next, it was determined if a polymer-modified cement composite extruded by a nozzle system could satisfy the required properties. The aspects essential to printing 3D concrete structures include extrudability and buildability, which are significantly affected by flowability and open time; the hardened properties include compressive and flexural strengths [4]. Thus, this study considered the feasibility of a polymer-modified cement composite based on the properties required for 3D concrete printing material.

# 2 Background

## 2.1 3D Printing Material

3D printing technology, also known as additive manufacturing, produces a 3D object using an additive method. Layers accumulate vertically until the entire object is printed out [5]. There are many types of 3D printers, which can be categorized by the type of material used, such as liquids, solids, powders, or sheets [6]. Concrete, which

**Fig. 3.1** 3D printed house [7]

**Fig. 3.2** Geometric building component [8]

is a representative construction material, is of the solid type. The examples presented below demonstrate that 3D concrete printing technology can be applicable in construction projects such as residences or geometric building components, as shown in Figs. 3.1 and 3.2.

## 2.2 Applicable 3D Concrete Printing Technologies

### 2.2.1 Contour Crafting by the University of Southern California, USA

Behrokh Khoshnevis, of the University of Southern California, developed the contour crafting process in 1988. Contour crafting is controlled by a computer-based gantry system that prints out the walls of a house without formwork. Rather, a cement-based paste is employed according to a layer-by-layer approach [1].

### 2.2.2 Freeform Construction by Loughborough University, UK

Freeform Construction was developed by a research team at Loughborough University. This 3D printing technology uses cement mortar to build large-scale concrete components for construction [3].

## 2.3 Properties of Polymer-Modified Concrete (PMC)

Polymer latexes or dispersions are widely used in the development of polymer-modified concrete. Specifically, the most representative polymeric modifiers are styrene–butadiene rubber, polyacrylic ester, and ethylene–vinyl acetate copolymers [9].

### 2.3.1 Workability

The workability of PMC is better than the workability of ordinary Portland cement because the size of the surfactant's particles in the polymer latex is more consistent [9].

### 2.3.2 Water Retention

The PMC's water retention ability is better than that of Portland cement because of the hydrophilic colloidal properties of the polymer latex [9].

### 2.3.3 Setting Behavior

PMC usually takes more time to set than Portland cement. However, this set time can be controlled by the type of polymer and polymer-to-cement ratio [9].

### 2.3.4 Strength

PMC's tensile and flexural strength levels are higher than they are for Portland cement because the bond strength between the cement and aggregates is enhanced by the polymer [9].

### 2.3.5 Adhesion

PMC has a stronger adhesion than Portland cement. Hence, it binds tightly to existing structures [9].

### 2.3.6 Deformability and Shrinkage

PMC usually has a larger ductility than does Portland cement. Also, PMC's shrinkage tends to decrease when the polymer-to-cement ratio is increased [9].

## 3 Experimental Program

### *3.1 Materials and Mix Design*

To investigate the feasibility of using polymer-modified cement composite as a 3D concrete printing material, test substances were prepared by adding admixtures to Portland cement, fly ash, B F slag, and silica fume. From an economic perspective, 60% to 75% of concrete volume is filled with aggregates because the unit cost of cement is higher [10]. However, inexpensive Portland cement must be used as the primary material for printing out concrete structures because aggregates cannot be added to smoothly extruded fresh concrete through the nozzle of a 3D printer; the concrete layer will not have a smooth surface. Therefore, the properties of the concrete must be modified to develop a new concrete material appropriate for 3D printers. In this study, while Portland cement was used as the primary material, a flow rate that was hoped to develop optimal flowability was applied as a mixing reference. Tests were repeated to determine the flow rate at which optimal flowability could be developed. The mixing ratios of the materials used in the experiments are shown in Table 3.1.

**Table 3.1** Mix Proportions of Polymer-Modified Cement Composites (by $kg/m^3$)

| P/C ratio | Polymer | Cement | Fly ash | B F slag | Silica fume | Sand | Water |
|---|---|---|---|---|---|---|---|
| 0 | 0 | 748 | 187 | 224 | 97 | 673 | 371 |
| 5 | 37 | 748 | 187 | 224 | 97 | 673 | 334 |
| 10 | 75 | 748 | 187 | 224 | 97 | 673 | 295 |
| 15 | 112 | 748 | 187 | 224 | 97 | 673 | 258 |

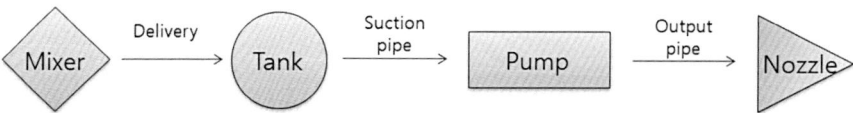

**Fig. 3.3** Printing system

## 3.2  3D Concrete Printer

Gantry and robotic arm systems are representative types of 3D concrete printers that extrude fresh concrete from a nozzle. The nozzle of a 3D printer is moved along $x$-, $y$-, and $z$-axes to build a concrete structure based on a 3D model created by a designer and 3D modeling software. However, each part of the 3D concrete printer may be handled separately. For example, a concrete mixer, tank, pump, and nozzle may all be attached to 3D concrete printers. The primary elements are the concrete pump and nozzle; the pump is usually of either a peristaltic or screw type, and the diameter of the nozzle is between 4 mm and 20 mm [4, 11–13]. Figure 3.3 illustrates a printing sequence for the 3D concrete printer used in this study, wherein the cross-sectional area of the nozzle hole was 3.5 $cm^2$.

## 3.3  Test Procedures

### 3.3.1  Flowability

The flowability was tested according to ASTM C1437-15 (the standard test method for the flow of hydraulic cement mortar) [14].

### 3.3.2  Extrudability

The extrudability was evaluated by measuring the length of the filament extruded in 1 min.

### 3.3.3  Open Time

The open time was assessed by measuring the initial and final setting times, according to ASTM C191-13 (the standard test methods for the time of setting of hydraulic cement by Vicat needle) [15].

### 3.3.4 Buildability

The buildability was estimated by measuring the average number of layers and their height, while keeping the layer length at 30 cm.

## 4 Results and Discussion

### *4.1 Flowability*

Flowability [4] is an indicator of the level of ease of moving a printing material mixed in a mixer from the tank to the nozzle. It is also known as workability [11] or printability [6]. Methods for testing the flowability include slump, flow, compacting factor, and Vebe time tests. However, according to BS 1881: Part 102, 1983 [16], the flow test method is the most appropriate means of examining high workability. Therefore, in this study, the flowability of the polymer-modified cement composite was evaluated according to the flow test method. Figure 3.4 shows the experimental procedures applied to determine the appropriate flowability. The results indicate that the most suitable flowability of polymer-modified cement composite used as a 3D printing concrete material had a flow rate of 70%.

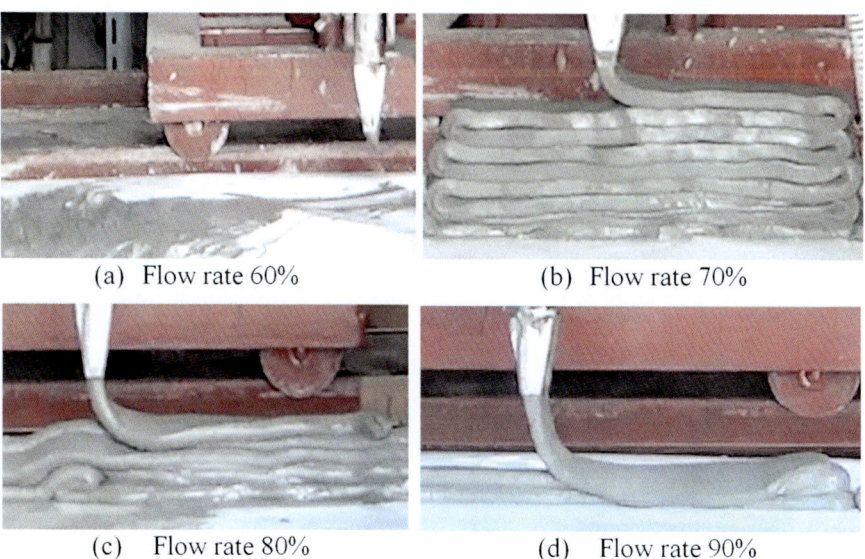

**Fig. 3.4** Finding an appropriate test for flow rate: (**a**) flow rate 60%, (**b**) flow rate 70%, (**c**) flow rate 80%, (**d**) flow rate 90%

## 4.2 Extrudability

Extrudability [4, 11], also referred to as pumpability [17], represents the degree of smooth extrusion of a printing material from a nozzle. A 3D printed work that uses a filament requires that the melting filament be continuously extruded as fresh material flows smoothly through the pump, from the tank to the nozzle [13]. In this study, the extrudability of the polymer-modified cement composite was evaluated by measuring the concrete extrusion velocity through a nozzle with a cross-sectional area of 3.5 $cm^2$. The extrusion velocity was calculated by dividing the continuous length of ten concrete layers (wherein a single layer was 30 cm long) by the time taken to complete the layering work. The results indicate that the extrusion velocity decreased when a polymer was added, which may have been because of the viscosity of the added polymer.

## 4.3 Open Time

Open time refers to the time the extruded and layered filament requires to harden [4]. The open time of cementitious materials, closely related to the setting time, is generally measured using the Vicat needle test method. However, open time may be evaluated by measuring the workability change over time in a slump test or via the shear vane test [13]. Hence, in this study, the Vicat needle test method was employed to evaluate the open time of the polymer-modified cement composite. This method is most extensively used for setting the test times of cementitious materials. The results indicate that both the initial and final setting times increased as the polymer content rose.

## 4.4 Buildability

Buildability is the number of layers of the filament extruded from the nozzle; it represents how high the filament may be laminated without deformation of the lower layers [4]. In this study, while the layer length of the polymer-modified cement composite was fixed at 30 cm, it was layered repeatedly and continuously over the layer length until the stacked layers collapsed. Figure 3.5 shows the buildability test. The results indicate that the surface was smooth and the filament thickness was uniform when the polymer content was high. In addition, 13–15 layers could be formed prior to collapse, to a layer height of 16–20 cm. This result indicates that the buildability of the polymer-modified cement composite increased as the polymer content increased.

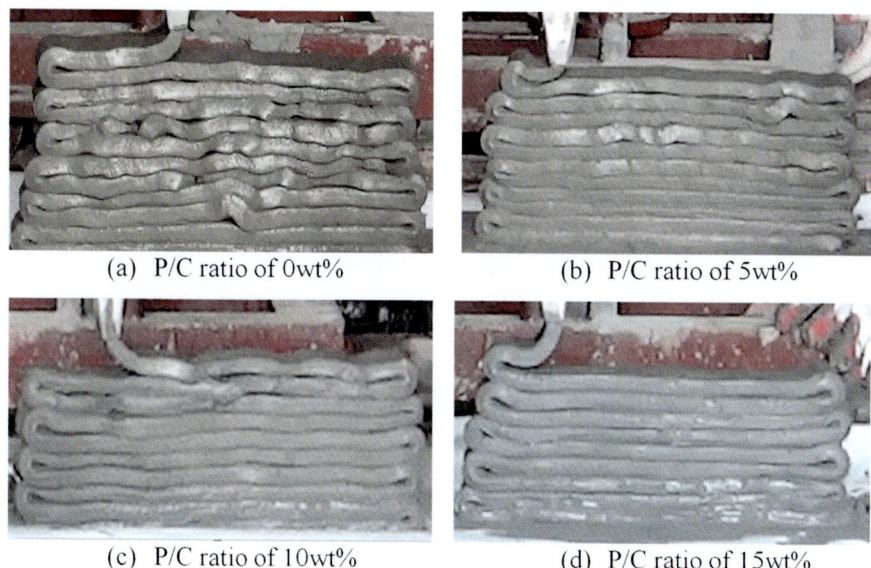

**Fig. 3.5** Buildability tests: (**a**) P/C ratio of 0 wt. %, (**b**) P/C ratio of 5 wt. %, (**c**) P/C ratio of 10 wt. %, (**d**) P/C ratio of 15 wt. %

## 5 Conclusions

This research, as a fundamental study of the development of polymer-modified cement composites for 3D concrete printing technology, was conducted by focusing on fresh properties such as flowability, extrudability, open time, and buildability. The results of this work indicate that the addition of a water-soluble polymer may improve the performance of concrete printing material. Further studies are needed on the hardened properties of this type of composites, such as the compressive, flexural, and adhesive strengths, in order to verify the effect of adding a polymer to concrete printing material. Finally, this study found that polymer-modified cement composite has the potential to be used as a 3D concrete printing material.

**Acknowledgments** This research was supported by a national R&D program via the National Research Foundation of Korea (NRF), and funded by the Ministry of Science & ICT (NRF-2016R1D1A1B03930763).

## References

1. Hwang, D., & Khoshnevis, B. (2005). An innovative construction process: Contour crafting. *22th international symposium on automation and robotics in construction,* Ferrara, Italy.

2. Khoshnevis, B., & Bekey, G. (2002). Automated construction using contour crafting: Applications on earth and beyond. *19th international symposium on automation and robotics in construction,* Gaithersburg, MD, USA.
3. Buswell, R. A., Soar, R. C., Gibb, A. G. F., & Thorpe, A. (2007). Freeform construction: Mega-scale rapid manufacturing for construction. *Automation in Construction, 16,* 224–231.
4. Malaeb, Z., Hachem, H., Tourbah, A., Maalouf, T., Zarwi, N. E., & Hamzeh, F. (2015). 3D concrete printing: Machine and design. *International Journal of Civil Engineering and Technology, 6*(6), 14–22.
5. Beck, J., Fritz, B., Siewiorek, D. & Wiess, L. (1992). Manufacturing mechatronics using thermal spray shape deposition. *Proceedings of the 1992 Solid Freeform Fabrication Symposium,* (pp. 272–279) Austin, TX, USA.
6. Gosselin, C., Duballet, R., Roux, P., Gaudillière, N., Dirrenberger, J., & Morel, P. (2016). Large-scale 3D printing of ultra-high performance concrete: A new processing route for architects and builders. *Materials and Design, 100,* 102–109.
7. Tess (2017). US army 3D prints 512-sq-ft barracks using concrete AM tech developed in collaboration with NASA. http://www.3ders.org/articles/20170824-us-army-3d-prints-512-sq-ft-barracks-using-concrete-am-tech-developed-in-collaboration-with-nasa.html
8. Bensoussan, H. (2016). Interviewing XtreeE: 3D printing concrete to push the limits of construction. https://www.sculpteo.com/blog/2016/12/07/interviewing-xtreee-3d-printing-concrete-to-push-the-limits-of-construction/
9. Chandra, S., & Ohama, Y. (1994). *Polymers in concrete* (pp. 111–135). CRC Press.
10. Ford, G. S., & Souwak, L. J. (2017). *The economic impact of the natural aggregates industry: A national, state, and county analysis* (pp. 1–24). Phoenix Center for Advanced Legal & Economic Public Policy Studies.
11. Hager, I., Golonka, A., & Putanowicz, R. (2016). 3D printing of buildings and building components as the future of sustainable construction. *Procedia Engineering, 151,* 292–299.
12. Cesaretti, G., Dini, E., Kestelier, X. D., & Colla, V. (2014). Building components for an outpost on the lunar soil by means of a novel 3D printing technology. *Acta Astronautica, 93,* 430–450.
13. Le, T. T., Austin, S. A., Lim, S., Buswell, R. A., Law, R., Gibb, A. G. F., & Thorpe, T. (2012). Mix design and fresh properties for high-performance printing concrete. *Materials and Structures, 45,* 1221–1232.
14. ASTM C1437-15. (2015). *Standard test method for flow of hydraulic cement mortar.* West Conshohocken, PA: ASTM International.
15. ASTM C191-13. (2013). *Standard test methods for time of setting of hydraulic cement by vicat needle.* West Conshohocken: ASTM International.
16. BS. (1881). *Part 102 (1983). Method for determination of slump.* London: BS Standard.
17. Warszawski, A., & Navon, R. (1998). Implementation of robotics in buildings: Current status and future prospects. *Journal of Construction Engineering and Management, 124,* 31–41.

# Chapter 4
# Experimental Analysis and Micromechanics-Based Prediction of the Elastic and Creep Properties of Polymer-Modified Concrete at Early Ages

Luise Göbel, Bernhard Pichler, and Andrea Osburg

Polymer-modified concrete (PCC) has been used since the 1980s mainly for repair and restoration. Nowadays, it is also increasingly applied in construction. The desirable future integration of PCC into guidelines and standards requires a reliable mathematical description of the mechanical behavior of PCC. Notably, PCC exhibits less elastic stiffness and a more pronounced creep activity compared to conventional concrete. This contribution presents a combined experimental-computational study concerning early-age mechanical properties of PCC. Experimental characterization comprised 3 min-long creep tests which were performed every hour, spanning material ages from 1 day after production up to 8 days. This allowed for a quasi-continuous quantification of the early-age evolutions of the elastic stiffness and of the non-aging creep properties. As for computational modeling, an existing multiscale model for the elastic stiffness of concrete is extended toward the consideration of polymers. It is shown that the extended model can reliably describe the elastic stiffness of PCC, provided that entrapped air is adequately considered.

L. Göbel (✉)
Bauhaus-Universität Weimar, Research Training Group 1462, Weimar, Germany
e-mail: luise.goebel@uni-weimar.de

B. Pichler
TU Wien – Vienna University of Technology, Institute for Mechanics of Materials and Structures, Vienna, Austria

A. Osburg
Bauhaus-Universität Weimar, F. A. Finger-Institute of Building Engineering Materials, Weimar, Germany

# 1 Introduction

The increasing use of polymer-modified concrete (PCC) in construction motivates its future integration into guidelines and standards. To this end, the mechanical behavior must be not only well understood, but it must be also predictable. Studies concerning the mechanical properties of well-hydrated (and, hence, *mature*) PCC have revealed a lower stiffness and a more pronounced creep activity in comparison to conventional concrete [1, 2]. Less information is available concerning the behavior of PCC at early ages, during which material properties evolve significantly because of the ongoing hydration process, i.e., the chemical reaction between cement and water. The present contribution focuses on the early-age elastic and creep behavior of PCC. A recently developed test protocol comprising ultrashort and hourly repeated creep tests [3] is applied to PCC. It provides quasi-continuous access to the evolution of elastic and creep properties. Experimental data are used to test the predictive capability of a multiscale elasticity model for PCC.

# 2 Experimental Campaign

## 2.1 Materials

Laboratory concretes were produced with Portland cement CEM I 42.5 N provided by Lafarge (Austria), distilled water, and oven-dried quartz gravel "Pannonia Kies" (Austria) with a maximum diameter of 8 mm. The initial water-to-cement mass ratio ($w/c$) amounted to 0.40 and the initial aggregate-to-cement mass ratio ($a/c$) to 3.0. Three different concrete mixes were produced: one reference mix without polymers and two PCC. The latter exhibited an initial polymer-to-cement mass ratio ($p/c$) amounting to 0.10. The two PCC differed in terms of the polymer type; see Table 4.1.

**Table 4.1** Characteristics of the polymers

| Characteristic | Polymer P1 | Polymer P2 |
|---|---|---|
| Delivery form | Dispersion | Dispersion |
| Main constituents | Styrene, acrylic acid ester | Styrene, butadiene, acrylonitrile |
| Minimum film formation temperature [°C] | 30 | 16 |
| Mean particle size [μm] | 0.19 | 0.20 |

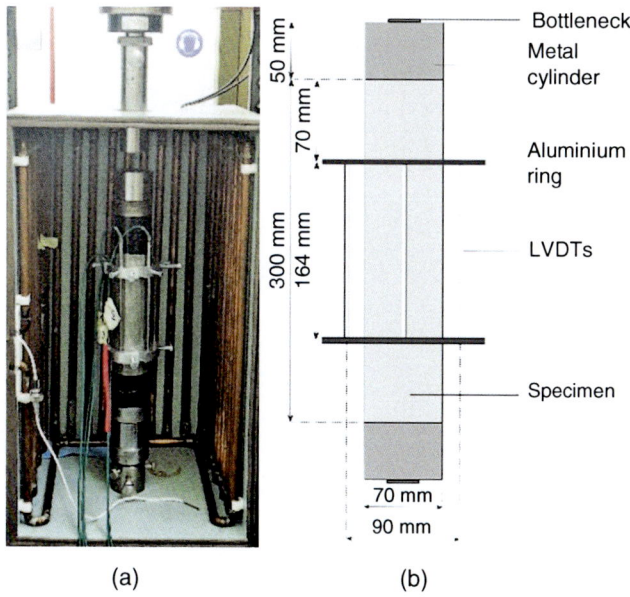

**Fig. 4.1** Test setup placed inside a temperature-controlled chamber (**a**) and schematic sketch (**b**) including the specimen, LVDTs, and metal cylinders

## 2.2 Ultrashort Creep Tests

Ultrashort creep tests were performed following the test protocol developed by Irfan-ul-Hassan et al. [3]. Cylindrical specimens with a diameter of 7 cm and a height of 30 cm were used. They were subjected – once per hour – to uniaxial compressive creep tests, from the first day after production up to material ages of 8 days. In total, 168 creep tests were carried out on each specimen. A testing machine of type Zwick/Roell Z050 (Germany) was used. It was placed inside a temperature-controlled chamber in order to guarantee a constant temperature of 20 °C; see Fig. 4.1. Five inductive displacement transducers (LVDT) were mounted to the specimens. They measured length changes of the specimens. Metal cylinders were put on top and below the specimens. They ensured a central (and, hence, *uniform*) loading of the specimens.

Focusing on *linear* creep of *undamaged* PCC specimens motivated the restriction of the applied load levels to a maximum of 7.5% of the estimated compressive strength at the time instant of testing. The applied load plateaus were updated regularly to account for the ongoing hydration, and the corresponding gain of the compressive strength, of the specimens. The duration of each creep test was 3 min, excluding the very short loading and unloading phases; see Fig. 4.2. This implies that the microstructure remained in very good approximation unaltered during every single creep test. Subsequent creep tests, in turn, referred to different microstructures because of the ongoing hydration process.

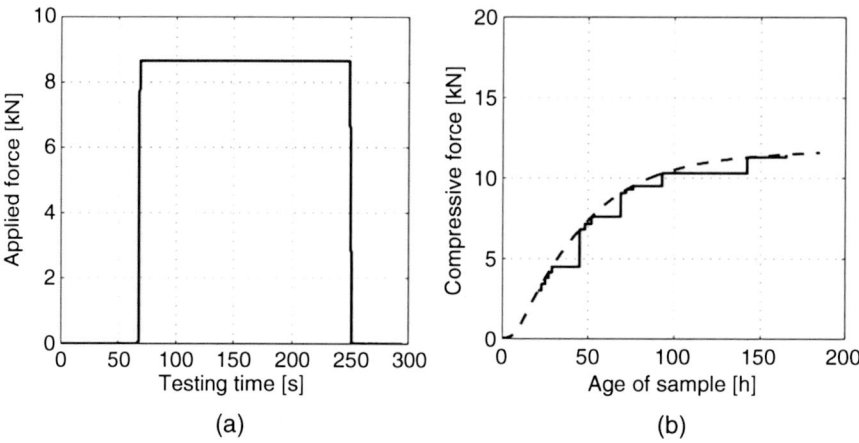

**Fig. 4.2** Loading regime of a single creep test applied to the PCC at sample ages of 75 h (**a**) and regularly updated load plateau values (**b**)

The test evaluation follows Irfan-ul-Hassan et al. [3], i.e., the measured deformation behavior of the tested materials is represented using a power-law creep function of the form

$$J^{\text{exp}}(t-\tau) = \frac{1}{E^{\text{exp}}} + \frac{1}{E_c^{\text{exp}}}\left(\frac{t-\tau}{t_{\text{ref}}}\right)^{\beta^{\text{exp}}} \quad (4.1)$$

where $t$ is the chronological time, $\tau$ the time instant of loading, and $t_{\text{ref}}$ a reference time amounting to 86,400 s, $E^{\text{exp}}$ the elastic modulus, $E_c^{\text{exp}}$ the creep modulus, and $\beta^{\text{exp}}$ the power-law exponent. At first, the elastic modulus is derived from the unloading path of the measured readings; see [3] for further details. Subsequently, the creep modulus and the power-law exponent are identified simultaneously by minimizing the sum of the squared errors between the experimentally measured and the modeled creep strain evolutions.

Quasi-isothermal differential calorimetry is used for quantifying the time evolution of the hydration degree. The latter is defined as the mass of hydrated cement over the mass of the initially used cement. This way, a specific value of the hydration degree could be assigned to each of the creep tests.

## 2.3 Elastic and Creep Properties of Polymer-Modified Concretes

The results obtained from the evaluation of the ultrashort creep tests involving the unmodified reference concrete and two polymer-modified concretes are summarized in Fig. 4.3. Young's moduli of all three tested concretes evolve in very good

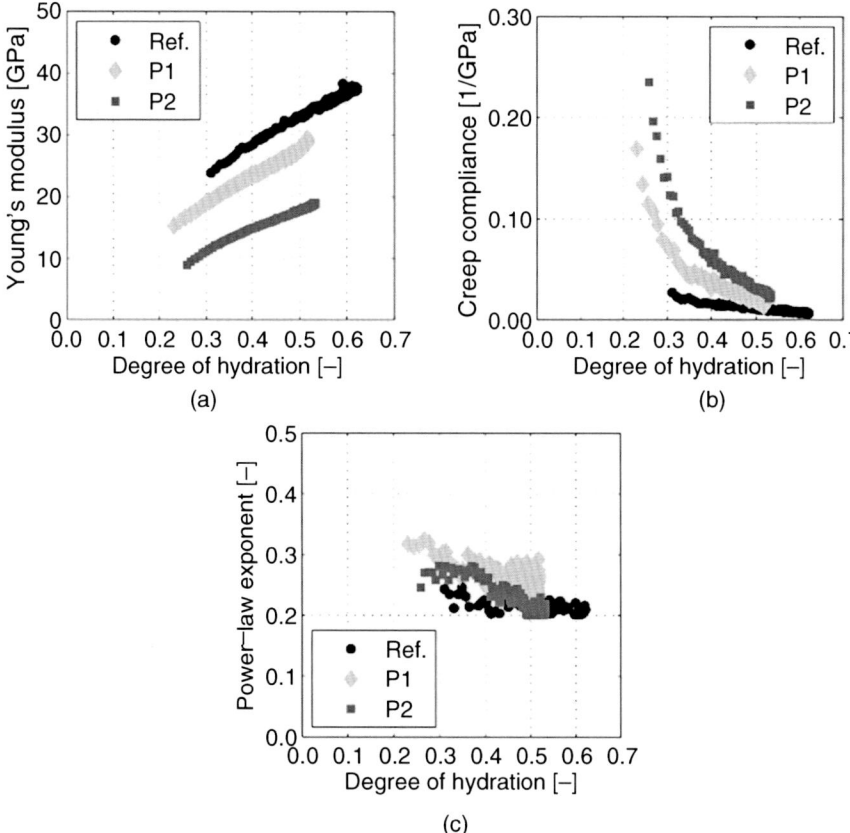

**Fig. 4.3** Evolution of Young's modulus (**a**), the creep modulus (**b**), and the power-law exponent (**c**) of different concrete mixtures with $w/c = 0.40$

approximation linearly with the degree of hydration. The polymer-modified concretes P1 and P2 exhibit a smaller stiffness than the unmodified reference material. This is a consequence of (i) the retarded hydration process in comparison to conventional concrete [4], (ii) the low stiffness of the polymers, and (iii) possibly also an increased entrapped air volume fraction. Concrete P1 exhibits a larger Young's modulus than the concrete P2. This suggests that the entrapped air porosity of P2 is significantly larger than that of P1.

The creep compliance, i.e., the inverse of the creep modulus, of the two PCC is larger than that of the unmodified reference concrete. These results quantify the more pronounced creep activity of the PCC. This expected effect results (i) from possible sliding effects of the polymers within the binder matrix and (ii) from the increased stress concentration into cement hydrates resulting from the entrapped air. The power-law exponent is virtually the same for all three concretes.

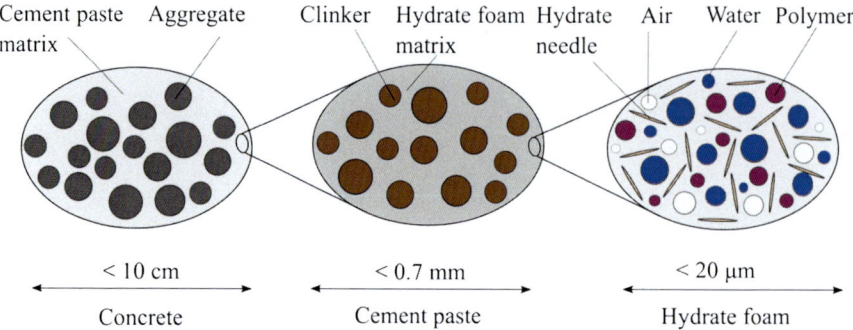

**Fig. 4.4** Micromechanical representation of polymer-modified concrete

## 3 Multiscale Modeling Based on Continuum Micromechanics

### 3.1 Homogenization of Elastic Stiffness of PCC

Multiscale modeling of the early-age elastic stiffness of PCC is carried out in the framework of continuum micromechanics. This is appealing because essential microstructural characteristics of hierarchically organized microheterogeneous materials can be accounted for in a straightforward manner. In addition, the modeling is based on representative volume elements complying with the concept of separation of scales [5].

Following Pichler and Hellmich [6], concrete is represented by a three-scale homogenization scheme; see Fig. 4.4. Concrete is considered to be a matrix-inclusion composite, where aggregates are embedded in a homogenized cement paste matrix. Cement paste is considered – at the next smaller scale of observation – to be a matrix-inclusion composite, where cement clinker particles are embedded in a homogenized hydrate foam matrix. At the next smaller scale of observation, the hydrate foam is considered to exhibit a polycrystalline microstructure containing spherical water and air pores (capillary pores) as well as hydrate gel needles. The latter are isotropically oriented in all space directions.

As for PCC, the described microstructural representation of concrete is extended. Due to the small sizes of the polymer particles, they are introduced as an additional spherical phase at the microscale of the hydrate foam; see Fig. 4.4. Consideration of polymers at the same scale as the hydrate gel needles is also motivated by the fact that polymers adsorb on the surfaces of hydration products.

Upscaling of the elastic stiffness from the microstructure of the hydrate foam all the way up to the scale of PCC is performed according to the approach described in [6]. In this context, it is noteworthy that the multiscale model requires four types of input parameters: quantitative information on the *shape* of the individual material phases and on their *interaction*, as well as quantitative information on their *elastic*

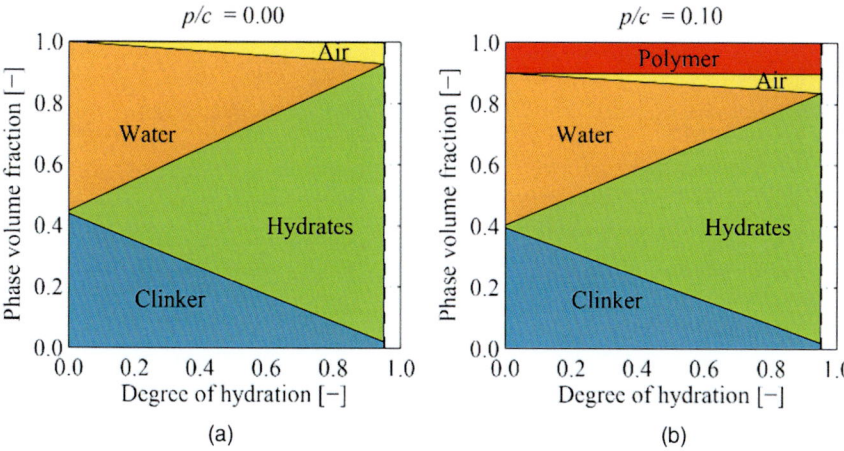

**Fig. 4.5** Volume fractions of cement pastes with $w/c = 0.40$ as a function of the hydration degree: (**a**) cement paste without polymers, (**b**) polymer-modified cement paste with $p/c = 0.10$

*stiffness constants* and on the hydration-induced evolutions of their *volume fractions*. As for the latter, Powers' model [7] is extended toward the consideration of a constant polymer volume fraction; see Fig. 4.5. The elastic stiffness constants of the hydrate gel needles, the clinker grains, and the aggregates can be found in the literature; see, e.g., [6]. The elastic stiffness of the used polymers, in turn, was determined by means of ultrasonic measurements [8].

## 3.2 Validation

Model-predicted evolutions of Young's modulus of the two tested PCC are compared with their experimentally measured counterparts; see Fig. 4.6. The multiscale model predicts the stiffness evolution of the concrete modified with polymer P1 quite reliably, both qualitatively and quantitatively. However, Young's modulus of the concrete modified with polymer P2 is overestimated significantly.

The stiffness evolution of concrete P2 is overestimated because the model does not consider entrapped air pores. The latter were very likely introduced into the specimens during mixing of the raw materials. This entrapped air porosity is estimated to amount to some 11%, based on the measured mass density of the tested PCC specimen and on its particle density. Analogously, the total porosities of the reference concrete specimen and the specimen P1 amounted to 10.5% and 11.3%, respectively. Thus, the additional entrapped air porosity in concrete P1 is almost as low as in the reference mix and does not require to be modeled explicitly. However, consideration of different entrapped air porosities from 5% to 20% yields improved

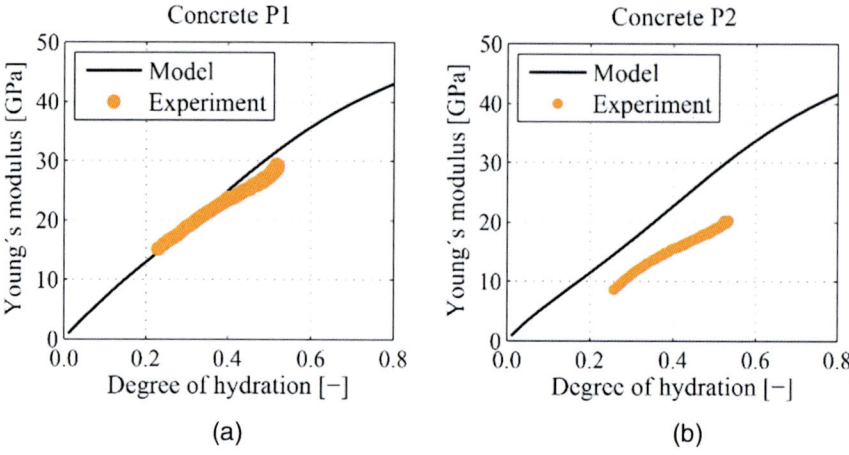

**Fig. 4.6** Comparison of model-predicted and experimentally measured stiffness evolutions of concrete P1 (**a**) and concrete P2 (**b**)

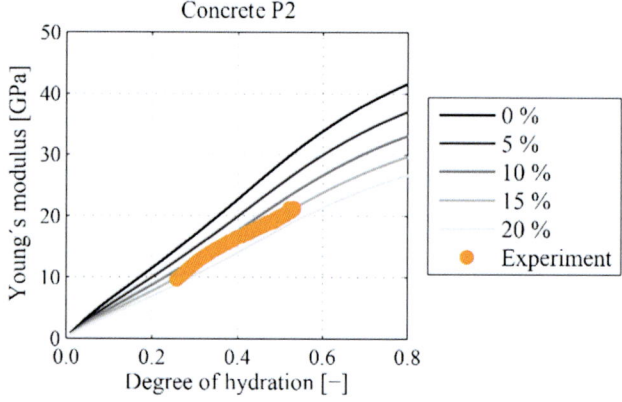

**Fig. 4.7** Consideration of an additional entrapped air porosity for concrete P2

model-predicted stiffness evolutions of concrete P2; see Fig. 4.7. The higher the additional air void content, the lower is the predicted Young's modulus of concrete.

## 4 Discussion

Concerning the micromechanical model of PCC, illustrated in Fig. 4.4, it is important to note that further ways of introducing the polymer phase are conceivable. First, the scale at which the polymers are inserted could be altered. When the polymers are dispersed, they exhibit particle sizes in the range of a few hundreds of nanometers.

4 Experimental Analysis and Micromechanics-Based Prediction of the... 45

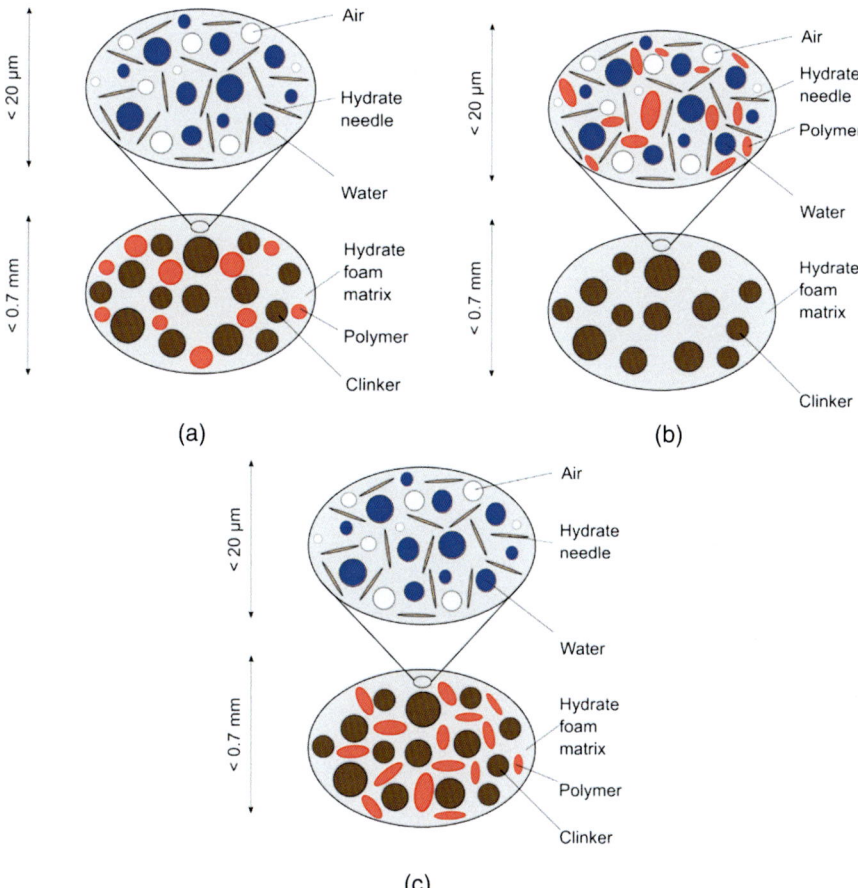

**Fig. 4.8** Possible micromechanical representations of PCC: polymers as spheres at the cement paste scale (**a**), polymers as cylindrical inclusions at the cement paste scale (**b**) as well as at the hydrate foam scale (**c**)

However, within the hardened cementitious matrix, the polymers may coalesce and assemble to larger compounds [9]. This motivates the consideration of the polymers at the scale of cement paste, one scale above the hydrate foam (Fig. 4.8a).

Furthermore, the variation of the form of the polymer inclusions resulting from the film formation process could be accounted for. For that, the polymers are modeled as ellipsoidal inclusions instead of spheres; see Fig. 4.8b, c.

The modeling of polymers as spheres and cylinders refers to two limit cases: A sphere represents a prolate form with an aspect ratio of one, while cylinders can be seen as prolates with an infinite aspect ratio; see also [10]. Thus, considering the polymers as both cylinders and spheres ensures to capture these two bounds.

The multiscale model predictions for PCC are derived using the micromechanical representations illustrated in Figs. 4.4 and 4.8. The model-predicted stiffness

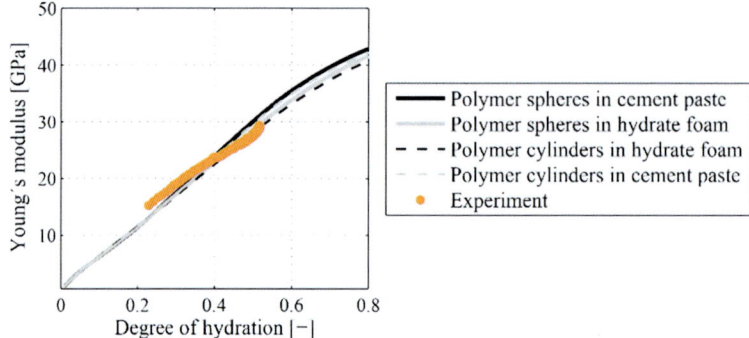

**Fig. 4.9** Model-predicted stiffness evolutions considering different micromechanical representations of PCC, compared with experimental results

evolutions are compared to the experimentally determined Young's moduli of specimen P1; see Fig. 4.9.

The influences of both the shape of the polymer and the scale at which the polymer is introduced on the elastic Young's modulus of concrete are rather small. Particularly, the very early hydration stages are hardly affected by the micromechanical representation. Small differences between the model predictions are observable for hydration degrees larger than 0.5.

## 5 Conclusions

Homogenization methods of continuum micromechanics were shown to be useful when it comes to the estimation of the early-age evolution of the elastic stiffness of PCC. Herein, it was found (i) that the *same* dosage of *different* polymers may result in significantly different early-age stiffness evolutions and (ii) that these stiffness differences may be explained by different amounts of entrapped air.

It can be concluded that future inclusion of PCC into standards and recommendations shall take the specific polymer type into consideration. Special attention deserves the interaction of the polymers with the other raw materials, which may result in entrapping a significant amount of air during the production of concrete.

**Acknowledgement** This research is supported by the German Research Foundation (DFG) via the Research Training Group 1462, which is gratefully acknowledged. This work was further supported by a short-term scientific mission (STSM) grant from COST Action TU1404 "Towards the Next Generation of Standards for Service Life of Cement-based Materials and Structures," which is gratefully acknowledged.

# References

1. Al-Zahrani, M. M., Maslehuddin, M., Al-Dulaijan, S. U., & Ibrahim, M. (2003). Mechanical properties and durability characteristics of polymer- and cement-based repair materials. *Cement and Concrete Composites, 25*, 527–537.
2. Wang, R., Wang, P.-M., & Li, X.-G. (2005). Physical and mechanical properties of styrene–butadiene rubber emulsion modified cement mortars. *Cement and Concrete Research, 35*, 900–906.
3. Irfan-ul-Hassan, M., Pichler, B., Reihsner, R., & Hellmich, C. (2016). Elastic and creep properties of young cement paste, as determined from hourly repeated minute-long quasi-static tests. *Cement and Concrete Research, 82*, 36–49.
4. Kong, X., Emmerling, S., Pakusch, J., Rueckel, M., & Nieberle, J. (2015). Retardation effect of styrene-acrylate copolymer latexes on cement hydration. *Cement and Concrete Research, 75*, 23–41.
5. Drugan, W. R., & Willis, J. R. (1996). A micromechanics-based nonlocal constitutive equation and estimates of representative volume element size for elastic composites. *Journal of the Mechanics and Physics of Solids, 44*, 497–524.
6. Pichler, B., & Hellmich, C. (2011). Upscaling quasi-brittle strength of cement paste and mortar: A multi-scale engineering mechanics model. *Cement and Concrete Research, 41*, 467–476.
7. Powers, T. C., & Brownyard, T. L. (1948). Studies of the physical properties of hardened Portland cement paste. *Research Laboratories of the Portland Cement Association Bulletin, 43*, 101–132.
8. Göbel, L., Pichler, B., & Osburg, A. (2017). Early-age experimental characterization and semi-analytical modeling of elasticity and creep of polymer-modified cement pastes. In *Second international RILEM conference on early-age cracking and serviceability in cement-based materials and structures*; 12.09.–14.09.2017. Brussels.
9. Afridi, M., Ohama, Y., Demura, K., & Iqbal, M. Z. (2003). Development of polymer films by the coalescence of polymer particles in powdered and aqueous polymer-modified mortars. *Cement and Concrete Research, 33*, 1715–1721.
10. Pichler, B., Hellmich, C., & Eberhardsteiner, J. (2009). Spherical and acicular representation of hydrates in a micromechanical model for cement paste: Prediction of early-age elasticity and strength. *Acta Mechanica, 203*, 137–162.

# Chapter 5
# Durability and Long-Term Performance of Fiber-Reinforced Polymer as a New Civil Engineering Material

Brahim Benmokrane and Ahmed H. Ali

In recent years, fiber-reinforced polymer (FRP) composites have been increasingly used for civil engineering applications such as columns, beams, and slabs to all-composite bridge decks. However, the durability of FRP, especially under harsh environmental conditions, is now recognized as the most critical topic of research. The lack of a comprehensive database on durability of FRP materials makes it difficult for the practicing civil engineer and designer to use FRP composites on a routine basis. The current paper presents the most significant research work conducted and published on durability performance of FRPs, as internal reinforcement, in the concrete members. Its durability has been extensively investigated in the last two decades. A comprehensive review of the literature, including degradation mechanisms, accelerated tests for long-term performance, and the effects of environment parameters on the durability of FRPs will be presented and discussed. In addition, proposed service-life prediction models for FRP materials will be reviewed.

## 1 Introduction

Reinforced concrete (RC) structures often deteriorate due to the corrosion of steel reinforcement when exposed to corrosive media, as shown in Fig. 5.1a. Repair and restoration costs in the USA, Canada, and the majority of European countries account for a high percentage of total national expenditures on infrastructure. The United States (US) Federal Highway Administration (FHWA) reported that eliminating the nation's bridge deficient backlog by 2028 would require an investment of

---

B. Benmokrane (✉) · A. H. Ali
University of Sherbrooke, Sherbrooke, QC, Canada
e-mail: Brahim.Benmokrane@usherbrooke.ca

© Springer International Publishing AG, part of Springer Nature 2018
M. M. Reda Taha (ed.), *International Congress on Polymers in Concrete (ICPIC 2018)*, https://doi.org/10.1007/978-3-319-78175-4_5

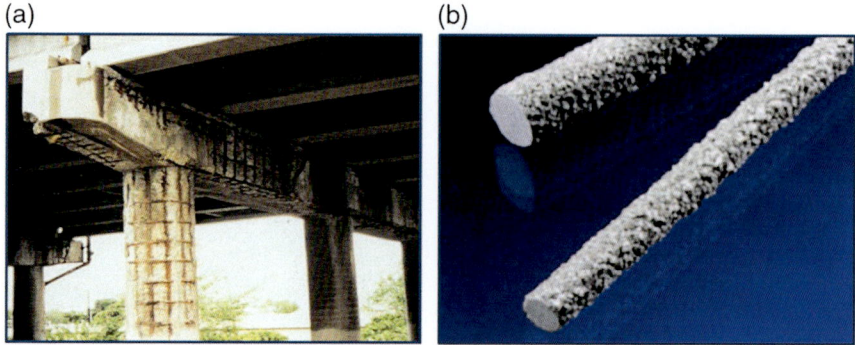

**Fig. 5.1** (a) Steel corrosion in concrete bridge and (b) sand-coated GFRP bars

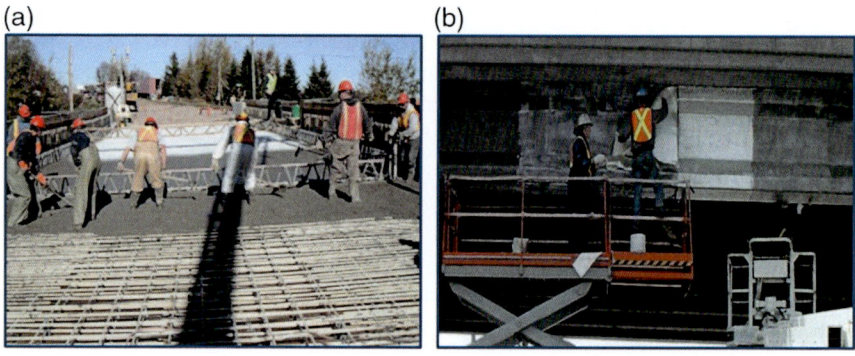

**Fig. 5.2** (a) Deck slab reinforced with GFRP. (b) Installation of GFRP sheet

$20.5 billion annually because of corroded steel reinforcement [1]. Canada's deficit for its municipality infrastructure, which represents 70% of the country's total infrastructure, was estimated to be $60 billion in 2004 and is expected to grow at $2 billion per year. A relatively recent solution, to overcome the corrosion of steel reinforcement, is the fiber-reinforced polymer (FRP) reinforcement as conventional construction materials. FRPs have been increasingly used in civil engineering for the last 20 years, due to their corrosion resistance, light weight, and high strength [2]. FRP composite materials, consisting of strong and stiff fibers—such as glass, carbon, and aramid—impregnated in polymeric resins (such as polyesters, vinyl esters, and epoxies), are being increasingly used in civil infrastructure applications. Additionally, FRP reinforcement has the durability qualities and long-term performance especially in harsh environments which encourage designers and civil engineers to use the FRP composite materials in concrete structures. Figures 5.1b and 5.2 show typical GFRP reinforcement and its use in a concrete deck slab, respectively. In fact, despite certifying the structural integrity of composite bars, the industry still needs long-term durability data on the material to gain great potential for the integration of the FRP into the infrastructure applications. Considerable work has been conducted on the durability of FRP composites originating from chemistry, aerospace, and military industries. Nevertheless, limited amount of research work was conducted on the durability of FRP composites in civil infrastructure applications. The current paper presents the recent research studies on the durability performance of FRP composites used as internal and external reinforcement for concrete structures. The main investigated topics are related to the effects of various environmental conditions, such as alkalinity, moisture creep, and relaxation, on FRP reinforcement. Most of these parameters have been extensively investigated and the results can be found elsewhere [1, 2, 6].

## 2 Durability Concerns

In spite of the fact that FRP materials are noncorrodible in nature and have higher tensile strength relative to conventional steel reinforcing bars, FRP performance can degrade under the effect of harsh environments, leading to the reduction of the strength properties of FRP materials. The literature (Robert [3]; Robert and Benmokrane [4]; Benmokrane [5]) indicates that FRP performance deteriorates due to the effect of sustained loading, moisture, variations of temperatures, or alkalinity. The degree of deterioration is dependent on different parameters such as type of fiber and resin, manufacturing process (curing rate, thermal microcracks, porosity, non-impregnated fibers, void content), and exposure environments [1]. Furthermore, adding FRP composites to concrete structures complicates the durability performance of FRP-reinforced concrete structures due to the combined effect of FRP composites, interface, concrete, and various environmental and mechanical conditions. Therefore, the durability assessment of the FRPs in the concrete structures is a very complex and multidimensional task. The following sections are presented as a summary of the various durability aspects of FRP composites, concrete members reinforced with FRP, and bond behavior of FRP in concrete.

## 3 Main Parameters Affecting the Durability of Internal FRP Reinforcement

Different types of fibers are used in manufacturing FRP bars as internal reinforcement, such as glass FRP, carbon FRP, aramid FRP, and basalt (BFRP). The degree of damage/deterioration to internal FRP reinforcement depends primarily on fundamental factors such as fiber type and volume fraction, resin type, morphology and adhesion of the fiber-matrix interface, severity of the exposure environments, and the fabrication process. Most FRP bars—susceptible to degradation by harsh environments such as moisture and alkalinity—are E-glass fibers; in contrast, carbon fibers are relatively inert in such environments. Aramid fibers, on the other hand, are highly resistant to abrasion and impact, but are sensitive to creep, moisture, and ultraviolet light. Appropriate performance depends on having a suitable resin to protect the fibers. The resin system's durability depends on many parameters such as resin components, their individual properties and proportions, and curing time and conditions.

## 3.1 Effect of Moisture

In recent decades, one of the most studied topics related FRP durability was conducted on the fluids effect on the performance of FRP composites. Ben Daly [6] showed that the moisture-diffusion process in pultruded composites and the saturation level attained could be related to the presence of fillers and additives in the matrix. It has been proven that the sorption rate is controlled by the matrix's chemical structure, interface, and manufacturing process.

Consequently, extensive studies have been conducted to control the diffusion process by using resin matrices with lower permeability [7], improving the interface zone by using suitable sizing chemistry, or selecting an appropriate molding process to reduce void content. Moreover, moisture ingress can degrade the resin through chemical attack or a drop in the $Tg$. For this reason, fluids affect dominant matrix properties, such as the transverse strength of FRP, and these properties decrease more with increasing exposure time and temperature. Due to the chemical and physical attacks, glass fibers are sensitive to fluid ingress. Considerable investigations have been conducted in this research topic [1, 2, 6, 7, 9–11]. Carbon fibers are not affected by fluid ingress, but the resin matrix is usually affected, so the performance of this composite is nevertheless affected. In the case of unidirectional carbon composites, this usually leads to reduced compressive and shear strength, with little impact on tensile strength, since it is especially dominated by the fibers, which are not affected by fluids [11]. At a higher temperature, aramid fibers are mostly affected by fluids. AFRP composites saturated in water have been reported to lose 35% of their flexural strength at room temperature and up to 55% if stressed and under wet/dry thermal cycles.

## 3.2 Degradation Mechanism of GFRP Bars in Alkaline Environments

The concrete environment is characterized by high alkalinity, with a pH range between 12 and 13. This alkaline environment damages glass fibers through loss in toughness and strength and increased embrittlement. Carbon fibers have a good resistance to alkaline, followed by aramid and glass fibers. Glass fibers are deteriorated due to the combination of two processes: the chemical attack and concentration of hydration products between individual filaments. The embrittlement of fibers is due to the nucleation of calcium hydroxide on the surface of fiber. Fiber surface pitting and roughness occurred due to hydroxylation, reducing fiber properties in the presence of moisture. Moreover, the calcium, sodium, and potassium ions found in the concrete pore solution are highly aggressive against the glass fibers. Therefore, glass fiber degradation not only depends on the high pH level but also due to the combination of alkali salts, pH, and moisture. On the other hand, the strength of aramid fibers degrades in alkaline environments. Its strength loses 74%, when Kevlar 29 was subjected to 10% sodium hydroxide solution for 1000 h. Carbon fibers are not supposed to be affected by alkaline solutions at any concentration or by water temperatures up to boiling [2, 12]. However, Judd [1] reported that they were resistant to alkaline solutions at all concentrations and different temperatures up to boiling. Carbon tows immersed for 6168 hours in a 50% sodium hydroxide solution showed that the variations in strength properties are only about 15%. Although an appropriate resin matrix (vinylester, epoxy) provides certain level of protection to fibers from alkaline degradation, migration of high pH solutions and alkali salts through resin (or through void, crack, interface between fiber/matrix) to the fiber surface is possible. Chu [13] conducted studies on characterizing and modeling the effects of moisture and alkalis on E-glass/vinyl ester composite strips at various temperatures (22 °C, 40 °C, 60 °C, and 80 °C). The degradations of tensile strength were between 35% and 62% of the initial strength.

Kim [14] conducted a durability investigation on two GFRP types (E-glass/vinyl ester) under various environmental conditions (moisture, chloride, alkali, and freeze-thaw cycling) at 25, 40, and 80 °C for up to 132 days. It can be concluded that an alkaline environment had greater impact on the degradation of the tensile strength of GFRP bars than the other environmental conditions. Robert [4] studied the mechanical, durability, and microstructural characterization of unstressed GFRP bars subjected to alkaline and saline solutions. This conditioning was used to simulate the effect of seawater and deicing salts on GFRP bars. The results showed that there is no significant effect on the durability of the concrete-wrapped GFRP bars immersed in saline solution or tap water; the GFRP bars in saline solution evidenced very high long-term durability. Benmokrane [1] conducted recent studies on the durability performance of carbon-fiber-composite cable (CFCC) tendons exposed to elevated temperatures and an alkaline environment. Specimens were exposed to alkaline solutions for 1000, 3000, 5000, and 7000 h elevated temperatures (22 °C, 40 °C, 50 °C, and 60 °C). The test results revealed a 7.17% reduction in tensile strength after 7000 h of conditioning at 60 °C. Recently, Benmokrane [2] assessed the physical and mechanical properties of GFRP bars made with three types of reins (vinyl ester, isophthalic polyester, or epoxy). The investigation was conducted by immersing the GFRP bars in an alkaline solution at 60°C for 1000, 3000 and 5000 h. The test results reveal that the vinyl ester and epoxy GFRP bars had the best strengths and lowest degradation rate after conditioning, while the polyester GFRP bars experienced the lowest strength properties after conditioning.

## 3.3 Effect of Freeze and Freeze/Thaw Cycles

Freezing and freeze/thaw exposures do not affect fibers, although such exposure can affect the resin and the fiber/resin interface. Most researches on this subject were conducted on aerospace materials. The literature indicates that freezing and thawing have very limited impact on pultruded FRP composites [15]. In general, at low temperatures, complex residual stress arises within FRP composites as a result of matrix stiffening and mismatch of thermal expansion coefficients of the matrix and resin as well as of the FRP and concrete. The presence of deicing salts under wet conditions with subsequent freeze/thaw cycling can cause microcrack formation and gradual degradation due to crystal formation and increased salt concentration. Alves [16] investigated the durability of FRP bonding to concrete elements subjected simultaneously to 250 freeze/thaw cycles and sustained load. The test results reveal that the combined conditions increased the GFRP-concrete bond strength.

## 4 Durability of GFRP-Reinforced Concrete in Field Structures

In 2004, ISIS Canada, a Canadian Network of Centres of Excellence, launched a major study to obtain field data on the durability of GFRP in concrete exposed to natural environments. The objective of the research was to provide performance data on a GFRP that has been used in several structures across Canada. Concrete cores containing the GFRP were removed from several 5- to 8-year-old exposed structures, and the GFRP's physical and chemical compositions were analyzed at the microscopic level. Five ISIS Canada field demonstration structures built in five different provinces (Hall's Harbor Wharf in Nova Scotia, Joffre Bridge in Quebec, Chatham Bridge in Ontario, Crowchild Trail Bridge in Alberta, Waterloo Creek Bridge in British Columbia), exposed to a wide range of environmental conditions, were chosen (Fig. 5.3). Their selection reflects a wide range of environmental conditions that are representative of the Canadian climate. Three research teams from four Canadian universities independently performed microanalyses of the

**Fig. 5.3** Pictures from the five field demonstration projects considered

GFRP and surrounding concrete. They used different analytical methods to (a) investigate whether or not the GFRP in the concrete field structures had been attacked by alkalis and (b) compare the composition of the GFRP removed from in-service structures to that of control specimens, which were saved from the projects and not exposed to the concrete environment. Direct comparisons were carried out with "virgin" GFRP rods preserved under controlled laboratory conditions. The results indicate that no deterioration of the GFRP was observed in any of the field demonstration structures included in this study and that no chemical degradation processes occurred within the GFRP due to concrete alkalinity.

## 5 Environmental Reduction Factors ($C_E$) for GFRP Bars

The upgrading of ACI 440 [17] comes accompanied by the introduction of design recommendation for shear design, indirect deflection control, and flexural crack control of concrete structures reinforced with FRP. The design tensile strength of GRPP bars, according to the ACI 440 design code [17], is given by Eq. 5.1:

$$f_{fu} = C_E.f_{fu}^* \quad f_{fu} = C_E.f_{fu}^* \tag{5.1}$$

where $f_{fu}$ and $f_{fu}^*$ are design tensile strength and guaranteed tensile strength of the GFRP bar, respectively. We note that for the GFRP bars in the ACI 440 [17] design guideline, the environmental reduction factor ($C_E$) is recommended to be 0.7 or 0.8 for concrete element exposed to and not exposed to earth and weather, respectively. In order to achieve the ACI committee requirements about the $C_E$ values for the long-term durability of GFRP bar in concrete, the $C_E$ values were calculated based on long-term durability database from the literature [1, 6, 7, 10] by using Eq. 5.2:

$$C_E = f_{pred} / f_{fu}^* \tag{5.2}$$

where $f_{pred}$ = predicted tensile strength of GFRP bar at 75 or 100 years which was calculated based on Arrhenius relationship. Based on the results of experimental and predicted tensile strengths in the literature, the factor ($C_E$) was assessed using Eq. 5.2. Consequently, a new value for the factor ($C_E$) value was calculated based on the guaranteed tensile strength (Eq. 5.2). It can be observed that the calculated values of the reduction factor ($C_E$) are higher than those values in ACI 440 [17]. Therefore, we recommend that the reduction factor ($C_E$) should be used as *0.90*.

# 6 Life Prediction Approaches for Long-Term Performance of FRP Bars

## 6.1 Arrhenius Relation

The popular Arrhenius relation has been adopted by researchers [1, 2, 7] to recast the long-term performance of FRP bars in a civil engineering environment based on accelerated short-term aging test data. The degradation rate is expressed as

$$k = A\exp\left(\frac{-E_a}{RT}\right) \quad (5.3)$$

$$\ln\left(\frac{1}{k}\right) = \frac{E_a}{R}\frac{1}{T} - \ln(A) \quad (5.4)$$

where $k$, $A$, $E_a$, $R$, and $T$ are the degradation rate (1/time), the material and degradation process constant, the activation energy, the universal gas constant, and the temperature.

## 6.2 Degradation Laws

Different mathematical equations describing the relationship between the strength retention and aging time have been proposed by researchers and are based on different theoretical foundations. Davalos [18] stated that there are generally four types of FRP strength models and the prediction procedures for those models are all based on the Arrhenius equations presented in Eqs. (5.3) and (5.4). Serbescu [21] claimed that there are mainly two approaches for the performance prediction of FRP bars: measuring either "strength retention" or "moisture absorption." The following is a brief description of the four widely used mathematical models present in the literature. Tannous [19] proposed the "moisture absorption" model:

$$Y = 100\left(1 - \frac{\sqrt{2DCt}}{r_o}\right) \quad (5.5)$$

where $Y$ is the strength retention (%) in this and all other equations presented in this paper, $t$ is the exposure time, $D$ is the diffusion coefficient, $C$ is the concentration of the solution, and $r_o$ is the radius of the FRP bar. This model assumes that the affected area is completely degraded and unable to carry any load, which may not be entirely true. Additionally, the determination of the coefficients $D$ and $C$ from moisture absorption tests makes its use rather complicated. In addition, this equation cannot be used when the solution is distilled water, as the value of $C$ would be zero. The second model adopted an exponential relationship between strength retention and aging time. Debonding at the fiber-matrix interface is assumed to be the major degradation mechanism in this model as is described via the following equation:

$$Y = 100\exp\left(\frac{-t}{\tau}\right) \quad (5.6)$$

where $\tau$ is a fitted coefficient using the least squares method. The tensile strength retention (%) at an infinite exposure time is assumed to be zero in this model. This model was originally used to predict the flexural strength retention of composite laminates and had been adopted by many scholars [8, 18] to predict the long-term performance of FRP bars. The third model adopted a linear relationship between the strength retention and the logarithm of the aging time via:

$$Y = a.\log(t) + b \quad (5.7)$$

where $a$ and $b$ are regression constants. Litherland [20] first developed this model and successfully predicted the residual strength of glass fiber using this model. It is worth noting that Eq. (5.7) is a widely used degradation model but does not hypothesize the degradation mechanism. Some researchers have found, however, that the degradation lines at different temperatures in a single logarithmic scale from Eq. (5.7) are not parallel. Serbescu [21] used a double logarithmic scale in his study to plot the experimentally obtained tensile strength percentages in the fourth model described here:

$$\log(Y) = a.\log(t) + b \quad (5.8)$$

Based on Eq. (5.8), an approach for the calculation of the environmental strength reduction factor ($\eta_{env,t}$, which corresponds to $1/C_E$ in the ACI 440 [17]) was established in fib bulletin 40. For detailed steps, the reader is referred to fib bulletin 40. The aforementioned second, third, and fourth models all belong to the "strength retention" approach. Recently, Dong [22] proposed a refined prediction method for the long-term behavior of FRP bars that considers the effects of service year, concrete wrap, environmental humidity, and seasonal temperature fluctuations. According to the available accelerated aging tests data in the literature, the environmental reduction factors (ERFs) for an FRP bar used as a concrete internal reinforcement in a field environment are predicted. The ERF under a specific design lifetime in a year (DL), service ambient temperature, and humidity for FRP bars embedded in concrete may be written as

$$ERF = [1 - \Delta_1 + \rho.\log(DLTSF)\eta] \quad (5.9)$$

where DL is the design lifetime in a year, $\Delta_1$ is the strength deterioration after 1 year (365 days) of experimental exposure at $T_0$, a correction factor ($\eta$) is the ratio of water in concrete at different RHs, $\rho$ is a linear regression's slope, and TSF is the time shift factor for temperature $T$ and temperature $T_1$, which can be obtained based on Eq. 5.10.:

$$\text{TSF} = e^{[B/(T1+273.15)]} - e^{[B/(T+273.15)]} \tag{5.10}$$

## 7 Conclusions

Application of FRP reinforcement in different structures has proven very successful to date. Based on the investigated study, the durability performance of FRP materials is generally very good in comparison with other, more conventional, construction materials. However, the long-term durability of FRPs is still not fully understood. Therefore, more research is required on the durability of FRPs to fill the gaps of knowledge.

## References

1. Benmokrane, B., Ali, A. H., Mohamed, H. M., Robert, M., & ElSafty, A. (2016). Durability performance and service life of CFCC tendons exposed to elevated temperature and alkaline environment. *Journal of Composites for Construction*, 04015043. https://doi.org/10.1061/(ASCE)CC.1943-5614.0000606.
2. Benmokrane, B., Ali, A. H., Mohamed, H. M., ElSafty, A., & Manalo, A. (2017). Laboratory assessment and durability performance of vinyl-ester, polyester, and epoxy glass-FRP bars for concrete structures. *Composites Part B Engineering, 114*, 163–174.
3. Robert, M., Cousin, P., & Benmokrane, B. (2009). Durability of GFRP reinforcing bars embedded in moist concrete. *Journal of Composites for Construction, 13*, 66–73.
4. Robert, M., & Benmokrane, B. (2013). Combined effects of saline solution and moist concrete on long-term durability of GFRP reinforcing bars. *Construction and Building Materials, 38*, 274–284.
5. Benmokrane, B., Robert, M., Mohamed, H. M., Ali, A. H., & Cousin, P. (2017). Durability assessment of glass FRP solid and hollow bars (rock bolts) for application in ground control of Jurong rock caverns in Singapore. *Journal of Composites for Construction, 21*(3), 1–4.
6. Ben Daly, H., Ben Brahim, H., Hfaied, N., Harchay, M., & Boukhili, R. (2007). Investigation of water absorption in pultruded composites containing fillers and low profile additives. *Polymer Composites, 28*(3), 355–364.
7. Benmokrane, B. (2000). *Improvement of the durability performance of glass fiber reinforced polymer (GFRP) reinforcements for concrete structures* (p. 50). Technical Report, Civil Engineering Department, University of Sherbrooke, Sherbrooke, Quebec, Canada.
8. Chen, Y., Davalos, J. F., Ray, I., & Kim, H. Y. (2007). Accelerated aging tests for evaluation of durability performance of FRP reinforcing bars reinforcing bars for concrete structures. *Composite Structures, 78*(1), 101–111.
9. Robert, M., & Benmokrane, B. (2010). Effect of aging on bond of GFRP bars embedded in concrete. *Cement and Concrete Composites Journal, 32*(6), 461–467.
10. Wang, J., GangaRao, H., Liang, R., & Liu, W. (2016). Durability and prediction models of fiber-reinforced polymer composites under various environmental conditions: A critical review. *Journal of Reinforced Plastics and Composites, 35*(3), 179–211.
11. Hancox, N. L., & Mayer, R. M. (1994). *Design data for reinforced plastics* (pp. 202–204). New York: Chapman & Hall.

12. Judd, N. C. W. (1971). The chemical resistance of carbon fibers and a carbon fiber\polyester composites. *The first international conference of carbon fibers, plastics institute*, London (pp. 1–8).
13. Chu, W., Wu, L., & Karbhari, V. (2004). Durability evaluation of moderate temperature cured E-glass/vinylester systems. *Composite Structures, 66*, 367–376.
14. Kim, H., Park, Y., & You, Y. (2008). Short-term durability test for GFRP rods under various environmental conditions. *Composite Structures, 83*, 37–47.
15. Lord, H. W., & Dutta, P. K. (1988). On the design of polymeric composite structure for cold regions application. *Journal of Reinforced Plastics and Composites, 7*, 435–459.
16. Alves, J., El-Ragaby, A., & El-Salakawy, E. (2010). Bond strength of glass FRP bars in concrete subjected to freeze-thaw cycles and sustained loads. *The 5th international conference on FRP composites in civil engineering*, September 27–29, Beijing.
17. ACI (American Concrete Institute). (2015). *Guide for the design and construction of structural concrete reinforced with FRP bars*. Farmington Hills : ACI 440.1R-15.
18. Davalos, J. F., Chen, Y., & Ray, I. (2012). Long-term durability prediction models for GFRP bars in concrete environment. *Journal of Composite Materials, 46*(16), 1899–1914.
19. Tannous, F. E. (1998). Environmental effects on the mechanical properties of E-glass FRP rebars. *ACI Materials Journal, 95*(2), 87–100.
20. Litherland, K. L., Oakley, D. R., & Proctor, B. A. (1981). The use of accelerated ageing procedures to predict the long term strength of GRC composites. *Cement and Concrete Research, 11*(3), 455–466.
21. Serbescu, A., Guadagnini, M., & Pilakoutas, K. (2014). Mechanical characterization of basalt FRP rebars and long-term strength predictive model. *Journal of Composites for Construction, 19*(2), 04014037.
22. Dong, Z., Wu, G., Zhao, X. L., & Wan, Z. K. (2017). A refined prediction method for the long-term performance of BFRP bars serviced in field environments. *Construction and Building Materials, 155*, 1072–1080.

# Chapter 6
# Nano-modified Polymer Concrete: A New Material for Smart and Resilient Structures

Mahmoud M. Reda Taha

Polymer concrete (PC) is a type of concrete where polymer binder replaces cement. PC has been used in field applications since the 1950s. Specifically, PC has been widely used for precast architectural facade, underground utilities, manholes, machine foundations, bridge deck overlays and closures, and other applications. PC is characterized by high compressive and tensile strengths and superior durability compared with conventional Portland cement concrete. However, the use of PC is limited due to its relatively higher cost than conventional concrete. This paper suggests that a significant change in PC can take place by incorporating a very limited amount of nanomaterials (below 2% of carbon nanotubes or alumina nanoparticles by weight of the resin). Incorporating such small amount of nanomaterials will result in significant improvement in mechanical properties of PC including strength, ductility, and fracture toughness. We also show that incorporating 2.0 wt.% of pristine carbon nanotubes can result in significant improvement in ductility and can allow self-sensing capabilities of PC. The significance of using nanomaterials is fundamental and will open doors for using PC as a material for smart and resilient structures under extreme loadings.

## 1 Introduction

Two types of polymer concrete (PC) are widely used in the construction field including polymer-modified concrete (PMC) and polymer concrete (PC) [1]. In PC, no cement is used and polymer is used as a binder to replace cement. PC has a number of advantages over conventional cement concrete such as improved

---

M. M. Reda Taha (✉)
Department of Civil Engineering, University of New Mexico, Albuquerque, NM, USA
e-mail: mrtaha@unm.edu

cracking strength, improved bond strength to existing concrete, and improved durability characteristics [1, 2]. The significant improvement of both PMC and PC compared with conventional cement concrete is attributed to the significant enhancement in the PC microstructure and its strong impact on permeability that typically governs concrete durability and longevity [2]. PC has been preferred in construction projects involving highway surfacing, bridge deck overlays, wall facades, wastewater pipes and tanks, manholes, machine foundations, dams and reservoirs [3–7]. To compare, Portland cement concrete with compressive strength of 40–80 MPa provides flexural strength of 4–12 MPa, a direct tensile strength in the range of 2.0–4.0 MPa, and strain at failure of 0.01–0.05%. On the other hand, PC compressive strength is in the range of 50–130 MPa, flexural strength in the range of 10–45 MPa, and tensile strength up to 8.0–12 MPa [8–11]. However, the commercial use of both types of polymer concrete is lower than that of conventional concrete due to the limited improvements in mechanical properties compared with cost. To further improve PC mechanical properties, researchers examined the use of synthetic chopped glass and carbon fibers at 0.5–7.5% by weight [12] and reported limited improvements in mechanical characteristics of PC [12]. Incorporating fibers, however, decreased PC workability and thus limits applicability, and the improvements associated with crack resistance, ductility, and deformability induced by fibers were not substantial [13]. We suggest significant improvement in tensile strength and tension strain at failure (ductility) and energy absorption (toughness) and that crack resistance (fracture toughness) can take place by incorporating nanomaterials such as carbon nanotubes (CNTs) or alumina nanoparticles (ANPs) in the polymer resin prior to the fabrication of PC.

## 2 Nanomaterials for Altering Polymers

In recent years, researchers have examined the use of nanomaterials to alter concrete and polymer composite behavior. Nanomaterials provide significantly high surface area than microfibers and additives and thus interact at several dimensions altering the performance at the mesoscale [14, 15]. The chemical polymeric nature of the PC binder maximizes utilization of nanoparticles by creating a chemical interaction platform via the polymerization process. However, the efficiency of nanoparticles in altering the mechanical performance is greatly influenced by the level of dispersion [16, 17]. Dispersion techniques vary in effectiveness with the type (e.g., geometry being spherical, tubular, or flakey) and the content of the nanomaterials. Nanomaterials with high aspect ratio such as carbon nanotubes (CNTs) are more challenging to disperse than other nanomaterials with spherical geometry such as silica nanoparticles as their high aspect ratio hinders the dispersion process [18, 19]. Moreover, surface functionalization can play a major role with any of the above techniques to enable improved dispersion. Most common dispersion techniques are ultrasonication, ball milling, stirring, shear mixing, and extrusion [18, 19].

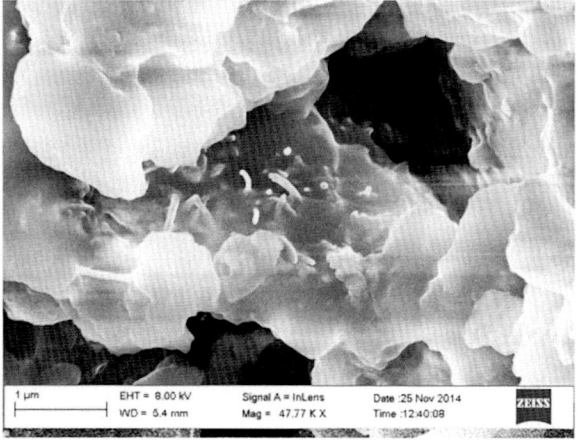

**Fig. 6.1** SEM micrograph of MWCNTs dispersed in epoxy matrix

CNTs are tubular materials made from single or multiple graphite sheets. Carbon arc discharge and chemical vapor deposition (CVD) are the two typical methods used to produce CNTs. Based on the quality of the source materials, the production method and temperature of single-walled carbon nanotubes (SWCNTs) or multi-walled carbon nanotubes (MWCNTs) can be produced [20]. SWCNTs have been proved to be stronger, stiffer, and thus more expensive than MWCNTs. MWCNTs typically are produced at a lower temperature of 450 °C compared with 700 °C typically used to produce SWNCTs [21]. While surface functionalization is a well-known technique to improve dispersion by enabling repulsion of similar charge nanomaterials, such functionalization can also be used to improve the interaction of the nanomaterials with the base polymer matrix. For instance, it was shown that significant improvement in electrical and mechanical characteristics of unsaturated polyester was made possible by incorporating MWCNTs that are surface functionalized with silane group [22]. Moreover, exceptional mechanical properties have been associated with improvements in strength, fracture toughness, fatigue, glass transition temperature, durability, and viscoelastic properties of polymers [23–25]. The above discussion proves that incorporating functionalized nanomaterials specifically MWCNTs can open a new window for producing a new class of PC with a set of pre-engineered mechanical properties. A scanning electron micrograph of MWCNTs dispersed in a polymer matrix that was later used to produce PC is shown in Fig. 6.1.

## 3 Nano-modified PC

### 3.1 Materials

The most typically used polymer binders for PC are epoxy, polyester, vinyl ester, and methacrylate. It is proposed to alter the polymer matrix with nanomaterials

**Table 6.1** Mix proportions by weight of neat PC, kg/m³

| Resin | Hardener | Filler powder | Coarse aggregate |
|---|---|---|---|
| 288 | 128 | 1570 | 320 |

**Table 6.2** Properties of MWCNTs

| | |
|---|---|
| Outer diameter/inner, nm | 20–30/5–10 |
| MWCNTs ash, % wt. | <1.5% |
| Purity, % | >95% |
| Length, μm | 10–30 |
| Specific surface area, m²/g | 110 |
| Bulk density, g/cm³ | 0.28 |
| COOH content % wt. | 1.2% |

including CNTs, ANPs, and other nanomaterials. This section demonstrates the ability of two types of nanomaterials, specifically MWCNTs and ANPs, to provide a new class of PC with superior mechanical properties. The work described here is focused on PC produced using epoxy resin. The epoxy used is low modulus polysulfide siloxane epoxy typically used to manufacture PC. The epoxy consists of two components: bisphenol-based epoxy resin and phenol-based epoxy hardener. The resin was mixed with a fine aggregate mix of crystalline silica (quartz) and bauxite-based aggregate with nominal maximum size of 4.75 mm and a fineness modulus of 3.2. Neat PC mix proportion is presented in Table 6.1. Nano-modification presented here was conducted using two types of nanomaterials: MWCNTs and ANPs. Cheap Tubes, Inc., supplied the MWCNTs. The MWCNTs were produced using the catalyzed chemical vapor deposition (CCVD) method at a relatively low temperature of 450 °C with purity above 95 wt.%. MWCNTs, used here, were functionalized by carboxylic (COOH) group of 1.23 wt.% of the MWCNTs weight. The MWCNTs used have 20–30 nm outer diameter and 10–30 μm length. Table 6.2 presents the main characteristics of MWCNTs. Furthermore, ANPs from Sigma Aldrich, Inc., with particle size of <50 nm were used separately to modify the PC with weight content below 3.0 wt.% as weight of the resin.

## 3.2 Experimental Methods

Neat PC mix (Table 6.1) and PC mixes incorporating 0.5, 1.0, and 1.5% COOH-functionalized MWCNTs were produced. The PC mixes with MWCNTs were tested for compressive strength, flexural strength, and impact resistance following ASTM standards as shown in Fig. 6.2 [26–28].

Furthermore, tension tests were conducted for PC mixes incorporating ANPs. Tensile strength of PC incorporating ANPs was performed on dog bone specimens following ASTM-D638-14 [29]. The stress-strain of PC under tension stress was observed using the test setup in Fig. 6.3a. Finally, fracture toughness tests of the PC

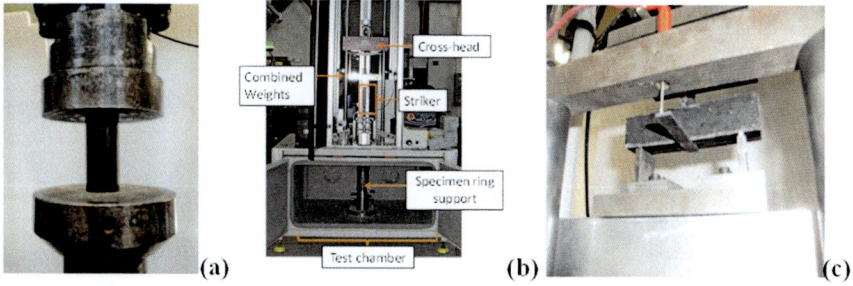

**Fig. 6.2** Mechanical testing of PC incorporating MWCNTs: (**a**) compressive strength test, (**b**) impact testing equipment, (**c**) flexural strength test

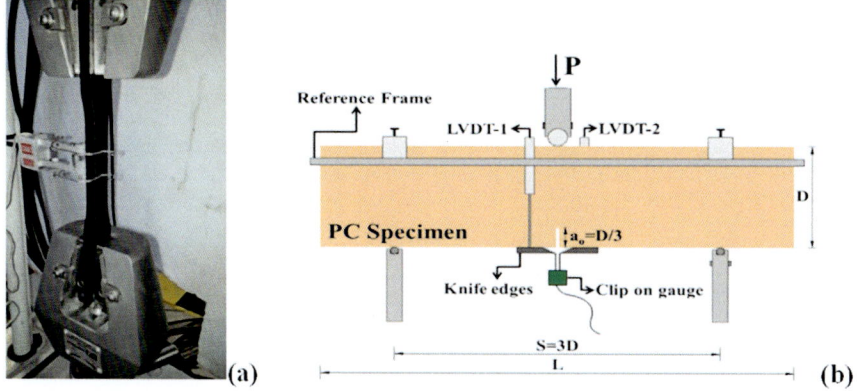

**Fig. 6.3** (**a**) Tension test of PC incorporating ANPs following ASTM D638–14 [30] (**b**) Fracture toughness tests of PC incorporating MWCNTs or ANPs

mixes incorporating either MWCNTs or ANPs were also performed. Figure 6.3b shows a schematic of the fracture toughness test setup.

A fracture test of PC specimens incorporating MWCNTs or ANPs was conducted in three-point bending mode using notched and unnotched specimens following the guidelines provided by ACI 446 [30]. Three-point bending specimens with dimensions of 300 mm length and 25 mm$^2$ cross section were tested. Notched prism samples with initial crack length of 8.4 mm ± 0.25 mm depth and 0.75 mm width were loaded with 75 mm span length. Two knife edges were attached to both sides of the notch where a crack mouth opening displacement (CMOD) gage was used. A reference frame with two linear variable differential transformers (LVDTs) recorded mid-span displacement accurately. The specimens were loaded in displacement control protocol with a loading rate ranging from 0.20 to 0.45 mm/min [31]. Quasi-brittle fracture mechanics (QBFM) approach using the effective elastic crack modulus theory was used due to the high ductility and nonlinear behavior of PC [32]. Critical stress intensity factor ($K_{IC}$), critical elastic energy release rate ($G_{IC}$), and critical plastic energy release rate (J-integral ($J_{IC}$)) are calculated in Eqs. (6.1), (6.2), (6.3), (6.4), and (6.5).

$$K_{IC} = g_1(\alpha)\sigma_c\sqrt{\pi a_c} \qquad (6.1)$$

$$G_{IC} = \frac{K_{IC}^2(1-v^2)}{E_{UN}} \qquad (6.2)$$

$$J_{IC} = \frac{2}{H_c b}(A_N - A_{UN}) \qquad (6.3)$$

$$A_N = \sum_{i=0}^{n_p} \frac{(\Delta_{i+1} - \Delta_i)(P_{i+1} + P_i)}{2}; n_p = \frac{P_c}{N_c} \qquad (6.4)$$

$$A_{UN} = \sum_{i=0}^{n_p} \frac{(\Delta_{i+1} - \Delta_i)(P_{i+1} + P_i)}{2}; n_p = \frac{P_c}{N_c} \qquad (6.5)$$

where $a$, crack depth; $a_c$, critical crack depth; $A_N$, area under the load displacement of notched specimen up to the peak load of that specimen; $A_{UN}$, area under the load displacement of unnotched specimen up to the peak load of similar notched specimen; $b$, specimen's width; $d$, specimen's depth; $E_{UN}$, elastic modulus of uncracked sample; $G_{IC}$, critical energy release rate; $H_c$, critical ligament length; $J_{IC}$, critical J-integral; $K_{IC}$, critical stress intensity factor; $N_c$, number of steps up to peak load of notched specimen; $P_c$, peak load of notched specimen; $P_i$, load at given time step; $\alpha$, notch to depth ratio; $\alpha_i$, initial notch to depth ratio; $\alpha_c$, critical notch to depth ratio; $\Delta_i$, displacement at given time step; $\sigma_c$, critical stress of notched prism; and $v$, Poisson's ratio, 0.35. Details of the fracture toughness test method and analysis can be found elsewhere [31].

## 3.3 Experimental Results

While incorporating MWCNTs did not result in any improvement of the compressive strength of PC as apparent in Fig. 6.4a, it resulted in significant improvement in the flexural strength of PC as shown in Fig. 6.4b. It is apparent that the optimal content of COOH-MWCNTs is within the range of 0.5–1.0 wt.% of the polymer resin. Further increase in MWCNTs content affects the epoxy and PC flowability, which in its turn results in entrapping air and thus reducing the flexural strength of PC.

Examining the impact test results, it is apparent that incorporating MWCNTs resulted in increasing the impact strength and energy with an optimal content within 0.5–1.0 wt.% similar to that reported above. Figure 6.5a shows the significance of MWCNTs on the impact strength of PC, and Fig. 6.5b shows the effect of MWCNTs on the total impact energy showing a 17% increase in impact energy for 1.0 wt.%

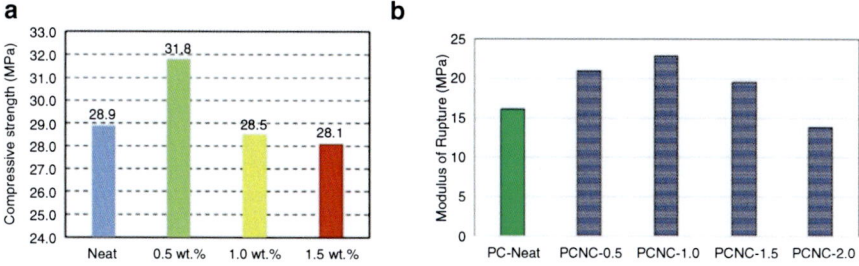

**Fig. 6.4** Mechanical properties of PC incorporating COOH-MWCNTs: (**a**) compressive strength, (**b**) flexural strength

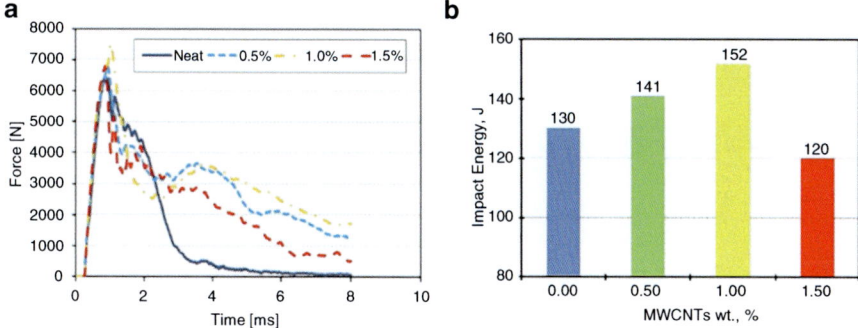

**Fig. 6.5** Effect of MWCNTs on impact resistance of PC. (**a**) Force-time response. (**b**) Impact energy of PC with MWCNTs

MWCNTs. Figure 6.6 shows the ability of MWCNTs to improve energy dissipation in PC and thus to alter the cracking pattern of PC producing four cracks under impact striking load compared with three cracks for neat PC. Details on yield line analysis for cracking pattern to proving the superior ability of PC with MWNCTs to absorb impact energy can be found elsewhere [33]. Finally, COOH-MWCNTs showed the ability to significantly improve fracture toughness of PC. Figure 6.7 shows that adding 0.5, 1.0, 1.5, and 2.0 wt.% MWCNTs resulted in improved fracture toughness of PC by 128, 94, 83, and 113%, respectively. It is apparent that COOH-MWCNTs can double the fracture toughness of PC with as low as 0.5 wt.% of the polymer resin. On the other hand, incorporating ANPs resulted in significant improvement in ductility of PC. Figure 6.8 shows the stress-strain of PC incorporating ANPs. It is apparent that ANPs can improve the failure strain of PC by an order of 200% compared with neat PC. Strains at failure ranging of 4.0–5.0% are unprecedented. Furthermore, Fig. 6.9 shows the significant increase of fracture toughness of PC incorporating ANPs.

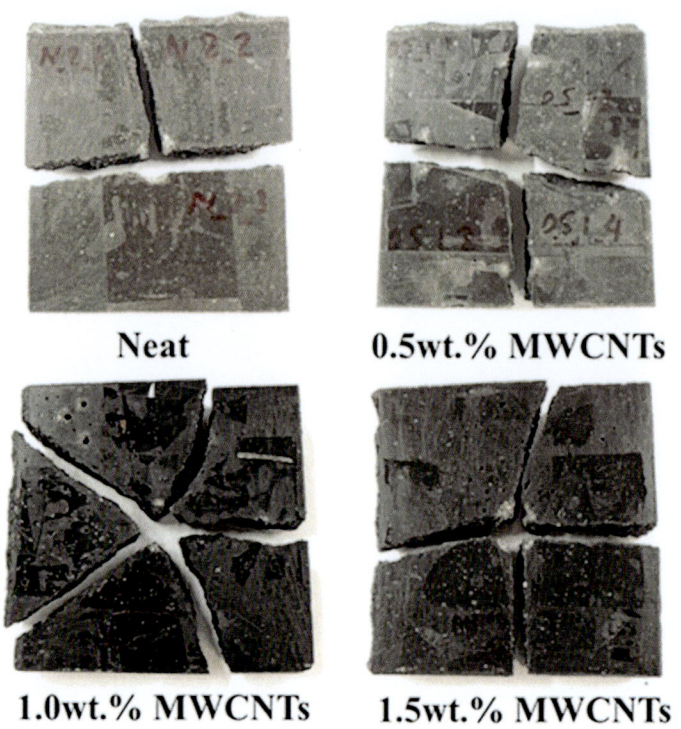

**Fig. 6.6** Cracking pattern of PC incorporating MWCNTs showing the significance of MWCNTs of altering the number cracks in PC due to impact load

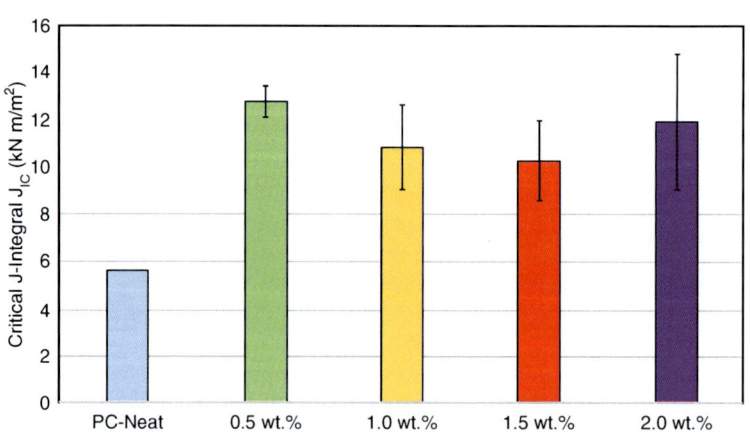

**Fig. 6.7** Significance of COOH-MWCNTs on nonlinear fracture toughness $J_{IC}$ of PC showing the ability of MWCNTs to double fracture toughness of PC

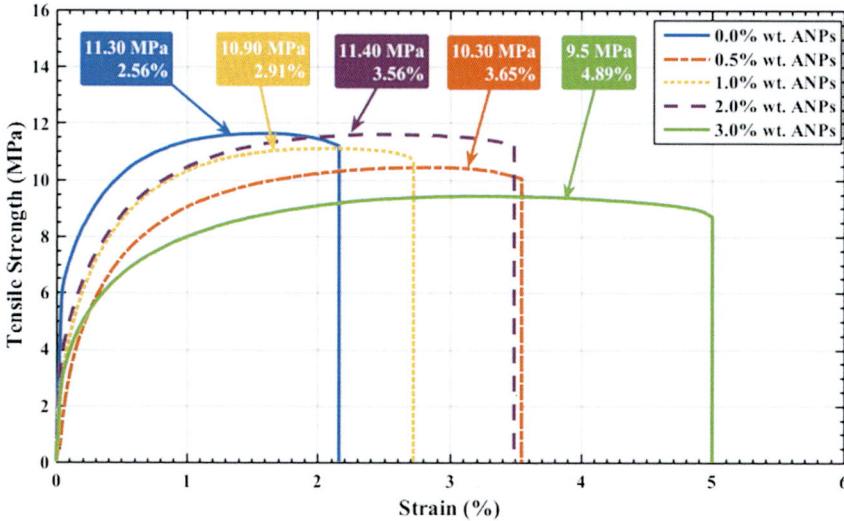

**Fig. 6.8** Stress-strain curves of PCs incorporating ANPs showing superior ductility (failure strain) of PC with 2.0% wt. and 3.0% wt. ANPs [34]

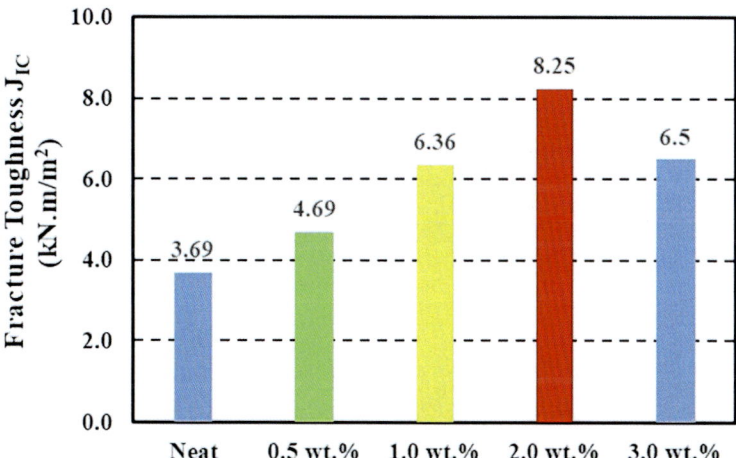

**Fig. 6.9** Effect of incorporating ANPs on the nonlinear fracture toughness $J_{IC}$ of PC showing the ability of ANPs to double fracture toughness of PC

Microstructural investigations of PC incorporating MWCNTs and ANPs showed that the significant improvement in the mechanical properties and fracture toughness of PC is attributed to the ability of both materials to alter the epoxy polymerization in PC. Investigations using dynamic mechanical analyzer (DMA) showed the ability of ANPs to significantly reduce the cross-linking density while that of COOH-MWCNTs to increase the cross-linking density. The two materials thus act differently, while ANPs improve PC ductility and COOH-MWCNTs improve the strength of PC. Further details on the significance of ANPs on cross-linking of PC are discussed elsewhere [34].

## 4  PC for Smart and Resilient Structures

The above research shows the ability of MWCNTs and ANPs to significantly improve the mechanical properties of PC specifically strength, ductility, and fracture toughness. Furthermore, the research has shown that pristine MWCNTs can also alter the electrical conductivity of PC enabling a conductive material in addition to significant increase in ductility. Figure 6.10a shows testing of a PC beam incorporating pristine MWCNTs where significant ductility of PC is apparent. Furthermore,

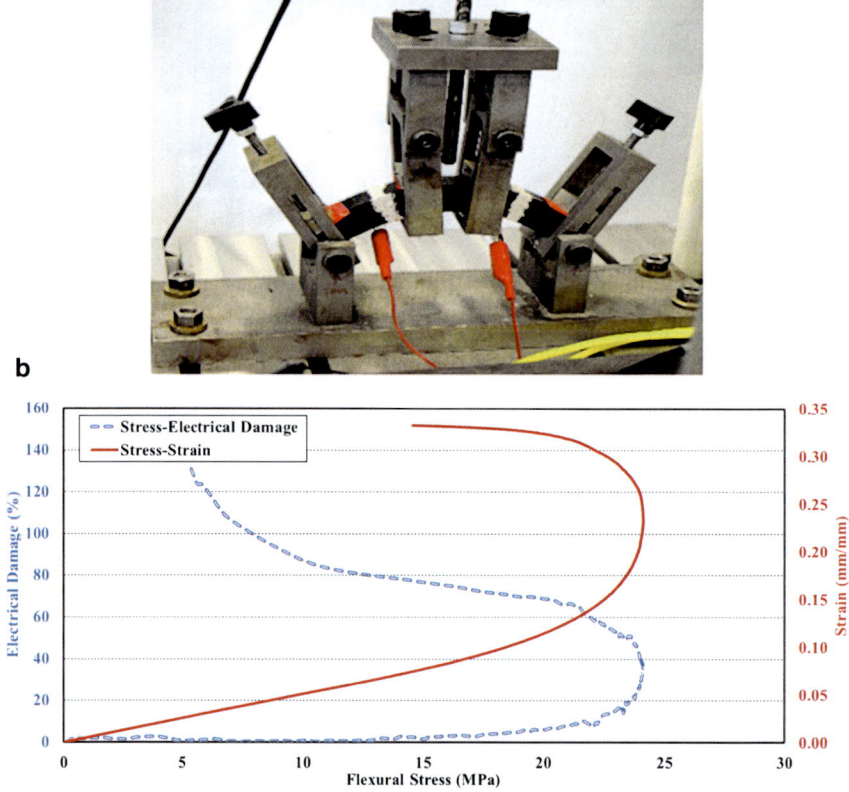

**Fig. 6.10** PC incorporating pristine MWCNTs: (**a**) flexural testing showing extreme ductility, (**b**) self-sensing PC incorporating MWCNTs showing the possible sense of damage using electrical resistance measurements of PC

the beam is connected to a source meter to measure its electrical resistance. Figure 6.10b shows the change in electrical resistance of the PC as compared to the stress-strain of PC. Further research is warranted to examine the fatigue strength of the new PC and to examine the use of the new PC with nanomaterials in strengthening of structures to improve structural resilience to extreme loading.

## 5 Conclusion

This paper presents details on the development of a new type of PC with significantly improved strength, superior ductility, high fracture toughness, and self-sensing capabilities. It is apparent that incorporating MWCNTs or ANPs in polymer resin prior to fabricating PC can result in significant improvement in the mechanical characteristics of PC. The new PC presents an attractive alternative for smart and resilient structures where the improved mechanical behavior is needed.

**Acknowledgments** The author cordially thanks the financial support over the last 10 years of numerous funding agencies including NSF, USDOT, US DOE, NMDOT, Transpo Industries, and Epoxy Chemicals. Special thanks to all post-doctors and graduate students who participated in this research over the years.

## References

1. ACI Committee 548. (2009). Guide for the Use of Polymers in Concrete, ACI 548.1R-2009, American Concrete Institute, Farmington Hills, MI, USA.
2. Ohama, Y. (1995). *Handbook of polymer-modified concrete and mortars: Properties and process technology*. New York: William Andrew.
3. Laredo dos Reis, J. M. (2005). Mechanical characterization of fiber reinforced polymer concrete. *Materials Research, 8*, 357–360.
4. Wheat, D. L., Fowler, D. W., & Negheimish, A. I. (1993). Thermal and fatigue behavior of polymer concrete overlaid beams. *ASCE Journal of Materials in Civil Engineering, 5*, 460–477.
5. Yeon, K.-S., Yeon, J. H., Choi, Y.-S., & Min, S.-H. (2014). Deformation behavior of acrylic polymer concrete: Effects of methacrylic acid and curing temperature. *Construction and Building Materials, 63*, 125–131.
6. Elalaoui, O., Ghorbel, E., Mignot, V., & Ouezdou, M. B. (2012). Mechanical and physical properties of epoxy polymer concrete after exposure to temperatures up to 250 °C. *Construction and Building Materials, 27*, 415–424.
7. Shokrieh, M., Rezvani, S., & Mosalmani, R. (2015). A novel polymer concrete made from fine silica sand and polyester. *Mechanics of Composite Materials, 51*(5), 571–580.
8. Agavriloaie, L., Oprea, S., Barbuta, M., & Luca, F. (2012). Characterization of polymer concrete with epoxy polyurethane acryl matrix. *Construction and Building Materials, 37*, 190–196.
9. Reis, J. M. L. (2006). Fracture and flexural characterization of natural fiber-reinforced polymer concrete. *Construction and Building Materials, 20*, 673–678.
10. Son, S.-W., & Yeon, J. H. (2012). Mechanical properties of acrylic polymer concrete containing methacrylic acid as an additive. *Construction and Building Materials, 37*, 669–679.

11. Reis, J. M. L., & Ferreira, A. J. M. (2004). Assessment of fracture properties of epoxy polymer concrete reinforced with short carbon and glass fibers. *Construction and Building Materials, 18*, 523–528.
12. Martínez-Barrera, G., Vigueras-Santiago, E., Martínez-López, M., Ribeiro, M. C. S., Ferreira, A. J. M., & Brostow, W. (2013). Luffa fibers and gamma radiation as improvement tools of polymer concrete. *Construction and Building Materials, 47*, 86–91.
13. Toufigh, V., Toufigh, V., Saadatmanesh, H., & Ahmari, S. (2013). Strength evaluation and energy-dissipation behavior of fiber-reinforced polymer concrete. *Advances in Civil Engineering Materials, 2*, 622–636.
14. Ganguli, S., Aglan, H., Dennig, P., & Irvin, G. (2006). Effect of loading and surface modification of MWCNTs on the fracture behavior of epoxy nanocomposites. *Journal of Reinforced Plastics and Composites, 25*, 175–188.
15. Theodore, M., Hosur, M., Thomas, J., & Jeelani, S. (2011). Influence of functionalization on proper-ties of MWCNT–epoxy nanocomposites. *Materials Science and Engineering A, 528*(3), 1192–1200.
16. Zhu, J., Peng, H., Rodriguez-Macias, F., Margrave, J., Khabashesku, V., Imam, A., & Barrera, E. (2004). Reinforcing epoxy polymer composites through covalent integration of functionalized nanotubes. *Advanced Functional Materials, 14*, 643–648.
17. David, O. B., Banks-Sills, L., Aboudi, J., Fourman, V., Eliasi, R., Simhi, T., & Raz, O. (2014). Evaluation of the mechanical properties of PMMA reinforced with carbon nanotubes experiments and modeling. *Experimental Mechanics: International Journal, 54*, 175–186.
18. Yu, N., Zhang, Z. H., & He, S. Y. (2008). Fracture toughness and fatigue life of MWCNT/ epoxy composites. *Materials Science and Engineering A, 494*, 380–384.
19. Tang, L.-C., Wan, Y.-J., Peng, K., Pei, Y.-B., Wu, L.-B., Chen, L.-M., & Lai, G.-Q. (2013). Fracture toughness and electrical conductivity of epoxy composites filled with carbon nanotubes and spherical particles. *Composites Part A: Applied Science and Manufacturing, 45*, 95–101.
20. Zhou, D., & Chow, L. (2003). Complex structure of carbon nanotubes and their implications for formation mechanism. *Journal of Applied Physics, 93*(12), 9972–9976.
21. Sharma, R., & Iqbal, Z. (2004). In situ observations of carbon nanotube formation using environmental transmission electron microscopy. *Applied Physics Letters, 84*, 990.
22. Swain, S., Sharma, R. A., Patil, S., Bhattacharya, S., Gadiyaram, S. P., & Chaudhari, L. (2012). Effect of allyl modified/silane modified multiwalled carbon nano tubes on the electrical properties of unsaturated polyester resin composites. *Transactions on Electrical and Electronic Materials, 13*(6), 267–272.
23. Thakre, P., Lagoudas, D., Riddick, J., Gates, T., Frankland, S. J., Ratcliffe, J., & Barrera, E. (2011). Investigation of the effect of single wall carbon nanotubes on interlaminar fracture toughness of woven carbon fiber-epoxy composites. *Journal of Composite Materials, 45*, 1091–1107.
24. Wetzel, B., Rosso, P., Haupert, F., & Friedrich, K. (2006). Epoxy nanocomposites - fracture and toughening mechanisms. *Engineering Fracture Mechanics, 73*, 2375–2398.
25. Salemi, N., & Behfarnia, K. (2013). Effect of nano-particles on durability of fiber-reinforced concrete pavement. *Construction and Building Materials, 48*, 934–941.
26. ASTM C109. (2008). Standard Test Method for Compressive Strength of Hydraulic Cement Mortars, ASTM International, West Conshohocken, PA, USA.
27. ASTM D7136. (2007). Standard Test Method for Measuring the Damage Resistance of a Fiber-Reinforced Polymer Matrix Composite to a Drop-Weight Impact Event. ASTM International, West Conshohocken, PA, USA.
28. ASTM C78. (2002). Standard Test Method for Flexural Strength of Concrete (Using Simple Beam with Third-Point Loading), ASTM International, West Conshohocken, PA, USA.
29. ASTM D638. (2014). Standard Test Method for Tensile Properties of Plastics, ASTM International, West Conshohocken, PA, USA.

30. ACI 446. (2009). *Fracture toughness testing of concrete*. Farmington Hills: American Concrete Institute.
31. Douba, A. E., Emiroglu, M., Tarefder, R., Kandil, U. F., & Reda Taha, M. M. (2017). Improving fracture toughness of polymer concrete using carbon nanotubes. *Journal of TRB, 2612*, 96–103.
32. Emiroglu, M., Douba, A. E., Tarefder, R., Kandil, U. F., & Reda Taha, M. M. (2017). New polymer concrete with superior ductility and fracture toughness using alumina nanoparticle. *ASCE: Journal of Materials in Civil Engineering, 29*(8), 04017069.
33. Reda Taha, M. M., Xiao, S., Yi, J., & Shrive, N. G. (2002). Evaluation of flexural fracture toughness for quasi-brittle structural materials using a simple test method. *Canadian Journal of Civil Engineering, 29*(4), 567–575.
34. Daghash, S. M., Soliman, E., Kandil, U. F., & Reda Taha, M. M. (2016). Improving impact resistance of polymer concrete using CNTs. *International Journal of Concrete Structures and Materials, 10*(4), 539–553.

# Part II
# Polymer Materials

# Chapter 7
# Bio-Based Superplasticizers for Cement-Based Materials

**Stephan Partschefeld and Andrea Osburg**

The additives to regulate the fresh and solid properties of mortar and concrete are of enormous importance to the building materials industry. The superplasticizers with a proportion of 85% of all admixtures are the most frequently used type of additives. With a share of 60%, the polycarboxylate ethers dominate the market for building chemical additives. Starch, which can be modified by appropriate chemical modification in such a way that fluidizing properties are produced, represents an alternative basic material. For this purpose, it is necessary first to reduce the molecular weight of the natural polysaccharide and then to introduce anionic charges into the degraded starch molecules. The influence on the rheological properties of Portland cement pastes was determined by means of rotational viscosimeters and compared with commercially available superplasticizers. To determine the interaction mechanism with Portland cement, adsorption experiments and calorimetric studies were carried out. At the same concentration, the synthesized starch superplasticizers show comparable flow characteristics to the PCE superplasticizers. The interaction mechanism is based on adsorption at the first hydration products of the cement. A low degree of polymerization and a high amount of anionic charges showed the most intensive liquefaction effect.

## 1 Introduction

In 2016, the world consumption of cement was approximately 4.2 billion tons, with 7 kg per ton of cement used in admixtures [1]. In this case, 85% of all additives were flow agents in which 60% of them were polycarboxylate ether (PCE), so

---

S. Partschefeld (✉) · A. Osburg
Bauhaus-Universität Weimar, Faculty of Civil Engineering, F. A. Finger-Institute for Building Material Engineering, Chair of Building Chemistry and Polymer Materials, Weimar, Germany
e-mail: stephan.partschefeld@uni-weimar.de

© Springer International Publishing AG, part of Springer Nature 2018
M. M. Reda Taha (ed.), *International Congress on Polymers in Concrete (ICPIC 2018)*, https://doi.org/10.1007/978-3-319-78175-4_7

approximately 15 million tons of PCE superplasticizer was needed [2]. Looking at the forecast for the next few years, cement consumption is going to continue to increase, and therefore the demand for superplasticizers is likely to rise disproportionately. For this reason, it is necessary to look for alternative solutions and to use renewable resources to replace petrochemical products like polycarboxylate ethers. If it is possible to develop superplasticizers from renewable raw materials for cement-based construction materials, then a much bigger application field than now will open up. While starch derivatives have previously been used only in division areas like plaster and tile adhesives systems, the application as additives in concrete may now be possible [3]. Driving power in this project is the preservation of fossil resources and the sustainable use of natural products and renewable resources, i.e., the creation of a large environmental benefit.

## 2  Materials and Methods

A Portland cement CEM I 52.5 N Milke classic® from HeidelbergCement was used for the following tests. A maize starch (Ingredion), a wheat starch (Roth), and a potato waste starch (Ablig), which is a residue in the production of potato products, were used as basic materials for the synthesis of the starch superplasticizers. The synthesis was carried out in two steps. First, the molecular weight of the starch was reduced by acid hydrolysis with HCl and, in the second step, anionic charges were introduced by sulfoethylation with sodium vinylsulfonate under alkaline conditions [4]. Figure 7.1 shows the reaction pattern for the synthesis of flow agents from starch.

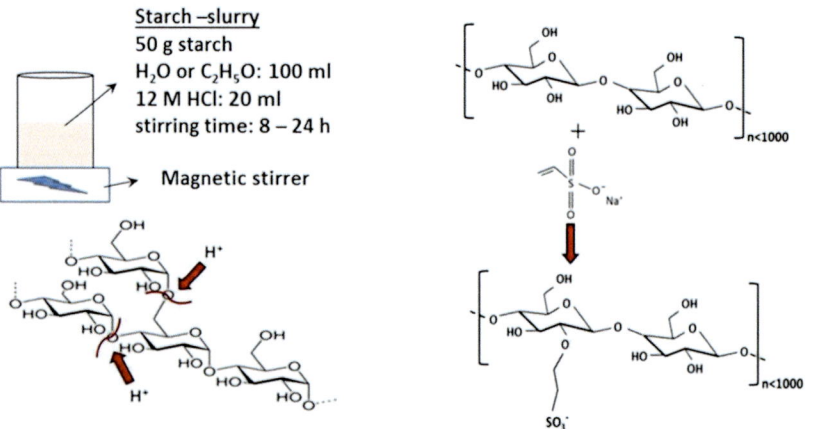

**Fig. 7.1** Reaction patterns for the synthesis of superplasticizers from starch

**Table 7.1** Molecular parameters of the generated starch flow agents

| Admixture | Type of starch | Molecular weight $M_n$ [g/mol] | Anionic charge [C/g] |
|---|---|---|---|
| Crystal-1 | Maize | 10,257 | 11.18 |
| Ro-1 | Wheat | 50,418 | 9.78 |
| Ro-2 | Wheat | 50,418 | 6.62 |
| Ab-1 | Potato (waste starch) | 20,920 | 6.85 |

By varying the process parameters (time, concentration, and slurry medium) during synthesis, different starch flow agents, with varying molecular weight and anionic charges, were produced. Table 7.1 shows the molecular parameters of the synthesized starch superplasticizers determined by means of size exclusion chromatography and charge titration.

The dispersing performance of the starch superplasticizers was determined by means of rheological investigations with a rotation viscometer (Rheotec, Brookfield DV III-ultra) and compared with commercially available superplasticizers polycarboxylate ether (PCE) and polycondensate (PC). Deionized water-to-cement ratio (W/C) of the cement pastes was approximately 0.35. The influence of the starch flow agents on early hydration was determined by calorimetric studies with an isothermal calorimeter (C3-Prozesstechnik) at 20 °C. The interaction principle was carried out by adsorption experiments using the phenol-sulfuric acid method [5]. Suspensions consisting of starch flow agent and 0.02 M Ca(OH)$_2$ solution were stirred with cement (liquid/solid ratio of 20) for 2 h and then centrifuged at 6000 rpm. The concentration of the starch flow agents in the supernatant was determined by means of the resulting coloration, as a result of the reaction with 5% phenol solution and 96% sulfuric acid at 490 nm wavelength with a spectral photometer (UV line, Zeiss). The adsorption rate was determined by comparison with the initial concentration [6].

# 3 Results and Discussion

Figure 7.2 showed that the rheological properties of the cement pastes were significantly modified by the addition of the superplasticizers. While the liquefying effect of the polycondensate flow agent is relatively low at 0.5 wt.%, the starch flow agent exhibits a similarly strong dispersing performance as the PCE superplasticizer. The comparison of the different starch superplasticizers showed that the samples with the highest amount of anionic charges, Crystal-1 and Ro-1, had the highest dispersing performance in the cement pastes. The comparison of the sample Ro-2 and Ab-1 showed that a low molecular weight had a higher dispersing performance at similarly the same amount of anionic charges. Generally, a low molecular weight ($M_n$ < 100,000 Da) and high anionic charges (>5 C/g) are the key parameters for a high dispersing performance. In addition to influencing the rheological properties, the starch flow agents have a significant influence on the early hydration of the Portland cement. Compared to commercially available PCE and PC, which also

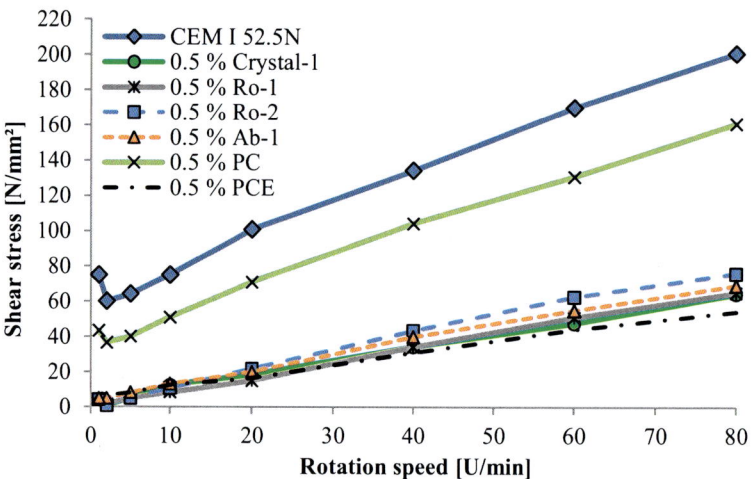

**Fig. 7.2** Influence of the starch superplasticizers and marketable flow agents on the flow curves of the Portland cement ($W/C = 0.35$)

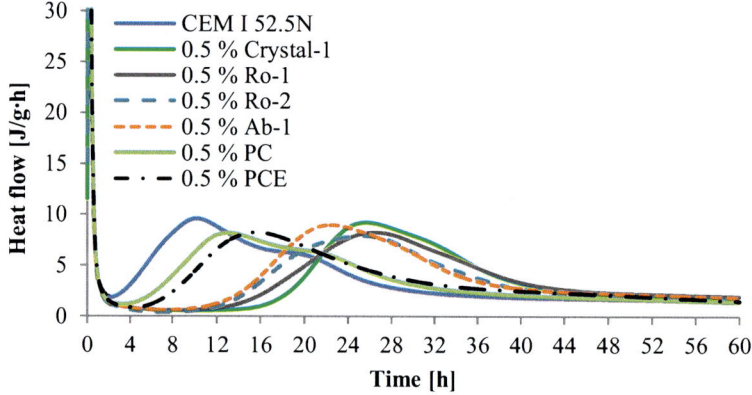

**Fig. 7.3** Influence of the starch superplasticizers and marketable flow agents on the early hydration of the Portland cement ($W/C = 0.35$) measured isothermal at 20 °C

delay the hydration, the starch flow agents show an even higher delay in cement hydration. Figure 7.3 shows the influence of the superplasticizers on early hydration by calorimetric measurements.

It is apparent that especially the pastes containing the starch superplasticizers with high anionic charges showed the highest delay of the hydration maximum. Especially the samples Crystal-1 and Ro-1 delay the hydration maximum up to 26 h. The starch flow agents Ro-2 and Ab-2 with lower anionic charges raise the maximum of hydration to 22 h.

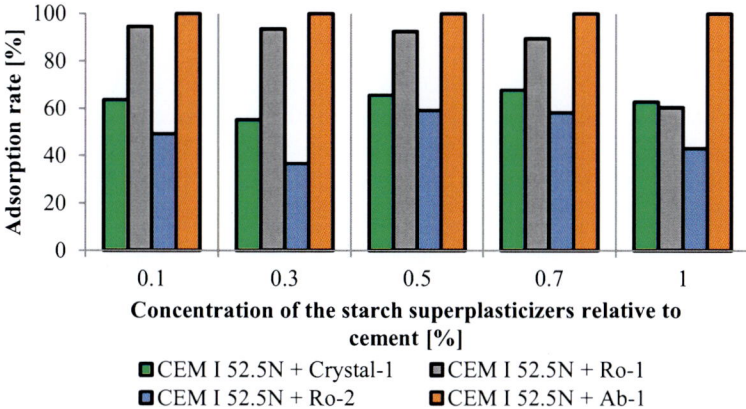

**Fig. 7.4** Adsorption rate of the superplasticizers in dependence of their concentration relative to the cement

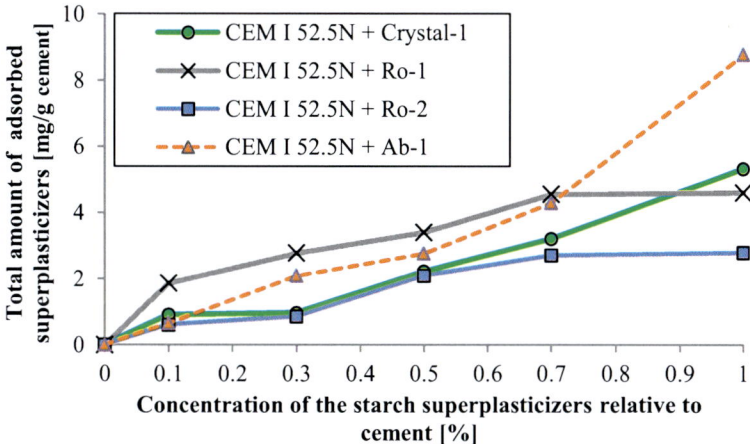

**Fig. 7.5** Total amount of the adsorbed starch superplasticizers relative to cement

Moreover, it was found that the dormant period is lengthened up to 9 h by the starch flow agents. The measurements indicate that the mechanism of interaction is based on the adsorption on the first hydration products. The results of the adsorption experiments are represented in Figs. 7.4 and 7.5. It is apparent that the interaction mechanism is based on the adsorption of the starch flow agents on the first hydration products.

The comparison of the adsorption rates of the samples Ro-1 and Ro-2 suggests that the higher the amount of anionic charge, the more superplasticizer is adsorbed on the cement surface.

The comparison of samples Ro-2 and Ab-1 with similarly the same amount of anionic charge showed that the lower the molecular weight of the starch flow agents, the more flow agent can adsorb on the cement surface. The calculation of the total amount of adsorbed starch superplasticizer showed that especially the samples Ro-1 and Ro-2 saturate the cement surface at limit concentrations of 0.7%. After that, a further addition of the samples does not lead to more adsorption on the cement surface. However, measurements at higher concentrations up to 5% could show whether there is a formation of double layers or more layers of superplasticizers on the cement surface.

## 4 Conclusions

Based upon this study, it can be concluded that by relatively simple chemical modification, superplasticizers based on native or waste starch can be produced. One of the main conclusions of this study is that the key parameters for creating superplasticizers from starch are the molecular weight and the amount of anionic charges introduced. The results demonstrated that the reduction of the molecular weight to DP < 400 and the introduction of anionic charges are crucial to attaining a high dispersing performance. It was noted that the higher the amount of anionic charges, the lower the yield point and, accordingly, the viscosity of the cement pastes is. An important conclusion, related to the mechanism of interaction, are the synthesized starch superplasticizers adsorbed on the surface of the first hydration products of the cement after mixing. This led to an extension of the dormant period in cement hydration and delayed the maximum of hydration up to 26 h. Therefore, this type of superplasticizers is suitable for the use in transport concrete.

## References

1. Global Cement Report 12th Edition, International Cement review. https://www.cemnet.com/Publications/Item/176633/the-global-cement-report-12th-edition.html.
2. Plank, J. Concrete admixtures – Where are we now and what can we expect in the future.19. Ibausil Proceedings, Weimar, 16.09.2015–18.09.2015, 32–41.
3. Crépy, L., Petit, J.-Y., Wirquin, E., Martin, P., & Joly, N. (2014). Synthesis and evaluation of starch based polymers as potential dispersants in cement pastes and self leveling compounds. *Cement and Concrete Composites, 45*, 29–38.
4. Vieira, M. C., Klemm, D., Einfeldt, L., & Albrecht, G. (2005). Dispersing agents for cement based on modified polysaccharides. *Cement and Concrete Research, 35*, 883–890.
5. Pourchez, J., Ruot, B., Debayle, J., Pourchez, E., & Grosseau, P. (2010). Some aspects of cellulose ethers influence on water transport and porous structure of cement-based materials. *Cement and Concrete Research, 40*, 242–252.
6. Peschard, A., Govin, A., Pourchez, J., Bertrand, L., Maximilien, S., & Guilhot, B. (2006). Effects of polysaccharides on the hydration of cement suspension. *Journal of the European Ceramic Society, 26*, 1439–1445.

# Chapter 8
# Screening Encapsulated Polymeric Healing Agents for Carbonation-Exposed Self-Healing Concrete, Service Life Extension, and Environmental Benefit

Philip Van den Heede, Bjorn Van Belleghem, Maria Adelaide Araújo, João Feiteira, and Nele De Belie

By incorporating encapsulated polymers in concrete, cracks can be healed autonomously upon occurrence. This is of high value for steel reinforced concrete structures subject to carbonation-induced corrosion. This paper presents the results of a rapid colorimetric screening test to assess the carbonation resistance of self-healing concretes containing encapsulated polymer-based healing agents. Four systems were tested for inhibition of further carbonation near artificially induced cracks (width: 300 μm). Next, the time to steel depassivation was assessed probabilistically in comparison with cracked concrete. With an adequately working pressurized PU precursor, the concrete would remain repair-free for at least 100 years. Subsequent life cycle assessment in SimaPro showed a potential environmental benefit (72–78%) for the ten CML-IA baseline impact categories which is mainly due to the service life extension possible with a properly working self-healing concrete.

---

P. Van den Heede · B. Van Belleghem
Magnel Laboratory for Concrete Research, Department of Structural Engineering, Ghent University, Ghent, Belgium

Strategic Initiative Materials (SIM vzw), project ISHECO within the program 'SHE', Ghent, Belgium

M. A. Araújo
Magnel Laboratory for Concrete Research, Department of Structural Engineering, Ghent University, Ghent, Belgium

Polymer Chemistry and Biomaterials Group, Department of Organic and Macromolecular Chemistry, Ghent University, Ghent, Belgium

J. Feiteira · N. De Belie (✉)
Magnel Laboratory for Concrete Research, Department of Structural Engineering, Ghent University, Ghent, Belgium
e-mail: nele.debelie@ugent.be

# 1 Introduction

Giving concrete self-healing properties is seen as a relevant goal since it has a low tensile strength and is therefore quite susceptible to cracking. These cracks count as preferential pathways for corrosion-inducing substances. One possibility of ensuring autonomous crack healing would consist of embedding encapsulated polymeric healing agents (HAs) that break upon crack occurrence, upon which the HA is released from the capsules into the crack and hardens [1]. Until now, the healing performance was mainly assessed on a proof-of-concept level by subjecting the self-healing concrete to sorptivity/permeability tests and quantifying the reduced water uptake after crack healing [2]. Although the output of such tests is certainly relevant, it cannot be linked directly with the service life extension that can be achieved. The latter information is considered imperative to convince engineers and contractors to use self-healing concrete on a larger scale. Therefore, the service life extension possible with an encapsulated in-house-developed high-viscosity and a commercial low-viscosity polyurethane (PU) precursor has been investigated recently for marine, chloride-exposed environments [3]. Nevertheless, corrosion can also be initiated by carbonation as $CO_2$ can penetrate the concrete, reach the rebar, and depassivate it. Carbonation-exposed self-healing concrete will only be effective if the HA in the crack can inhibit further ingress of $CO_2$ gas which may be more difficult than creating a barrier against a chloride-bearing liquid. The preference may go to a different HA then. In this paper, four types of polymeric HAs were screened for carbonation resistance. Service life and life cycle assessment (LCA) were also done.

# 2 (Self-Healing) Concrete Manufacturing and Preconditioning

Carbonation tests should be done on concrete for carbonation-exposed areas, e.g., exposure class XC3 (moderately humid, while sheltered from rain). The preference would then go to an ordinary Portland cement (OPC) concrete with a cement content of 300 kg/m$^3$ and a water-to-cement (W/C) ratio of 0.55 (cf. NBN B15-001). However, carbonation of an OPC system would require a substantial exposure time. Therefore, 50% of the cement was replaced with pozzolanic class F fly ash. Six test series of nine cylindrical test specimens were cast in the same mold setup as in [2]: an uncracked reference (UN), a cracked reference (CR), and self-healing concrete with four different encapsulated HAs: HA_1 being a noncommercial high-viscosity PU precursor, HA_2 being a commercial low-viscosity PU precursor, HA_3 being HA_2 with addition of 5 wt.% of accelerator and 5 wt.% of benzoyl peroxide (BPO) to pressurize the HA inside capsules and ensure a better flow-out once broken, and HA_4 being an in-house-developed two-component, elastic, and pH-sensitive acrylate end-capped polymer. More information on the precise composition of each of the abovementioned HAs and their capsules for encapsulation can be found in [2]. Cracks with a 25 mm depth were induced by putting thin brass plates with a thickness of 0.3 mm in the molds just before casting. For HA_1, HA_2, and HA_3, the plates contained three holes with a 20 mm spacing. The capsules containing the PU precursor were put through these holes and fixed with nylon threads to the sides of the molds. For HA_4, the configuration was a little bit different. Two coupled capsules were going through the plates with a 25 mm spacing. For the cracked reference, thin brass plates without holes were used. After casting, capsule breakage and crack healing were triggered by pulling out the thin brass plates from the hardened concrete. This was done after 7 days of optimal curing at 20 °C and 95% relative humidity (RH). Once the HA became visible near the crack mouth (after around 1 h), the samples were cured further on for another 7 days.

## 3 Carbonation Test and Healing Performance

After curing, the samples were covered with Al-tape on all sides except for one 100-mm-diameter exposure surface. Then, they were stored in a chamber with 1% $CO_2$, 20 °C, and 60% RH. Fourteen days later, the samples were split. Subsequently, the carbonation front was visualized with phenolphthalein which leaves the carbonated area colorless and the non-carbonated zone purple. In an unhealed crack, it extends beyond the $\pm 9$ mm carbonation depth for uncracked concrete (Fig. 8.1a, b).

When measuring the average carbonation depth inside the crack starting from the crack wall and tip onward, values of 1–3 mm were recorded. The fact that the latter carbonation depths were substantially less than the carbonation depth from the uncracked sample surface may be related to the humidity inside the crack. There, the RH most probably locally differs from the 60% that is generally maintained in the carbonation chamber which is ideal for carbonation to occur to its fullest extent. The carbonation rate from the uncracked concrete surface can clearly not be assumed equal to the carbonation rate from the crack tip onward. Thus, simply assuming a reduced concrete cover on top of the steel reinforcement at the crack location without modifying the carbonation rate accordingly would be an incorrect approach in service life prediction. The carbonation rates from the uncracked surface and from

 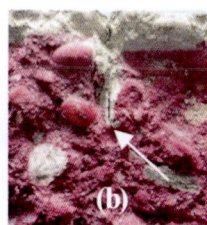  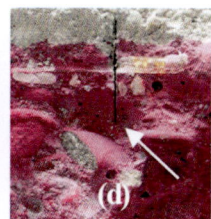

**Fig. 8.1** Carbonation front: (un)cracked (**a**, **b**), partially (**c**) and fully (**d**) healed

**Table 8.1** Crack healing performance of the four encapsulated HAs

| Test series | HA_1 | HA_2 | HA_3 | HA_4 |
|---|---|---|---|---|
| No. of samples healed | 0/9 | 5/9 | 8/9 | 2/9 |

the crack tip differ with a factor $3 \pm 1$. This factor was applied to the $R_{ACC,0}^{-1}$ value of OPC concrete when estimating the time to carbonation-induced steel depassivation (see subsequent section). For the test series with self-healing properties, two options exist. Either crack healing was incomplete, meaning that the carbonation front (see Fig. 8.1c) was not that different from the one that was recorded for the cracked test series (see Fig. 8.1b), or there was full crack healing with no carbonation visible at the crack wall and crack tip (see Fig. 8.1d). For incomplete healing, service life prediction also requires a factor to be applied to the $R_{ACC,0}^{-1}$ value in combination with a reduced concrete cover on top of the reinforcing steel (the concrete cover minus the 25 mm crack depth). However, unlike for the CR test series, this factor should not equal 3 as the carbonation depth from the crack is usually only around 1/8 of the carbonation depth from the surface away from the crack. Thus, for incomplete healing, the factor was assumed to be $8 \pm 2$ instead of $3 \pm 1$.

For now, rather disappointing results were obtained for HA_1 as none of the nine samples showed a fully healed crack (Table 8.1). Most probably, the high viscosity of the HA prevented a full flow-out from the capsule upon breakage and therefore complete crack filling. Using a low-viscosity commercial PU precursor (HA_2) on the other hand already proved a lot more beneficial. For this test series, already five out of nine samples behaved as if uncracked. The addition of small dosages of accelerator and BPO to the low-viscosity PU precursor (HA_3) led to even more satisfactory results with eight out of nine samples healed. Pressurizing the HA inside the capsules indeed seems to ensure a better release from the capsules and better closure near the crack mouth. Finally, the use of the two-component acrylate end-capped polymer (HA_4) as HA yielded a rather poor crack healing with only two out of nine samples healed. For that HA to work properly, not only an adequate release from the capsules but also a proper mixing of the two components is needed. In that perspective, the encapsulated HA_4 had a disadvantage in competing with the other HAs. The authors are fully aware of that. From the start, the goal was to develop a 1-component HA_4. The 2-component one that was tested now must be seen as an intermediate product in that development process. Further research is ongoing on how the two components can become one using a latent cross-linker.

## 4 Service Life Prediction and Life Cycle Assessment

The time to carbonation-induced steel depassivation was estimated using the limit state function mentioned in fib Bulletin 34 without weather function (Eq. 8.1).

$$g(d, x_c(t)) = d - \sqrt{2 \cdot k_e \cdot k_c \cdot \left(k_t \cdot R_{ACC,0}^{-1} + \varepsilon_t\right) \cdot C_S} \cdot \sqrt{t} \qquad (8.1)$$

with $d$, concrete cover [35 / 10 ± 8 mm, lognormal]; $x_c(t)$, carbonation depth at time t [years]; $k_e$, environmental function; $k_c$, execution transfer parameter; $k_t$, regression parameter [1.25 ± 0.35, normal]; $R^{-1}{}_{ACC,0}$, accelerated inverse effective carbonation resistance [(mm$^2$/years)/(kg/m$^3$), normal]; $\varepsilon_t$, error term [315.5 ± 48 (mm$^2$/years)/ (kg/m$^3$), normal]; and $C_S$, $CO_2$ concentration [0.00082 ± 0.0001 kg/m$^3$, normal]. $k_e$ takes into account that the RH of the environment [RH$_{real}$: 79 ± 9%, beta] differs from the reference RH imposed during the accelerated carbonation test [RH$_{ref}$: 60%, constant]. It depends on two regression parameters $f_e$ [5.0, constant] and $g_e$ [2.5, constant]. $k_c$ accounts for the curing effect and depends on the curing period $t_c$ [28 d, constant] and an exponent of regression $b_c$ [−0.807 ± 0.050, normal] (Eq. 8.2).

$$k_e \cdot k_c = \left(\left(1 - \left(\frac{RH_{real}}{100}\right)^{f_e}\right) / \left(1 - \left(\frac{RH_{real}}{100}\right)^{f_e}\right)\right)^{g_e} \cdot \left(\frac{t_c}{7}\right)^{b_c} \qquad (8.2)$$

Three scenarios were considered. For the first one, full crack healing was assumed, meaning that the concrete was modeled as if uncracked, with a full concrete cover of 35 mm and the inverse effective carbonation resistance valid for the concrete surface away from the crack (2684 ± 303 (mm$^2$/years)/(kg/m$^3$)) [4]. Secondly, incomplete crack healing was accounted for by reducing d to only 10 mm (= 35–25 mm) and using a $R_{ACC,0}^{-1}$ value to which the earlier mentioned factor 8 was applied (= 336 (mm$^2$/years)/(kg/m$^3$)). Finally, the cracked condition was simulated by assuming the same concrete cover of only 10 mm in combination with a $R_{ACC,0}^{-1}$ value of 895 (mm$^2$/years)/(kg/m$^3$) which is 1/3 of the value for the uncracked condition. Time-dependent reliability indices ($\beta \geq 1.3$) and probabilities of failure ($P_f \leq 0.10$) were calculated using the first-order reliability method of Comrel. When comparing the healed/uncracked, the partially healed, and the cracked condition, it is clear that the presence of a crack should be avoided at all cost. While it easily takes more than 100 years for the carbonation front to reach the rebar at a depth of 35 mm for the healed/uncracked condition, the presence of a 25 mm deep crack with a 0.3 mm width shortens this time to only 24 years. One should always aim for 100% healing because for a partially healed crack, it would only take about 24 years more – which is still less than 100 years – for the carbonation front to reach the rebar.

The LCA was very similar to the one described in [3] for a steel reinforced concrete slab with a design service life of 100 years as functional unit. Only now, the aspect of service life is related to carbonation instead of chloride-induced corrosion. For the cracked concrete slab, it was assumed that the entire 35 mm concrete cover is replaced as soon as the carbonation front reaches the rebar. Thus, when aiming at a design service life of 100 years, the original concrete cover volume plus four concrete repair volumes need to be considered in the LCA. On the other hand, there is the self-healing concrete slab which normally contains one layer of 4150 PU

**Table 8.2** CML-IA baseline impact indicators while assuming full healing

| Parameter | CR | HA_1 (%) | HA_2 (%) | HA_3 (%) | HA_4 (%) |
|---|---|---|---|---|---|
| Abiotic depletion ($\times 10^3$ MJ fossil) | 1.2 | −73 | −72 | −72 | −68 |
| Global warming ($\times 10^2$ kg $CO_2$ eq) | 2.6 | −78 | −78 | −78 | −77 |
| Ozone depletion ($\times 10^{-6}$ kg CFC-11 eq) | 8.2 | −78 | −78 | −77 | −71 |
| Human toxicity ($\times 10^1$ kg 1,4-DB eq) | 2.4 | −77 | −76 | −76 | −73 |
| Freshwater ecotoxicity ($\times 10^1$ kg 1,4-DB eq) | 1.5 | −75 | −75 | −74 | −67 |
| Marine ecotoxicity ($\times 10^4$ kg 1,4-DB eq) | 5.7 | −77 | −76 | −76 | −68 |
| Terrestrial ecotoxicity ($\times 10^{-1}$ kg 1,4-DB eq) | 3.8 | −78 | −77 | −77 | −70 |
| Photochemical oxidation ($\times 10^{-2}$ kg $C_2H_4$ eq) | 1.7 | −75 | −74 | −74 | −66 |
| Acidification ($\times 10^{-1}$ kg $SO_2$ eq) | 4.7 | −76 | −75 | −75 | −70 |
| Eutrophication ($\times 10^{-1}$ kg $PO_4$ eq) | 1.3 | −75 | −74 | −74 | −67 |

precursor filled capsules in case of HA_1 and HA_2 [3]. The same number of capsules can be assumed for HA_3, while 1.3 times this value (5394) can be taken into account for the coupled HA_4. In case of full crack healing, only the original concrete cover volume with inclusion of the required number of capsules filled with HA should be assessed in terms of environmental impact. The life cycle inventory data related to the concrete constituents (only this time with a 100% OPC binder system (300 kg/m³) and W/C: 0.55), the glass capsules HA_1 and HA_2, were identical to those used in [3]. The renowned Ecoinvent database was used as main data source. Modifications were done to existing datasets for the two HAs to make them more representative for the actual materials that were used. A similar approach was adopted for HA_3 and HA_4 with the use of additional stoichiometric information. All LCA calculations were performed in SimaPro 8 using the CML-IA method with ten baseline indicators (Table 8.2). Under the theoretical assumption that each of the proposed HAs could guarantee a 100-year repair-free service life, substantial environmental benefits seem possible for all ten impact indicators. For HA_1, HA_2, HA_3, and HA_4, the reductions in impact would amount to 73–78%, 72–78%, 72–78, and 66–77%, respectively. Apparently, a well-working HA_4 would impose a slightly higher overall environmental impact onto the slab than the PU-based HAs.

# 5 Conclusions

The presence of one layer of 4150–5394 capsules filled with polymeric HA does not impose substantial environmental burdens onto a steel reinforced concrete slab if this intervention eliminates cover replacements within the envisaged 100-year service life. However, carbonation screening tests revealed that the fully healed condition can only be achieved on a regular basis for a commercial low-viscosity PU precursor with addition of accelerator and BPO (HA_3) to pressurize the HA in the capsules for better flow-out and crack closure. For that specific HA, the reduction in impact would amount to 72–78% for the 10 CML-IA baseline impact indicators.

**Acknowledgments** This research under the program SHE (Engineered Self-Healing materials), project ISHECO (Impact of Self-Healing Engineered materials on steel COrrosion of reinforced concrete) was funded by SIM (Strategic Initiative Materials in Flanders) and VLAIO (Flanders Innovation & Entrepreneurship). Their financial support as well as the one from the EU 7th Framework Program (FP7/2007–2013) under grant agreement n° 309451 (HEALCON) is gratefully acknowledged.

# References

1. Van den Heede, P., et al. (2016). Neutron radiography based visualization and profiling of water uptake in (un)cracked and autonomously healed cementitious materials. *Materials, 9*, 1–28.
2. Van den Heede, P., et al. (2017). Screening of different encapsulated polymer-based healing agents for chloride exposed self-healing concrete using chloride migration tests. In *Proceedings of the 6th international conference on non-traditional cement and concrete* (pp. 1–7). Brno.
3. Van Belleghem, B., et al. (2017). Quantification of the service life extension and environmental benefit of chloride exposed self-healing concrete. *Materials, 10*, 1–22.
4. Van den Heede, P., & De Belie, N. (2015). Durability based life cycle assessment of concrete with supplementary cementitious materials exposed to carbonation. In *Proceedings of the international conference on sustainable structural concrete I* (pp. 13–24). La Plata.

# Chapter 9
# Synthesis and Characterization of Superabsorbent Polymer Hydrogels Used as Internal Curing Agents: Impact of Particle Shape on Mortar Compressive Strength

Stacey L. Kelly, Matthew J. Krafcik, and Kendra A. Erk

Superabsorbent polymer hydrogels have proven to be effective internal curing agents for high-performance concrete because of their ability to absorb and release large amounts of water during hydration and thus mitigate autogenous shrinkage. In this study, the impact of hydrogel particle shape on the microstructure and compressive strength of internally cured mortar was experimentally determined. Inverse suspension polymerization was used to synthesize spherical poly(sodium-acrylate acrylamide) hydrogel particles, while solution polymerization was used to create similarly sized angular particles with identical chemical composition. The hydrogels were characterized with swelling tests in water and cement pore solution. Particle shape did not impact the swelling behavior, and micrographs confirmed that the particles maintained their shape during mixing and placement. Despite the introduction of spherical- and angular-shaped voids from the swollen hydrogel particles, there were no significant differences observed between the compressive strengths of the control mortar and the mortars containing either the spherical or angular hydrogel particles.

## 1 Introduction

High-performance concrete is produced with a low water-to-cement (w/c) ratio resulting in higher strength and a very dense microstructure. However, autogenous shrinkage typically occurs in the early stages of curing, leading to cracking and compromised strength [1]. Shrinkage can be alleviated by the addition of internal curing agents, which are additives with high water storage capacity that provide the

S. L. Kelly · M. J. Krafcik · K. A. Erk (✉)
School of Materials Engineering, Purdue University, West Lafayette, IN, USA
e-mail: erk@purdue.edu

cement grains with a source of water during the curing process. Superabsorbent polymer (SAP) hydrogels have proven to be effective internal curing agents because of their ability to absorb and release large amounts of water [2].

Although internal curing agents have many benefits, the mechanical strength of the concrete can be negatively affected by the hydrogels. The mechanical strength is expected to increase when internal curing is employed. The increase in hydration during curing reduces the amount of autogenous shrinkage and results in a strong, dense microstructure. However, the increase in strength of the mortar can be counteracted by the formation of larger macropores caused by the swollen hydrogel particles. At w/c $\leq$ 0.35, the strength loss due to the macropores is usually overcome by the increased degree of hydration. However, the concentration of internal curing agents added is also important. Hydrogel dosages between 0.3% and 0.6% by weight of cement (bwoc) are typically used when studying the impact of internal curing. In general, increasing the amount of internal curing agents causes a decrease in strength, eventually outweighing the effects of the increased hydration [3].

Researchers have examined the benefits that the hydrogels impart on internally cured concrete but the impact of hydrogel particle shape on mortar strength has not been directly investigated. A few studies have examined the impact of hydrogels available in industry, some of which were angular and some that were spherical. However, in these cases, the exact chemistry of the polymers were not known or the size distributions were not the same [4, 5]. Cross-linking density, hydrogel particle size, and other factors affect the shrinkage and strength of mortar. so the impact of particle shape could not be determined. In this study, hydrogels were synthesized using solution polymerization and inverse suspension polymerization to produce angular and spherical particles. The hydrogel chemistry and particle size were controlled and quantified. These custom-synthesized hydrogels were previously found to effectively reduce autogenous shrinkage in mortar [6, 7]. In this paper, results from mortar compression testing are presented and the impact of particle shape on mortar strength is discussed.

## 2 Experimental Methods

Random poly(sodium-acrylate acrylamide)(PANa-PAM) copolymer hydrogels of two concentrations (17 wt.% PANa-83 wt.% PAM and 83 wt.% PANa-17 wt.% PAM) were synthesized for this study. In the rest of the document, the hydrogels will be referred to by their sodium acrylate concentrations. Angular hydrogel particles were synthesized via solution polymerization following the procedure in Krafcik et al. [6]. Spherical particles were synthesized using inverse suspension polymerization. The exact proportions for this method are listed in Table 9.1. The continuous phase, consisting of 225 ml of cyclohexane and 0.75 g of Span-80, was prepared in a round-bottom flask with an overhead mixer and nitrogen inlet. The aqueous phase was prepared by mixing water, acrylic acid (AA), acrylamide (AM), sodium hydroxide (NaOH) solution, and N,N'-methylenebisacrylamide (MBAM). The first

# 9 Synthesis and Characterization of Superabsorbent Polymer Hydrogels Used...

**Table 9.1** Exact proportions used for inverse suspension polymerization

| Hydrogel type | AA  | AM  | Water | NaOH Soln. | MBAM | NaPS |
|---------------|-----|-----|-------|------------|------|------|
| 17 wt.% PANa  | 0.4 | 2.0 | 11.4  | 0.6        | 0.05 | 1    |
| 83 wt.% PANa  | 2.0 | 0.4 | 8.8   | 3.2        | 0.05 | 1    |

All items are in milliliters except for AM and MBAM, which are in grams

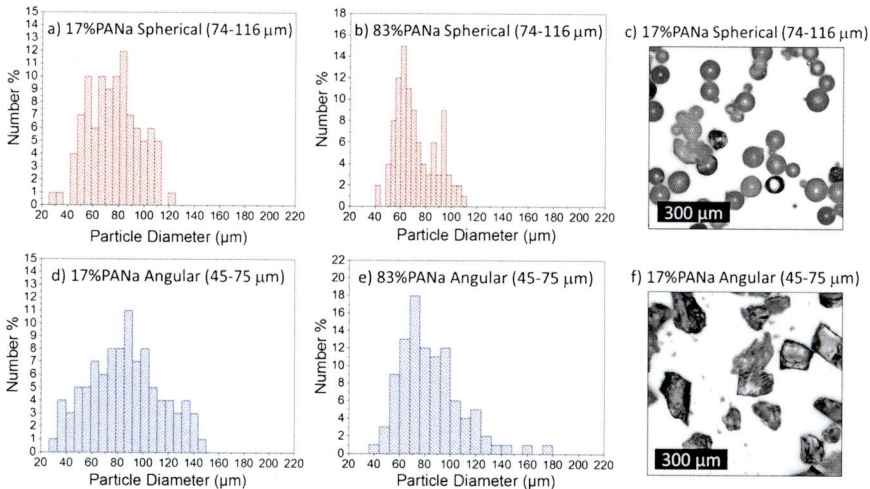

**Fig. 9.1** PSD and optical microscopy images of spherical and angular particles

initiator, sodium persulfate (NaPS), was added after 30 min of mixing. The aqueous solution was then transferred to the flask containing the continuous phase and the mixture was given 2 h to equilibrate. The polymerization was then catalyzed by the addition of 1.3 ml N,N,N',N'-tetramethylethylenediamine (TMED) and heated to 65 °C. After the polymerization was complete, the hydrogel particles were filtered from the continuous phase and rinsed with reverse osmosis (RO) water, ethanol, and acetone.

The hydrogel particles were sieved and particles were collected between 74 and 116 μm mesh for the spherical samples and 45 and 75 μm mesh for the angular samples. The particles' swelling behavior was analyzed in RO water and pore solution following the tea-bag method reported in Krafcik et al. [6]. Optical microscopy paired with ImageJ was used to determine the hydrogel particle size [7]. When the angular particles were optically measured, the longest dimension was recorded as the particle diameter. The hydrogel particle size distributions (PSD) are shown in Fig. 9.1. Note that despite the use of different mesh sieves, the PSD are very similar for the spherical and angular hydrogel particle samples.

The mortars for compression testing were cast into molds using the following mixture proportions: 666 kg cement, 233 kg water, 1440 kg sand, 0.2% hydrogels, and 4.7% water reducer (WRA) bwoc. Mortars were stored in an environmentally controlled chamber at (23.5 ± 0.5) °C and (50 ± 1)% relative humidity for 24 h.

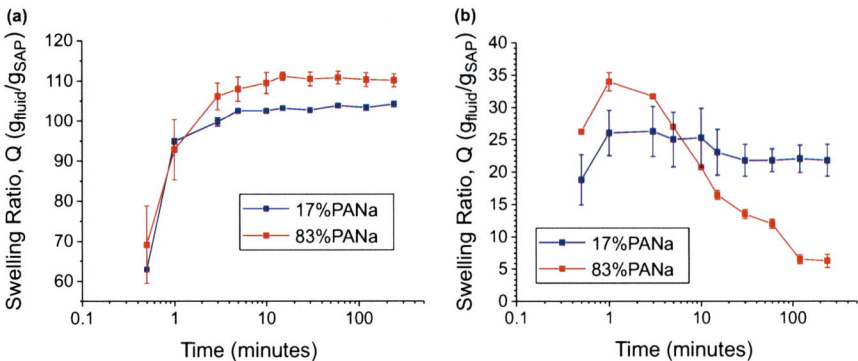

**Fig. 9.2** Swelling behavior of spherical 17 wt.% PANa and 83 wt.% PANa particles in (**a**) RO water and (**b**) pore solution

The samples were removed from the molds, sealed in airtight bags and returned to the chamber. Mortar samples were tested at 1, 3, 7, and 28 days after casting. Compressive strength was determined in accordance with ASTM-C109 [8].

## 3 Results and Discussion

Figure 9.2 reports the swelling behavior of spherical hydrogel particles in RO water and pore solution; results are comparable to the swelling kinetics of angular hydrogel particles reported previously [6]. As seen in Fig. 9.2a, the 83 wt.% PANa hydrogels displayed a higher equilibrium swelling ratio in RO water compared to the 17 wt.% PANa particles (i.e., 110 $g_{fluid}/g_{SAP}$ compared to 100 $g_{fluid}/g_{SAP}$). This was expected based on prior work [6, 9] due to higher concentration of sodium acrylate in the 83 wt.% hydrogels. The sodium acrylate groups along the polymer backbone deprotonate in alkaline solutions, and the repulsion of the resulting negative charges along the polymer backbone causes an increase in the osmotic pressure. As a result, liquid diffuses into the polymer network, and the hydrogel particle increases in volume.

As shown in Fig. 9.2b, the maximum swelling ratio of the hydrogels in pore solution was found to be significantly lower than the swelling ratio in RO water. The reduced swelling capacity and deswelling of the hydrogel particles is explained by the presence of cations in the pore solution. Sodium, potassium, and calcium are the cations with the highest concentration in pore solution [9]. The anionic sites on the polymer backbone electrostatically bond with these cations. Monovalent cations shield the anionic sites. Divalent and trivalent cations can complex with several sites on the polymer backbone. As reported previously, shielding and complexation reduce the volumetric expansion of the polymer network and cause the hydrogel to expel water from within the network [6, 9]. Deswelling is more prominent in

# 9 Synthesis and Characterization of Superabsorbent Polymer Hydrogels Used...

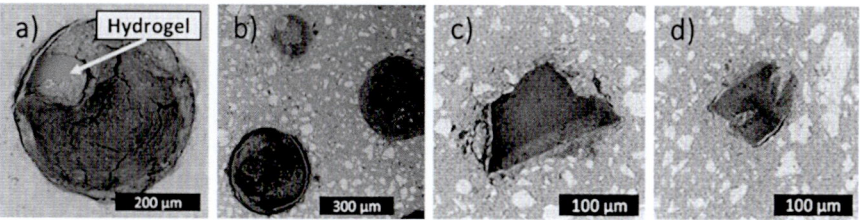

**Fig. 9.3** BSE images of cement paste containing spherical 17 wt.% PANa in (**a**) and (**b**) and angular 83 wt.% PANa in (**c**) and (**d**)

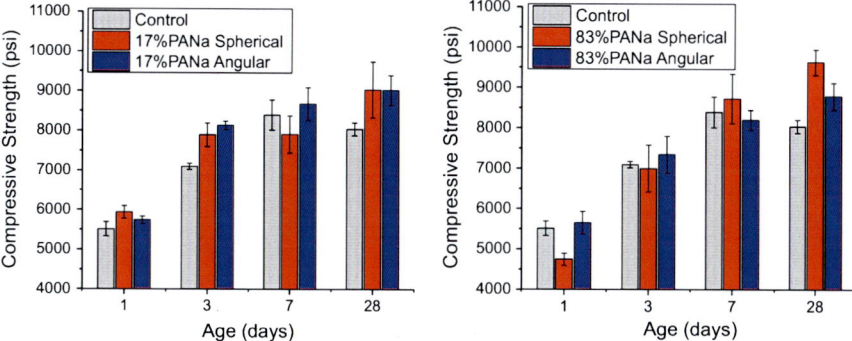

**Fig. 9.4** Compressive strength of mortar with and without hydrogel particles Each bar represents the average compressive strength of three samples

hydrogels containing higher amounts of sodium acrylate due to the higher amount of anionic sites on the polymer backbone.

Figure 9.3 displays backscattered electron (BSE) images of polished cement paste sections containing spherical and angular hydrogel particles. Because swollen hydrogels consist primarily of water, they typically have low mechanical strength [10]. Investigation of the paste void structure indicates that the mechanical forces present during batching, mixing, and placement did not result in deformation of the hydrogel particles, which Fig. 9.3 shows maintained their shape during processing.

The development of compressive strength for control mortar specimens (no hydrogels) and mortar containing hydrogel particles are shown in Fig. 9.4. Compressive strength of all mortar samples increased with time. It was expected that the mortar samples containing spherical hydrogels would have increased compressive strength since they experienced less autogenous shrinkage and should have a stronger microstructure. It was also expected that the irregularly shaped voids resulting from the angular particles would introduce stress concentrations where cracks could propagate, while the structure of the spherical hydrogels voids allow the compressive stresses to be transferred by dome action [4]. However, for the hydrogel dosage of 0.2% bwoc used in this study, the compressive strength of the mortars containing spherical and angular hydrogels was comparable at all ages.

This unexpected result could be due to the relatively small size and concentration of hydrogels used in the mortar samples. Mortar contains primarily aggregates, which are also irregular in shape and can reach 1 cm in diameter. It is possible that the effect of aggregate shape influences the compressive strength much more than the hydrogel particles. In inhomogeneous materials, such as mortar, the stress is defined as the average of the microstress over a certain volume of material. To obtain accurate data for the compressive strength, the cross section of this volume should ideally be much larger than any inhomogeneities, that is, the size of the aggregates [11]. Thus, in larger mortar samples or mortar containing a greater dosage of hydrogels, differences in compressive strength might become apparent.

The mortars containing hydrogels had similar strengths to that of the control mortar at 1 and 7 days, except the mortar with 83 wt.% PANa spherical hydrogels at 1 day of curing. The reason for the low strength of this sample is not precisely known. If the hydrogels were deswelling in the first 24 h, the strength of the mortar should be similar to that of the control with no internal curing. Alternatively, if the hydrogels were retaining liquid, it would be expected to display comparable strength with the 17 wt.% PANa samples. The reasoning behind the low strength should be examined further. The mortars containing 83 wt.% PANa demonstrated equivalent strength after 3 days but 17 wt.% PANa had higher strength than the control. This suggests that the 83 wt.% PANa hydrogels are releasing their curing water before the mortar has set, as it performs the same as the control mortar. Most importantly, the mortars containing hydrogel particles all experienced an increase in strength compared to the control mortar after 28 days. This is a significant result because it demonstrates that despite the increase in porosity caused by the addition of hydrogels the compressive strength is not reduced and could possibly be increased by the presence of these hydrogel-based internal curing agents.

This study was supported by the National Science Foundation (award #1454360).

## References

1. Mignon, A., Snoeck, D., Dubruel, P., Vlierberghe, S. V., & Belie, N. D. (2017). Crack mitigation in concrete: superabsorbent polymers as key to success? *Materials, 10*(3), 237.
2. Jensen, O. M., & Lura, P. (2006). Techniques and materials for internal water curing of concrete. *Materials and Structures, 39*, 817–825.
3. Snoeck, D., Schaubroeck, D., Dubruel, P., & Belie, N. D. (2014). Effect of high amounts of superabsorbent polymers and additional water on the workability, microstructure and strength of mortars with a water-to-cement ratio of 0.50. *Construction and Building Materials, 72*, 148–157.
4. Snoeck, D. (2015). *Self-healing and microstructure of cementitious materials with microfibres and superabsorbent polymers.* Dissertation, Ghent University, Ghent, Belgium.
5. Mechtcherine, V., et al. (2014). Effect of internal curing by using superabsorbent polymers (SAP) on autogenous shrinkage and other properties of a high-performance fine-grained concrete: results of a RILEM round-robin test. *Materials and Structures, 47*(3), 541–562.

6. Krafcik, M. J., & Erk, K. A. (2016). Characterization of superabsorbent poly(sodium-acrylate acrylamide) hydrogels and influence of chemical structure on internally cured mortar. *Materials and Structures, 49*, 4765.
7. Kelly, S. L. (2017). *Inverse Suspension Polymerization of Superabsorbent Polymer (SAP) Hydrogels for Internally Cured Concrete*. Master's thesis, Purdue University, West Lafayette, Indiana.
8. ASTM C109–16a. (2016). *Standard test method for compressive strength of hydraulic cement mortars*. West Conshohocken, PA: ASTM International.
9. Zhu, Q., Barney, C. W., & Erk, K. A. (2015). Effect of ionic crosslinking on the swelling and mechanical response of model superabsorbent polymer hydrogels for internally cured concrete. *Materials and Structures, 48*, 2261–2276.
10. Sun, J. Y., Zhao, X., Illeperuma, W. R. K., Chaudhuri, O., Oh, K. H., Mooney, D. J., & Suo, Z. (2012). Highly stretchable and tough hydrogels. *Nature, 489*(7414), 133–136.
11. Bažant, Z. P., & Oh, B. H. (1983). Crack band theory for fracture of concrete. *Materials and Structures, 16*(3), 155–177.

# Chapter 10
# Evaluation of Microencapsulated Corrosion Inhibitors in Reinforced Concrete

Reece Goldsberry, Jose Milla, Melvin McElwee, Marwa M. Hassan, and Homero Castaneda

Reinforced concrete (RC) structures are vital to the US infrastructure due to their relatively low cost, durability, and high strength. However, when concrete cracks, the steel reinforcement is susceptible to corrosion due to the ingress of harmful agents. The resulting corrosion products occupy a greater volume than the steel reinforcement, leading to an expansion that creates tensile stresses within the concrete matrix, causing further deterioration through cracking, delamination, and spalling. In order to mitigate corrosion, this study aims to develop and characterize the performance of encapsulated calcium nitrate tetrahydrate embedded in reinforced concrete beams for corrosion inhibition at varying concentrations (as a percentage by weight of cement). The corrosion and inhibitor penetration mechanism between microcapsules/concrete and rebar/concrete interfaces was preliminarily characterized using an electrochemical impedance spectroscopy (EIS) technique. In addition, the effects of the microcapsules on the concrete properties, such as the compressive strength and surface resistivity, were evaluated.

## 1 Introduction

Steel reinforcement in concrete is generally in a passive state due to the surface oxide layer in a high alkalinity environment. However, several factors can cause the breakdown of the passive layer and the corrosion of steel material, such as aggressive agents (e.g., chlorides ions), the alkalinity reduction of the concrete, and the

R. Goldsberry · H. Castaneda
Texas A&M University, College Station, TX, USA

J. Milla · M. McElwee · M. M. Hassan (✉)
Louisiana State University, Baton Rouge, LA, USA
e-mail: marwa@lsu.edu

© Springer International Publishing AG, part of Springer Nature 2018
M. M. Reda Taha (ed.), *International Congress on Polymers in Concrete (ICPIC 2018)*, https://doi.org/10.1007/978-3-319-78175-4_10

existence of cracks in concrete. To mitigate the damage caused by the corrosion of steel reinforcement in concrete, researchers have proposed embedding microencapsulated corrosion inhibitors in concrete. Zuo et al. prepared polymer/metal hydroxide microcapsules that steadily released the encapsulated materials over time. Using calcium hydroxide and barium hydroxide as corrosion inhibitors, the results showed the microcapsules successfully delayed the decline on the pH values of concrete, thus decreasing the corrosion rate of rebar [1]. Dong et al. prepared polystyrene microcapsules containing sodium monofluorophosphate. The results showed strong corrosion inhibition in a simulated concrete environment, measured with an electrochemical impedance spectroscopy (EIS) technique [2]. In this study, the authors propose embedding microcapsules containing a corrosion inhibitor, calcium nitrate tetrahydrate, within the concrete matrix. This mechanism is activated during a cracking event, where the microcapsules are ruptured and release the corrosion inhibitor. Therefore, this study proposes a technology with potential to influence surface properties and indirectly the transport properties, the rebar corrosion rate, and the mechanism of reaction at the steel/concrete interface.

## 2 Experimental Methods

### 2.1 Microcapsule Preparation

The microencapsulation procedure was adapted from a previous study [3], albeit with a few modifications. The process is based on a water-in-oil, suspension polymerization reaction of polyurea-formaldehyde. Kerosene was selected as an organic solvent as it is more economical than hexane. Moreover, the calcium nitrate solution concentration was increased from 16.7% to 25% to increase the amount of encapsulated core material. The suspension polymerization reaction was enabled by heating at an elevated temperature (40 °C) in the presence of an acid catalyst (sulfonic acid) for 2 h. The sulfonic acid concentration used was 0.25% by wt. of organic solvent, and no secondary emulsifier was used.

Once the microcapsules were synthesized, any trace of sulfonic acid on the microcapsules' surface was neutralized by washing the microcapsule slurry with a 1% sodium bicarbonate solution. The microcapsules were then recovered through a vacuum filtration. The samples were observed using a scanning electron microscope (SEM) as shown in Fig. 10.1, revealing a smooth outer surface and a spherical shape. The mean microcapsule diameter based from 250 readings was 60 μm.

### 2.2 Concrete Mix Design

A water-cement ratio of 0.42 was selected through preliminary laboratory tests. The maximum aggregate size was 19 mm for the coarse aggregate and 4.76 mm for the

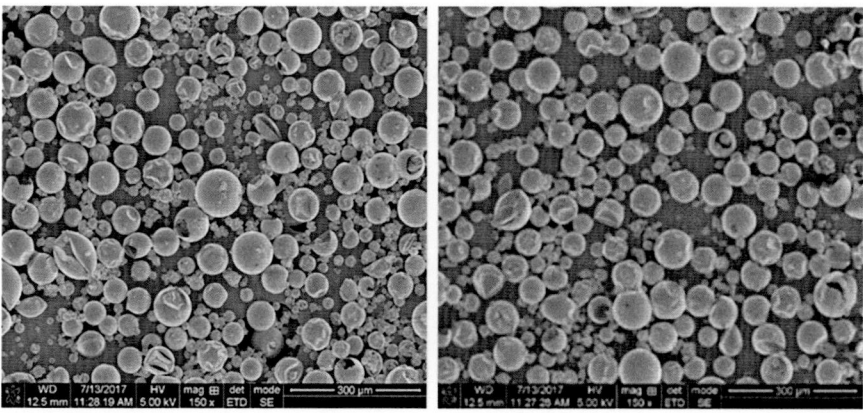

**Fig. 10.1** Electron microscope images of microcapsules synthesized

**Table 10.1** Description of specimen groups

| Sample | Corrosion | Microcapsule concentration |
|---|---|---|
| Control | None | 0.00% |
| CN-0.25 | Calcium nitrate | 0.25% |
| CN-0.50 | Calcium nitrate | 0.50% |
| CN-2.00 | Calcium nitrate | 2.00% |

fine aggregate, respectively. The microcapsules were embedded at varying concentrations (0.25%, 0.50%, and 2.00%) by weight of cement to determine the minimal dosage required to mitigate corrosion considerably. A superplasticizer was added to increase workability to the concrete mix design. A defoaming agent was also introduced to counter the increases in air voids caused by the addition of microcapsules in concrete.

## 2.3 Concrete Testing

Concrete cylinders were made for compressive strength (ASTM C39), and surface resistivity tests (AASHTO TP 95), while concrete beams were made for corrosion testing (ASTM G109). The concrete samples were cast and cured in laboratory settings per ASTM C192 guidelines. A total of three 100 mm × 200 mm cylinders and three 150 mm × 150 mm × 280 mm beams were poured per specimen group. Table 10.1 shows the characteristics of the specimens used in this study. All beam specimens used were subject to a three-point loading system, where a slow strain rate (0.005 in/s) was applied until a crack was induced. The crack sizes induced ranged from 0.2 to 0.45 mm.

The ASTM G109 standard test method is a suitable and convenient way to investigate the rebar corrosion and the effects of chemical admixtures. The core of

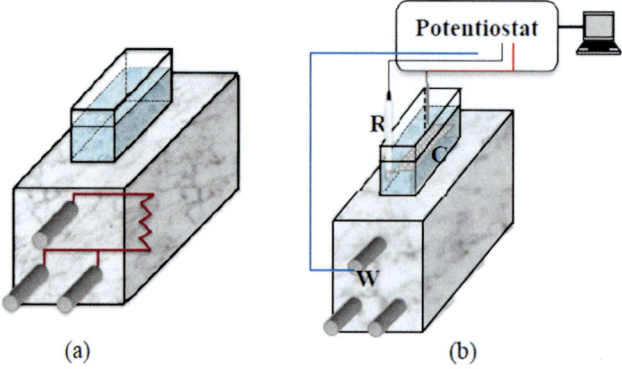

**Fig. 10.2** Electrochemical setup of the reinforced concrete samples

this method is measuring the voltage across a resistor connecting the top bar (as anodic rebar) and the bottom bar (as cathodic rebar). Therefore, the changes of the current can be monitored and calculated. However, to gain insight into the corrosion and inhibitor penetration mechanism, implementation of electrochemical techniques is necessary. Electrochemical impedance spectroscopy (EIS) as an AC-based technique can characterize, quantify, and monitor the state of the rebar/concrete interfaces and provide insight into the corrosion mechanisms occurring at the substrate/electrolyte level. The wide frequency range allows for the detection and characterization of interfacial phenomena occurring at different frequencies. Moreover, the application of a small-amplitude AC signal shows no significant influence on the chemical reactions and electrode processes [4, 5]. Therefore, the dynamics of surface changes can be continually characterized as illustrated in Fig. 10.2. The samples include a typical three electrode system where working electrode is the rebar, counter electrode is a metallic platinum mesh, and the reference electrode is the saturated calomel electrode. The electrochemical testing includes the open circuit potential followed by EIS for each day of sodium chloride exposure.

## 3 Results

The compressive strength results shown in Table 10.2 indicate that an increase in microcapsule concentration has a negative impact on strength, where the highest microcapsule concentration (2% by wt. of cement) resulted in an 18% strength reduction. Furthermore, the surface resistivity tests showed that the addition of microcapsules dropped the chloride permeability level from "low" to "moderate" for the tested mix design.

A statistical test using Fisher's least squares difference (LSD) was used to determine if the addition of microcapsules had a significant effect on the compressive strength of concrete. The analysis shows that the control specimen group is

**Table 10.2** Description of concrete samples produced

| Sample ID | Concentration (% by wt. of cement) | Compressive strength (MPa) | Surface resistivity (kΩ-cm) | Chloride penetrability |
|---|---|---|---|---|
| Control | N/A | 70.8 | 21.9 | Low |
| CN-0.25 | 0.25 | 66.4 | 20.1 | Moderate |
| CN-0.50 | 0.50 | 53.3 | 18.7 | Moderate |
| CN-2.00 | 2.00 | 58.1 | 15.3 | Moderate |

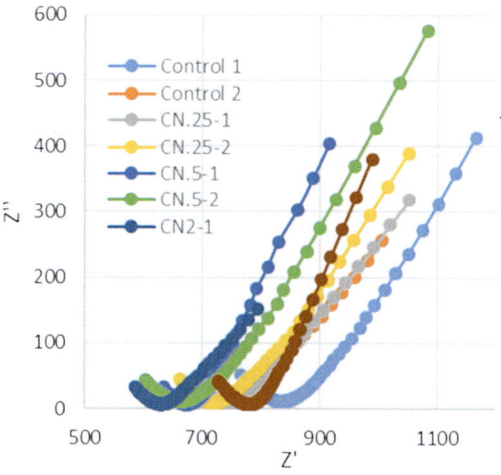

**Fig. 10.3** EIS results for different concrete samples

significantly different from those samples admixed with microcapsules at 0.5% and 2.0%. Even though the defoaming agent was shown to improve dispersibility of the microcapsules in water, it is possible that with the defoamer concentration used it was still insufficient to adequately disperse the capsules throughout the concrete matrix. Thus, any clusters of microcapsules would weaken the cementitious matrix and increase the variability of the compressive strength considerably with respect to the control specimens.

EIS was performed to provide information about the mechanisms of corrosion degradation of the different concrete specimens. The electrochemical results were reported from all specimen groups after 1 day of ponding. The initial exposure of the Nyquist representation for the concrete specimens during the continuous electrolyte exposure is displayed in Fig. 10.3.

According to the EIS signal, the impedance spectra for the concrete specimens showed one capacitive (not a complete loop) behavior which could be related to the properties of the concrete matrix, while the low-frequency shows linear behavior that is characteristic of a mass transfer process. The EIS complex diagram represents the presence of a passive film on the steel surface in each sample. The capacitive loop observed at high frequencies for these specimens suggests the water permeation at the initial, pre-cracked conditions allowed the ingress of aggressive ions. The comparison in impedance between the control specimens and the microcapsule-containing specimens shows that the microcapsules had a minimal effect in providing different mechanisms to the steel interface in the presence of cracks, since this trend was observed in every sample. However, crack widths over 0.1 mm have a significant effect on the chloride diffusion coefficients of concrete [6], and thus it is possible that any corrosion inhibition effect would have been insufficient in such high diffusion coefficients. In addition, it is also possible that a higher calcium nitrate concentration is required to mitigate corrosion adequately.

## 4 Conclusions

The use of microcapsules containing calcium nitrate tetrahydrate as a corrosion inhibitor was explored in this study. The concentration of microcapsules added had a significant effect on the compressive strength of concrete when added at 0.5% and 2.0% by weight of cement, leading up to 18% strength reduction. Surface resistivity tests also indicated that a slight increase in chloride penetrability was attributed to the addition of microcapsules. With respect to corrosion tests, the crack widths obtained allowed a significant amount of chloride to penetrate the concrete cover in a shorter period of time. The EIS results indicate that the release of the microcapsule's corrosion inhibitor was negligible due to the exposure time on the specimen to the corrosive environment (to induce the formation of a protective passive layer or modification on the mass transfer process) and nitrate interacting with the electrolyte uptake. Further testing is ongoing to ensure that the cracks induced are smaller than 0.1 mm, in order to successfully rupture and release the corrosion inhibitor in the microcapsules, and prevent a substantial increase in the chloride diffusion coefficient. This can be achieved by using a lower strain rate in the third-point loading test and the addition of polymer fibers to have more control on the crack width obtained.

## References

1. Zuo, J., Zhan, J., Dong, B., Luo, C., Qiuhong, L., & Chen, D. (2017). Preparation of metal hydroxide microcapsules and the effect on pH value of concrete. *Construction and Building Materials, 155*, 323–331.
2. Dong, B., Wang, Y., Ding, W., Li, S., Han, N., Xing, F., & Lu, Y. (2014). Electrochemical impedance study on steel corrosion in the simulated concrete system with a novel self-healing microcapsule. *Construction and Building Materials, 56*, 1–6.
3. Milla, J., Hassan, M. M., & Rupnow, T. (2017). Evaluation of self-healing concrete with microencapsulated calcium nitrate. *Journal of Materials in Civil Engineering, 29*(12), 04017235.
4. Wittmann, F. H. (2007). Effective chloride barrier for reinforced concrete structures in order to extend the service-life. In *Advances in construction materials 2007* (pp. 427–437). Berlin/Heidelberg: Springer.

5. Otieno, M., Beushausen, H., & Alexander, M. (2016). Chloride-induced corrosion of steel in cracked concrete – Part I: Experimental studies under accelerated and natural marine environments. *Cement and Concrete Research, 79*, 373–385.
6. Jang, S. Y., Kim, B. S., & Oh, B. H. (2011). Effect of crack width on chloride diffusion coefficients of concrete by steady-state migration tests. *Cement and Concrete Research, 41*(1), 9–19.

# Chapter 11
# The Use of Polymer Additions to Enhance the Performance of Industrial and Residential Decorative Concrete Flooring

Michelle Sykes and Deon Kruger

This paper provides an overview of an investigation into the use of various polymeric additions into ordinary concrete formulations in order to enhance the performance of such materials when used for decorative flooring in industrial, commercial and residential applications.

Using the ACI 302.1R-04-Guide for Concrete Floor and Slab Construction as the basis for this research, various polymer additions were investigated to optimise the durability and aesthetics of decorative flooring when exposed to extreme service conditions such as high-level UV exposure, cyclic temperature variations, high-impact shock loading, chemical spillage and high-level foot and light vehicle traffic [1].

Material properties that are enhanced include strength, crack resistance, prevention of discoloration and flaking, maintaining of lustre as well as long-term durability.

The paper will conclude with a summary of the findings and a recommendation of the most appropriate type and ratio of polymer addition for optimum enhancement of decorative concrete flooring.

## 1 Introduction

This study is unique since it identifies whether the addition of polymers to conventional concrete formulations will enhance the overall product of decorative flooring and dominate the industry of concrete flooring. This study will delve into the examination of which type of polymer and loading will enhance the overall aesthetic

---

M. Sykes · D. Kruger (✉)
Department of Civil Engineering Science, University of Johannesburg, Johannesburg, South Africa
e-mail: dkruger@uj.ac.za

properties of the decorative concrete flooring. The addition of polymers to concrete is a fairly new concept with minimal guides, textbooks or codes to direct designers on this material. Therefore, for this research, ACI 302.R-04-Guide for Concrete Floor and Slab Construction serves as the basis to investigate high-quality decorative polymer-modified concrete slabs for various service classes in terms of concrete materials and concrete mixture proportions. Although the service class and ideal qualities of the floor design depends on both the basic concrete mix design and the type and dosage of polymer modifier, there is no fixed or universal rule for specifying the most correct type or ratio of polymer; in turn, the goal of this study was to clarify this uncertainty.

There are three categories of concrete-polymer materials: polymer concrete (PC), polymer-modified concrete (PMC) and polymer-impregnated concrete (PIC). This study will focus solely on polymer-modified concrete, due to its practical properties. PMC is a Portland cement concrete which is modified with polymer or monomer systems which are added when the concrete is mixed [2]. PMC is attractive because it improves the adhesion, imperviousness, flexibility and shrinkage resistance properties of concrete materials [3].

Polymers enhance the concrete system through two phases known as cement hydration and polymer film formation while binding. When the PMC material cures, the polymer particles coalesce into a network structure that is contained by the hydrated cement paste [6]. This will enhance the durability of the PMC material through the resistance of moisture transmission, increased waterproofness and freeze-thaw resistance. Factors such as exposure conditions, formulation of the PMC system, level of polymer loading, polymer type and water-to-cement ratio will influence the degree of microstructural integration of the polymer system as well as the Portland cement that is essential for the long-term durability of PMC material [5].

## 2 ACI Guide Requirements

The ACI guide provides information of the material for users during the planning, designing and execution of construction. Based on the guide, a floor classification of type 4 single course was selected; further specifications of design considerations, materials, batching, mixing, placing, finishing, curing and protection were followed, in accordance with the guide. A common Portland cement of ASTM C150 Type I was suggested for floor class 4. A coarse aggregate of 19 mm was assumed, as the guide mentioned, since coarse aggregates are favourable to ensure shrinkage reduction because of a lower water content requirement. Unless the main aim is to achieve a high flexural strength, it is optional to use a natural sand as a fine aggregate. The minimum cementitious content required for maximum aggregate size of 19 mm is specified as 320 kg/m$^3$ and was thus used in this study [1]. Admixture usage should conform to the requirements of ASTM C494 based on the type of admixture. Addition of a polymer to the mix may act as a water-reducing admixture that is advantageous to compressive and flexural strength escalation; however, it may delay the initial setting time as well as the final finishing time in hot climates, while retardation could cause plastic shrinkage cracking and surface crusting [1].

## 3 Latexes Tested and Mix Designs

For the experimental tests, performed in this study, three commercially available types of polymer systems were selected for evaluation. They were polyacrylic ester emulsion (PAE) with the trade name Texicryl 13-072; styrene-butadiene rubber latex (SBR), available as Savinex 29Y40; and vinyl acetate/ethylene copolymer (VAE) available as Vinnapas 5010 N. The Savinex 29Y40 SBR and Texicryl 13-072 PAE are both water-based emulsions with a milky white physical appearance which are slightly odorous. The SBR latex is a styrene acrylic copolymer emulsion with 45% polymer solids by weight, a pH rating of 9.0–10.0 and relative density of 1.0 at 20 °C. The PAE emulsion (Texicryl 13-072) is a dispersed emulsion copolymer based on acrylic esters containing 60% by weight of polymer solids. It has a boiling point of approximately 100 °C, a relative density of 1.10 at 20 °C, a pH rating of 8.0–9.0 and fully miscible to water. The VAE (Vinnapas 5010 N) is a copolymer white powder resin which disperses freely in water. This powder resin has a high ethylene content and glass transition temperature which can reach below freezing point. It is odourless with a pH rating of 7. Its particle size is max. 2% over 400 μm with a 99 ± 1% by weight of polymer solids [6].

### 3.1 Design of Polymer Modified Concrete Mixes

The polymer-modified concrete mix proportion consisted of 1 (cement):1 (sand):1 (stone) by weight with varying water-cement ratios based on the type of polymer and the quantity of polymer loading [4]. The control mixture with 0% polymer solid was used as the basis to which the latex-modified specimens were evaluated against. The first test specimen of styrene-butadiene rubber latex (SBR1) of 5% polymer solid was designed with a water-cement ratio of 0.625. The second specimen SBR2 had the same water-cement ratio, but it was clear while mixing the water content was too high, since the mixture was not cohesive. This suggests that a higher polymer loading reduces the water requirement of the mix design. The last styrene butadiene rubber latex specimen with 15% polymer solid had a lower w/c ratio of 0.366. The first polyacrylic ester (PAE1)-modified specimen of 5% polymer solid had a w/c ratio of 0.45, the second polyacrylic ester (PAE2) specimen with 10% polymer solid had a w/c ratio of 0.307, and the last polyacrylic ester (PAE3) specimen with 15% polymer solid had a lower w/c ratio of 0.294. The first vinyl acetate/ethylene copolymer (VAE1) specimen with 5% polymer solid had a w/c ratio of 0.45, the second vinyl acetate/ethylene copolymer (VAE2) specimen with 10% had the lowest overall w/c ratio of 0.2616, and the last vinyl acetate/ethylene copolymer (VAE3) specimen with 2.5% polymer solid had a w/c ratio of 0.3013.

## 4 Test Results and Discussion

Results of compressive strength based on the Schmidt hammer and tensile strength from the tensile pull-off test for each test specimen were recorded in Table 11.1.

On review of the results, the unmodified control mix CON exhibited a high compressive strength with a reading 83% higher than the concrete substrate. The first and second modified mixes SBR1 and SBR2 had a 32% lower compressive strength than the control mix. SBR3 had the highest compressive strength of all the modified mixes which was 18% higher than the unmodified control mix. The first polyacrylic ester emulsion specimen, PAE1, had the same compressive strength as the control mix. The PAE2 mix had a 9% lower compressive strength than PAE1 mix. The last polyacrylic ester emulsion-modified mix, PAE3, had the highest compressive strength of 23 MPa, 4% higher than the unmodified control mix compressive strength but 11% lower than the SBR3 mix. Vinyl acetate/ethylene copolymer-modified mixes, VAE1 and VAE2, had the same compressive strength even though the polymer loadings were 5% and 10% respectively. The final modified mix, VAE3, had the second highest compressive strength (8% lower than the SBR3 mix) but contained the lowest polymer loading of 2.5%.

The compressive strength highlighted the importance of proper cement hydration together with the polymer film formation process during the initial stages of the PMC materials strength development; for this reason, initial wet curing is a fundamental stage for ensuring long-term stability of the PMC materials. Both the control and VAE3 mixes exhibited high compressive strength. It was only mix SBR3 that outperformed the other mixes significantly.

**Table 11.1** Compressive strength (MPa) and tensile strength (kN)

| Mix type | Compressive strength, MPa | Tensile strength, kN |
|---|---|---|
| Base | 12 | 1 |
| CON | 22 | 1 |
| SBR1 | 15 | 0.8 |
| SBR2 | 15 | 1.2 |
| SBR3 | 26 | 1 |
| PAE1 | 22 | 1 |
| PAE2 | 20 | 1 |
| PAE3 | 23 | 1 |
| VAE1 | 18 | 1 |
| VAE2 | 18 | 1 |
| VAE3 | 24 | 1 |

The tensile strength of all the specimens had basically comparable results of 1 kN. These results may be inconclusive as it could in fact be due to the limitation of the superglue used to bond the test bolt to the concrete surface, since some of the test did show concrete coating failure. Specimens SBR1, SBR2, PAE2, VAE1 and VAE2 pulled off the concrete surface to which the bolt was bonded/glued, thus indicating concrete coating failure. The rest of the specimens, SBR3, PAE1, PAE3, VAE3 and CON, showed no sign of concrete surface detachment and, hence, no concrete coating failure.

Regarding the visual performance, specimens VAE1 and VAE2 exhibited amplified amounts of cracks compared to the lower vinyl acetate/ethylene copolymer-modified mix VAE3 as well as the other polymer-modified specimens. Conversely, the VAE2 mix maintained the best lustre of all the specimens, while the SBR1 and CON specimens had moderate lustres.

## 5 Conclusion

The main focus of the study was to identify the most appropriate type and ratio of polymer addition for optimum enhancement of decorative flooring PMC coatings with a thickness of around 30 mm.

From the results of the study, the SBR3 mix had the highest compressive strength of 26 MPa, about 18% higher than the strength of the unmodified mix. The second highest compressive strength of 24 MPa, 8% lower than the SBR3 result and 9% higher than the control mix, VAE3 had the lowest polymer solid of 2.5% by weight of cement which was 83% lower polymer solid than for SBR3 mix. The PAE3 had compressive strength results of 4% which was higher than the unmodified mix.

The compressive strength development is interconnected to the polymer film formation. The compressive strength results also indicate that the long-term stability of the PMC material is very dependent on the critical development of the cement hydration during the initial wet curing phase.

The tensile strengths were inconclusive; however, the detachment of the concrete surface bonded to the bolt, as it was pulled off, was noticed in some specimens, indicating coating failure. Specimens SBR3, PAE1, PAE2, VAE3 and CON had no concrete surface sheared off; therefore, the concrete did not fail. Mix VAE2 maintained the best lustre of all the specimens during the evaluation period.

The ACI 302.1R-04 Guide for Concrete Floor and Slab advises that water-reducing admixtures, abiding to ASTM C494 requirements, will increase the compressive strength and flexural strength, but will not essentially improve the finishing characteristics nor will it reduce shrinkage [1].

It can therefore be concluded, based on the findings described above, that the addition of polymers to the concrete formulation may improve the strength and aesthetics of decorative flooring, when compared to unmodified coatings. Additionally, the best performing polymer in this trial was the styrene-butadiene rubber latex, when used as a PMC with a polymer loading of 15%.

# References

1. ACI 302.1 R-04. (2004). Guide for concrete floor and slab construction. *American Concrete Institute* (pp. 1–77).
2. ACI Committee. (1995). State-of-the-art report on polymer modified concrete. *American Concrete Institute, ACI, 548*, 1–47.
3. Depuy, G. W. (1996). Polymer modified concrete-properties and applications. *Construction Repair, 10*(2), 63–67.
4. Owens, G. (2012). *Fundamentals of concrete, section edition* (p. 134). Midrand: Cement and Concrete Institute.
5. Ramli, M., & Swamy, R. N. (1996). Development of polymer-modified cement systems for durable concrete construction. In M. W. Hussin (Ed.), *Proceedings $3^{rd}$ Asia Pacific conference on structural engineering and construction* (APSEC '96), Johor Bahru, Malaysia, (pp. 375–382).
6. Ramli, M., & Swamy, R. N. (1997). A rational mix design methodology for latex modified concrete. *3rd Southern African Conference on polymers in concrete*. Johannesburg, South Africa.

# Chapter 12
# The Effect of Glucose on the Properties of Cement Paste

**Samantha Mirante and Ali Ghahremaninezhad**

This study presents the results of an investigation of the effect of glucose on the hydration, microstructure, and properties of cement pastes. The hydration was studied using non-evaporable water content measurement and thermogravimetric analysis (TGA). Electrical resistivity as a measure of the transport property of cement pastes was evaluated using electrochemical impedance spectroscopy (EIS). It was found that all cement pastes showed a similar degree of hydration at late ages. The cement paste with a low concentration of glucose (0.05%) showed a slightly higher compressive strength and electrical resistivity compared to the control cement paste. This was most likely due to improved workability and dispersion of cement particles in this cement paste. However, the cement paste with 0.25% glucose showed a lower electrical resistivity compared to the control cement paste.

## 1 Introduction

Chemical admixtures are widely used to impart specific properties to concrete mixtures. Retarders are a class of admixtures used to increase the setting time of the concrete, thereby allowing adequate time for concrete transport, placement, and finishing. Sugars have been shown to act as hydration retarders [1–5]. Several prior studies have examined the influence of glucose on the hydration process of cement [6–8]. It is generally understood that delayed hydration as a result of glucose addition is due to the poisoning of nucleation sites reducing the rate of hydration product formation [8]. In spite of many prior investigations on the hydration behavior of cementitious material with glucose, studies focused on the impact of

---

S. Mirante · A. Ghahremaninezhad (✉)
Department of Civil, Architectural and Environmental Engineering, University of Miami, Coral Gables, FL, USA
e-mail: a.ghahremani@miami.edu

glucose on the properties of cementitious materials at late ages are scarce. Therefore, this paper aims to evaluate the hydration, compressive strength, electrical resistivity, and microstructure of cement pastes with added glucose at late ages.

## 2 Experiments

### 2.1 Materials and Sample Preparation

The samples tested in these experiments were made of a type I/II Portland cement and were mixed with a water/cement ratio of 0.5 to create a cement paste. Cement paste with no added glucose was used as control, and specimens with glucose additions of 0.05% and 0.25% by cement mass were used for testing. These three cement paste mixes were cast into cubes (50 × 50 × 50 mm) according to ASTM C 109.

### 2.2 Non-evaporable Water Content

Cement pastes were ground, passed through the sieve #60, and dried in a furnace for roughly 1 day at 105 °C. They were then placed back into the furnace for about 3 h at 1000–1100 °C. The mass of the powder after initial drying over 24 h ($M^1$) and the mass after being subjected to ignition for 3 h ($M^2$) were used to calculate $Wn$, normalized per cement mass, as follows:

$$Wn = \frac{M^1 - M^2}{M^2} - \text{LOI} \qquad (12.1)$$

where LOI is the loss on ignition of the cement. The average value of two samples for each cement paste is reported.

### 2.3 TGA Analysis

TGA was utilized to measure the content of calcium hydroxide (CH) in the samples. A Netzsch TG was used for testing at a heating rate of 20 °C/min in a temperature range of 25 °C–1000 °C. The change in mass in the range between 400 °C and 500 °C, normalized per cement mass, was used to calculate CH content.

## 2.4 Compressive Strength

The compressive strengths of the cubic specimens were measured using the SATEC material testing instrument. The compressive strengths of five replicate cubes were measured and averaged to produce the final value.

## 2.5 Electrical Resistivity

Electrical resistivity is determined by the morphological characteristics of pores and the chemistry of the pore fluid in cement-based materials [9]. Thus, electrical resistivity was measured to provide information about the transport properties of the cement paste samples. A Gamry Reference 600 analyzer with AC signals of 250 mV and a frequency range of $10^6$ to 10 Hz measured the electrical resistivity of the cement paste cubes. The average value of five replicate specimens was reported.

## 2.6 Scanning Electron Microscopy

The microstructure of each of the mixtures at 28 days was examined with a JEOL JSM-6010PLUS/LA scanning electron microscope. The sample preparation followed the standard procedure.

# 3 Results and Discussion

## 3.1 Degree of Hydration

The results of non-evaporable water content ($Wn$) measurement at 3 days, 14 days, and 28 days of curing of the cement pastes are presented in Fig. 12.1. An increase in $Wn$ with time is observed in all cement pastes, which is related to the progression of hydration with time. A noticeable increase in the $Wn$ of the cement paste with 0.05% glucose compared to both the control cement paste and cement paste with 0.25% glucose can be noted. This increase was not expected and can be attributed to a more uniform dispersal of cement particles in the mixture as a result of the improved workability of this cement paste. In a mixture with glucose, the sugar molecules can adsorb onto the surface of unhydrated cement particles or hydration products and can cause repulsion between the solid particles. It appears that in the cement paste with a low concentration of glucose the effect of improved rheology outweighed the retarding effect of glucose on hydration. It is noted that with continued hydration all cement pastes showed a similar degree of hydration at 14 days and 28 days.

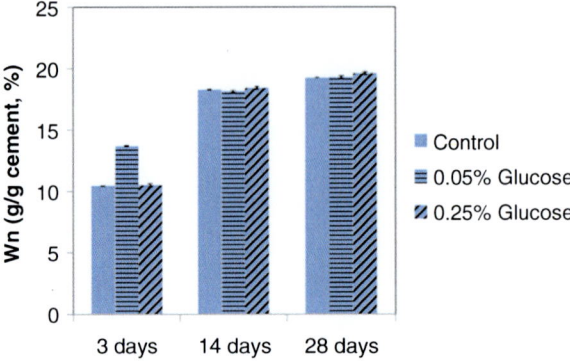

**Fig. 12.1** Non-evaporable water content results

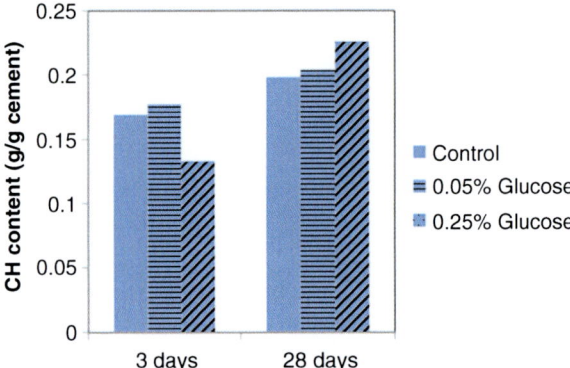

**Fig. 12.2** Ca(OH)$_2$ content of the cement pastes at 3 days and 28 days of age

## 3.2 TGA Analysis

The CH content per cement mass of the cement pastes at 3 days and 28 days of curing is shown in Fig. 12.2. It is seen that the CH content of the cement paste with 0.25% glucose is lower than other pastes at 3 days. However, at 28 days this trend is reversed, and this cement paste showed the highest CH content. By comparing the CH content results with the non-evaporable water content $Wn$ results of the cement paste containing 0.25% glucose, it can be noted that the rate of CH production from 3 days until 28 days is greater than the rate of increase of other hydration products mainly consisting of calcium silicate hydrate (C-S-H). The rate of increase of CH in the control cement paste and the cement paste with 0.05% glucose follows a similar trend as seen from the figure.

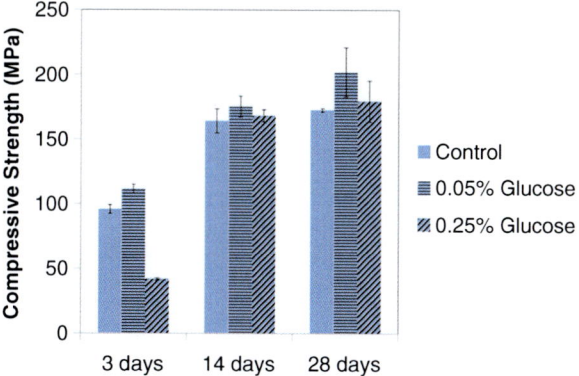

**Fig. 12.3** Compressive strength of the control paste and modified pastes

## 3.3 Compressive Strength

The results of compressive strength tests are shown in Fig. 12.3. A noticeable reduction in compressive strength in the cement paste with 0.25% glucose can be seen as compared to the other cement pastes at 3 days. However, the compressive strength of this cement paste improved at later ages and showed a similar value to the control cement paste. The improvement in the compressive strength of the cement paste with 0.25% glucose at later ages can be explained in light of the similar degree of hydration of this cement paste to that of the control cement paste at later ages, as seen from Fig. 12.1. However, the lower compressive strength of this cement paste at 3 days is not consistent with the degree of hydration of this paste at 3 days, where a reduction is not observed. The lower compressive strength of the cement paste with 0.25% glucose could be due to the effect of glucose on the microstructural morphology or interfacial binding between various phases of hydration products in this cement paste at 3 days. More studies are required to validate this hypothesis and shed more light on the underlying mechanisms affecting the compressive strength of this cement paste. A small increase in the compressive strength of the cement paste with 0.05% glucose at all ages examined in this study can be observed. This increase could be attributed to a more uniform microstructure of this cement paste as a result of increased workability.

## 3.4 Electrical Resistivity

The electrical resistivity of the control cement paste and the cement pastes modified with 0.05% and 0.25% glucose at various ages is shown in Fig. 12.4. The cement paste with 0.25% glucose exhibited a noticeably lower electrical resistivity compared to the control cement paste at all ages studied here. A slightly lower electrical resistivity of the cement paste with 0.05% glucose compared to the control cement paste at 3 days is observed. It is interesting to note a higher increase in electrical

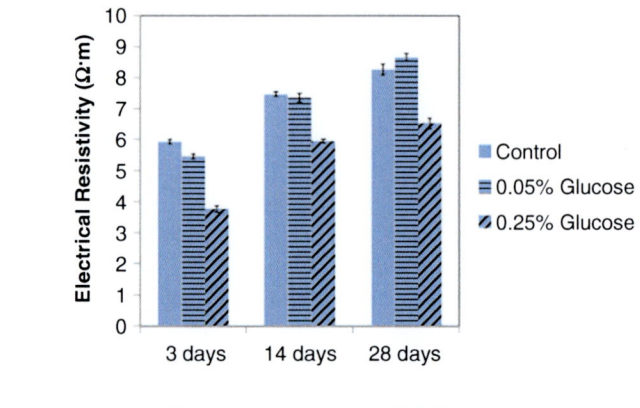

**Fig. 12.4** Electrical resistivity of the cement pastes at varied ages

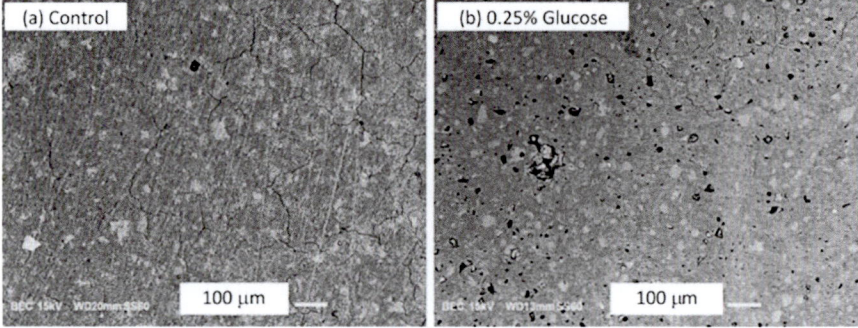

**Fig. 12.5** SEM images of the cement pastes at 28 days

resistivity with time in the cement paste with 0.05% leading to a slightly higher value at 28 days compared to that of the control cement paste. This increase in electrical resistivity is consistent with the improved compressive strength of this cement paste as shown in Fig. 12.3. Since electrical resistivity is strongly influenced by the densification of the microstructure, the results of electrical resistivity measurements can provide insight into the effect of glucose on the transport properties of cement pastes. Electrical resistivity is also dependent on the pore solution chemistry. The effect of pore solution chemistry on electrical resistivity is not studied in this paper.

## 3.5 Scanning Electron Microscopy

Figure 12.5 shows the SEM images corresponding to the control cement paste and the cement paste with 0.25% glucose. A large number of microvoids in the range of 20–70 μm are evident in the cement paste with glucose as seen from the micrograph. The mechanism of microvoid formation in this cement paste is currently under investigation and will be reported in future contributions.

## 4 Conclusions

The degree of hydration, compressive strength, electrical resistivity, and microstructure of the control cement paste and the cement pastes with 0.05% and 0.25% additions of glucose were studied in this paper. The following conclusions can be drawn from the results of this investigation:

- All cement pastes showed a similar degree of hydration at late ages. An increase in the early age (3 days) degree of hydration of the cement paste with 0.05% glucose was examined, which can be related to increased workability and improved dispersion of mixture in this cement paste.
- The cement paste with 0.05% glucose showed an increased compressive strength and electrical resistivity compared to the control cement at 28 days. However, the cement paste with 0.25% glucose exhibited a reduction in electrical resistivity compared to the control cement paste at 28 days. The compressive strength of this paste was similar to that of the control paste at 28 days.
- Microscopic examination showed formation of a large number of microvoids in the cement paste with 0.25% glucose.

## References

1. Thomas, N. L., & Birchall, J. D. (1983). The retarding action of sugars on cement hydration. *Cement and Concrete Research, 13*(6), 830–842.
2. Hubler, M. H., Thomas, J. J., & Jennings, H. M. (2011). Influence of nucleation seeding on the hydration kinetics and compressive strength of alkali activated slag paste. *Cement and Concrete Research, 41*(8), 842–846.
3. Juenger, M. C. G., & Jennings, H. M. (2002). New insights into the effects of sugar on the hydration and microstructure of cement pastes. *Cement and Concrete Research, 32*(3), 393–399.
4. Peterson, V. K., & Juenger, M. C. G. (2006). Hydration of tricalcium silicate: Effects of CaCl2 and sucrose on reaction kinetics and product formation. *Chemistry of Materials, 18*(24), 5798–5804.
5. Bishop, M., & Barron, A. R. (2006). Cement hydration inhibition with sucrose, tartaric acid, and lignosulfonate: Analytical and spectroscopic study. *Industrial & Engineering Chemistry Research, 45*(21), 7042–7049.
6. Yasuda, S., Ima, K., & Matsushita, Y. (2002). Manufacture of wood-cement boards VII: Cement-hardening inhibitory compounds of hannoki (Japanese alder, Alnus japonica Steud.) *Journal of Wood Science, 48*(3), 242.
7. Kochova, K., Schollbach, K., Gauvin, F., & Brouwers, H. J. H. (2017). Effect of saccharides on the hydration of ordinary Portland cement. *Construction and Building Materials, 150*, 268–275.
8. Na, B., Wang, Z., Wang, H., & Lu, X. (2014). Wood-cement compatibility review. *Wood Research, 59*(5), 813–826.
9. Neithalath, N., Weiss, J., & Olek, J. (2006). Characterizing enhanced porosity concrete using electrical impedance to predict acoustic and hydraulic performance. *Cement and Concrete Research, 36*(11), 2074–2085.

# Chapter 13
# Microstructured Polymers and Their Influences on the Mechanical Properties of PCC

Alexander Flohr, Luise Göbel, and Andrea Osburg

Concretes are modified with polymers in order to improve their durability and adhesive strength. However, polymer-modified cementitious materials exhibit lower elastic moduli and higher viscous and plastic deformations in comparison to unmodified systems. The macroscopic properties are governed by microstructural changes in the binder matrix, which consists of both cementitious and polymer components. Herein, different pure polymer specimens and microstructured polymers were characterized using specific load tests and nanoindentation in order to better understand the microscopic origin of the macroscopic deformation behavior and ultimately the mechanical properties of PCC. The link between the micromechanical and macroscopic properties is established using a continuum micromechanics approach. Multiscale models aiming at the prediction of the deformation behavior of polymer-modified cementitious materials are developed with input parameters that are partially obtained from the experimental investigations. The comparison of the modeling results with the experimentally determined deformations is satisfactorily good, underlining the predictive capability of the modeling approach. The improvement of prediction models is crucial for the application of PCC for construction purposes and will encourage their integration into guidelines.

## 1 Introduction

Polymer-modified cement concrete (PCC) is primarily used to repair and rehabilitate damaged or corroded concrete structures. Its durability, adhesion, shrinkage, and cracking processes have been studied intensively [1–3] during the last few decades.

---

A. Flohr (✉) · L. Göbel · A. Osburg
Bauhaus-Universität Weimar, Department of Civil Engineering, F. A. Finger-Institute of Building Materials Science, Weimar, Germany
e-mail: alexander.flohr@uni-weimar.de

The investigation of the PCC-microstructure was another focus [4, 5], especially in terms of durability. A lot of research activities are also concerned with the effects of polymer-modifications on the mechanical properties of concrete [6, 7]. They reveal that the PCC have a clearly decreased elastic modulus and compressive strength, but increased tensile strength and creep deformations compared to normal concrete. However, there is a lack of general statements about the structural behavior. In particular, the time- and stress-dependent material properties of the PCC have been explored only partially, especially considering microstructure. But the knowledge of micromechanical properties is not only essential for the improvement of the performance of the material at the macroscopic scale; it is also required for the development of prediction models that incorporate microstructural information.

Nanoindentation is an established technique for the determination of material properties at the microscale, which has been refined during the past years and successfully applied in studies of hydrated cement pastes [8–10]. Furthermore, statistical nanoindentation techniques have been employed on cementitious materials in order to investigate creep effects at the microscopic scale [11]. Wang et al. investigated polymer-modified cement pastes and determined the average indentation modulus of the cement pastes as a function of the polymer content in order to identify the finer-scale origin of the macroscopic properties [12].

The target of the study presented here is the quantification of the elastic properties on the microscale of polymer-modified cement pastes. Particularly, correlations between the micromechanical properties, the arrangement of the microstructural components, and the macroscopic mechanical behavior are derived. Results from different load tests on microstructured polymers and polymer-modified concretes as well as nanoindentation tests on polymer films are presented. Furthermore, the experimental data are coupled with a multiscale model, which aims at the prediction of the elastic properties of polymer-modified cementitious materials. A continuum micromechanics approach is applied to highlight the significance of the micromechanical properties.

## 2 Materials

The investigations were performed with ordinary Portland cement CEM I. For the polymer modification, a re-dispersible powder (polymer 1) and a dispersion (polymer 2) on the basis of styrene/acrylate as well as a dispersion (polymer 3) on the basis of styrene/butadiene were used. The polymer–cement ratio differed between 0.05 and 0.20. With given processing and storage conditions (20 °C, 65% rel. humidity) the polymers 1 and 3 form continuous films in the hardened concrete. Polymer 2 does not form films under these conditions, because its minimum film formation temperature is 30 °C. The polymer particles do not merge, but they have an adhesive bond to the aggregates and to the hardened cement matrix. Furthermore, for the investigations on microstructured polymers, specimens of concrete flour and

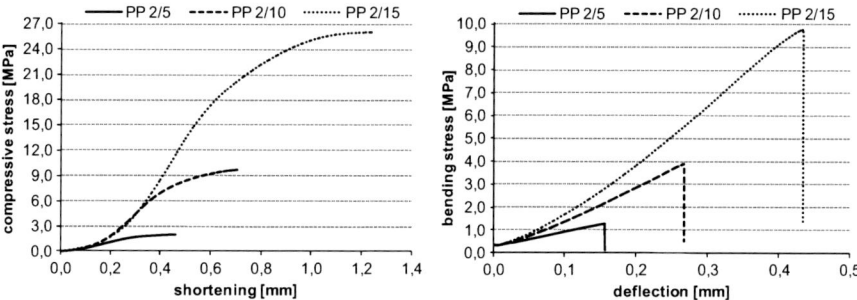

**Fig. 13.1** Influence of the polymer content on the compressive stress (left) and the bending stress (right) of concrete flour mixes

polymers were produced. Aggregates were used in the middle range of grading curve AB 16 (DIN 1045-2) for the tests with concrete.

## 3 Methods and Results

To establish an elementary understanding on how polymers influence the load-bearing and deformation behavior of concrete, tests on so-called microstructured polymers were performed. Therefore, specimens of concrete flour and polymers were produced with dimensions of $10 \times 10 \times 60$ mm$^3$. The particles of the concrete flour were agglutinated by the polymers. The result was a polymeric network that is similar to that in PCC. For determining the load-bearing and deformation behavior, three-point bending tests and also compressive tests were conducted and evaluated as a function of the polymer content and type. Figure 13.1 shows the mechanical behavior of the samples depending on the polymer content for the example of polymer 2 (PP 2) Beside the reference, which only consists of concrete flour, specimens with a polymer content of 5% (PP 2/5), 10% (PP 2/10) and 15% (PP 2/15) in relation to the flour weight were tested. The reference did not exhibit a measurable compressive or bending strength. So, it is verified that the concrete flour is inert. Figure 13.2 points out how different polymer types (PP 1/10, PP 2/10, PP 3/10) influence the strength and deformations of the samples.

At first it can be stated that a film formation is not compulsory for a load transmission. Even the polymers that do not form load-bearing films (PP 2) are capable to increase the strength due to an adhesive bond between the single constituents. Furthermore, the potential to transmit load increases with increasing density of the polymeric network, which goes along with increasing polymer content. Then, the differing deformation capacities of the single polymers should be considered, which can vary from a ductile (PP 1 and PP 3) to a brittle (PP 2) behavior. A direct transfer of these results to the mode of operation of the polymers in the structure of mortar or concrete is not possible because of the testing conditions

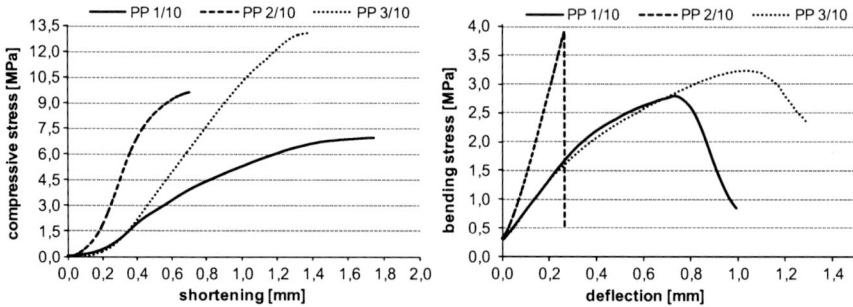

**Fig. 13.2** Influence of the polymer type on the compressive stress (left) and the bending stress (right) of concrete flour mixes

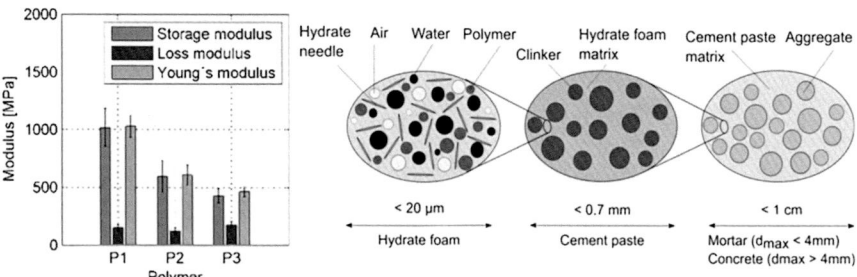

**Fig. 13.3** Micromechanical properties of the polymers (left) and morphological model of PCC (right)

and the complex processes during load subjections. However, it allows first conclusions about the way they function.

For a more profound investigation on the relationship between the microstructure and macroscopic properties, experimental data are required, which can also be coupled with a semi-analytical multiscale model that aims at the prediction of the elastic properties of PCC. To this end, nanoindentation tests on the pure polymer films were carried out (see Fig. 13.3; left). The polymer specimens had the dimensions of $10 \times 10 \times 1$ mm$^3$. The viscoelastic properties are described by means of the complex modulus $E^*$:

$$E^* = E' + iE'' \tag{13.1}$$

In Eq. 13.1, $E'$ represents the storage modulus, which describes the amount of elastic energy stored in the material, and $E''$ is the loss modulus, which describes the amount of energy dissipated in the material. That provides access to the Young's modulus $E$ of the polymers according to Eq. 13.2:

$$E = \sqrt{E'^2 + iE''^2} \qquad (13.2)$$

Beside the elastic deformation resistance of the polymers, the elastic properties of the components of the cement paste and of the aggregates are necessary for the calculation of the elastic properties of PCC. The properties of the cementitious phases as well as of the aggregates were taken from literature [13].

The computational approach correlating the micromechanical properties to the macroscopic properties of PCC is based on continuum micromechanics [14]. Essential is the definition of representative volume elements (RVE). This allows the hierarchically organized morphological representation of PCC (see Fig. 13.3; right). At each scale of interest, quasi-homogeneous material phases are defined. To every phase, the elastic properties, the shape, the type of interaction, and a time-depending phase volume fraction are assigned. The volume of the binder phase depends on the hydration degree, which is related to time instants of 2, 7, and 28 days in this study. The hydration degrees of the cement pastes were determined experimentally. After that, the volume fractions were calculated with an appropriate hydration model [15]. The homogenized elastic properties of the single RVE were computed, starting at the lowest level of observation according to Eq. 13.3 [16].

$$\mathbb{C}_{\mathrm{hom}} = \sum_{i=1}^{n} f_i \mathbb{C}_i : \mathbb{A}_i \qquad (13.3)$$

where $\mathbb{C}_{\mathrm{hom}}$ is the homogenized stiffness tens or, $f_i$ the phase volume fraction, $\mathbb{C}_i$ the phase stiffness, and $\mathbb{A}_i$ the phase strain concentration tensor.

To validate the model, the Young's moduli of concretes, modified with the already presented polymers, were determined (see Fig. 13.4; left). The concretes were prepared with an aggregate–cement ratio of approximately 5.5, a water–cement ratio of 0.4, and polymer–cement ratios of 0.05 and 0.20. Cylindrical specimens with a height of 30 cm and a diameter of 15 cm were produced. The dynamic moduli of the concrete cylinders were determined according to DIN EN 12504–4 using the ultrasonic pulse velocity at testing ages of 2, 7, and 28 days. The model results, that is, the elastic moduli derived from the homogenized stiffness, are compared to the

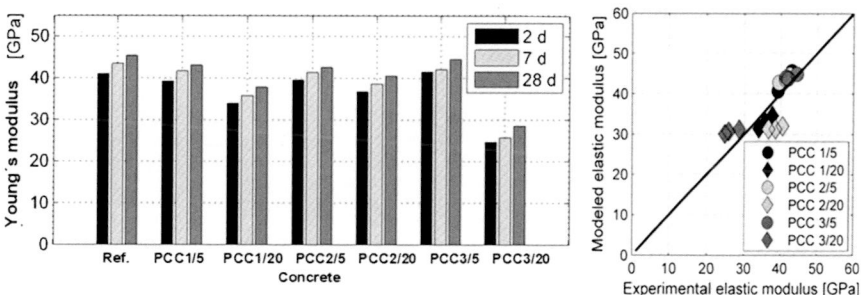

**Fig. 13.4** Young's moduli of the PCC at testing age of 2, 7, and 28 days (left) and validation of the model prediction (right)

results of the experimental multiscale analysis. The comparison between model-prediction and the experimental results shows a good agreement (see Fig. 13.4; right), revealing that the multiscale approach is able to reliably predict the elastic behavior of PCC.

## 4 Conclusions

The strong link between microscopic and macroscopic mechanical properties is evaluated using a continuum micromechanics approach. Nanoindentation tests on polymer films provide information about the elastic properties of the polymers, which serve as input parameters in a computational model. A multiscale homogenization model is adapted by the introduction of spherical polymer particles at the scale of a few microns. The model, which is based on the principles of continuum micromechanics, aims at the prediction of the elastic behavior of cementitious materials. For validation purposes, a multiscale experimental study comprising mechanical tests was carried out. The comparison between model predictions and experiments shows that the model is able to reliably predict the elastic behavior of polymer-modified cementitious materials. The results will provide further steps toward the establishment of correlations between the microstructure and macroscopic properties. The development of prediction models for polymer-modified cementitious materials represents an important step toward the use for construction purposes and accelerates their implementation in guidelines.

**Acknowledgement** This research has been supported by the German Research Foundation (DFG) through Research Training Group 1462, which is gratefully acknowledged.

## References

1. Dimmig, A. (2002). Einflüsse von Polymeren auf die Mikrostruktur und die Dauerhaftigkeit kunststoffmodifizierter Mörtel (PCC), PhD Dissertation, Bauhaus-Universität Weimar, Germany, 2002.
2. Swamy, R. N. (1995). Durability properties concrete composites with polymers. *8th international congress of polymers in concrete* (pp. 21–34). Oostende.
3. Bode, K. A. (2008). *Untersuchungen zu den kohäsiven und adhäsiven Eigenschaften von PCC* (Doctoral dissertation, Dissertation). Bauhaus-Universität Weimar.
4. Dimmig-Osburg, A. (2005). New model for the formation of the microstructure of polymer-modified mortar. *BFT, 10*, 27–36.
5. Beeldens, A., & Van Gemert, D. (2004). Integrated model of microstructure building in polymer cement concrete. In *11th international congress on polymers in concrete* (pp. 1–10). Berlin.
6. Lohaus, L., & Anders, S. (2004). Effects of polymer modification on the mechanical and fracture properties of high and ultra-high strength concrete, *11th international congress on polymers in concrete* (pp. 183–190). Berlin.

7. Mangat, P. S. (1982). Creep characteristics of polymer modified concrete under uniaxial compression. *3rd international congress on polymers in concrete* (pp. 193–208). Koriyama.
8. Vandamme, M., Ulm, F.-J., & Fonollosa, P. (2010). Nanogranular packing of CSH at substochiometric conditions. *Cement and Concrete Research, 40*(1), 14–26.
9. Nemecek, J., Kralik, V., & Vondrejc, J. (2013). Micromechanical analysis of heterogeneous structural materials. *Cement and Concrete Composites, 36*, 85–92.
10. Sebastiani, M., Moscatelli, R., Ridi, F., Baglioni, P., & Carassiti, F. (2016). Highresolution high-speed nanoindentation mapping of cement pastes - Unravelling the effect of microstructure on the mechanical properties of hydrated phases. *Materials & Design, 97*, 372–380.
11. Vandamme, M., & Ulm, F.-J. (2013). Nanoindentation investigation of creep properties of calcium silicate hydrates. *Cement and Concrete Research, 52*, 38–52.
12. Wang, R., Lackner, R., & Wang, P.-M. (2011). Effect of styrene-butadiene rubber latex on mechanical properties of cementitious materials highlighted by means of nanoindentation. *Strain, 47*, 117–126.
13. Velez, K., Maximilien, S., Damidot, D., Fantozzi, G., & Sorrentino, F. (2001). Determination by nano-indentation of elastic modulus and hardness of pure constituents of Portland cement clinker. *Cement and Concrete Research, 31*(4), 555–561.
14. Zaoui, A. (2002). Continuum micromechanics: A survey. *Journal of Engineering Mechanics, 128*(8), 808–816.
15. Powers, T. C., & Brownyard, T. L. (1948). Studies of the physical properties of hardened portland cement paste. *Research Laboratories of the Portland Cement Association Bulletin, 22*, 101–992.
16. Pichler, B., & Hellmich, C. (2011). Upscaling quasi-brittle strength of cement paste and mortar: A multiscale engineering mechanics model. *Cement and Concrete Research, 41*(5), 467–476.

# Chapter 14
# Stability of Latex in Cement Paste: Experimental Study and Theoretical Analysis

**Dongdong Han, Weideng Chen, and Shiyun Zhong**

To clarify the key factors affecting the stability of latex in cement paste, five latexes with varied hard and soft monomer contents were prepared, and the relationship between the stability of latex in cement paste and the amount of surfactant was investigated through flowability test and residue on sieve test. Besides, the amounts of surfactant required for covering the bare surface of polymer particles and the coefficient of restitution of polymer particles during collision process were calculated. Results show that the stability of latex in cement paste decreases with the increase of soft monomer content. Instead of the coverage ratio of surfactant on polymer particle, the coefficient of restitution of polymer particles determines the stability of latex in cement paste, and the higher coefficient of restitution contributes to the higher stability of latex in cement paste. These results provide some suggestions for the development of latex used for modification of cement-based materials.

## 1 Introduction

Polymer latex has been widely used for the modification of cement-based materials because it can improve various properties [1]. However, most of the latexes coagulate when mixed with cement due to lack of required stability [1–4], especially when the glass transition temperature ($T_g$) of latex is lower than 0 °C [5]. $T_g$ is determined by the hard and soft monomer content during emulsion polymerization [6], which implies that hard and soft monomer content affects the stability of latex in

---

D. Han · S. Zhong (✉)
Key Laboratory of Advanced Civil Engineering Materials, Ministry of Education, Tongji University, Shanghai, China

W. Chen
Fujian Academy of Building Research, Fuzhou, China

cement paste. However, few studies have investigated why it affects the stability of latex in cement paste.

In this paper, five latexes with different styrene (hard monomer)/butyl acrylate (soft monomer) ratio were synthesized, and the amounts of surfactant required for keeping latex stable in cement paste were investigated. Besides, the bare ratio of polymer surface, the theoretical amounts of surfactant required for keeping latex stable in cement paste, and the coefficient of restitution of polymer particles were calculated theoretically, so the influencing mechanism of hard and soft monomer contents on the stability of latex in cement paste could be analyzed.

## 2 Experimental

### 2.1 Materials

P•I 42.5 Portland cement and five self-made styrene-acrylate (SA) latexes were used. The compositions and properties of latexes were described in reference [6]. The latexes were marked with BA××, BA represents butyl acrylate, and the following value indicates the weight percentage of BA compared with total monomers.

### 2.2 Stability of Latex in Cement Paste

The stability of latex in cement paste was evaluated by flowability test and residue on sieve test, and the specific process of these tests can be found in the work by Han et al. [7]. Water/cement ratio of 0.35 and polymer/cement ratio of 0.1 were chosen to prepare the paste. The dosage of surfactant OP10 was given in millimole (mmol) per gram of polymer.

### 2.3 Surfactant Titration

When the surface of polymer particle is not completely covered by surfactant, the surface tension of latex will decrease with the increase of surfactant concentration until reaching equilibrium. At inflection point, the concentration of free surfactant in latex reaches critical micelle concentration (CMC). The concentration of surfactant at inflection point of titration curve is considered as the amount of surfactant required in titration. The mass concentration of latex for titration test was 10%, and the latex was titrated with mixed surfactant. The mixed surfactant was same as that used in emulsion polymerization, which was prepared by mixing sodium lauryl sulfate ($C_{12}H_{25}SO_4Na$) and alkyl alcohol polyoxyethylene ether AEO9 ($C_{12-14}H_{25-29}O(CH_2CH_2O)_9H$) at a mass ratio of 1:1.5.

**Table 14.1** Flowability and residual ratio on sieve of cement paste

| Paste type | Amount of OP10/ (mmol/g) | Flowability of paste/mm | | | Residual ratio on sieve (0 min)/(%) |
|---|---|---|---|---|---|
| | | 0 min | 60 min | 120 min | |
| CP | 0 | 50 | 40 | 40 | 0.2 |
| BA49 | 0 | 112 | 115 | 110 | 1.6 |
| BA54 | 0 | 65 | 83 | 43 | 1.9 |
| BA59 | 0.02 | –[a] | –[a] | –[a] | 80.4 |
| | 0.04 | 100 | 128 | 145 | 1.5 |
| BA64 | 0.06 | 23[b] | 40 | 60 | 13 |
| | 0.08 | 57 | 100 | 89 | 2 |
| BA69 | 0.12 | 23[b] | 23[b] | 23[b] | 10.7 |
| | 0.16 | 108 | 120 | 115 | 1.7 |

Note: [a]The symbol "–" means the paste is not a real paste but cohesionless particles
[b]The flowability of 23 mm means that the paste does not flow because it is the diameter of the tube for the test

## 3 Results and Discussion

### *3.1 Experimental Study on Stability of Latex in Cement Paste*

The flowability and residual ratio on sieve of different pastes are shown in Table 14.1. It is found that when BA monomer content is more than 54 percent, the latexes cannot be used for modifying cement paste when without extra OP10, latex modified cement pastes (LMCP) show low flowability and high residual ratio on sieve. With the increase of OP10 amount, the flowability of LMCP increases significantly, and the residual ratio on sieve decreases accordingly. All latexes can be stable in cement paste only if there is enough surfactant available.

If the latex is considered to be stable in cement paste when the flowability of LMCP is higher than that of pure cement paste (CP), then the actual amounts of OP10 required for keeping latex stable in cement paste are 0, 0, 0.04, 0.08, and 0.16 mmol/g for BA49, BA54, BA59, BA64, and BA69 latex, respectively. The stability of latex in cement paste decreases with the increase of BA monomer in latex.

### *3.2 Theoretical Analysis*

#### 3.2.1 Surfactant Amount Required for Covering Bare Surface of Polymer Particle

It is generally recognized that the stability of latex is maximum when the surfaces of polymer particles are completely covered by surfactant [8]. From the experimental fact that all latexes can be stable in cement paste only if there is enough surfactant

**Table 14.2** $C_{titration}$, $\mu$, $B_{OP10}$, and $A_{OP10}$ of different latexes

| Latex type | $C_{titration}$/(mg/L) | $\mu$ | $B_{OP10}$/(mmol/g) | $A_{OP10}$/(mmol/g) |
|---|---|---|---|---|
| BA49 | 1600 | 0.38 | 0.048 | 0 |
| BA54 | 1240 | 0.32 | 0.041 | 0 |
| BA59 | 900 | 0.26 | 0.036 | 0.04 |
| BA64 | 940 | 0.25 | 0.032 | 0.08 |
| BA69 | 680 | 0.20 | 0.028 | 0.16 |

available, we speculated that the polymer particle surface that is not covered by surfactant (bare surface) may be a reason for the destabilization of latex in cement paste. The bare surface divided by the total surface of the polymer particles is called bare ratio of polymer particles ($\mu$), which could be calculated by Eq. 14.1. The calculation results are shown in Table 14.2.

$$\mu = \frac{C_{titration} - CMC}{C_{adsorbed}} \qquad (14.1)$$

where $C_{titration}$ is the amount of surfactant required in surfactant titration test [mg/L], which is listed in Table 14.2; CMC is the critical micelle concentration of mixed surfactant [40.5 mg/L]; and $C_{adsorbed}$ is the total amount of surfactant adsorbed on polymer particles [mg/L], which can be calculated by Eq. 14.2.

$$C_{adsorbed} = C_{titration} + C_{initial} - CMC \qquad (14.2)$$

where $C_{initial}$ is the initial surfactant amount in polymerization process [2550 mg/L].

It was assumed that the coverage area of an OP10 molecule on SA polymer was equal to that on polystyrene particle, which is 0.54nm² [9]. Then the amounts of OP10 required for covering the bare surface of SA particles ($B_{OP10}$) were calculated as indicated in Table 14.2. It is found that with the increase of BA monomer content, $\mu$ and $B_{OP10}$ decrease. Comparing $B_{OP10}$ and the actual required OP10 for keeping latex stable in cement paste ($A_{OP10}$, Table 14.2), it is found that the latexes with BA monomer content less than 54 percent can be very stable in cement paste even if part of the surface of polymer particles is bare, while the latexes with BA monomer content more than 59 percent cannot be stable in cement paste even if the surface of polymer particles are covered by surfactant completely, which indicates that the bare ratio of polymer particles is not a critical factor affecting stability of latex in cement paste.

### 3.2.2 Collision Between Polymer Particles During Mixing Process

Polymer particles will collide with each other during the mixing process of cement paste. Whether the particles aggregate together or separate from each other depends on the elastic force and Van der Waals force during collision process. It is reported that with the restitution coefficient decrease, the decaying rate of elastic force increases; as a result, the particles tend to aggregate together after collision.

As reported by Roy [10], coefficient of restitution ($e$) can be calculated by Eq. 14.3:

$$e = \frac{1.9 H^{5/8}}{E_{eff}^{1/2} \rho_{polymer}^{1/8} V^{1/4}} \quad (14.3)$$

where $H$ is the hardness of polymer particles [MPa], $\rho_{polymer}$ is the density of the polymer particles [1 g/cm$^3$], $V$ is the relative velocity of particles [m/s], and $E_{eff}$ is the effective modulus of the particles [MPa], which can be written as Eq. 14.4:

$$\frac{1}{E_{eff}} = \frac{1-v_1^2}{E_1} + \frac{1-v_2^2}{E_2} \quad (14.4)$$

where $E_1$ and $E_2$ are the elastic modulus of colliding polymer particles [MPa], and $v_1$ and $v_2$ are the Poisson ratio of polymer particles. In this study, the colliding particles are the same to each other, so the elastic modulus of polymer ($E$) $= E_1 = E_2$ and $v_1 = v_2 = 0.499$.

The hardness/elastic modulus ($H/E$) ratio of polymer is between 0.05 and 0.1, and $H/E$ is near 0.1 for the polymer of high elasticity [11]. Tg of polymers used in this study is lower than room temperature (20 °C), so the polymers are in high elastic state, and the relationship between H and E of these polymers can be expressed as Eq. 14.5:

$$H/E = 0.1 \quad (14.5)$$

It is assumed that $V$ is the same in different LMCP systems. Based on Eqs. 14.3, 14.4, and 14.5, the relationship between $eV^{1/4}$ and $E$ can be obtained as Eq. 14.6:

$$eV^{\frac{1}{4}} = 0.55 E^{\frac{1}{8}} \quad (14.6)$$

The $E$ of polymers was obtained by tensile stress-strain curves of polymer films. $E$ and $eV^{1/4}$ of different polymer particles are shown in Table 14.3. It is found that $eV^{1/4}$ decreases with the increase of BA monomer content, which means that the polymer particles with higher content of BA monomer are more inclined to aggregate together after collision. This explains well why the latex with higher content of BA monomer is more difficult to maintain stable in cement paste.

**Table 14.3** The elastic modulus and $eV^{1/4}$ of polymer particles

| Polymer particle type | Elastic modulus ($E$)/MPa | $eV^{1/4}$ |
|---|---|---|
| BA49 | 18.05 | 0.800 |
| BA54 | 2.12 | 0.604 |
| BA59 | 0.59 | 0.515 |
| BA64 | 0.36 | 0.484 |
| BA69 | 0.16 | 0.437 |

## 4 Conclusions

The stability of latex in cement paste decreases with the increase of BA monomer content in polymer. The coefficient of restitution of polymer particles decreases significantly with the increase of BA monomer content, which determines the stability of latex in cement paste rather than the coverage ratio of surfactant to polymer particles. These results provide some guidance for the application of latex in cement-based materials.

## References

1. ACI 548.3R-09. (2009). *Report on polymer-modified concrete*. Farmington Hills: American Concrete Institute.
2. Cai, Y., Wang, P. M., & Zhong, S. Y. (2015). Influence of coagulation of polymer dispersion on the properties of polymer-modified mortar. *Advanced Materials Research, 1129*, 162–168.
3. Merlin, F., Guitouni, H., Mouhoubi, H., Mariot, S., Vallée, F., & Van, D. H. (2005). Adsorption and heterocoagulation of nonionic surfactants and latex particles on cement hydrates. *Journal of Colloid and Interface Science, 281*(1), 1–10.
4. Zhao, W., Zhang, Y., Zhao, Y., & Zhang, H. (2014). Effects of twain-20 on performance of PB-g-PSG latex modified cement mortar and its mechanism of action. *Journal of Building Materials, 17*(4), 566–571.
5. Xu, J., Feng, Z. J., Fu, L. F., & Meng, W. K. (2015). A low Tg acrylate latex for cement-based materials and its preparation method. *CN104530304A*. Patent filled.
6. Han, D. D., Chen, W. D., & Zhong, S. Y. (2017). Physical retardation mechanism of latex polymer on the early hydration of cement. *Advances in Cement Research*. https://doi.org/10.1680/jadcr.17.00055.
7. Han, D. D., Chen, W. D., Zhong, S. Y., & He, Y. (2017). Effects of nonionic surfactants and external factors on stability of latex in cement paste. *Journal of Applied Polymer Science*. https://doi.org/10.1002/app.45946.
8. Zaccone, A., Wu, H., Marco Lattuada, A., & Morbidelli, M. (2008). Correlation between colloidal stability and surfactant adsorption/association phenomena studied by light scattering. *Journal of Physical Chemistry B, 112*(7), 1976–1986.
9. Kronberg, B., Käil, L., & Stenius, P. (1981). Adsorption of nonionic surfactants on latexes. *Journal of Dispersion Science and Technology, 2*(2–3), 215–232.
10. Roy, M. (1999). Dynamics of impact of sphere on flat surfaces of polymer matrix composites. *Bulletin of Materials Science, 22*(6), 1009–1012.
11. Amitay-Sadovsky, E., & Wagner, H. D. (1998). Evaluation of young's modulus of polymers from knoop microindentation tests. *Polymer, 39*(11), 2387–2390.

# Chapter 15
# Experimental Verification of Use of Secondary Raw Materials as Fillers in Epoxy Polymer Concrete

Rostislav Drochytka and Jakub Hodul

Generally, fillers make up more than 70% of the volume in polymer concrete (PC). Silica sand of various fractions with optimal round-shaped grains is one of the most widely used fillers in PC. Silica sand is the primary raw material and its replacement by some progressive secondary raw materials could be appropriate if PC shows the same or better physical and mechanical properties. In this paper, the possibility of using secondary raw materials such as waste container glass (WCG), waste auto glass (WAG), waste foundry sand (WFS), waste slag (WS), and fly ash contaminated by denitrification process (FAD) are examined to see if they can replace the commonly used round-grain pure crystal silica sand (REF) in epoxy PC. Based on the compressive strength and three-point flexural strength test results, the best PC formulations were selected, which were subsequently tested for the pull-off bond strength. Samples with optimum amount of the filler were also monitored for the microstructure using a high-resolution optical microscope. It was found that the most suitable fillers of the secondary raw materials tested were waste foundry sand (WFS) and waste container glass (WCG). It was also determined that the optimal amount of these progressive fillers of suitable granulometry is 75% by weight, when the PC showed the best physical and mechanical parameters. It was confirmed that by replacing the currently used primary natural material, silica sand, with the secondary raw materials, it is possible to achieve better properties of PC and at the same time improve the environmental and economic benefits of PC production.

---

R. Drochytka · J. Hodul (✉)
Brno University of Technology, Faculty of Civil Engineering, AdMaS Centre, Brno, Czech Republic
e-mail: hodul.j@fce.vutbr.cz

## 1 Introduction

Polymer concrete (PC) is a composite material in which aggregates are bonded together with polymer resins [1]. PC is being extensively used as a suitable substitute for cement concrete in a variety of applications such as construction and structural repairs, pavements, floors, and dams [2, 3]. The properties of PC depend on many factors including the type and content of filler or resin; type and particle size of aggregates; and also dosage of components in the PC mixture [4]. The grading of aggregates in the PC is nonstandardized till date and varies widely from system to system. In addition to the coarse and fine aggregates, microfillers are also added sometimes to the PC mainly with an aim to fill the microvoids [5]. At present, crystal quartz sand is the most commonly used filler for PC in the Czech Republic. It is washed several times, free from humus-like substances and after drying, the round-grain sand is graded to a 0–1.5 mm granulometry. However, this is the primary raw material and its replacement with a suitable treated secondary raw material could be appropriate. Crushed and sorted transparent container waste glass seems to be the progressive substitute [6]. The objective of this experiment was to achieve as high a filling of the selected secondary raw materials (WCG, WAG, WFS, WS, FAD) as possible while achieving the best PC properties that would result in the reduction of epoxy resin (ER) volume and an efficient use of the by-products. All these materials are cumulated in the Czech Republic in a large quantity and it is very probable that it will be possible to use them as substitute for primary raw materials, such as silica sand.

## 2 Materials

**Binder** As a polymer binder, a low-viscosity, double-component, solvent-free material on epoxy base was used. The A component (epoxy resin (ER)) is based on bisphenol A and the B component (hardener) is based on aromatic amines (aniline, 2,4,6-tris (dimethylaminomethyl)phenol). This material is mainly designed for the preparation of PC floors in industrial building with high loads, where there is a high demand for resistance against humid temperature and mechanical load. The recommended temperature for the application is +20 °C and during the application and polymerization, water or any other aggressive medium may not enter the material. During the samples production, the A component was first of all properly and slowly mixed with a slow-rotating mixing machine together with the selected type and amount of filler, while later the hardener was added (B component) and the mixture was homogenized again. The A:B mixing ratio was 3.2:1 and the treatability of the mixed material was 15 min.

**Filler** The tested fillers were: waste container glass (WCG), waste auto glass (WAG), waste foundry sand (WFS), waste slag (WS), and filter fly ash contaminated by flue gas denitrification (FAD) (see Tables 15.1 and 15.2 and Fig. 15.1). The

**Table 15.1** Chemical composition of the fillers (representative oxides) [% of dry]

| Filler | SiO$_2$ | CaO | Fe$_2$O$_3$ | MgO | Al$_2$O$_3$ | MnO | K$_2$O | Na$_2$O |
|---|---|---|---|---|---|---|---|---|
| WCG | 75.35 | 7.45 | 0.17 | 4.23 | 0.53 | – | 0.19 | 11.87 |
| WAG | 70.32 | 9.23 | 0.15 | 3.65 | 0.71 | – | 0.32 | 12.02 |
| WFS | 94.0 | 0.17 | 0.39 | 0.06 | 1.72 | 0.01 | 0.73 | 1.99 |
| WS | 41.5 | 24.8 | 9.89 | 7.23 | 11.6 | 0.50 | 0.44 | 0.27 |
| FAD | 53.4 | 4.13 | 9.45 | 1.97 | 17.8 | 0.15 | 1.78 | 0.65 |
| REF | 99.5 | 0.04 | 0.03 | 0.01 | 0.21 | – | 0.12 | 0.03 |

**Table 15.2** Density of the fillers

|  | WCG | WAG | WFS | WS | FAD | REF |
|---|---|---|---|---|---|---|
| Density [kg/m$^3$] | 2530 | 2540 | 2680 | 2890 | 2530 | 2662 |

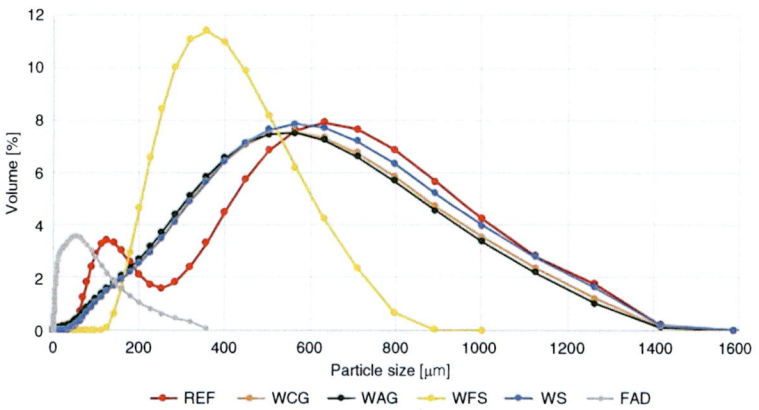

**Fig. 15.1** Particle size distribution of the fillers

chosen waste glasses and slag were ground before their application and sorted into the granulometry, as in the case of the reference pure silica sand (REF) (see Fig. 15.1). During the grinding process, the foil had to be removed from the auto glass, which would have caused nonhomogeneity of the resulting material. Chosen secondary raw materials can be obtained in the Czech Republic free of charge. However, when using waste glass and slag, it is necessary to include the costs for cleaning, crushing, and sorting, which makes about $0.4 per kg. Therefore, from the economic point of view, the use of fly ash is most appropriate as it does not require further treatment. The reference sand price is around $0.8 per kg. FAD comprises the same volume of ammonia ions (NH$_3$), more specifically 3.79 mg/kg of the dry matter, which is caused by the flue gas denitrification impact – the injection of a water solution of urea into a combustion boiler under high temperatures. The requirement for NO$_x$ emission limits reduction for industrial resources within the Czech Republic is based on the European Parliament and Committee directive 2010/75/EU.

**Fig. 15.2** Determination of flexural (**a**) and compressive strength (**b**)

**Formulations** In this experiment, 23 formulations were tested in total, while the filling volume was selected to be within a range of 50–85% (percentage of proportion in PC). FAD could be added in a smaller volume as in the case of coarse-grained fillers, as the finer the filler particles, the more the viscosity of the ER.

## 3 Methods

**Compressive and Flexural Strength** The determination of the compressive strength (Fig. 15.2b) and three-point flexural strength (Fig. 15.2a) was performed in line with the EN 13892-2 standard. For the compressive strength test, fragments of the samples (20 × 20 × 100 mm) after the flexural strength test were used and the area of the supporting metal pressing jigs was 400 mm$^2$ (Fig. 15.2b).

**Pull-Off Bond Strength** The measurement of the pull-off bond strength was performed according to the EN 1542 standard. A penetrated concrete vibro-pressed pavement block of 30 × 30 cm size was used as a base for the application of the PC. A fresh PC of 5 mm thickness was applied on the entire concrete surface. This test was performed with the use of metal targets of 50 mm in diameter, stuck to the hardened surface of the PC by epoxy glue (Sikadur 31 CF). After 24 h, the adhesion was tested by the DYNA Proceq pull-off tester.

**Microstructure** The microstructure of the chosen samples was observed by the Leica DFC450 C optical microscope with a high display resolution up to a magnification of 500×. The distribution of filler particles was mostly monitored within the epoxy matrix and if the selected samples showed homogeneity in the entire cross section of 20 × 20 × 10 mm in size. The distribution of filler particles has a significant effect on the resulting physical and mechanical parameters of PC. Silica sand has a higher density and therefore higher susceptibility to sedimentation can be observed.

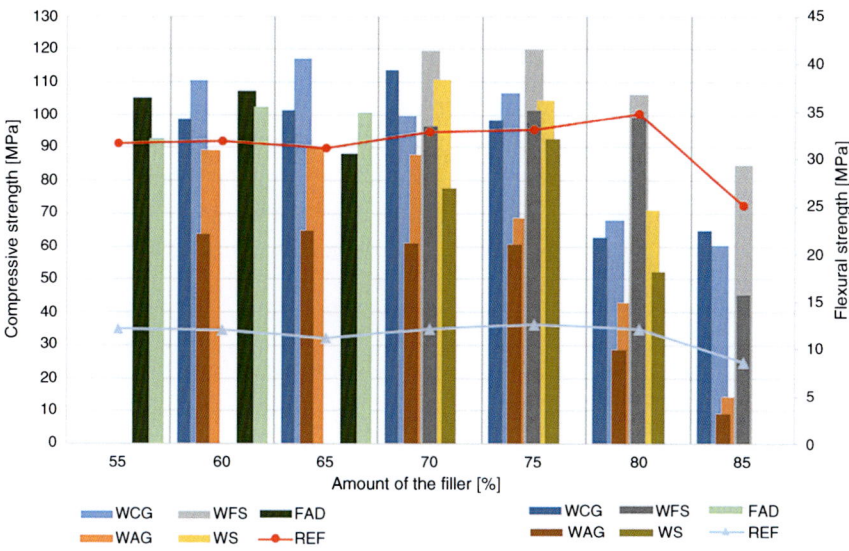

**Fig. 15.3** Results of the compressive and flexural strength

## 4 Results and Discussion

**Compressive and Flexural Strength** The results of the compressive and flexural strength tests are displayed in Fig. 15.3. The highest compressive strength was found in the sample of PC with 75% WFS (120 MPa). In the case of REF samples, the maximum compressive strength was achieved below 80% of filling, while this value is 20 MPa lower than the 75% WFS sample. Thanks to its shape index, WFS was probably incorporated into the epoxy matrix, preferably from all the fillers. From the results it is clear that the WFS contamination has no negative impact on the final mechanical parameters of the PC. Based on the results, it is clearly visible that it is only in the case of PC samples with WAG that no higher compressive strength values were achieved in comparison with the REF samples. The highest flexural strength was achieved by the PC sample with 70% WCG (39.4 MPa), while this value is 3 MPa higher than in the case of the 75% REF sample, which showed the highest flexural strength. Samples that had the highest compressive strength (75% WFS) showed a flexural strength of 35.1 MPa, which in practice is the same value as the REF samples. It was proved that in the case of coarse-grained fillers during filling higher than 75%, the strength started to markedly decrease. This fact is the result of the problematic incorporation of the filler under high filling into the epoxy matrix and it is impossible to cover all the filler particles. It is very probable that due to the problematic homogenization of the fresh mixture, the hardened PC is nonhomogeneous and no strong contact zone of filler-binder is secured when the main damage occurs. A filling of 75% seems to be the optimum for coarse-grained fillers, while for fine-grained fillers, such as fly ash (FAD), a 60% filling is optimum.

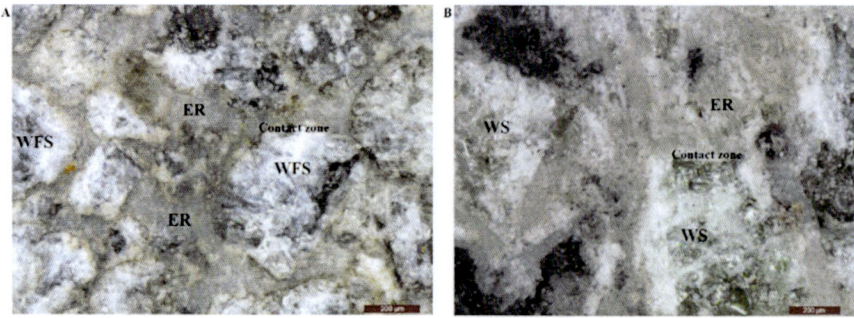

**Fig. 15.4** Photomicrographs of 75% WFS (**a**) and 75% WS (**b**) sample (100×)

**Pull-Off Bond Strength** Break in the base concrete (2–5 mm) was reported for all the tested formulas, including the reference ones, and the pull-off bond strength value was higher than 4 MPa. On the basis of such results, it is clear that the adhesion of tested PC samples with the base concrete is exceptional. The minimum required adhesion of repair materials with a static function (R4 class) with the base concrete is more than 2.0 MPa, pursuant to the EN 1504-3 standard.

**Microstructure** In Fig. 15.4, it is possible to see the photomicrographs from the optical microscope under a magnification of 100×. It is clear that the filler particles (WS, WFS) are ideally covered by the epoxy binder (ER) and are equally distributed within the matter. At the contact zone point, there is no separation visible between the used filler and epoxy binder.

## 5 Summary

In the performed experiments it was found that the most appropriate substitution for the silica sand of the 0–1.5 mm granulometry in the PC seems to be waste foundry sand (WFS) and waste container glass (WCG). The most optimal proportion of these secondary fillers is 75% wt. (300% from the ER binder amount), when the highest compressive strength was achieved (120 MPa) and the same flexural strength as in the case of REF samples (35 MPa).

**Acknowledgments** This paper has been written under project no. LO1408 "AdMaS UP – Advanced Materials, Structures and Technologies", supported by the Ministry of Education, Youth, and Sports under "National Sustainability Programme I".

## References

1. Mebarkia, S., & Vipulanandan, C. (1995). Mechanical properties and water diffusion in polyester polymer concrete. *Journal of Engineering Mechanics, 121*, 1359–1365.
2. Vipulanandan, C., & Dharmarajan, N. (1987). Flexural behaviour of polyester polymer concrete. *Cement and Concrete Research, 17*, 219–230.
3. Czarnecki, L., Garbacz, A., & Kurach, J. (2000). On the characterization of polymer concretefracture surface. *Cement and Concrete Composites, 23*, 399–409.
4. Bărbuță, M., Harja, M., & Baran, I. (2010). Comparison of mechanical properties for polymer concrete with different types of filler. *Journal of Materials in Civil Engineering, 22*, 696–701.
5. Bedi, R., Chandra, R., & Singh, S. P. (2013). Mechanical properties of polymer concrete. *Journal of Composites, 2013*, 1–11.
6. Hodul, J., Hodná, J., Drochytka, R., & Vyhnánková, M. (2016). Utilization of waste glass in polymer concrete. *Materials Science Forum, 865*, 171–177.

# Chapter 16
# Effects of Anionic Asphalt Emulsion on Early-Age Cement Hydration

Jinxiang Hong, Wei Li, and Kejin Wang

Cement-asphalt mortar (CAM) is a cement-based, asphalt-modified, inorganic-organic composite. In this study, the effects of anionic asphalt emulsion on early-age cement hydration of CAM pastes are investigated. The setting behavior, heat of hydration, and electrical resistivity of CAM pastes with different asphalt-to-cement (A/C) ratios are evaluated. The results indicated that the set time of cement paste was prolonged and the heat released from cement hydration was delayed with increased A/C. During the early state of cement hydration, the resistivity of the CAM paste studied was higher than that of the corresponding Portland cement paste, indicating that the asphalt emulsion hindered the ion dissolution of the cement in the CAM paste. However, during the acceleration period of cement hydration, the resistivity of the CAM paste studied was lower than that of the corresponding Portland cement paste, implying that the formation of microstructure of the CAM paste was delayed. This retardation phenomenon could be attributed to two mechanisms: (i) the active sites of cement particles that could have been occupied by asphalt emulsion through anionic emulsifier via electrostatic interaction and (ii) the most surfaces of the cement particles that might have been covered by asphalt membrane due to the demulsification of asphalt emulsion.

---

J. Hong (✉)
Jiangsu Research Institute of Building Science Co., Ltd, Nanjing, China

Department of Civil, Construction, and Environmental Engineering, Iowa State University, Ames, IA, USA
e-mail: hongjinxiang@cnjsjk.cn

W. Li
College of Civil Science and Engineering, Yangzhou University, Yangzhou, China

K. Wang
Department of Civil, Construction, and Environmental Engineering, Iowa State University, Ames, IA, USA

# 1 Introduction

Cement-asphalt mortar (CAM) is a cement-based, asphalt-modified, inorganic-organic composite, which consists of Portland cement (hereafter PC), asphalt emulsion, water, and other related admixtures. This hybrid material combines the high compressive strength of cement paste with the high flexibility of asphalt, and the resulting composite possesses not only desirable strength, stiffness, temperature resistance but also much improved flexibility and ductility [1]. CAM has been extensively used as a grouting material in the cushion layer of the slab track system of high-speed railways (HSR) in China, primarily due to its excellent damping property.

In a CAM mixture, asphalt is often introduced as an emulsion form. During mixing, the droplets of asphalt often adsorb on the surfaces of cement particles. As a layer of asphalt droplet coats the cement particles, it influences the CAM mixture rheological property, cement hydration, and set behavior. As cement hydrates, water in the mixture reduces due to cement hydration and evaporation. This result in a thin membrane of asphalt that lines capillary pores and cracks interweaves with cement hydration products and wraps unhydrated cement and aggregate particles. Consequently, the hardened CAM has excellent mechanical and durability properties.

The properties of CAM can be strongly affected by the type of asphalt emulsion used. The anionic asphalt emulsion is primarily adsorbed onto the positively charged surface of aluminate phases, such as $C_3A$ and some small parts of silicate phases [2]. Therefore, the anionic asphalt emulsion could only exert some slight hindrances to the hydration of $C_3S$ and $C_2S$, which are the main contributors to the strength development of PC paste. Therefore, one could deduce that anionic asphalt emulsion is more suitable for formulating CAM in the applications where high strength is required. However, little study has been reported to confirm this concept.

In the present study, an anionic asphalt emulsion was used to formulate a type II CAM, where a relatively high strength was required under the current Chinese standard for high-speed railway. The effects of this anionic asphalt emulsion on the early-age cement hydration process were characterized by set time, isothermal calorimetry, and electrical resistivity of the CAM. Based on these experimental test results, three possible mechanisms by which the anionic asphalt emulsion influences cement hydration in CAM were then discussed.

# 2 Raw Materials, Sample Preparation, and Test Methods

## 2.1 Raw Materials

Portland cement (PC), type P.I 42.5, complying with the Chinese standard GB8076-2008, was used in this study. An anionic asphalt emulsion (with a solid content of 60%) was supplied by Jiangsu Bote New Materials Co., Ltd.

## 2.2 Sample Preparation

Paste samples were used for all tests conducted in the present study. The CAM pastes had a fixed water-to-cement (W/C) ratio of 0.41 and different asphalt-to-cement (A/C) ratios, namely, 0, 0.16, 0.24, and 0.32. Based on their A/C, the CAM mixes studied were named as PC (i.e., CA-0, CA-0.16, CA-0.24, and CA-0.32). All these pastes were tested for both set time and isothermal calorimetry. However, the electrical resistivity tests were carried out only for two CAM pastes, with A/C of 0 and 0.24.

To prepare samples, water and asphalt emulsion was mixed first. (The amount of water in the anionic asphalt emulsion was considered in the calculation of W/C.) Cement was then added into the solution. The mixture was then mixed according to the standard practice for mechanical mixing of hydraulic cement paste.

## 2.3 Test Methods

### 2.3.1 Set Time Test

Set time was measured according to the standard of ISO 9597:2008 Cement-Test Method, determination of set time and soundness, NEQ. Both the initial and final set times of CAM pastes with different anionic asphalt emulsion were measured, and the test temperature was kept at 20 °C.

### 2.3.2 Calorimetry Test

The fresh CAM pastes were put into a plastic bottle within 10 min after mixing, and the heat released from the paste was measured by a self-regulated isothermal conduction calorimeter. The testing temperature was equilibrated at 20 °C between the specimen and the instrument before conducting the measurement, and the test lasted about 72 h.

### 2.3.3 Electrical Resistivity Measurement

This measurement was performed using a noncontact impedance measurement (NCIM CCR-II). In the test, a fresh CAM paste was cast in a ring-shaped plastic mold. AC currents with different frequencies were applied to the sample via a transformer core. The current going through the tested sample was recorded, and the resistivity of the sample was then computed.

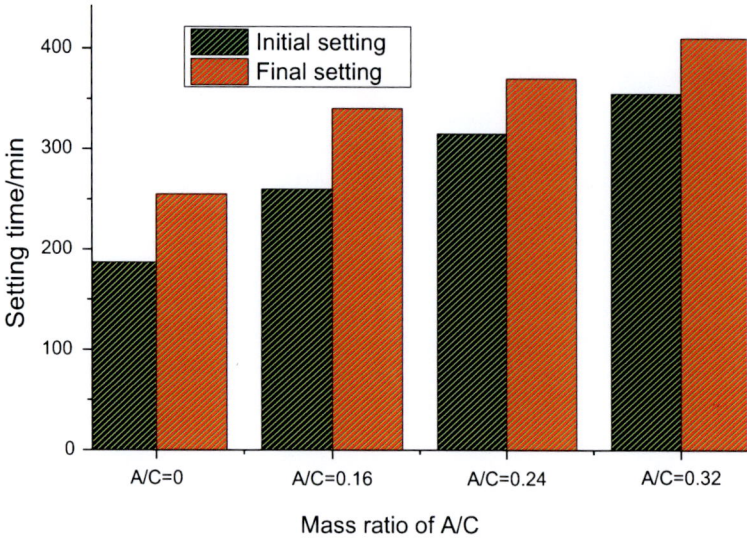

**Fig. 16.1** Influence of anionic asphalt emulsion on set time [3]

## 3 Results and Discussion

### 3.1 Influence of Anionic Asphalt Emulsion on Cement Set Time

Figure 16.1 shows that both the initial and the final set times were prolonged with the increase of the A/C from 0 to 0.32, which implies that the PC hydration was retarded by the anionic asphalt emulsion. As mentioned previously, during CAM mixing, a layer of asphalt droplet often coats cement particles, which could prevent the cement particles to contact with water, thus retarding cement hydration and delaying the paste set times.

### 3.2 Influence of Anionic Asphalt Emulsion on the Hydration Heat

Figure 16.2a presents the heat flow during the first 70 h of PC hydration at different A/C, and Fig. 16.2b shows the details of heat evolution during the first 10 h.

From Fig. 16.2b, it can be seen that the induction period of heat flow curve was prolonged with the increase of A/C. It is generally agreed that the surface of aluminate phase and certain part of silicate phases show positive charge after mixing with water, which is mainly due to the discrepancy of the migration rate of different kinds of ions from the cement particles into solution [4]. Hence, the anionic asphalt emulsion can be

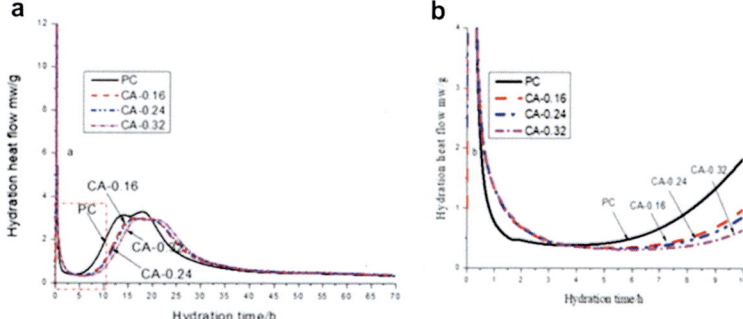

**Fig. 16.2** Influence of asphalt emulsion on heat of cement hydration [3] (**a**) heat flow curve, (**b, c**) the initial part of (**a**) (time ≤ 10 h)

adsorbed onto the positively charged sites of cement particles through electrostatic interaction. Consequently, the surface of cement particles could be covered by an asphalt membrane, which would hinder the further dissolution of cement particles and resulted in the slow increase of the cumulative concentration of $Ca^{2+}$ with the increase of A/C during the induction period. Therefore, the duration of the induction period was prolonged due to the lower concentration of $Ca^{2+}$ in the presence of anionic asphalt emulsion, i.e., the PC hydration process was retarded by the anionic asphalt emulsion.

### 3.3 Influence of Anionic Asphalt Emulsion on the Electrical Resistivity

In this study, the electrical resistivity was employed to investigate the evolution of the early-stage hydration process of CAM paste. Figure 16.3 shows the evolution of electrical resistivity of the CAM paste at an A/C of 0.24 in comparison with the pure PC paste within the first 24 h.

It is evidenced that during the dissolution period, the electrical resistivity of both the CAM paste and the pure PC paste reduced quickly to a minimum value, which lasted for about 100 min and 50 min, respectively. Additionally, it can be seen that the electrical resistivity of the CAM paste was higher than that of the pure PC paste during the dissolution and induction periods, while the opposite can be observed during the acceleration period.

It is well known that, during the early dissolution stage, the electrical resistivity of fresh PC paste is mainly determined by the ion concentration in the fresh paste which, in turn, is controlled by the dissolution process. That is, the electrical resistivity of the fresh PC paste decreases with the increase of the ion concentration in the solution.

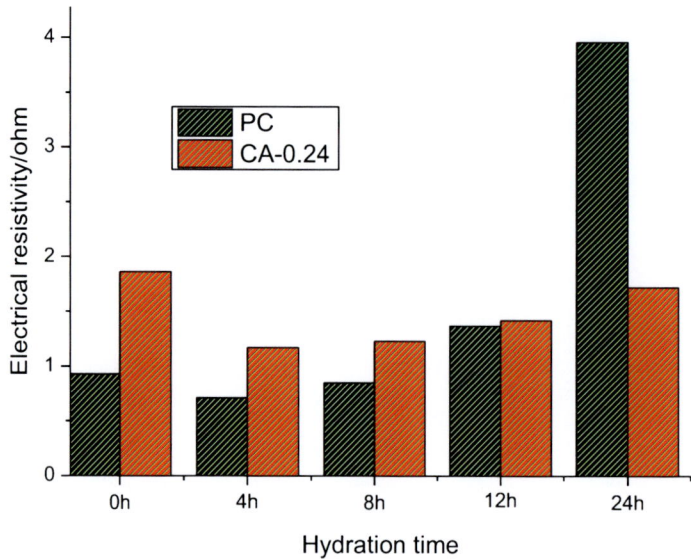

**Fig. 16.3** Influence of anionic asphalt emulsion on electrical resistivity [3]

On the contrary, during the acceleration hydration period, the electrical resistivity of the CAM paste was lower than that of the PC paste, which would indicate that the microstructure formed in the CAM paste was more porous than that of the PC paste. This, again, could be attributed to the barrier formed by the asphalt membrane on the surface of cement grains which could have adversely affected the dissolution process of cement and, hence, retarded the hydration process and the formation of the microstructure of the CAM paste.

## 3.4 The Retardation Mechanism of Asphalt Emulsion on PC Hydration

**Mechanism I: Selective adsorption of anionic emulsifier via electrostatic attraction**
Anionic asphalt emulsion can be selectively adsorbed onto those positively charged sites on the surface of cement grains, such as those sites occupied by aluminate phase or certain part of silicate phases, through anionic emulsifier via electrostatic adsorption. This layer of adsorbed anionic asphalt emulsion would then hinder further dissolution of chemical ions from cement particles and subsequent deposition of hydration products on the surface of cement particles, thus retarding cement hydration.

**Mechanism II: Formation of coating from demulsification of anionic asphalt emulsion**

Once the counterions, such as $Ca^{2+}$, are dissolved from cement particles into the double electrode layers of asphalt emulsion, along with the consumption of water during PC hydration process, the asphalt emulsion would demulsify. This would result in the formation of aggregation of asphalt particles or the formation of asphalt membrane. Due to the surfaces of cement particles covered by asphalt as a membrane, the ions releasing from the cement particles and the deposition reactions of the cement particles were hindered.

## 4 Conclusions

1. The set time of PC was prolonged with increased A/C.
2. The heat of the cement hydration was delayed with increased A/C, and so did set time of the CAM paste.
3. During the early hydration period, the electrical resistivity of fresh CAM paste was higher than that of the pure PC paste; but later, at the accelerated hydration period, it was lower.
4. The possible retardation mechanisms by which the anionic asphalt emulsion influences the PC hydration are (i) the active sites of aluminate phase and some silicate phases could have been occupied by asphalt emulsion particles through anionic emulsifier via electrostatic interaction and (ii) the surface of cement particles might have been covered by a layer of asphalt membrane due to the demulsification of asphalt emulsion.

**Acknowledgments** The present study is a part of the research project supported by the National Natural Science Foundation of China (No.51708483). This paper is modified from a part of publication of Ref. [3].

## References

1. Lu, C.-T., Kuo, M.-F., & Shen, D.-H. (2009). Composition and reaction mechanism of cement-asphalt mastic. *Construction and Building Materials, 23*, 2580–2585.
2. Plank, J., & Hirsch, C. (2007). Impact of zeta potential of early hydration phases on superplasticizer adsorption. *Cement and Concrete Research, 37*, 537–542.
3. Li, W., Hong, J., Zhu, X., Bai, Y., Liu, J., & Miao, C. (2018). Retardation mechanism of anionic asphalt emulsion on the hydration of Portland cement. *Construction and Building Materials, 163*, 938–948. (in press).
4. Li, W., Zhu, X., Hong, J., & Zuo, W. (2015). Effect of anionic emulsifier on cement hydration and its interaction mechanism. *Construction and Building Materials, 93*, 1003–1011.

# Chapter 17
# Polymer Solutions for Protection of Concrete Exposed to Strong Alkaline or Acid Effluent on Industrial Installations

Nicolas Roche and Hervé Davaux

Reinforced concrete (RC) structures of industrial installations may be exposed to various chemical solutions due to accidental or common causes (tank failure, piping leaks, control mishandling, etc.). Besides chemical tanks specifically designed for long-term storage, surrounding concrete retention basins are generally not protected properly against unexpected exposure to aggressive solutions. This raises concerns for environmental protection against potential effluent leaks. This work presents a general industrial case study of protective coatings' chemical resistance assessment and an economically optimal strategy to enhance RC structures' chemical protection. An experimental study of strong acid resistance evaluation has been conducted on various polymer coatings. The results led to a multi-material approach for chemical resistance retrofitting of existing RC structures.

## 1 Introduction

Industrial facility structures are commonly built with reinforced concrete and steel. Many industries (chemical, petroleum, water treatment, energy provider, etc.) use aggressive chemicals in their transformation processes. These chemicals are stored in dedicated tanks in which properties correspond to specified chemicals [1, 2]. However, the mandatory retention spill basins under these tanks are often built in reinforced concrete which is notoriously weak versus many aggressive chemicals especially sulfuric acid [3, 4]. In fact, it is often assumed that a sufficiently reinforced

---

N. Roche (✉)
EDF Ceidre TEGG, Aix-en-provence, France
e-mail: nicolas-n.roche@edf.fr

H. Davaux
EDF CNEPE, Tours, France

concrete will not crack; thus, the concrete itself will satisfy the chemical tightness in case of major spillage. Experience showed that these structures are often overlooked during the construction phase, leading to underperforming RC against crack formation and low thickness RC slabs. This raises concerns for staff safety and environmental protection regarding potential effluent leaks. This work focuses on improving aggressive chemical management in contact with RC structures while complying with industrial constraints. Different polymer coating solutions have been tested with highly concentrated sulfuric acid, and an optimized RC structure retrofitting strategy has been developed and tested on a running industrial facility.

## 2 Experimental Approach

### 2.1 Materials

Concrete protection can be successfully realized using various polymer coatings. Leaching and corrosion can be prevented with watertight coatings consisting of thermoplastic or elastomeric membranes, but also thermoset coatings. However, elastomeric materials commonly used as concrete protection have generally poor resistance against highly aggressive chemicals, and thus were not considered in this work. Table 17.1 presents the selected polymer references. Vinyl ester resins have not been included because of health safety issues linked to styrene monomer (registered in REACH) use during the wet lay-up process. Epoxy coatings already proved to be efficient against low concentration sulfuric acid [5], but no literature was found concerning 96% concentrated sulfuric acid.

The epoxy resin coupon is composed of a glass fiber-reinforced stratification of DGEBA (bisphenol A diglycidyl ether) monomer-based epoxy with the protective outer layer being based on novolac epoxy, a higher grade than common DGEBA type (Fig. 17.1).

**Table 17.1** Materials

| Polymer type | Apply mode | Coupons |
|---|---|---|
| Epoxy resin | Manual wet lay-up stratification | 200 × 100 × 40 mm coated concrete specimens |
| PVDF[a] | Prebuild membranes, mortar embedded and welded | 200 × 100 × 4 mm specimens with welding ribbon |
| ECTFE[a] | Prebuild membranes, mortar embedded and welded | 200 × 100 × 4 mm specimens with welding ribbon |
| HDPE[a] | Prebuild membranes, mortar embedded and welded | 200 × 100 × 4 mm specimens with welding ribbon |

[a]*PVDF* polyvinylidene fluoride, *ECTFE* ethylene chlorotrifluoroethylene, *HDPE* high-density polyethylene

# 17 Polymer Solutions for Protection of Concrete Exposed to Strong Alkaline...

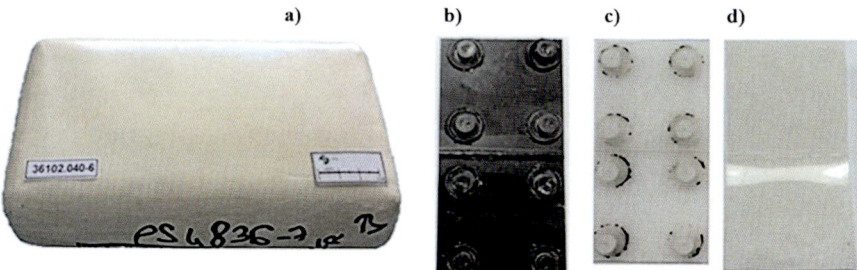

**Fig. 17.1** Samples (**a**) epoxy coating, (**b**) HDPE, (**c**) PVDF, (**d**) ECTFE

## 2.2 Exposure Conditions

This study is focused on the protection of RC retention spill basin associated with 96 w% sulfuric acid storage tanks. A minimum of two coupons per material were fully immersed in a 96% sulfuric acid solution, thermoregulated at 30 °C. Aging duration varied from 1 to 180 days depending on the coating.

## 2.3 Characterization

Weight and dimensional modifications were followed during aging for all coupons. Cross sections of the novolac epoxy coatings on concrete substrate were characterized with binoculars in order to evaluate the aging damage. Tensile testing was performed on the thermoplastic coupons. A minimum of four 5A-type dumbbell test samples were carved from the aging coupons and tested at ambient temperature according to ISO 527-2 standards with a Wolpert TT1220 dynamometer with a 50 mm/min crosshead speed. The welding ribbon was also tested according to NF T 54-122 on a minimum of two parallelepipedic test samples.

# 3 Results and Discussion

## 3.1 Epoxy Coatings

Three different coupons were sampled at different immersion durations. The outer protection layer thickness of the stratified coating was measured every centimeter alongside the cross-section perimeter and reported on Fig. 17.2.

Although the concrete roughness and manual wet-lay process makes it difficult to generate a constant thickness, it is shown that the novolac resin was attacked by sulfuric acid solution, gradually decreasing the outer layer thickness. This evolution is represented on the cross-section photographs on Fig. 17.3.

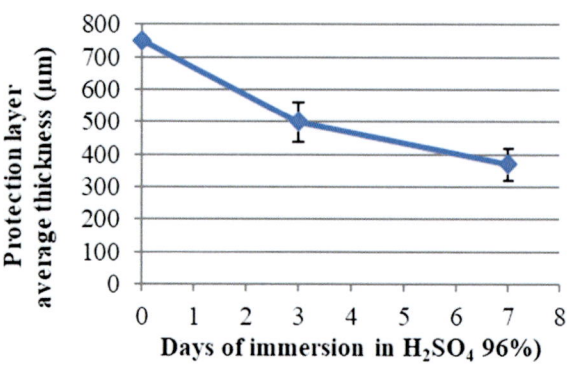

**Fig. 17.2** Protection layer thickness versus immersion time

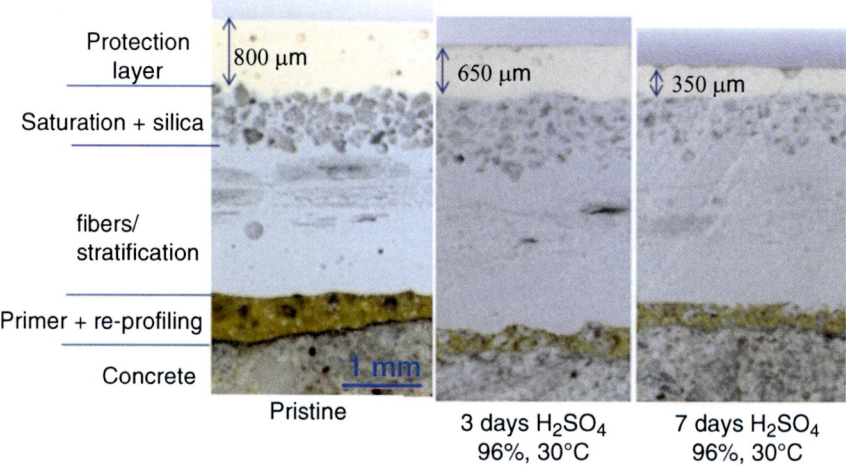

**Fig. 17.3** Binocular cross-section observations of novolac epoxy stratified coatings after immersion in 96 w% sulfuric acid

These first results enabled to make a gross estimate of the protection layer erosion kinetic at 50 μm/day in contact with 96 w% concentrated sulfuric acid.

## 3.2  Thermoplastics Membranes

Visual observations of the thermoplastic coupons after 7 months of immersion did not pick up any morphological nor weight change for HDPE, PVDF, and ECTFE samples. Table 17.2 summarizes the HDPE tensile property evolution during aging. HDPE yield properties do not appear to be affected by the contact with highly concentrated sulfuric acid, and the welding ribbon is also unaffected. The yield properties of concrete embedded thermoplastic membrane are linked to its capacity

**Table 17.2** Tensile testing results on HDPE samples

| Reference | Aging | $E$ (MPa) | $\sigma_y$ (MPa) | $\varepsilon_y$ (%) | $\varepsilon_b$ (%) |
|---|---|---|---|---|---|
| HDPE | Pristine | 679 ± 41 | 20 ± 2 | 10 ± 1 | 67 ± 19 |
| HDPE | 180 days 96 w% $H_2SO_4$ | 897 ± 112 | 19 ± 2 | 9 ± 1 | 51 ± 7 |
| HDPE (welding ribbon) | Pristine | 660 ± 51 | 23 ± 1 | 5 ± 0 | 90 ± 9 |
| HDPE (welding ribbon) | 180 days 96 w% $H_2SO_4$ | 773 ± 84 | 20 ± 1 | 6 ± 1 | 57 ± 8 |

to maintain the anchor knobs integrity and the membrane thickness without plastic deformation, thus maintaining the products function. A tensile modulus increase and elongation at break decrease are noted. Chemical resistance manufacturers' database [6] predicted the high chemical resistance of fluorinated thermoplastics (ECTFE, PVDF), but the HDPE was not known for having equivalent performance. Additional testing after 270 days will be performed in order to confirm these results.

## 4 Industrial Application

Concrete protection coatings have been applied since the 1970s, among which the stratified epoxy was considered having the best performance and durability in terms of watertightness and chemical resistance. Other advantages included the manual *wet lay-up* process allowing adapting to site-specific RC structure morphology and surrounding equipment such as pipe crossing or anchor plates. This work experimental result showed that epoxy coating solutions were not suited for prolonged contact with 96 w% sulfuric acid. Thermoplastic membranes, however, proved to be very resistant to permanent contact with 96 w% sulfuric acid solutions. Based on these findings, the best solution for RC structures protection against 96 w% sulfuric acid incidental spillages would be by applying a thermoplastic membrane on top of the exposed RC surfaces. However, thermoplastic membrane installations are twice as expensive as epoxy coatings on simple geometries and approximately ten times costlier for complex instrumented structures. Thus, industrial and economic constraints led to a multi-material approach with epoxy coatings combined with thermoplastic membranes. Figure 17.4 shows the concrete protection design of 96 w% sulfuric acid storage tank building. Low slope retention basins are coated with epoxy because of the nonpermanent nature of incidental effluent flow. The effluent flow leak would be collected in a PVDF sump pit where permanent contact requires the best chemical protection of concrete ensured by a thermoplastic membrane. Key point was insuring the continuity between epoxy coatings and the PVDF membrane. Considering the low adherence of most adhesive to thermoplastics, other bonding techniques have been implemented.

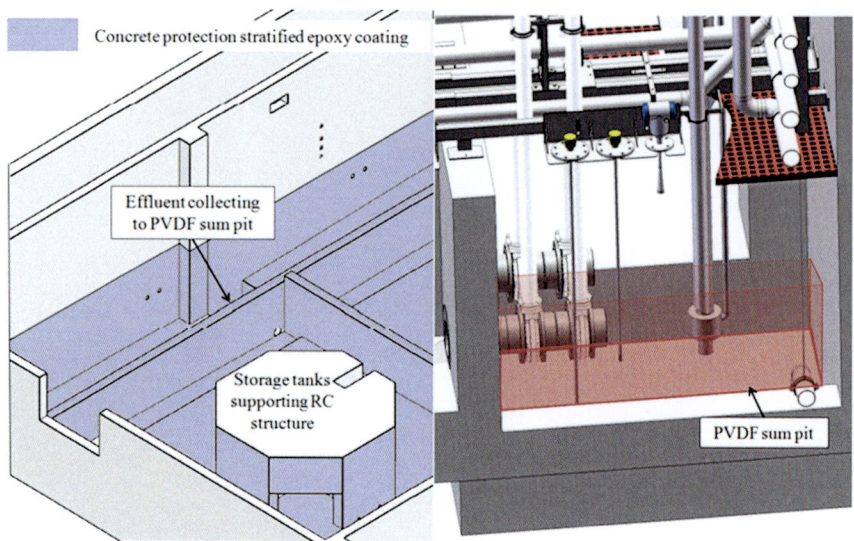

**Fig. 17.4** Design adaptation of a sulfuric acid storage installation

## 5 Conclusion

This work showed that an efficient and durable concrete protection strategy against aggressive chemicals is based on in-depth material behavior knowledge. Indeed, studying the performance and field of application of various polymer-based RC protections allowed obtaining an efficient yet economically viable solution. Also, despite manufacturers' chemical resistance ranking, experimental work on HDPE resin showed outstanding chemical resistance against highly concentrated sulfuric acid.

## References

1. European standard NF EN 13121. (2003). GRP tanks and vessels for use above ground – Part 2: composite materials – Chemical resistance.
2. American Society of Mechanical Engineers. (2011). *Reinforced thermoset plastic corrosion-resistant equipment*. New York: American Society of Mechanical Engineers.
3. Yuan, H., Dangla, P., Chatellier, P., & Chaussadent, T. (2013). Degradation modelling of concrete submitted to sulfuric acid attack. *Cement and Concrete Research, 53*, 267–277.
4. Kawai, K., Yamaji, S., & Shinmi, T. (2005). Concrete deterioration caused by sulfuric acid attack. In *International conference on durability of building materials and components. Lyon* (pp. 17–20).
5. Vipulanandan, C., & Liu, J. (2002). Glass-fiber mat-reinforced epoxy coating for concrete in sulfuric acid environment. *Cement and Concrete Research, 32*(2), 205–210.
6. Bürkle GmbH. (2018). Chemical resistance. Version 3.6 (15.12.2017), https://www.buerkle.de/files_pdf/wissenswertes/chemical_resistance_en.pdf.

# Chapter 18
# Lightweight Filled Epoxy Resins for Timber Restoration

Torben Wiegand and Andrea Osburg

Currently used methods for the restoration of damaged timber constructions involve the removal of huge parts of the intact wood, although it is often desired to largely keep it. Furthermore, it is not useful to strain the remaining timber with the high weight of compactly filled mortar. Thus, the application of a high-performance, lightweight filled epoxy resin can increase the amount of remaining historic structures and, in addition, strengthen it. Fillers like foam glass are comparatively cheap, decrease the density (<1 g/cm$^3$), and increase the workability of the hardened material. A further reduction of the density without an appreciable reduction of the material's stability was possible by using high-performance hollow glass microspheres as filling material. The mechanical properties of numerous mortars have been determined, and materials with a density as low as 0.44 g/cm$^3$ but also with high strengths have been developed. It was possible to increase the mechanical properties by using additives, and epoxy resins based on renewable sources have been used as binders. Also, the thermal behaviour of chosen mortars has been investigated by TGA, coupled with FTIR, in order to identify the degradation products.

## 1 Introduction

Timber constructions exhibit numerous advantages like relatively high strength, lightweight, abundance of material, and workability. Due to its biological origin, it is susceptible to deterioration, e.g. by insect attack, moisture, fungi, and soft rot [1]. In spite of the increasing use of solid ceilings since the 1940s, it can be assumed

---

T. Wiegand · A. Osburg (✉)
Bauhaus-Universität Weimar, Faculty of Civil Engineering, F. A. Finger-Institute for Building Material Engineering, Chair of Building Chemistry and Polymer Materials, Weimar, Germany
e-mail: andrea.osburg@uni-weimar.de

that in approximately 40% of the existing residential buildings, wood beams are used for ceiling construction [2]. Established methods for the restoration of damaged timber constructions (e.g. the BETA method) need to remove huge parts of the intact wood as well [3]. Application of the BETA method or other ones, using compactly filled polymer concretes, e.g. [4, 5], usually leads to heavier construction material, which can be hardly worked after curing. For example, it is almost impossible to penetrate it with nails or screws.

The application of a high-performance, lightweight filled material seems to be a suitable method to increase the amount of remaining historic structures, to increase the strength of the resulting construction elements as well, and to improve the workability of the hardened material.

## 2 Choices of Binder and Filling Material

Materials for timber restoration require good adhesion to wood, high durability, and low density. Possible binders for these mortars are polyurethane, polyacrylate, unsaturated polyester, and epoxy resin [6]. Best suitable for the design of a restoration material for harmed wood constructions seemed to be epoxy (EP) resin because of its comparatively little odour development, its excellent adhesion to timber, and its adjustable processing time achieved by varying the amine hardener. Numerous combinations of epoxy resins and amine hardeners have been tested to identify one with comparably good mechanical properties and a pot life of approx. 0.5 h. The combinations tested were filled with foam glass, which is a comparatively cheap, lightweight filler made from glass waste. Several grain sizes of this material, bearing diameters from 0.04 to 2 mm, have been used. This led to a material with a density of $0.88$ $g/cm^3$, a flexural strength of $22.5 \pm 0.8$ $N/mm^2$, and a compressive strength of $47.0 \pm 2.3$ $N/mm^2$, respectively. Due to the fact that the density and the strength of the resulting material as well are dependent on the maximum grain size used for the mortar, a further reduction of the density without an appreciable reduction of the material's strength required a (partial) substitution of the filling material. High-performance hollow glass microspheres with grain sizes of 12–115 μm showed appropriate properties for this application. So it was possible to develop materials with a density as low as $0.44$ $g/cm^3$ and a higher strength-to-density ratio. Figure 18.1 shows a CT (computer tomography) image of foam glass-filled epoxy resin (left) and a hollow glass microsphere-filled epoxy resin (middle). As can be seen, there are more dark areas (which indicate low density; in this case air) in the hollow glass microsphere-filled resin. It is possible to vary the colour of the material by adding different kinds of rock flour or pigments as well (Fig. 18.1, right).

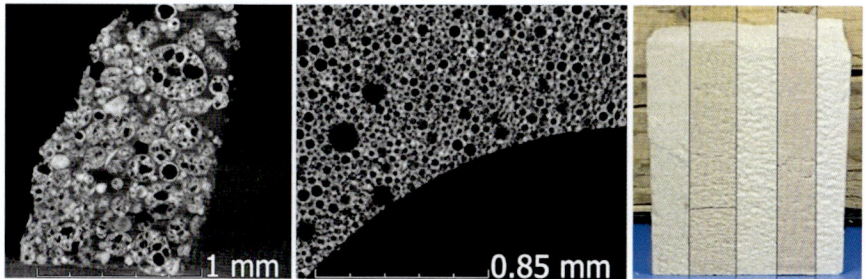

**Fig. 18.1** Left: CT image of foam glass-filled epoxy resin; middle: CT image of hollow glass microsphere-filled epoxy resin; right: possible colours of the mortars (choice)

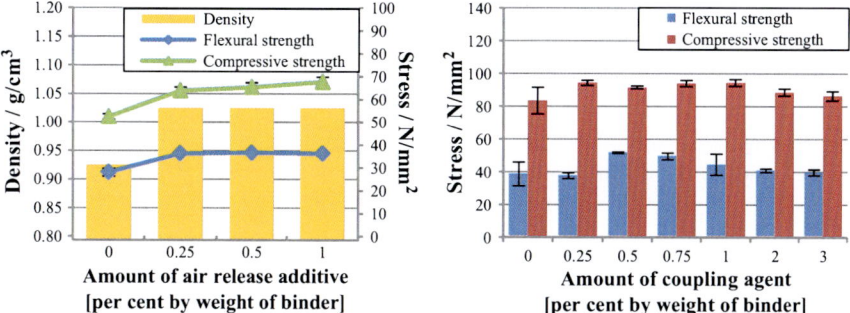

**Fig. 18.2** Left: influence of air release additive to the density and strengths of EP-mortar; right: influence of a coupling agent to the strengths of EP-mortar

## 3 Uses of Additives

During the mixing process, air is embedded in the mortar. This effect is increased due to the low density of the filling materials. For repeatable results, it is necessary to obtain constant air content in the hardened mortar. Therefore, the use of different, commercially available air release additives had been examined. The results for the most effective one (a solution of foam destroying polymers) are shown in Fig. 18.2 (left). The calculated theoretical density of the chosen mortar is 1.13 g/cm$^3$. The density of the hardened material, ascertained of material without any additives, is 0.93 g/cm$^3$. Addition of 0.25 weight per cent of the additive led to an increase of the density up to 1.03 g/cm$^3$, which means an enhancement of about 10%. The flexural strength increased by about 25% and the compressive strength increased by about 20% as well. Further addition of the air release additive did not lead to higher densities or improved strengths. In mortars with low binder content, no influence of the additive was observed.

For a further improvement of the mechanical properties, while maintaining the constant densities of the material, the suitability of coupling agents has been examined. Actually, these coupling agents are used for glass fibre-reinforced EP and form covalent bonds between the binder and the filling material. They can be simply added during the mixing process. As Fig. 18.2 (right) shows, there is a significant improvement of the flexural strength of the material (71.4 m% of EP and 28.6 m% of hollow glass microspheres [12–30 µm grain size], density: 0.77 g/cm$^3$), observed after the addition of 0.5 weight per cent of the coupling agent. With further addition of the agent, the effect decreases. The compressive strength did not change remarkably. In mortars with low binder content, no effect after addition of the coupling agent was observed after addition of the coupling agent. Another method, which deals with the preparation of the filling material before the mixing process [7], showed no influence either.

## 4 Adhesive Tensile Strengths of Chosen Mortars on Timber

Six chosen mortars (Table 18.1) in the shape of an approximately 1 cm thick circle with a diameter of approx. 9 cm have been put on timber (softwood).

The adhesive tensile strength has been determined based on DIN 18555–6 after 7 days of storage in a laboratory climate and after 6 weeks of a different kind of storage. A part of the samples has been stored in laboratory climate for the whole 6 weeks and another part has been stored for 4 of the 6 weeks in a climate chamber at temperatures from −20 °C to +60 °C and in changing humidity as well.

The adhesive tensile strengths obtained from the mortars are shown in Fig. 18.3 alongside the tensile strengths of pure timber for comparison. No significant differences of the adhesive tensile strengths exist, with the exception of mortar H3, which has lower binder content, and in general the values are in the same range as they are for pure wood. This is also confirmed by most of the fracture patterns obtained (Fig. 18.4, middle), which predominantly show cohesion fractures of the wood. As shown in Fig. 18.4. (right), the fracture patterns obtained from determination the adhesive strength of the very flexible mortar H6, which is made completely from renewable sources, differ completely from the others and show only fractures near

**Table 18.1** Differences in the composition of the investigated mortars

| Mortar | H1 | H2 | H3 | H4 | H5 | H6 |
|---|---|---|---|---|---|---|
| Binder | Petroleum-based | | | Petroleum-based, more flexible | From partially renewable sources | From renewable sources, very flexible |
| Content of binder [m%] | 54.0 | 47.4 | 54.0 | 54.0 | 54.0 | 54.0 |
| Air release additive and coupling agent | Yes | Yes | No | Yes | Yes | Yes |
| Filling material | Foam glass and hollow glass microspheres | | | | | |
| Content of filling material [m%] | 46.0 | 52.6 | 46.0 | 46.0 | 46.0 | 46.0 |

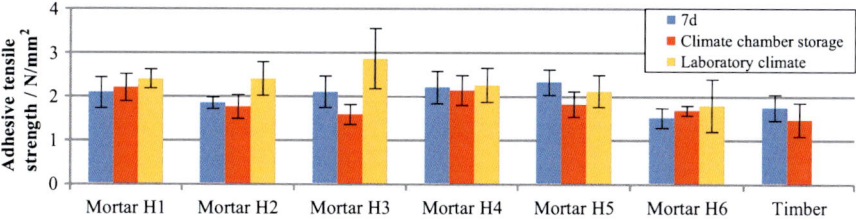

**Fig. 18.3** Adhesive tensile strengths of chosen mortars on timber (softwood)

**Fig. 18.4** Left: setup for determination of the adhesive tensile strength of chosen mortars; middle: fracture patterns obtained when petroleum-based EP is used as binder; right: fracture patterns obtained when EP from renewable sources is used

the surface. Nevertheless, the obtained values obtained from mortar H6 are also in the same range as they are for pure wood.

## 5 Thermal Behaviours of Chosen Mortars and FTIR Analysis of Gaseous Degradation Products

Mortars H1 till H6 have also been characterised through thermal analysis and the gaseous degradation products have been investigated by FTIR analysis using a coupled SDT-FTIR system. The inflection point of the mass reduction and the maximum of the detected IR intensity occur almost at the same temperature. A comparison of the IR spectrum obtained at the maximum with the database of the IR software used showed good matches between the IR spectra of 4-propylphenol and 4-isopropyl phenol. Both of them can be possible thermal degradation products of bisphenol A-based epoxy resin. In the case of mortar H6, the IR spectrum obtained matches long-chained alkanes. This indicates that the constitution of mortar H6 is different from the others and can be an explanation for its deviating behaviour during the adhesion tensile strength tests.

## 6 Conclusions

It was shown that it is possible to develop several epoxy resin-bonded materials with low densities, but high mechanical strengths and excellent adhesion to timber. Foam glass and hollow glass microspheres are suitable lightweight fillers, and their use led to densities as low as 0.44 g/cm$^3$. Furthermore, it was possible to improve the mechanical properties of the materials by using small amounts of degassing additives, as well as coupling agents. The amount of room temperature-cured epoxy resins made from renewable sources is very limited, but one made to 56% from renewable sources is also suitable as binder for timber reconstruction, despite a remarkable reduction of the compressive strengths obtained. The adhesion of chosen mortars to softwood has been examined and is not lower than the cohesion within the timber. The mortars showed similar behaviour at the characterisation through thermal analysis. The FTIR analysis of the gaseous degradation products is a possibility to attain indicators to the composition of the material analysed.

## 7 Acknowledgement

The authors wish to acknowledge the German Federal Ministry of Education and Research for financial funding.

## References

1. Lißner, K., & Rug, W. (2000). *Holzbausanierung: Grundlagen und Praxis der sicheren Ausführung.* Berlin/Heidelberg: Springer.
2. Schober, K.-U. (2008). *Untersuchungen zum Tragverhalten hybrider Verbundkonstruktionen aus Polymerbeton, faserverstärkten Kunststoffen und Holz [Doctoral thesis].* Weimar: Bauhaus-Universität Weimar.
3. Paul, O. (1989). Das Beta-Verfahren. *Bautenschutz Bausanierung, 12*(1), 17–22.
4. Schober, K.-U. (2010). Hochleistungskunststoffe für die Tragwerkserhaltung von Holzkonstruktionen im Bestand. *Bausubstanz, 1*(4).
5. Kim, Y. J., Hossain, M., & Harries, K. A. (2013). CFRP strengthening of timber beams recovered from a 32 years old quonset: Element and system level tests. *Engineering Structures, 57,* 213–221.

6. Merkblatt E-1-7 - Holzergänzungen. *WTA – Wissenschaftlich-Technische Arbeitsgemeinschaft für Bauwerkserhaltung und Denkmalpflege e.V.* 2012, 09.2012/D.
7. Schönherr, P. (2012). *Multiple Oberflächenfunktionalisierung von Mischgläser- und Siliciumoxidpartikeln als Komponenten für Kompositmaterialien [Doctoral thesis]*. Chemnitz: Technische Universität Chemnitz.

# Chapter 19
# Effect of Methyl Methacrylate Monomer on Properties of Unsaturated Polyester Resin-Based Polymer Concrete

Kyu-Seok Yeon, Nan Ji Jin, and Jung Heum Yeon

A study was conducted to experimentally investigate fresh and hardened properties of concrete produced by using a methyl methacrylate (MMA) monomer as a diluent to improve the workability of unsaturated polyester (UP) resin-based polymer concrete. The results showed that as the MMA content increased, the workability and working life were largely improved, while the setting shrinkage was rather reduced. In addition, an increase in MMA content reduced the elastic modulus and compressive strength, but the reductions of these properties were much smaller than the one produced using a styrene monomer (SM). In conclusion, the use of MMA as the diluent to highly viscous UP resin was more beneficial than the use of SM to ensure the various properties of polymer concrete.

## 1 Introduction

Unlike conventional cement concrete, polymer concrete is made using a liquid resin instead of cement and water. The liquid resins most widely used to fabricate polymer concrete include unsaturated polyester (UP) [1–4] and acrylic resins [5, 6]. The polymer concrete incorporating UP resin has several advantages, such as rapid hardening, easy working life control, high strength, and excellent durability

---

K. -S. Yeon
Department of Regional Infrastructure Engineering, Kangwon National University, Chuncheon, South Korea

N. J. Jin
Dongil Engineering Consultants Co. Ltd., Seoul, South Korea

J. H. Yeon (✉)
Gachon University, Department of Civil and Environmental Engineering, Seongnam, South Korea
e-mail: jyeon@gachon.ac.kr

[7]. However, since the UP resin is highly viscous, its workability is quite limited, along with a large amount of setting shrinkage occurring at early ages [8]. In particular, when the highly viscous UP resin is not diluted, a greater amount of resin needs to be used to achieve a certain level of workability. The high viscosity of the UP resin may be decreased by adding a styrene monomer (SM), but it tends to significantly decrease the concrete strength.

To resolve this problem, studies have been conducted to modify the binder by adding a methyl methacrylate (MMA) monomer instead of SM [8]. Addition of the MMA monomer to UP resin not only decreases the viscosity and improves the workability of the concrete but also reduces the binder consumption, saving the material cost.

Considering the findings described above, the present experimental study was conducted to investigate the effect of MMA monomer on the properties of UP resin-based polymer concrete.

## 2 Materials and Methods

1. Binder
   UP resin, MMA monomer, catalyst, and accelerator
2. Aggregate and filler
3. Determination of mixing proportions (Table 19.1)
4. Flow test: ASTM C 1437 (standard test method for flow of hydraulic cement mortar)
5. Working life test: KS F 2484 (standard method of measuring working life of polyester resin concrete)
6. Setting shrinkage test: Ohama-Demura method
7. Elastic modulus test: ASTM C 469M (standard test method for static modulus of elasticity and Poisson's ratio of concrete in compression)
8. Compressive strength test: ASTM C 39M (standard test method for compressive strength of cylindrical concrete specimens)

Table 19.1 Binder formulation and mix proportion

| Binder content (wt.%) | Binder formulation | | | Filler (wt.%) | Fine aggregate (wt.%) |
| --- | --- | --- | --- | --- | --- |
| | UP:MMA (wt.%) | MEKPO (phr[a]) | DMA (phr[a]) | | |
| 10.0 | 100:0 | 1.0 | 0.5 | 24.0 | 66.0 |
| | 90:10 | | | | |
| | 80:20 | | | | |
| | 70:30 | | | | |
| | 60:40 | | | | |

[a]Parts per hundred parts of resin

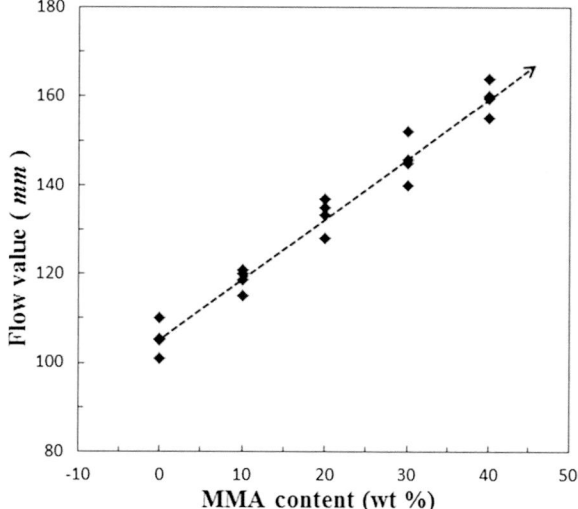

**Fig. 19.1** MMA content vs. flow value

## 3 Results and Discussion

### 3.1 Workability

Since commercially available UP resins are highly viscous, an expensive binder is often added to obtain required workability, and the workability is almost lost at lower temperatures as the viscosity becomes higher at a lower temperature [9]. Figure 19.1 shows the workability of the UP resin-based polymer concrete prepared by adding MMA to resolve the viscosity problem. Figure 19.1 shows that the workability estimated by the flow test increased as the MMA content increased. The flow value was 105 mm in the absence of MMA, but it significantly increased to 159.6 mm with an MMA content of 40 wt.%. This result confirms that the addition of MMA significantly improved the workability of polymer concrete.

### 3.2 Working Life

The working life of polymer concrete is affected by various factors, including the type and amount of the catalyst and accelerator added as well as the curing temperature. Therefore, the working life can be controlled relatively easily by appropriately choosing these factors. However, since the present study was conducted to investigate the effect of MMA content on the working life, only the MMA content varied while keeping the other factors constant. Figure 19.2 shows the results of the test.

As shown in Fig. 19.2, the working life increased as the MMA content increased. The working life in the absence of MMA was 26 min but increased by about

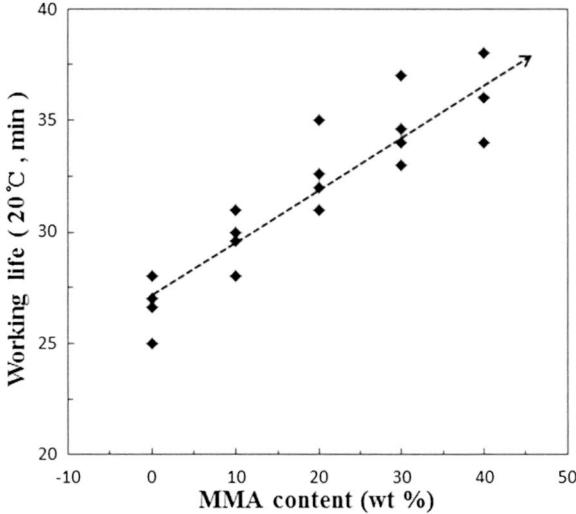

**Fig. 19.2** MMA content vs. working life

10–36 min when the MMA content was 40 wt.%. This indicates that the addition of MMA is one of the effective factors that increases the working life.

## 3.3 Setting Shrinkage

The setting shrinkage of polymer concrete has a similar phenomenon to the autogenous shrinkage of cement concrete. The difference is that polymer concrete is hardened by polymerization while cement concrete is hardened by cement hydration. Generally, the setting shrinkage of polymer concrete at an age of 3 h is over 90% of the ultimate setting shrinkage [10].

The factors affecting the setting shrinkage of polymer concrete include the type and amount of the catalyst and accelerator added, as well as the curing temperature, as in the case of working life. The setting shrinkage may be controlled by adding a shrinkage-reducing agent [10]. However, since the present study was conducted to investigate the effect of MMA content on the setting shrinkage, only the MMA content varied in the testing program. Figure 19.3 shows the results of the test. As shown in the result, the setting shrinkage was $73 \times 10^{-4}$ in the absence of the MMA but decreased to $23 \times 10^{-4}$ when the MMA content was 40 wt.%. This result indicates that the setting shrinkage of the UP resin-based polymer concrete may be decreased by adding MMA.

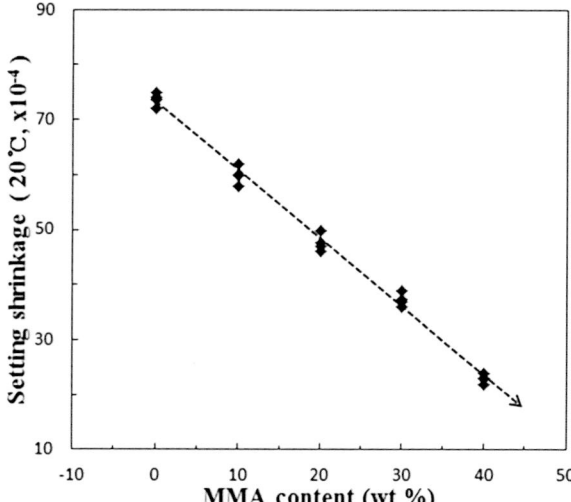

**Fig. 19.3** MMA content vs. setting shrinkage

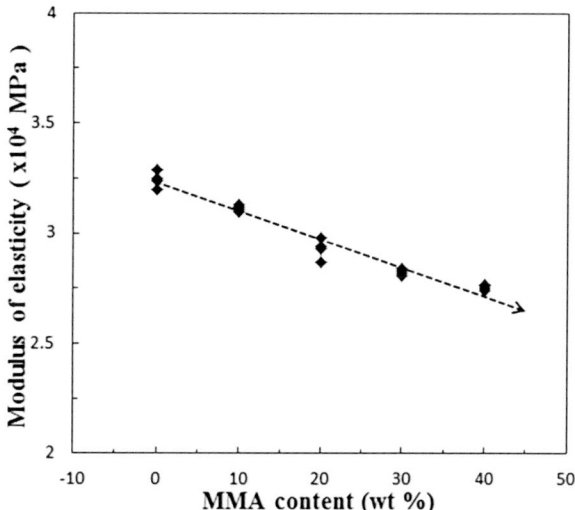

**Fig. 19.4** MMA content vs. MOE

## 3.4 Elastic Modulus

The elastic modulus was evaluated from the compressive stress-strain curves. The elastic modulus of polymer concrete is different from that of cement concrete because the elastic modulus of the polymer concrete is linear, from origin to about 65% of the ultimate stress. Figure 19.4 shows the elastic modulus measured by varying the MMA content. As shown in Fig. 19.4, the elastic modulus in the absence of MMA was $3.24 \times 10^4$ MPa but was reduced by 15.1% when the MMA content was 40 wt.%. However, a previous study on the elastic modulus as a function of SM content [11] showed that the elastic modulus in the absence of SM decreased much more (by 22.8%) when the SM content was 40%. Comparison with the results of previous studies on the elastic modulus indicates that the elastic modulus of the polymer concrete in the present study was lower than that of cement concrete with a compressive strength of 60 MPa ($3.6 \times 10^4$ MPa) but was similar to that of other types of polymer concrete [4, 11].

## 3.5 Compressive Strength

Various factors, including the type of binder, mixture proportions, and curing temperature, affect the compressive strength of polymer concrete. In the present study, the compressive strength was measured by varying only the MMA content while keeping the other factors constant. Figure 19.5 shows the results of the test.

As shown in Fig. 19.5, the compressive strength of the polymer concrete was 82.1 MPa in the absence of MMA but decreased by 6.3% when the MMA content was 40 wt.%. This result was consistent with the compressive strength test results on UP-MMA polymer concrete reported by Hyun et al. [8]. However, the compressive strength of the UP resin-based polymer concrete depending on the styrene monomer content showed that the compressive strength with a styrene content of 60 wt.% was about 23.0% lower than the mixture with no styrene [12], indicating the addition of MMA is better than the addition of SM.

The compressive strength was reduced more by the addition of SM than by the addition of MMA, because macrophase separation may have occurred when the SM originally included in the UP resin was combined with the added SM.

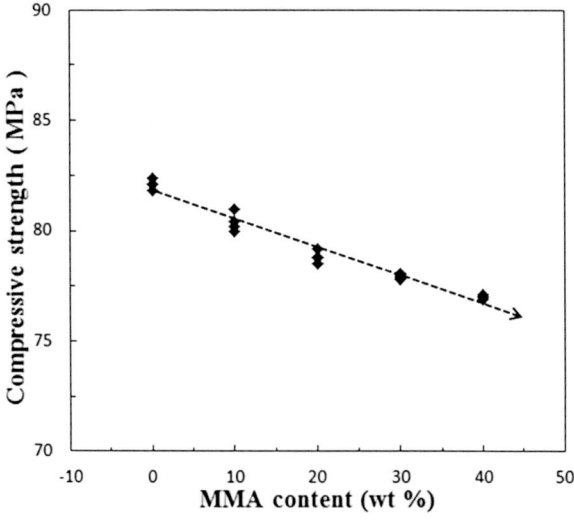

**Fig. 19.5** MMA content vs. compressive strength

## 4 Conclusions

1. The workability and working life were significantly improved as the MMA content increased, while the setting shrinkage rather decreased.
2. The elastic modulus and the compressive strength were reduced with an increased MMA content. The decreases in those properties were much smaller than the ones resulted from adding SM.
3. In sum, the addition of MMA as a diluent to a highly viscous UP resin is more beneficial than the addition of SM to enhance the overall properties of polymer concrete. It is recommended to add a proper dosage of MMA depending on the usage.

**Acknowledgment** This research was supported by Basic Science Research Program through the National Research Foundation of Korea (NRF) funded by the Ministry of Science, ICT & Future Planning (No. NRF-2015R1C1A1A01052102). This research was also supported by National R&D Program through the National Research Foundation of Korea (NRF) funded by the Ministry of Science & ICT (NRF-2016R1D1A1B03930763).

## References

1. Vipulanandan, C., Dharmarajan, N., & Ching, E. (1988). Mechanical behaviour of polymer concrete system. *Journal of Materials and Structures, 21*, 268–277.
2. Ignacio, C., Ferraz, V., & Orefice, R. L. (2008). Study of the behavior of polyester concrete containing ionomers as curing agent. *Journal of Applied Polymer Science, 108*(4), 2682–2690.
3. Martinez-Barrera, G., Villarruel, U., Vigueras-Santiago, E., Hernandez-Lopez, S., & Brostow, W. (2008). Compressive strength of gamma-irradiated polymer concrete. *Journal of Polymer Composites, 29*(11), 1210–1217.
4. Gorninski, J. P., Dal Molin, D. C., & Kazmierczak, C. S. (2004). Study of the modulus of elasticity of polymer concrete compounds and comparative assessment of polymer concrete and portland cement concrete. *Journal of Cement and Concrete Research, 34*, 2091–2095.
5. Kobayashi, T., & Ohama, Y. (1984). Low-temperature curing of polymethyl methacrylate polymer concrete. *Journal of Transportation Research Record, 1003*, 15–18.
6. Fowler, D. W., Meyer, A. H., & Paul, D. R. (2007). Low temperature curing of polymer concrete. *Proceedings of the 3rd international congress on polymers in concrete, Koriyama, Japan* (pp. 421–434).
7. Chandra, S., & Ohama, Y. (1994). *Polymers in concrete* (pp. 189–191). CRC Press, USA.
8. Hyun, S. H., & Yeon, J. H. (2012). Strength development characteristics of UP-MMA based polymer concrete with different curing temperature. *Construction and Building Materials, 37*, 387–397.
9. Ohama, Y., & Terata, O. (1979). Determination methods for consistency of polyester resin concrete, *Proceedings of the Twenty-Second Japan Congress on Materials Research, Kyoto, Japan* (pp. 352–355).
10. Yeon, K. S., & Yeon, J. H. (2012). Setting shrinkage, coefficient of thermal expansion, and elastic modulus of UP-MMA based polymer concrete. *Journal of the Korea Concrete Institute, 24*(4), 491–498.
11. Demura, K. (1982). *Development of resin concrete for construction works* (Ph.D. dissertation). Nihon University, Japan, (in Japanese).
12. Neville, A. M. (1995). Elasticity, shrinkage and creep. In *Properties of concrete* (4th ed., pp. 412–481). Pearson Education Ltd., England, UK.

# Chapter 20
# Analysis of Mechanical Behavior and Durability of Coatings for Use as Flooring in the Petroleum Industry

Jane Proszek Gorninski and Jéssica Maiara de Freitas

This work analyzes the properties of four compositions of polymer mortar through flexural strength, water absorption, and abrasive wear. For this study, the epoxy resin type used as binder was used due to its high chemical and corrosion resistance, hardness, flexibility, good mechanical strength, and outstanding adhesion to most substrates. The binder was used in different combinations, producing four coating chemical compositions; they are (1) Quartz M10, Quartz 307, fly ash, and epoxy resin; (2) sand crushing, gravel dust, and epoxy resin filler; (3) gravel dust without the 2.4 mm and 4.8 mm fractions, fly ash filler, and epoxy resin; and (4) gravel dust without the 2.4 mm and 4.8 mm fractions and with gravel dust filler and epoxy resin. Through analysis of the results, it was observed that the studied floors reached values of up to 40 MPa for flexural strength. The water absorption was 0.17%, showing the effectiveness of the compositions, since a coating is considered as high performance coating when absorption is 0.5%. In addition, the results also showed high abrasive resistance.

## 1 Introduction

Petroleum is of paramount importance in the economy, accounting for various geopolitical and socioeconomic changes worldwide. The recent discovery of the pre-salt layer in Brazilian territory is, if the operation is conducted in a balanced manner, a great opportunity for social and economic development for the country. From this discovery, Brazil now becomes one of the countries with large oil reserves, this status generating future investment in research and development of new

---

J. P. Gorninski (✉) · J. M. de Freitas
Federal University of São Paulo, São Paulo, Brazil

technologies in the country, increasing and optimizing the infrastructure of all existing refineries [1, 2].

The polymeric mortars have been an alternative in industrial flooring, providing high mechanical and chemical resistance, essential qualities for flooring that constantly support loads, machine movement, and/or that of chemicals, besides presenting high adhesion to substrates and resistance to high temperatures [3, 4].

This applied research has developed composites from matrices formed by thermoset resins of the bisphenol epoxy novalac type. These resins have high cross-link density, besides methylene groups introduced between the bisphenol A aromatic rings, increasing the adhesion and flexibility of these resins.

The polymer composites do not corrode in the presence of hydrocarbons, such as acetone, gasoline, ethanol, methanol, petroleum, diesel, and vegetable oils. The objective of this work is to develop floor coatings for petroleum and petrochemical industries, using thermosetting polymers and inorganic components as aggregates and additions. The composites must show resistance to corrosive agents, as well as mechanical strength and impermeability.

## 2 Materials and Methods

### 2.1 *Materials*

The epoxy resins used in this study were prepared using Araldite GY 279, a type of DGEBA, combined with two types of hardeners, Aradur 450 and Aradur 2963.

The coatings were produced using the following aggregates: gravel dust, gravel dust without the 2.4 mm fraction and without the 4.8 mm fraction, and M10 industrial quartz and 307 industrial quartz. The gravel dust has a sustainable origin and is replacing the natural aggregates in civil construction. Table 20.1 shows the particle size of the gravel dust obtained from crushing.

There are significant differences in the particle size distribution of the two sands used in the compositions of this work. It was found that the quartz sand had a higher amount of grain in the 1.2 mm nominal aperture sieve. The gravel dust has a more

**Table 20.1** Compositions of polymeric coatings used

| Flooring | Composition |
|---|---|
| M.Q.E. | M10 Quartz, [a]307 Quartz (8%), [b]fly ash (24%), epoxy resin |
| M.A.E. | Gravel dust, [c]gravel dust filler (14%), epoxy resin |
| M.P.C.E. | Gravel dust without 2.4 mm and 4.8 mm [d]fly ash (4%), epoxy resin |
| M.P.F. | Gravel dust without 2.4 mm and 4.8 mm, [e]gravel dust filler (4%), epoxy resin |

[a]Percentage by mass in relation to the M10 Quartz aggregate
[b]Percentage by mass relative to the sum of the two aggregates (M10 Quartz +307 Quartz)
[c]Percentage by mass, relative to the gravel dust aggregate
[d]Percentage by mass, relative to the gravel dust aggregate without the 2.4 mm
[e]Percentage by mass, relative to the gravel dust aggregate without the 2.4 mm and 4.8 mm fractions

uniform particle size distribution. The maximum size of the quartz is 2.4 mm and the gravel dust is 4.8 mm, but the gravel dust adopted in this work had the maximum size 1.2 mm, however de 0.6 mm and <0.15 were the in higher amount 21 and 22 percent, respectively.

The fillers used in this work were fly ash and gravel dust filler, which is the fraction of fines extracted from the gravel dust aggregate. Fly ash, suggested in this study as filler, is a residue generated by the combustion of coal in thermoelectric power plants; it is separated and collected in electrostatic precipitators or mechanical collectors [5].

According to its particle size distribution, it could be seen that the fly ash has a higher concentration of grains in the range between 10 and 100 μm, which characterizes it as a powdery material. Fly ash had its specific mass determined according to the norm NM 23 [6], whose value is 2.11 g/cm$^3$.

## 2.2 Methods

Molding followed the specifications of NBR 5738 [7]. The flexural strength tests followed the NBR norm 12142 [8].

The water absorption test was performed according to NBR 9778-09, 2009 [9]. For the abrasive wear, samples were molded in the dimensions of $5 \times 5 \times 5$ cm and tested by the CIENTEC method, cited in Gorninski [10].

Table 20.1 shows the polymeric mortar compositions adopted for this study. It should be noted that for all four compositions, the same hardener resin system was adopted 1:10 (resin/hardener); as aggregates, quartz and gravel dust were used and as fillers gravel dust and fly ash. The following abbreviations are adopted for the compositions: M.Q.E. (M10 Quartz, 307 Quartz, fly ash, and epoxy resin), M.A.E. (gravel dust, fillers of gravel dust, and epoxy resin), M.P.C.E. (gravel dust without the 2.4 mm and 4.8 mm fractions, fly ash, and epoxy resin), and M.P.F. (gravel dust without the 2.4 mm and 4.8 mm fractions, gravel dust, and epoxy resin filler). The resin system content was 20% by weight, relative to the sum of all inorganic components, aggregates, and fillers.

## 3 Results and Analysis

### 3.1 Flexural Strength

Table 20.2 shows the individual values and the average of these values obtained in the flexural strength test.

The flexural strength property has great influence on the quality and durability of the polymeric flooring. It can be seen that the values obtained in the tests, showed in Table 20.2, are above those required for industrial flooring, given that, for coating

**Table 20.2** Individual and average flexural strength values for each of the four coating compositions

|  | Result [MPa] | | | Average [MPa] | S [MPa] | CV [%] |
| --- | --- | --- | --- | --- | --- | --- |
|  | 1 | 2 | 3 | | | |
| M.Q.E. | 36. 00 | 34. 50 | 32. 00 | 34. 17 | 2. 02 | 5. 91 |
| M.A.E. | 33. 00 | 31. 50 | 30. 00 | 31. 50 | 1. 50 | 4. 76 |
| M.P.C.E. | 36. 00 | 36. 00 | 37. 00 | 36. 33 | 0. 58 | 1. 59 |
| M.P.F. | 37. 00 | 39. 00 | 38. 00 | 38. 00 | 1. 00 | 2. 63 |

**Table 20.3** Average water absorption, in percentage, for the four types of polymeric coatings

| Coating type | Absorbed water (%) |
| --- | --- |
| M.Q.E. | 0.12 |
| M.A.E. | 0.17 |
| M.P.C.E. | 0.07 |
| M.P.F. | 0.08 |

Type 1 to be considered high performance, it must have a flexural strength greater than or equal to 20 MPa according to NBR norm 14050 [11]. Thus, all four coating compositions tested were efficient regarding this property, since they presented flexural strength values between 31.5 and 38 MPa.

The compositions that had the highest flexural strength values strength were the M.P.F. and M.P.C.E. This strength increase is due, probably, to the grain size, since these compositions consisted mainly of gravel dust fractions of up to 1.2 mm and fly ash.

Similar to that occurred in the compression behavior, in this test, the three compositions having the gravel dust as aggregate showed excellent flexural strength values, with the M.P.F. containing the gravel dust aggregate with a particle size up to 1.2 mm and gravel dust fillers was that which reached 38.00 MPa flexural strength. Thus, it can be seen that the composition M.Q.E., which is comprised of virgin industrial aggregate, was not more efficient as to this property than coatings having all recycled inorganic components.

## 3.2 Water Absorption

Table 20.3 shows the water absorption percentages for each of the four polymer compositions. These values were obtained from an average of six individual samples tested for each of the four compositions.

The M.P.C.E. coating, made up of gravel dust without the 2.4 mm and 4.8 mm fractions and fly ash, at 4% in relation to the aggregate, presented the lowest water absorption, around 0.07%. This behavior may have arisen due to the maximum compactness of the inorganic compounds. Moreover, the fly ash significantly contributes to a water absorption reduction. In this sense, Gorninski et al. [12] state that one of the reasons to use fly ash as a fine aggregate is the increase in the mechanical properties and reduced water absorption in polymeric concrete. The fly ash improves the workability of fresh concrete and the strength of the hardened material, resulting in a material with an excellent surface finish.

The M.A.E. composition, that contains gravel dust and gravel dust filler at a 14% in relation to the aggregate, had the highest water absorption. This result suggests that the use of gravel dust aggregate in its original particle size contributed to the existence of empty spaces, which possibly were not properly filled with the filler and the binder.

The other two compositions also showed good results for this test. The M.P.F. coating obtained 0.08% water absorption, while the M.Q.E. coating showed absorption near 0.12%. These results also demonstrate the feasibility of using recycled aggregates in polymeric coatings, since the composition containing virgin industrial aggregate (M.Q.E.) behaved similar to or worse than the compositions containing the recycled aggregates. Therefore, the four tested polymer compositions fulfill the necessary requirements, regarding water absorption, for a coating to be considered high performance, because they showed values between 0.07 and 0.12% absorption, not exceeding the maximum allowed of 1% (for the flooring Type 1), according to NBR norm 14050 (2004) [12].

## 3.3 Abrasive Wear

The result of this test is expressed in terms of wear, which is the average of the differences between the initial and final heights of five points defined on the sample. Through these results, it was found that there was little wear of the samples of the four compositions tested (Table 20.4).

The results presented on Table 20.4 showed that the M.Q.E. composition containing industrial M10 Quartz, industrial 307 Quartz, and fly ash was the one that had the best performance among the four polymeric coatings, presenting an abrasive wear of 1.40 mm. Its components contributed in the grain packaging, providing greater compactness to this composition.

The M.A.E. polymer composition, formed by gravel dust and the gravel dust fillers adopted in 14% content in the aggregate, presented the worst performance, undergoing an abrasive wear of 2.99 mm. This behavior probably was due to the geometry and size of the aggregate grains, which had not undergone enough packaging. Furthermore, the resin may have been insufficient for bonding all the aggregates.

**Table 20.4** Mean values of abrasive wear obtained for each of the four polymer compositions

| Coating type | Abrasive wear [mm] |
|---|---|
| M.Q.E. | 1.40 |
| M.A.E. | 2.99 |
| M.P.F. | 2.85 |
| M.P.C.E. | 2.75 |

The M.P.F. composition which contains no 2.4 mm and 4.8 mm gravel dust fractions and the gravel dust filler at 4% in relation to the aggregate and M.P.C.E consisting of gravel dust without the 2.4 mm and 4.8 mm fractions and fly ash content of 4% relative to the aggregate presented similar wear, 2.85 mm for the M.P.F. and 2.75 mm for the M.P.C.E. Through the analysis of these two compositions, it is clear that, adopting the same aggregates and changing only the filler, the composition containing the fly ash showed better performance than that containing gravel dust fillers. Possibly the sphericity of the fly ash contributed to the increase of this property at the expense of the more angular character of gravel dust fillers.

# 4 Conclusions

The compositions that presented the highest flexural strength values were the M.P.F. and M.P.C.E. This resistance increase is due, probably, to the grain size, since these compositions consist mainly of gravel dust fractions up to 1.2 mm and fly ash. Therefore, it was concluded that the type of aggregate had influence on flexural strength, since the mortars produced with gravel dust obtained higher flexural strength values.

The water absorption was insignificant, demonstrating the impermeability of the polymer mortar. The highest absorption was obtained in the M.A.E. composition, 0.17%, meeting one of the requirements to be considered a high performance coating, the value for which can be no greater than 1% absorption.

The mortars exhibited good resistance to abrasive wear, undergoing a maximum wear of 2.99 mm for the M.A.E. composition.

# References

1. Petróleo Brasileiro S.A. – Petrobras. (2014). Incorporação de Reservas e Novos Recordes de Produção no Pré-sal, Rio de Janeiro. Available in: http://www.investidorpetrobras.com.br/pt/destaques/destaques-mobile/incorporacao-de-reservas-e-novos-recordes-de-producao-no-pre-sal.htm. Accessed in: jun/2014.
2. Callister, W. D., Jr., & Rethwisch, D. G. (2012). *Fundamentals of materials science and engineering: An integrated approach*. John Wiley & Sons, New York.
3. Gorninski, J. P., Dal Molin, D. C., & Kazmierczak, C. S. (2004). Study of the modulus of elasticity of polymer concrete compounds and comparative assessment of polymer concrete and portland cement concrete. *Cement and Concrete Research, 34*(11), 2091–2095.
4. Dikeon, J. T., & Kaeding, A. O. (1992). U.S. and other specifications and standards for polymer concretes. In *International congress on polymer in concrete, 7, 1992* (pp.9–25). Moscow.
5. Silva, N. I., Calarge, L. M., Chies, F., Mallmann, J. E., & Zwonok, O. (1999). Caracterização de cinzas volantes para aproveitamento cerâmico. *Cerâmica, 45*(296), 184–187.
6. Associação Brasileira de Normas Técnicas. (2000). Cimento Portland e outros materiais em pó – determinação da massa específica: NBR NM 23. Rio De Janeiro, Local.

7. Associação Brasileira de Normar Técnicas. (2003). Moldagem e cura de corpos-de-prova cilíndricos ou prismáticos de concreto: NBR 5738. Local.
8. Associação Brasileira de Normar Técnicas. (2010). Determinação da resistência à tração na flexão em corpos-de-prova prismáticos: NBR 12142. Rio de Janeiro.
9. Associação Brasileira de Normas Técnicas. (2009). Argamassa e concreto endurecidos – Determinação da absorção de água por imersão – Índice de vazios e massa especifica. NBR 9778-09. Rio de Janeiro.
10. Gorninski, J. (2002). Study on the influence of isophtalic and ortophtalic polymer resins and fly ash concentration on PC mechanical properties and durability. Thesis (PhD Engineering) 167p. Graduate Program in Civil Engineering - Universidade Federal do Rio Grande do Sul, Brasil. Porto Alegre.
11. Associação Brasileira de Normar Técnicas. (2004). Sistemas de revestimentos de alto desempenho, à base de resinas epoxídicas e agregados minerais – Projeto, execução e avaliação do desempenho – Procedimento: NBR 14050. Rio de Janeiro.
12. Gorninski, J. P. (1996). Investigação do comportamento mecânico do concreto polímero de resina poliéster. Porto Alegre, p 103 (Doctoral dissertation, Thesis (MSC Engineering)). Programa de Pós-Graduação em Engenharia de Minas, Metalurgia e de Minas, Universidade Federal do Rio Grande do Sul, Brazil.

# Chapter 21
# Evaluation of the Performance of Engineered Cementitious Composites (ECC) Produced from Local Materials

Gabriel Arce, Hassan Noorvand, Marwa Hassan, and Tyson Rupnow

Engineered cementitious composites (ECC) are steady-state multiple-cracking, strain-hardening cementitious materials that significantly enhance ductility of traditional cement-based materials. This investigation focuses on the development of ECC utilizing locally available materials in the state of Louisiana. The influence of Class C and Class F fly ash at different dosages and low fiber content (1.5% volume fraction) were investigated for cost-effectiveness of the composite. Compressive and bending tests were conducted to characterize the mechanical properties of ECC mixes. Results showed a trade-off between strength and ductility. Increments in fly ash content as well as utilization of Class F ash (which lowered strength) favored ductility of the composites.

## 1 Introduction

Engineered Cementitious Composites (ECC) are a special type of high-performance, fiber-reinforced cementitious composites (HPFRCC). In contrast to HPFRCC, ECC are micromechanically designed to exhibit a highly ductile behavior at low fiber contents (between 1% and 2% volume fraction), which make these materials cost-effective and practical to implement using traditional construction techniques [1, 2]. In order to achieve a metal-like strain-hardening ductile behavior, ECC design theory requires the fulfillment of two fundamental criteria: strength and energy [3].

---

G. Arce (✉) · H. Noorvand · M. Hassan
Department of Construction Management, Louisiana State University, Baton Rouge, LA, USA
e-mail: garcea1@lsu.edu

T. Rupnow
Louisiana Transportation Research Center, Baton Rouge, LA, USA

The strength criterion guarantees that adequate fiber-bridging capacity exists after the first-cracking stress of the material is reached according to the following equation [2]:

$$\sigma_0 \geq \sigma_{fc} \tag{21.1}$$

where $\sigma_0$ is the maximum fiber-bridging strength, and $\sigma_{fc}$ is the matrix first-cracking strength.

On the other hand, the energy criterion (or steady-state cracking criterion) provides for the occurrence of steady-state (constant crack width) multiple-cracking formation (in contrast to localized-cracking formation on plain or traditional fiber-reinforced concrete) [4]. The energy criterion is satisfied if the crack tip matrix toughness is equal or lower than the complementary energy of the fiber bridging relation, as presented in the following equation [2]:

$$J'_b \geq J_{tip} \approx \frac{K_m^2}{E_m} \tag{21.2}$$

where $J'_b$ is the complementary energy of the fiber bridging relation, $J_{tip}$ is the crack tip matrix toughness, $K_m$ is the matrix fracture toughness, and $E_m$ is the matrix Young's modulus.

Consistently with Eqs. 21.1 and 21.2, the pseudo strain-hardening (PSH) performance indexes, PSH strength ($\sigma_0/\sigma_{fc}$) and PSH energy ($J'_b/J_{tip}$), are presented as the parameters controlling the performance of ECC materials [4]. If both ratios are higher than 1, then the strength and the energy criteria will be met, and ductile strain-hardening behavior of the composite is possible; otherwise, if any of these ratios is less than 1, the tensile-softening behavior of typical fiber-reinforced concrete (FRC) will govern [2, 4]. It is important to note that the equality signs on Eqs. 21.1 and 21.2 assume a perfectly homogeneous material; thus, in reality the need for PSH indexes higher than one is required for robust strain-hardening performance [2, 4]. Experimental evidence suggests that a PSH strength of 1.45 and PSH energy of 3 correlate to robust strain-hardening of polyvinyl alcohol (PVA) fiber-reinforced cementitious composites [4].

## 2 Objectives

This study focuses on the development of PVA-ECC utilizing locally available materials in the state of Louisiana. The implementation of local materials and the utilization of low fiber content and high replacements of cement with fly ash throughout this investigation aim to make ECC cost-effective. To this end, the main objectives of this study were: (1) to develop PVA-ECC mixes utilizing locally available materials; (2) to evaluate and correlate compressive strength, flexural

**Table 21.1** Chemical composition of cement and fly ash (weight %)

| Material | $SiO_2$ | $Al_2O_3$ | $Fe_2O_3$ | CaO | MgO | $SO_3$ | $K_2O$ | $TiO_2$ |
|---|---|---|---|---|---|---|---|---|
| Cement | 19.2 | 4.8 | 3.4 | 65.8 | 2.2 | 3.6 | 0.5 | 0.2 |
| Class-C | 20.3 | 12.2 | 11.8 | 41.3 | 4.7 | 2.5 | 0.6 | 1.7 |
| Class-F | 45.7 | 17.2 | 16.6 | 10.7 | 1.6 | 0.6 | 3.2 | 1.7 |

**Table 21.2** ECC mix design proportions by weight

| ID | Cement | Fly ash | Water | Sand | HRWR[a] | Fibers (Vol%) | W/B | S/B |
|---|---|---|---|---|---|---|---|---|
| M1 | 1 | 1.2 | 0.60 | 0.80 | 0.625 | 1.5 | 0.27 | 0.36 |
| M2 | 1 | 2.2 | 0.87 | 1.16 | 0.286 | 1.5 | 0.27 | 0.36 |
| M3 | 1 | 4.4 | 1.47 | 1.96 | 0.227 | 1.5 | 0.27 | 0.36 |

[a]HRWR dosage by weight of cement

strength, and ductility of PVA-ECC mixes to different levels of matrix/interface tailoring (by varying types and contents of fly ash in the ECC mix designs).

## 3 Experimental Program

### 3.1 Materials

The materials utilized for the development of ECC mixes came from local sources (excepting the fibers). The materials utilized are the following: Type I Portland cement, river sand with a maximum particle size of 1.18 mm and a fineness modulus of 1.96, Class C and Class F fly ash, REC 15 polyvinyl alcohol (PVA) fibers from Kuraray Co. Ltd. in Japan (12 mm in length, 39 μm in diameter, 40 GPa Young's modulus, and 1456 MPa tensile strength), and a polycarboxylate-based high-range water reducer (HRWR). Chemical composition of cement and fly ash are detailed in Table 21.1.

### 3.2 Specimen Preparation

Three replicas of cylindrical (101.6 × 203.2 mm) and prismatic (101.6 × 101.6 × 355.6 mm) specimens were cast to evaluate the compressive and flexural performance of ECC mixtures. Class C and Class F fly ash, at three levels of replacement of cement (55%, 69%, and 81% replacement by weight), were evaluated. Moreover, water-to-binder ratio (W/B), sand-to-binder ratio (S/B), and fiber content were maintained constant at 0.27, 0.36, and 1.5%, respectively (as shown in Table 21.2).

Specimens were cast and demolded after 24 h (specimens were covered with a plastic sheet to prevent moisture loss). Subsequently, specimens were allowed to cure for 28 days under water-submerged conditions at a 23 ± 2 °C temperature, according to ASTM C192.

## 3.3 ECC Testing

Specimens were tested in compression and flexion. Compression test was performed according to ASTM C 39. Moreover, the bending test was performed by three-point bending according to ASTM C 1609.

# 4 Results and Analysis

## 4.1 Compressive Strength

The 28-day average compressive strength results of Class C and Class F ECC mixtures M1, M2, and M3 are presented in Fig. 21.1a. As expected, increments in fly ash content from 55% replacement (M1) to 69% (M2) and 81% (M3) produced a progressive reduction in the compressive strength of the ECC materials for Class C and Class F ash. Furthermore, Class C specimens exhibited higher compressive strengths compared to Class F specimens.

It is important to notice that, except for M3 mix with Class F fly ash, all the materials evaluated in this study exhibited higher compressive strengths than that of normal concrete (30 MPa).

## 4.2 Flexural Performance

After 28 days of curing, the flexural performance of ECC beams was assessed by three-point bending test. Figure 21.1b–d presents the first-cracking strength, peak strength, and deflection at peak strength for M1, M2, and M3 with Class C and Class F fly ash, respectively (determined according to ASTM C 1609 guidelines). Except for M1, which exhibited a strain-softening behavior with localized-cracking (FRC like), all the mixes evaluated in this study exhibited ductile strain-hardening behavior with the formation of multiple cracking (as shown in Fig. 21.2). M1 performance occurred likely due to excessive strength of the cementitious matrix, which produced an excessive first-cracking strength (as shown in Fig. 21.1b) and crack tip fracture toughness. Consequently, PSH strength and/or PSH energy performance indexes decreased to the level of not meeting the necessary requirements for pseudo strain-hardening behavior.

From the mixes with successful strain-hardening performance (M2 and M3), M2 was the mix with the highest flexural strength (as shown in Fig. 21.1c). Yet, M3

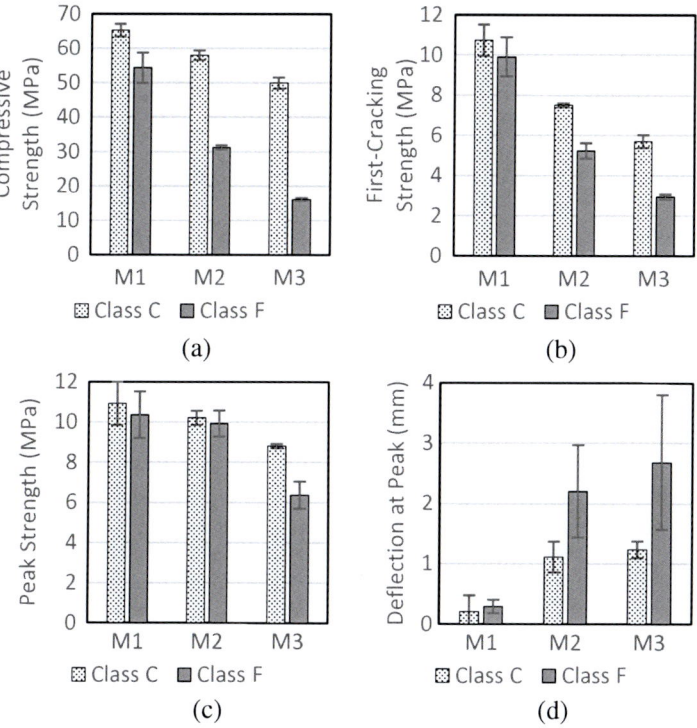

**Fig. 21.1** Mechanical properties of ECC: (**a**) compressive strength, (**b**) first-cracking strength, (**c**) peak strength, (**d**) deflection at peak strength

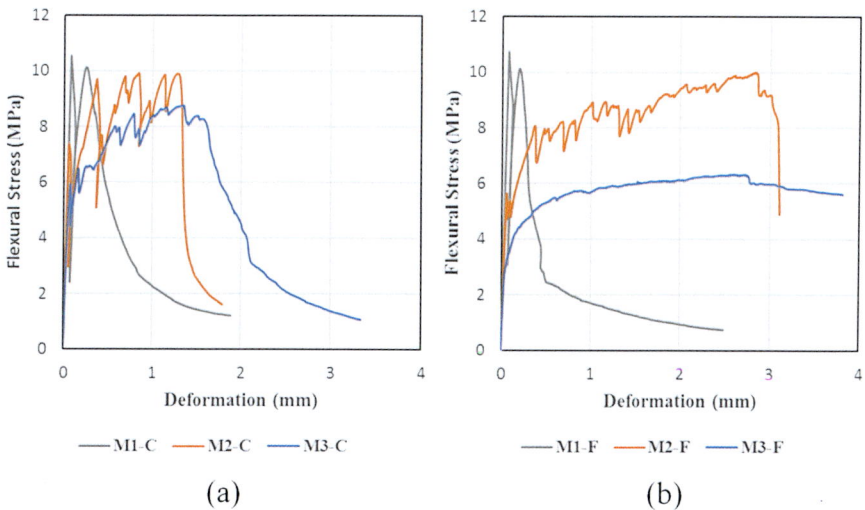

**Fig. 21.2** Flexural stress versus deformation curves: (**a**) Class C fly ash, (**b**) Class F fly ash

possessed the highest ductility as shown in Fig. 21.1d. Moreover, for all mixes, specimens containing Class C ash exhibited higher first-cracking strength, higher flexural strength (peak strength), and lower ductility (deflection), as shown in Fig. 21.1b–d. The trade-off between strength and ductility observed between M2 and M3 mixes as well as between Class C and Class F series is typical of ECC materials and highlights the importance of tailoring ECC materials for specific applications.

## 5  Conclusions

PVA-ECC utilizing locally available materials in the state of Louisiana was successfully developed. Furthermore, the correlation between compressive strength, flexural strength, and ductility of PVA-ECC mixes to different levels of matrix/interface tailoring was investigated. The following conclusions can be drawn:

- The compressive strength of ECC mixtures was significantly affected by the type and content of fly ash. Higher fly ash contents produced lower strengths. Moreover, Class C ash produced higher compressive strengths than Class F ash specimens.
- Three-point bending test results demonstrated that the increase in cement replacement with fly ash as well as the utilization of Class F ash was in favor of multiple-cracking and strain-hardening. Yet, a trade-off between flexural strength and ductility was observed.
- M1 mixture, having the lowest fly ash content, did not exhibit strain-hardening due to failure to meet the strength and/or energy criteria allowing strain-softening behavior similar to that of FRC to prevail. On the other hand, M2 and M3 exhibited a robust strain-hardening behavior for Class C and Class F ash series.

## References

1. Lepech, M. D., & Li, V. C. (2008). Large-scale processing of engineered cementitious composites. *ACI Materials Journal, 105*(4), 358–366.
2. Li, V. C. (2008). Engineered Cementitious Composites (ECC) – Material, structural, and durability performance. In *Concrete construction engineering handbook* (Vol. 78). Boca Raton: CRC Press.
3. Li, V. C. (2012). Tailoring ECC for special attributes: A review. *International Journal of Concrete Structures and Materials, 6*(3), 135–144. https://doi.org/10.1007/s40069-012-0018-8.
4. Yang, E. (2008). Designing added functions in engineered cementitious composites. PhD Dissertation, Department of Civil Engineering, The University of Michigan, 2008, 276 p.

# Chapter 22
# Application of Phase Change Material (PCM) in Concrete for Thermal Energy Storage

Nengfu Tao and Hai Huang

Phase Change Material (PCM) has the ability to absorb and to release a large amount of latent heat during its temperature-constant phase change process. This characteristic makes PCM an ideal candidate for building thermal energy storage (TES). The incorporation of phase change materials (PCMs) in building materials has attracted a lot of research interest due to the concern on energy efficiency. PCM-concrete can be used for reducing the building energy consumption and enhancing the comfort of the building. Significant research showed that PCM-concrete has better latent heat storage and thermal performance. In this paper, butyl stearate (BS), calcium chloride hexahydrate, and polyethylene glycol 600 were investigated as PCMs incorporating concrete. The PCMs consisting of butyl stearate and calcium chloride hexahydrate were successfully encapsulated by PMMA polymerization. The encapsulated PCMs were successfully incorporated into cement mortar. The thermal properties of encapsulated PCMs were studied by differential scanning calorimetry (DSC), transmission electron microscopy (TEM), and thermal conductivity analyzer. The mechanical properties of PCM-based concrete were tested according to Singapore Standards.

## 1 Introduction

The buildings sector accounts for about 38% of all energy consumption in Singapore [1]. Thermal energy storage (TES) is one of the most effective methods of increasing energy efficiency in building. Phase change materials (PCMs) have

---

N. Tao
School of Architecture & the Built Environment, Singapore Polytechnic, Singapore, Singapore
e-mail: cnftao@sp.edu.sg

H. Huang (✉)
Advanced Materials Technology Centre, Singapore Polytechnic, Singapore, Singapore
e-mail: Huang_Hai@sp.edu.sg

been used for thermal energy storage in buildings for several decades [2–12]. - PCM-based concrete tremendously enhanced the energy storage capacity compared to normal concrete [13–17]. Therefore, PCM-concrete has attracted a lot of research interest worldwide.

There are two types of phase change materials (PCMs) that are commonly used in concrete: inorganic and organic. Inorganic PCM has high volumetric heat storage capacity and good thermal conductivity. Moreover, it is cheap and nonflammable. The most common inorganic PCMs are hydrated salts. On the other hand, the disadvantages of inorganic PCMs are very high volume change and super cooling. Inorganic PCMs may remain in a liquid state below its freezing point, which is called super cooling. Super cooling makes the PCMs ineffective for building energy storage. Organic PCMs comprise paraffin and non-paraffin types. Most of the organic PCMs are chemically stable, safe, and nonreactive. In addition, they have no phase segregation during phase change cycles and problems of super cooling because of self-nucleating properties. Paraffin wax (PAR) is a hydrocarbon and regarded as one of the most popular PCMs used in concrete, because of its chemical stability and low cost. The majority of organic non-paraffin PCMs are fatty acids. Butyl stearate is the most appropriate material because of its excellent properties, that is, low cost, high latent heat storage, low volume change, and inflammable nature.

## 2 Experimental Section

### 2.1 PCMs Formulation

The PCMs formulation investigated in this study include:

1. 100% butyl stearate
2. 100% calcium chloride hexahydrate
3. 100% polyethylene 600
4. 50% butyl stearate + 50% calcium chloride hexahydrate
5. 30% butyl stearate + 70% calcium chloride hexahydrate
6. 70% butyl stearate + 30% calcium chloride hexahydrate
7. 50% butyl stearate + 50% PE 600
8. 70% butyl stearate + 30% PE 600
9. 50% polyethylene glycol E 600 + 50% calcium chloride hexahydrate
10. 70% polyethylene glycol E 600 + 30% calcium chloride hexahydrate

### 2.2 Encapsulation of PCMs

Butyl stearate (CAS Number: 123-95-5) and calcium chloride hexahydrate were used as phase change materials and obtained from Tee Hai Chem Pte ltd and VWR chemical together with methylmethacrylate (CAS number: 80-62-6) and *allyl*methacrylate (CAS number: 96-05-9), used as shell forming monomers. The polymerization initiator used was tertbutylhydroperoxide (CAS number: 75-91-2). NP9 was used as a surfactant. Ferrous sulfate (CAS number: 7782-63-0), sodium thiosulfate (CAS number: 7772-98-7), and ammonium persulfate (CAS number: 7727-54-0) were used without further purification.

PMMA-PCM capsules were prepared by emulsion polymerization. A reaction flask containing 200 mL deionized water, 50 g PCM, and 4 g of NP9 (surfactant) was mechanically mixed for 30 min before polymerization; 50 mL methylmethacrylate, 5 g allyl methyl acrylate, 2 mL freshly prepared $FeSO_4$-$7H_2O$ solution (0.3 g of $FeSO_4$-$7H_2O$ in 200 mL water), and 0.5 g of ammonium persulfate were added. Addition of allyl methacrylate as cross-linking agent during polymerization produced uniformly sized and shaped capsules. Moreover, the incorporation of a secondary initiator during polymerization further helped to produce uniformly sized and shaped microcapsules in high yield without a subsequent process. The resultant mixture was vigorously mixed at 750 rpm for 30 min. An extra 0.5 g $Na_2S_2O_3$ and 2.00 g 70% tertbutyl hydroperoxide solution were added at this point and the reaction medium was heated to 90 °C and mixed for an extra hour. The colloidal emulsion concentrated by casting water. The precipitate was dried under vacuum at 40 °C for 24 h.

## 2.3 Preparation of PCM-Based Concrete

The PCM-concrete samples were prepared according to SS EN 206-1:2014. PCM was 5% of mortar and replaced standard sands.

## 3 Results and Discussions

Polyethylene glycol E600 and its combinations with other PCMs were not able to be encapsulated in PMMA due to its chemical properties. The encapsulated PCMs consisting of butyl stearate and calcium chloride hexahydrate were successfully developed.

## 3.1 Morphology Investigation by SEM

Surface morphology of obtained encapsulated PCMs was studied by using SEM. Figure 22.1 shows the micrograph of PMMA-butyl stearate capsules. The PMMA-butyl stearate capsules have spherical shape without edges or sharp dents. The particle sizes were in the range between 50 um and 120 um.

**Fig. 22.1** SEM micrograph of PMMA-butyl stearate capsules

Table 22.1 Thermal conductivity of some encapsulated PCMs

| Encapsulated PCMs | Average thermal conductivity (W/mK) |
|---|---|
| 100% polyethylene glycol 600 | 0.254 |
| 100% calcium chloride hexahydrate | 5.067 |
| 100% butyl stearate | 0.090 |
| 100% decanoic acid | 0.131 |
| 50% butyl stearate +50% calcium chloride hexahydrate | 2.531 |

## 3.2 Thermal Conductivity

Thermal conductivity is a material property describing the ability to conduct heat. A thermal conductivity analyzer was used to measure the thermal conductivity of the PCMs. Table 22.1 shows the thermal conductivity of PCMs. Calcium chloride hexahydrate and 50% butyl stearate +50% calcium chloride had very high value of thermal conductivity. The high thermal conductivity would be very important in building insulation.

## 3.3 DSC Measurement of Encapsulated PCMs

Thermal properties of encapsulated PCMs such as melting points and latent heats were measured by DSC. The analyses were carried out at 20 °C/min heating rate under a constant stream of argon at flow rate of 60 mL/ min.

DSC curve of encapsulated butyl stearate is shown in Fig. 22.2. The first peak in the graph is where the PCM is going through a phase change, melting. The

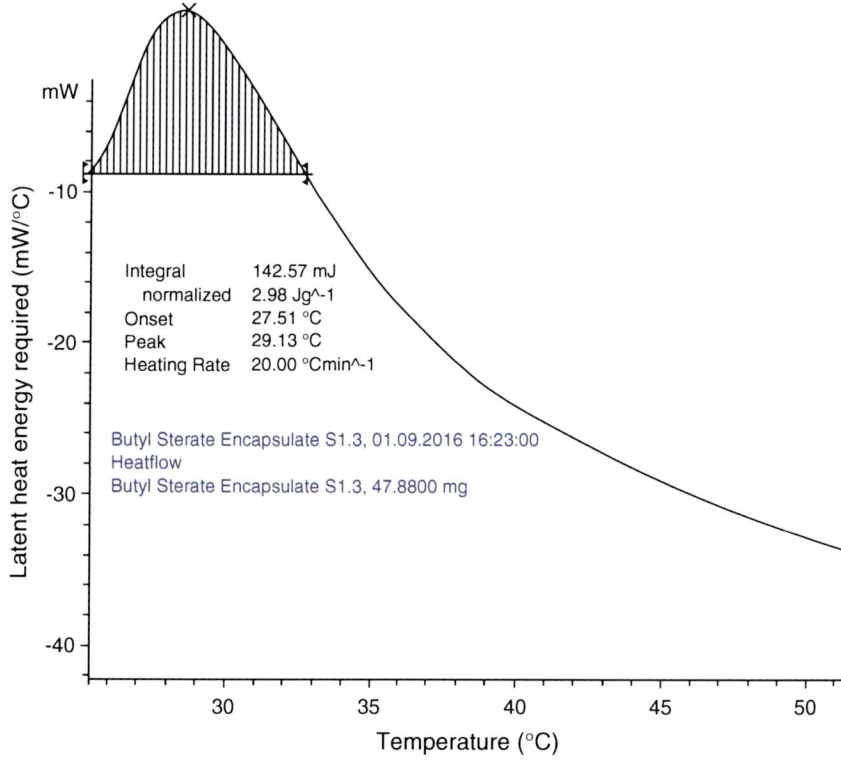

**Fig. 22.2** DSC graph of encapsulated butyl stearate

temperature at which this phase change occurs is 29 °C. This is hence suitable for Singapore weather climate as it falls between the range of 25 and 31 °C. The initial increase in the graph shows that the material is going through an endothermic reaction. The area under the graph represents the change in h, latent heat energy required for melting, 142.57 J/g.

Figure 22.3 showed the DSC results of encapsulated butyl stearate/calcium chloride hexahydrate mixture. The temperature at which this phase change occurs is 32.59 °C. Latent heat of fusion is 145.39 J/g.

## 3.4 Mechanical Properties of PCM Concrete

Table 22.2 shows the compressive strength results which indicate that the incorporation of PCM had reduced the compressive strength of concrete. However, it is still in the tolerable range.

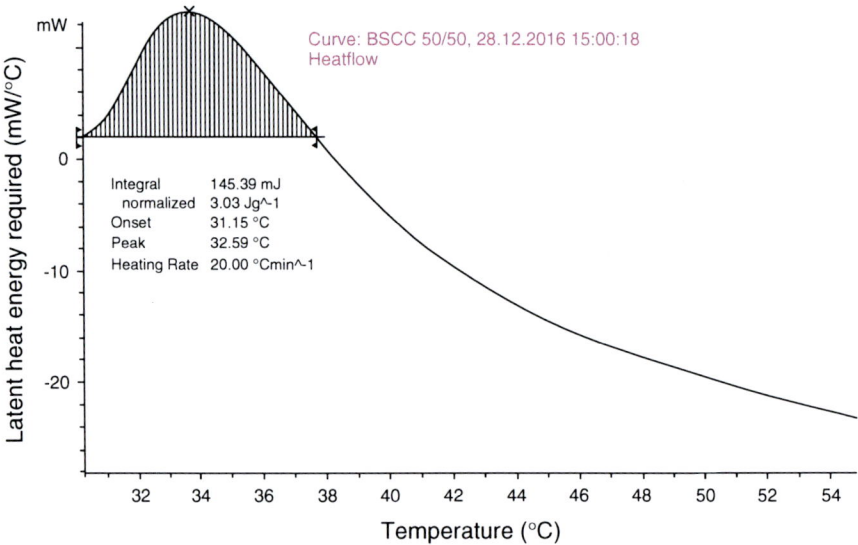

**Fig. 22.3** DSC graph of encapsulated butyl stearate/calcium chloride hexahydrate

**Table 22.2** Mechanical properties of PCM concrete

| Age | Compressive strength (Mpa) of control | Compressive strength (Mpa) of PCM concrete |
|---|---|---|
| 2 day | 12.5 | 10.0 |
| 7 day | 28.0 | 21.3 |
| 30 day | 42.2 | 31.7 |

## 4 Conclusions

The following conclusions were achieved in this research:

1. The PCMs consisting of butyl stearate and calcium chloride hexahydrate are suitable for building insulation in tropical climate. The phase change temperatures are in the range of 25–33 °C. The PCMs had high latent heat, which was around 152 J/g and high thermal conductivity, that is, 5 W/mK.
2. The PCMs consisting of butyl stearate and calcium chloride hexahydrate were successfully encapsulated by PMMA polymerization and incorporated into cement mortar.
3. Polyethylene glycol E600 is not suitable for tropical building insulators. The phase change temperature is very low, 22 °C. This chemical was also unable to form a stable mixture with butyl stearate or calcium chloride hexahydrate.
4. Polyethylene glycol 600 cannot be encapsulated into PMMA beads.

# References

1. Building & Construction Authority. (n.d.). Retrieved September 29, 2017, from https://www.bca.gov.sg/
2. Yuan, Y., Zhang, N., Tao, W., Cao, X., & He, Y. (2014). Fatty acids as phase change materials: a review. *Renewable and Sustainable Energy Reviews, 29*, 482–498.
3. Aydın, A. A., & Aydın, A. (2012). High-chain fatty acid esters of 1-hexadecanol for low temperature thermal energy storage with phase change materials. *Solar Energy Materials and Solar Cells, 96*, 93–100.
4. Madessa, H. B. (2014). A review of the performance of buildings integrated with Phase change material: Opportunities for application in cold climate. *Energy Procedia, 62*, 318–328.
5. Zhu, N., Ma, Z., & Wang, S. (2009). Dynamic characteristics and energy performance of buildings using phase change materials: a review. *Energy Conversion and Management, 50*(12), 3169–3181.
6. Wang, X., Zhang, Y., Xiao, W., Zeng, R., Zhang, Q., & Di, H. (2009). Review on thermal performance of phase change energy storage building envelope. *Chinese Science Bulletin, 54*(6), 920–928.
7. Khudhair, A. M., & Farid, M. M. (2004). A review on energy conservation in building applications with thermal storage by latent heat using phase change materials. *Energy Conversion and Management, 45*(2), 263–275.
8. Pérez-Lombard, L., Ortiz, J., & Pout, C. (2008). A review on buildings energy consumption information. *Energy and Buildings, 40*(3), 394–398.
9. Ismail, K. A. R., & Castro, J. N. C. (1997). PCM thermal insulation in buildings. *International Journal of Energy Research, 21*(14), 1281–1296.
10. Tyagi, V. V., & Buddhi, D. (2007). PCM thermal storage in buildings: a state of art. *Renewable and Sustainable Energy Reviews, 11*(6), 1146–1166.
11. Kuznik, F., Virgone, J., & Noel, J. (2008). Optimization of a phase change material wallboard for building use. *Applied Thermal Engineering, 28*(11), 1291–1298.
12. Pasupathy, A., Velraj, R., & Seeniraj, R. V. (2008). Phase change material-based building architecture for thermal management in residential and commercial establishments. *Renewable and Sustainable Energy Reviews, 12*(1), 39–64.
13. Hawes, D. W., & Feldman, D. (1992). Absorption of phase change materials in concrete. *Solar Energy Materials and Solar Cells, 27*(2), 91–101.
14. Hawes, D. W., Banu, D., & Feldman, D. (1992). The stability of phase change materials in concrete. *Solar Energy Materials and Solar Cells, 27*(2), 103–118.
15. Lee, T. (1998). *Latent and sensible heat storage in concrete blocks* (Doctoral dissertation, Concordia University).
16. Hadjieva, M., Stoykov, R., & Filipova, T. Z. (2000). Composite salt-hydrate concrete system for building energy storage. *Renewable Energy, 19*(1), 111–115.
17. Zhang, D., Li, Z., Zhou, J., & Wu, K. (2004). Development of thermal energy storage concrete. *Cement and Concrete Research, 34*(6), 927–934.

# Chapter 23
# Mortars with Phase Change Materials (PCM) and Stone Waste to Improve Energy Efficiency in Buildings

**Mariaenrica Frigione, Mariateresa Lettieri, Antonella Sarcinella, and José Barroso de Aguiar**

The main objective of this contribution is the study of mortars with the incorporation of polymer-based phase change materials (PCM) for the improvement of energy efficiency in buildings. The mortars are intended for an indoor thermal comfort in the typical climatic conditions of the Southern European countries. Production waste, such as stone powder from quarry, will also be incorporated in the mortars. The finer powder is proposed as mortar aggregate and, at the same time, as support for the PCM. Firstly, different procedures aimed at effectively introducing the selected polymeric material (PEG) into the Lecce Stone have been performed. The chemical and thermal characterization of these compounds has been carried out. The LS/PEG composites have been, then, added to a mortar. Experiments are in progress in order to characterize from chemical, physical, and thermal point of view the mortars with and without PCM, following the recommendations of the international standards in this field. In addition, the studied materials will be used to build laboratory-scale prototypes that will be tested in real environmental conditions.

---

M. Frigione (✉) · A. Sarcinella
Innovation Engineering Department, University of Salento, Lecce, Italy
e-mail: mariaenrica.frigione@unisalento.it

M. Lettieri
Institute of Archaeological Heritage, Monuments and Sites, CNR – IBAM, Lecce, Italy

J. B. de Aguiar
Civil Engineering Department, University of Minho, Guimarães, Portugal

# 1 Introduction

The world consumes a huge amount of fossil fuels and energy for heating and cooling buildings. This represents a serious global problem because a large use of fossil fuels increases climate change. Furthermore, these materials have high costs and the oil fields are running out rapidly [1]. For this reason, in the last two decades, research has been concentrated to find a possible way to reduce the negative environmental impacts through the study of environmentally friendly and renewable energy technologies [2].

The energy efficiency of buildings is one of the main objectives of international policy, because buildings are one of the largest energy consumers for heating and cooling necessities. Systems that can reduce the cooling/heating energy demand to maintain the internal thermal comfort are, then, constantly explored. The use of phase change materials (PCM) in building applications has been found as a suitable method to stabilize the indoor temperature [3]. The operating principle of PCM consists of a change in their status, according to the environment temperature [4]. During daytime, when the temperature is higher, the PCM melts and retains part of the heat involved in the melting process; then, this material possesses the capability to release the previously stored energy when the temperature decreases, changing its status from liquid to solid [5]. A wide variety of materials with different melting point ranges is employed as PCM. Based on their chemical composition, these materials are classified as organic, inorganic, and eutectic [2, 6].

Suitable phase change materials for thermal energy storage of buildings must have, above all, melting temperature lying in the practical range of operation, large latent heat of fusion/crystallization, and high thermal conductivity. Chemical stability, small volume variation during solidification/melting, low toxicity, and low costs are also required [3]. The recent research in this field depicts a rising interest on the use of PCM in mortars [7]. Different methods of PCM incorporation in mortars make it possible to obtain systems with different characteristics and different energy efficiency [8].

Starting from previous experiences [9], a research aimed at incorporating a polymer-based phase change material in a mortar has been undertaken. Materials intended for indoor thermal comfort in climatic conditions typical of the Southern European countries have been considered. Stone powder from waste production was adopted as support for the PCM, and the same powder will be used as mortar aggregate. The use of natural stone waste represents an element of novelty. This choice allows exploiting the waste materials, pursuing the environmental protection strategies. The chemical and thermal characterization of both the PCM and the stone material was first performed. The same investigations were repeated on the PCM-stone composites obtained after mixing under vacuum for different time spans.

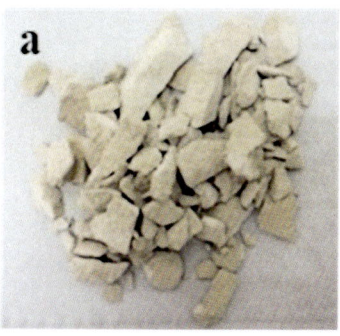

**Fig. 23.1** Lecce Stone in flakes (**a**) and sieved Lecce Stone (**b**)

**Table 23.1** Sample of PCM-stone composites

| Sample | LS [g] | PEG 1000 (g) | Vacuum time (min) | Impregnation time (min) |
|---|---|---|---|---|
| LS/PEG-10 | 10.0 | 1.0 | 30 | 10 |
| LS/PEG-30 | 10.0 | 1.0 | 30 | 30 |
| LS/PEG-60 | 10.0 | 1.0 | 30 | 60 |

## 2 Materials and Methods

Lecce Stone (LS) was used as supporting matrix. This is a typical stone of South Italy (quarries located near Lecce, in the Apulia region), composed of $CaCO_3$ (92–95%) and in a smaller percentage of glauconite, quartz, feldspar, muscovite, phosphate, and clay materials. The material provided by a local company was in the form of flakes (Fig. 23.1a), and it was sieved to obtain little granules (Fig. 23.1b) with granulometry between 1.0 mm > x > 2.0 mm.

Polyethylene glycol 1000 (PEG1000) selected as PCM was purchased from Sigma-Aldrich company (Germany). PEG1000 was chosen since its melting temperature is in the range 37–40 °C and it is non-toxic and eco-friendly.

PEG1000 was incorporated into the stone granules using a shape-stabilization principle, due to its lower costs of production. The chosen amount (i.e., 10 g) of Lecce Stone was placed in a flask. Air removal was performed at the vacuum pressure of 0.1 MPa for 30 min. Then, PEG1000, previously heated in oven at 80 °C, was added to the stone, in order to obtain LS/PEG composites, using different mixing times (from 10 to 60 min).

Composition and details of production for each system are reported in Table 23.1.

A FT-IR Thermo Nicolet Nexus spectrometer, equipped with a deuterated triglycine sulfate (DTGS) detector, was used for the chemical characterization of the produced systems as well as the pristine materials. The samples, mixed with KBr, were analyzed in transmission mode. The spectra were acquired in the range of 4000–400 cm$^{-1}$, with a resolution of 4 cm$^{-1}$ and 32 scans per measurement; the background spectrum was collected on a pellet made of KBr only.

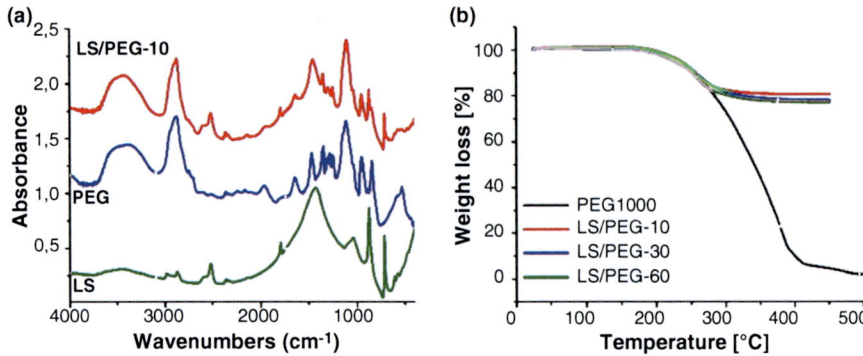

**Fig. 23.2** (**a**) FT-IR spectra of LS, pure PEG and PCM-stone composite. (**b**) TGA curves of pure PEG and PCM-stone composites

**Table 23.2** TGA analysis: degradation temperatures and percentage composition of pure PEG and PCM-stone composites

| Sample | Onset (°C) | Endset (°C) | Residual mass (%) | Amount of PEG (%) |
|---|---|---|---|---|
| PEG1000 | 289.3 | 401.1 | 1.7 | 100 |
| LS/PEG-10 | 219.1 | 286.0 | 80.5 | 19.5 |
| LS/PEG-30 | 218.2 | 296.1 | 78.0 | 22.0 |
| LS/PEG-60 | 219.1 | 292.6 | 77.0 | 23.0 |

The thermal properties were measured by a DSC1, and the thermal degradation temperature was measured by a TGA/DSC1, both instruments by Mettler Toledo.

## 3 Results and Discussion

The FT-IR spectra of the composite materials are very similar to one another, irrespective of the impregnation time. The absorbance bands of PEG are the main peaks (Fig. 23.2). The absence of new signal proves that no chemical interaction occurred between PEG and the stone material. In fact, slight shifts (e.g., the peak at 1114 cm$^{-1}$ shifted to 1104 cm$^{-1}$), observed in the spectra of the composites, indicate some physical attractions, including hydrogen bonds [10, 11], between the two components. The hydrogen bonds occurring in the composite can contribute to the stability of the PCM material because the PEG molecules are tied to the stone and lose their freedom of motion.

To evaluate thermal resistance and the degradation temperature, TGA analysis was used. The related results are shown in Fig. 23.2b and briefly summarized in Table 23.2.

Thermal resistance is one of the most important properties for a selected PCM for a thermal energy storage (TES) application. In this case, the selected PCM and the

**Fig. 23.3** DSC curves of the pure PEG and PCM-stone composites

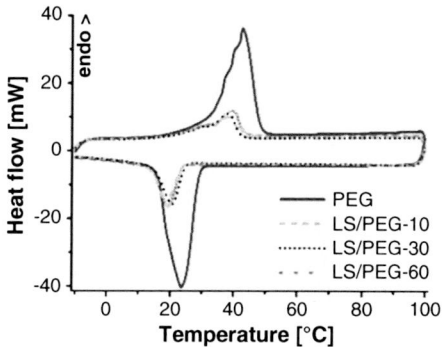

**Table 23.3** LHTES properties of PCM and of PCM-stone composites

| Sample | Heating | | Cooling | |
|---|---|---|---|---|
| | Tm (°C) | ΔHm (J/g) | Tc (°C) | ΔHc (J/g) |
| PEG1000 | 42.8 | 129.3 | 23.6 | 129.8 |
| LS/PEG-10 | 38.4 | 28.2 | 18.5 | 28.7 |
| LS/PEG-30 | 38.2 | 31.1 | 20.4 | 31.3 |
| LS/PEG-60 | 39.3 | 27.7 | 19.4 | 26.2 |

produced PCM-stone composites were characterized between 25 and 450 °C. The same range of temperatures was used to test the PCM-stone composites, to ensure the complete degradation of PEG inside. As shown in Fig. 23.2b, PEG1000 begins to lose weight at 290 °C, and the degradation process is complete just above 400 °C. For this reason, the selected PCM can be regarded as a material with high thermal resistance and good thermal stability, as reported in literature [11].

The complete degradation of PEG in the PCM-stone composites occurs at a lower temperature than the pure PEG, irrespective of the amount of PCM into the stone.

TGA analysis was also employed to evaluate the amount of PEG truly absorbed by the granules of LS. The percentage of PEG contained in each PCM-stone composite is around 20% by weight. It was, then, concluded that the granules of stone were effectively impregnated by PEG1000 using the procedure under vacuum, increasing the amount of PEG by increasing the impregnation time.

To evaluate latent heat thermal energy storage (LHTES) properties of PCM-stone composites, the melting temperature and latent heat capacity were measured. Figure 23.3 shows the DSC thermograms of pure PEG1000 and of PCM-stone composites, including heating and cooling cycles. Table 23.3 summarizes melting (Tm) and crystallization (Tc) temperatures with the relative latent heat of melting (ΔHm) and crystallization (ΔHc) for each system.

As seen in Fig. 23.3 and in Table 23.3, PEG1000 and PCM-stone composites show endothermic (melting) peaks at around 38–40 °C during the heating stage and crystallization peaks at about 20–25 °C during cooling. The reduced values of melting and crystallization enthalpies found for LS/PEG samples are related to the lower amount of PEG into the PCM-stone composites.

## 4 Conclusions

The production of LS/PEG composites under vacuum has proved to be successful in effectively impregnating the granules of stone. All the obtained materials exhibit high thermal resistance and good stability. No significant differences in properties are observed changing the impregnation time. However, LS/PEG-60 can be judged more favorable than others, since the longer impregnation time can insure a deeper penetration of PEG and, in turn, a further increased stability.

The measured properties confirm that these materials can be regarded as promising candidates for building applications. Starting from this characterization, tests of addition of LS/PEG to mortars are in progress in order to evaluate the behavior of the obtained materials as PCM. The developed mortars, in fact, should present adequate characteristics in fresh and hardened states.

**Acknowledgments** The authors wish to thank Ing. L. Pascali and staff of S.I.PRE. S.r.l. (Cutrofiano, Lecce, Italy) for the technical support and Pitardi Cavamonti Company (Melpignano, Lecce, Italy) for supplying the Lecce Stone flakes.

## References

1. Jeon, J., Lee, J. H., Seo, J., Jeong, S. G., & Kim, S. (2013). Application of PCM thermal energy storage system to reduce building energy consumption. *Journal of Thermal Analysis and Calorimetry, 111*, 279–288.
2. Kalnæs, S. E., & Jelle, B. P. (2015). Phase change materials and products for building applications: A state-of-the-art review and future research opportunities. *Energy and Buildings, 94*, 150–176.
3. Cabeza, L., Castell, A., Barreneche, C., Gracia, A., & Fernández, A. (2011). Materials used as PCM in thermal energy storage in buildings: A review. *Renewable and Sustainable Energy Reviews, 15*, 1675–1695.
4. Cunha, S., Aguiar, J., & Pacheco-Torgal, F. (2015). Effect of temperature on mortars with incorporation of phase change materials. *Construction and Building Materials, 98*, 89–101.
5. Zalba, B., Marín, J., Cabeza, L., & Mehling, H. (2003). Review on thermal energy storage with phase change: Materials, heat transfer analysis and applications. *Applied Thermal Engineering, 23*, 251–283.
6. Akeiber, H., Nejat, P., Majid, M. Z. A., Wahid, M. A., Jomehzadeh, F., Famileh, I. Z., Calautit, J. K., Hughes, B. R., & Zaki, S. A. (2016). A review on phase change material (PCM) for sustainable passive cooling in building envelopes. *Renewable and Sustainable Energy Reviews, 60*, 1470–1497.
7. Venkateswara Rao, V., Parameshwaran, R., & Vinayaka Ram, V. (2018). PCM-mortar based construction materials for energy efficient buildings: A review on research trends. *Energy and Buildings, 158*, 95–122.
8. Cunha, S., Aguiar, J., Ferreira, V., & Tadeu, A. (2015). Mortars based in different binders with incorporation of phase-change materials: Physical and mechanical properties. *European Journal of Environmental and Civil Engineering, 19*, 1–18.
9. Kheradmand, M., Castro-Gomes, J., Azenha, M., Silva, P. D., Aguiar, J., & Zoorob, S. E. (2015). Assessing the feasibility of impregnating phase change materials in lightweight

aggregate for development of thermal energy storage systems. *Construction and Building Materials, 89*, 48–59.

10. Guo-Qiang, Q., Cheng-Lu, L., Rui-Ying, B., Zheng-Ying, L., Wei, Y., Bang-Hu, X., & Ming-Bo, Y. (2014). Polyethylene glycol based shape-stabilized phase change material for thermal energy storage with ultra-low content of graphene oxide. *Solar Energy Materials and Solar Cells, 123*, 171–177.

11. Sarı, A. (2016). Thermal energy storage characteristics of bentonite-based composite PCMs with enhanced thermal conductivity as novel thermal storage building materials. *Energy Conversion and Management, 117*, 132–141.

# Chapter 24
# Physical and Mechanical Properties of Cement Mortars with Direct Incorporation of Phase Change Material

Sandra Cunha, José Aguiar, Victor Ferreira, and António Tadeu

The world economic evolution causes a significant intensification in the energy consumption. So, the energy efficiency of buildings is very important because this is one of the principal areas, responsible of the energy consumption in developed countries. The phase change materials (PCM) are smart materials, with the capacity to control the temperature variation, due to their capacity of absorbing and releasing energy. The PCM can be integrated into construction materials using different techniques, such as encapsulation, shape stabilization, direct incorporation, and immersion. The most common form of this material utilization is through encapsulation. It should be noted that, currently, there are still high production costs involved with PCM encapsulation. Thus, it is imperative to develop mortars with PCM incorporation based on raw materials and techniques with lower costs, such as PCM direct incorporation, to address the high production costs of macro- or microencapsulated PCM. The main objective of this work was the study of the physical and mechanical properties of mortars with direct incorporation of PCM, with a melting temperature of about 22 °C. Based on the results obtained, it can be concluded that the addition of this material caused some changes in the fresh and hardened properties of cement mortars.

---

S. Cunha (✉) · J. Aguiar
University of Minho, Guimarães, Portugal

V. Ferreira
University of Aveiro, Aveiro, Portugal

A. Tadeu
University of Coimbra, Rua Luís Reis Santos – Pólo II da Universidade, Coimbra, Portugal

© Springer International Publishing AG, part of Springer Nature 2018
M. M. Reda Taha (ed.), *International Congress on Polymers in Concrete (ICPIC 2018)*, https://doi.org/10.1007/978-3-319-78175-4_24

# 1 Introduction

The energy efficiency of buildings is now one of the main objectives of national and international energy policy [1]. The residential sector is one of the leading sectors in energy consumption in developed countries, representing 40% of energy consumption and $CO_2$ emissions to the atmosphere, in the European Union.

The thermal energy storage systems, based in PCM such as an energy storage technology, have a high potential for increasing the energy efficiency of buildings. The PCM has the ability to reduce the temperature difference, due to their capability to regulate the temperature, based in their capacity to absorb and release the energy to the environment. The working procedure is activated with the change in its properties, in accordance with the surrounding temperature, absorption, and release of stored energy to the environment [2, 3].

The PCM can be integrated into construction materials using diverse techniques, such as encapsulation, shape stabilization, direct incorporation, and immersion [4]. The PCM is generally used encapsulated. There are two types of encapsulation: macroencapsulation and microencapsulation. The use of PCM encapsulated in building materials presents some limitations, such as high costs of the PCM, due to the encapsulation processes and the decrease of the mechanical strengths of the materials doped with PCM microcapsules. Thus, the development of new construction materials based in techniques or raw materials is crucial, with high thermal performance and low cost. In the construction sector, the utilization of mortars for interior coating is a technique very used to finish walls and ceilings. Thus, it becomes imperative the development of these materials, with direct incorporation of nonencapsulated PCM. Thus, it is possible to reduce the high costs of the materials doped with macro- or microencapsulated PCM.

The main objective of this work is to study the influence of direct incorporation of PCM in interior mortars. Thus, the main physical and mechanical properties were studied. The evaluated properties were the workability, density, water absorption by capillarity, water absorption by immersion, microstructure, hydration process, flexural strength, and compressive strength. All the properties were determined based on three different ambient temperatures (10 °C, 25 °C, and 40 °C). Four different mortar compositions were also developed, with different content of PCM (0%, 2.5%, 5%, and 7.5%).

# 2 Materials, Compositions, and Test Procedures

The used cement was CEM II B-L 32.5 N with density of 3030 kg/m$^3$. The sand used has an average particle size of 439,9 μm and a density of 2600 kg/m$^3$. Based in the granulometric distribution, the parameters D50, D10, and D90 were obtained. The D10 corresponds to 150 μm, D50 corresponds to 310 μm, and the D90 corresponds to 480 μm. Synthetic polyamide fibers with a length of 6 mm, thickness of 22.3 μm,

Table 24.1 Mortar compositions (kg/m$^3$)

| Composition | Cement | Sand | PCM | SP | Fibers | Water | Liquid/binder |
|---|---|---|---|---|---|---|---|
| CEM-0PCM | 500 | 1279 | 0 | 15 | 5 | 325 | 0.65 |
| CEM-2.5PCM | 500 | 1290 | 32.2 | 15 | 5 | 275 | 0.61 |
| CEM-5PCM | 500 | 1244 | 62.2 | 15 | 5 | 250 | 0.62 |
| CEM-7.5PCM | 500 | 1204 | 90.3 | 15 | 5 | 225 | 0.63 |

and density of 1380 kg/m$^3$ were used. The superplasticizer used was a polyacrylate, with a density of 1050 kg/m$^3$. Finally, the PCM used is paraffin with temperature transition between 20 and 23 °C, enthalpy of 200 kJ/kg, and densities of 760 kg/m$^3$ and 700 kg/m$^3$, respectively, in solid and liquid states.

Based on these materials, four compositions were developed. The principal objective was to evaluate the possibility of using PCM direct incorporation into mortars for interior coating of building walls. The compositions were evaluated during all the curing process, from the fresh state up to 28 days. In the fresh state, the workability was studied. In hardened state, the water absorption by capillarity, microstructure, flexural strength, and compression strength were evaluated. The developed mortars are presented in Table 24.1.

The hardened state tests were performed after the specimen's submission, during 24 h, to three different ambient temperatures: 25 °C (reference temperature), 10 °C, and 40 °C. The selection of the testing temperatures took into account the PCM transition temperature. The tests performed at 10 °C evaluate the properties of the cement mortars with the PCM in the solid state. On the other hand, the tests performed at 25 °C and 40 °C evaluate the mortars with the PCM in the transition state and in the liquid state, respectively.

The workability tests were performed based on the flow table method stated by the European standard EN 1015-3 [5]. The resulting value within the test was only considered when equal to 160 ± 5 mm.

The observation of the microstructure of the developed mortars was performed using a scanning electron microscope; cylindrical specimens with diameter and height of 1 cm were prepared.

The water absorption by capillarity tests was performed based on the European standard EN 1015-18 [6].

The flexural and compressive behaviors were determined based on the standard EN 1015-11 [7]. The flexural and compressive tests were performed at loading rates of 50 N/s and 150 N/s, respectively.

## 3 Results and Discussion

Regarding the workability results, it was possible to observe that the incorporation of 2.5% of nonencapsulated PCM leads to a reduction superior to 15% in the water content. However, it should be noted that the liquid material/binder ratio remains

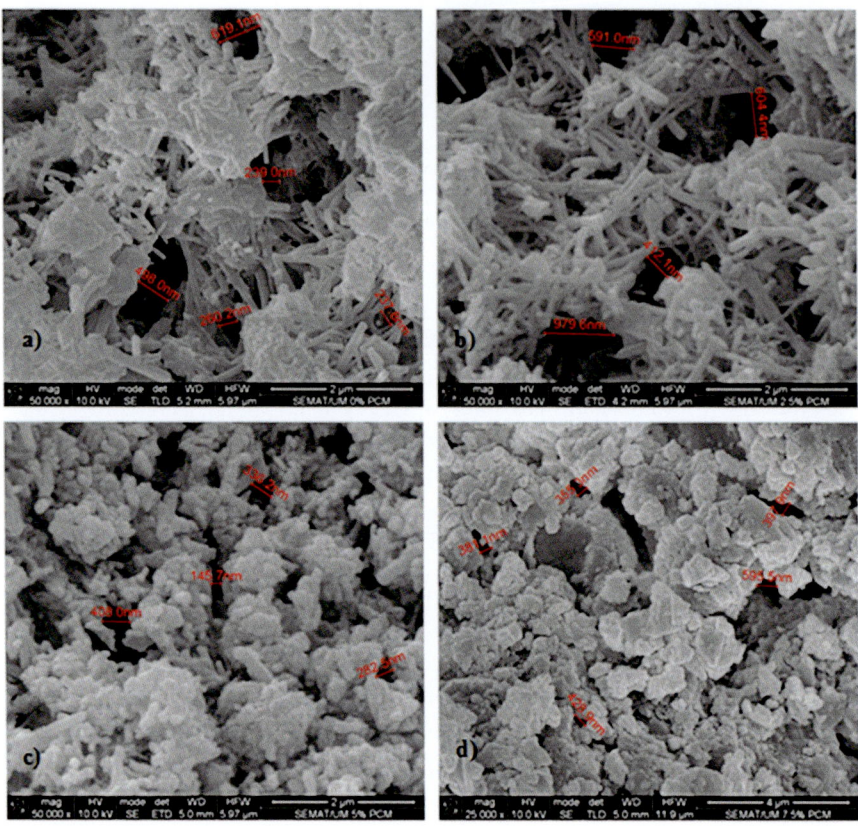

**Fig. 24.1** Microscope observation: (**a**) 0% of PCM; (**b**) 2.5% of PCM; (**c**) 5% of PCM; (**d**) 7.5% of PCM

constant. This behavior can be justified by the introduction of the PCM in mortars in the liquid state, which in part can operate as agent for the development of a homogeneous mortar, substituting part of the water.

The electron microscope observations showed a good correlation between different materials evidenced by the absence of cracks in the microstructure of the mortars which reveals an adequate process of mixing, application, and curing of the mortars. Other observations were performed in order to evaluate the micropore distribution in the developed mortars. In the mortars doped with PCM, a reduction in the pore amount and in its size was observed (Fig. 24.1). The mortar with 7.5% of nonencapsulated PCM incorporation shows a more compact microstructure when compared to the mortar without PCM. This behavior can be related with the reduction of the water content in the mortars with PCM incorporation. It was also possible to observe that the mortars without PCM incorporation present a more crystalline microstructure, resulting from the increase cement hydration, when compared to the mortars doped with PCM. This behavior can be justified by the

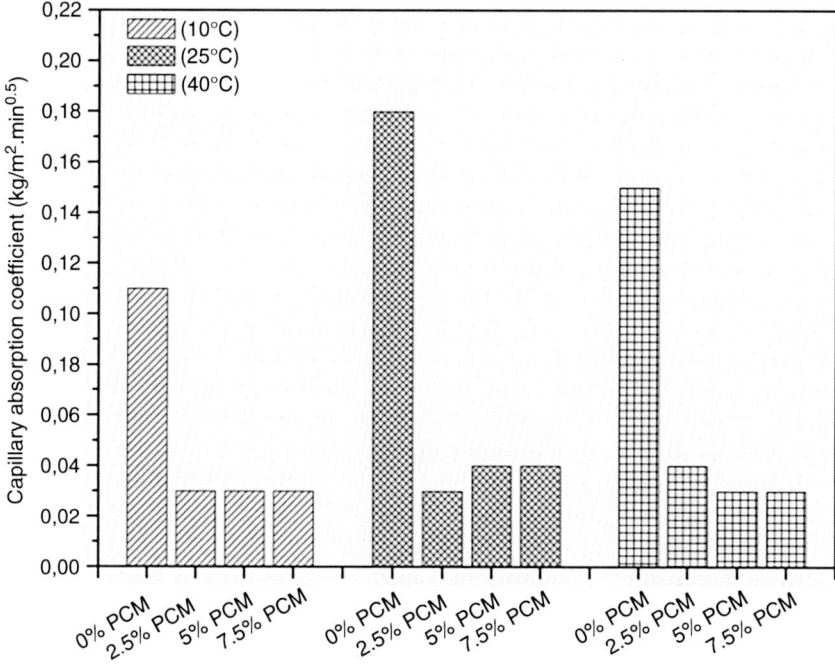

**Fig. 24.2** Capillary absorption coefficient of the developed mortars

porosity decrease, since there is less space in the PCM mortar microstructure to the cement hydration reactions.

Concerning the water absorption by capillarity (Fig. 24.2), it was possible to verify that the nonencapsulated PCM incorporation leads to a reduction in the capillary absorption coefficient superior to 73%. This behavior is associated with the partial or total occupation of the mortar pores by the PCM. On the other hand, it was also possible to observe that the mortars with PCM incorporation showed a similar capillary absorption coefficient when submitted to different ambient temperatures. Thus, it was possible to conclude that even in different states (solid, transition, and liquid), the PCM did not move outside the mortar pores. Regarding the mortars without PCM incorporation, it was possible to verify a higher capillary absorption coefficient, due to the presence of empty pores in the mortars.

Figure 24.3 presents the flexural and compressive strengths of the developed mortars when submitted to different temperatures. It was possible to verify that the incorporation of nonencapsulated PCM did not cause significant changes in the flexural and compressive strengths. This behavior can be justified by the PCM contained inside the pores, not weakening the mechanical strength. On the other hand, a liquid material/binder ratio similar for all developed compositions was observed.

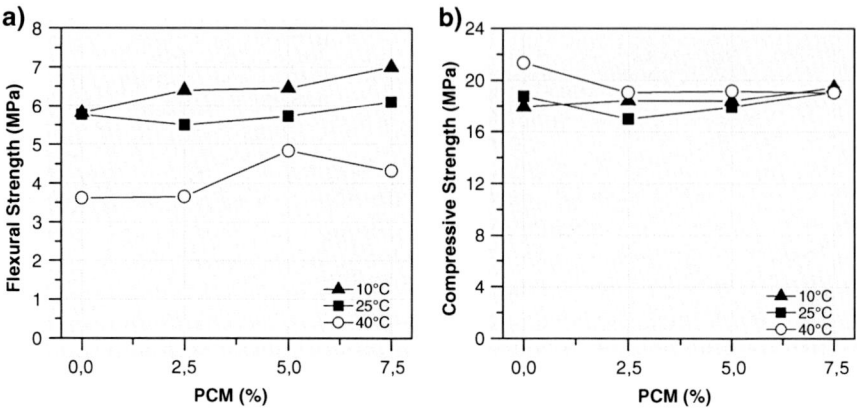

**Fig. 24.3** Mechanical behavior of the developed mortars

## 4 Conclusions

Based on the results obtained, it can be concluded that the PCM direct incorporation causes some changes to the properties of cement mortars in fresh and hardened states. The amount of water added to the mixtures decreased with the PCM incorporation. However, the liquid material/binder ratio of the mortars is similar. This indicates that all mortars generally require the same amount of liquid (water and PCM) to form a homogeneous paste. Concerning the water absorption by capillarity tests, it was detected that the direct incorporation of nonencapsulated PCM led to a reduction in the capillary absorption coefficient, due to partial or total occupation of the mortar pores by the PCM. According to the flexural and compressive strengths, it can be concluded that the direct incorporation did not cause significant changes in the mechanical behavior of the mortars. Thus, it is possible to conclude that the utilization of nonencapsulated phase change materials can be seen as a good and more economical solution for the energy efficiency of the buildings.

**Acknowledgments** The authors acknowledge the Portuguese Foundation for Science and Technology (FCT) for the financial support of PhD scholarship SFRH/BD/95611/2013.

## References

1. Soares, N., Costa, J., Gaspar, A., & Santos, P. (2013). Review of passive PCM latent heat thermal energy storage systems towards buildings energy efficiency. *Energy and Buildings, 59*, 82–103.
2. Zhang, Y., Zhou, G., Lin, K., Zhang, K., & Di, H. (2007). Application of latent heat thermal energy storage in buildings: State-of-the-art and outlook. *Building and Environment, 42*, 2197–2209.

3. Zalba, B., Marín, J., Cabeza, L., & Mehling, H. (2003). Review on thermal energy storage with phase change: Materials, heat transfer analysis and applications. *Applied Thermal Engineering, 23*, 251–283.
4. Memon, S. A. (2014). Phase change materials integrated in building walls: A state of the art review. *Renewable and Sustainable Energy Reviews, 31*, 870–906.
5. European Committee for Standardization. (2004). EN 1015-3, Methods of test for mortar for masonry – Part 3: Determination of consistence of fresh mortar (by flow table). European Standards.
6. European Committee for Standardization. (2002). EN 1015-18, Methods of test for mortar for masonry – Part 18: Determination of water absorption coefficient due to capillary action of hardened mortar. European Standards.
7. European Committee for Standardization. (1999). EN 1015-11, Methods of test for mortar for masonry – Part 11: Determination of flexural and compressive strength of hardened mortar. European Standards.

# Chapter 25
# Mechanical Performance of Fly Ash Geopolymeric Mortars Containing Phase Change Materials

M. Kheradmand, Z. Abdollahnejad, and F. Pacheco Torgal

European Union (EU) aims to achieve nearly zero-energy (public) building (NZEB) by the end of 2018. This very ambitious target would be more easily fulfilled if high-thermal performance materials like phase change materials (PCMs) are to be used. This paper reports experimental results on the mechanical properties of geopolymeric mortars containing PCMs at ambient temperature and after exposure to high temperature. The results show that the inclusion of PCMs is responsible for a reduction of the mechanical strength of the mortars. Several mixtures showed an increase in compressive strength after being exposed to high temperatures. Since PCMs are made of flammable materials, geopolymeric mortars are more advantageous than Portland cement-based mortars for PCM incorporation.

## 1 Introduction

Global warming is considered one of the worst problems faced by our planet, and the production of energy is considered to be the first responsible for that problem [1].

Buildings are a major contributor for carbon dioxide emissions, and higher energy efficiency levels are required [2]. Building energy efficiency will not only be able to cut carbon dioxide emissions but also create new jobs [3]. In this context the European Energy Performance of Buildings Directive 2002/91/EC represents a step forward by defining very ambitious requirements for buildings [4]. Accordingly

---

M. Kheradmand · Z. Abdollahnejad
University of Minho, Gumarães, Portugal

F. Pacheco Torgal (✉)
University of Minho, Gumarães, Portugal

University of Sungkyunkwan, Suwon, Republic of Korea
e-mail: torgal@civil.uminho.pt

by January 1 of 2021, all new private buildings must be of the nearly zero-energy type. For public buildings the requirement is even more ambitious and sets January 1 of 2019 as the deadline. For such targets to be met, new and improved materials will be needed. This is the case of PCMs [5] that use chemical bonds to store or release heat hence reducing energy consumption. PCMs can absorb heat inside buildings avoiding excessive heating and reducing cooling needs. Also the European agenda regarding resource efficiency requires that waste is to be managed as a resource [6]. This is a very important goal in the European context of a circular economy and zero-waste target [7]. Thus materials that have the ability for the reuse of several types of wastes such as geopolymers are to deserve a special attention from the scientific community [8, 9]. This is the case of fly ash [10]. In this context this paper reports experimental results on the properties of fly ash geopolymeric mortars containing PCMs and different ratios of activator/binder and sodium silicate/sodium hydroxide.

## 2 Experimental Programme

The binder precursor was composed by 90% of fly ash and 10% calcium hydroxide. This is because previous investigations show that calcium hydroxide is crucial for the durability of geopolymers [11] and that some authors reported that 10% is an optimum amount [12]. Solid sodium hydroxide, which was obtained from commercially available product of Ercros, SA, Spain, was used to prepare the 12 M NaOH solution. The chemical composition of the sodium hydroxide was 25% $Na_2O$ and 75% $H_2O$. The sodium silicate liquid was supplied by MARCANDE, Portugal. The chemical composition of the sodium silicate was of 13.5% $Na_2O$, 58.7% $SiO_2$ and 45.2% $H_2O$. The fly ash was obtained from the PEGO Thermal Power Plant in Portugal, and it was classified as class F according to ASTM-C618 [13] standard. It was used as the base material for the production of the geopolymers. The chemical composition of the fly ash is presented in Table 25.1.

The adopted for the mortars are Portland cement type I class 42.5R from Secil, Portugal, and calcium hydroxide from Lusical H100. In terms of chemical components, OPC contains 63.3% CaO, 21.4%$SiO_2$, 4.0%$Fe_2O$, 3.3%$Al_2O_3$, 2.4%MgO and other components. The calcium hydroxide used in this study contains more than 99% CaO. The sand was used as inert filler provided from the MIBAL, Minas de Barqueiros, SA, Portugal, in which they are passing from 4.75 mm sieve and remaining on 0.6 mm sieve. The altered sieve was used in order to remove dusts from the sand particles. The superplasticizer was commercially available in polyacrylate from Acronal series, with a density of 1050 kg.m$^3$ from BASF. One

Table 25.1 Major oxides in fly ash

| $SiO_2$ | $Al_2O_3$ | $Fe_2O_3$ | CaO | MgO | $Na_2O$ | $K_2O$ | $TiO_2$ |
|---|---|---|---|---|---|---|---|
| 60.8 | 22.7 | 7.6 | 1.0 | 2.2 | 1.5 | 2.7 | 1.5 |

Table 25.2 Properties of PCMs

| Operating temperature range | Latent heat of fusion (J/g) | Melting point (°C) | Apparent density at solid state (kg/m$^3$) | Particle size distribution range (µm) |
|---|---|---|---|---|
| 60.8 | 22.7 | 7.6 | 1.0 | 2.2 |

type of organic microencapsulated PCM was considered: BSF26 with a melting temperature of 26 °C. The properties of the selected PCM for this study are provided by the manufacturer and are presented in Table 25.2.

Geopolymer specimens were prepared with respect to the following steps: (i) homogenization of sodium silicate and NaOH solution (12 M) for 1 min, (ii) mixing all the solid materials together by using standard mortar mixture upon speed I (65 rpm) for 3 min and (iii) addition of solution into the blends and mixture for 1 min with speed I (65 rpm) and another 1 min with speed II (90 rpm). Then the mixture was transferred to metallic moulds. After nearly 4 h, the specimens were demoulded and kept sealed with the plastic wrap and then left in the same curing conditions until the date of testing. The specimens were cured in laboratory conditions (25 °C and 65%RH). Compressive strength testing was carried out at 7, 14 and 28 days with respect to the recommendations of the European standard EN1015–1. In order to determine the compressive strength of the mortars, a total number of six specimens were used for each mortar mix in which two specimens were tested at proposed ages of testing. The specimens used for compressive strength had 50 mm × 50 mm × 50 mm. A second series of the specimens with 28 days of curing were tested for compressive strength after being exposed to high temperature (200 °C and 600 °C) during 4 h. Next, the specimens were cooled down lasting about 12 h. at room temperature, and immediately the compressive tests were performed (following the above curing/testing procedure). Then the specimens were kept sealed with a plastic wrap under at laboratory environment until testing date of 7 days, 14 days and 28 days. After that, the compressive strength measurements were carried out through a compressive strength apparatus model LLOYD-LR50KPlus with the capacity of 50kN.

## 3 Results and Discussion

The results of the compressive strength are shown in Fig. 25.1. The increase of PCM incorporation resulted in a decrease of compressive strength from ≈16 MPa to ≈4 MPa at 28 days. It can also be observed that there is a noticeable difference in changes of compressive strength with addition of different amount of PCM when compared to reference specimens (without PCM), particularly at 14 and 28 days. PCMs exhibit little effect of compressive strength at 7 days.

Previous studies [14] show that the PCM incorporation in the mortars increases porosity of specimens. Furthermore, increasing of curing ages has always enhanced

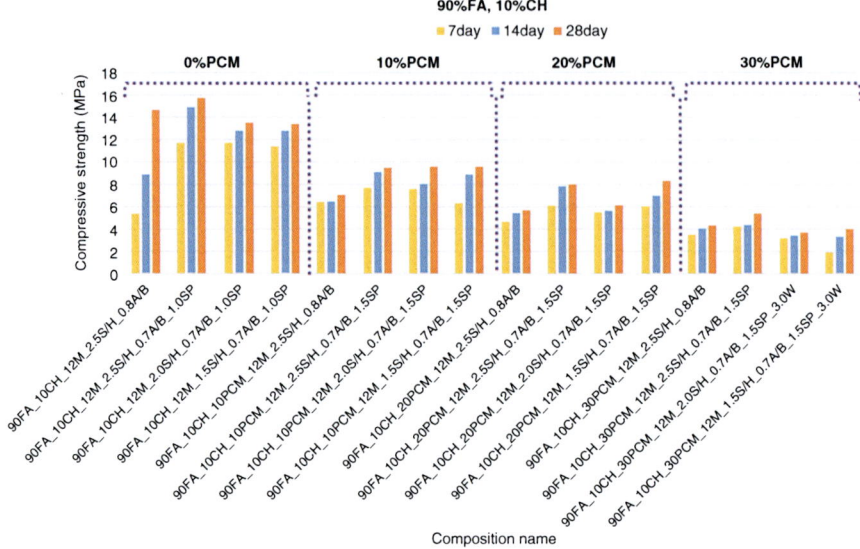

**Fig. 25.1** Compressive strength of geopolymeric mortars

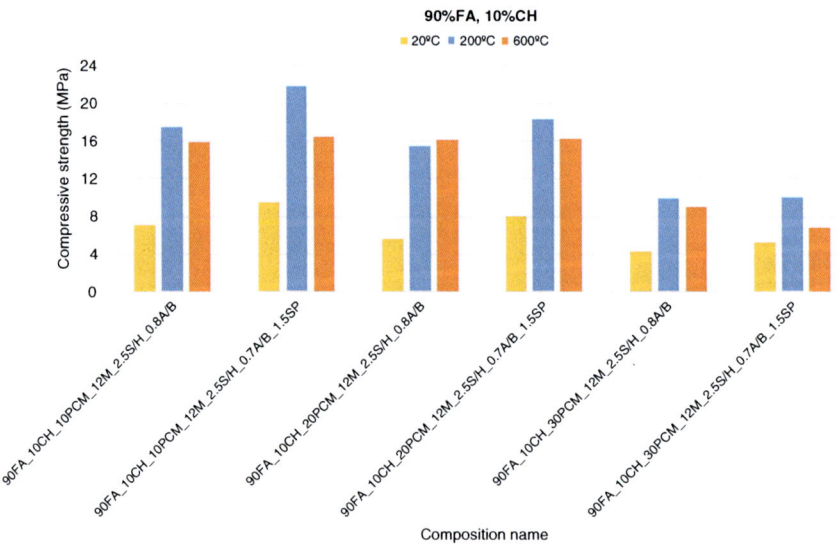

**Fig. 25.2** Compressive strength of geopolymeric mortars exposed to high temperature

the compressive strength. Interestingly, compressive strength at 14 and 28 days is almost similar in most of the cases which is typical of sodium silicate-based geopolymeric mixtures. Figure 25.2 shows the compressive strength results of PCM-based mortars after being exposed to high temperatures. The difference between the strength at temperature of 20 °C and temperature of 200 °C is quite significant. In mixtures with 90%FA and 10%CH, the compressive strength is 60% higher than that of the strength of specimens at temperature of 20 °C. The mixtures with 90%FA and 10%CH when exposed to a temperature of 600 °C still showed an improvement in its compressive strength higher than 50%; however, it is slightly lower than the compressive strength of reference specimens when exposed to a temperature of 600 °C. Several authors present different explanations for the strength increase of geopolymers exposed to a high temperature. Some state that the strength increase is attributed to a combination of polymerization reaction and the sintering reactions of unreacted fly ash particles [15, 16]. More recently others attributed the strength increase to promotion of polycondensation between chain-like geopolymer gels [17]. As to the compressive strength reduction of mixtures exposed to a temperature of 600 °C, it can be maybe due to thermal incompatibility arising from nonuniform temperature distribution as suggested by others [18, 19]. Geopolymeric mortars with PCM do not show any destruction of the specimens after exposing to the high temperatures. This finding constitutes an important advantage of the PCM geopolymeric mortar when compared with conventional Portland cement-based PCM mortars because other authors [20] noticed that the PCM-based mortars can be destroyed upon high temperature expositions. Since PCMs are made by flammable materials, this means that geopolymeric mortars are in fact preferable to Portland cement-based mortars for PCM incorporation.

## 4 Conclusions

The inclusion of PCMs is responsible for a serious reduction of the mechanical strength of the geopolymeric mortars from ≈16 MPa to ≈4 MPa. Several mixtures showed an increase in compressive strength after being exposed to high temperatures. This strength increase may be attributed to a combination of polymerization reaction and the sintering reactions of unreacted fly ash particle or due to the polycondensation between chain-like geopolymer gels. Since PCMs are made of flammable materials, geopolymeric mortars are more advantageous than Portland cement-based mortars for PCM incorporation.

## References

1. Hook, M., & Tang, X. (2013). Depletion of fossil fuels and anthropogenic climate change-a review. *Energy Policy, 52*, 797–809.
2. COM. (2011). 885/2, Energy Roadmap 2050. European Commission, Brussels.
3. Lund, H., & Hvelplund, F. (2012). The economic crisis and sustainable development: The design of job creation strategies by use of concrete institutional economics. *Energy, 43*, 192–200.

4. European Union. (2010). Directive 2010/31/EU of the European Parliament and of the Council of May 19th, 2010 on the energy performance of buildings (recast). Official Journal of the European Union.
5. Pacheco-Torgal, F. (2014). Eco-efficient construction and building materials research under the EU Framework Programme Horizon 2020. *Construction and Building Materials, 51*, 151–162.
6. European Commission. (2011). Roadmap to a resource efficient Europe. *COM (2011) 571*. EC, Brussels.
7. COM. (2014, July 2). 398 final. Towards a circular economy: A zero waste programme for Europe. Communication from the commission to the European Parliament, *The council, the European economic and social committee and the committee of the regions*. Brussels
8. Payá, J., Monzó, J., Borrachero, M., & Tashima, M. (2014). Reuse of aluminosilicate industrial waste materials in the production of alkali-activated concrete binders. In F. Pacheco-Torgal, J. Labrincha, A. Palomo, C. Leonelli, & P. Chindaprasirt (Eds.), *Handbook of alkali-activated cements, mortars and concretes* (pp. 487–518). Cambridge: WoddHead Publishing.
9. Bernal, S., Rodríguez, E., Kirchheim, A., & Provis, J. (2016). Management and valorisation of wastes through use in producing alkali-activated. Cement materials. *Journal of Chemical Technology & Biotechnology, 91*, 2365–2388.
10. American Coal Ash Association. (2016). https://www.acaa-usa.org/Publications/Production-Use-Reports
11. Van Deventer, J. S., Provis, J. L., & Duxson, P. (2012). Technical and commercial progress in the adoption of geopolymer cement. *Minerals Engineering, 29*, 89–104.
12. Pacheco-Torgal, F., Castro-Gomes, J. P., & Jalali, S. (2008). Investigations on mix design of tungsten mine waste geopolymeric binder. *Construction and Building Materials, 22*(9), 1939–1949.
13. ASTM C618 – 15. (2015). Standard specification for coal fly ash and raw or calcined natural Pozzolan for use in concrete, ASTM International, West Conshohocken, PA. www.astm.org
14. Zhang, H., Xing, F., Cui, H.-Z., Chen, D.-Z., Ouyang, X., Xu, S.-Z., Wang, J.-X., Huang, Y.-T., Zuo, J.-D., & Tang, J.-N. (2016). A novel phase-change cement composite for thermal energy storage: Fabrication, thermal and mechanical properties. *Applied Energy, 170*, 130–139.
15. Kong, D., Sanjayan, J., & Sagoe-Crentsil, K. (2007). Comparative performance of geopolymers made with metakaolin and fly ash after exposure to elevated temperatures. *Cement and Concrete Research, 37*, 1583–1589.
16. Kong, D., & Sanjayan, J. (2008). Damage behavior of geopolymer composites exposed to elevated temperatures. *Cement and Concrete Composites, 30*, 986–991.
17. Luna-Galiano, Y., Cornejo, Y., Leiva, C., Vilches, L., & Fernández-Pereira, C. (2015). Properties of fly ash and metakaolín based geopolymer panels under fire resistance tests. *Materiales de Construcción, 65*(319), e059.
18. Pan, Z., Sanjayan, J. G., & Rangan, B. V. (2009). An investigation of the mechanisms for strength gain or loss of geopolymer mortar after exposure to elevated temperature. *Journal of Materials Science, 44*, 1873–1880.
19. Guerrieri, M., & Sanjayan, J. (2010). Behavior of combined fly ash/slag-based geopolymers when exposed to high temperatures. *Journal of Fire Mater, 34*, 163–175.
20. Cunha, S., Aguiar, J., & Pacheco-Torgal, F. (2015). Effect of temperature on mortars with incorporation of phase change materials. *Construction and Building Materials, 98*, 89–101.

# Part III
# Polymer Concrete

# Chapter 26
# Are Polymers Still Driving Forces in Concrete Technology?

**Lech Czarnecki, Mahmoud Reda Taha, and Ru Wang**

In less than one century, concrete has become the most widely used construction material worldwide. Today it is difficult to imagine a concrete totally without polymers. An implantation of polymers into concrete has taken effect in the form of concrete-polymer composite [C-PC = PMC + PCC + PIC + PC]. On the way of C-PC development, several milestones are recognized and discussed here with particular emphasis on innovative milestones that shaped the use of polymers in concrete. As the difference between polymer-cement concrete and ordinary concrete diminishes, the question: "Are polymers still driving forces in concrete technology?" arises. This question is about the future of the concrete technology. The authors consider various optional answers. One of the most promising options seems to be that the routes of nanotechnology and the organic-inorganic chemically bound composites will open the gate for the development of a new generation of sustainable polymer. The new polymer concrete will have superior performance characteristics that will enable its use for 3D-printed sustainable concrete structures.

---

L. Czarnecki (✉)
Instytut Techniki Budowlanej (ITB), Warszawa, Poland
e-mail: l.czarnecki@itb.pl

M. Reda Taha
University of New Mexico, Albuquerque, NM, USA

R. Wang
Tongji University, Shanghai, China

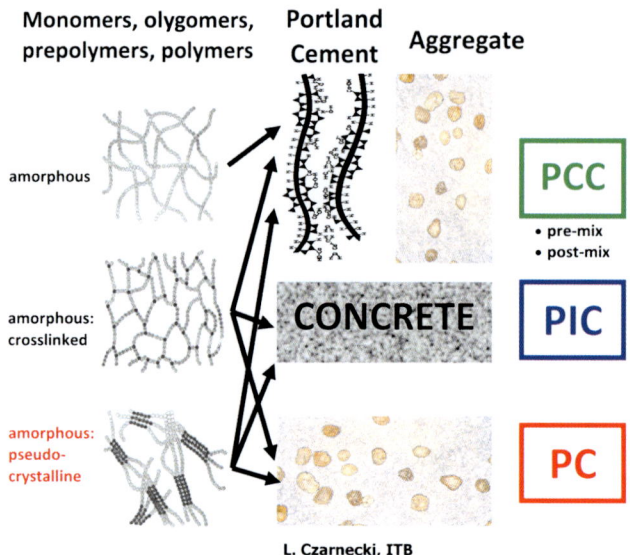

**Fig. 26.1** The fascinating concept of C-PC

## 1 Introduction

In less than one century, concrete has become the most widely used construction material worldwide. In less than half of a century, it is difficult today to imagine a concrete without polymers. The driving forces are those factors (external and internal) that have the greatest influence on the given activity outcomes. The question: "Are polymers still driving forces in concrete technology?" is a question about the future of concrete technology. The proper recognition of driving forces is always verified by the future, and, due to that, it is uncertain and risky. This paper does not bring a decisive answer but rather discusses this important question and tries to foresee the future of the use of polymers in concrete in a very fast-changing and technology-driven construction industry.

## 2 Polymers in/on and for Concrete: Lesson from the Past

Decades ago, an implementation of polymers into concrete has taken effect in the form of concrete-polymer composites (C-PC=PMC + PCC + PIC + PC) (Fig. 26.1).

The original scope was polymers in concrete which at that time included polymer (less than 5% cement mass)-modified concrete, PMC; polymer (more than 5% cement mass)-cement concrete, PCC; polymer-impregnated concrete, PIC; and cementless polymer concrete, PC. There are three ways of technological implementation of polymer into concrete: into fresh concrete mixture (PMC, PCC),

**Table 26.1** Scope of concrete-polymer composites (C-PC) [1]

| Polymers | | |
|---|---|---|
| *in* | *on* | *for* |
| **Concrete** | | |
| Polymer concrete, PC<br>Polymer-modified concrete, PMC<br>Polymer-cement concrete, PCC<br>Polymer-impregnated concrete, PIC<br>Polymer fibers in concrete, PFiC<br>Polymer aggregate in concrete | Polymer overlays<br>Polymer coatings and waterproofing materials<br>Polymer used for bonding materials to concrete<br>Fiber-reinforced polymers for strengthening Concrete, FRP-C | Polymer repair mortars<br>Polymer crack repair |

polymerization after mixing (post-mix) or before mixing (pre-mix); into hardened concrete (PIC), polymer "forced" under pressure into "concrete"; and into fresh concrete mixture without Portland cement, solely polymer as a binder. The 5% of cement mass which divided the PMC and PCC is not just an arbitrary number. Above this limit polymer is able to create a continuous network and acts as co-binder together with the Portland cement. The use of polymers in, on, and for concrete is summarized in Table 26.1.

The International Congress on Polymers in Concrete (ICPIC) – Congresses – established the milestones for C-PC. The aphorisms characterized each Congress are presented in Table 26.2. The history of C-PC is well documented elsewhere [2]. After the material models were elaborated [3, 4], the PCC research and application abruptly rose up. Example models include PCC-pre-mix [3, 4], PCC-post-mix [5], PC [6], and PIC [7]. Areas of application where C-PC is irreplaceable included:

- PCC – in typical concrete repairing, overlays, industry floors
- PC – as above performing under aggressive chemical environment
- PIC – as the mean of preserving the monuments and old building

## 3 Polymer Driving Forces and Shaping New Trends

The aim is always to produce *better concrete*. However right now it is obvious that the *better concrete* means *sustainable concrete*. The sustainable development requirements should overrule any technological activity. It has been shown that the compatibility between polymers and Portland cement as well as other concrete components is an overriding requirement. What has been fascinating from the beginning that small amount of polymers, sometimes very small, can make a significant effect on concrete. It is also important to realize that each type of polymer has its unique characteristics in cement modification. For example, recent findings

**Table 26.2** 15 ICPIC's aphorisms 1975–2015

| No | Year, city | Motivation | Remarks |
|---|---|---|---|
| I | 1975, London | The use of polymers in concrete as a new technology | Precast elements and bridge deck overlays |
| II | 1978, Austin | C-PC versatility proved and applications widespread | Focused on the use of polymers in concrete |
| III | 1981, Koriyama | C-PC is used in different area (repair and overlays) | Shrinkage, creep and durability of C-PC |
| IV | 1984, Darmstadt | "Material model" appears as the category | Epoxy cement concrete and GRPC panels |
| V | 1987, Brighton | Tailoring C-PC properties for various applications | Materials models for performance control |
| VI | 1990, Shanghai | Worldwide use of polymers in concrete comes true | Interaction between polymers and concrete |
| VII | 1992, Moscow | How good C-PC is good enough? | Evaluation, simulation and optimization |
| VIII | 1995, Oostende | Mathematical modeling and design for durability | High-strength PC and polymer fiber concrete |
| IX | 1998, Bologna | Micro-/macrostructure relations for new C-PC | Nanotechnology and recycled plastics In PC |
| X | 2001, Hawaii | How polymers develop sustainable C-PC | Self-repair PCC; metallic monomer PC |
| XI | 2004, Berlin | Integrated model PCC, hardener-free epoxy | Water-soluble polymer in PCC |
| XII | 2007, Chuncheon | Sustainable C-PCs and nanotechnology in C-PC | Extensive studies of C-PC performance |
| XIII | 2010, Madeira | High-performance CPC (HPC-PC) | Interfacial zone of polymer in concrete |
| XIV | 2013, Shanghai | Modeling of cement and polymer hardening | Polymer film forming thermodynamics |
| XV | 2015, Singapore | C-PC potential aging | Self-sensing PC and nano-modified PC |

by van Gemert and Knapen [8] pointed out the valuable use of water-soluble polymer (WSP) instead of the liquid polymer. Due to the thermodynamic conditions of WSP, such polymer will be placed in nano-area. As the result of changing the polymer position from micro-area into nano-area, the polymer will not bridge between the crack edges of concrete but between the hexagonal plates of portlandite: $Ca(OH)_2$. In such situation, even the addition of polymer within 1% of cement mass will improve the tensile strength by 50% higher than with 10% polymer content in ordinary situation. Furthermore, chemical interaction has not been the point of focus in many cases compared with physical interaction. However, the important role of chemical interaction has been only very rarely denied [9, 10]. Chemical interaction could result in the formation of complex structures, changes in the hydrated cement phases morphology, composition, and quantities. For example, it is generally believed that vinyl acetate group contained in EVA copolymer suffers hydrolysis when dispersed in an alkaline medium [11] that is called saponification. When EVA is dispersed in a Ca(OH)$_2$ saturated solution, as in the case of pore solution of cement pastes, the acetate anion CH$_3$COO released in the alkaline hydrolysis of EVA combines with Ca$^{2+}$ released in the dissolution of cement grains and the product of this interaction is calcium acetate, Ca(CH$_3$COO)$_2$. The interaction between acrylic polymers and cement was also studied in the literatures [12, 13]. Chemical interactions at the nanoscale have also been reported. Soliman et al. [14] showed the rule of the chemical interaction of the COOH functional group with multi-walled carbon nanotubes (MWCNTs) in reacting with styrene-butadiene rubber (SBR) polymer latex-modified cement. It was shown that such chemical interaction enables MWCNTs to alter the polymerization process of the SBR film in the existence of cement and to form a new polymer-cement composite with significantly improved mechanical and fracture properties.

## 4 C-PC: Roadmap for the Way Ahead

Existing research works have demonstrated many evidences of chemical interactions between polymer and cement in concrete-polymer composites through various analytical methods including infrared (IR) spectroscopy, thermal analysis, nuclear magnetic resonance (NMR) microscopy, etc. [15]. Some of the reactions explain the failure mode of this type of materials such as the hydrolysis of ester group. Others explain the strengthening mechanism such as the formation of chelates in it. Chemical and physical interactions between cement and polymers are the two sides of one coin. Only when the chemical interactions between polymer and cement are clearly understood, we will be able to better explain the micro and macro structure relationship in concrete-polymer composites. The authors believe that new developments into polymer concrete will happen through investigations that target producing new classes of polymer concrete with superior mechanical and durability performances and with characteristics not attainable in cement concrete such as self-healing, self-leveling, and self-sensing using nanotechnology. Example research work investigating this route has recently been published [16]. Another route of development is through investigating the organic-inorganic chemically bound composites [17, 18] and the possible production of new generation of polymer concrete using such compounds with new capabilities never observed in polymer concrete. Finally, we believe that polymer concrete will be a major player in the new world of 3D-printed concrete or what is known today as additive construction by extrusion [19]. The versatility of polymer concrete and the possible manipulation of its thixotropic, hardening, and mechanical properties through the use of polymers will enable new opportunities for polymer concrete beyond many other concretes available today for field applications.

## 5 Conclusions

Progress in concrete-polymer composites technology has been driven mainly by material modification. Polymers in concrete as modifiers until now play significant function. The role of polymers in concrete could not be overestimated. The question: "Are polymers still driving forces in concrete technology?" about the future of the concrete technology is still open. In our review, we show that future concrete-polymer composites should materialize the main contemporary ideas: sustainable development in civilization. We discussed the role of chemical interaction on concrete-polymer composites, and we postulated that feature will shape the future of concrete-polymer composites. A roadmap for the use of concrete-polymer composites in the future is discussed including the use of nanotechnology to produce concrete-polymer composites with superior characteristics. The future of the organic-inorganic chemically bound composites and the potential use of polymer concrete for 3D-printed concrete structures are suggested.

## References

1. Czarnecki, L., Fowler, D., Kruger, D., Ping, L. J., Van Gemert, D., & Weichold, O. (2013). Defining scope of ICPIC. The ICPIC unpublished report. Personal Communique.
2. Czarnecki, L. (1985). Status of polymer concrete. *Concrete International, 7*(7), 47–53.
3. Ohama, Y. (1987). Principle of latex modification and some typical properties of latex modified mortars and concretes. *ACI Materials Journal, 84*(6), 511–518.
4. Van Gemert, D., & Beeldens, A. (2013). Evolution in modeling microstructure formation in polymer-cement-concrete. *Restoration of Buildings and Monuments, 19*(2/3), 97–108.
5. Łukowski, P. (2016). Studies on the microstructure of epoxy-cement composites. *Archives of Civil Engineering, 62*, 101–113.
6. Czarnecki, L. (1984). Introduction to material model of polymer concrete. *Proceedings of ICPIC 1984*, Darmstadt (pp. 50–64).
7. Chen, C. H., Huang, R., & Jiann Kuo, W. (2007). Preparation and properties of polymer impregnated concrete. *Journal of the Chinese Institute of Engineers, 30*(1), 163–168.
8. Knapen, E., & Van Gemert, D. (2009). Cement hydration and microstructure formation in the presence of water-soluble polymers. *Cement and Concrete Research, 39*(1), 6–13.
9. Riley, V., & Razi, I. (1974). Polymers additive for cement composites: a review. *Composites, 5*(1), 27–33.
10. Zeng, S., Short, C., & Page, C. (1996). Early-age hydration kinetics of polymer-modified cement. *Advances in Cement Research, 8*(29), 1–9.
11. Ohama, Y. (1998). Polymer-based admixtures. *Cement and Concrete Composites, 20*, 189–212.
12. Tian, Y., Jin, X., Jin, N., et al. (2013). Research on the microstructure formation of polyacrylate latex modified mortars. *Construction and Building Materials, 47*, 1381–1394.
13. Wang, R., Yao, L. J., & Wang, P. M. (2013). Mechanism analysis and effect of styrene-acrylate copolymer powder on cement hydrates. *Construction and Building Materials, 41*, 538–544.
14. Soliman, E., Kandil, U. F., & Reda Taha, M. M. (2012). The significance of carbon nanotubes on styrene butadiene rubber (SBR) and SBR modified mortar. *Materials and Structures, 45*(6), 803–816.

15. Wang, R., Li, J., Zhang, T., & Czarnecki, L. (2016). Chemical interaction between polymer and cement in polymer-cement concrete. *Bulletin of the Polish Academy of Sciences, Technical Sciences, 64*, 785–792.
16. Emiroglu, M., Douba, A. E., Tarefder, R., Kandil, U. F., & Reda Taha, M. M. (2017). New polymer concrete with superior ductility and fracture toughness using alumina nanoparticle. *ASCE Journal of Materials in Civil Engineering, 29*, 8.
17. Czarnecki, L. (2013). Sustainable concrete; is nanotechnology the future of concrete polymer composites? *Advanced Materials Research, 687*, 3–11.
18. Czarnecki, L., & Schorn, H. (2007). Nanomonitoring of polymer cement concrete microstructure. *Restoration of Buildings and Monuments, 13*(3), 141–152.
19. Reda Taha, M. M., & Soliman, M. M. (2018). Nano-modified polymer concrete and composites – A future prospective. Conc. Int., In review.

# Chapter 27
# Mechanical Properties of Polymer Cement-Fiber-Reinforced Concrete (PC-FRC): Comparison Based on Experimental Studies

Tomasz Piotrowski, Piotr Prochoń, and Alice Capuana

The paper presents results of the comparative tests on OC and PCC modified with a constant volume of steel, macro-, and micro-polypropylene fibers. The influence of these fibers on FRC and PC-FRC mechanical properties was evaluated by flexural and compressive strength tests, as well as bending strength and breaking load of a concrete pavement slab in accordance with EN 1339. Significant differences in results were observed between FRC and PC-FRC. Increase of flexural strength in PC-FRC beams made with steel and macro-polypropylene fibers in comparison with reference PCC samples has been shown. Breaking load of PC-FRC slabs was close to reference PCC ones. The enhancement of compressive strength on cubic samples of PC-FRC in relation to PCC was also observed. Further research with different polymer additives and fibers with other volume contents is considered.

## 1 Introduction

When comparing the ordinary concrete (OC) to the polymer cement concrete (PCC), the second one is usually characterized by improved tensile and flexural strength, adhesion, and durability [1, 2]. Those effects are highly dependent on the polymer-cement ratio, water-cement (water-binder) ratio, type of polymer used, and curing conditions, especially humidity and temperature [3]. PCC are used commonly as repair material [4, 5]. Fiber-reinforced concretes (FRC) provide improvement of technical properties in relation to OC and change the post-cracking behavior of concrete [6]. This allows to use FRC in several applications, such as offshore structures, architectural panels, road and bridge engineering, and structures in

---

T. Piotrowski (✉) · P. Prochoń · A. Capuana
WUT, Faculty of Civil Engineering, Warsaw, Poland
e-mail: t.piotrowski@il.pw.edu.pl

seismic regions [7]. The fibers are generally made of steel, glass, and natural or synthetic material. Depending on the aim of modification, they can improve concrete characteristics in different ways [8]. This paper deals with the study of effects of different types of fibers on mechanical properties of OC and PCC.

## 2 Materials and Methods

### 2.1 Cement

Blast furnace cement (CEM III/A 42.5 N) with specific gravity 2.97 g/cm$^3$, initial and final setting time of 224 min and 292 min, respectively, in accordance with EN 197-1, was used as a binding material.

### 2.2 Aggregates

River sand with density 2.45 g/cm$^3$ was used as fine aggregate. Coarse aggregate granite fractions 2/8 and 8/16 with volumetric mass density 2.64 g/cm$^3$ were applied. Both aggregates comply with EN 12620 + A1.

### 2.3 Fibers

In this work, effects on mechanical properties of cement composites of waved steel, micro-, and macro-polypropylene fibers were studied. Description of used fibers is shown in Table 27.1.

### 2.4 Polymer

For PCC concrete samples, carboxylated styrene-butadiene latex was used with 47% containment of solid particles. The dispersion was alkaline (pH = 10).

**Table 27.1** Description of different fibers

| Fiber | Description |
| --- | --- |
| Steel (ST) | 50 mm long, 2 mm wide, and 0.5 mm thick, density 7.85 g/cm$^3$ |
| Micro-polypropylene (MP1) | 12 mm long, diameter 38 µm, density 0.91 g/cm$^3$ |
| Macro-polypropylene (MP2) | 39 mm length, diameter 0.78 mm, density 0.91 g/cm$^3$ |

Table 27.2 Concrete mix compound

| | Cement kg/m³ | Water | Sand | Coarse aggregate | Polymer dispersion | Steel fibers | Polypropylene fibers |
|---|---|---|---|---|---|---|---|
| OPC | 305 | 133 | 712 | 1323 | 0 | 0 | 0 |
| FRC_S | 305 | 133 | 712 | 1323 | 0 | 25 | 0 |
| FRC_MP1 | 305 | 133 | 712 | 1323 | 0 | 0 | 0.6 |
| FRC_MP2 | 305 | 133 | 712 | 1323 | 0 | 0 | 3 |
| PCC | 305 | 133 | 712 | 1323 | 20 | 0 | 0 |
| PC-FRC_ST | 305 | 133 | 712 | 1323 | 20 | 25 | 0 |
| PC-FRC_MP1 | 305 | 133 | 712 | 1323 | 20 | 0 | 0.6 |
| PC-FRC_MP2 | 305 | 133 | 712 | 1323 | 20 | 0 | 3 |

## 3 Mixture Proportions

The mixture proportions are presented in Table 27.2. The water-to-binder ratio was kept constant as 0.43. In accordance with producer's information card, the addition of polymer dispersion in PCC samples should be 15–30 kg/m³. All the batches were mixed in a stationary mixer; the samples were molded and left for 24 h before demolding. The OC samples were put then into a climatic chamber up to 28 days with the humidity 95% and the temperature $18 \pm 2\,^{\circ}\mathrm{C}$. The PCC samples were cured for 5 days in the climatic chamber and for the following 23 days in the ambient conditions.

## 4 Experimental Methodology

### 4.1 Test of Fresh Concrete

Workability is an important fresh concrete property described by its consistency which impacts quality, strength, and appearance of the hardened construction or element. It is said that cement composite is workable when it can be easily placed and compacted homogeneously. The aim is to assure proper consistency of concrete mixture and avoid its bleeding or segregation. The workability should be determined in accordance with the type of construction, used formworks, and way of transport. In this work the slump test according to EN 12350-2 was conducted for evaluation of the concrete mixes workability. The result is presented in Fig. 27.1.

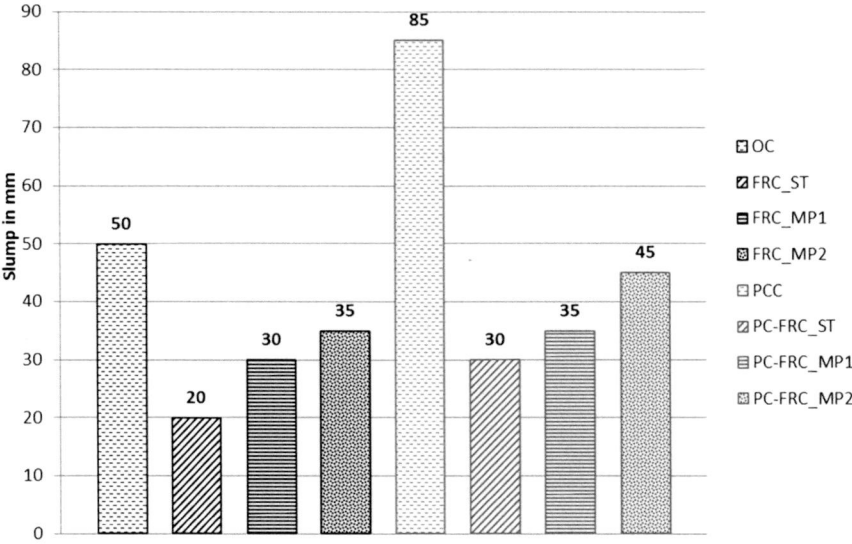

**Fig. 27.1** Effects of fibers on workability of concrete

Higher slump value for PCC samples is observed due to extra amount of water from polymer dispersion. Addition of fibers in FRC and PC-FRC decreased the slump of the fresh concrete. It was observed that addition of steel fibers in both types of concrete minimize workability of concrete compared to other fibers. Comparison between synthetic fibers, with the same volume fraction, shows that microfibers have decreased workability of concrete less than macro-fibers. The presence of fibers in OC and PCC blocks more cement paste and increases the fresh concrete viscosity with a lower slump as a consequence.

## 4.2 Test on Hardened Concrete

In this research flexural strength on slabs (400 × 400 × 50 mm) and beams (75 × 75 × 275 mm) was measured. They were taken after 28 days in accordance with EN 1339 App. F and EN 12390-5 for three-point flexural strength of slabs and beams, respectively. Additionally the compressive strength on cubic samples (100 × 100 × 100 mm) according to EN 12390-4 was performed after 28 days.

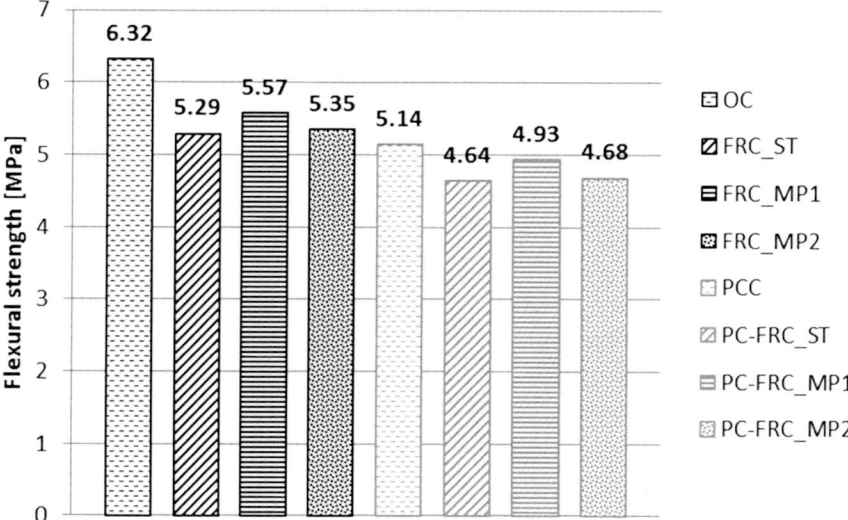

**Fig. 27.2** Results of flexural strength tests for slabs

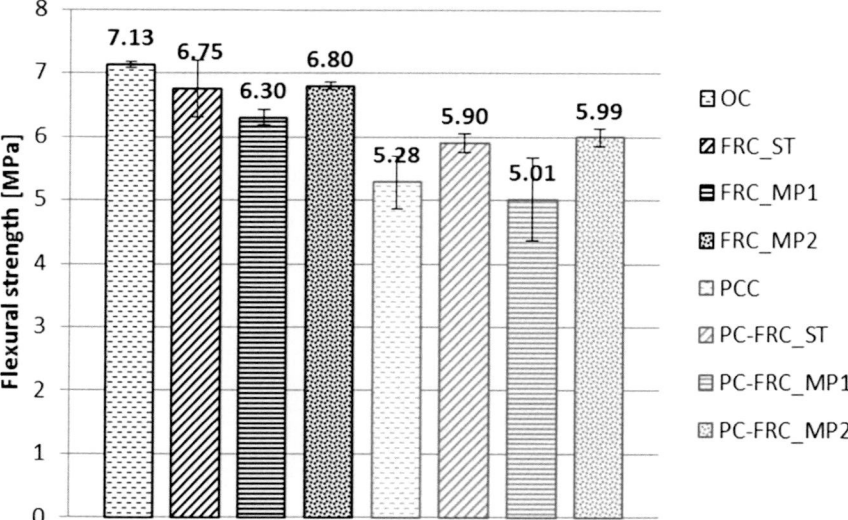

**Fig. 27.3** Results of flexural strength tests for beams

## 5 Experimental Results and Discussions

Results of flexural strength for examined concretes on beam and slabs specimen with different fibers are shown in Figs. 27.2 and 27.3. It is observed that for slab specimens adding any fibers to concrete decreased flexural strength. In case of

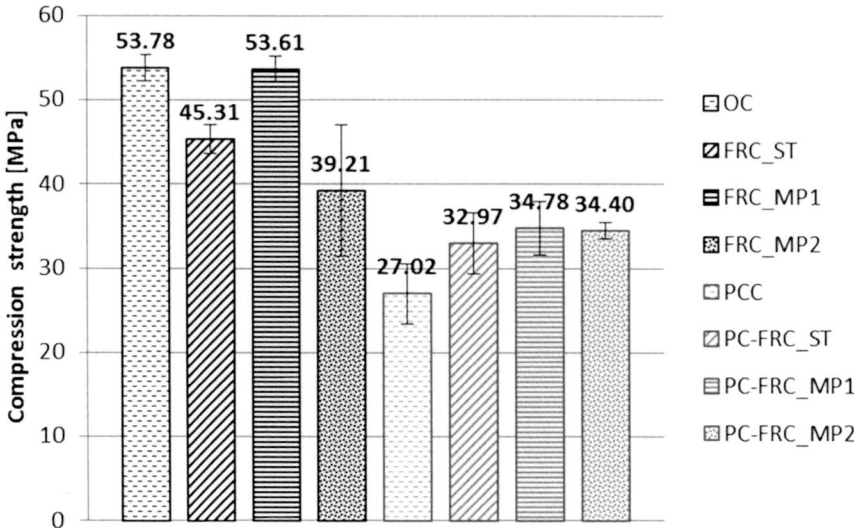

**Fig. 27.4** Results of compressive strength tests for cubes

OC, the addition of steel fiber reinforcement (FRC_ST) decreased the flexural strength by 16%. Similarly in PCC steel fibers (PC-FRC_ST) are reducing the flexural strength close to 10%. In both types of composites (FRC and PC-FRC), the best results on slabs were obtained for polypropylene microfibers.

Comparing the results presented in Fig. 27.2 for slabs with values in Fig. 27.3 representing the flexural strength on beams, it is observed that results for beams are higher and a decrease in strength is lower. For PC-FRC specimens, steel and macro-polypropylene fibers increased PCC flexural strength more than 13%. In both types of specimens, the PCC mixes show lower flexural strength than OC. This difference may result from compatibility problems between cement matrix, polymer dispersion, and fibers.

Figure 27.4 presents the results of compressive strength test on cube samples with different fiber reinforcement of OC and PCC. For OC the best compressive strength value is obtained for FRC with micro-polypropylene fibers (FRC-MP1). The macro-polypropylene fibers (FRC-MP2) caused a reduction of this parameter by more than 27%. In case of PCC, both steel and polypropylene fibers (micro and macro ones) gained better results than reference PCC.

## 6 Conclusion

Obtaining certain flexural strength value depends not only on type of fiber reinforcement used but also on the type of specimens. Using fibers in PCC can boost both flexural and compressive strength. Further analysis on different fibers content in OC and PCC mixes is needed.

## References

1. Sasse, H. R. (1986). *Adhesion between polymers and concrete*. Dordrecht: Springer Science + Business Media.
2. Cao, Q., Sun, W., Guo, L., & Zhang, G. (2012). Polymer-modified concrete with improved flexural toughness and mechanism analysis. *Journal of Wuhan University of Technology--Materials Science Edition, 27*(3), 597–601.
3. Łukowski, P. (2016). Material concrete modification (polish). *SPC Cracow*, p. 355.
4. Czarnecki, L., Łukowski, P., & Garbacz, A. (2017). Repair and protection in concrete constructions (polish). *PWN, Warsaw*, p. 272.
5. Bissonnette, B., Courard, L., & Garbacz, A. (2016). *Concrete surface engineering* (p. 255). Boca Raton: T&F Group.
6. Bentur, A., & Mindess, S. (2007). *Fibre reinforced cementitious composites* (p. 601). London: T&F Group.
7. Vairagade, V. S., & Kene, K. S. (2013). Strength of normal concrete using metallic and synthetic fibers. *Procedia Engineering, 51*, 132–140.
8. Masuelli, M. A. (2013). *Fiber reinforced polymers—The technology applied for concrete repair*. San Luis: InTech. *CC BY, 3*.

# Chapter 28
# Environmental Temperature and Humidity Adaptability of Polymer-Modified Cement Mortar

Ru Wang, Shaokang Zhang, and Peiming Wang

It is well known that polymer can improve the performance of cement mortar, e.g., the workability, flexural strength, tensile bond strength, flexibility, and durability. But environmental conditions such as temperature and humidity variation influence the hydration of cement and the morphology of polymer in cement mortar. Therefore, the service environment will impact the performance of polymer-modified cement mortar. In recent years, researches on the influence of curing temperature and humidity variation on the properties of polymer-modified cement mortar are increasing, which refer to a wide range of temperature and humidity changes. This chapter summarizes the research progress in our laboratory on the effect of curing temperature and humidity on the properties of polymer-modified Portland cement mortar and polymer-modified calcium sulfoaluminate cement mortar.

## 1 Introduction

Due to the diversity of the world climate and application, polymer-modified cement mortar has to adapt to different environmental conditions. For example, in China, there are different climates, i.e., high temperature/high humidity, high temperature/low humidity, low temperature/high humidity, and low temperature/low humidity, except normal temperature and humidity. Recently, research on the influence of curing conditions on the properties of polymer-modified cement mortar is on the increase. At the beginning, people care about the influence of water curing, water-air mixed curing, and air-curing and try to find an appropriate curing method for polymer-modified cement mortar [1–5]. People almost reached a consensus that an

---

R. Wang (✉) · S. Zhang · P. Wang
School of Materials Science and Engineering, Tongji University, Shanghai, China
e-mail: ruwang@tongji.edu.cn

appropriate water-air mixed curing is beneficial to strength development. But air-curing has to be actually used in a lot of practical engineering. So the modified mortars have to face different air-curing conditions in practical applications because of the changed climate. So the influence of humidity and temperature on the properties becomes necessary and relative research is increasing. Previous researches showed that the curing regime used within 28 days is critical to the bond strength of polymer-modified mortar [6]. A reduction in the mechanical strength of polymer cement mortar was observed with temperature increasing from 20 to 60 °C [7]. 50 °C/90%RH has an adverse effect on the strength of acrylic emulsion polymer-modified mortar compared to normal condition [8]. Muthadhi et al. [9] found that styrene-butadiene rubber (SBR)-modified concrete can be exposed to 400 °C for 3 h without any adverse effect on strength properties. Adding polymer into cement mortar is one of the effective methods to improve the tensile bond strength of the mortar on substrates. The factors affecting the tensile bond strength are the types and dosages of polymer and cement, water-cement ratio, curing system, etc. Up to now, people failed to give sufficient attention to the curing system. Therefore, this chapter will focus on the effect of curing temperature and humidity on the properties of polymer-modified cement mortar, especially the tensile bond strength.

## 2 Effect of Temperature and Humidity on the Tensile Bond Strength of Polymer-Modified Portland Cement Mortar

Hydroxyethyl methyl cellulose (HEMC) and ethylene-vinyl acetate (EVA) copolymer were adopted to modify Portland cement mortar. The influence of curing humidity (40%, 65%, and 90% RH) and temperature (0, 5, 10, and 20 °C) on the tensile bond strength at 28 days of the modified mortar on concrete was studied [10]. The tensile bond strength at 28 days of the modified mortar on concrete was tested according to JGJ/T70-2009. The results show that the addition of HEMC in cement mortar is beneficial to the development of tensile bond strength under low but not high humidity conditions at 20 °C. Either only adding EVA or adding both EVA and HEMC to modify cement mortar, the tensile bond strength increases along with the increase of the dosage of EVA under low humidity conditions, while under high humidity conditions, the high dosage of EVA is not good to the development of the tensile bond strength (Fig. 28.1a). Low temperature has little effect on the tensile bond strength of cement mortar without adding polymers. 5 °C curing is suitable for the development of the tensile bond strength of cement mortar modified with only HEMC. The tensile bond strength of EVA-modified cement mortar decreases with decreasing temperature. When the temperature rises, the tensile bond strength of EVA and 0.3% HEMC-modified cement mortar basically increases, but 10 °C is not suitable for the development of the tensile bond strength (Fig. 28.1b). When the

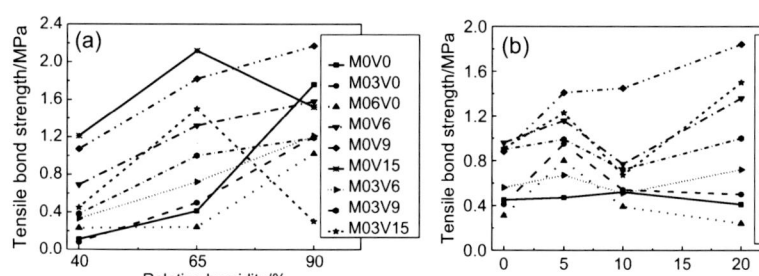

**Fig. 28.1** Relation of tensile bond strength with curing (**a**) humidity and (**b**) temperature of different polymer-modified cement mortars (M, HEMC; V, EVA; number $n$, $n\%$; number $0n$, $0.n\%$)

dosage of HEMC is high, the curing temperature has little effect on the tensile bond strength of EVA-modified cement mortar with HEMC.

The adhesive strength of paving cement mortar seems to be a more important property for the ceramic tiles with very low water absorption rate which developed recently. The influence of humidity (30%, 55%, and 80% RH) on the tensile bond strength of HEMC- and EVA-modified cement mortar on this kind of ceramic tile at 40 °C and 80 °C was studied and compared with that on concrete [11]. Under all researched curing conditions, the tensile bond strength of all kinds of mortars on ceramic tile is much lower than that on concrete. The addition of HEMC into EVA-modified cement mortar can reduce the difference in tensile bond strengths caused by different substrate materials. Whether mixed with HEMC or not, the high humidity (80% RH) makes the tensile bond strength of 15% EVA-modified cement mortar to decrease at 80 °C. When EVA is added at a dosage below 15% together with HEMC, the tensile bond strength of the mortar at 40 °C increases with the increase of moisture. At the same humidity, increasing temperature is unfavorable to the development of the tensile bond strength of the mortar on ceramic tile generally. In addition, the condition of 80 °C with 80% RH is the most unfavorable condition for the tensile bond strength (Fig. 28.2).

The influence of curing humidity (40%, 65%, and 90% RH) and low (0, 5, 10, and 20 °C) and high (70 °C) temperatures on the tensile bond strength of cement mortar modified by HEMC and EVA on steel, glass, and tile substrates at 28 days was studied [12]. The development of the tensile bond strength of cement mortar under different conditions varies with the types of polymer and substrate. For steel, glass, and tile substrates, the tensile bond strength of cement mortar increases with the increase of humidity when only HEMC is added. 65% RH is beneficial to the development of the tensile bond strength of cement mortar when adding only EVA or both HEMC and EVA. The development of the tensile bond strength of cement mortar varies with the types and dosages of polymer under low temperature, but in general 20 °C is beneficial to the development of the tensile bond strength. When adding only HEMC or EVA, the tensile bond strength of cement mortar on steel and glass decreases with the increase of pre-curing days at high temperature, while that on tile increases. However, when adding both HEMC and EVA, the

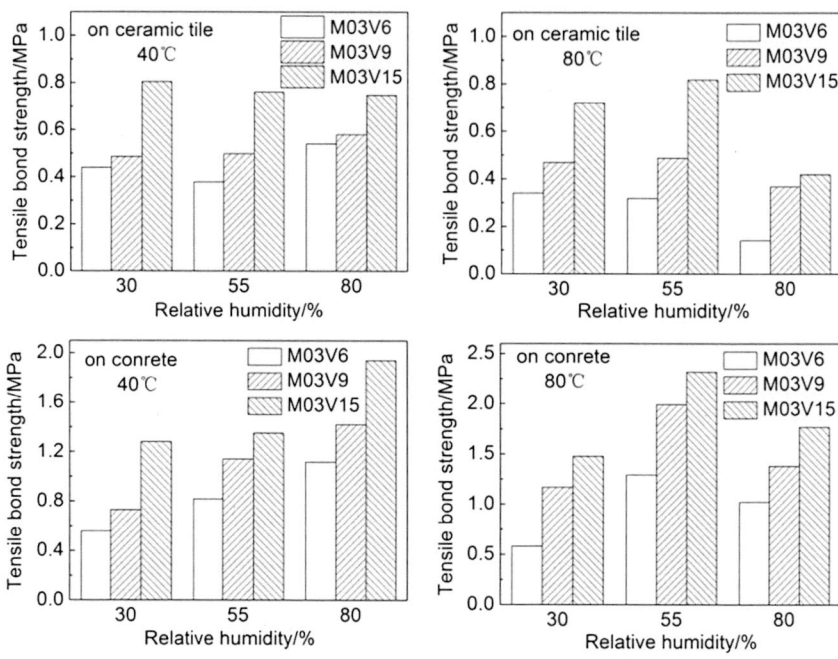

**Fig. 28.2** Relation of tensile bond strength with curing humidity of different polymer-modified cement mortars (M, HEMC; V, EVA; number $n$, $n\%$; number $0n$, $0.n\%$)

tensile bond strength of cement mortar on steel, glass, and tile increases with the increase of pre-curing days.

## 3 Effect of Temperature and Humidity on the Alkali Efflorescence Resistance of Polymer-Modified Portland Cement Decorative Mortar

Decorative mortar with hydroxypropyl methylcellulose (HPMC) and ethylene/vinyl chloride/vinyl lauryl redispersible terpolymer (E/VL/VC) as polymer additives was developed. Research on the influence of curing temperature and humidity on its properties showed that, except the effect on mechanical strength, the influence of temperature and humidity on the alkali efflorescence resistance, tested according to JC/T 1024-2007, of the decorative mortar has certain regularities. The effect of humidity is greater than temperature. At 90% RH, the alkali efflorescence resistance of the mortar is very good, showing no visible trace of efflorescence; the mortar modified by only HPMC has the best alkali efflorescence resistance. At 60% RH, the alkali efflorescence resistance of the mortar is generally poor, showing moderate and severe efflorescence. The main product of alkali efflorescence of Portland cement decorative mortar is $CaCO_3$; $Ca^{2+}$ is the most important basic ion which causes the efflorescence. But only the content of $Ca(OH)_2$ cannot predict the efflorescence situation under different curing temperature and humidity conditions. The positive correlation existing between opening porosity and efflorescence of plain mortar has not been found for the mortar modified with HPMC and E/VL/VC [13].

## 4 Effect of Temperature and Humidity on the Properties of Polymer-Modified Sulfoaluminate Cement Mortar

Styrene-butadiene rubber (SBR) copolymer dispersion was used to modify calcium sulfoaluminate (CSA) cement mortar. The effect of curing temperature (low, 0, 5 °C; room, 20 °C; high, 40 °C) and humidity (30%, 60%, and 90% RH) on the properties of the SBR/CSA mortar was analyzed [14]. The compressive and flexural strength of the SBR/CSA mortar was tested according to GB/T 17671-1999. It is found that the SBR dispersion improves the properties of the CSA mortar at different curing temperatures and humidities. The compressive, flexural, and tensile bond strengths firstly decrease and then increase with the increase of SBR dosage. The flexural and tensile bond strengths increase, the flexibility improves, and the water absorption rate decreases in the case of appropriate addition of SBR dispersion. The compressive, flexural, and tensile bond strengths increase with curing temperature, but the effect level is different at different ages and SBR dosages (Fig. 28.3). Curing temperature has different effects on the shrinkage rate of the control and modified mortars, the control mortar expands in early age at high temperature, but the modified mortars don't expand at all curing temperatures. SBR dispersion can improve the performance of the CSA mortar at all the researched temperatures. The increase of humidity has benefit to the performance of the SBR/CSA mortar. The flexural, compressive, and tensile bond strengths of the mortar increase with the increase of humidity. The higher humidity leads to lower shrinkage and better

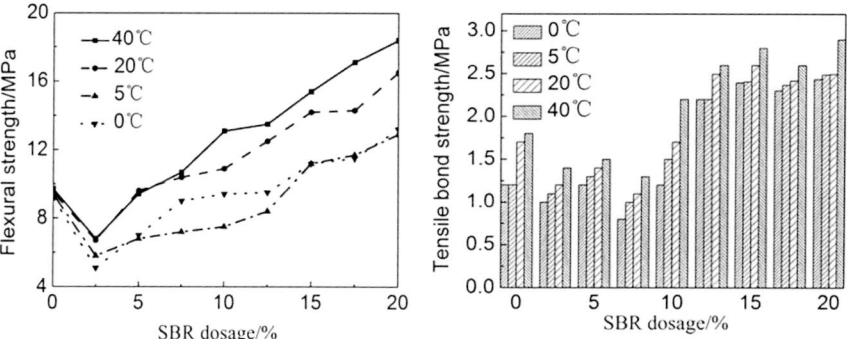

**Fig. 28.3** Flexural and tensile bond strengths of the SBR/CSA mortar at 28 days and different curing temperatures

impermeability. But the effect of the humidity is not so significant as SBR dosage and temperature.

## 5 Conclusions

The effect of temperature and humidity on the properties of polymer-modified cement mortars is complicated, which depends on not only the kinds and dosages of polymers in cement mortar, e.g., the polymers that have no good water tolerance are not suitable for use under high humidity conditions, but also the applied substrate for the tensile bond strength. The effect mechanism is also not monotonous. More research is needed to confirm this theory. In practical applications, people should choose polymers according to the climate carefully.

**Acknowledgements** The authors acknowledge the financial support by the National Natural Science Foundation of China (51572196) and Sino-German Center for Research Promotion (GZ 1290).

## References

1. Wang, P. M., Zhang, G. F., & Wu, J. G. (2004). Effect of polymer dry powders on bond strength of cement mortar under different maintenance conditions. *Dry Mortar, 12*, 37–39.
2. Li, F. (2011). Effect of curing condition on the performance of cement-based ceramic tile adhesive and its mechanism. *Guangdong Building Materials, 10*, 97–99.
3. Li, F., Yang, Y. Y., & Wu, J. G. (2004). Effect of curing conditions on mechanical properties of polymer modified cement mortar for bridge deck. *Journal of Building Materials, 12*, 37–39.
4. Ji, Y., & Zhang, J. (2009). Analysis of influence factors on adhesion between polymer modified mortar and EPS board. *Proceedings of the 3rd national conference on commercial mortar*, Wuhan, China, (pp. 73–78).
5. Wang, R., & Wang, P. M. (2009). Physical properties of SBR latex-modified mortar under different curing conditions. *Journal of the Chinese Ceramic Society, 37*, 2118–2123.
6. He, D. H. (2010). Effect of curing regime on bonding strength of EVA modified mortar to tile. *Journal of Wuhan University of Technology, 25*, 346–348.
7. Rashid, K., Zhang, D., Ueda, T., et al. (2016). Investigation on concrete-PCM interface under elevated temperature: At material level and member level. *Construction and Building Materials, 125*, 465–478.
8. Kwon, H. M., Nguyen, T. N., & Le, T. A. (2009). Improvement of the strength of acrylic emulsion polymer-modified mortar in high temperature and high humidity by blast furnace slag. *KSCE Journal of Civil Engineering, 13*, 23–30.
9. Muthadhi, A., & Kothandaraman, S. (2014). Experimental investigations on polymer-modified concrete subjected to elevated temperatures. *Materials and Structures, 47*, 977–986.
10. Chen, C. Y. (2013). *The influences of curing regime on the tensile bond strength of cement mortar*. Tongji University, Shanghai, China.
11. Shou, M. J. (2017). *Influence of temperature and humidity on the tensile bond strength of polymer modified cement mortar*. Tongji University, Shanghai, China.

12. Xia, F. (2015). *Influence of curing conditions and substrates on the tensile bond strength of cement mortar*. Tongji University, Shanghai, China.
13. Du, D. (2017). *Effect of temperature and humidity on performance of cement-based decorative mortar*. Tongji University, Shanghai, China.
14. Xu, Y. D. (2017). *Early performance of styrene-butadiene rubber latex/calcium sulphoaluminate cement composite cementitious material under different conditions of temperature and humidity*. Tongji University, Shanghai, China.

# Chapter 29
# Innovative Polymer-Modified Pervious Concrete

Aly M. Said and Oscar Quiroz

The United States generates a tremendous amount of waste latex paint annually, which is a significant challenge to recycle as it contains volatile organic compounds. However, waste latex paint can be utilized to produce an economic latex-modified pervious concrete that is superior to regular pervious concrete. This study evaluates pervious concrete produced using waste latex paint in comparison to regular pervious concrete as well as that produced using commercially manufactured styrene-butadiene rubber. Waste latex paint added to concrete results in characteristics comparable to polymer-modified concrete made with commercial latex products.

## 1 Introduction

Pervious concrete is a special type of concrete which contains 15–25% air voids. This particularly high void ratio is accomplished by minimizing or eliminating fine aggregate as well as reducing the water-to-cement ($w/c$) ratio. Pervious concrete was initially used in 1852 but has recently adopted in many applications due to its capacity to recharge groundwater, capture rainwater, reduce runoff, and help meet EPA stormwater regulations [1]. It can also contribute to LEED credits in some projects.

The structure of pervious concrete relies on a robust cementitious paste to prevent aggregate separation. Huang et al. [2] studied the ability of polymers to enhance the strength of pervious concrete. Their investigation underlined the importance of

A. M. Said (✉)
The Pennsylvania State University, University Park, PA, USA
e-mail: aly.said@engr.psu.edu

O. Quiroz
Slater Hanifan Group, Las Vegas, Nevada, USA

balancing strength and permeability. In addition to polymers, the investigators created mixtures with sand and fibers. They found improved mechanical properties in terms of compressive and tensile strength for concrete mixtures containing latex and for mixtures containing fibers. Nonetheless, the latex mixture exhibited a decrease of 5–10% in porosity volume. Their findings showed that pervious concrete mixture was capable of providing adequate strength and permeability with the inclusion of polymer and sand. Some studies showed the feasibility of using waste latex paint (WLP) in concrete [3, 4] and may enhance the cementitious paste component in terms of strength.

Pervious concrete was identified as a target application for WLP recycling in concrete for multiple reasons. First, latex, as an additive, has been shown to improve the mechanical properties of pervious concrete. Accordingly, the study aims to identify the strength enhancement from WLP. Second, pervious concrete is increasingly adopted as an aspect of sustainable construction. Lastly, WLP can possibly produce a stronger pervious concrete with a recycled material. A WLP-based pervious concrete mixture should have the capacity to meet or surpass conventional pervious concrete strength characteristics.

## 2 Research Significance

With the enormous amounts of waste latex paint generated in the United States and worldwide, there is a significant need for an approach to recycle it. The major part of this waste paint is left to dry then discarded in landfills. The current investigation proposes a mean to rid the environment of waste latex paint as a pollutant, in a very stable way, while providing an added value to concrete produced.

## 3 Experimental Program

The materials used in this study were provided by regional ready-mix concrete producer and conformed to specifications for pervious concrete production. The binder used was a type V cement (sulfate resistant), which is used throughout the region since sulfates are present in the soil. The cement used conformed to ASTM C150 [5] requirements and had a specific gravity of 3.15. Class F fly ash was used as a cementitious replacement. Fly ash has the ability to enhance sulfate resistance of concrete and is accordingly used locally in concrete slabs on grade and pavements. The fly ash used conformed to the requirements set by ASTM 618 [6] and had a specific gravity of 2.3. Coarse aggregate used in the study had a 2.84 specific gravity, a 0.6% absorption, a 98.6 dry rodded unit weight, and a gradation of No. 67 per AASHTO standards. The paint used in this study was obtained from domestic waste in a closed container. After inquiry, it was indicated by the local hardware store clerk that this is an average product in terms of quality and pricing. The admixtures used

# 29 Innovative Polymer-Modified Pervious Concrete

Table 29.1 Proportions of the studied mixtures

| Mixture | NC | 5% WLP | 10% WLP |
|---|---|---|---|
| Cement, lb/yd³ | 500 | 500 | 500 |
| Fly ash, lb/yd³ | 125 | 125 | 125 |
| Coarse aggregate, lb/yd³ | 2500 | 2380 | 2250 |
| Fibers, lb/yd³ | – | – | – |
| Water, lb/yd³ | 190 | 160 | 130 |
| WLP, lb/yd³ | 0 | 60 | 120 |
| Total water, lb/yd³ | 190 | 190 | 190 |
| Designed air content | 24% | 24% | 24% |
| Solid content of polymer | – | 54% | 54% |
| w/c ratio | 0.30 | 0.30 | 0.30 |
| p/c ratio | – | 0.05 | 0.10 |

Fig. 29.1 View of the fresh unit weight of concrete test

were high-range water reducing admixtures conforming to ASTM C494 [7] and with a density of 8.8 lb/gal. The mix proportions of the studied mixtures are presented in Table 29.1.

A battery of tests was performed on the mixes to assess the improvements imparted by the addition of the waste latex paint to the pervious concrete mixes. These tests include fresh unit weight (Fig. 29.1), compressive strength, a modified freeze-thaw test, porosity, and usable void percentage.

The compressive strength test of concrete cylinders was conducted in accordance with ASTM C39 standards. Due to the amount of voids in pervious concrete, a modified ASTM C666 [8] freeze-thaw testing was conducted by using the mass loss as the indicator for freeze-thaw resistance instead of the typical ultrasonic wave measurements. Finally, the porosity of the cylindrical specimens was assessed by evaluating the weight of water that a typical $4'' \times 8''$ cylinder can hold, whereas the percentage of useable voids was determined as the volume of that water divided by the volume of the entire cylinder.

## 4 Results and Discussion

Results show that normal concrete and 5% WLP experienced an increase in unit weight from the fresh to the hardened density, whereas the 10% WLP exhibited a decrease in hardened unit weight. The reduction in hardened unit weight may be a sign of the weakness of the cementitious paste connecting the aggregate. The weaker paste can permit aggregate to be lost between fresh density and hardened density measurements. Tennis et al. [1] indicate that typically hardened densities of pervious concrete should be within the range of 100 lb/ft$^3$ to 125 lb/ft$^3$, which was accomplished by the tested mixtures.

The tested compressive strength of the WLP mixtures at 28-days displayed a reduction with increasing percentage of latex paint addition as illustrated in Fig. 29.2. Increasing the WLP content from 5% to 10% resulted in a decrease in the 7-day and 28-day compressive strength considerably. The incorporation of 5% WLP resulted in a decrease in compressive strength for 7-day and 28-day testing by less than 10% with reference to normal concrete. Regarding the modified freeze-thaw resistance evaluation, the mixture containing 5% WLP was not deemed failing until after 250 cycles. It is noteworthy that normal concrete exhibited the best resistance to freezing and thawing performance.

Results also indicated that pervious concrete exhibited an increase in porosity with the increase of waste latex paint content. Such result was expected, given the decrease in unit weight discussed earlier. Regarding the usable void percentage, an increase in void content was associated with the increase in WLP content as shown in Fig. 29.3. Normal concrete has usable void percentage that falls between the 5% WLP and 10% WLP content.

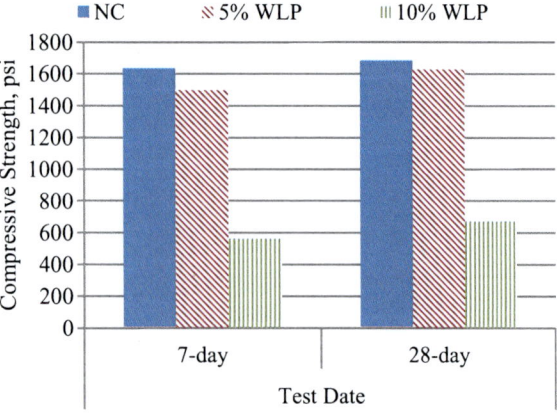

**Fig. 29.2** Compressive strength testing of the studied pervious concrete mixtures at 7 and 28 days

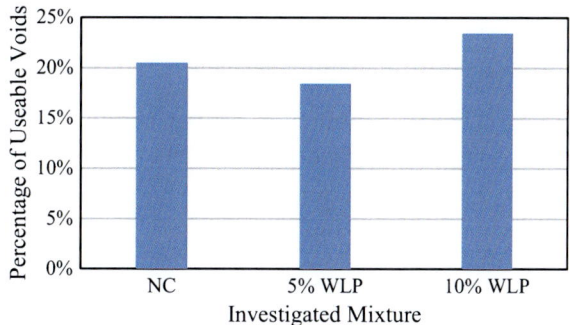

**Fig. 29.3** Percentage of useable voids for the studied mixtures

## 5 Conclusions

The presented study showed that the use of waste latex paint in the production of pervious concrete is feasible. The addition on a 10% of waste latex paint can positively enhance the porosity of the pervious concrete. However, maintaining the amount of waste latex paint at 5% minimizes the impact of the paint on the compressive strength of the tested pervious concrete mixtures in terms of porosity, strength, and freeze-thaw resistance.

## References

1. Tennis, P. D., Leming, M. L., & Akers, D. J. (2004). *Pervious concrete pavements*. Skokie: Portland Cement Association and National Ready Mixed Concrete Association.
2. Huang, B., Wu, H., Shu, X., & Burdette, E. G. (2005). Laboratory evaluation of permeability and strength of polymer-modified pervious concrete. *Construction and Building Materials, 24*(5), 818–823.
3. Quiroz, O. I., & Said, A. (2010). Properties of Economical Latex-Modified Concrete Using Recycled Paint. Special Publication on Frontiers in the Use of Polymers in Concrete, 16p.
4. Said, A. M., Quiroz, O. I., Hatchett, D. W., & Elgawady, M. (2016). Latex-modified concrete overlays using waste paint. *Construction and Building Materials, 123*, 191–197.
5. ASTM C150. (2011). *Standard specification for Portland cement*. West Conshohocken: American Society for Testing and Materials.
6. ASTM C618. (2008). *Standard specification for coal fly ash and raw or calcined natural pozzolan for use in concrete*. West Conshohocken: American Society for Testing and Materials.
7. ASTM C494. (2010). *Chemical admixtures in concrete*. West Conshohocken: American Society for Testing and Materials.
8. ASTM C666. (2008). *Standard test method for resistance of concrete to rapid freezing and thawing*. West Conshohocken: American Society for Testing and Materials.

# Chapter 30
# Combined Methods to Investigate the Crack-Bridging Ability of Waterproofing Membranes

**Marius Waldvogel, Roger Zurbriggen, Alfons Berger, and Marco Herwegh**

Polymer modification of dry mortars with redispersible polymer powders is a common approach to increase the mortar's mechanical properties. The use of polymer-modified mortars (PMMs) as waterproofing membranes requires a well-performing crack-bridging ability (CBA) to avoid infiltration of water into the construction owing to shrinkage or expansion. In order to evaluate the performance of PMMs, we present below a combination of crack-bridging tests with optical analysis. The latter is exerted with digital image analysis on photos acquired during the tensile tests and pictures of CBA-samples, which have been fixated after a certain displacement. Additionally, thin sections of fixated CBA-samples are investigated using transmitted light, UV light, and SEM microscopy. These methods are developed to understand the (micro)mechanical initiation and evolution of cracks in order to improve the crack-bridging ability of waterproofing membranes.

## 1 Introduction

Polymer-modified mortars (PMMs) are important products in the building industry as they have increasing tensile and flexural strength [1–3] as well as adhesion [1–3] and show the ability to bridge (micro)cracks [3] which formed during dimensional changes (e.g., shrinkage, expansion) [4]. Due to their crack-bridging ability (CBA), protective coatings of concrete substrates are often polymer-based [4, 5]. Taking advantage from the CBA of polymers as protective parameter [4, 5], PMMs can be

---

M. Waldvogel (✉) · A. Berger · M. Herwegh
Institute of Geological Sciences, University of Bern, Bern, Switzerland
e-mail: marius.waldvogel@geo.unibe.ch

R. Zurbriggen
RD&I Department, Akzo Nobel Chemicals AG, Sempach Station, Switzerland

used as waterproofing membranes. Investigations of mechanical properties of polymer-modified mortars are often narrowed down to specific relations such as strength to polymer-cement ratio [2, 3], stress-temperature [3], stress-strain [4, 6], or stress-displacement relations [6]. In order to gain a general understanding of the (micro)mechanical behavior and CBA of PMMs used as waterproofing membrane, a combination of established methods with suggested new approaches will be presented in the following. Finally, numerical models of the CBA of waterproofing membranes are verified with the results.

## 2 Methods

The basic *composition* of the investigated waterproofing membrane consists of a one-component [1] dry mix composed of quartz sand (0.1–0.3 mm grain size), ordinary Portland cement (OPC, CEM I 52.5 N), and a redispersible polymer powder (RPP; on the base of a vinyl acetate/ethylene copolymer). The composition is kept simple to determine the influence of the RPP on the CBA of the waterproofing membrane. Therefore, the RPP content is increased systematically (*T1*). Simultaneously, the cement-to-water ratio is varied to maintain a similar viscosity and workability [6, 7] (*T1*), parameters which are fundamental for the application on the construction site. Due to the amount of water needed for redispersion of the polymer powder, water content raises with increasing RPP content from 17 wt.% to 22 wt.% (*T1*). Exceptions are given for 3 wt.% and 5 wt.% RPP samples which need higher amounts of water to be workable (*T1*) (Table 30.1).

As the CBA is not only investigated as a function of RPP content but also of temperature, waterproofing membranes of fixed RPP content are produced and tested in 5 °C intervals from +20 °C down to −20 °C (*T2*) (Table 30.2).

In order to gain as many information about the CBA as possible, the waterproofing membrane is applied in two geometries: Dog-bone geometry (hereafter referred

**Table 30.1** Composition of RPP, OPC, quartz sand, and water (all in wt. %) in waterproofing membranes of varying RPP content

| Component\wt. % | *C1* | *C2* | *C3* | *C4* | *C5* | *C6* | *C7* | *C8* |
|---|---|---|---|---|---|---|---|---|
| RPP | 3 | 5 | 10 | 15 | 20 | 25 | 30 | 35 |
| OPC | 19 | 19 | 19 | 19 | 19 | 19 | 19 | 19 |
| Quartz sand | 78 | 76 | 71 | 66 | 61 | 56 | 51 | 46 |
| Water | 17 | 17 | 16 | 16 | 17 | 19 | 20 | 22 |

**Table 30.2** Temperatures used during uniaxial testing of waterproofing membranes with fixed RPP content

|  | *T1* | *T2* | *T3* | *T4* | *T5* | *T6* | *T7* | *T8* | *T9* | *T10* |
|---|---|---|---|---|---|---|---|---|---|---|
| RPP [wt. %] | 25 | 25 | 25 | 25 | 25 | 25 | 25 | 25 | 25 | 25 |
| Temperature [°C] | 23 | 20 | 15 | 10 | 5 | 0 | −5 | −10 | −15 | −20 |

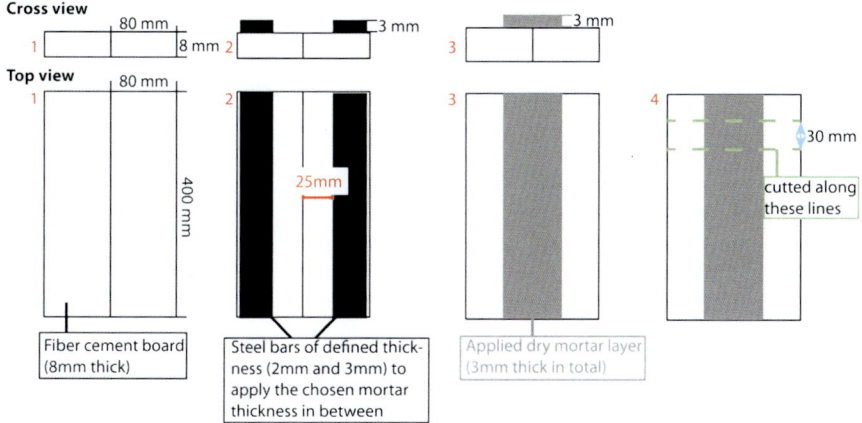

**Fig. 30.1** Sketch of application of waterproofing membrane on substrate

to as dog-bone) and crack-bridging ability sample geometry (Fig. 30.1; hereafter referred to as CBA-sample).

*Dog-bone* geometry is produced by creating a 3 mm thick layer of the waterproofing membrane, which is dried at 23 °C and 50% r.h. for 28 days. Later dog-bones of defined dimensions are punched out [8, type 5]. With the dog-bone geometry, mechanical properties are determined by tensile testing.

*CBA-samples* are produced by covering two fiber cement boards (each 400 mm × 80 mm × 8 mm), juxtaposed along their long edge to form a pre-crack in the substrate, with a 50-mm-wide and 400-mm-long layer of the waterproofing membrane of a total thickness of 3 mm. This layer is applied in two steps: 2 mm thick layer followed by 1 mm thick one with a resting time of 5 h between the two applications. After a drying period of 24 h, the same procedure is done on the back side. After that, the prepared plates are dried at 23 °C and 50% r.h. for 28 days. Prior to testing the individual CBA-samples are produced by cutting 30-mm-wide strips parallel to the short edge (Fig. 30.1). The width and thickness of the waterproofing membrane are specified by an EU norm [9]. With the CBA-sample geometry, the CBA of the waterproofing membrane is tested.

*Uniaxial tensile testing (UTT)* is performed using a static materials testing machine (Z020/TH2S by Zwick/Roell). For temperature-dependent measurement, the machine is encased by a heating/cooling system (RS-Simulatoren TEE65/40X). Besides the mandatory temperature steps to test the CBA [9], dog-bones and CBA-samples are tested in additional steps (*T2*) for a more detailed investigation of the temperature-dependent deformation behavior. Testing speed is 100 mm/min for dog-bones [8] and 0.15 mm/min for CBA-samples [9].

*Time-lapse photos* are shot during CBA-sample testing in a fixed time interval of 4 seconds. With the given speed of 0.15 mm/min, the pictures represent continuous displacement increments of 10 μm.

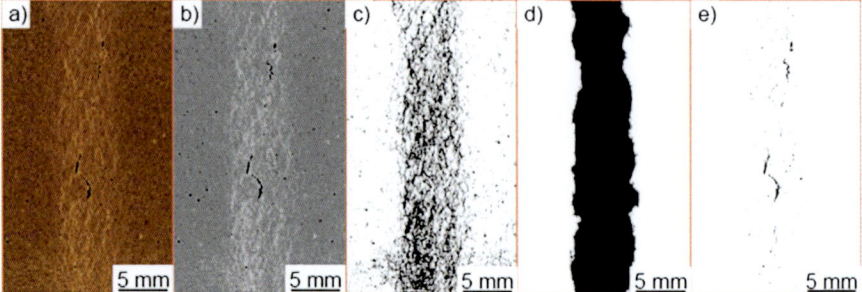

**Fig. 30.2** Steps of image analysis ranging from (**a**) acquiring a color picture and (**b**) its reduction first to a gray scale and then to (**c**) a binary picture of strain whitening, (**d**) the area of strain whitening, and (**e**) the location of cracks and holes

Individual CBA-samples are stopped and fixated at specific displacements for further microstructural investigations and their development as function of displacement.

*Orthographic pictures* are taken of the surface of the fixated deformed CBA-samples for a subsequent image analysis.

With a *binocular* the surfaces of fixated deformed CBA-samples are investigated. Furthermore, the continuation of the cracks and holes into depth can be traced by varying the focus level. With an attached camera, pictures can be taken for image analysis.

*Transmitted light* and *UV-light microscopy* is done on thin sections of fixated, deformed CBA-samples. With an attached camera, photos and mosaics of (parts of) the thin sections can be taken.

*SEM microscopy* is used to investigate thin sections of fixated, deformed CBA-samples and dog-bone surfaces with SE, BSE, and EDX signals. Surfaces from the dog-bones are the rupture surface created by the tensile test and created non-ruptured surfaces by cutting and ion polishing out of the dog-bones. The investigations made by the SEM can be documented with an attached camera.

The photos acquired as described above are processed and analyzed using ImageJ 1.51j8. The processing depends on the purpose of the analysis. Figure 30.2c, d shows the processing result for a later analysis of the stress whitening, while Fig. 30.2e is processed to analyze the cracks and holes.

## 3 Results

*Uniaxial testing* provides stress-strain data for the dog-bone geometry. These data are used to assign material properties in numerical models. The load-displacement data gained from crack-bridging tests are the basis for a qualitative and quantitative evaluation of the numerical models.

*Image analysis*, exerted on orthographic and time-lapse pictures, determines the initiation and evolution of strain whitening depending on displacement. Furthermore, these pictures reveal data on the size, orientation, and geometrical descriptors of holes and cracks providing fundamentals to understand the evolution from holes to cracks.

*Time-lapse photos* are evaluated after image processing (Fig. 30.2). First, the width of the strain localization zone and its evolution can be mapped with increasing displacement. Furthermore, the high temporal and spatial evolution of the time-lapse photos allows unraveling precisely the initiation of both holes and cracks and their changes in space and time.

Using the *binocular* reveals information about the 3D continuation of holes and cracks into CBA-sample depth. Additionally, the interaction of crack tips with holes or RPP/cement matrix on the surface is examinable providing important insights to understand crack growth.

*Optical and UV microscopy* of thin sections visualizes crack initiation, crack growth, and paths of crack growth in the vertical section to the lateral observations made by the prior mentioned tools. Besides that, pore geometry, pore size, and porosity can be mapped and analyzed.

*SEM microscopy* on thin sections reveals details about possible adhesive failure between waterproofing membrane and substrate. Furthermore, it leads to a high-resolution investigation of the co-matrix by revealing the hydration stage of the cement particle and possible lateral or vertical changes in the composition of the co-matrix. Exerted on dog-bones, *SEM microscopy* reveals insights in the behavior of the RPP film during deformation by investigating the rupture surface, which can be compared to a non-deformed surface.

## 4 Discussion

Previous studies applied different mechanical tests to either investigate the initiation of cracks in protective coatings [4, 5], the strength of polymer-modified mortars [6, 7], or adhesion strength [1, 3] of polymer-modified mortars. The combination of these tests with (time-lapse) photography of the CBA-sample's surface transmitted light, UV light, and SEM microscopy on thin sections, and processing of these pictures with image analysis adds high-resolution spatial information on the onset and evolution of strain localization and cracking. The deformation behavior of the polymer matrix can be traced by comparing undeformed and deformed areas of dog-bone samples using SEM microscopy, which was generally used so far to either investigate polymer film formation [1, 10], polymer film-cement particle interaction [10, 11], or rupture surfaces [7]. In this way continuous spatial information on the evolution of microcracks and their propagation to macrocracks is available, which can be correlated with the bulk mechanical properties.

## 5 Conclusion and Outlook

In this study, the methods are refined to monitor and study the displacement-driven deformation processes as well as associated structures over the whole experiment. In this way, we gain new insights into the crack-bridging behavior of waterproofing membranes (Fig. 30.2c, d) as well as the initiation and propagation of individual crack tips. Such information is crucial to detect the crack-bridging (micro)mechanisms within the waterproofing membranes, which is an important step in order to enhance CBA in future products. First results show an increase in the CBA with increasing RPP content [3] and increasing temperature [12] due to the capability to accommodate an enhanced amount of deformation by elastoplastic deformation prior to the onset of macroscopic fracturing. Furthermore, the width of strain whitening increases nonlinearly with increasing displacement showing a strong initial strain whitening and a decreasing widening with increasing displacement.

**Acknowledgments** Josef Kaufmann, Frank Winnefeld, and Nikolajs Toropovs from EMPA are acknowledged for fruitful discussions. We thank Sebastian Dettmar for his careful sample preparation. CTI is gratefully thanked for financial support of project No. 18797.1 PFIW-IW.

## References

1. Schulze, J., & Killermann, O. (2001). Long-term performance of redispersible powders in mortars. *Cement and Concrete Research, 31*, 357–362.
2. Ohama, Y. (1998). Polymer-based admixtures. *Cement and Concrete Composites, 31*, 189–212.
3. Ohama, Y. (1995). *Handbook of polymer-modified concrete and mortars properties and process technology* (p. 236). Noyes Publications, New Jersey, U.S.A.
4. Delucchi, M., Barbucci, A., Temtchenko, T., Poggio, T., & Cerisola, G. (2002). Study of the crack-bridging ability of organic coatings for concrete: Analysis of the mechanical behaviour of unsupported and supported films. *Progress in Organic Coatings, 44*, 261–269.
5. Delucchi, M., Barbucci, A., & Cerisola, G. (2004). Crack-bridging ability of organic coatings for concrete: Influence of the method of concrete cracking, thickness and nature of the coating. *Progress in Organic Coatings, 49*, 336–341.
6. Pascal, S., Alliche, A., & Pilvin, P. (2004). Mechanical behaviour of polymer modified mortars. *Materials Science and Engineering A, 380*, 1–8.
7. Bureau, L., Alliche, A., Pilvin, P., & Pascal, S. (2001). Mechanical characterization of a styrene – Butadiene modified mortar. *Materials Science and Engineering A, 308*, 233–240.
8. European Committee for Standardization; EN ISO 527-1-3; Plastics - Determination of tensile properties Parts 1-3; German version EN IS0 527-1-3; 1996; Brussels.
9. European Committee for Standardization; FprEN 14891 - Liquid-applied water impermeable products for use beneath ceramic tiling bonded with adhesives - Requirements, test methods, evaluation of conformity, classification and designation; 2011; Brussels.
10. Afridi, M. U. K., Ohama, Y., Demura, K., & Iqbal, M. Z. (2003). Development of polymer films by the coalescence of polymer particles in powdered and aqueous polymer-modified mortars. *Cement and Concrete Research, 33*, 1715–1721.
11. Ollitrault-Fichet, R., Gauthier, C., Clamen, G., & Boch, P. (1998). Microstructural aspects in a polymer-modified cement. *Cement and Concrete Research, 28*, 1687–1693.
12. Callister, W. D., Jr. (2007). *Materials science and engineering an introduction* (p. 722). Wiley, New York, U.S.A.

# Chapter 31
# Polymer Concrete for a Modular Construction System: Investigation of Mechanical Properties and Bond Behaviour by Means of X-Ray CT

**Franziska Vogt, Alexander Gypser, Florian Kleiner, and Andrea Osburg**

In the presented paper, the results of a BMWi-supported project of applied research are reported. The main objective of this work is the development of a system of PC elements for the application as roof or floor plates. Based on the determination of the creep behaviour of the PC at different temperatures, reinforcement techniques for the PC elements were developed. As reinforcement, materials from renewable resources such as bamboo, Tonkin cane and Chinese silvergrass were tested. The challenge was to ensure a sufficient residual loading capacity of the elements despite the already high bending tensile strength of the PC. It was found that especially Tonkin cane performed sufficiently well as reinforcement with regard to loading capacity. However, X-ray CT investigations revealed insufficient bond between reinforcement and PC especially within the vicinity of the bearings, which means the anchoring has to be improved further.

## 1 Introduction

In earlier studies [1, 2], a polyester resin-based PC for a modular construction system for low-cost housing was developed together with the research partner PolyCare Research Technology GmbH. In the developed PC, desert sands have been used as fillers. Desert sand is a locally available resource which hitherto has not been considered suitable for the use in construction due to its specific properties (e.g. particle shape, particle size distribution, chemical composition and impurities, organic components). It could be shown that the PC elements containing desert sand fulfil all requirements concerning structural stability and durability.

---

F. Vogt (✉) · A. Gypser · F. Kleiner · A. Osburg
Bauhaus-Universität Weimar, F. A. Finger Institute for Building Material Engineering (FIB), Weimar, Germany
e-mail: franziska.vogt@uni-weimar.de

**Table 31.1** Formulation and selected properties of the PC at 20 °C

| Binder | | | PC | |
|---|---|---|---|---|
| Formulation | | Content in PC M.-% | Compressive strength N/mm$^2$ | Bending tensile strength N/mm$^2$ |
| Resin | UP | 13.35 | 108 | 29 |
| Initiator | | 1.70%[a] | | |
| Accelerator | | 1.35%[a] | | |

[a]In relation to the resin content

The focus of this study of applied research is the development of a technology to fabricate self-supporting PC construction elements using natural materials as reinforcement. The aim of the application of reinforcement is to ensure a certain residual load-bearing capacity of the elements in case of failure of the PC matrix. This guarantees multifunctionality of the PC elements and enables also their application in regions with increased risk of earthquakes.

## 2 Materials

### 2.1 Polymer Concrete

The binder of the PC is a medium reactive orthophthalic polyester resin which was cured by copolymerisation with styrene. Methyl ethyl ketone peroxide was used as accelerator; the organic peroxide was combined with a cobalt accelerator to start the polymerisation at room temperature [1].

Originally the PC formulation was based on desert sands as aggregates. Due to difficulties with the acquisition of the material for research purposes (e.g. customs regulations), natural sand from the Upper Rhine Plain was used instead (Table 31.1).

### 2.2 Reinforcement Materials

In relation to its compressive strength, PC has a considerably higher tensile bending strength than conventional cement concrete. Therefore, the reinforcement materials need to have a very high tensile strength in order to be able to sustain the occurring tensile force in case of failure of the PC matrix under bending stress.

As reinforcement materials for the PC elements, natural materials such as bamboo and Tonkin cane, hemp rope and Chinese silvergrass were used. Natural materials as reinforcement were applied as they are locally available renewable resource and thus contribute to the sustainability of the modular construction system. The said materials were tested for their suitability as reinforcement in polymer concrete by preliminary tests. Aspects such as sufficient straightness, comparable cross-sectional

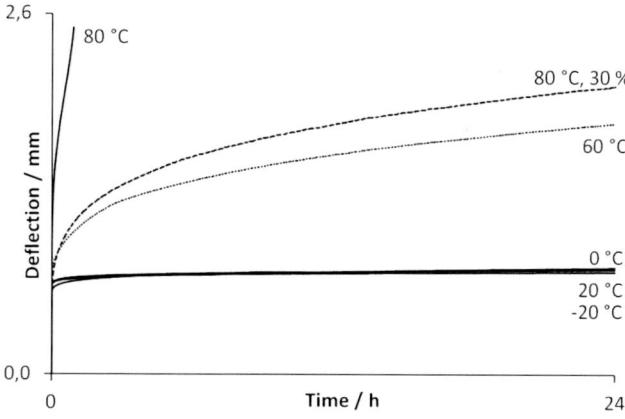

**Fig. 31.1** Deflection of the PC under three-point loading

area with the use of several reinforcing elements and elimination of possible adhesions that interfere with the bonding to the polymer concrete were the main focus.

## 3 PC: Determination of Creep Behaviour

The short-term flexural creep behaviour of the PC was tested by three-point loading in accordance with DIN EN ISO 899-2 at different temperatures. Unreinforced PC samples (420 mm × 60 mm × 25 mm) were subjected to a constant test load over a period of 24 h. The test load was defined as 50% of the mean value of the breaking load during short-term fracture test at the respective temperature. Test temperatures were in the range of −20 °C to 80 °C. Over the entire test period, the deformations of the prisms were recorded continuously between the supports with a displacement transducer.

The diagram in Fig. 31.1 shows the examined deflection of the PC samples at respective temperatures. The applied test force and the mean values of the deflections after 1 h and after 24 h are listed in Table 31.2.

In Fig. 31.1 and Table 31.2, it can be seen that the short-term creep of the polymer concrete depends to a great extent on the applied load and test temperature. At 20 °C and a continuous load of about 860 N, the deflection after 1 h was about 0.66 mm and increased only slowly to a value of 0.77 mm after 24 h. Similar results of deflection were obtained at 0 and −20 °C although the applied test load was higher. At 60 °C, the test load was reduced by about 30%, but the deformation after 1 h increased by more than 50% compared to the test at room temperature. Furthermore, a significantly steeper increase in the curve was observed at 60 °C over 24 h, and the deflection increased by more than 70%. The greatest difference was found at 80 °C. Although the test force was reduced by almost half, the deflection increased almost

**Table 31.2** Results of three-point loading test

| Test temperature °C | Test load N | Deflection after 1 h mm | Deflection after 24 h mm | Change of deflection % |
|---|---|---|---|---|
| −20 | 1048 | 0.69 | 0.74 | 7 |
| 0 | 899 | 0.70 | 0.76 | 9 |
| 20 | 864 | 0.66 | 0.77 | 17 |
| 60 | 621 | 1.04 | 1.81 | 74 |
| 80 | 490 | 2.42 | – | – |
| 80[a] | 294 | 1.11 | 2.08 | 87 |

[a]With test load defined as 30% of the mean value of the breaking load during short-term fracture test

fourfold compared to the test at room temperature. All three prisms were broken at 80 °C and 50% of the test load of the short-term fracture test after a little over an hour under this load. A second test at 80 °C with reduced test load (30% of the mean value of the breaking load during short-term fracture test) was performed. It could be shown that the PC can withstand a certain amount of load at 80 °C. Nevertheless, in order to ensure a required load-bearing capacity of the PC elements also at higher temperatures, certain measures have to be taken. One possibility is the application of reinforcement.

## 4 Flexural Tensile Strength of Reinforced PC Elements

The flexural tensile strength and the deformation behaviour of polymer concrete with reinforcement were examined at 20 °C on test specimens with the dimensions 700 mm × 150 mm × 35 mm; in case of bamboo cane, the depth of the samples was increased to 70 mm. All reinforcement materials were installed in such a way that the largest possible effective depth is achieved. The results are plotted in Fig. 31.2. It can be seen that the main bearing capacity of the samples is provided by the PC matrix ("fracture of matrix"). The generally lower bending tensile strength of the PC matrix of the samples with reinforcement can be explained with the reduction of the cross-sectional area of the PC matrix due to insertion of the reinforcement.

The load was removed immediately after the failure of the matrix, and the samples were loaded again in order to determine the residual load-bearing capacity of the reinforcement ("fracture of reinforcement"). The best results were achieved with bamboo and Tonkin canes where the remaining load-bearing capacity of the reinforcement was equal to or higher than that of the PC matrix. Chinese silvergrass proved to be not suitable because of insufficient bonding due to the alternating leaves on the stems. Tonkin canes can on one hand be used to increase the load-bearing capacity of PC elements at higher temperatures (above 20 °C), and on the other hand, the material PC can be saved with regard to sustainability.

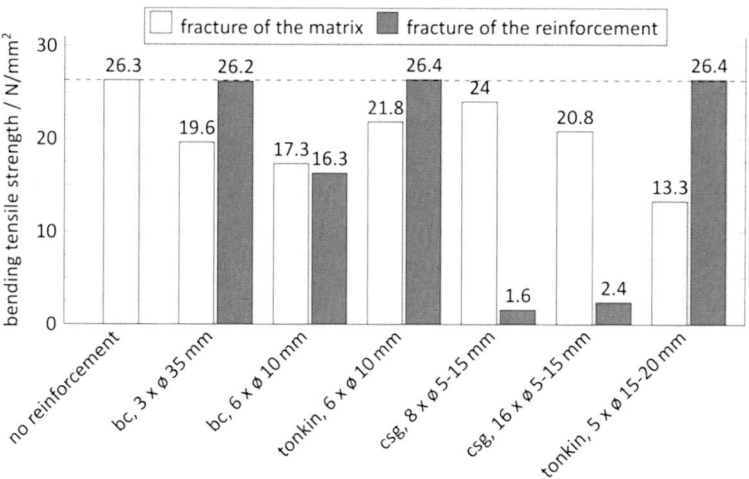

**Fig. 31.2** Results of bending tensile tests with reinforcement (selection); bc bamboo cane, csg Chinese silvergrass

## 5 Reinforced PC Elements: Investigation of Bond Behaviour by Means of X-Ray CT

X-ray computer tomography (X-ray CT) is a fairly recently established method for the non-destructive investigation of the microstructure of building materials. In this study it was used to visualise and investigate the bond between the reinforcement materials and the PC as well as the development of cracks due to loading. As the object size for CT scans is limited, samples were cut from the test specimens used for the determination of the flexural tensile strength. These samples of reinforced PC elements were scanned in a "GE Phoenix nanotom m research|edition" X-ray CT at 160 kV and 315 μA. The focal spot size was approx. 2 μm. The acquisition and reconstruction software allows for the correction of ring artefacts as well as of artefacts due to random movement of the sample and Feldkamp artefacts. After reconstruction the 2D cross sections were analysed with 3D-software "VG Studio Max 3.0" [3]. The challenge of scanning multiphase materials such as PC is the combination of low- and high-absorbing materials. High voltage is needed to penetrate the high-absorbing phase (PC filler) fully, which is disadvantageous for the contrast of the CT images. Low-absorbing materials (resin matrix, bamboo cane) become less distinguishable.

In Fig. 31.3 a CT-slice and a 3D image of a PC element with bamboo reinforcement are shown. High-absorbing materials, i.e. materials with higher densities, are displayed in lighter greyscale values, low-absorbing materials appear darker and air voids are black. In the CT-slice image on the left-hand side, the crack appearing in the PC matrix due to loading continues around the bamboo cane's surface and again

**Fig. 31.3** Greyscale image of CT-slice (left) and 3D image (right) of PC element with bamboo reinforcement

into the PC matrix. Debonding of PC and bamboo cane is clearly visible, which means that the bond between the PC matrix and the reinforcement is not sufficient.

The samples were taken in the vicinity of the bearing. Debonding of the reinforcement in that area typically means that the anchoring of the reinforcement bars in the PC element is inadequate. Although the bending tensile strength of the PC matrix and the tensile strength of the reinforcement are sufficient for the intended purposes, the bond between the two components needs to be improved in order to increase the bearing capacity of the PC elements further.

## 6 Summary

The results of the study show that materials from renewable resources can be used as reinforcement in PC construction elements in order to ensure certain residual load-bearing capacity in case of failure of the PC matrix. Tonkin and bamboo cane proved to be suitable. Installation of the reinforcement and especially anchoring of the same still require optimization, which is the subject of our current research.

## References

1. Osburg, A., Gypser, A., & Ulrich, M. (2015). Development of polymer concrete with non-standardised fillers for innovative building materials. In *Advanced materials research* (Vol. 1129, pp. 484–491). Switzerland: Trans Tech Publications.
2. Dimmig-Osburg, A. (2014). Polymerbeton aus Wüstensand – Ein Projekt der angewandten Forschung, polymer concrete produced with desert sand – A project of applied research. *Restoration of Buildings and Monuments, 20*(5), 361–370. ISSN 1864-7251.
3. Vogt, F., Hadlich, C., & Schirmer, U. (2017). Visualisation and verification of polymers in cementitious materials using computer tomography – A case study. *3rd international conference on tomography of materials and structures*, 26.–30.06.2017, Lund.

# Chapter 32
# Bending and Crack Characteristics of Polymer Lattice-Reinforced Mortar

Brian Salazar, Ian Williams, Parham Aghdasi, Claudia Ostertag, and Hayden Taylor

As the construction industry moves toward precast structures, a more mechanically robust concrete is required. To provide joints with higher mechanical toughness, we propose reinforcing the mortar with a polymeric lattice. This study considers the mechanical effects of reinforcing concrete with a polymeric lattice and compares this new technique against the standard fiber-reinforced method. We investigate the octet lattice, which is notable for its high specific strength. Lattices with 23.4 mm unit cells were prototyped out of polylactic acid using a thermoplastic extrusion-based 3D printer. These lattices were placed in a rectangular mold, infiltrated with concrete, and vibrated. The resulting specimens were tested for flexural strength in four-point bending on a hydraulic testing machine. Bending test results show that the lattice-reinforced beams achieve a net deflection at peak load that is 2.5 times greater than that of the fiber-reinforced beams. Further, the lattice-reinforced beams obtain toughness and peak load values comparable to the fiber-reinforced beams while allowing for easier processing. Fabrication of these lattice structures for use in construction can readily be scaled, as the polymeric lattices can be manufactured through injection molding.

B. Salazar (✉) · H. Taylor
University of California, Berkeley, Department of Mechanical Engineering, Berkeley, CA, USA
e-mail: brian10salazar@berkeley.edu

I. Williams · P. Aghdasi · C. Ostertag
University of California, Berkeley, Department of Civil Engineering, Berkeley, CA, USA

**Fig. 32.1** A polymeric octet lattice beam for use as concrete reinforcement

# 1 Introduction

In 1961, Buckminster Fuller patented his vision for building walls, roofs, and floors constructed out of struts in an octahedron-tetrahedron arrangement [1]. Since then, the octet lattice (shown in Fig. 32.1) has been experimentally shown to have a high specific strength [2]. Other research has focused on the optimizing the octet shape for bending performance [3], and Chiras has found that members in metallic octet lattices fail due to material yielding (in tension) or buckling (in compression) [4]. O'Masta's experiments with metallic lattices show that, after a critical member thickness, the octet's fracture toughness increases as the member diameter increases [5].

Polymer fibers are usually added to concrete beams for improved crack resistance. However, the amount of polymeric fibers that can be incorporated into a mortar is limited to about 5% by volume due to practical reasons. We demonstrate the ability to use lattice reinforcement with a polymeric volume fraction up to 30.5% and compare the lattice reinforcement with randomly dispersed polymer-fiber reinforcement and unreinforced beams.

# 2 Materials and Methods

We 3D-printed polymeric lattices out of polylactic acid (PLA) on a Type A Pro fused-deposition modeling machine using a 0.4 mm nozzle and 0.2 mm layer height. Though the octet lattice is very complex, its optimal angles and size allowed it to be

**Table 32.1** Mortar mix weight proportions for the two mortars

| Reinforcement type | Cement | Fly ash | Water | Fine aggregates | Superplasticizer | Fibers |
|---|---|---|---|---|---|---|
| Plain and lattices | 1.00 | 0.18 | 0.60 | 2.05 | 0.0024 | 0 |
| 8 mm PVA fibers | 1.00 | 0.10 | 0.55 | 1.01 | 0.0018 | 0.05 |

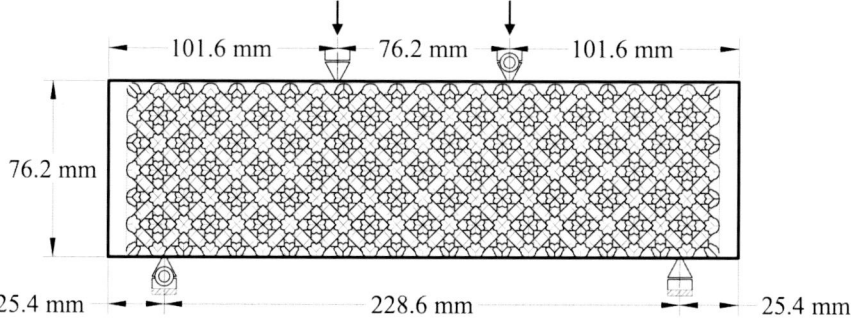

**Fig. 32.2** Four-point beam bending testing setup, with the internal lattice visible

printed without support material. We printed two lattices with a 3.175 mm member diameter and three lattices with a 4.62 mm member diameter. Variation in the 3D printing process resulted in up to a 2.5% weight variation in lattices of a given member diameter.

High flowability is required for the concrete to infiltrate the dense polymer lattices. This is achieved by using a high water-to-cementitious materials ratio of 50% and a high-range, water-reducing admixture (ADVA Flex superplasticizer). The mortar mix (shown in Table 32.1) replaces 15% of the portland cement with ASTM C618 Class F fly ash [6] and contains fine aggregates (P.W. Gillibrand #90 Silver Sand, SG = 2.6). Once mixed, the mortar reached a flow diameter of 25 mm, based on the ASTM C1437-07 flow test [7]. A second mortar, shown in Table 32.1, was made and included 8 mm polyvinyl alcohol (PVA) fibers at a 2.88% fiber volume fraction. This mix was so viscous that it was not possible to perform the ASTM C1437-07 flow test [7].

The lattice reinforcement was placed into 76.2 mm × 76.2 mm × 279.4 mm rectangular beam molds. The mortar mixture was then poured into the molds and vibrated until it completely infiltrated the lattices. To make the unreinforced beams, the mortar was simply poured into the mold and vibrated; the fiber-reinforced beams were made in a similar fashion, accompanied by rodding. The beams were stored in a fog room for 7 days.

We tested the beams in four-point bending, as shown in Fig. 32.2, on a Universal Testing Machine (Baldwin Southwark Tate-Emery Testing Machine) in accordance with ASTM C78M [8]. We used LabVIEW for data capture from the force sensor and two linear variable differential transformers (LVDTs). One LVDT was placed on the beam's front side (shown in Fig. 32.2), and the other LVDT was placed on the back; the displacement was taken as the average of the LVDT readouts.

## 3 Results and Discussion

Whereas the unreinforced beams revealed a flat fracture surface, the lattice-reinforced beams developed the textured fracture surfaces shown in Fig. 32.3. Mortar beams with polymeric lattices formed multiple cracks. The crack propagation was restricted, as shown in Fig. 32.4, by the polymeric lattice.

We analyzed the test data in accordance with ASTM 1609 [9]; toughness is defined as the area under the load-displacement curve up until a displacement of $L/150$, where $L$ is the beam span. We used MATLAB's "trapz" function to calculate the toughness for our experiments, where $L = 228.6$ mm. The lattice-reinforced beams exhibit multiple peaks (shown in Fig. 32.5), where each peak corresponds to a

**Fig. 32.3** Fracture surfaces of plain mortar, fiber, and lattice-reinforced beams

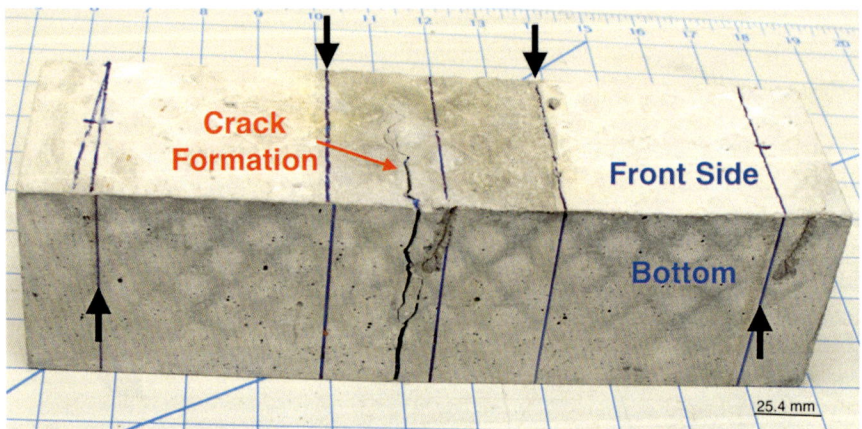

**Fig. 32.4** Crack propagating through a PLA-reinforced mortar beam

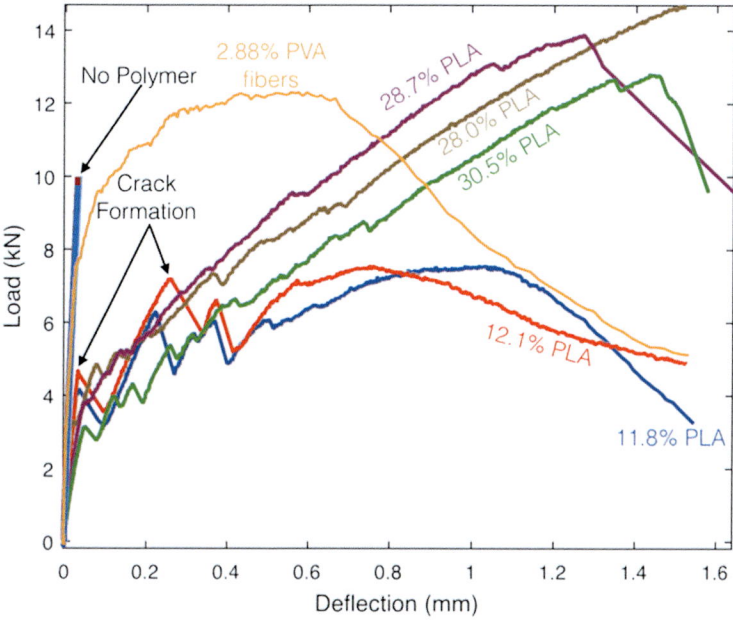

**Fig. 32.5** Force vs deflection profiles for plain, fiber, and lattice beams

**Table 32.2** Material characteristics of tested beams

| Beam description | Toughness | First peak load | Peak load | Deflection at peak load |
|---|---|---|---|---|
| No polymer | 0.185 J | 10.0 kN | 10.0 kN | 0.0345 mm |
| No polymer | 0.166 J | 10.0 kN | 10.0 kN | 0.0293 mm |
| No polymer | 0.158 J | 9.74 kN | 9.74 kN | 0.0298 mm |
| 2.88% PVA fibers | 14.3 J | 6.33 kN | 12.3 kN | 0.557 mm |
| 11.8% PLA lattice | 9.07 J | 4.17 kN | 7.56 kN | 0.985 mm |
| 12.1% PLA lattice | 9.29 J | 4.67 kN | 7.56 kN | 0.750 mm |
| 28.0% PLA lattice | 14.9 J | 3.31 kN | 14.6 kN | 1.52 mm |
| 28.7% PLA lattice | 16.4 J | 5.25 kN | 13.9 kN | 1.27 mm |
| 30.5% PLA lattice | 13.6 J | 3.15 kN | 12.8 kN | 1.43 mm |

new crack forming. As such, we report the first peak load and the overall (max) peak load in Table 32.2. For plain beams, the first peak load and peak load are equal, since they did not experience multiple crack formations.

## 4 Conclusion

While the polymeric lattice provides initiation sites for cracks through the mortar, it also serves to deflect the cracks and prevents them from quickly propagating through the beam. The three lattice-reinforced beams with the highest amount of PLA attain toughness values 88 times higher than the toughness of plain beams; they also reach a peak load that is 1.4 times greater than the peak load reached by the plain mortar beams. The reduced stiffness displayed by lattice-reinforced beams is to be expected, based on the rule of mixtures, due to the higher polymeric ratio and may not be relevant in some applications. While the lattice-reinforced beams obtain toughness and peak load values comparable to the fiber-reinforced beams, the lattice-reinforced beams achieve a net deflection at peak load that is 2.5 times greater than that of the fiber-reinforced beams. Furthermore, the fiber-reinforced mortar presents workability issues, due to its low flowability, while the lattice reinforcement can be quickly injection molded from polymers and then easily infiltrated with non-fiber-containing mortar or concrete. Manufacturing from a recycled polymer would make this method inexpensive and would provide a new usage stream for used polymers.

We can adjust the polymeric lattice to concrete weight ratio to achieve desired mechanical properties and are exploring other lattice materials and geometries. These mechanical properties indicate that lattice-reinforced concrete composite beams are promising for construction applications. We envisage lattice reinforcement for use in nonbearing members, such as in building façade systems, where the polymeric structure can be designed to support bending loads.

## References

1. Buckminster, F. R. (1961). Synergetic building construction, *US2986241 A* .Washington, DC: U. S. Patent and Trademark Office.
2. Deshpande, V. S., Fleck, N. A., & Ashby, M. F. (2001). Effective properties of the octet-truss lattice material. *Journal of the Mechanics and Physics of Solids, 49*(8), 1747–1769. https://doi.org/10.1016/S0022-5096(01)00010-2.
3. Wicks, N., & Hutchinson, J. W. (2001). Optimal truss plates. *International Journal of Solids and Structures, 38*(30–31), 5165–5183. https://doi.org/10.1016/S0020-7683(00)00315-2.
4. Chiras, S., Mumm, D. R., Evans, A. G., Wicks, N., Hutchinson, J. W., Dharmasena, K., Wadley, H. N. G., & Fichter, S. (2002). The structural performance of near-optimized truss core panels. *International Journal of Solids and Structures, 39*(15), 4093–4115. https://doi.org/10.1016/S0020-7683(02)00241-X.
5. O'Masta, M. R., Dong, L., St-Pierre, L., Wadley, H. N. G., & Deshpande, V. S. (2017). The fracture toughness of octet-truss lattices. *Journal of the Mechanics and Physics of Solids, 98* (October 2016), 271–289. https://doi.org/10.1016/j.jmps.2016.09.009.
6. ASTM International. (2015). *ASTM C618-15 standard specification for coal fly ash and raw or calcined natural pozzolan for use in concrete*. https://doi.org/10.1520/C0618-15.
7. ASTM International. (2015). *ASTM C1437-15 standard test method for flow of hydraulic cement mortar*. Retrieved from https://doi.org/10.1520/C1437-15.
8. ASTM International. (2016). *ASTM C78/C78M-16 standard test method for flexural strength of concrete (using simple beam with third-point loading)*. https://doi.org/10.1520/C0078_C0078M-16.
9. ASTM International. (2012). *ASTM C1609/C1609M-12 standard test method for flexural performance of fiber-reinforced concrete (using beam with third-point loading)*. https://doi.org/10.1520/C1609_C1609M-12.

# Chapter 33
# The Influence of Specimen Shape and Size on the PCC Compressive Strength Values

Joanna J. Sokołowska, Tomasz Piotrowski, and Iga Gajda

The paper discusses the effect of the specimen shape and size on the result of compressive strength tests performed on polymer cement concrete (PCC). The shapes of tested specimens included cylinders and cubes of dimensions as recommended in EN-206 when determining the compressive strength class of cement concrete. Moreover, apart from cubes of standardized size of 150 mm, the research included tests of cubes of size 71 mm and 100 mm. The tests were performed on PCC with two different polymer aqueous dispersions: styrene-acrylic copolymer and carboxylated styrene-butadiene latex. As a reference, the tests were performed on the (polymer-less) ordinary concrete (OC). The results showed that the general rule – the smaller is the specimen, the higher values of strength it obtains – cannot be directly applied to PCC. Also the relation between compressive strength measured on cylinder and cube specimens (both for OC and PCC) did not show the conformity with reference ratios for OC.

## 1 Introduction

The most commonly done concrete tests include the compressive strength test. Results of compressive strength measurement depend on many factors. Important factors are size and shape of the specimen, as the stress concentrates differently, depending on the geometry of the object being loaded. Following the procedures set out in the standard documents, to determine concrete strength class, tests are performed on cubic and/or cylindrical specimens. According to the American standard ASTM C39, testing is done using cylindrical specimens of size 6 ×

J. J. Sokołowska · T. Piotrowski (✉) · I. Gajda
Warsaw University of Technology, Faculty of Civil Engineering, Warsaw, Poland
e-mail: t.piotrowski@il.pw.edu.pl

© Springer International Publishing AG, part of Springer Nature 2018
M. M. Reda Taha (ed.), *International Congress on Polymers in Concrete*
*(ICPIC 2018)*, https://doi.org/10.1007/978-3-319-78175-4_33

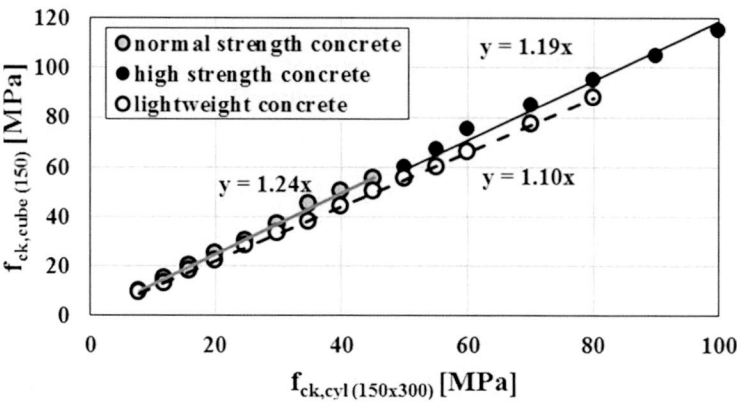

**Fig. 33.1** Relation $f_{ck,\text{cube}\,(150)}/f_{ck,\text{cyl}\,(150\,\times\,300)}$ [on basis of EN 206]

12 inch (150 × 300 mm). Though in ASTM C470, one can find information on specimen molds having diameters from 2 to 36 inch (50–900 mm). If the ratio between specimen length ($L$) and diameter ($D$) is 1.75 or less, the result of compressive strength value obtained using such specimen should be multiplied by the proper correction factor (ranging from 0.87 for $L/D = 1.00$–0.98 for $L/D = 1.75$). The European Standard EN 206 determines as concrete compressive strength class the minimal characteristic strength obtained for cylinder (150 mm diameter and 300 mm height) or cube (150 mm edge length). EN 206 defines characteristic strength as that value below which not more than 5% of the test results will fall. In other words, 95% of all specimens will have strengths in excess of the design characteristic strength. The relation calculated on the basis of compressive characteristic strengths for cubic and cylindrical specimens $f_{ck,\text{cube}}/f_{ck,\text{cyl}}$ varies for different concrete strength classes. For example, it is 1.24 in the case of ordinary concrete class C45/55, 1.19 in case of high-strength concrete of class C50/60, or 1.10 in case of lightweight concrete of class LC45/55 (Fig. 33.1).

PN-B-06265:2004 (Polish national supplement to standard PN-EN 206-1:2003) states that in case of normal weight and heavyweight high-strength concrete, compressive strength obtained on cubes of size 100 mm can be used to estimate the strength that would be obtained on cubes of size 150 mm (i.e., used for determining concrete compressive class). In such case the values of strength are converted by multiplier of 0.95 for each single result (as it was expected to obtain lower value using larger specimen). Meanwhile empirical relations between the values of compressive strength of concrete determined on specimens of different sizes elaborated by Jamroży [1] are described in Eq. 33.1. (the graphical representation of those findings is presented on Fig. 33.2):

$$f_{c28,\text{cube}(150)} = 1.05 f_{c28,\text{cube}(200)} = 0.90 f_{c28,\text{cube}(100)} = 1.25 f_{c28,\text{cyl}(150 \times 300)} \quad (33.1)$$

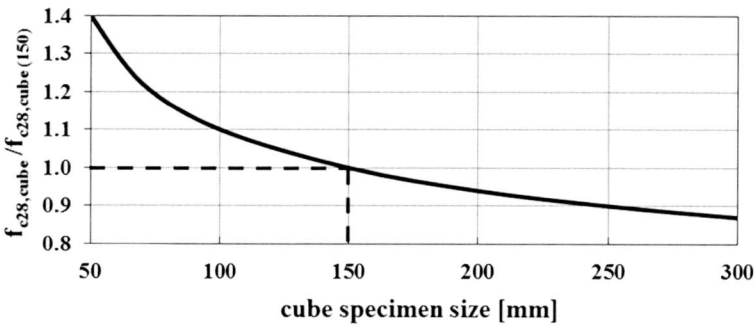

**Fig. 33.2** Relation $f_{c28,cube\,(d)}/f_{c28,cube\,(150)}$ [1]

Taking into consideration above, one can say that the influence of specimen shape and size on the compressive strength values of ordinary concrete is thoroughly recognized and described. This issue was also investigated for many types of "not ordinary" concretes, and results were not consistent with the literature and standards [2–5]. The aim of the paper was to discuss the same effects on the result of compressive strength tests performed on polymer cement concrete (PCC). According to EN 934-2, PCC contains at least 5% of polymer in relation to cement mass, but previous investigations showed that the copolymer for PCC is added in larger amounts, typically about 10% [6, 7] but even up to 20% [8].

## 2 Materials

The subjects of the research were two types of composites: OC and PCC. There were two experiments performed. The used materials were the following:

- *Cement.* Experiment 1 concerned concretes contained blast-furnace slag cement CEM III/A 42.5 N LH/HSR/NA of normal early strength, low heat of hydration, and high resistance to sulfate and alkali corrosion. Experiment 2 concerned concretes containing CEM I 42.5 R (Portland cement of high early strength).
- *Water.* In case of all composites, tap water (fulfilling requirements formulated in standard EN 1008) was used as the mixing water.
- *Aggregate.* All concretes contained the same aggregate. As the finest fraction (0/2 mm), a local river sand was used. Apart from sand, crushed granites of fractions 2/8 mm and 8/16 mm were used.
- *Polymer binder.* Two aqueous dispersions of polymer were used as co-binders. The first dispersion (referred later as D1) was a dispersion of styrene-acrylic copolymer; the content of the polymer in the dispersion was 33%. The second one (D2), carboxylated styrene-butadiene (XSB) latex, was more reach in the polymer concentration was 47% of solid polymer.

**Table 33.1** Concrete composition [kg/m$^3$]

| Symbol | Cement | w/c | Water | Water in D | Polymer in D | Dispersion D | 0/2 mm | 2/8 mm | 8/16 mm |
|---|---|---|---|---|---|---|---|---|---|
| Experiment 1 | | | | | | | | | |
| OC1 | 390 | 0.45 | 175 | – | – | – | 656 | 563 | 656 |
| PCC1-D1-10 | 390 | 0.45 | 97 | 78 | 39 | 117 | 656 | 563 | 656 |
| PCC1-D2-10 | 390 | 0.45 | 131 | 44 | 39 | 83 | 656 | 563 | 656 |
| Experiment 2 | | | | | | | | | |
| OC2 | 345 | 0.43 | 150 | – | – | – | 730 | 445 | 695 |
| PCC2-D1-5 | 345 | 0.43 | 115 | 35 | 17 | 52 | 730 | 445 | 695 |
| PCC2-D2-5 | 345 | 0.43 | 130 | 20 | 17 | 37 | 730 | 445 | 695 |
| PCC2-D2-8 | 345 | 0.43 | 119 | 31 | 28 | 59 | 730 | 445 | 695 |

The mix compositions are presented in Table 33.1. When adding polymer, the amount of cement was not reduced – polymer was a co-binder introduced into the mix in contents of 10%, 5%, and 8% of cement mass. Each time the water present in the polymer dispersion was taken into consideration when calculating total water content and keeping w/c constant 0.45 in Experiment 1 and 0.43 in Experiment 2. The OC specimens were put then into a climatic chamber up to 28 days with the humidity 95% and the temperature 18 ± 2 °C. The PCC specimens were cured for 5 days in the climatic chamber and for the remaining 23 days in the ambient conditions.

## 3 Testing

In case of OC1, cubes (150 mm) and cylinders (150 × 300 mm) were prepared. For PCC1 additional cubes (71 mm) and (100 mm) were cast. In Experiment 2 only cubes (100 mm) and cubes (150 mm) were prepared. Compressive strength test after 28 days was performed according to EN 12390-4. The average values were used to find relations between the shape and size of the specimen and the value of compressive strength for OC and PCC.

**Table 33.2** Compressive strength results

| Symbol | Shape | Size d [mm] | $f_{cm}$ [MPa] | $f_{cmin}$ [MPa] | SD [MPa] | CV [%] | $f_{cm,cube}$ (150) / $f_{cm,cyl}$ (150 × 300) | $f_{cm,cube}$ (150) / $f_{cm,cube}$ (d) | Class |
|---|---|---|---|---|---|---|---|---|---|
| **Experiment 1** | | | | | | | | | |
| OC1 | Cube | 150 | 45.22 | 44.32 | 0.80 | 1.8 | 1.49 | | C30/37 |
| | Cylinder | 150 × 300 | 30.38 | 27.30 | 4.36 | 14.3 | | | C25/30 |
| PCC1-D1-10 | Cube | 71 | 43.85 | | 2.40 | 5.5 | | 1.06 | |
| | | 100 | 42.37 | | 3.91 | 9.2 | | 1.10 | |
| | | 150 | 46.53 | 44.39 | 0.98 | 2.1 | 1.35 | | C30/37[a] |
| | Cylinder | 150 × 300 | 34.51 | 31.94 | 3.00 | 8.7 | | | C30/37[a] |
| PCC1-D2-10 | Cube | 71 | 31.01 | | 2.98 | 9.6 | | 1.18 | |
| | | 100 | 36.09 | | 1.57 | 4.4 | | 1.01 | |
| | | 150 | 36.52 | 35.40 | 0.98 | 2.7 | 1.24 | | C25/30[a] |
| | Cylinder | 150 × 300 | 29.40 | 28.28 | 0.92 | 3.1 | | | C25/30[a] |
| **Experiment 2** | | | | | | | | | |
| OC2 | Cube | 100 | 66.66 | | 2.32 | 3.5 | | 0.94 | |
| | | 150 | 62.95 | | 2.24 | 3.6 | | | C45/55 |
| PCC2-D1-5 | Cube | 100 | 46.84 | | 0.56 | 1.2 | | 0.93 | |
| | | 150 | 43.58 | | 5.62 | 12.9 | | | C30/37[a] |
| PCC2-D2-5 | Cube | 100 | 44.05 | | 5.17 | 11.8 | | 1.04 | |
| | | 150 | 45.91 | | 1.91 | 4.2 | | | C30/37[a] |
| PCC2-D2-8 | Cube | 100 | 25.14 | | 1.13 | 4.5 | | 1.24 | |
| | | 150 | 31.12 | | 2.05 | 6.6 | | | C20/25[a] |

[a]Compressive strength class evaluation analogically to procedure used in OC technology

# 4 Experimental Results and Discussion

The compressive strength characteristics, i.e., $f_{cm}$ and $f_{c,min}$, of all tested concretes are given in Table 33.2. As expected, because of the higher w/c ratio, OC1 concrete compressive strength and – as a consequence – its compressive strength class were lower than in case of OC2. Taking into consideration that analyzed populations of test results were described in most of the cases by low CV (and the lowest obtained result $f_{c,min}$ and mean $f_{cm}$ adopted close values), determination of concrete class depended in general on $f_{cm}$ (and indeed on "$f_{cm}$ – 4 MPa" according to EN 206 procedure). Therefore, it seemed reasonable to analyze ratio between the mean values, i.e., $f_{cm,cube\ (150)}/f_{cm,cyl\ (150\times 300)}$, rather than characteristic values $f_{ck,cube}/f_{ck,cyl}$. In case of OC1 $f_{cm,cube\ (150)}/f_{cm,cyl\ (150\times 300)}$ was 1.49, so the compressive strength class determined using cylindrical specimens results (C25/30) turned out to be lower than the one determined on the basis of cube specimens results (C30/37). In Experiment 1 $f_{cm,cube\ (150)}/f_{cm,cyl\ (150\times 300)}$ ratios were 1.35 for PCC1-D1-10 and 1.24 for PCC1-D2-10. Although application of dispersion D2 resulted in formation of polymer cement concrete with significantly lower properties than D1, the evaluation (done analogically to procedure used in case of OC technology) based on cube and cylinder specimens in both cases led to the same "class" of concretes – C30/37 or C20/25 – regardless the geometry of the specimens (compare Table 33.2).

The main aim of Experiment 2 was to focus on the size of tested specimen and its influence on the result of compressive strength test. The obtained data (Table 33.2) enabled to determine the empirical ratio between mean values of compressive strength of ordinary concrete (OC2) and three types of polymer cement concrete tested using cubes of size 150, 100, and 71 mm. The empirical ratio $f_{cm,cube\ (150)}/f_{cm,cube\ (100)} = 0.94$ determined for OC2 was close to the value 0.95 (given in PN-B 06265). Surprisingly this ratio determined for PCC2-D1-5 containing only 5% of polymer was nearly the same – 0.93. In case of other polymer cement concretes, the ratio was higher than 1.00 (see Table 33.2), which shows the opposite tendency describing the effect of size of specimen on its compressive strength: the smaller the size of the cube, the lower the compressive strength values in measurements. This finding was also noted in case of relation between results obtained for cubes 150 and 71 mm ($f_{cm,cube\ (150)}/f_{cm,cube\ (71)} > 1.00$).

## 5 Conclusions

The aim of the paper was to verify if the relation of compressive strength measured on specimens of different shapes and sizes given for OC could be applied to PCC. Results of first experiment show that in case of OC but also PCC the known relation describing effect of specimen shape on compressive strength value was confirmed but the effect was generally stronger. On the other hand, there was no clear influence of the cube size on compressive strength results. OC and PCC containing 5% of styrene-acrylic copolymer, which according to EN 934-2 is the maximum value for considering polymer as a admixture, are characterized with $f_{cm,cube\ (150)}/f_{cm,cube\ (100)} < 1.00$ as expected. However the results of compressive strength of PCC with the same and higher amount of carboxylated styrene-butadiene latex showed opposite. This could be an effect of impact of polymer phase on the fracture mechanism. Analysis of this phenomenon and shape and size dependence on compressive strength test results for PCC mixes need to be continued using a wider variety of PCC specimens.

## References

1. Jamroży, Z. (2015). *Beton i jego technologie/Concrete and its technology*. Warsaw: Wydawnictwo Naukowe PWN (in Polish).
2. Tokyay, M., & Özdemir, M. (1997). Specimen shape and size effect on the compressive strength of higher strength concrete. *Cement and Concrete Research, 27*(8), 1281–1289. https://doi.org/10.1016/S0008-8846(97)00104-X.
3. Hamad, A. J. (2017). Size and shape effect of specimen on the compressive strength of HPLWFC reinforced with glass fibres. *Journal of King Saud University – Engineering Sciences, 29*(4), 373–380. https://doi.org/10.1016/j.jksues.2015.09.003.
4. Dehestani, M., Nikbin, I. M., & Asadollahi, S. (2014). Effects of specimen shape and size on the compressive strength of self-consolidating concrete (SCC). *Construction and Building Materials, 66*, 685–691. https://doi.org/10.1016/j.conbuildmat.2014.06.008.
5. Sim, J.-I., Yang, K.-H., Kim, H.-Y., & Choi, B.-J. (2013). Size and shape effects on compressive strength of lightweight concrete. *Construction and Building Materials, 38*, 854–864. https://doi.org/10.1016/j.conbuildmat.2012.09.073.
6. Jaworska, B., Sokołowska, J. J., Łukowski, P., & Jaworski, J. (2015). Waste mineral powders as a components of polymer-cement composites. *Archives of Civil Engineering, 61*(4), 199–210. https://doi.org/10.1515/ace-2015-0045.
7. Piotrowski, T., & Gawroński, P. (2015). Chemical resistance of concrete-polymer composites – Comparison based on experimental studies. *Advanced Materials Research, 1129*, 123–130. https://doi.org/10.4028/www.scientific.net/AMR.1129.123.
8. Sokołowska, J. J., Piotrowski, T., Garbacz, A., & Kowalik, P. (2013). Effect of introducing recycled polymer aggregate on the properties of C-PC composites. *Advanced Materials Research, 687*, 520–526. https://doi.org/10.4028/www.scientific.net/AMR.687.520.

# Chapter 34
# Long-Term Investigation on the Compressive Strength of Polymer Concrete with Fly Ash

Joanna J. Sokołowska

For the past years, a lot was said about the great potential of reuse of the fly ashes in the building materials industry. Fly ashes have gained recognition as a good addition to various concretes, including ordinary concrete, polymer cement concrete, and polymer concrete. Many research was conducted on the positive effect of fly ash on the technological and technical properties; however, most of them included tests performed after several days or weeks. Not much published data concerned tests conducted after a longer period of time, confirming the durability of polymer composites with fly ashes. The paper presents the results of compressive strength of polymer concretes with fly ashes tested after longer time. Composites with vinyl ester resin, standard sand, and fluidized fly ash were tested after 18 months, and composites with the same resin, river sand, and fluidized or siliceous fly ash were tested after 7 years. The results were compared with the results of the respective specimens tested after 14 days. Research results indicated that there was no reduction in compressive strength. Moreover, in all cases, a significant improvement of strength was noted. The research has demonstrated the usefulness of fly ash as a component for the production of durable polymer concrete.

## 1 Introduction and Research Scope

Starting from the very beginning, i.e., development of the concept of polymer concrete (PC) over 50 years ago and later through subsequent improving that concept and PC compositions, the property indicated as a great advantage in contradiction to ordinary concrete (OC) was the excellent durability [1–3]. Despite

J. J. Sokołowska (✉)
Warsaw University of Technology, Faculty of Civil Engineering, Warsaw, Poland
e-mail: j.sokolowska@il.pw.edu.pl

© Springer International Publishing AG, part of Springer Nature 2018
M. M. Reda Taha (ed.), *International Congress on Polymers in Concrete (ICPIC 2018)*, https://doi.org/10.1007/978-3-319-78175-4_34

the modifications of polymer binder or aggregate, polymer concrete is considered to be resistant to water penetration and consequently to freezing and thawing or to aggressive substances, including acids and alkaline media [2, 4, 5] (acid resistance has always been a major drawback of composites containing cement [6, 7]). Polymer concrete is not resistant to high temperatures, but it has been selected for applications that do not require this ability. Nevertheless, it is generally considered to be durable material and resistant to most types of chemical corrosion.

One could say a lot about the potential chemical resistance and mechanical properties of polymer concretes, even the ones containing industrial wastes or by-products, such as fly ashes. Their high compressive, flexural, and tensile strength and elasticity modulus have been confirmed by laboratory tests [4, 8] and on site [2, 9], including industrial environments [1]. However, most of testing was performed after 14, 28, or 90 days. Not much research has been done on what happens after long-term use of these composites and whether their ability to maintain required level of properties over time is really as high as one might expect.

The author was particularly interested in the stability of PC containing siliceous and fluidized fly ashes since the development of such pro-ecological composite has been a subject of her research work in the past years [10–12]. Presented paper concerns the compressive strength of vinyl ester concretes with two kinds of fly ash tested after 1.5 and 7 years. The results were compared with the results of the respective specimens tested after 14 days. Results indicated that there was no reduction in compressive strength for that time. Moreover, in all cases an improvement of strength was noted. The research has demonstrated the usefulness of fly ashes as components of durable polymer concrete.

## 2  Materials, Design, and Testing

Presented research is the continuation of the work realized in years 2010–2011. All compositions of tested PC, namely, vinyl ester concrete (Table 34.1), were designed and prepared according to statistic design (for details, see [10]) and were a small section of a larger experimental program. Presented research included 4 compositions of PC with standard sand (fulfilling requirements of EN 196-1) and fluidized fly ash, $FA_F$, tested after 14 days and 1.5 years (compositions No 1 ÷ 4 according to Table 34.1) and 10 compositions of PC with river sand and same fluidized fly ash (compositions No 5 ÷ 9 according to Table 34.1) or with siliceous fly ash, $FA_S$ (compositions No 10 ÷ 14 according to Table 34.1), tested after 14 days and 7 years. Composites differed in proportions of constituents, such as binder and aggregate, but also fly ash content in microfiller or in total PC mass.

Specimens tested in presented research were the halves of beams of dimensions 40 × 40 × 160 mm remaining after flexural test. They were stored in the laboratory conditions for periods of 14 days (time of PC curing recommended in standard EN 1542) and 18 or 84 months. The change in the compressive strength over time was considered as the measure of the long-term durability of PC with fly ash.

**Table 34.1** PC compositions per 1 m$^3$

| No | VE | Q | FA$_F$ | FA$_S$ | S | G | FA/PC, % | FA/M, % |
|---|---|---|---|---|---|---|---|---|
| 1 | 231 | 327 | 87 | 0 | 546 | 1109 | 3.8 | 21.0 |
| 2 | 231 | 413 | 109 | 0 | 513 | 1031 | 4.7 | 21.0 |
| 3 | 271 | 271 | 271 | 0 | 496 | 992 | 11.8 | 50.0 |
| 4 | 326 | 123 | 462 | 0 | 463 | 926 | 20.1 | 79.0 |
| 5 | 271 | 333 | 8 | 0 | 496 | 992 | 0.4 | 1.5 |
| 6 | 231 | 413 | 109 | 0 | 513 | 1031 | 4.7 | 21.0 |
| 7 | 326 | 462 | 123 | 0 | 463 | 926 | 5.3 | 21.0 |
| 8 | 271 | 271 | 271 | 0 | 496 | 992 | 11.8 | 50.0 |
| 9 | 326 | 123 | 462 | 0 | 463 | 926 | 20.1 | 79.0 |
| 10 | 271 | 333 | 0 | 8 | 496 | 992 | 0.4 | 1.5 |
| 11 | 231 | 413 | 0 | 109 | 513 | 1031 | 4.7 | 21.0 |
| 12 | 326 | 462 | 0 | 123 | 463 | 926 | 5.3 | 21.0 |
| 13 | 271 | 271 | 0 | 271 | 496 | 992 | 11.8 | 50.0 |
| 14 | 326 | 123 | 0 | 462 | 463 | 926 | 20.1 | 79.0 |

*VE* vinyl ester, *Q* quartz powder, *FA$_F$* fluidized fly ash, *FA$_S$* siliceous fly ash, *S* sand, *G* gravel, *M* microfiller

**Table 34.2** Granulometry of fluidized and siliceous fly ashes used in tested PC

| Property | Fluidized fly ash | Siliceous fly ash |
|---|---|---|
| Size range, μm | 0.12 ÷ 77.34 | 0.17 ÷ 200.00 |
| Mode/average size, μm | 0.25/4.05 | 4.96/36.78 |
| D$_{50}$/D$_{90}$ μm | 0.25/12.41 | 29.91/101.46 |
| Specific surface area, cm$^2$/cm$^3$ | 14,825 | 18,294 |

**Fig. 34.1** SEM observations of fluidized (left) and siliceous (right) fly ash

Both used fly ashes presented different granulometry (Table 34.2) and morphology (Fig. 34.1) and differently influenced the workability of the mix. It was expected that PC with siliceous fly ash of nearly perfect spherical particles, despite the larger specific area of that filler, would perform better. The consistence of mixes with

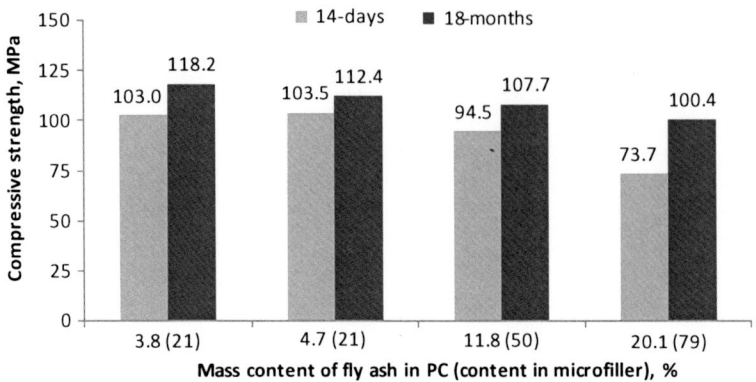

**Fig. 34.2** Compressive strength of PC with fluidized fly ash and standard sand (compositions No 1 ÷ 4 according to Table 34.1) after 14 days and 18 months

siliceous fly ash was more liquid, and their microstructure was more homogeneous than in the case of analogous composites with fluidized fly ash.

## 3 Results and Discussion

The compressive test results obtained for vinyl ester concrete with standard sand and fluidized fly ash after 14 days and 18 months are given on Fig. 34.2. Firstly, it is worth noting that irrespectively of the concrete composition no decrease in strength in time was observed. Moreover, the values of strength increased in time. This is an interesting observation, as it is generally believed that PC achieved mechanical properties after short time of curing [2]. Meanwhile, after 18 months, the increase by 8.6–36.2% (on average by 18.4%) was noted. This may be due to the fact that introducing fly ash, a by-product of fluidized coal combustion using calcium sorbents, containing lots of calcium compounds (including $Ca(OH)_2$) into the polymer matrix, could initiate additional cross-linking and the whole process could had been extended (effect observed in case of epoxies [13]). Secondly, both after 14 days and 18 months, it was noticed that the more fluidized fly ash was in the microfiller (and in total composite mass), the lower was the compressive strength value. However, after 18 months all PC, even the richest in fly ash (79% of $FA_F$ in the microfiller, corresponding to 20.1% of the total composite mass) presented compressive strength no lower than 100 MPa.

The compressive strength of PC with river sand and fluidized fly ash obtained after 14 days and 7 years is given on Fig. 34.3. The strength of analogous composites with siliceous fly ash is given on Fig. 34.4. Despite the expectation that using siliceous fly ash would result in better mechanical strength of PC, it was not explicitly confirmed in the laboratory. The results obtained for 14-day-old composites of the same quantitative compositions but with various fly ashes differed from

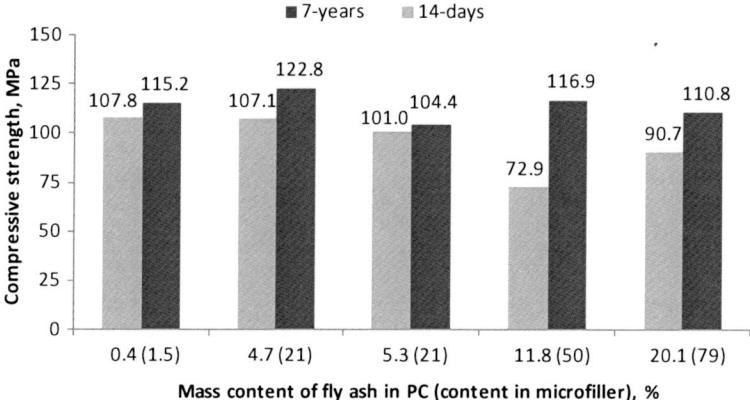

**Fig. 34.3** Compressive strength of PC with fluidized fly ash and river sand (compositions No 5 ÷ 9 according to Table 34.1) after 14 days and 7 years

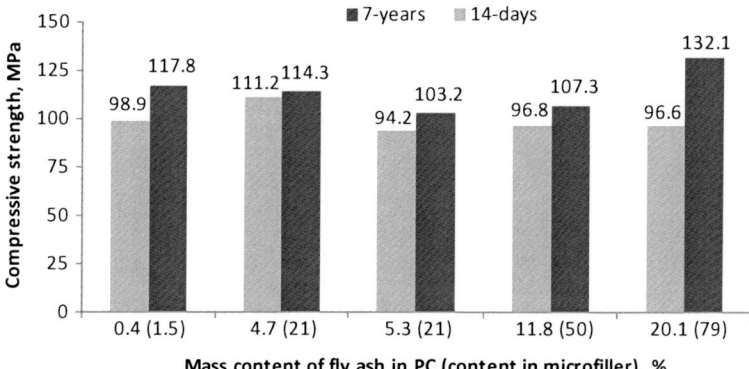

**Fig. 34.4** Compressive strength of PC with siliceous fly ash and river sand (compositions No 10 ÷ 14 according to Table 34.1) after 14 days and 7 years

3.7 to 24.7% in plus or minus. After 7 years the differences have decreased and ranged from 1.2% to 16.2% in plus or minus. Nevertheless, in each case, an increase in strength in time was observed. In case of PC with siliceous fly ash, the increase was from 2.8 ÷ 36.7% (on average about 16%). In the case of PC with fluidized fly ash, the increase was from 3.4 to even 60.4% (on average about 22%).

There was no clear relation between fly ash content and compressive strength of 14-day-old nor 7-year-old composites. The results obtained after 14 days seemed to indicate that compressive strength of PC was decreasing with the increase of fluidized fly ash content in the microfiller fraction (and total composite mass). However, after 7 years even in the case of the composition richest in this fly ash (79% of $FA_F$ in the microfiller, corresponding to 20.1% of the total composite mass), the strength value stabilized at 110.8 MPa, which is higher than the initial value of the PC with negligible amount of fly ash (0.4% of microfiller, which corresponds to 1.5% of total composite mass) (compare first and last bar on Fig. 34.3).

Similar effect was observed in case of concrete with siliceous fly ash. After 7 years, in the case of the composition richest in $FA_S$ (79% of microfiller, corresponding to 20.1% of the total composite mass), the compressive strength value stabilized at 132.1 MPa. This is about 1/3 higher than the initial values of the concrete with negligible or essential amounts of fly ash (compare data on Fig. 34.4).

# 4 Conclusions

The presented results indicated that there was no reduction in compressive strength of vinyl ester concrete with fly ashes after 1.5 nor after 7 years. Moreover, in all analyzed cases, an improvement (often significant) of strength in time was noted. It can be concluded that the cross-linking of vinyl ester in analyzed composites might lasted longer than it was expected on the basis of general knowledge. This could be an effect of the fly ashes chemical composition that, in contradiction to pure quartz powder, included noticeable amounts of calcium compounds that could initialize the additional cross-linking. On the other hand, introducing into the polymer matrix, fluidized fly ash particles may had caused difficulties in polymer setting, elongating the process. Nevertheless, the research has demonstrated the usefulness of considered fly ashes – both fluidized and siliceous – as components for the production of durable polymer concrete retaining or even getting better properties in time.

## References

1. Guide for use of polymers in concrete. (2009). Report of ACI Committee 548, 1R-09, Detroit.
2. Czarnecki, L. (2010). Polymer concretes. *Cement-Lime-Concrete, 15*(2), 63–85.
3. Fowler, D. W. (2013). Application of PC and PMC in industry and industrial environment. *Advanced Materials Research, 687*, 21–25.
4. Gorninski, J. P., Dal Molin, D. C., & Kazmierczak, C. S. (2007). Strength degradation of polymer concrete in acidic environments. *Cement & Concrete Composites, 29*, 637–645.
5. Sokołowska, J. J., Woyciechowski, P., Łukowski, P., & Kida, K. (2015). Effect of perlite waste powder on chemical resistance of polymer concrete composites. *Advanced Materials Research, 1129*, 516–522.
6. Shi, C., & Stegemann, J. A. (2000). Acid corrosion resistance of different cementing materials. *Cement and Concrete Research, 30*(2000), 803–808.
7. Sokołowska, J. J., & Woyciechowski, P. (2013). Effect of acidic environments on cement concrete degradation. In *Proceedings of the 3rd international conference on sustainable construction materials & technologies – SCMT3*. Kyoto. (paper No 85).
8. Hwang, E. H., & Kim, J. M. (2015). Characteristics of polyester polymer concrete using spherical aggregates from industrial by-products(II) (use of fly ash and atomizing reduction steel slag). *Korean Chemical Engineering Research, 2*(53), 364–371.

9. Yeon, K. S., Kawakami, M., Choi, Y. S., Hwang, J. Y., Min, S. H., & Yeon, J. H. (2013). Remodeling of deteriorated irrigation aqueducts using precast polymer concrete flume. *Advanced Materials Research, 687*, 35–44.
10. Czarnecki, L., Garbacz, A., & Sokołowska, J. J. (2010). Fly ash polymer concretes. In *Proceedings of 2nd int. conference on sustainable construction materials and technologies* (pp. 28–30). Ancona.
11. Garbacz, A., Sokołowska, J. J., Lutomirski, A., & Courard, L. (2012). Fly ash polymer concrete quality assessment using ultrasonic method. In *Proceedings of the 7th Asian symposium on polymers in concrete ASPIC 2012* (pp. 573–580). Istanbul.
12. Garbacz, A., & Sokołowska, J. J. (2013). Concrete-like polymer composites with fly ashes – Comparative study. *Construction and Building Materials, 38*, 689–699.
13. Katsuta, T., Ohama, Y., & Demura, K. (2001). Investigation of microcracks self-repair function of polymer-modified mortars using epoxy resins without hardeners. In *Proceedings of the 10th international congress on polymers in concrete, 1, CD-Proceedings*.

# Chapter 35
# Overlays: A Great Use for Polymer Concrete

David W. Fowler and David P. Whitney

Polymer concrete (PC) overlays have been successfully used since the 1970s primarily by transportation agencies on bridges. Polymers used for overlays have seen significant improvements, and the current overlays have proven to be durable, skid resistant, and cost-effective. Over 2400 overlays had been installed by 2009, and overlays are a standard bridge protection system by departments of transportation based on an extensive survey of US and Canadian provinces. Slurry, multiple-layer, and premixed overlays are the most widely used. Epoxies are the most widely used resins. Many states have published specifications on the construction of overlays.

## 1 Introduction

Polymer concrete (PC) has been used for years for many applications, but none has been more appropriate than for overlays. The first overlays were single layers of coal tar epoxy broomed onto the substrate and seeded with fine aggregate; however, they were permeable and not very durable under traffic. By the mid-1970s, polyester-styrene and methyl methacrylate monomer systems were being placed using the broom-and-seed method [1]. Eventually premixed PC that was screeded into place began to be used. Initially, many of the thicker, more brittle layers delaminated because of the thermal incompatibility of the overlay and substrate. Through increased attention to monomer and resin formulations and improved understanding of the causes of delamination, the performance of thin-bonded polymer concrete overlays (TPOs) has improved significantly.

---

D. W. Fowler (✉) · D. P. Whitney
University of Texas, Austin, TX, USA
e-mail: dwf@mail.utexas.edu

The use of TPOs has increased significantly in recent years. Sprinkel [2] reported that before 1990, 139 TPOs were placed. Between 1990 and 1999, 416 more had been placed. The results of a more recent survey will be discussed in this paper.

## 2 Uses of Thin Polymer Overlays

TPOs are an excellent means of protecting decks that are in fairly good condition but are at risk for chloride and water penetration; they are not a repair method. Multiple-layer overlays are preferred for use on decks that have a good ride quality since the overlays follow the contours of the deck surface. Slurry and premixed overlays are best for decks that have many surface irregularities.

According to Sprinkel [2] the most likely candidates for TPOs are bridges that "are in need of a skid resistant surface but have peak-hour traffic volumes that are so high that it is not practical to close a lane to apply the surface except during off-peak traffic periods" and "are those in which increases in dead load, reductions in overhead clearance, and modifications to joints and drains must be held to a minimum." According to Carter [3] TPOs can provide service lives of up to 20 years although some maintenance will be required. TPO installation requires specialized expertise. The primary failures observed have been the result of workmanship or contractor-related errors. Polymer wearing surfaces may lose toughness and ductility with time.

Wide cracks should be filled with a gravity-fill polymer that is compatible with the primer and overlay to prevent future delaminations. Concrete with a chloride ion content in excess of $0.77$ kg/m$^3$ at the level of steel reinforcement should be removed and replaced before placing the overlay in order to minimize the likelihood of steel corrosion causing cracking and delamination of the overlay [2].

## 3 Results of Surveys

Surveys of all states and Canadian provinces were conducted to determine the use of TPOs in their jurisdiction, reasons for using TPOs, polymers used, failures, and other pertinent information. Responses were received from 40 states and 7 provinces. Some of the findings were [4]:

- At the time of the survey (2009), approximately 2400 TPOs had been installed by states and 147 by provinces reporting. California placed the most (520), but 7 other states and 1 province each reported placing 100 or more.
- Three states and one province no longer use TPOs for various reasons including no more problems with decks, administrative problems in enforcing specifications, poor performance, and difficulty in achieving adequate inspection accompanied by wet conditions.

- Most states indicated more than one reason for using TPOs including improving skid resistance (most common reason), extending life of bridge decks, waterproofing the deck, repairing spalled and crack decks, and restoring a uniform appearance.
- Nearly all states use epoxy resins, and the most common type was the multiple layer. California was the exception, having installed over 500 premixed polyester resin overlays.
- The majority of states use contractors for installation, but ten states use their own forces for at least some installations.
- New Mexico, Georgia, and LaGuardia Airport required warranties ranging from 1 to 10 years.
- Many states have published specifications for TPOs.
- Three states reported overlay costs: $60/m$^2$ for epoxy multiple-layer TPO, $114/m$^2$ for an epoxy-urethane TPO, and $66/m$^2$ for epoxy multiple-layer TPOs that included shot blasting.

Respondents indicated many reasons for failures that included:

- Poor condition of deck
- Repaired areas not sufficiently dry or roughened
- Inadequate surface preparation
- Cool, damp weather during installation
- Deck too damp at the time of overlay installation
- Construction problems
- Inadequate quality control
- Use of snow chains

Many states have specifications that can be accessed through their websites. Reference [4] contains a list of available specifications and the links.

Nine contractors having considerable experience with TPOs and seven material supplier representatives were interviewed by telephone. The problems they encountered tended to be the same as those mentioned by the agencies. Contractors reported problems with cure times related to traffic openings caused by resins curing too slowly to permit opening as required by contract. They also reported that plans always indicate less quantities of materials than required. Material suppliers complained that agencies require that they must provide technical support on site.

## 4 Proven Practices

Proven practices are available from the literature and from the practices adopted by state agencies. These practices have produced overlays that have been durable and long lasting [4].

Experience has shown that the bridge decks that are the most likely candidates are those that need a skid-resistant surface and have peak-hour traffic volumes so high that it is not practical to close a lane, in which increases in dead load, reductions in overhead clearance, and modifications to joints and drains must be kept to a minimum [5]. TPOs properly constructed should last at least 20 to 25 years [3, 5].

Evaluation, repair, and surface preparation of the substrate are essential in order to achieve a satisfactory overlay. The deck needs to be sounded for delamination and evaluated for corrosion activity using appropriate methods. Concrete with high levels of chlorides needs to be removed. The concrete must have an acceptable minimum tensile rupture strength or else be removed and replaced. Materials for repair or replacement should be of low shrinkage and placed in accordance with manufacturer's instructions. Cracks wider than 1 mm should be filled with a gravity-fill resin that is compatible with the resin in the overlay system. The surface must be cleaned, usually by shot blasting, just prior to overlay application. The surface should be dry and free of oil or other deleterious materials [4].

The most commonly used polymer systems for TPOs include epoxies (including modified epoxy-urethanes), polyester-styrenes, and methyl methacrylates (MMAs). The polymers that exhibit the best performance have low modulus and high elongation. Three overlay methods are used to apply the overlays. Care should be taken to ensure that the concrete repair materials and/or primers used on the decks are compatible with the monomers and resins used in the overlay.

**Multiple-Layer Overlay**

One or more applications of the resin system are placed, each followed by an application of aggregate onto the concrete (Figs. 35.1 and 35.2). If more than one application is required to produce the required thickness, the first layer is allowed to cure prior to applying the second application that is immediately followed by the application of aggregate. The recommended binder is epoxy. The thickness can vary from 6 to 12 mm.

**Fig. 35.1** Applying epoxy to clean surface

**Fig. 35.2** Blowing aggregate onto surface

**Fig. 35.3** Placement of polyester-styrene PC

## Premixed Overlay

The premixed TPO is typically used for thicker (75 mm and greater) overlays. It works well on uneven surfaces where the thicker, lower modulus TPO can resist the stress and impact of tire chains and can accommodate more wear. Epoxy and polyester-styrene are the recommended resins. High molecular weight methacrylate is often used as a primer for polyester-styrene overlays to improve bond. The polymer resins and aggregates are premixed and placed on the deck using paving machines or screeds (Figs. 35.3 and 35.4). Aggregates are often broadcast on the surface to provide more skid resistance.

## Slurry Overlay

The slurry incorporates aggregates into the binder before placement on the deck. The lower viscosity of the binder, usually epoxy or methacrylate, requires a well-graded fine filler to support and more uniformly disperse the larger sand particles and the desired depth. They are normally thinner than premixes and about the same

**Fig. 35.4** Completed premixed with paver P-S overlay

thickness as multi-layer TPOs. Slurry overlays require a primer and usually a seal coat to help bind the aggregate that is broadcast on the surface.

## 5 Repairs

Failures, often in the form of delaminations, are repaired by first sounding the area to determine the extent of failure. The perimeter of the delaminated area is saw-cut in a rectangular pattern. The overlay inside the saw-cut is chipped out down to the concrete substrate. The concrete surface is cleaned, typically by shot blasting. The same or similar materials and application methods are used to make the repair.

## 6 Future Needs

Continued use and success of polymer concrete overlays will require a research to improve their performance and have a long life including [4]:

- Development of improved resins to perform under high and low temperatures, heavy traffic loads, and heavy traffic volumes
- Development of resins that will self-heal if cracks develop
- Development of recycled resins that perform well in overlays to lower costs and to improve sustainability

- Improved analytical techniques to select material properties and thicknesses of overlays to perform satisfactorily under the anticipated conditions
- Development of tests and procedures to more accurately predict long-term performance under a wide range of expected conditions
- Development of nondestructive test methods to accurately determine if the concrete substrates are suitable for applying overlays

# 7 Conclusions

Polymer concrete overlays have proven to be one of the most effective and useful applications of concrete polymer materials. State and provincial transportation agencies have used PC overlays in increasing numbers with very good results. Many have their own specifications, and there are numerous contractors who are very proficient in applying them. Multiple-layer, slurry, and premixed overlays are the most common. Overlays give the best performance when applied to decks that are still in good condition. Adequate surface preparation, dry and warm weather, an experienced application, and good workmanship coupled with resins that have low modulus and high elongation generally ensure a successful application.

# References

1. ACI 548.5R. (1994, January). *Guide for polymer concrete overlays*. American Concrete Institute, Farmington Hills, MI. Reapproved 1998, 26 pp.
2. Sprinkel, M. M. (2003). Twenty-five-year experience with polymer concrete bridge deck overlays. *Polymers in concrete: The first thirty years,* ACI Special Publication No. 214, pp. 51–62.
3. Carter, P. D. (1993). Thin polymer wearing surfaces for preventative maintenance of bridge decks. *Polymer concrete*, ACI Special Publication No. 137 pp. 29–48.
4. Fowler, D. W., & Whitney, D. P. (2011). *Long-term performance of polymer concrete for bridge decks* (p. 63). Washington, D.C: NCHRP Synthesis 423, National Cooperative Highway Research Program, Transportation Research Board.
5. Sprinkel, M. M. (1997). Nineteen year performance of polymer concrete bridge decks. *In-place performance of polymer concrete overlays*, ACI Special Publication No. 169, pp. 42–74.

# Chapter 36
# PC with Superior Ductility Using Mixture of Pristine and Functionalized Carbon Nanotubes

AlaEddin Douba and Mahmoud Reda Taha

Polymer concrete (PC) replaces cement hydration with polymerization by utilizing epoxy polymer binders as a substitute of cement creating impermeable and highly durable concrete. It has been reported that the tensile properties of PC can be significantly improved by incorporating multi-walled carbon nanotubes (MWCNTs) at 2.0 wt.% content. In this work, pristine and carboxyl functionalized MWCNTs are mixed together to investigate the effect of functionalization as well as to engineer PC mechanical performance. Results show that hybrid mixes outperform plain mixes of either type of MWCNTs achieving strains at failure up to 5.5% and improvements in toughness up to 184%. The resultant PC also offers tensile strength in the range of 10 MPa. Microstructural analysis using scanning electron microscope (SEM) and dynamic modulus analyzer (DMA) reveals proper dispersion of MWCNTs and increased cross-linking density due to functionalization. Furthermore, it is evident that functional group concentration of 0.001–0.018 wt.% maximizes PC ductility. The proposed PC is an attractive alternative for concrete joints under extreme loading events such as earthquakes and hurricanes.

---

A. Douba
Department of Civil Engineering and Engineering Mechanics, Columbia University, New York, NY, USA

M. Reda Taha (✉)
Department of Civil Engineering, University of New Mexico, Albuquerque, NM, USA
e-mail: mrtaha@unm.edu

# 1 Introduction

Polymer concrete (PC) is one of several polymer-based concretes in which the cement binder is replaced by polymers. The cement hydration process is hence transformed to polymerization in which its intensity can be described through cross-linking density. PC has become a primary material for overlays, machine foundations, manholes, drainage pipes, and architectural panels [1]. This is due to PC's excellent weather and chemical resistance and its good mechanical properties. Structural applications of PC require PC to have superior mechanical behavior to justify its relatively high cost compared with Portland cement concrete.

Nanomaterials offer superior mechanical properties improvements at low contents enabled by their extreme geometries. Multi-walled carbon nanotubes (MWCNTs) offer a significantly large surface area [2]. This enables a larger interactive interface within the host matrix enabling multiple influence pathways using mechanical and chemical interactions. Previous research showed that maximum improvements can be attained at 2.0 wt.% MWCNTs [3]. MWCNTs are often functionalized to improve dispersion in which functional groups are attached throughout their surface. The functional group might also enable a chemical interaction interface between MWCNTs and the host matrix. Thus, functionalized MWCNTs (F-MWCNTs) can be considered chemically reactive, while pristine MWCNTs (P-MWCNTs) are nonreactive. Other scholars reported improvements in strength, fracture toughness, fatigue, glass transition temperature, and viscoelastic recovery using varying contents of either P-MWCNTs or F-MWCNTs [3–9]. We hypothesize that utilizing a mix of F-MWCNTs and P-MWCNTs simultaneously can result in superior improvement utilizing different interactive mechanisms of each type of MWCNTs with epoxy. PC samples containing different ratios of P-MWCNTs and F-MWCNTs were tested in tension. The results were analyzed for their tensile properties in conjunction with the functional group content.

# 2 Experimental Methods

## 2.1 *Materials*

P-MWCNTs and F-MWCNTs, of similar geometrical properties of 5–10 nm inner diameter, 20–30 nm outer diameter, and an aspect ratio of 1000, were used. Carboxyl groups (COOH) were used in MWCNTs functionalization and was performed using acid treatment chemistry of $H_2SO_4/HNO_3$ at 1.23 wt.% content of MWCNTs. All MWCNTs were supplied by Cheap Tubes Inc. The polymer binder and the aggregate were obtained from Transpo Industries, Inc. utilizing low modulus polysulfide epoxy consisting of bisphenol A/epichlorohydrin with silane resin and a hardener of diethylenetriamine (DETA), phenol, 4,4′-(1-methylethylidene)bis-, and tetraethylenepentamine. A combination of ceramic microsphere powder and crystalline silica (quartz) was used as aggregate.

## 2.2 Nanocomposite Synthesis and PC Production

MWCNTs were dispersed in the epoxy resin prior to PC production using a sequence of magnetic stirring and ultrasonication at elevated temperature. MWCNTs were first added to the resin; the resin was then heated to 110 °C in an oil bath and magnetically stirred at 800 rpm for 2 h. The epoxy solution was then placed in a degassed distilled water at 60 °C, and ultrasonic waves were applied for another 2 h. No additional protocol was followed for mixes containing both P-MWCNTs and F-MWCNTs (hybrid) to eliminate discrepancy between dispersion energy of plain and hybrid mixes.

After the epoxy nanocomposite was left to cool down to room temperature, the PC production process started. The epoxy resin (nanocomposite or neat) was mixed with the hardener at low speed for 2 min using a hand mixer. The aggregate filler was then added slowly and mixed with the polymer for another 2 min. Flowability test was performed immediately after mixing PC slurry after ASTM C1437 [10]. In this test, PC was placed in a 70/100 mm cone of 50 mm height in two layers each compacted 20 times. The cone was then left, and 25 drops were applied in which PC spread over the flow table. The sum of 4 diameters represents the flowability of PC. Tension specimen was cast filling dog bone-shaped specimen satisfying ASTM D638 [11] and left to harden for 16 h. All specimens were then heat treated at 60 °C for 24 h to accelerate polymerization minimizing testing errors. Table 36.1 summarizes all mixes performed with their designation.

## 2.3 Tension Test

Tension test was performed using ASTM D638 [11] type III specimens. The test utilized an extensometer attached to the mid third of the specimen's height. Loading was performed at 3.75 mm/min according to ASTM standard. The results were analyzed for tensile strength, strain at failure, and toughness which was calculated by calculating the area under the stress-strain curves of each specimen. For each mix, a minimum of five specimens were tested where the mean is reported for mechanical performance and the median was used for the stress-strain plots.

Table 36.1 PC mixes designation with respect to their MWCNTs content

| Content (wt.%) | PC-neat | PCNC1 | PCNC2 | PCNC3 | PCNC4 | PCNC5 | PCNC6 |
|---|---|---|---|---|---|---|---|
| P-MWCNTs | 0.0 | 2.0 | 1.9 | 1.5 | 0.5 | 0.1 | 0.0 |
| F-MWCNTs | 0.0 | 0.0 | 0.1 | 0.5 | 1.5 | 1.9 | 2.0 |

## 2.4 Scanning Electron Microscope (SEM)

SEM images were collected for cured epoxy polymer specimen containing no aggregate filler. Mixes containing 2.0 wt.% P-MWCNTs and 1.9 P-MWCNTs with 0.1 wt.% F-MWCNTs were selected to represent all mixes since P-MWCNTs is harder to disperse than F-MWCNTs. Scans were collected using Hitachi S-5200 Nano SEM with a range of $100\times$–$2{,}000{,}000\times$.

## 2.5 Dynamic Modulus Analyzer (DMA)

Three-point bending test of epoxy polymer specimen with no aggregate filler was performed using DMA. The test utilized a protocol of temperature ramp of 3 °C/min at 1 Hz. Specimen of this test was significantly larger than SEM measuring 20 mm × 12 mm × 3 mm. The test was critical for measuring the cross-linking density of all PC mixes which was performed using Eq. 36.1:

$$v_e = \frac{E'}{3RT} \qquad (36.1)$$

where $v_e$ is cross-linking density, $E'$ is the tensile storage modulus in the rubbery plateau, $T$ is the temperature in Kelvins, and $R$ is the gas constant taken as 8.3145 J/(K mol).

## 3 Results and Discussion

Figure 36.1 shows SEM micrographs collected for PCNC1 and PCNC2. The SEM micrographs show good dispersion of MWCNTs in the epoxy matrix indicating good dispersion. No agglomeration of MWCNTs is visible. Flowability of all PC mixes was measured immediately after producing PC slurry. Results show

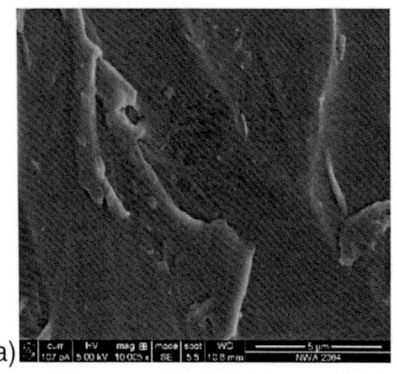
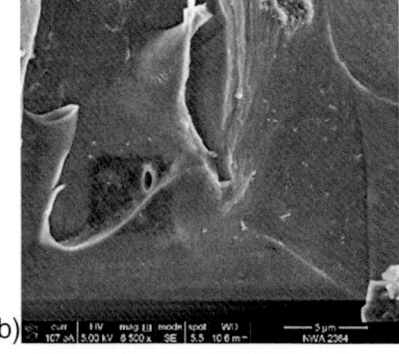

**Fig. 36.1** SEM micrographs of (**a**) PCNC1 and (**b**) PCNC2

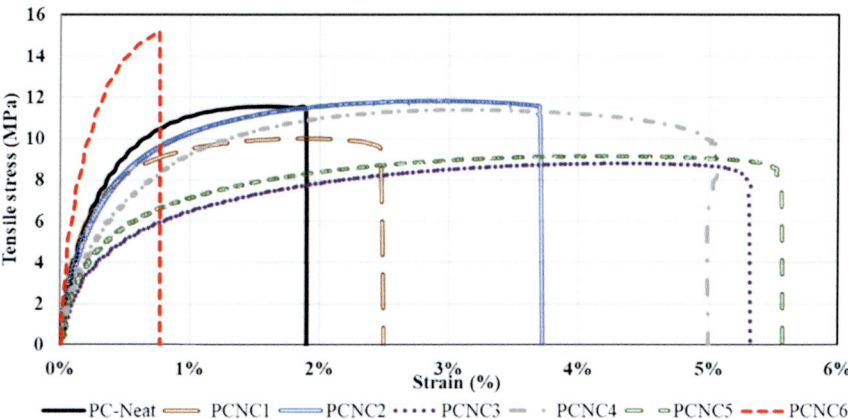

**Fig. 36.2** Stress-strain curves of all PC mixes

insignificant difference between all PCNC mixes with an average decrease of 21% from that of PC-Neat. Nevertheless, the produced PC shows high level of flowability, and incorporating MWCNTs does not affect PC ability to be cast. Stress-strain curves of all PC mixes are presented in Fig. 36.2. All mixes containing P-MWCNTs show significant improvement in ductility represented by both strain at failure and toughness. This is accompanied by a reduction in tensile strength up to 25% from that of PC-Neat at 11.8 MPa where PCNC mixes strength ranged between 9 and 11 MPa. In fact, PCNC2 (containing 1.9 wt.% P-MWCNTs and 0.1 wt.% F-MWCNTs) shows unprecedented levels of strain at failure of 5.5%. On the other hand, PCNC6 (the only mix containing only F-MWCNTs) shows significant improvement in tensile strength reaching 15.4 MPa. However, this improvement in strength comes with a reduction in strain at failure of 0.8% compared with 1.85% of PC-Neat. To further examine the effect of changing the mix content of P-MWCNTs to F-MWCNTs, the tensile properties were normalized by that of PC-Neat as shown in Fig. 36.3. It is evident that the reduction in tensile strength limited by 25% is outweighed by the significant improvement in ductility measuring up to 200% increase in strain at failure and 184% increase in toughness compared with PC-Neat. More importantly, it is interesting to observe the relationship of COOH content and the mechanical performance of PC incorporating MWCNTs. It is apparent that between 0.001 wt.% and 0.180 wt.% COOH functionalization is able to improve the ductility of PC when using 2.0 wt.% MWCNTs content. Our experimental observations confirm our hypothesis that combining both F-MWCNTs and P-MWCNTs can combine their individual effect offering improved tensile properties over mixes containing only one type of MWCNTs. Furthermore, it is important to realize that MWCNTs also offer mechanical reinforcing effect like that of microfibers. This effect occurs due to the high aspect ratio of MWCNTs which improves crack arrest and prolongs ductility. This is evident in Fig. 36.2 where all mixes containing P-MWCNTs (PCNC1 to PCNC5) show considerably high strain at failure.

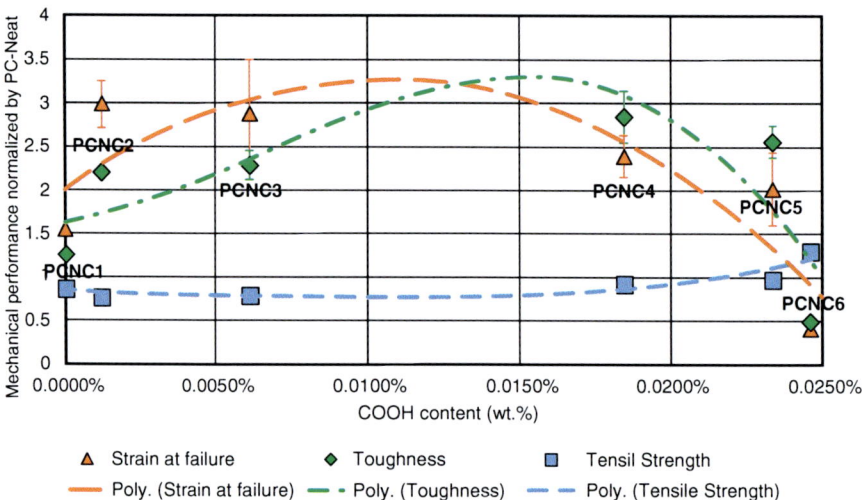

**Fig. 36.3** Normalized tensile properties of all PC mixes with their trend lines

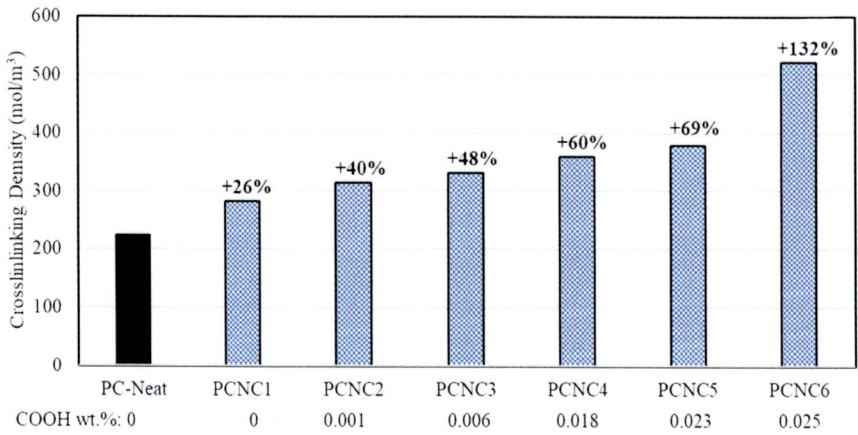

**Fig. 36.4** Cross-linking density of all PC mixes measured using DMA

The chemical effect of MWCNTs was examined via measuring cross-linking density using DMA. The results are presented in Fig. 36.4. The cross-linking density increases with increasing the content of functional group up to 132%. While COOH content between PCNC5 and PCNC6 did not vary significantly, introducing P-MWCNTs even at low content in PCNC5 is sufficient to prevent excessive increase in cross-linking density. This allows PCNC5 to maintain high ductility while PCNC6 to reach high strength. Evidently, using F-MWCNTs only results in significant improvement in tensile strength. This is explained by the embedment of the MWCNTs in the polymer matrix due to its chemical reaction with the polymer matrix.

## 4 Conclusion

This research examined the effects of incorporating a mix of P-MWCNTs and F-MWCNTs at different mix ratios with a fixed 2.0 wt.% total MWCNTs content. Tension and DMA tests were performed. SEM micrographs were observed for epoxy polymer with no aggregate filler. SEM images confirm good dispersion of all mixes. PC with mixed pristine and functionalized MWCNTs reaches unprecedented levels of ductility with failure strain of 5.5%. Results also show that increasing the content of COOH functional group results in increased polymer cross-linking. The presence of P-MWCNTs even at low content prevents increase in cross-linking and thus improved ductility. On the other hand, PC incorporating F-MWCNTs only reaches high tensile strength of 15.4 MPa. The proposed PC with MWCNTs offers a very ductile concrete suitable for extreme loading events.

**Acknowledgment** This work was supported by Transpo Industries, Inc., NY. The authors greatly acknowledge this support.

## References

1. Hosseinali, M., Toufigh, V., & Shirkhorshidi, S. M. (2017). Experimental investigation and constitutive modeling of polymer concrete and sand interface. *International Journal of Geomechanics, 17*(1), 04016043–04016043.
2. Ganguli, S., Aglan, H., Dennig, P., & Irvin, G. (2006). Effect of loading and surface modification of MWCNTs on the fracture behavior of epoxy nanocomposites. *Journal of Reinforced Plastics and Composites, 25*, 175–188.
3. Genedy, M., Daghash, S., Soliman, E., & Reda Taha, M. M. (2015). Improving fatigue performance of GFRP composites using carbon nanotubes. *Fibers, 3*, 3–29.
4. Yu, N., Zhang, Z. H., & He, S. Y. (2008). Fracture toughness and fatigue life of MWCNT/epoxy composites. *Materials Science and Engineering A, 494*, 380–384. https://doi.org/10.1016/j.msea.2008.04.051.
5. Tang, L.-C., Wan, Y.-J., Peng, K., Pei, Y.-B., Wu, L.-B., Chen, L.-M., & Lai, G.-Q. (2013). Fracture toughness and electrical conductivity of epoxy composites filled with carbon nanotubes and spherical particles. *Composites Part A: Applied Science and Manufacturing, 45*, 95–101. https://doi.org/10.1016/j.compositesa.2012.09.012.
6. Thakre, P., Lagoudas, D., Riddick, J., Gates, T., Frankland, S. J., Ratcliffe, J., & Barrera, E. (2011). Investigation of the effect of ingle wall carbon nanotubes on interlaminar fracture toughness of woven carbon fiber-epoxy composites. *Journal of Composite Materials, 45*, 1091–1107.
7. Wetzel, B., Rosso, P., Haupert, F., & Friedrich, K. (2006). Epoxy nanocomposites - fracture and toughening mechanisms. *Engineering Fracture Mechanics, 73*, 2375–2398.
8. Salemi, N., & Behfarnia, K. (2013). Effect of nano-particles on durability of fiber-reinforced concrete pavement. *Construction and Building Materials, 48*, 934–941.
9. Shokrieh, M. M., Kefayati, A. R., & Chitsazzadeh, M. (2012). Fabrication and mechanical properties of clay/epoxy nanocomposite and its polymer concrete. *Materials and Design, 40*, 443–452.
10. ASTM C1437-13. (2013). *Standard test method for flow of hydraulic cement mortar*. West Conshohocken: ASTM International.
11. ASTM D638-14. (2014). *Standard test method for tensile properties of plastics*. West Conshohocken: ASTM International.

# Chapter 37
# Contribution of Concrete-Polymer Composites and Ancient Mortar Technology to Sustainable Construction

Dionys Van Gemert, Lech Czarnecki, Ru Wang, and Özlem Cizer

Recent findings on microstructure formation in polymer-cement concrete are presented. There is strong evidence that interactions between polymers and cement minerals are not only of physical nature but also of chemical nature. Chemical interactions and bonds are nanoscale phenomena. They can increase adhesion between phases and improve mechanical and physical properties of the composite material. The combination of polymers with cement concrete only dates back to the beginning of the twentieth century. However, natural polymers were already used in ancient times to enhance the properties and durability of plasters, mortars, and concretes. Analysis of the composition of ancient binders and mortars with proven durability reveals the hardening activation methods. These methods are applied to improve the hydration mechanisms of pozzolans and industrial by-products to develop more sustainable binders for construction industry.

## 1 Introduction

The yearly global Conferences of Parties (COP) on climate change deal with the worldwide concern of a sustainable development of societies and nations. The first COP took place in Berlin in 1995, but the best known is the COP 3 Kyoto Climate Change Conference of 1997, which adopted the Kyoto Protocol [1]: 37 countries

---

D. Van Gemert (✉) · Ö. Cizer
KU Leuven, Heverlee, Belgium
e-mail: dionys.vangemert@triconsult.be

L. Czarnecki
Building Research Institute (ITB), Warsaw, Poland

R. Wang
Tongji University, Shanghai, China

committed themselves to a reduction of 5.2% each below the 1991 production level of four greenhouse gases (GHG) (carbon dioxide, methane, nitrous oxide, sulfur hexafluoride) and of two groups of gases (hydrofluorocarbons and perfluorocarbons). As of June 2017, 195 members of United Nations Framework Convention on Climate Change signed the Paris Agreement of December 2015. But already in June 2017, the USA withdrew from this Paris Agreement, which cast a shadow over the latest COP 23 in Bonn, November 2017. In the meantime, $CO_2$ content in the air continuously increased from 325.03 ppm in January 1970 to 354.87 ppm in January 1991 to 397.80 ppm in January 2014 and 406.82 ppm in December 2017 [2].

The combination of hydraulic cement and polymer in concrete creates opportunities for beneficial synergies, which result from the intended interactions between cement and polymer particles in the fresh mix and between cement hydration and polymer hardening systems during curing and from the interactions between hardened cement hydrates and hardened polymer structures. Performance enhancement and a wide range of innovative properties and applications become possible. Additionally, the study of ancient binders and mortars reveals opportunities to develop modern, durable inorganic binders, not only for restoration and revitalization projects but also as alternatives for cement. Both PCC (polymer-cement concrete) and inorganic binders contribute to the reduction of the carbon footprint of construction and cement industry [3].

## 2 Milestones in PCC Microstructure Formation Modeling

Several milestones can be observed on the road to building the PCC microstructure formation model, parallel with the history of ICPIC Congresses [4]. At the first ICPIC Congress in 1975, H. R. Sasse [5] already presented the interaction between polymer admixtures and cement hydrates. He assumed that the polymers formed extremely thin resin films or netlike structures on the hydrate surfaces. During hydration, these films are penetrated and swallowed up by newly formed hydration products, thus losing their effectiveness. However, that assumption is only valid for the low ratio polymer admixtures in his study. At first, the models only envisaged the interaction of polymer, cement paste, and aggregates in the hardened state [6, 7]. The original three-step model proposed by Y. Ohama [8] took into account the hardening process of the polymer phase. Afterward, numerous specifications and modifications to this model have been presented [4]. Beeldens et al. [9] proposed an integrated model, in which the interaction between cement hydration and polymer hardening is integrated. Dimmig-Osburg [10] included adsorption of polymer on cement particles, whereas Ye [11] considered possible flocculation effects of the polymer particles, leading to discontinuous distribution of polymer throughout the microstructure. Enhancement of SEM resolution and magnification capabilities also enabled to study the effect of very low amounts of polymer on the microstructure at the nanoscale, e.g., to study the positioning and influence of water-soluble

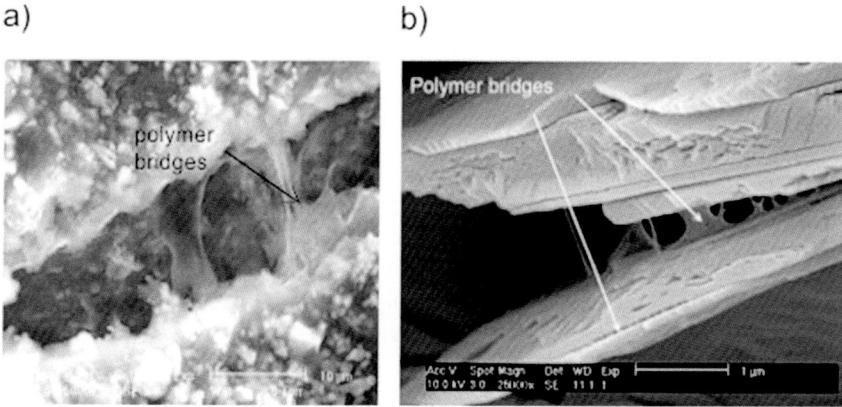

**Fig. 37.1** (a) Polymer bridging between crack edges of concrete on microlevel [9]. (b) Polymer bridging between portlandite plates on nano-level [12]

polymer in between hexagonal portlandite plates [12]. The difference between micro- and nano-interaction is clearly represented in Fig. 37.1. It is obvious that the above presented polymer-cement concrete technology development and microstructure models only and solely involve physical mechanisms and physical interactions by which the synergy phenomena are obtained. Compared to physical interaction, chemical interaction is not considered in many cases. The authors are convinced that in PCC technology, the next stage will be organic-inorganic composites in which some components are chemically bound, additionally to their physical interactions. That should open new expectations and opportunities for further progress in material design [13].

## 3 Chemical Interactions Between Polymer and Cement

Chemical bonding (ionic, covalent, or metallic bonds) as an aspect of bonding of polymer materials to concrete at molecular scale was already considered by Sasse and Fiebrich in 1983 [14], but they attributed bonding primarily to van der Waals forces (internal dipoles originating from dispersion, induction, orientation effects) and to micromechanical interlocking mechanisms. However, recent studies show evidence of chemical interactions between polymers and hydrating Portland cement. Chemical interaction may result in the formation of complex structures, as well as in changes in the morphology, composition, and quantities of hydrated cement phases [15]. Various studies have been conducted to understand the chemical interactions between polymer and cement. Typically two approaches are used in the research process: one is searching chemical interactions in simulated pore water solution, and the other is searching in cement or cement-containing paste. The available research reports on chemical interaction between polymer and cement specifically deal with EVA and acrylic polymers.

Ethylene-vinyl acetate (EVA) copolymer is a very popular polymer, used to modify cement mortar and concrete, because of its balanced performance to cost ratio. Research has been conducted to observe the interaction at both early and late stages of cement hydration through various analysis methods. It is generally believed that the vinyl acetate group contained in EVA copolymer suffers hydrolysis when dispersed in an alkaline medium, known as saponification. When EVA is dispersed in a $Ca(OH)_2$ saturated solution, simulating the pore solution of cement pastes, the acetate anion $CH_3COO^-$ released in the alkaline hydrolysis of EVA combines with $Ca^{2+}$ released in the dissolution of cement grains as shown in Eq. 37.1, and the product of this interaction is calcium acetate, $Ca(CH_3COO)_2$.

$$Ca(OH)_2 + 2CH_3COO^- \Rightarrow Ca(CH_3COO)_2 + 2OH^- \tag{37.1}$$

Equation 37.1 indicates that the $Ca(OH)_2$ content decreases. It was observed that the ettringite crystals are well formed and that many Hadley's grains (hollow hydrated cement particles) are formed. A porous, calcium-rich hexagonal structure phase was found, probably caused by an acetic acid attack on $Ca(OH)_2$ crystals. This chemical interaction can decrease the EVA flexibility and the increase of the modulus of elasticity of the EVA modified cement material.

The reaction scheme between $Ca^{2+}$ and polyacrylate chain is shown in Fig. 37.2 [16].

In 1981, Chandra and Berntsson [17] presented research on the chemical interaction between calcium hydroxide and styrene-acrylic ester copolymer, as later confirmed by Knapen and Van Gemert [12].

The ion bond between a calcium ion and a polycarboxylate enhances the bond strength inside and between particles. Other examples of chemical interactions are listed in [13].

Strengthening mechanisms may be explained by the formation of chelates, chemical compounds in the form of heterocyclic rings, containing a metal ion attached by coordinate covalent bonds to at least two nonmetal ions. This is similar to the complexes formed by natural polymers in ancient lime- and pozzolana-based mortars.

## 4 Ancient Mortar Technology for Sustainable Construction

Some ancient Roman mortars are found to have a remarkably durable behavior in aggressive environments. The Roman concrete is made by mixing burnt lime with volcanic ash and volcanic tuff [18]. By using seawater at mixing underwater concrete, the mixing water instantly triggered a hot chemical reaction. The lime was hydrated and reacted with the ash to cement the whole structure together. Vitruvius (± 85–20 BC) reported that the best maritime concrete was obtained

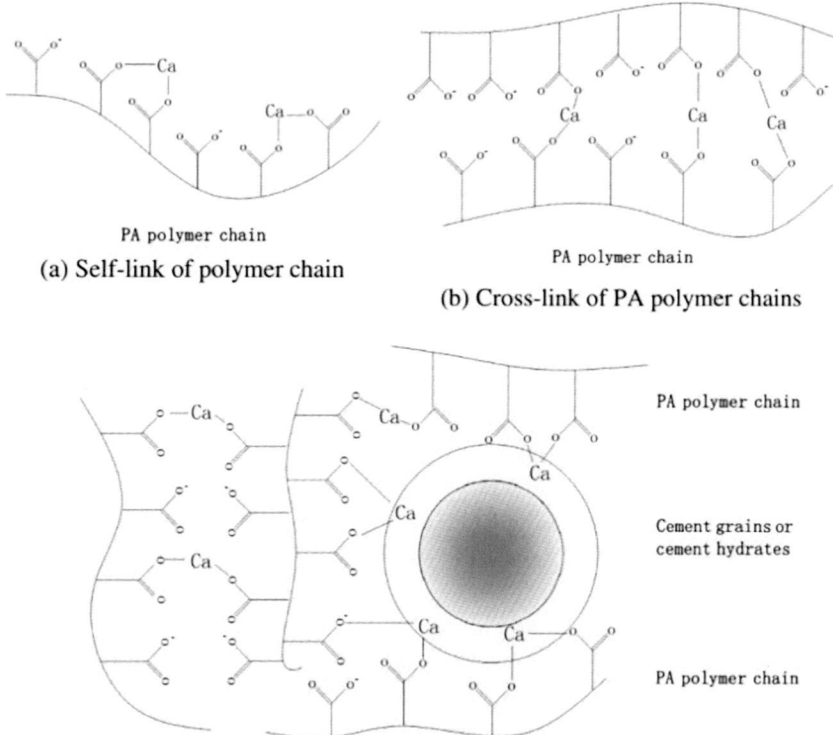

**Fig. 37.2** Reaction between $Ca^{2+}$ and polyacrylate polymer chain

with ash from volcanic regions of the Gulf of Naples, originating from eruptions of, e.g., Mt. Vesuvius. Plinius the Elder (23 or 24 AD – 25th August 79 AD) reported on the benefits of volcanic ashes, especially those from sites near the town of Pozzuoli. Strangely enough, Plinius died in the eruption of Mt. Vesuvius on 25th August 79 AD. Ashes with similar properties, found in many parts of the world, are called "pozzolan" as well. Monteiro et al. [18] investigated maritime concrete from Pozzuoli Bay. Portland cement paste is a compound of calcium, silicates, and hydrates (CSH). On the contrary, Roman concrete contains compounds with less silicon and added aluminum. The resulting CASH binder reveals to be very stable. The CSH phases in modern concrete resemble a combination of naturally occurring layered minerals, called tobermorite and jennite, but never with ideal crystalline structures. The strength and stiffness of ancient seawater concrete are found to be much higher, thanks to the aluminum-tobermorite combination.

Environmentally friendlier modern cements already include volcanic ash or fly ashes as partial substitute for Portland clinker. Roman cement only contained about 10% of lime (burnt at 900 °C compared to Portland cement at 1450 °C), mixed with aluminum-rich pozzolan ash and seawater. Montciro estimates that about 40% of

Portland cement could be replaced by pozzolans. Pozzolans are found all over the world.

Natural polymers (e.g., milk products, oils, blood, etc.) were used in ancient mortars. All these natural products contain various proteins, cellulose, polysaccharides, and fat. Proteins are built up by amino acid residues arranged in long chains, linked by so-called peptide bond $CONH_2$ [19]. Different hydrophobic and hydrophilic segments can be present along the chain of a single protein molecule. The hydrophobic parts go to the air phase (pores and capillaries), and the hydrophilic parts go to the water phase. At high pH values, hydrolysis of the chain takes place. Parts of the chain form salts with alkali and complexes through cross-linking with metal ions.

## 5 Conclusions

The collection of more specific evidence of chemical interactions and the quantification of the influence of these chemical interactions on the performance of polymer-modified cementitious materials are needed. Only a clear understanding of the chemical interactions between polymers and cements can help to correctly explain the micro- and macrostructure relationships in concrete-polymer composites. Exploiting the mineralogical properties of pozzolans in combination with polymers and modern cements is a promising way to reduce the ecological impact of worldwide cement production.

## References

1. United Nations, Kyoto Protocol to the United Nations Framework Convention on Climate Change, UN Doc FCCC/CP/1997/7/Add.1, Dec. 10, 1997; 37 ILM 22 (1998).
2. https://www.esrl.noaa.gov/gmd/ccgg/trends/index.html
3. Van Gemert, D., & Cizer, Ö. (2015). Combining mineral and polymer binder material science for sustainability in construction and restoration. *Restoration of Buildings and Monuments, 21*, 149–164.
4. Van Gemert, D., & Beeldens, A. (2013). Evolution in modeling microstructure formation in polymer-cement concrete. *Restoration of Buildings and Monuments, 19*(2–3), 97–108.
5. Sasse, H. R. (1975). Water-soluble plastics as concrete admixtures. *Proceedings of first international congress on polymer concretes* (pp. 168–173). The Construction Press.
6. Bareš, R. (1985). A conception of a structural theory of composite materials. In A. Brandt & I. Marshall (Eds.), *Brittle matrix composites I* (pp. 25–48), Elsevier Applied Science Publishers, London and New York.
7. Van Gemert, D., Czarnecki, L., & Bareš, R. (1988). Basis for selection of PC and PCC for concrete repair. *International Journal of Cement Composites and Lightweight Concrete, 10*, 121–123.
8. Ohama, Y. (1987). Principle of latex modification and some typical properties of latex modified mortars and concretes. *ACI Materials Journal, 84*, 511–518.

9. Beeldens, A., Van Gemert, D., Schorn, H., Ohama, Y., & Czarnecki, L. (2005). From microstructure to macrostructure: An integrated model of structure formation in polymer modified concrete. *RILEM Materials and Structures, 38*(280), 601–607.
10. Dimmig-Osburg, A. (2007). Microstructure of PCC – Effects of polymer components and additives. *Proceedings of 12th ICPIC, Chuncheon* (pp. 239–248).
11. Tian, Y., Li, Z., Ma, H., & Jin, N. (2011). An investigation on the microstructure formation of polymer modified mortars in the presence of polyacrylate latex. In C. Leung, & K. Wan (Eds.), *Proceedings of international RILEM conference on advances in construction materials through science and engineering* (pp. 71–77), 3–5 September 2011, Hong-Kong.
12. Knapen, E., & Van Gemert, D. (2009). Cement hydration and microstructure formation in the presence of water-soluble polymers. *Cement and Concrete Research, 39*(1), 6–13.
13. Wang, R., Li, J., Zhang, T., & Czarnecki, L. (2016). Chemical interaction between polymer and cement in polymer-cement concrete. *Bulletin of the Polish Academy of Sciences. Technical Sciences, 64*, 785–792.
14. Sasse, H. R., & Fiebrich, M. (1983). Bonding of polymer materials to concrete. *RILEM Materials and Structures, 16*(94), 293–301.
15. Wang, R., Wang, G., & Wang, P. (2015). Status of research and application of C-PC in China. *Advanced Materials Research, Vol. 1129, Proceedings of ICPIC XV Singapore* (pp. 59–68).
16. Tian, Y., Jin, X., Jin, N., et al. (2013). Research on the microstructure formation of polyacrylate latex modified mortars. *Journal Construction and Building Materials, 47*, 1381–1394.
17. Chandra, S., & Berntsson, L. (1981). Behavior of calcium hydroxide with styrene acrylates polymer dispersion. *Journal Cement and Concrete Research, 11*, 125–129.
18. Monteiro, P. (2013). *Roman seawater concrete holds the secret to cutting carbon emissions.* Berkeley Lab. http://newscenter.lbl.gov/2013/06/04/roman-concrete/
19. Chandra, S. (2003). *History of architecture and ancient building materials in India.* New Delhi: Tech Books International.

# Chapter 38
# Smart Monitoring of Movement and Internal Temperature Changes Within Polymer Modified Concrete Repair Patches

Johannes Bester, Jacques G. Engelbrecht, and Michael Grobler

Polymer modified concrete repair materials (PMCRMs) offer significant protection to concrete reinforcement, if the PMCRM is compatible with the concrete substrate, and the instructions given by the manufacturer are followed. However, if the PMCRM is not compatible with the concrete substrate, or the manufacturer's instructions are not followed, volume changes within concrete repair materials (CRMs) caused by shrinkage may result in the repair system failing by means of delamination or cracking. This paper demonstrates the use of an optical fibre strain and temperature sensor to provide valuable information regarding the movement and internal temperature changes which occur within concrete repairs, which may help extend the service life and provide a better understanding of PMCRMs. A strain measuring (fibre Bragg grating) FBG sensor was implemented to measure strain within a polymer modified concrete repair patch (PMCRP). The optical fibre monitoring system consisted of an internal temperature measuring FBG sensor and a strain measuring FBG sensor, which monitored the temperature and movement changes within a PMCRP. The device allowed for easy installation of the strain measuring FBG, and allowed for the optical fibre sensor to be reused, thus it was possible to measure both the internal temperature and the movement of the PMCRP soon after placement of the repair material.

---

J. Bester (✉) · J. G. Engelbrecht
Department of Civil Engineering Science, University of Johannesburg, Johannesburg, South Africa
e-mail: jannesb@uj.ac.za

M. Grobler
Department of Electrical and Electronic Engineering, University of Johannesburg, Johannesburg, South Africa

# 1 Background

## 1.1 Concrete Patch Repair Failure

Concrete is the most widely used building material, but because of poor design, incorrect building procedures and underestimating environmental conditions, concrete's durability may be affected negatively [1]. Concrete is therefore susceptible to failure caused by deterioration, making concrete repair work a necessity [2].

A variety of PMCRM are available on the market, such as polymer latexes, liquid resins, re-dispersible polymer powders, monomers and water-soluble polymers [3]. Common failure modes in composite materials are delamination and cracking. Delamination is the separation of one or more layers within the material, which results in the material being made up of separate layers [4]. This separation will result in the loss of mechanical toughness, which in turn could result in repairing a repair. This obviously has financial implications for the owner of the structure. Cracking may be because of restrained shrinkage [2]. Hence, the accurate measuring of movement and thermal changes of patch repairs is paramount to counteracting concrete repair patch failure.

## 1.2 Optical Fibre Sensors

The measurement of movement as soon as possible within the PMCRM is important, as a good deal of the shrinkage could occur during the first few hours after placement. Some of the conventional instruments used to measure strain in concrete and concrete repair materials include: the ring test, Baenziger block, photogrammetric systems, non-contact lasers, Moiré interferometry and strain gauges.

Considerable research on the use of optical fibre sensors for structure condition monitoring is focused on the long-term health of structures and investigates the properties of hardened concrete by measuring strain. Banerji et al. [5] used optical fibre strain sensors to monitor a concrete box-girder bridge. In this work, optical fiber strain gauges were clamped to the surface of the bridge using grouted studs and shows that an optical fibre strain sensor system provides adequate concrete surface strain information under loaded conditions. Lai et al. [6] implemented FBG sensors to monitor real-time strain in the liner concrete of a tunnel structure. It showed that during in situ monitoring, FBG-based sensors can provide safety and stability information of a tunnel liner structure. Yazdizadeh et al. [7] used electrical strain gauges and FBG strain sensors to explore shrinkage and creep of concrete and found FBG sensors a suitable alternative method to study time-dependant properties of concrete. The results from Banerji et al. [5–7] show that FBG sensors can be effectively used for quantifiable measurement of concrete structure strain to assess structure health in the civil engineering industry. Wong et al. [8–10] show that FBG sensors can be employed during the early age of various cement and concrete combinations to measure shrinkage and curing temperature.

FBGs are intrinsic fibre sensors in optical fibres where the refractive index of the fibre core is periodically modulated by UV laser illumination [11]. When light from a broadband source is launched into an optical fibre and propagates through this modulated structure, Bragg reflection causes one wavelength to be selectively reflected, so that it satisfies Eq. 38.1. The reflected Bragg wavelength is given by:

$$\lambda_b = 2n_{eff}\Lambda, \qquad (38.1)$$

where $n_{eff}$ indicates the effective refractive index with respect to the core propagating mode and $\Lambda$ [pico-meter] is the grating period of the FBG.

As can be seen from Eq. 38.1, any change in the periodicity or refractive index of the FBG will result in a change in the reflected Bragg wavelength and any strain or temperature-induced effect on the FBG will produce a shift in the reflected Bragg wavelength.

## 2 Experimental Design

### 2.1 Concrete Patch Repair

There were four samples used in this experiment. Product A is a polymer modified concrete repair material which is fibre-reinforced and meets the requirements of class-R4 of EN 1504-3 [12]. Product B and C are polymer modified concrete repair materials that contain silica fume which enhances the compressive strength, bond strength and abrasion resistance, whilst reducing bleeding of a cementitious material. A very dry cementitious-based concrete mix, containing no polymers, was used as the control, referred to as product D. A 250 × 250 × 70 mm repair area was done on a 3-year-old ready-mix supplied 25 MPa (floor slab) concrete substrate of dimensions 1000 × 500 × 180 mm.

### 2.2 Optical System

FBG sensors embedded into the concrete repairs allow for the early age monitoring of the movement and temperature changes. However, these sensors may also be attached to the surface of composites. Figure 38.1 shows the optical setup to monitor the strain and temperature of the PMCRM. An FBG interrogator (Smartscan SBI) was used to collect the wavelength information for the strain and temperature FBG measurements.

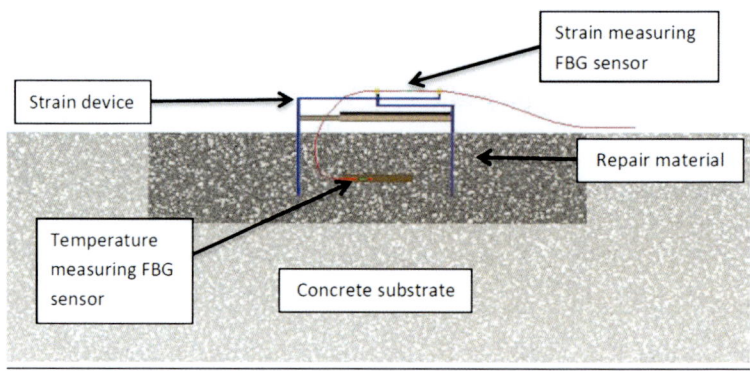

**Fig. 38.1** Strain and temperature FBG measurement

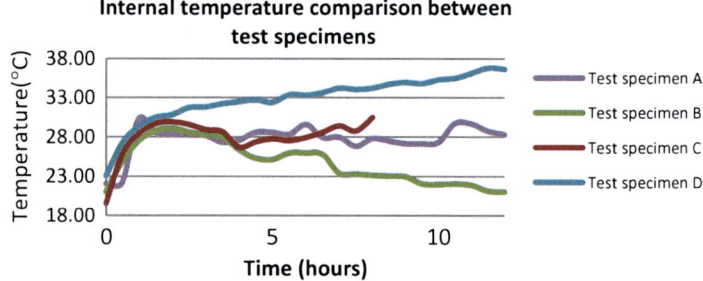

**Fig. 38.2** Internal temperature measurement

## 3 Results

### 3.1 Temperature Measurement and Movement Measurement

The internal changes in temperature for the four specimens are shown in Fig. 38.2. It shows that the four specimens experienced a temperature increase during the first 2–3 h due to the exothermic hydration process. Test specimens A, B and C experienced a decrease in internal temperature after the second hour, which corresponds to shrinkage shown in Fig. 38.3, starting at the third hour. The internal temperature of test specimen A increased by 6.4 °C during the first 2 h, whereas specimens B and C increased by 8.1 °C and 10.4 °C during this period. The internal temperature of specimen D increased during the entire 12 h period. Test specimen B was at a temperature of 21 °C at 12 h, which was the same as the ambient starting temperature.

Figure 38.3 displays the change in length of each test specimen during the first 12 h. It can be seen that test specimen B experienced expansion during the first 3 h, whereafter shrinkage was experienced by specimens B, C and D. Test specimen A experienced a maximum expansion of 0.053 mm, and little to no movement occurred

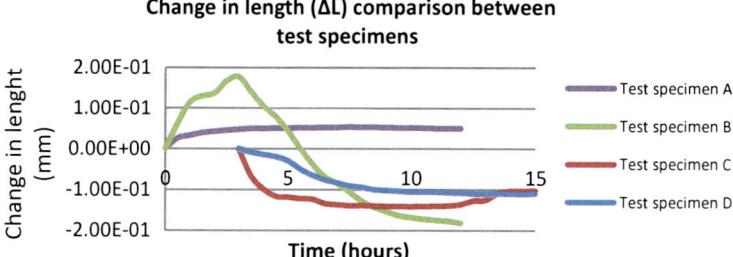

**Fig. 38.3** Strain measurement

after approximately 3 h. Test specimens C and D were only measured for strain starting from the third hour due to difficulties experienced with the optical fibres; as a result, the movement during the first 3 h was not measured.

Test specimen B experienced a higher change in length throughout the 12 h (0.177 mm expansion and shrinkage thereafter of 0.35 mm) when compared to specimens A, C and D. This is most likely due to the difference in composition between the specimens. The movement which occurred in test specimens B and C over 12 h is much larger than that of the test specimens A and D. Specimen B and C were the only test specimens which cracked on the concrete/repair material interface. It was observed that at approximately 8 h into the testing, cracks were forming on the interface between the concrete substrate and the PMCRM of specimen B. This is expected when looking at the amount of shrinkage that took place in the period of 3–7 h.

## 4 Conclusion

The FBG strain measuring device, and the FBG temperature sensor was able to measure the movement and internal temperature of the different PMCRM specimens from a very early age. The movement and internal temperature results taken during the first 12 h of the different PMCRM specimens gave general indications of the relationship between the change in internal temperature and the change in length. This application of FBGs allows for the accurate measurement of both temperature and movement from repairs directly after placement. This, in turn, could assist in measuring excessive movement, which could result in delamination and cracking of the PMCRM resulting in a failed CPR.

Future work would entail the development of a remote sensing strain and temperature optical fibre sensor system to monitor early age parameters which could be left on a structure for condition monitoring.

**Acknowledgements** This work was supported by Telkom, CBI, NLC and the URC of UJ.

# References

1. Owens, G. (Ed.). (2012). *Fundamentals of concrete* (2nd ed.pp. 115–128). Midrand: Cement & Concrete Institute.
2. Ballim, Y., Alexander, M., & Beushausen, H. (2009). Durability of concrete. In *Fultons concrete technology* (pp. 155–188). Midrand: Cement & Concrete Institute.
3. Ohama, Y. (1995). *Handbook of polymer-modified concrete and mortars: Properties and process technology*. New Jersey: Noyes Publications.
4. Cantwell, W. J., & Morton, J. (1991). The impact resistance of composite materials – A review. *Composites, 22*(5), 347–362.
5. Banerji, P., Chikermane, S., Grattan, K., Sun, T., Surre, F., & Scott, R. (2011). Application of fiber-optic strain sensors for monitoring of a pre-stressed concrete box girder bridge. In *Sensors, 2011 IEEE*, Limerick, Ireland (pp. 1345–1348).
6. Lai, J., Qiu, J., Fan, H., Zhang, Q., & Hu, Z. (2016). Fiber bragg grating sensors-based in situ monitoring and safety assessment of loess tunnel. *Journal of Sensors*, (pp. 1–10).
7. Yazdizadeh, Z., Marzouk, H., & AliHadianfard, M. (2017). Monitoring of concrete shrinkage and creep using Fiber Bragg Grating sensors. *Construction and Buiding Materials, 137*, 505–512.
8. Coetzee, W., Van Eck, S., Grobler, M., Vannucci, M., Marinez Manuel, R., & Bester, J. (2015). Embedded Fibre Bragg Gratings to measure shrinkage during the early age of concrete. *Proceedings of the 12th IEEE 2015 Africon international conference*. Addis Ababa, Ethiopia.
9. Wong, A., Childs, P. B. R., Macken, T., Peng, G., & Gowripalan, N. (2007). Simultaneous measurement of shrinkage and temperature of reactive powder concrete at early-age using fibre Bragg grating sensors. *Cement & Concrete Composites, 29*, 490–497.
10. Dong, L., Ibrahim, Z., & Ismail, Z. (2013). The use of fibre Bragg grating to monitor the early age curing temperature of cement paste. *Electronic Journal of Structural Engineering, 13*(1), 56–59.
11. Spammer, S., & Fuhr, P. (1998). Concrete embedded optical fibre Bragg grating strain sensors. *IEEE international symposium on industrial electronics. Pretoria*.
12. European Standards. (2005). EN 1504-3, *Products and systems for the protection and repair of concrete structures – Definitions, requirements, quality control and evaluation of conformity – Part 3: Structural and non-structural repair*. British Standards Institution, London, UK.

# Chapter 39
# PIC: Does It Have Potential?

David W. Fowler

Polymer-impregnated concrete (PIC) was developed in the 1970s by Brookhaven National Laboratory and the Bureau of Reclamation. Fully impregnated concrete had excellent strength and durability properties. Partially impregnated concrete had very good resistance to water and chloride penetration and to abrasion. Neither of these processes is used to any significant degree. The reasons for the lack of use and the potential for future use are discussed including potential applications to make use of the very good material properties that are possible.

## 1 Introduction

Polymer-impregnated concrete (PIC) was developed in the 1970s by the collaborative research by Brookhaven National Laboratory (BNL) and the Bureau of Reclamation (BuRec) [1]. PIC is produced by impregnating portland cement concrete with a low-viscosity monomer, e.g., methyl methacrylate (MMA), that is subsequently polymerized in situ. Full-depth impregnation was shown to result in a three- or fourfold increase in compressive and tensile strength, 50% or higher modulus of elasticity, excellent resistance to freezing and thawing and acid attack, and excellent resistance to abrasion. Partial-depth impregnation, which does not require the extensive drying, vacuum, and pressure required for full depth, was developed to provide resistance to intrusion of chlorides and water to bridge decks and parking garages in field conditions. Considerable research was performed at BNL, BuRec, and universities in the USA and abroad. The research led to many potential applications of PIC that caught the attention of the concrete and related industries. During the 1970s,

D. W. Fowler (✉)
University of Texas, Austin, TX, USA
e-mail: dwf@mail.utexas.edu

**Fig. 39.1** Pretensioned PIC bridge slab (BuRec)

lofty predictions were made for both types of PIC to take advantage of their excellent mechanical and durability properties.

Some of the applications for full-depth impregnation that were made and tested included water and sewer pipe, tunnel liners, precast, pretensioned bridge deck slabs (Fig. 39.1), quarter-scale posttensioned beams, piling, form liners, and sanitary ware. Full impregnation never received traction, and a Japanese firm is the only known company that is producing PIC products in the form of stay-in-place form liners. Partial-depth impregnation applications include bridge deck surfaces, parking garage floors, dam inlet tunnel walls and stilling basin floors, statuary, floor tile, and murals. One firm is using partial-depth impregnation to treat garage floors and bridge decks using a vacuum impregnation process. The process has also been used to treat statuary and wall murals that are subject to deterioration.

Numerous bridge decks were treated in Texas and other states. At that time partial depth was potentially the best method for waterproofing a deck without adding substantial weight. The first partial-depth impregnation that was one of the most complex ever performed was on the inlet walls and stilling basin floor of Dworshak Dam in Idaho. The inlet walls had been badly eroded due to cavitation of the water flowing through the dam. The walls were repaired with high-strength concrete covered with steel fiber-reinforced concrete. The surface was then impregnated to a depth of about 25 mm. The floor of the stilling basin into which the water from the inlets flowed was eroded through a 1.2 m-thick high-strength concrete and into the rock below to a depth of about 1.5 m. The floor was repaired with high-strength concrete covered with FRC and then impregnated under some very hostile environmental conditions. Later inspections showed that both walls and floor had no significant scouring or erosion (Fig. 39.2).

**Fig. 39.2** Applying monomer (USACOE) on stilling basin floor

## 2 What Happened to PIC?

PIC had such an impressive beginning. For portland cement concrete with 28-day compressive strength of 35 MPa, the full-depth PIC from which it is made has approximately:

- 140 MPa compressive strength
- 10 MPa tensile strength
- 16 MPa modulus of rupture
- 20,000 MPa modulus of elasticity
- 0.6% water absorption
- 3500 cycles of freezing and thawing resistance
- 10 times the acid resistance as control concrete
- 5 times the abrasion resistance of control concrete
- Very low creep

The potential applications were endless, many of which were listed in the previous section. One of the most intriguing is posttensioned beams. The Bureau of Reclamation produced and tested posttensioned precast bridge panels shown in Fig. 39.1. However, the pretensioning was done before impregnation, and the large increase in compressive strength of PIC was not taken advantage of. Posttensioning is done after impregnation, and the advantages of much higher compressive strength, tensile strength, modulus of elasticity, and low creep can be utilized.

At the University of Texas at Austin, we made and tested posttensioned PIC I-beams that were 200 mm deep and 2.29 m long (Fig. 39.3) [2]. The beams were fully impregnated with polymer and had a compressive strength of 104 MPa. The beams were dried at up to 130C, cooled to room temperature, and evacuated in a

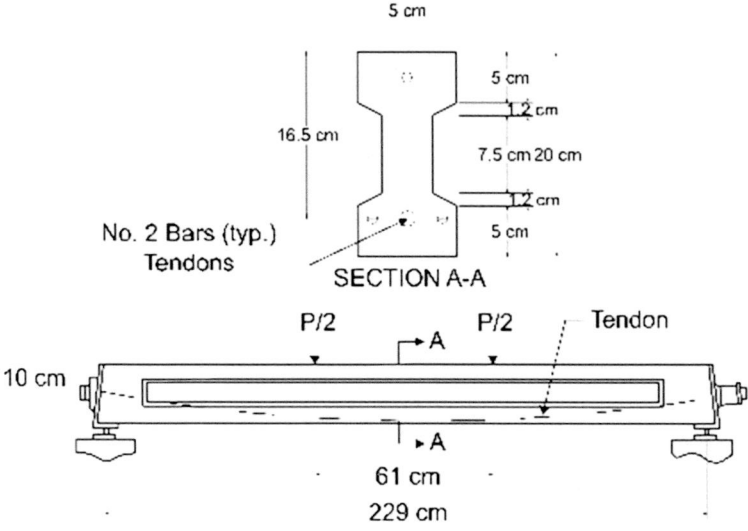

**Fig. 39.3** Dimensions of PIC beam

chamber at a vacuum of 750 mm Hg. Methyl methacrylate monomer system was introduced under a pressure of 340 kPa and was polymerized by immersion in water at 82C. The vacuum and pressure were required to achieve full impregnation in the pores of the concrete. Unbonded 6 mm wires, ASTM grade 240 (1650 MPa), were used for the tendons. Two, four, six, and eight wires were used in the four PIC beams tested; the control could accommodate only two wires based on tensile stress as a result of posttensioning. The higher tensile strength of the PIC permitted the use of more wires. The wires were stressed to approximately 70% of their ultimate strength.

The test results (Fig. 39.4) indicate that the PIC beams could carry considerably more load because of the larger number of wires. The one with two wires carried the same load as the control with about 50% more deflection. The one with eight wires carried three times as much load as the control with about the same deflection. Combined with the impermeability of PIC that prevents water and chloride intrusion, PIC posttensioned beams that can carry three times the load of non-PIC beams of the same size seemed to have a tremendous future for highway girders, particularly in coastal areas or northern climates where chlorides pose a problem for normal portland cement concrete. But it did not happen. Why?

The cause was the process that is required to achieve polymer impregnation. (1) Monomers were new to the concrete industry, and PIC was considered to be an "exotic" material that was too advanced for the construction industry at that time. (2) The process of drying the concrete, subjecting it to a vacuum, immersing the beam in monomer, and applying pressure and applying heat to polymerize the monomer was considered too complex and, as a result, too expensive for the industry. It would have taken an entrepreneur willing to invest in an impregnation facility large enough to produce full-size beams to prove the process and determine

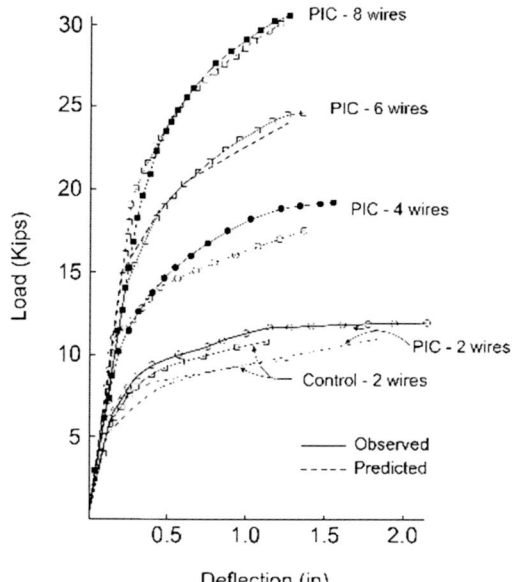

**Fig. 39.4** Load-deflection curve

the cost. But until there was a reasonable assurance that PIC would be acceptable to DOTs and other owners, no one would make the investment. It became a Catch-22.

Partial-depth impregnation was developed to permit concrete bridge decks to be treated. At that time there was considerable interest in developing procedures to prevent chloride and water intrusion into decks. The process developed at UT Austin consisted of placing a 10 mm layer of sand on the deck that served to prevent thermal shock during drying and later to hold the monomer in place during soaking, drying the deck using infrared heaters for up to 6 h to achieve a concrete temperature at the desired depth of polymer impregnation of about 130C, cooling the concrete, applying about 1 kg/sq. m of monomer system and letting it soak for 4–6 h to achieve a penetration depth of at least 20 mm, and polymerizing the monomer using steam heat inside shallow enclosures on the deck [3]. The process required about 2 days for each section treated. The process produced impregnation depths of 18–25 mm that gave a durable, abrasion-resistant surface that was essentially impermeable. Again, why were they not continued to be used?

First, the process was too time-consuming. Second, and most importantly, polymer concrete overlays began to be developed, and they were much simpler, faster, and easier to apply. The cost was usually less. As a result, a very good process for producing durable, strong, and impermeable concrete surfaces did not continue to be used.

## 3  Is There a Future for PIC?

**Full-Impregnation PIC**

Full-impregnation produces PIC with excellent properties for structural applications including posttensioned beams, slabs, and other members particularly when durability is important. No other concrete material has the same combination of strength, stiffness, durability, and impermeability. It can be used to significantly improve the properties of low-strength materials. If the material can be produced at a cost that is competitive with other concretes, it can be successful.

What is needed? The process needs simplifying. The time required needs to be reduced. A more rapid drying procedure needs to be found, perhaps a form of vacuum drying. Situating the impregnation facility near a source of waste heat, e.g., cement kiln or brick masonry kiln, could lower the cost of drying and perhaps polymerizing the monomer.

**Partial-Depth Impregnation**

Partial-depth impregnation has no future in bridge decks because of the time and cost; polymer concrete overlays are a better solution from the standpoint of cost, time, and simplicity and are a widely used method. However, the process has considerable potential for other products that require surface strengthening, improved durability, or improved wear resistance including:

- Building tile. Absorptive, low-strength cementitious tile can be improved significantly. Figure 39.5 shows low-quality tile before and after impregnation and after a sandblast abrasion test had been performed. The tile were dried, cooled, immersed in monomer for 5–15 min, and polymerized. The flexural strength was increased by a factor of three, the water would bead on the surface, and the abrasion resistance was improved significantly.
- Sewer pipe. The inside face of sewer pipe can be impregnated to provide acid resistance and reduce abrasion and scaling.
- Floors and statutory using a vacuum process involving placing a plastic membrane over the floor or around the statue for drying, applying a vacuum, and

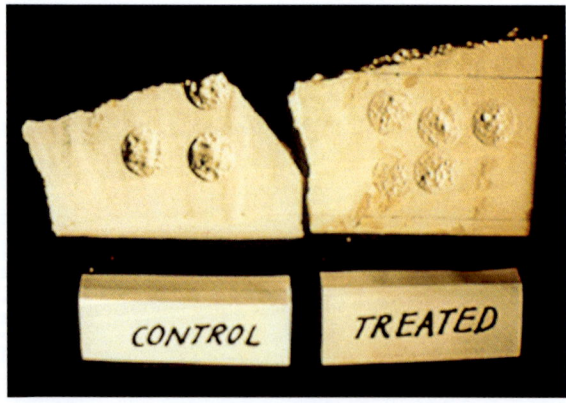

**Fig. 39.5** Tile after partial depth

**Fig. 39.6** Vacuum impregnation process impregnation

applying the monomer. This process has considerable potential for treating low-strength floors and unusually shaped concrete members (Fig. 39.6). The method is commercially available in the USA but is not well known.

## 4 Will PIC Ever Find a Use?

The excellent mechanical and durability properties make full-depth impregnation very attractive. If it becomes a viable material, it will require advancements in the technology, including drying and polymerization in situ. It would be an excellent material for posttensioned and other structural concrete members.

Partial-depth impregnation has some intriguing possibilities for improving the properties of products that require stronger or more durable or more abrasive-resistant surfaces. The available technology, including vacuum impregnation, can be adapted to floors, parking garages, and unusual shapes such as statues.

## References

1. Dikeou, J. T. (1978). Polymers in concrete: new construction achievements on the horizon. *Proceedings of the second international congress of polymers in concrete*, Austin (pp. 1–8).
2. Limsuwan, E., Fowler, D. W., Paul, D. R., & Burns, N. H. (1978). Flexural behavior of post-tensioned polymer-impregnated beams. *Proceedings of the second international congress of polymers in concrete*, Austin (pp. 361–380).
3. Bartholomew, J., Fowler, D. W., & Paul, D. R. (1978). Current status of bridge deck impregnation. *Proceedings of the second international congress of polymers in concrete*, Austin (pp. 399–411).

# Chapter 40
# A Perspective on 40 Years of Polymers in Concrete History

Albert O. Kaeding

This paper presents a brief history of the author's experience with using polymers in concrete beginning in 1976. The manufacture of precast underground utility structures are a significant application of polymer concrete. As the use of these products grew, standards were developed and applied to make the products more consistent from the various suppliers. There are some significant differences among the standards in use. These differences are discussed, and some suggestions for improvements to make the standards agree are presented. Drainage structures are also being successfully manufactured and used, and their use is discussed. Both underground utility structures and drainage structures are typically made using polyester-based polymer concrete with vinyl ester being used where corrosive environments are present.

## 1 Underground Utility Enclosures

The use of precast polymer concrete (PC) for underground utility structures began in the 1970s. Product development took place with the involvement of companies who would become producers and several West Coast utilities who would use the products. Utilities and producers undertook early development of standards for these precast PC products. The products and the standards evolved over the years from four basic enclosure sizes to the several hundred sizes offered by today's producers. Use of the enclosures expanded from use by a few electrical and telephone utilities in the western United States to today's use by utilities across the United States and including water distribution utilities, wind energy producers, and

---

A. O. Kaeding (✉)
Consulting Structural Engineer, Ormond Beach, FL, USA
e-mail: aok@iag.net

**Fig. 40.1** Typical enclosure

state highway departments. Additional applicable standards have been developed over the years by the American Concrete Institute, ASTM International, the Western Underground Committee, and ANSI. The focus of the specifying standards is to provide a design basis, to provide provisions for testing enclosures with simulated structural loads, and to provide material tests to assure the products will be suitable for the varying environments in which they will be used.

Underground utility enclosures (Fig. 40.1) [1] are used for housing underground outside plant equipment by the electrical, telecommunications, transportation, and water utilities. Electric utilities house equipment such as cable junctions and underground transformers. The telecommunications industry uses underground enclosures for fiber optic splices and for cables connected to their switching equipment. The transportation industry makes traffic signal and street lighting connections in underground enclosures and splices cable runs for smart highways. Water utilities house water meters and backflow preventers in the enclosures. These utilities also use precast PC pads to support surface-mounted equipment such as transformers and switchgear.

Underground utility enclosures have historically been made from precast Portland cement concrete and, more recently, from precast PC. Polyester-based PC is normally used. Where wheel loads are not anticipated, enclosures may also be made from thermoplastics. Precast PC was first used around 1975 to manufacture underground utility enclosures. Today about four manufacturers produce underground enclosures and distribute them nationally. There are a number of smaller regional and local producers as well. The enclosures vary in size from about 6 × 8 × 6 inches to 8 × 8 × 8 feet. The smaller enclosures are typically referred to as handholes. The larger units are considered manholes. The advantage of PC enclosures compared to concrete is the lower weight for the same strength and better durability during handling.

Typical construction of an enclosure includes a four-sided box with a cover. Normally, the enclosures do not have a bottom. Sometimes an extension is included that can be installed below the box to produce a deeper enclosure. The covers are fiberglass-reinforced or steel and fiberglass-reinforced PC. The boxes are either fiberglass-reinforced PC or a composite of fiberglass-reinforced polymer (FRP) and fiberglass-reinforced PC. In the composite construction, the upper section is of fiberglass-reinforced PC to provide rigidity and abrasion resistance. The lower wall portion of the box is FRP, providing very high strength and low weight. Fiberglass-reinforced PC in composite with FRP provides a very lightweight, strong enclosure and is preferred by many utilities.

## 2 Design Standards for Underground Utility Structures

Underground utility enclosures are subject to loads from pedestrian foot traffic up to loaded, multi-axle trucks. Wheel loads may be partially on the enclosure or fully on the cover at locations that result in maximum shear and moments. In addition, wheel loads on the pavement or soil next to the enclosure produce a resulting pressure on the walls of the enclosure. Manufacturers rate various sizes and strength enclosures according to one or more standards. Design accounts for the standing weight or slowly moving weight of the vehicle, either on the cover or on the surrounding soil or pavement. PC enclosures have not been designed to withstand the heavy, high-speed traffic encountered in primary and secondary roadways.

The earliest standard developed was Western Underground Committee Guide 3.6, *Nonconcrete Enclosures* (Guide 3.6) [2]. This standard was developed by a number of utility companies on the West Coast of the United States. The standard is performance-based and describes common sizes and features for underground enclosures used by these West Coast utilities. Included in the performance requirements are resistance to chemicals commonly found in and near roadways and parking lots, minimum performance when exposed to ultraviolet light, maximum water absorption, resistance to projectile impact, and flammability limits. The latest edition of this standard was issued in May 1988. In that revision, one loading is described with test requirements to show compliance but without a corresponding design load.

Another standard is ANSI SCTE 77, *Specification for Underground Enclosure Integrity* (ANSI SCTE 77) [3]. This standard is nationally recognized. It essentially contains the same material performance requirements as Guide 3.6. This standard has five different load ratings with a design load and test load for each rating. The design loads and test loads include both minimum vertical and minimum horizontal loadings on the enclosures.

A third standard is ASTM C857, *Standard Practice for Minimum Structural Design Loading for Underground Precast Concrete Utility Structures* (ASTM C857) [4]. While the standard states that it applies to precast concrete utility structures in the scope, the loading described can be applied to underground utility structures made from other materials. This standard specifies four specific vehicle and pedestrian load designations and describes how to apply the loading to the top of the underground structure. It also contains formulas for calculating applicable side loading on an underground structure.

**Table 40.1** Applicable load standards for underground enclosures

| Standard | Load designation | Design load, lbs | Load + impact, lbs | Test load, lbs | Safety factor |
|---|---|---|---|---|---|
| WUC 3.6 | Incidental | n/a | n/a | 10,400 | n/a |
| ANSI SCTE 77 | Tier 5 | 5000 | n/a | 7500 | 1.5 |
| ANSI SCTE 77 | Tier 8 | 8000 | n/a | 12,000 | 1.5 |
| ASTM C857/C497 | A-8 | 8000 | 10,400 | 22,568 | 2.8 |
| AASHTO/T33 | H-15, HS-15 | 12,000 | 15,600 | 33,852 | 2.8 |
| ASTM C857/C497 | A-12 | 12,000 | 15,600 | 33,852 | 2.8 |
| ANSI SCTE 77 | Tier 15 | 15,000 | n/a | 22,500 | 1.5 |
| AASHTO/T33 | H-20, HS-20 | 16,000 | 20,800 | 45,136 | 2.8 |
| ASTM C857/C497 | A-16 | 16,000 | 20,800 | 45,136 | 2.8 |
| ANSI SCTE 77 | Tier 22 | 22,500 | n/a | 33,750 | 1.5 |

In the United States, standardized truckloads are specified by AASHTO, *Highway Bridge Specification* [5]. The bridge specification provides vehicle design loads that correspond to the loads specified in ASTM C857. Proof of compliance testing can be performed using ASTM C497, *Standard Test Methods for Concrete Pipe, Manhole Sections, or Tile* [6], ANSI SCTE 77, or AASHTO T33, *Standard Method of Test For Determining Physical And Chemical Properties of Culvert Pipe, Sewer Pipe, and Drain Tile* [7]. The American Concrete Institute has published a standardized testing procedure, ACI 548.7, *Test Method for Load Capacity of Polymer Concrete Underground Utility Structures* [8], that can be used to compare the strength of various underground enclosures using a standardized test method.

Design loads imposed on underground structures are generally governed by the code or standard that applies to the installation location and the utility doing the installation. For power-generating and distribution utilities, the governing code is the National Electric Safety Code [9]. Section 323 of this code describes design loading that is the same as AASHTO load designation H-20 and ASTM C857 load designation A-16. Underground power distribution in locations such as shopping centers and other private applications is subject to the National Electric Code [10]. The Electric Code specifies that enclosures comply with ANSI SCTE 77. Enclosures installed in highway right-of-ways are generally governed by state department of transportation standards.

There are significant differences in the loads the various standards specify. Table 40.1 compares some of the design loads and test loads from the standards and illustrates some of the differences. ASTM C857 specifies the same layout for a design truck and the same wheel loads as are specified in the Highway Bridge Specification. ANSI SCTE 77 has one load that matches the ASTM standard, designated tier 8, but other loads specified are different from the AASHTO and ASTM standards. The tier 15-load designation specifies a 15,000 lb design load but has nearly the same test load as the A-8 load, which has a design load of 8000 lbs.

ANSI SCTE 77 includes testing loads and test methods also. Note that ANSI SCTE 77 has compliance test loads that are 1.5 times the design load (safety factor = 1.5) where AASHTO and ASTM compliance test loads are 2.8 times the design load (safety factor = 2.8). ANSI SCTE 77 specifies a heavy truck wheel load of 15,000 lbs, while AASHTO and ASTM specify a heavy truck wheel load of 16,000 lbs. Finally, ANSI SCTE 77 includes a 22,500 lb design wheel load but relaxes the test requirement to the use of a $10 \times 20$ inch load plate rather than the $10 \times 10$ in plate used for all other tests. ANSI SCTE does not specify a factor for wheel impact.

WUC 3.6 and ANSI SCTE 77 both specify a vertical test load on the edge of the box with one half ($5 \times 10$ in. of the load plate on the box wall. This test, however, specifies that the same test load used on the cover is to be applied. This effectively doubles the load pressure on the box. In actual application, a wheel load applied in this manner would have the same unit force applied since one-half of the wheel load would be supported on the adjoining pavement. Therefore, this test requirement seems to be too high.

The ANSI SCTE 77 lateral or side loading specified is substantially higher than the other standards for small enclosures. For deeper and longer underground enclosures, the SCTE design load and test load are not conservative. They substantially understate the side loads expected when calculated using rational design procedures.

## 3 Drainage Structures

Drainage structures are typically in the form of a precast, sloped, trench drain with fiberglass, steel, or cast iron grating as a cover to permit liquid inflow (Fig. 40.2) [11]. The systems include provisions for turns and intersections and catch basins to connect to underground piping. Precast PC drainage systems became available in the United States in the last half of the 1970s. The first products on the market were a European design being manufactured in the United States by a subsidiary of a European company. A US company began manufacturing these structures by the end of the 1970s followed by the entry of several additional manufacturers into the market. Drainage structures are designed for the specified design loading for projects where they are used. In parking areas, this is normally ASTM C857 A-8 or A-16.

**Fig. 40.2** Trench drains

These products are typically cast using a polyester-based PC. Where corrosive fluids may be present, a vinyl ester-based PC is used.

## 4 Conclusions

Forty years ago, the use of PC as a precasting material was just beginning in the United States. Japan and some European countries had already established some precast products by that time. Precast PC is now being used extensively for a number of products in the United States. The most successful precast products over the past 40 years appear to be underground utility structures and drainage channels. Over this time, standards have evolved to describe minimum requirements for underground structures since they are now widely used by utilities and highway departments. The following changes should be considered to make these standards uniform in defining expected performance:

- ASTM C857 should incorporate PC into its scope of materials covered.
- ANSI SCTE 77 should remove the tier loading descriptions and reference ASTM C857 for applicable loads or describe the same loads.
- The ANSI SCTE and WUC 3.6 box vertical load should be revised to apply a pressure equal to that produced by the cover test.
- Box lateral loads should be determined by rational soil pressure analysis as in ASTM C857 with a minimum load to account for installation and backfilling practices.
- ANSI SCTE 77 should increase the safety factors for compliance test loads to match ASTM C857/ASTM C497.

Additional research into the fatigue behavior of enclosure covers and into the effects of the high deflection limits used in the standards would broaden the application areas for these structures.

## References

1. Underground enclosure. (2016). Retrieved from http://www.raiproducts.com/traffic/underground-enclosures.html.
2. Western Underground Committee Guide 3.6. (2015). *Nonconcrete enclosures*. Retrieved from http://www.westernunderground.org/
3. Society of Cable Telecommunications Engineers. (2010). *Specification for underground enclosure integrity*. Exton: Author.
4. ASTM C857-16. (2016). *Standard practice for minimum structural design loading for underground precast concrete utility structures*. West Conshohocken: ASTM International.
5. AASHTO. (2002). *Standard specification for highway bridges*. Washington, DC: American Association of State Highway and Transportation Officials.
6. ASTM C497-17. (2017). *Standard test methods for concrete pipe, manhole sections, or tile*. West Conshohocken, ASTM International.

7. AASHTO. (1982). *Standard method of test for determining physical and chemical properties of culvert pipe, sewer pipe, and drain tile*. AASHTO Designation T33. Washington, DC: American Association of State Highway and Transportation Officials.
8. ACI Committee 548. (2004). *Test method for load capacity of polymer concrete underground utility structures*. Farmington Hills: American Concrete Institute.
9. IEEE Standards Association. (2017). *National electric safety code*. Piscataway: Author.
10. National Fire Protection Association. (2017). *National electric code*. Quincy: Author.
11. Polymer concrete drain. (2016). Retrieved from http://www.tradekey.com/product-free/Polymer-Concrete-Drain-Trench-Drain-En1433-Liner-Channel-3992714.html

# Chapter 41
# Development Length of Steel Reinforcement in Polymer Concrete for Bridge Deck Closure

Moneeb Genedy, Rahulreddy Chennareddy, Michael Stenko, and Mahmoud Reda Taha

Accelerated bridge construction techniques are getting attention due to their ability to reduce construction time and thus lower cost of bridges. Precast concrete girders and bridge deck panels are typically used today in the construction of bridges. This construction method creates the need for flowable yet high strength and good bond for filling the joints between the precast elements. To achieve the required integrity and enable high bond strength, polymethyl methacrylate (PMMA) polymer concrete (PC) has been suggested for bridge deck closure. This paper examines the development length of steel reinforcement when embedded in PMMA-PC. The results of the experimental investigation show that PMMA-PC has a development length shorter than that used with ultra-high performance concrete (UHPC).

## 1 Introduction

To meet the growing construction demand, accelerated bridge construction (ABC) techniques are becoming more common worldwide. These techniques enable reducing construction time and thus lowering construction cost [1]. Typically, precast bridge girders and deck panels are cast offsite and transported to the bridge site. The use of ABC techniques requires developing special concrete for field joints, including deck closures, with relatively high flowability and high strength. A major challenge in cast-in-place deck closure is the insufficient space for developing the steel reinforcing bars extended from the precast bridge deck panels into the deck

M. Genedy (✉) · R. Chennareddy · M. Reda Taha
Department of Civil Engineering, University of New Mexico, Albuquerque, NM, USA
e-mail: moneeb@unm.edu

M. Stenko
Transpo Industries, New Rochelle, NY, USA

closure. Therefore, cast-in-place concrete with relatively high bond strength to existing concrete surface and steel reinforcing bars is required. Researchers have suggested using ultra-high performance concrete (UHPC) for its high performance including high bond strength with steel reinforcing bars compared with normal concrete [2]. However, field application of UHPC presents a true challenge because of the many requirements in mixing, placing, and curing of UHPC that are necessary to avoid cracking. Furthermore, UHPC has a relatively very high cost compared with all other types of concrete. Nevertheless, today about 32 bridges have been constructed in the US bridge deck closures using UHPC [3].

Polymers are known to have a significantly higher bond strength to existing surfaces compared with any cementitious material. Polymer concretes (PC) are therefore favorable when strong bond to existing concrete or reinforcing bars is necessary. Moreover, PC is corrosion resistant and has a very good resistance to cracking and thus is a good material for use in bridge decks. The above properties make PC a favorable material for bridge deck overlays [4].

In the current study, we investigate the potential use of polymethyl methacrylate (PMMA) polymer concrete (PC) denoted here as PMMA-PC as a potential material for bridge deck closures. The low viscosity of PMMA and the ease of mixing and relatively high workability of PMMA-PC are key features for its use for bridge deck closure. Here, we examine the minimum developing length of steel reinforcing bars when embedded in PMMA-PC. Pullout bond strength tests are performed, the development length is determined, and the results are compared to those of UHPC.

## 2 Experimental Methods

### 2.1 Materials

PMMA-PC was produced by mixing methyl methacrylate resin with benzoyl peroxide initiator and well-graded aggregate of 25.4 mm nominal maximum size. The mixing ratio was 6.2% resin to 93.8% solids by weight. PMMA-PC was mixed then left to cure for 1 week in standard lab temperature of 22 °C.

### 2.2 Strength Testing

Five cylinders of PMMA-PC were tested under compression following ASTM C39 [5] and splitting tension following ASTM C496 [6]. PMMA-PC showed a compressive strength and split tensile strength of 67.4 ± 0.9 MPa and 6.91 ± 0.4 MPa, respectively.

**Fig. 41.1** Pullout test setup

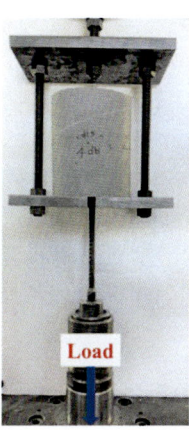

## 2.3 Pullout Test

The pullout test was conducted to determine the minimum development length of No. 4 (US designation) #13 (SI designation) with 0.5 in (12.7 mm) diameter and No. 5 (US designation) #16 (SI designation) with 0.625 in (15.9 mm) diameter steel reinforcing bars. The embedment length was measured as a multiplier of the reinforcing bar diameter ($d$). Three specimens were prepared with embedment lengths of $4d$, $6d$, $8d$, and $10d$. Pullout test specimens were tested at age of 7 days using the pullout test setup shown in Fig. 41.1. The steel bar was pulled out from the concrete cylinders with displacement control loading rate of 1 mm/min. The load and displacement were recorded continually with sampling rate of 1 Hz.

## 3 Results and Discussion

All specimens were tested up to failure. All specimens with embedment length of $4d$ failed due to de-bonding between the steel bar and concrete specimens. For steel bars with embedment of $6d$, failure occurred either due to bar rupture or de-bonding between the steel bar and the PMMA-PC after steel bar yield. For specimens with embedment length of $8d$ and $10d$, failure in all specimens took place due to steel bar rupture. For specimens with No. 4 (#13) reinforcing bars, all specimens with embedment length of $4d$ failed due to de-bonding between the steel bar and PMMA-PC. For specimens with embedment length of $6d$, the failure occurred either due to steel bar rupture or de-bonding between the steel bar and PMMA-PC after the yield of the bar. For specimens with embedment length $8d$ and $10d$, the failure in all specimens was due to steel bar rupture. All the failure modes observed are shown in Fig. 41.2. For No. 5 (#16) reinforcing bars, all specimens with embedment length of $4d$ failed due to de-bonding between the rebar and PMMA-PC after steel yield as shown in Fig. 41.3a. For specimens with embedment length $6d$, $8d$, and $10d$, failure

**Fig. 41.2** Failure due to (**a**) de-bonding of bar for embedment length 4$d$, (**b**) bar rupture for embedment of 6$d$, (**c**) de-bonding of bar for embedment length of 6$d$, (**d**) bar rupture for embedment length of 10$d$ (similar to 8$d$)

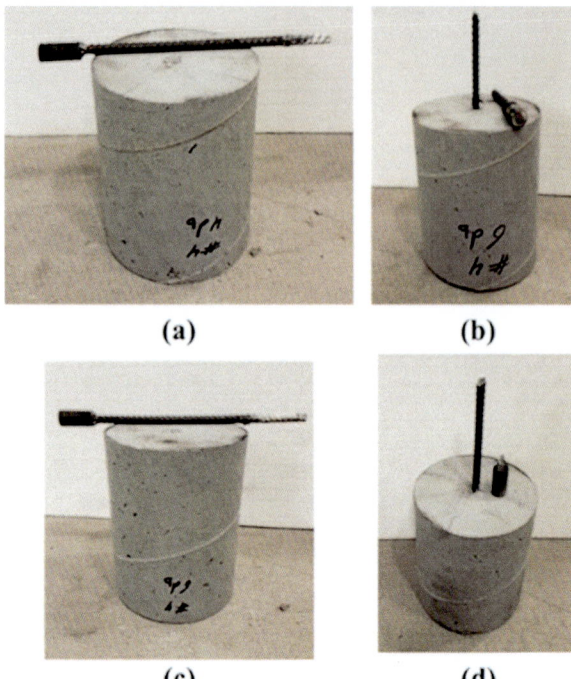

**Fig. 41.3** Failure due to (**a**) de-bonding of bar for embedment length 4$d$, (**b**) bar rupture for embedment of 6$d$ (similar failures were observed for 8$d$ and 10$d$ embedment lengths)

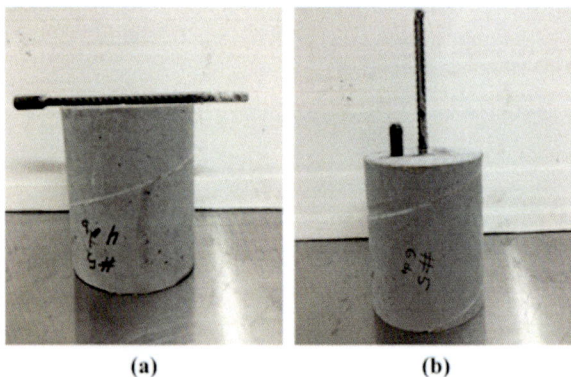

occurred due to steel bar rupture after passing the yield strength as shown in Fig. 41.3b. The maximum load developed in the steel bar for each embedment length for No. 4 (#13) and No. 5 (#16) reinforcing bars is shown in Figs. 41.4 and 41.5. Moreover, the maximum stress in each bar is also shown in Fig. 41.6.

The bond strength between the steel bars and PMMA-PC was calculated from the specimens that showed de-bonding as failure mode. The bond strength for No. 4 (#13) and No. 4 (#16) reinforcing bars was found to be 25.3 ± 0.9 MPa and

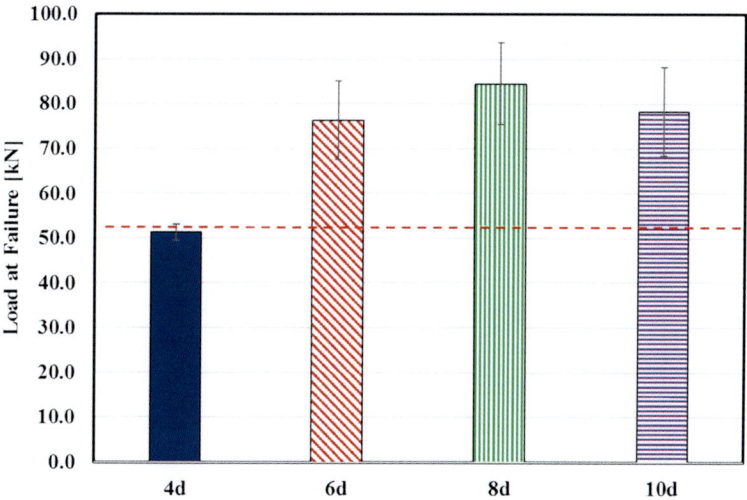

**Fig. 41.4** Load at failure for No. 4 (#13) reinforcing bar for all embedment length (red line represents the load required to achieve the design yield strength of the steel rebar "52.5 kN")

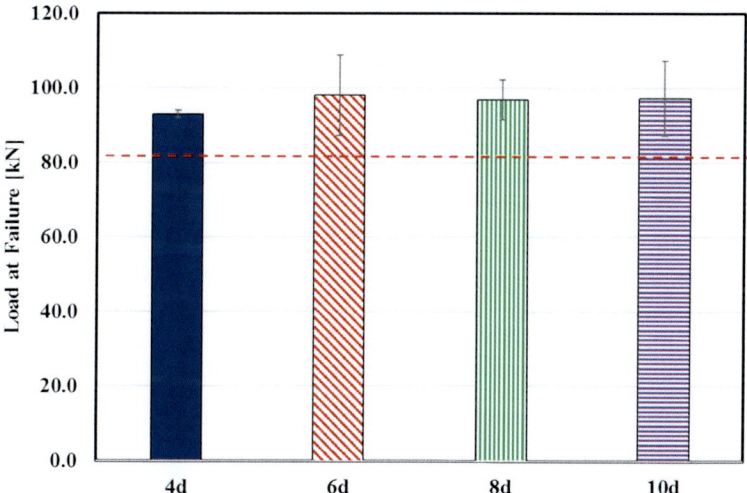

**Fig. 41.5** Load at failure for No. 5 (#16) reinforcing bar for all embedment length (red line represents the load required to achieve the design yield strength of the steel rebar "81.85 kN")

$28.9 \pm 0.3$ MPa, respectively. The development length of the rebar was determined using Eq. (41.1):

$$l_d = \frac{f_y \cdot d}{4 f_d} \quad (41.1)$$

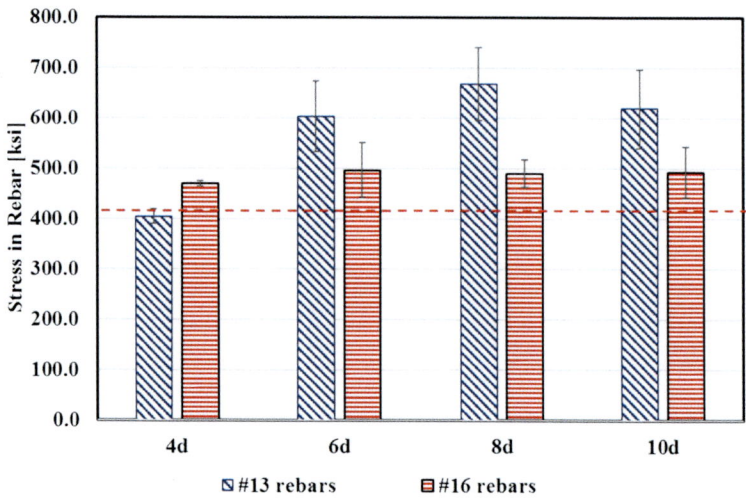

**Fig. 41.6** Maximum stress in No. 4 (#13) and No. 5 (#16) reinforcing bars for each embedment length (red line represents the design yield strength of the steel rebar "413 MPa")

where $l_d$ is the development length [mm], $f_y$ is the yield stress of steel [MPa], $d$ is the rebar diameter [mm], and $f_d$ is the bond strength between concrete and steel [MPa]. The minimum development length required for the steel bars to achieve yield strength was found to be $4.1d$ and $3.6d$ for #13 and #16 steel bars, respectively.

Based on the ACI 318, the required development length for steel bars embedded in normal concrete with similar compressive strength is at least 24 times the steel bar diameter [7]. For UHPC, the development length was found experimentally to range between 12 and 18 times the steel bar diameters [8]. On the other hand, the required development length of steel bars embedded in PMMA-PC was found to be less than 4.5 times the steel bar diameter. The development length necessary for embedding the steel reinforcement in bridge deck closures for PMMA-PC is almost half of that of UHPC. Our experimental investigation shows that PMMA-PC has superior bond strength with steel compared with normal concrete and UHPC. This makes the PMMA-PC a very suitable material for deck closures in precast bridges with relatively short closure gaps.

## 4 Conclusions

Pullout tests were conducted on steel reinforcing bars embedded in PMMA-PC to determine the required developing length. No. 4 (#13) and No. 5 (#16) steel reinforcing bars with embedment length of 4, 6, 8, and 10 times the bar diameters were tested. The minimum development length required for steel bars embedded in PMMA-PC was found to range between 3.6 and 4.1 times the reinforcing bar diameter. This development length is less than one-half of the minimum development length of 12 to 18 times the bar diameter required for use with UHPC. It is apparent that PMMA-PC is an excellent candidate for bridge deck closures and requires significantly narrower precast gap spacing compared with UHPC.

# References

1. Culmo, M. P. (2011). *Accelerated bridge construction-experience in design, fabrication and erection of prefabricated bridge elements and systems* (No. FHWA-HIF-12-013).
2. Yuan, J., & Graybeal, B. A. (2014). *Bond behavior of reinforcing steel in ultra-high performance concrete* (No. FHWA-HRT-14-090).
3. Graybeal, B. (2011). *Ultra-high performance concrete. US Department of Transportation, Federal Highway Administration* (p. 8). FHWA-HRT-11-038.
4. Whitney, D. P., & Fowler, D. W. (2015). New applications for polymer overlays. In *Advanced materials research* (Vol. 1129, pp. 277–282). Trans Tech Publications.
5. ASTM C39-17b. (2017). *Standard test method for compressive strength of cylindrical concrete specimens*. West Conshohocken: ASTM International.
6. ASTM C496-17. (2017). *Standard test method for splitting tensile strength of cylindrical concrete specimens*. West Conshohocken: ASTM International.
7. ACI Committee 318. (2015). *Building code requirements for structural concrete (ACI 318-14)*. Farmington Hills: American Concrete Institute.
8. Saleem, M. A., Mirmiran, A., Xia, J., & Mackie, K. (2012). Development length of high-strength steel rebar in ultrahigh performance concrete. *Journal of Materials in Civil Engineering, 25*(8), 991–998.

# Chapter 42
# Development of Ultrarapid-Hardening Epoxy Mortar for Railway Sleepers

Sunhee Hong, Jaehoon Lee, Duhyouk Kim, Junwoo Kim, and Yong Jeong

This study aims to develop epoxy mortar that is quick-setting and of high strength for emergency work on railway sleepers. Polymer mortars using commercial epoxy resin with hardeners were prepared with various epoxy equivalent weights (E.E.W.) and active hydrogen equivalent weights (A.H.E.W.). These mortars were mixed, cured, and tested for compressive and adhesive strength.

As a result, an ultrarapid-hardening epoxy mortar using an epoxy resin binder system was developed. This system is composed of a part A resin [E.E.W.:182.5 (g/eq)] and a part B hardener [A.H.E.W:33.9 (g/eq)]. This epoxy mortar developed higher adhesive strength than typical cementitious quick-setting materials. Also, this epoxy mortar showed superior workability and physical properties to existing products, and there were no defects such as brittleness and cracks, or poor adhesion in the field application. It is concluded from the results of this study that an excellent quality epoxy mortar for use on railway sleepers was developed.

## 1 Introduction

Special material is required for the replacement work of railway sleepers, which must be finished within 3–4 hours during late night hours after train service time. Ultrarapid-hardening cementitious material, which is used for emergency work in the construction field, can meet the required compressive strength and working time. However, the adhesive strength of cementitious ultrarapid-hardening material is not very high. Also, the nature of train service with its repeated impact compromises long-term durability and adhesive strength [1]. Therefore, ultrarapid-hardening cementitious material is not suitable as replacement material for railway sleepers.

S. Hong (✉) · J. Lee · D. Kim · J. Kim · Y. Jeong
R&D Center, SAMPYO Industry, Gyeonggi-do, South Korea

**Table 42.1** Properties of the epoxy resin

| E.E.W. (eq/g) | Viscosity (cps at 25 °C) | Appearance |
|---|---|---|
| 187 | 12,500 | Colorless liquid |

**Table 42.2** Properties of the reactive diluent

| E.E.W. (eq/g) | Viscosity (cps at 25 °C) | Appearance |
|---|---|---|
| 150 | 2.5 | Colorless liquid |

In contrast, epoxy resin has excellent properties such as strong adhesion to almost every material, abrasion resistance, chemical resistance, dimensional stability, and so on. Also, when applying epoxy resins to railway turnout section, it can be satisfied to the required electric insulation because of signal system on the railway. Therefore, this study intends to develop epoxy mortar which develops high strength and quick-setting for railway sleepers by using epoxy resin binder and sand.

## 2 Materials

### 2.1 Epoxy Resin

A commonly used bisphenol A type epoxy resin was used. Its properties are listed in Table 42.1.

### 2.2 Reactive diluent

The reactive diluent used in this study is monofunctional glycidyl ether, and its properties are listed in Table 42.2.

### 2.3 Hardener

Hardeners used in this study are amine compounds such as polyamide, aliphatic amine, and di-ethylene tri-amine, and the properties are listed in Table 42.3.

### 2.4 Accelerator

A tertiary amine was used as an accelerator in this study. The properties of accelerator are listed in Table 42.4.

**Table 42.3** Properties of the hardeners

| Type | | A.H.E.W. (eq/g) | Viscosity (cps at 25 °C) | Appearance |
|---|---|---|---|---|
| Hardener A | Polyamide | 70 | 2 | Yellow liquid |
| Hardener B | Di-ethylene tri-amine | 70 | 1,250 | Colorless liquid |
| Hardener C | Aliphatic amine | 21 | 4 | Brown liquid |

**Table 42.4** Properties of accelerator

| A.H.E.W. (eq/g) | Viscosity (cps at 25 °C) | Appearance |
|---|---|---|
| 15 | 200 | Yellow liquid |

## 2.5 Sand

Silica sand (Size: 0.3–1.0 mm) was used for filling the resinous binder system.

## 3 Testing Procedures

In this study, epoxy binder systems were prepared with proper weight ratios of resin to hardener (part A: part B = 3:1), and mixed together. Then, the epoxy binders were mixed with sand in a weight ratio of 1:5, and the resulting mortar's compressive strength was measured after curing at 20 °C.

First of all, part A (epoxy resin) of epoxy resin binders was tested for physical properties according to amount of reactive diluent. And part B (hardener) of epoxy resin binders was tested for physical properties according to the mixing ratio and component change. Then, selected mix proportion was tested for physical properties of epoxy resin binder and epoxy mortar, and the results were compared with the appropriate Korean Standard specifications.

## 4 Test Results and Discussions

### 4.1 Properties According to Reactive Diluent Content

Epoxy mortars were prepared with mix proportions of epoxy resin binders as listed in Table 42.5, and tested for compressive strength at different curing times.

As shown in Fig. 42.1, increasing the reactive diluent content decreased early strength of the epoxy mortars.

**Table 42.5** Mix proportions of epoxy resin binders

|  | Part A (wt.%) | | Part B (wt.%) | | |
|---|---|---|---|---|---|
|  | Epoxy resin | Reactive diluent | Hardener A | Hardener B | Accelerator |
| No.1 | 75 | 25 | 60 | 20 | 20 |
| No.2 | 80 | 20 | | | |
| No.3 | 85 | 15 | | | |
| No.4 | 90 | 10 | | | |

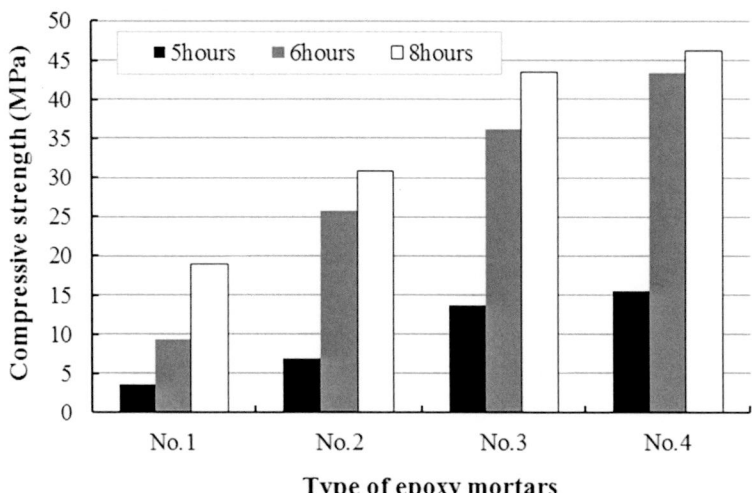

**Fig. 42.1** Compressive strength of epoxy mortars according to reactive diluent content

**Table 42.6** Mix proportions of epoxy resin binders

|  | Part A (wt.%) | | Part B (wt.%) | | |
|---|---|---|---|---|---|
|  | Epoxy resin | Reactive diluent | Hardener A | Hardener B | Accelerator |
| No.4 | 90 | 10 | 60 | 20 | 20 |
| No.5 | | | 50 | 20 | 30 |
| No.6 | | | 20 | 30 | 40 |

## 4.2 Properties According to Accelerator Content

Based on the highest strength achieved by mortar's binder No. 4 (shown in Fig. 42.1), the proportioning for Part A was established. For this part of the study, mix proportions for the Part B (hardener) of the binders were varied to account for increasing accelerator content, as listed in Table 42.6. And the result of compressive strength test was shown in Fig. 42.2. As a result, 5 h strength increased with increasing accelerator content. However, strengths of 1 day and 7 days tended to decrease.

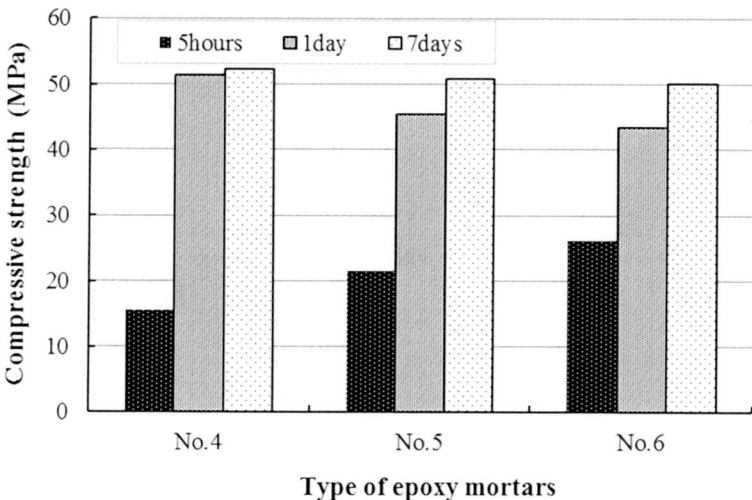

**Fig. 42.2** Compressive strength of epoxy mortars according to accelerator content

**Table 42.7** Mix proportions of epoxy resin binders

|        | Part A (wt.%) |                  | Part B (wt.%) |           |             |
|--------|---------------|------------------|---------------|-----------|-------------|
|        | Epoxy resin   | Reactive diluent | Hardener C    | Hardener B | Accelerator |
| No.7   | 90            | 10               | 90            | –         | 10          |
| No.8   |               |                  | 80            | –         | 20          |
| No.9   |               |                  | 80            | 10        | 10          |
| No.10  |               |                  | 70            | 20        | 10          |
| No.11  |               |                  | 60            | 30        | 10          |
| No.12  |               |                  | 50            | 40        | 10          |

## *4.3 Properties According to Changes in Hardener*

In order to accomplish the development of high strength in 3 h, the polyamide component of Hardener A was replaced with an aliphatic amine modified Mannich-based hardener. The mix proportions of the new epoxy binders are listed in Table 42.7.

The 2 h, 3 h, and 1 day compressive strengths of epoxy mortars with varying component changes in the hardener are shown in Fig. 42.3. The 2 h and 3 h strengths of epoxy mortars using No.10 and No.11 mix proportions were markedly higher than No.7 and No.8, which increased accelerator content. On the other hand, compressive strength of No.12 that the component of hardener B exceeded showed decreasing tendency. Therefore, it was ascertained that rapid-setting and high strength can be controlled by optimum content of part A and part B.

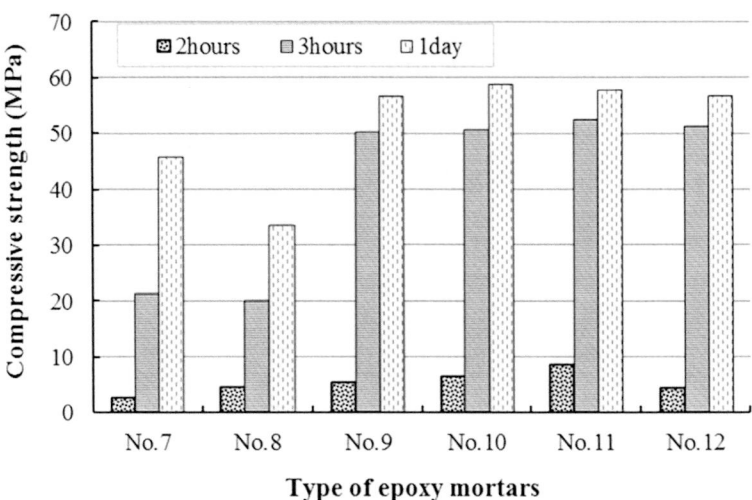

**Fig. 42.3** Compressive strength of epoxy mortars according to component change of hardener

**Table 42.8** Properties of epoxy resin binder

| Classification | Part A | Part B | |
|---|---|---|---|
| Appearance | Colorless liquid | Brown liquid | |
| Mixing ratio | 3 | 1 | (Weight ratio) |
| Specific gravity | 1.15±0.05 | 1.15±0.05 | (20±2°C) |
| Viscosity (cP) | 500±50 | | (20±2°C) |

## *4.4 Properties for the Optimum Mix Proportion*

Finally, No.11 was selected as the optimum binder system based on all the test results. The basic properties of No.11 epoxy resin binder were measured, and the results are listed in Table 42.8.

Properties of epoxy mortars using this epoxy resin binder were also evaluated, and compared with specified quality requirements in the related KS standard. The results are listed in Table 42.9. An evaluation of the properties of the optimum epoxy mortar developed in this study indicates that it satisfies the specified requirements of the KS standard, but it also surpasses the study's higher performance targets of faster set time and higher strength.

Table 42.9 Properties of the optimized epoxy mortar

|  |  | Evaluation result |  | Target performance | KS quality standard | Test method |
|---|---|---|---|---|---|---|
| Compressive strength (MPa) | 2h | 8.7 | 5↑ |  | KS F 4043 "Epoxy resin mortar for restoration in concrete structure" |
|  | 3h | 52.6 | 50↑ |  |  |
|  | 7days | 68.6 | 60↑ | 40↑ |  |
| Flexural strength (MPa) | 7days | 23.4 | 15↑ | 10↑ |  |
| Adhesive strength (MPa) | 1day | 4.7 | 4.0↑ | – |  |
|  | 7days | 4.8 | – | 1.5↑ |  |
| Length change (%) | 7days | 0.015 | ±0.050 | ±0.050 |  |

## 5 Conclusions

In this study, research to develop an ultrarapid-hardening epoxy mortar for railway sleepers was conducted. Through several experimental tests the effects of changes in proportions for the various epoxy binder components were evaluated. Epoxy mortars were made from specified variations in the binder system and sand. These mortar specimens were cured and tested. The following conclusions can be made from the test results:

1. The optimum reactive diluent content for the epoxy resin system greatly improves the viscosity of the epoxy binder and the early rate of strength development.
2. The proper accelerator content in the hardener portion of the epoxy binders is limited to less than 20%.
3. The study's target performance values for the epoxy mortar (2 h setting time and 3 h for high strength development) was accomplished with optimized hardener system composed of an aliphatic amine modified Mannich-based hardener (60 wt. %), di-ethylene tri-amine (30 wt.%), and accelerator (10 wt.%).
4. The epoxy mortar developed in this study not only satisfies the specified quality requirements of the KS standard, but it surpasses the study's own higher target performance requirements as well.

## References

1. Afia, S. H., & Shashikala, A. P. (2016). Suitability of rubber concrete for railway sleepers. *Perspectives in Science, 8*, 32–35.

# Chapter 43
# Contribution of C-PC to Resilience of Concrete Structures in Seismic Country Japan

Makoto Kawakami, Mikio Wakasugi, and Fujio Omata

Japan is a highly seismic country and there have been frequent damages to infrastructure such as bridges and buildings due to earthquakes. Therefore, securing the resilience of infrastructure corresponding to robustness, redundancy, resourcefulness, and rapidity is strongly required. In this study, the current status and practical countermeasures for concrete structures to enhance the resilience using concrete polymer composites(C-PC) were investigated and discussed. C-PC, which have high strength, high early strength development, and high durability, are effective for repair and strengthening to sustain the current performance of structures, to minimize damage, and to quickly recover design-level performance. Practical application of C-PC for the restoration and reconstruction in Japan is also presented.

## 1 Introduction

Nearly 20% of the world's earthquakes of magnitude 6 or greater have occurred in or around Japan. Japan has suffered great damages from the massive interplate earthquakes produced by plate subduction such as the Great East Japan Earthquake of 2011 and the inland crustal earthquakes caused by plate movements such as the Great Hanshin-Awaji Earthquake of 1995. In previous earthquakes, extensive unexpected damage and failure mode have been observed as shown in Fig. 43.1. These

M. Kawakami (✉)
Akita University, Akita-shi, Japan
e-mail: kawakami@gipc.akita-u.ac.jp

M. Wakasugi
Chemical Construction, Co., Ltd., Kobe, Japan

F. Omata
Maintenance Technology Japan Corporation, Co., Ltd., Tokyo, Japan

**Fig. 43.1** Collapse of bridges and buildings

range from collapse of mushroom-shaped slab bridges caused by the termination of longitudinal re-bars at mid-height, damage of piers of rectangular cross section, and unseating of superstructures and serious damage of RC buildings. These damages were attributed to underestimation of seismic force, lack of flexural capacity and shearing strength of the structural members, and large displacements between superstructures and substructures. Based on these serious damages, design specifications for bridges and buildings were quickly and intensively reviewed and construction codes were entirely revised [1, 2].

On the other hand, nowadays concrete–polymer composites (C-PC) are the indispensable material for repair and strengthening of concrete structures. Furthermore, basic research on C-PC is expected to promote their application in new fields [3–6]. However, in conventional studies, little information is known on the resilience of concrete structures using C-PC that could evaluate properly the structural performance before and after an earthquake disaster.

In this study, based on earthquake damage and the restoration of concrete structures in Japan, contribution of C-PC to maintenance and seismic strengthening of concrete structures is addressed. Specifically, current seismic measures using C-PC for RC piers and buildings, and bridge-falling prevention systems are presented. Furthermore, seismic retrofitting methods of RC building are also discussed.

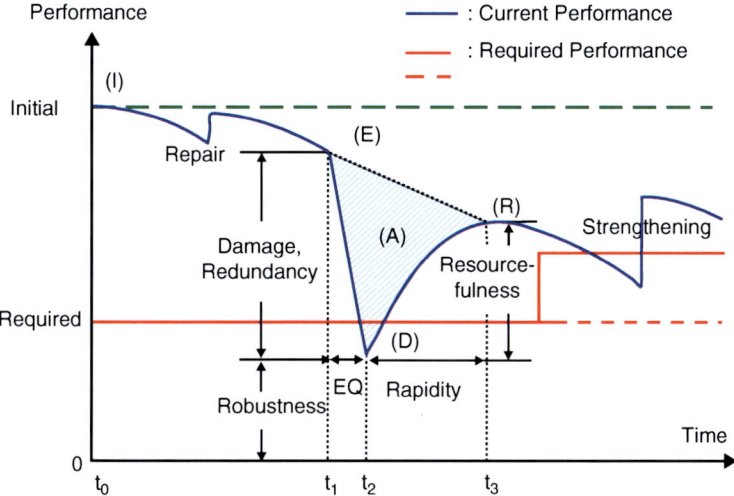

**Fig. 43.2** Conceptual life cycle and resilience of concrete structures

## 2 Resilience

Figure 43.2 is a conceptual illustration of life cycle and resilience of concrete structures.

### 2.1 From Completion to Earthquake Occurrence: (I) – (E)

The performance of structures gradually declines with time due to carbonation, chloride penetration, and frost damages of concrete. Therefore, suitable repair is important to restore and improve durability. Some representative repair methods [7] are surface protection, patching and crack repair, and C-PC and other similar materials applied for coating, plastering, shotcrete, filling, and impregnation.

### 2.2 Damage by Earthquake Occurrence: (E) – (D)

The performance of structures (D) drops in a short time depending on structural potential (E) upon earthquake occurrence. Seismic strengthening using C-PC and redundancy of structures are important to prevent collapse or decrease damage.

## 2.3 Recovery from Damage to Designated Performance: (D) – (R)

It is desirable to recover rapidly from the diminished structural performance (D) to the designated one (R). Just after an earthquake disaster, sufficient information as well as material and human resources are required in the affected areas of occasionally a large quantity debris. C-PC with their high strength and high early strength development are helpful to reduce the recovery time.

## 2.4 Resilience (A)

It is thus desirable to minimize the area A surrounded by the three points of the time-dependent structural performance: (E), (D), and (R). Therefore C-PC are considered to be extremely superior materials in preventing deterioration and damage of concrete structures, and reducing recovery time for early restoration.

# 3 Seismic Strengthening and Concrete–Polymer Composites

Sufficient dynamic bearing capacity of columns and walls, and reliable unseating prevention systems are required to prevent the collapse and falling of bridges [8]. In order to secure the seismic performance of bridges and building, polymers, especially epoxy resin, have been effectively applied in the following seismic strengthening methods.

## 3.1 Seismic Strengthening of Piers

Jacketing methods are applied for reinforcement of existing concrete piers in order to upgrade their flexural strength, shearing strength, and ductility. These methods are effective at the base part of the pier and at the cutoff sections of reinforcement, where the collapse usually begins.

### 3.1.1 Reinforced Concrete Jacketing Method: Fig. 43.3a

RC jacketing method, in which ordinary concrete is placed to the outside of existing RC piers after installing reinforcing steel, is adopted when sufficient construction duration and executing space are secured. By spraying PCM [9], which has high bond strength and low carbonation rate, the thickness of the placed concrete could be decreased to less than 20% of ordinary concrete.

(a) RC jacketing

(b) Steel jacketing

(c) FRP jacketing

(d-e) Bridge falling prevention systems

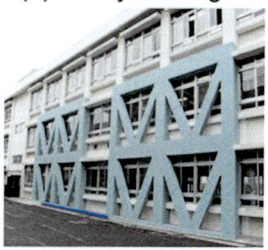
(f) Steel seismic brace

**Fig. 43.3** (a–f): Seismic strengthening methods

### 3.1.2  Steel Plate Jacketing Method: Fig. 43.3b

After installing steel sheet of thickness 6–12 mm around the existing RC pier, liquid epoxy resin or shrinkage compensated cement mortar is poured into the gap between steel jacket and RC pier. The base of pier is fixed to concrete footing by an intervening H-shaped steel. Then, epoxy resin is injected into the space of drilled holes in the concrete footing and anchor bars.

### 3.1.3  Carbon/Aramid Fiber Jacketing Method: Fig. 43.3c

Carbon/aramid fiber sheet having high modulus of elasticity is normally impregnated by epoxy resin and adhered to the existing RC pier [10]. Concentric application of the fiber sheet at the cutoff sections of reinforcement is particularly effective.

Furthermore, easy and quick retrofitting carbon/aramid fiber-based methods are developed in which hydraulic polyurethane resin is impregnated in aramid fiber sheet, and then water sprinkling and surface finishing are performed [11, 12].

## 3.2 Bridge-Falling Prevention Systems: Fig. 43.3d–e

The typical countermeasures to bridge-deck falling are extension of seat length, installation of unseating prevention devices including chain setting, bump prevention devices, and displacement control structures. Steel bracket is fundamentally utilized for those devices and structures, and there, anchor bolts are fixed to existing pier/abutment by use of epoxy resin. Injected liquid epoxy resin is effective in fixing the anchor bolts in horizontally bored hole.

## 3.3 Seismic Retrofitting Methods of RC Buildings: Fig. 43.3f

Seismic retrofitting with steel brace is most often used when a building structure has insufficient strength and ductility. Epoxy resin superior in adhesive property is effectively used to incorporate steel brace and frame in the existing beam and column. The rate of earthquake-resistance of school buildings retrofitted using this method is about 98% in Japan [13] and their seismic performance was confirmed by the fact that no school buildings collapsed in the 2016 Kumamoto Earthquake [14].

# 4 Conclusions

Contribution of C-PC to maintenance and resilience of concrete structures in seismic country Japan was investigated. The main conclusions obtained are summarized in the following:

1. In order to prevent the collapse of bridges, the earthquake-resistant performance of concrete piers has been improved. Epoxy resin, PCM, and hydraulic polyurethane are effectively used for jacketing methods.
2. Steel brackets are generally used for bridge-falling prevention devices, and those are fixed to pier and girder by anchor bolts and injected liquid epoxy resin.
3. Seismic retrofitting with steel brace for building is most often used and epoxy resin superior in adhesive property is effectively used to incorporate steel brace and frame in the existing beam and column.
4. Disaster resilience is evaluated by the product of the degree of damage multiplied by the recovery time. C-PC are powerful in their contribution in preventing disaster and in controlling damage with appropriate repair and reinforcement, and in the quick recovery of structural performance due to the high performance of materials. The development of excellent materials better than conventional C-PC obtained will be effective in enhancing the resilience of concrete structures.

# References

1. Japan Road Association. (2012). Specifications for Highway Bridges Part V Seismic Design 1–402.
2. Ministry of Land, Infrastructure, Transport and Tourism, Japan. (1981). Amendment of Order for Enforcement of the Building Standard Law.
3. Ohama, Y., & Ohta, M. (2013). Recent trends in research and development activities of polymer-modified paste, mortar and concrete. *Proceedings of 14th ICPIC*, Shanghai, China (pp. 26–34).
4. Aguiar, J., Ozkul, H., & Cunha, S. (2013). Report from 14th ICPIC and 7th ASPIC: New trends on concrete-polymer composites. *Proceedings of 14th ICPIC*, Shanghai, China (pp. 45–56).
5. Wang, R., Wang, G., & Wang, P. (2015). Status of research and application of concrete-polymer composites in China. *Proceedings of 15th ICPIC*, Singapore (pp. 59–68).
6. Okamoto, K., Tsuruta, K., & Naitou, T. (2015). Development and application of PIC form. *Proceedings of 15th ICPIC*, Singapore (pp. 159–161).
7. Japan Society of Civil Engineers. (2007). Standard specifications for concrete structures 2007 "Maintenance". 88–90.
8. Kawakami, M., Omata, F., & Matsuoka, S. (2010). Advanced Seismic Countermeasures for Concrete Bridges by Using Polymer in Japan. *Proceedings of 13th ICPIC*, Funchal-Madeira, Portugal (pp. 635–642).
9. Nakamura, S., Yamaguchi, K., Hino, S., & Sato, K. (2008). Seismic Retrofit of RC Pier with Termination of Main Reinforcements Using Mortar for Shotcrete. *Proceedings of the Japan Concrete Institute,* 30-3 (in Japanese), 1171–1176.
10. Japan Society of Civil Engineers. (2001). Recommendations for upgrading of Concrete Structures with Use of Continuous Fiber Sheets. 1–88.
11. Suzuki, M., Ojima, F., Itoh, M., & Katoh, Y. (2009). Basic properties and repair effect of an Emergency Retrofitting Method using TST-FiSH. *Proceedings of the Japan Concrete Institute,* 31–1(in Japanese), 2197–2202.
12. Suzuki, M., Itoh, M., Maki, T., & Katoh, Y. (2010). Influence of Physical Properties of Resin on Shear Capacity of RC Beams. *Proceedings of the Japan Concrete Institute,* 32–1(in Japanese), 1841–1846.
13. Disaster Risk Management Hub, et al. (2016). Making Schools Resilient at Scale: the Case of Japan 2016, 1–96.
14. Ministry of Education, Culture, Sports, Science and Technology, Japan. (2016). The Report of the Urgent Proposal for the Maintenance of School Facilities on Basis of Damage of Kumamoto Earthquake (in Japanese). http://www.mext.go.jp/component/b_menu/shingi/toushin/__icsFiles/afieldfile/2016/07/29/1374841_1_2_1.pdf

# Chapter 44
# Precast Polymer Concrete Panels for Use on Bridges and Tunnels

Michael S. Stenko

Portland cement concrete (PCC) has been used in the United States for over 100 years for various highway and bridge barrier shapes as well as tunnel bench walls and wall liners. For installations where corrosive materials and environments are not a factor, PCC is a viable option. However, in most uses exposure to moisture and road deicing chemicals has led to premature corrosion of embedded steel reinforcing results in spalling of the concrete cover. Rehabilitating existing deteriorated concrete elements is a significant issue both from a cost stand point as well as the time required for demolition and replacement.

Beginning in the late 1980s, a number of projects began using polymer materials based on their ability to resist the ingress of moisture and corrosive materials. Initially precast polymer concrete curb facing panels were introduced; however further developments in design and manufacturing capabilities saw the availability of a variety of shaped panels for infrastructure use. Today precast polymer concrete panels are manufactured for various shapes of highway and bridge barriers, tunnel bench walls, catwalks, and tunnel wall liners. The nonporous, chemical-resistant surface with strength three times that of PCC can be substituted for standard concrete. Because of the significant advantages over PCC barriers, this paper will discuss precast polymer concrete panels, manufacturing process, physical properties, size and shapes available, and various installation methods.

---

M. S. Stenko (✉)
Transpo Industries, Inc., New Rochelle, NY, USA
e-mail: mstenko@transpo.com

**Fig. 44.1** Night time and rainy condition visibility

# 1 What Are Precast Polymer Concrete Panels?

Polymer concrete uses a polymer-based binder instead of Portland cement. The binder is a thermosetting resin that develops high strength and good resistance to moisture, chemicals, and corrosive attack. It uses pre-dried blended aggregates to comprise the balance of the polymer matrix.

Panels are manufactured using specialized mixing equipment to blend unique filler components containing required pigments and unsaturated polyester resin to produce a matrix that is virtually void free and resistant to ingress of moisture. The exposed face of the panels is coated with a pigmented unsaturated polyester resin gel coat.

The gel-coated surface has several benefits: long-term resistance to degradation due to UV exposure, eliminates exposed surfaced aggregates that can attract hydrocarbon from vehicle exhaust causing discoloring, and low coefficient of friction on the exposed surface reduces vehicle tire run up which can lead to vehicle roll over. And, the panels will partially self-clean with rain, or they can be easily cleaned using environmentally friendly cleaning agents and help with nighttime visibility Fig. 44.1.

Each panel is cast with internal FRP reinforcing mesh that is designed to reduce spalling in the event of heavy vehicle impacts. Panels are manufactured in various sizes and shapes with a standard thickness of 1 1/2 inches (40 mm). Due to the high cost of polymer concrete materials, the panels are designed to be installed with lower-cost concrete backfill for new construction or with a flowable grout when retrofit is installed on existing concrete barriers.

# 2 Panel Manufacturing Process

Precasting can be accomplished using stainless steel or FRP molds with an application of mold release agent prior to casting.

The manufacturing process begins with an application of pigmented gel coat directly on the face of the panel molds. While the gel coat is still fluid, the surface is

**Fig. 44.2** Threaded anchor insert with J Bolt

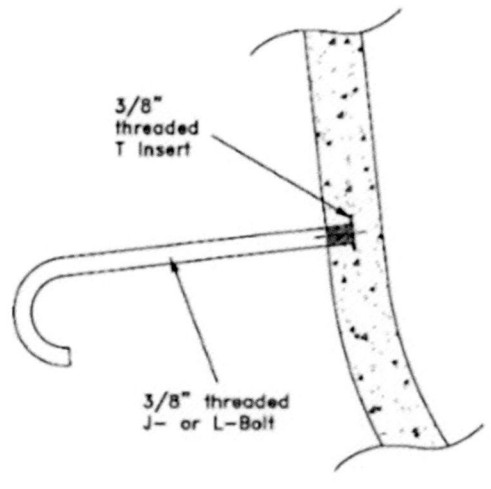

covered with a layer of 2.0 oz (57 gm) FRP reinforcing veil. The purpose of the veil is twofold, it protects the gel coat from penetration of coarse aggregates contained in the polymer matrix, and it reinforces the gel coat eliminating surface cracking during the curing process. After initial setting of the gel coat and prior to curing the specified panel, anchorage devices are installed. For new construction, anchorages are threaded inserts capable of accepting J bolts or L bolts which will permanently anchor the panels to the concrete backing (Fig. 44.2). Panels that are installed as a retrofit to existing concrete with chemical anchors utilize bolt hole blockouts with internally cast steel hole reinforcement Fig. 44.3.

Once anchorages have been installed, the molds are filled with the polymer matrix which may require the addition of shrinkage controlling chemicals to reduce dimensional inconsistencies and warping of oversized panels. Panels manufactured to tunnel applications will also require chemical additives that increase fire resistance and reduce flame spread and smoke development in case of exposure to a vehicle fire. Once the molds are filled to desired thickness, a sheet of FPR reinforcing mesh 1/8 in × 1/8 in. 4.5 oz/sy (3 mm × 3 mm, 150 gm/sm) is applied to increase resistance to flexural stress. After placement of FRP reinforcing pneumatic vibrators attached to the molds, consolidate the polymer matrix eliminating air voids.

Cast panels should be allowed to cure sufficiently so that they will not be damaged during demolding and handling. Curing can be accelerated using heat lamps or curing chambers, but caution should be taken when using either of these methods to maintain the panel flatness as these accelerated curing methods can cause panel warping.

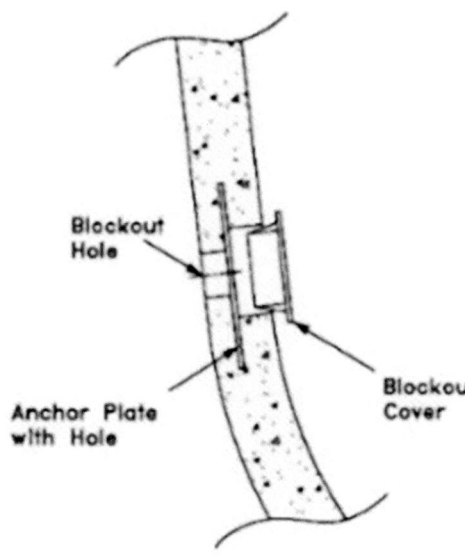

**Fig. 44.3** Bolt blockouts with internally cast steel hole

Table 44.1 Physical properties of polymer concrete panels

| Physical property | Value | Test method |
|---|---|---|
| Compressive strength | 14,000 psi (95 MPa) | ASTM C109 |
| Flexural strength | 3200 psi (22 MPa) | ASTM C293 |
| Impact strength | 100 ft.lb (135 N-m) | ASTM D2444 |
| Fire resistance | Class "A" | ASTM E84 |
| Flame spread | <25 | ASTM E84 |
| Smoke development | <75 | ASTM E84 |

## 3 Polymer Panel Physical Properties

The physical properties of the polymer matrix are far superior to those of standard Portland cement concrete and can be adjusted to meet project requirements [1–3]. Tunnel installations require testing for fire resistance, flame spread, and smoke development [4]. Most standard panels will exhibit the following physical properties (Table 44.1).

## 4 Panel Size, Shape, and Custom Options

Because the polymer matrix will undergo some shrinkage during curing, precast panel lengths have been limited to 12 ft (3.7 m). Sequential precasting allows panels to be manufactured in custom lengths (±0.5 in (12.5 mm)) that match existing

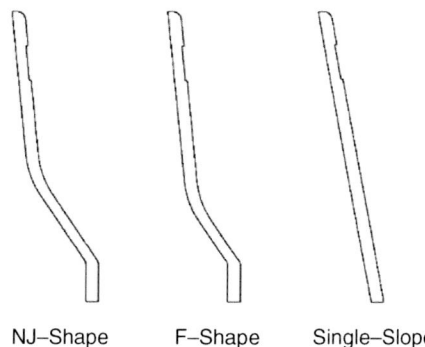

**Fig. 44.4** Common barrier shapes for roads and bridges

expansion joints and in tunnel applications air flue openings, wayfinding lights, and stair treads at catwalk openings. Panels can also be readily cut in the field with standard concrete saws which make custom lengths possible if detailed surveys are not available at the time of manufacturing.

Panels are currently manufactured in the three most commonly used shapes for highway and bridge use, New Jersey shape, F shape, and single slope (Fig. 44.4).

The overall height of each shape can be varied for future roadway overlays which reduces the need for milling of existing roadway surfaces prior to installation of new wearing surface material. Panel fabrication can incorporate a number of custom options to meet project requirements and specifications. The gel-coated panel surface color can be varied and remains consistent throughout production. Panels used on roadway and bridge applications are bright white, while tunnel panels specify a buff white to reduce glare from vehicle and tunnel lights. In addition to the variations in surface color, panels can be fabricated with a retroreflective safety stripe near the top. Because these safety stripes are on the vertical surface, they remain highly effective during almost all weather conditions, unlike pavement markings that can lose retroreflective properties if they become covered with debris, water, or snow. In addition, because the retroreflective stripe will not be exposed to any vehicle tires, they will not require replacement like standard pavement marking materials. After panels are cast, the surface to receive the retroreflective stripe is abrasive blasted to remove any mold release materials and lightly roughen the gel-coated surface.

A coat of pigmented unsaturated polyester resin is applied to the area primarily in white or yellow depending on the final installed location on the roadway. While the resin is still fluid highly retroreflective, glass beads are automatically applied and will become tightly bonded when the resin cures.

Tunnel applications require various special features to be incorporated into the panels. Precast stair treads allow pedestrian and workers to exit the roadway and retreat to the safety of the catwalk. During emergencies tunnels utilize wayfinding lights mounted on the barriers to direct pedestrians to exits or safety areas. Precast blockouts accommodate electrical box installation and assure proper light fixture location (Fig. 44.5).

**Fig. 44.5** Stair treads and precast blockouts

**Fig. 44.6** Anchor bolt holes

## 5 Panel Installation Options

For new construction or when existing concrete barriers are removed and concrete will be poured as a backing material, panels are fabricated with internally cast threaded inserts which allow for a variety of attachment hardware. The inserts can receive J bolts for composite bonding to the concrete backing. They can also be used for threaded rods and turnbuckles which are useful to maintain proper panel alignment during installation and concrete placement. Turnbuckles allow installation of multiple panels before placement of concrete backing which reduces installation cost.

For rehabilitation and retrofit installations, panels are fabricated with multiple blockouts for bolt connection to existing concrete. Each anchor bolt location is a recessed hole that can be used as a drill template for the installation of chemically bonded concrete anchors (Fig 44.6). Each anchor location is reinforced with a steel plate washer 4 in × 4 in × 1/8 in (100 mm × 100 mm × 3 mm) which is cast into the panels. Panels should be installed with a gap between the existing concrete surface and the panel of 1/2in (12 mm) minimum. After panels are installed, the gap should be filled with a flowable cementitious grout to achieve 100% contact between the panels and existing concrete. Plastic hole covers assure that dirt and moisture will not enter the anchor bolt blockout holes.

# 6 Conclusion

Over the last 10 years, precast polymer concrete panels have experienced increased use in the construction and rehabilitation of transportation infrastructure for a variety of reasons. Existing Portland cement concrete barriers are susceptible to moisture and corrosive materials causing steel reinforcement corrosion and concrete spalling. Products made with polymer matrix materials are virtually water and chemical resistant eliminating the deterioration experienced by standard concrete materials. With changes in safety standards, some existing safety barriers have become noncompliant primarily due to overall barrier height, and there is not a corrective option using convention materials. The ability of precast polymer concrete panels to be fabricated in various shapes and sizes, the ability to accomplish installation without the need for costly demolition of existing safety barriers, and the elimination of the need for concrete formwork reduce construction cost, project delivery time, and public inconvenience. The high visibility of the gel coat and the low surface friction offer increased safety to the traveling public. Incorporation of an all-weather retroreflective safety stripe increases driver awareness and can result in reduced accidents and injuries. Life cycle cost benefits associated with use of precast polymer concrete barrier panels make their use cost-effective for new construction as well as rehabilitation/repair projects.

All of these advantages and the need for sustainable infrastructure mean well-manufactured polymer concrete barriers will continue to be used in increasing number of projects worldwide.

# References

1. ASTM C109/C109M-16a. (2016). Standard Test Method for Compressive Strength of Hydraulic Cement Mortars (Using 2-in. or [50-mm] Cube Specimens).
2. ASTM C293/C293M-16. (2016). Standard Test Method for Flexural Strength of Concrete (Using Simple Beam With Center-Point Loading).
3. ASTM D2444-17. (2017). Standard Practice for Determination of the Impact Resistance of Thermoplastic Pipe and Fittings by Means of a Tup (Falling Weight).
4. ASTM E84-17a. (2017). Standard Test Method for Surface Burning Characteristics of Building Materials.

# Part IV
# Polymer Fiber Concrete

# Chapter 45
# The Effect of Combinations of Treated Polypropylene Fibers on the Energy Absorption of Fiber-Reinforced Shotcrete

Johannes J. Bester, Kulani D. Mapimele, and George Fanourakis

The dosage of fiber reinforcement in shotcrete increases the toughness and tensile stress and depends on the fibers' aspect ratio, toughness and energy absorption. Polypropylene (PP) fiber is hydrophobic, causing weak bonds to the cement matrix. One of the causes of the aforementioned is the "saturated hydrocarbon" classification of PP fibers. Their chemical structure is such that it is deprived of forming acid–base reactions needed at the interface of fiber–cement matrices. When surface treatment is applied, the bond can be increased, thus improving the load transfer mechanism. This increases the composite matrix's flexural, tensile strength and energy absorption characteristics. This research project compared the residual loads and energy absorption characteristics of two types of individual treated PP fibers (3000D and 450FT) to those yielded when combining the two fiber types, as described in the European Federation of National Associations Representing for Concrete (EFNARC) document. The fibers tested were surface treated by a process called Crypsination. The results indicate that the performance of the weaker single fiber was increased, when the PP treated fibers are combined. The combination of treated fibers yielded sufficient energy absorption to satisfy the toughness classification of EFNARC.

## 1 Introduction

Fibers are relatively small reinforcement agents, which come in varied materials such as glass, steel, synthetic, and natural types. The addition of these improves the concrete's tensile strength and reduces permeability, in turn increasing the concrete's

---

J. J. Bester (✉) · K. D. Mapimele · G. Fanourakis
University of Johannesburg, Faculty of Engineering and the Built Environment, Department of Civil Engineering Science, Johannesburg, South Africa
e-mail: jannesb@uj.ac.za

durability [1]. Fiber reinforcement is the preferred shotcrete reinforcement method because of its ability to increase the toughness and tensile strength of shotcrete. The dosage of the fibers to be used depends on the fiber aspect ratio, toughness, and energy absorption requirements. Fibers come in many different forms, shapes, and sizes and their usage is dependent on the application conditions and intended usage [2]. It is not surprising that steel fiber reinforcement is widely used in reinforced shotcrete, simply because steel has proven to perform relatively well in enhancing the tensile properties of steel-reinforced concrete [3].

## 2 Polypropylene Fibers

Polypropylene (PP) is a form of synthetic hydrocarbon polymer. It is hydrophobic, therefore causing weak bonds to the cement matrix, unless surface treatment is administered. They exhibit plastic stress–strain characteristics and a relatively high melting point compared to other synthetic fibers [2].

*Equivalent diameter and aspect ratio*—PP fibers have high aspect ratios (around 100), and are preferable as they increases the total surface area of the fibre–matrix bond even before modifications such as fibrillation and twisting [4], and have good interlocking abilities but have the tendency to ball during mixing [5].

*Relative density*—PP fibers are the lightest in their class at a relative density of 0.89–0.91. Therefore, it is quite favorable due to the lack of formwork and promotion of self-support existing in shotcrete [4].

*Tensile strength*—PP fibers have relatively higher tensile strengths ranging between 140 and 700 MPa [2] and exhibit high bending fatigue strengths [4].

*Elastic modulus*—PP fibers have a very low elastic modulus, at a value of 3.5–4.8 MPa. This primarily aids in the increased control of plastic shrinkage cracking and also, due to the fiber's ability to absorb large amounts of energy, increased impact resistance and toughness of the composite matrix [2].

*Ignition and melt temperature*—At 90 °C there exists no notable change in the strength of the fiber–concrete matrix. PP fibers can be installed permanently in areas heating up to 110 °C as well as temperatures as low as −20 °C [4].

## 3 Treatment of Polypropylene Fibers

Tests indicate that shear bond strengths between PP fibers and concrete matrices are of the range 0.2–1.3 MPa, which is very poor compared to the 3–4 MPa shown by cementitious matrices [6]. One of the causes of the aforementioned is the "saturated hydrocarbon" classification that PP fibers have been given, their chemical structure is such that it is deprived of forming acid–base reactions needed at the interface of fiber–cement matrices [7].The fibers tested in this report have been surface treated by an Oxyfiber process called Crypsination. Cryplon is applied on a polymer, thus

**Table 45.1** Mix design

| Material | Mass |
|---|---|
| Sand (kg) | 264 |
| Cement (kg) | 66 |
| C 40 superplasticizer (g) | 330 |
| Water (l) | 33 |

altering its surface properties. Another process is oxyfluorination, which entails the addition of a fluorine and oxygen mixture to alter the surface topography of hydrocarbon polymers to fluorocarbon polymers [4]. With sufficient interfacial bonding comes an improvement in the load transfer mechanism, this is the mechanism that transfers the load from the cement matrix to the fibers after cracking, which increases the composite matrix's flexural and tensile strength, then back to the cement matrix again through the shear surface between the fiber and cement interface [6].

## 4 Experimental Setup

The mix design (Table 45.1) does not resemble a typical shotcrete mix; however, the C40 superplasticizer aids in achieving properties that one would expect from shotcrete. This was done to reduce the usage of water. There were three samples each of the 3000D, 450FT, and 50:50 combination fibers, containing 770 g in total of the crypsinated fibers.

A power operated tilting rotating drum mixer was used. The panels were not sprayed, but were casted in panels by hand, and then compacted by slightly lifting and dropping the panels on the floor. Curing entailed spraying water on the samples and then covering them with blankets. This method of curing introduces a certain level of inconsistency, but since all the batches were cured this way the variations thereof are seen as negligible. The testing was conducted according to EFNARC energy absorption tests of the EFNARC EUROPEAN SPECIFICATION FOR SPRAYED CONCRETE:1996, Subsection (10.4) – Energy absorption class (plate test) [8].

## 5 Results

Sample 1 (Fig. 45.1) shows the results of the individual 770g 3000D, 770g 450FT, and the 50:50 combination load deformation graphs. The 3000D and 450FT residual loads are 27 kN and 6 kN, respectively, while the combined fiber curve is 17 kN, which approximated the actual average of the three individual fiber curves—the actual average is 16 kN.

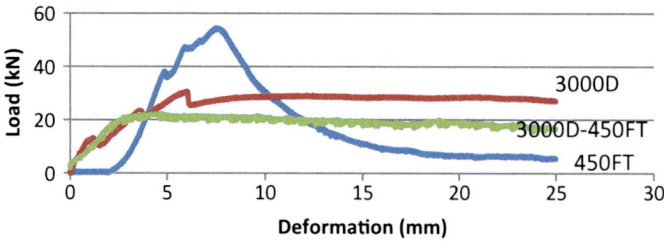

**Fig. 45.1** 3000D and 450FT fibers combination load deflection curves

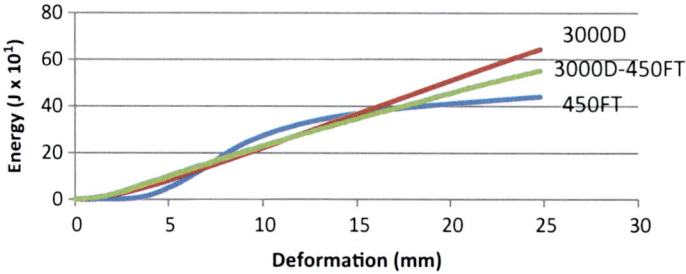

**Fig. 45.2** 3000D and 450FT fibers combination energy absorption curves

**Table 45.2** Residual load results

| Sample | Fiber type 3000D | Fiber type 450FT | Fiber type 3000D-450FT |
|---|---|---|---|
| 1 | 27 | 6 | 17 |
| 2 | 34 | 7 | 21 |
| 3 | 36 | 3 | 21 |
| Ave | 32.2 | 5.3 | 19.7 |

**Table 45.3** Energy absorption results

| Sample | Fiber type 3000D | Fiber type 450FT | Fiber type 3000D–450FT |
|---|---|---|---|
| 1 | 640 | 440 | 550 |
| 2 | 770 | 500 | 670 |
| 3 | 900 | 400 | 680 |
| Ave | 770 | 447 | 633 |

A similar trend was exhibited by the energy absorption results shown in Fig. 45.2, which show the 3000D and 450FT values being 440 J and 640 J, respectively, and the combined fiber value being 550 J, which is close to the actual average value of the individual fibers (540 J).

Due to space constraints, the load deflection and energy absorption curves for Sample 2 and Sample 3 cannot be shown graphically. However, the values of the residual loads and energy absorptions for all samples are shown in Tables 45.2 and 45.3.

With reference to Table 45.2, it is evident that the average residual load for the three fiber types for each sample was approximately the value of the 3000D–450FT (combined fibers). Furthermore, the residual loads of the 3000D–450FT (combined) fibers were an average of 40% (ranging from 37% to 41.7%) lower than the values yielded by the 3000D fibers. By contrast, the residual loads of the 3000D–450FT (combined) fibers were an average of 327,8% (ranging from 183.3% to 600%) higher than the value yielded by the 450FT fibers.

With reference to Table 45.3, as was the case in the residual loads (Table 45.2), the average energy absorption for the three fiber types for each sample was approximately the value for the 3000D–450FT (combined fibers). Furthermore, the energy absorption loads of the 3000D–450FT (combined) fibers were an average of 17.2% (ranging from 13% to 24.4%) lower than the values yielded by the 3000D fibers. The energy absorption of the 3000D–450FT (combined) fibers was an average of 36.3% (ranging from 25% to 50%) higher than the values yielded by the 450FT fibers.

## 6 Conclusions

The research proved to be a success in showing that the addition of two different fibers in one mix will result in the properties of the individual fibers working together as though one fiber type was used. The combining of the fibers (3000D and 450FT) resulted in residual loads and energy absorption values lower than those yielded when only including 3000D fibers, but higher than when including only the 450FT fibers. Although the combined fiber inclusion does not necessarily produce a panel that selected the best properties of the individual fibers, it rather appears as if the average of the two different fibers, or a best fit curve between the two fibers, is what is visible from test results of the combined fibers.

## References

1. Elvery, R. (1958). Effect of curing on permeability. In *Concrete practice* (pp. 105–106). London: Hastings and Folkestone.
2. Amtsbüchler, R., & van Heerden, H. (2009). Shotcrete. In *Fulton's concrete technology* (9th ed., pp. 349–357). Midrand: Cement & Concrete Institution.
3. Owens, G. (Ed.). (2012). *Fundamentals of concrete* (3rd ed., pp. 273–274). Midrand: Cement and Concrete Institute.
4. Tu, L. (1998). *Development of surface flourinated polypropylene fibres for use in concrete*. Johannesburg: University of Johannesburg.
5. ACI Committee 544. (2002). *State-of-the-art report on fiber reinforced concrete*. Farmington Hills: American Concrete Institute.
6. Bantur, A. (1990). *Fibre reinforced cementitious composites*. London/New York: Elsevier Applied Science. s.n.
7. Mittal, K. (1993). *Contact angle, wettability and adhesion*. Utrecht: VSP.
8. EFNARC. (1996). *European specification for sprayed concrete*. Surrey: Association House.

# Chapter 46
# Bond Performance of Steel-Reinforced Polymer (SRP) Subjected to Environmental Conditioning and Sustained Stress

Wei Wang and John J. Myers

Concrete beams reinforced with a steel-reinforced polymer (SRP) strengthening system were subjected to environmental conditioning with and without stress. They were subjected to environmental cycling in an environmental chamber that simulated the exterior weather of the Midwest United States. Two types of steel fibers (galvanized and brass-coated steel fibers) were investigated in this study. These SRP specimens experienced a series of freeze-thaw, varying temperature, and humidity cycles to experimentally investigate the influence of the accelerated aging environmental conditioning on bond performance between SRP strengthening system and concrete substrate. Flexural bending tests and direct pull-off bond tests were performed to evaluate the long-term bond performance of SRP-to-concrete interfaces. The flexural bending test results illustrated that the bond behavior between SRP and concrete was affected by the environmental cycling. The results of pull-off tests were scattered. This high variability was related to several issues including the nonhomogeneous characteristic of the concrete, applied load rate using a hand applied loading method, and the variable test method. The direct pull-off test was deemed to be a less reliable methodology to capture bond degradation than the flexural bending test.

## 1 Introduction

Currently, there are several composite application technologies to repair and retrofit deficient and aging concrete members in existing buildings and bridges. These technologies involve manual FRP lay-up, pre-cured laminate plates, near-surface-

---

W. Wang · J. J. Myers (✉)
Department of Civil, Architectural, and Environmental Engineering, Missouri University of Science and Technology, Rolla, MO, USA
e-mail: jmyers@mst.edu

mounted (NSM) bars, mechanically fastened FRP, and SRP. SRP is applied in a similar way as the FRP strengthening system. The non-galvanized configuration has been installed to an existing concrete bridge, Bridge P-0962, that is located on Highway B and spans Dousinbury Creek in Dallas County, Missouri, USA. The SRP strengthening system was utilized to reinforce the girders and deck of this bridge. This earlier configuration showed signs of rust in many locations after approximately 5 years in service. This was especially dominant in places that were able to drain from the deck to the girders or bents [1]. In this study, 3x2-G Hardwire® (RG) and 3x2 Hardwire® (RNG) were used to study the durability performance of concrete members reinforced externally with composite materials made from Hardwire® and epoxy (SRP). Sikadur 330 was used as epoxy. It is important to note, that the epoxy used in this study is a newer generation than that used previously with the non-galvanized system back in 2003 during the original demonstration application case. Flexural bending and pull-off tests were conducted in this study.

## 2 Experimental Programs

### 2.1 Flexural Bending Test

The dimension of the flexural specimen used in this study was 6 in. (width) × 6 in. (height) × 24 in. (length) (152.4 × 152.4 × 609.6 mm). A concrete grinder was utilized to roughen the concrete surface to improve the mechanical interlock between the concrete and SRP strengthening system which is common practice prior to field installation of the system. A saw cut with a width of approximately 0.125 in. (3.18 mm) and a depth of 2 in. (50 mm) was made on the tension side of the beam at mid-span consistent with the specified test method. Two layers of steel laminate strips were utilized with 1 in. (25.4 mm) width for the first ply and 0.75 in. (19.05 mm) width for second ply. Two steel squares were also staggered for direct pull-off testing on the side of the specimen. The reinforcement sheet for flexural testing was centered on the tension surface of concrete using Sikadur® 330 epoxy resin. The 12 in. (304.8 mm) long strips had a development length of 6 in. (152.4 mm) on each side of the saw cut based on previous research [2] and ACI 440.9R-15 [3]. The representative SRP specimens are shown in Fig. 46.1.

### 2.2 Direct Pull-Off Specimens

One layer of epoxy was brushed onto the roughened surface, and then a 5.5 × 5.5 in. (140 × 140 mm) steel sheet was applied. The second layer resin was applied to the first ply steel sheet and epoxy. Another 5.0 × 5.0 in. (127 × 127 mm) steel fabric

**Fig. 46.1** The typical SRP specimens

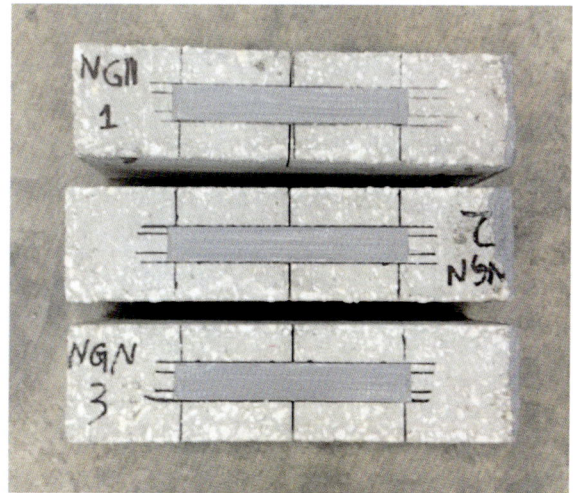

**Fig. 46.2** Spring-loaded fixture

square was performed in the exact same manner as the first steel ply. These two steel squares were also staggered to more easily accommodate direct pull-off testing.

## 2.3 Sustained Loading and Environmental Conditioning

RG and RNG series specimens were subjected to environmental conditions and sustained loading of 20% and 40% of ultimate tensile capacity, respectively, in an environmental chamber. Pairs of back-to-back specimens were subjected to the sustained three-point flexural load in the vertical orientation using a spring-loaded fixture to the desired sustained load. Spring-loaded fixture is shown in Fig. 46.2.

Data collected from the National Weather Service [4] and National Climatic Data Center [5] during a time frame from 1980 to 2013 was used to determine a suitable weather conditioning regime. In this study, one complete conditioning cycle was made up of 50 freeze-thaw cycles, 150 extreme temperature cycles, 150 relative humidity cycles, and 50 additional freeze-thaw cycles.

## 3 Flexural Bending Testing (Three-Point Load Testing)

The detailed dimension and tension surface of the specimen for the flexural bending test is illustrated in Fig. 46.3 per ACI 440.9R-15.

Load was applied at a constant rate of 0.005 in./min (0.127 mm/min) to cause failure of the specimens in 9–10 min. The deflections of both sides at the mid-span of the SRP specimen were measured by using two linear variable displacement transducers (LVDTs) with 0.5 in. (12.7 mm) extension. The failure loads and maximum deflections of conditioned and control specimens are summarized in Table 46.1.

It can be seen in Table 46.1 that the control specimens failed at higher loads. Meanwhile, they also showed greater deflections than those of conditioned samples. Therefore, the control specimens exhibited higher deformation ductility. The flexural

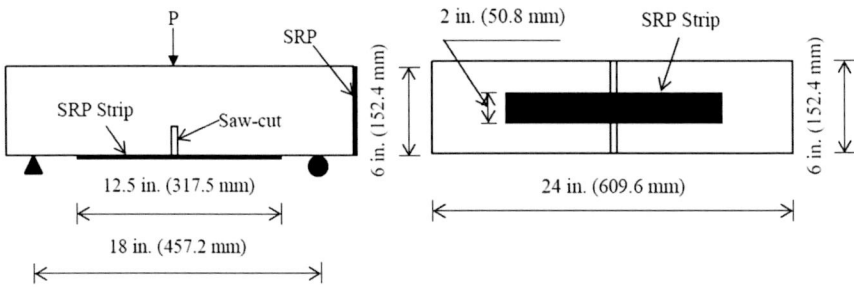

**Fig. 46.3** The setup of bending test

**Table 46.1** Summary of flexural bending tests

| Specimen | Avg. failure load (lb) | Avg. max. deflection (in.) |
|---|---|---|
| RG (40%) | 2998 | 0.0475 |
| RG (20%) | 3143 | 0.0593 |
| RG (unloaded) | 2831 | 0.0494 |
| RG control | 3471 | 0.0794 |
| RNG (40%) | 3178 | 0.0631 |
| RNG (20%) | 2968 | 0.0566 |
| RNG (unloaded) | 2890 | 0.0511 |
| RNG control | 3336 | 0.0748 |

Conversion Units: 1 in. = 25.4 mm, 1 lb. = 0.0044 kN

bending test results illustrated that the bond behavior between SRP and concrete was affected by the environmental cycling. However, the sustained load did not indicate any clear trend in the specimens studied.

## 4 Direct Pull-Off Testing

The concrete prisms with the SRP sides were maintained after the flexural bending tests to utilize for direct pull-off testing. Consistent with the testing standard, the diamond bit was used to drill the cores to separate the adhesion fixture from the surrounding SRP. The aluminum disk (dolly) with a 2 in. (50.8 mm) diameter was bonded to the SRP testing surface. Figure 46.4 illustrates a representative pull-off specimen and test setup. DYNA Z pull-off tester with digital manometer was used to pull the dolly with a pull pin off in direct tension.

According to ASTM D7522 [6], the pull-off bond strength (or concrete strength) was calculated by Eq. 46.1:

$$\sigma = \frac{4F_p}{\pi D^2} \qquad (46.1)$$

where $\sigma$ is the pull-off bond strength (psi [MPa]), $F_p$ is the pull-off force (lb [N]), and $D$ is the diameter of the loading fixture (in. [mm]). Table 46.2 illustrates the test results of pull-off specimens exposed to environmental chamber.

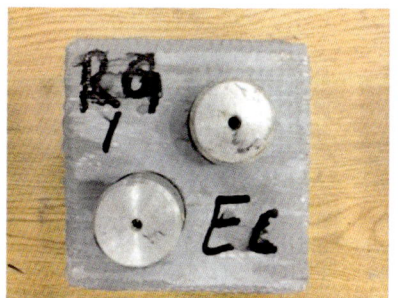

**Fig. 46.4** Pull-off specimen and test setup

**Table 46.2** Summary of pull-off tests

| SRP specimen | Average bond stress (psi) | Standard deviation | COV (%) |
|---|---|---|---|
| Conditioned RG | 369.5 | 88.9 | 24.1 |
| Conditioned RNG | 308.6 | 75.7 | 24.5 |
| Control RG | 318.2 | 61.7 | 19.4 |
| Control RNG | 229.0 | 132.4 | 57.8 |

Converse Units: 1 psi = 6.9 kPa

High standard deviations and coefficient of variations of the control and conditioned specimens were observed as shown in Table 46.2. For all of the pull-off test results illustrated a higher COV when compared to that of ASTM C 39 (10%) [7]. It is clear that the direct pull-off test was not able to capture any clear trends based on the observed standard deviation and COV obtained.

## 5 Conclusions

The results from the flexural bending test have suggested that the combination of freeze-thaw cycles, varying temperature, and humidity cycles resulted in some loss of the load-carrying capacity or bond shear strength between the concrete and the SRP strengthening system. It is also important to note that sustained loading did not indicate any clear degradation when compared to the specimen without any sustained loading. Through the results of this study, the flexural testing method specified in ACI 440.9R appears to be more effective to evaluate the long-term bond performance of SRP-to-concrete systems. In addition, the results of the pull-off test exhibited a large degree of scatter and variation indicating the variability of this test method is less than ideal. Therefore, while a direct pull-off test might be considered a technology to evaluate the long-term bond performance of SRP-to-concrete systems in field, it may not be an effective avenue. However, it may further be noted that the pull-off test method did satisfy the minimum ACI 440 required bond strength of 200 psi (1380 kPa).

## References

1. Myers, J. J., Holdener, D., Merkle, W., & Hernandez, E. (2008). Preservation of Missouri transportation infrastructures: Validation of FRP composite technology through field testing-in-situ load testing of bridges P-962, T-530, X-495, X-596 AND Y-298, *University of Missouri-Rolla*.
2. Gartner, A., Douglas, E. P., Dolan, C. W., & Hamilton, H. R. (2011). Small beam bond test method for CFRP composites applied to concrete. *Journal of Composites for Construction, 15* (1), 52–61.
3. ACI Committee 440.9R-15. (2015). *Guide to accelerated conditioning protocols for durability assessment of internal and external fiber reinforced polymer (FRP) reinforcement for concrete.* Farmington Hills: American Concrete Institute.
4. National Weather Service Forecast. (2013). http://www.weather.gov
5. NCDC (National Climatic Data Center). (2013). http://www.ncdc.noaa.gov
6. ASTM D7522/D7522M-15. (2009). Standard test method for pull-off strength for FRP bonded to concrete substrate, *American Society for Testing and Materials*.
7. ASTM C39/C39M-16b. (2016). Standard test method for compressive strength of cylindrical concrete specimens, *American Society for Testing and Materials*.

# Chapter 47
# High-Strength, Strain-Hardening Cement-Based Composites (HS-SHCC) Made with Different High-Performance Polymer Fibers

Marco Liebscher, Iurie Curosu, Viktor Mechtcherine, Astrid Drechsler, and Stefan Michel

This article presents an investigation on the tensile behavior of high-strength, strain-hardening cement-based composites (HS-SHCC) made with four different types of high-performance polymer microfibers. In particular, high-density polyethylene (HDPE), poly(p-phenylene-terephthalamide) (aramid), as-spun poly(p-phenylene-2,6-benzobisoxazole) (PBO), and high-modulus PBO fibers were examined in respect of their reinforcing effect in a high-strength, finely grained, cementitious matrix. Moreover, microscopic investigations were carried out to assess the fibers' ability to be wetted and to explain their interaction with the cementitious matrix. It was shown that the HS-SHCC made with PBO and aramid fibers yielded increased first crack stress and tensile strength, but also a considerably reduced crack width compared to the HS-SHCC reinforced with HDPE fibers. This was traced back to the considerably higher wettability of these fibers compared to the hydrophobic HDPE fibers, ensuring a stronger interfacial bond with the cementitious matrix, but also to their superior mechanical properties, such as tensile strength and Young's modulus.

## 1 Introduction

Despite the many advantageous properties of concrete, the material is also known for its poor tensile behavior and low tensile strength. By incorporating high-performance polymer microfibers and adapting the cementitious matrix, a special type of fiber-reinforced cementitious composite can be obtained, which not only

---

M. Liebscher (✉) · I. Curosu · V. Mechtcherine
Technische Universität Dresden, Institute of Construction Materials, Dresden, Germany
e-mail: marco.liebscher@tu-dresden.de

A. Drechsler · S. Michel
Leibniz-Institut für Polymerforschung Dresden e.V, Dresden, Germany

yields higher tensile strength compared to the nonreinforced matrix, but also a remarkable tensile quasi-ductility. When subjected to increasing tensile loading, such strain-hardening cement-based composites (SHCCs), also engineered cementitious composites (ECCs), yield high inelastic deformations ensured by the subsequent formation of multiple fine cracks prior to failure localization, that is, softening [1]. Such tensile behavior of SHCC is the result of a micromechanics-based design, which assumes a fine adjustment of the cementitious matrix in terms of its mechanical and physical properties in order to ensure a proper fiber anchorage on one side, and a sufficient energy absorption capacity in the crack bridging process on the other. Typical SHCCs are made with high-performance PVA fibers, usually in a volume content of up to 2% [2]. Another type of polymer fibers that are used in SHCC are those made of HDPE [3, 4]. Their superior mechanical properties and hydrophobic nature make them suitable for high-strength SHCC, which assumes a stronger and denser cementitious matrix. However, both the PVA fibers and the HDPE fibers possess a low temperature resistance. This leads to a dramatic degradation of the tensile behavior of SHCC when exposed to temperatures above 150 °C [5].

The present work is a comparative study of high-strength SHCC made with alternative high-performance polymer microfibers with different mechanical and thermomechanical properties. The work highlights the effect of fiber type on the tensile behavior of high-strength SHCC, and the reinforcing effect of the different fibers is explained according to the thermodynamically driven interactions between the fibers and the high-strength cementitious matrix.

## 2 Experimental Part

As-spun PBO (PBO-AS) and high-modulus (PBO-HM) Zylon® fibers were produced by Toyobo, Japan, representing the fibers with the highest tensile strength as well as modulus of elasticity in this study. The aramid fibers used are Technora® from Teijin. HDPE fibers Dyneema® SK62 were obtained from DSM, The Netherlands. A comparison of selected specific properties of the used fibers can be found in [6]. The SHCC preparation was performed by using a 5 L Hobart mixer. All samples contained 2 Vol.% of polymer fibers. The detailed sample composition of the fine-grained matrix was derived from an earlier study [4] and is given in Table 47.1.

Subsequently, dumbbell-shaped samples were produced in steel molds having a total length of 250 mm and cross-sectional dimensions in the narrowed middle region of 24 × 40 mm. Uniaxial tension tests were performed at 14 days in a 100 kN load capacity Instron universal testing machine in a deformation-controlled mode with a displacement rate of 0.05 mm/s. The tensile deformations of the specimens were measured on a gauge length of 100 mm in the middle of the specimens. Environmental Scanning Electron Microscopy (ESEM) was conducted by using a Quanta 250 FEG, FEI (The Netherlands). The pore solution was prepared by filtration of 2.5 kg cement in 13 kg tap water.

**Table 47.1** SHCC composition

|  | CEM I 52.5 R-SR3/NA [kg/m³] | Silica fume [kg/m³] | Quartz sand 0.06–0.2 mm [kg/m³] | Glenium ACE 460 [kg/m³] | Water [kg/m³] | Fibers (2% Vol.%) [kg/m³] |
|---|---|---|---|---|---|---|
| HDPE | 1460 | 292 | 145 | 35 | 315 | 20 |
| Aramid | 1460 | 292 | 145 | 35 | 315 | 28 |
| PBO-AS | 1460 | 292 | 145 | 35 | 315 | 31 |
| PBO-HM | 1460 | 292 | 145 | 35 | 315 | 31 |

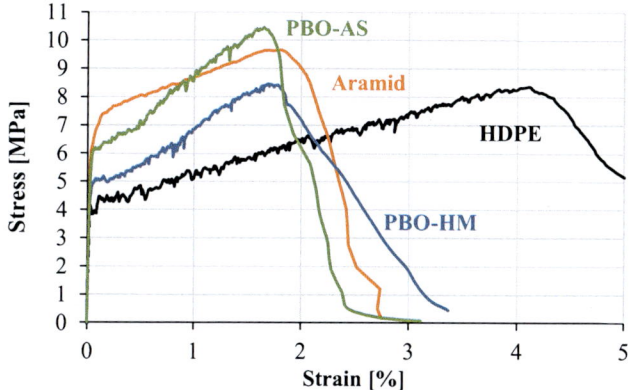

**Fig. 47.1** Representative stress–strain curves for HS-SHCC made with the four different polymer fibers. (Adapted from [6])

## 3 Results

For each material combination, a specific strain-hardening behavior can be observed, as shown in Fig. 47.1. The SHCC samples made with the nonpolar HDPE fibers possess the lowest first crack stress (~3.8 ± 0.4 MPa) by the highest ultimate strain (3.9% ± 0.9%), indicating a complete fiber pull-out behavior. In contrast, the SHCC made with aramid fibers yielded the highest first crack stress (6.3 ± 0.4 MPa) but the lowest strain capacity (1.4 ± 0.4%) as a result of the smallest crack width. Among all material combinations, the highest tensile strength (9.8 ± 0.9 MPa) was measured for the SHCC samples made with PBO-AS fiber. Samples mixed with PBO-HM show instead a lower first crack stress and also a lower tensile strength, by similar ultimate strain.

The differences in the deformation phenomena of the SHCC can be explained, on the one hand, by the specific mechanical properties of the fibers and, on the other hand, by their different wettability. Contact angle measurements with deionized water have shown that the most hydrophilic fibers are made of aramid and PBO-AS, followed by PBO-HM and the intrinsically hydrophobic HDPE

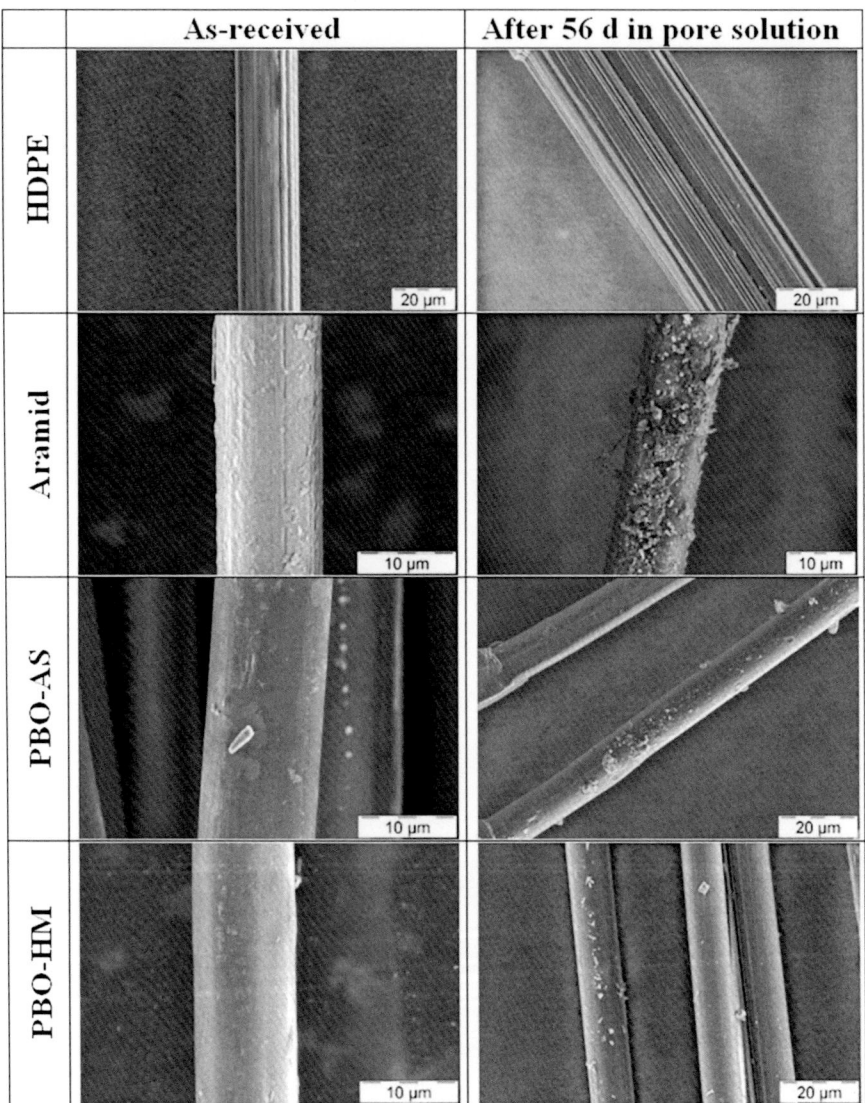

**Fig. 47.2** ESEM images of the different polymer fibers: as-received (left) and after storage in a pore solution

[6]. These differences determine the fibers' bonding behavior toward the cementitious matrix. By simulating a cementitious environment with a cement-based pore solution, ESEM analysis shows clearly more hydration products on the more hydrophilic fiber surfaces (Fig. 47.2). For the hydrophobic HDPE fiber, no residuals could be observed, due to its nonpolar molecular architecture. The slightly more hydrophilic PBO-HM fiber shows already in selected spots small mineral-based

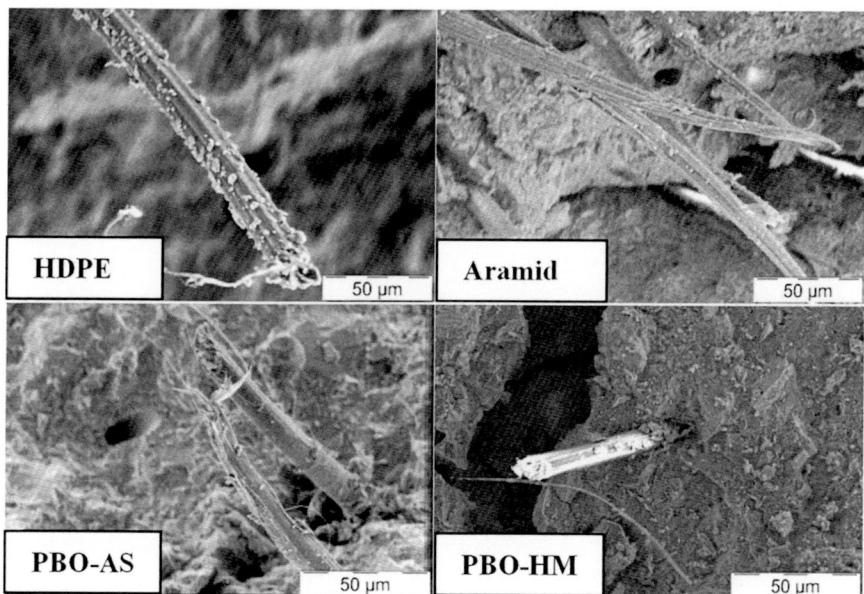

**Fig. 47.3** ESEM images of the damaged polymer fibers captured on the fracture surfaces (failure localization cracks) of tested SHCC specimens

features on the surfaces, which become even more prominent for the more hydrophilic PBO-AS fibers.

On the hydrophilic aramid fibers, big clumps of mineral-based products can be found, which explains the highest interfacial bond developed with the cementitious matrix [6], and consequently the highest first crack stress of the resulting SHCC. With decreased hydrophilic character, the fiber–concrete interface becomes weaker. The bonding behavior between the reinforcing fibers and the concrete matrix is determined more by a physical interaction, such as fiber friction and mechanical interlocking. This results in a lower capacity of the fibers to impede the formation and propagation of micro-cracks, and thus results in a reduction of the positive effect of the fibers on the first crack stress. Furthermore, a weaker interfacial bond leads also to a more pronounced fiber pull-out, which translates into wider cracks and lower tensile strength of the composite. It is noteworthy here that the SHCC made with HDPE fibers showed the highest average crack width (34.6 ± 9.6 μm), followed by the PBO-HM samples (20.3 ± 5.2 μm) [6]. For the hydrophilic aramid and PBO-AS fibers smaller average crack widths were measured, 14.6 ± 4.3 μm and 14.8 ± 4.5 μm, respectively [6].

Due to the different wetting behavior and their determined interfacial properties, also specific morphological features can be seen on the fibers captured on the fractured surfaces for the samples (Fig. 47.3). On the surface of the hydrophobic HDPE fibers concrete matrix residuals can be found, which are induced by the frictional damage during the extensive fiber pull-out. The stronger interfacial bond between the aramid fibers and the cementitious matrix, but also the slip-hardening pull-out behavior, result in a pronounced fiber rupture and surface damage. The tips of the hydrophilic fibers, which formed a chemical bond toward the cementitious matrix, are more deformed along the fiber axis than those of HDPE and PBO-HM, indicating a higher stress transfer during crack bridging, in particular seen for the aramid fibers. The fiber form fibrils and a gradual reduction of the fiber diameter can be observed.

## 4 Conclusions

The application of different high-performance polymer fibers as reinforcing agents in a finely grained cementitious matrix was systematically investigated and led to the following conclusions: The mechanical properties of SHCC are drastically influenced by the polymer fibers wettability and their mechanical properties. The higher the fiber polarity, the better is their wettability with water based materials such as cement pastes. This enables a chemical bond between concrete matrices and the polymer fibers resulting, for example, in a higher first crack stress of the SHCC, as seen in this study for aramid fibers. With nonpolar fibers, such as HDPE, no chemical bonding between concrete and the reinforcing agents occurs; the interface is determined by a physical bonding, for example, frictional forces or mechanical interlocking. This leads to smaller first crack stresses but to higher ultimate strains, due to the larger crack widths. Depending on fiber type, a targeted material design can be performed depending on the intended application of the composite and of the required mechanical properties.

**Acknowledgment** The authors acknowledge the financial support of the German Research Foundation (DFG) for funding the project ME 2938/16-1.

## References

1. Li, V. C. (2003). On Engineered Cementitious Composites (ECC). *Journal of Advanced Concrete Technology, 1*(3), 215–230.
2. Mechtcherine, V., Millon, O., Butler, M., & Thoma, K. (2011). Mechanical behaviour of strain hardening cement-based composites under impact loading. *Cement and Concrete Composites, 33*(1), 1–11.
3. Ranade, R., Li, V. C., Stults, M. D., Heard, W. F., & Rushing, T. S. (2013). Composite properties of high-strength, high-ductility concrete. *ACI Materials Journal, 110*(4), 413–422.
4. Curosu, I., Mechtcherine, V., & Millon, O. (2016). Effect of fiber properties and matrix composition on the tensile behavior of Strain-Hardening Cement-based Composites (SHCCs) subject to impact loading. *Cement and Concrete Research, 82*, 23–35.
5. Mechtcherine, V. (2013). Novel cement-based composites for the strengthening and repair of concrete structures. *Construction and Building Materials, 41*, 365–373.

6. Curosu, I., Liebscher, M., Mechtcherine, V., Bellmann, C., & Michel, S. (2017). Tensile behavior of High-Strength Strain-Hardening Cement-based Composites (HS-SHCC) made with high-performance polyethylene, aramid and PBO fibers. *Cement and Concrete Research, 98*, 71–81.

# Chapter 48
# Uniaxial Tensile Creep Behavior of Two Types of Polypropylene Fiber Reinforced Concrete

Rutger Vrijdaghs, Marco di Prisco, and Lucie Vandewalle

Structural polymeric macrofibers can be added to concrete to increase the residual capacity after matrix cracking. Polymeric fiber reinforced concrete (PFRC) can be designed according to the Model Code 2010, but no design guidelines are given to take creep behavior into account. In this work, the results of a multiscale experimental campaign into the crack-widening mechanisms of PFRC are detailed. Two different commercially available polypropylene fibers from the same manufacturer are tested. In the tests, individual fibers are subjected to long-term loading and the elongations are recorded. Furthermore, precracked PFRC cores are tested in a uniaxial tensile creep test at two load levels. The fiber creep tests highlight significant differences between the two fibers: the creep coefficient can differ an order of magnitude at similar load ratios. However, despite the much better performance at the individual fiber level, the FRC creep behavior does not vary to that degree. By comparing the single fiber performance with the FRC creep, it is found that pull-out creep and rupture can offset superior fiber creep performance.

## 1 Introduction

Since the introduction of fiber reinforced concrete (FRC) in the Model Code 2010 (MC2010) [1], designers can use this composite material in structural applications. While design rules are given to take the post-cracking tensile capacity of the material into consideration, long-term performance due to creep is not accounted for. Creep

---

R. Vrijdaghs (✉) · L. Vandewalle
Department of Civil Engineering, KU Leuven, Leuven, Belgium
e-mail: Rutger.Vrijdaghs@kuleuven.be

M. di Prisco
Department of Structural Engineering, Politecnico di Milano, Milan, Italy

© Springer International Publishing AG, part of Springer Nature 2018
M. M. Reda Taha (ed.), *International Congress on Polymers in Concrete (ICPIC 2018)*, https://doi.org/10.1007/978-3-319-78175-4_48

of FRC under tension is of high importance in structural design and the subject has been gaining attention in recent years. Different test methods have been proposed to study time-dependent crack width or deflections ranging from bending tests on prisms [2, 3] or panels [4] to uniaxial tension tests on prisms or cylinders [5, 6]. As fibers only take up forces after matrix cracking, all test methods use precracked specimens in which the fibers take up the forces in the cracked section. Creep of FRC in a cracked section is composed of up to three mechanisms: (1) compressive concrete creep in the case of bending tests, (2) time-dependent fiber pull-out, and (3) individual fiber creep in the case of polymeric FRC.In this work, uniaxial tension creep tests on two types of polypropylene (PP) FRC are discussed. Two different types of embossed PP fibers are used, and creep tests are performed on both individual fibers and the composite material.

## 2 Experimental Program

In the experimental program, two types of PP fibers are considered (designated as type A and B) and both fiber types are characterized according to EN 14889-2 [7]. The declared geometry (diameter $d$ and length $l$) and the mechanical results (fiber strength $f_t$ and cord modulus $E$) are summarized in Table 48.1.

Both fiber types are tested in a creep setup as described elsewhere [8]. The setup allows for different fibers to be tested independently in separate creep frames. For the creep test, three different load ratios are considered: 36%, 43%, and 53% of the fiber strength $f_t$.

In addition to the individual fiber creep tests, uniaxial tensile creep tests are performed on precracked PP FRC specimens. For both fiber types, 1 V% of fibers is added to a normal strength concrete with an average cube compressive strength of the concrete is 43 MPa as determined by EN 12390-3 [9]. The creep specimens are cored from a prism used in a characterization test according to EN 14651 [10]. The European Standard identifies the post-cracking tensile capacity of FRC in a displacement-controlled three-point bending test on a notched beam. The post-cracking tensile strength for specimens with type A and type B fibers is 1c and 2d, respectively. At the end of the test, cores are taken from these beams with a nominal diameter and height of 100 mm and 300 mm, respectively. At mid-height, the cylinders are notched to a diameter of 80 mm and the notched cores are precracked to an initial crack width of 0.2 mm in the notched section. The precracking is done in a custom designed precracking frame which allows eccentric load application in order to achieve uniform crack width growth after matrix cracking. The load is applied manually by three technicians and after the initial precrack width of 0.2 mm

**Table 48.1** Geometry and mechanical properties of the fibers

| Property | $d$ (mm) | $l$ (mm) | $f_t$ (MPa) | $E$ (MPa) |
|---|---|---|---|---|
| Type A | 0.9 | 45 | 451 | 4400 |
| Type B | 0.7 | 55 | 490 | 7900 |

is achieved, the load is gradually removed and the sample is placed in a cantilevered creep frame. The creep load is applied centrically and is expressed as a percentage of the residual post-cracking tensile strength measured during the precracking procedure. Two different load ratios (LR) are considered: 30% and 45% of the residual uniaxial strength. Each specimen is placed in a separate creep frame, and all testing is done in a climate controlled chamber at a constant 20 °C and 60% relative humidity. Further details about the precracking procedure and the creep setup can be found in literature [11].

## 3 Results and Discussion

The single fiber creep results are shown in Fig. 48.1 and highlight a significant difference between the two fibers. Note the different time scale in the figure. The stiffer fiber type B outperforms type A with respect to time to failure as well as total strain at failure. This is clearly indicated for the highest loaded samples, i.e., LR = 53%. In the case of type A, the fibers fail after 3 days at a total strain of 70%, while the first failure of a type B fiber at that LR was after 235 days at a strain of 40%. Similarly, the samples at 43% failed after 54 days at $\varepsilon = 98\%$ for type A, and no failure has been observed after 270 days at strains $\varepsilon < 16\%$.

The difference in single fiber creep behavior can be explained in terms of the creep compliance, i.e., the ratio between the creep strain and the applied stress. In an ideal viscoelastic material, the creep compliance is a function of time but not of the load ratio. Consequently, in a viscoelastic material, the compliance curves are superimposed and any deviation from the ideal behavior indicates the onset of plasticity in the fibers. The creep compliance is shown for the two fiber types in Fig. 48.2 on a double-logarithmic scale. For the type A fibers, a clear onset of plasticity is found at all load levels, with 53% samples already deforming plastically after 4 h and even the lowest loaded fibers exhibit plastic deformations within a week. In contrast, the creep compliance curves for type B remain superimposed until

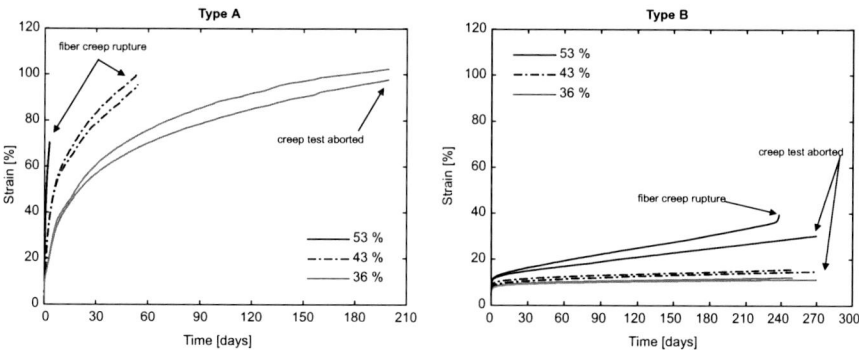

**Fig. 48.1** Single fiber creep for type A (left) and type B (right)

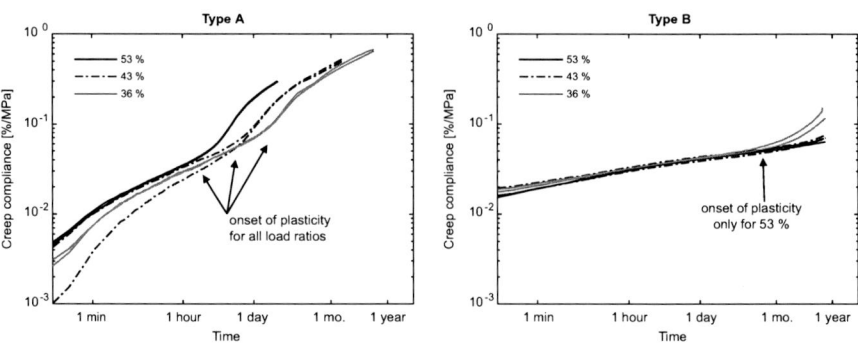

**Fig. 48.2** Creep compliance as a function of time for type A (left) and type B (right)

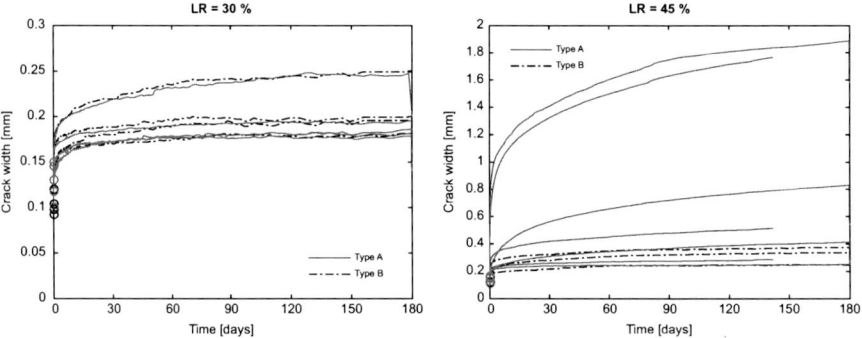

**Fig. 48.3** Crack width evolution for LR = 30% (left) and 45% (right)

the highest loaded samples deform plastically after 20 days. Furthermore, no plasticity is recorded for the samples at 43% and 36% after 270 days under load.

The results of the FRC creep tests are presented in Fig. 48.3 for the samples loaded at 30% and 45% after 180 days under sustained tensile loading. Note the different y-scales in the figure. The tests clearly show that at 30%, both fiber types exhibit a nearly identical crack width evolution and that in most cases, the average crack width remains below the 0.2 mm limit found in Eurocode 2 or MC2010. However, the samples exhibit very different behavior at 45%. In this case, the average crack width evolution of type B FRC corresponds to the best performing type A samples.

Comparing the single fiber creep with the FRC creep results, it is argued that part of the fibers in the cracked section deforms plastically for the type A FRC at 45%, causing the large crack widths observed for some specimens. Fiber type B exhibits practically no plasticity, thereby preventing excessive crack widths to form under creep loading. However, the differences in FRC creep are not as pronounced as in the single fiber creep results, leading to two conclusions. First, fiber creep alone cannot explain FRC creep, and time-dependent fiber pull-out should be considered as well. Tests on the short-term pull-out behavior have shown that type B fibers tend to rupture rather than pull out from the matrix owing to their smaller diameter. The smaller fiber diameter also increases the number of fibers in the cracked section with respect to type A at equal V%. The average fiber stress for type B fibers will be lower, but due to their higher tendency to rupture during pull-out, the better fiber creep behavior is partly offset by the increased chance of fiber rupture. A second consequence of the less pronounced FRC creep difference is that the fiber stress level is lower in the FRC specimens than the stress levels considered in the single fiber creep tests. At lower stresses, the type A fibers will not exhibit plastic deformations within a few hours or days but rather within several months. Since the fibers in the cracked section are not deforming plastically, the overall deformation and difference between the fibers will be smaller at lower load levels. Lastly, given the random distribution and orientation of fibers in FRC, large variations between different specimens can be expected. This is specifically an important issue for the specimens of this research as the cracked section is only 5000 mm$^2$, compared to nearly 19,000 mm$^2$ for the EN 14651 bending test.

## 4 Conclusion

In this chapter, the results of an experimental campaign into the creep behavior of normal-strength polymeric fiber reinforced concrete are presented in which two different polypropylene fiber types are considered. Owing to the different tensile strength and diameter of the fibers, the effect on the post-cracking tensile strength is significant. An increase from class 1c to 2d was observed when using the thinner, stronger fiber.

In the experimental program, creep tests were performed on individual fibers at three different load ratios: 36%, 43%, and 53% of the tensile strength. A clear difference in fiber creep performance is found, with the stiffer and stronger fiber creeping to a much lesser degree. Furthermore, the tendency for fiber creep failure was strongly reduced as well. Analysis of the creep compliance found that the difference between the two fiber types can be attributed to the onset of plastic deformations. Plasticity occurred within hours at 53% load for the weaker fiber and was only observed after 20 days in the stronger fiber.

In a second part of the experiments, notched FRC cores are subjected to sustained uniaxial tensile loads and the time-dependent crack widening is measured at loads of 30% and 45% of the residual tensile strength. No difference between the two fiber types was observed at the lower load level, but at the higher load, excessive crack widening was measured for the weaker fiber.

The comparison of the single fiber and FRC creep suggested that single fiber creep alone does not explain FRC creep and that time-dependent pull-out should be taken into account. For the samples considered in this research, it was found that the superior single fiber performance of one fiber type was partially offset by its higher tendency for fiber rupture during pull-out. Secondly, it is suggested that the fibers in the cracked sections are subjected to stress levels below the ones considered in the single fiber creep tests. Nevertheless, the coefficient of variation of all specimens was rather high and additional testing on more fibers is needed.

## References

1. Fédération internationale du béton (fib). (2010). *Model code 2010 first complete draft*. Lausanne: Fédération Internationale du Béton.
2. MacKay, J., & Trottier, J. F. (2004). Post-crack creep behavior of steel and synthetic FRC under flexural loading. In *Shotcrete: More engineering developments* (pp. 183–192). London: Taylor & Francis.
3. Kurtz, S., & Balaguru, P. (2000). Postcrack creep of polymeric fiber-reinforced concrete in flexure. *Cement and Concrete Research, 30*(2), 183–190.
4. Bernard, E. S. (2010). Influence of fiber type on creep deformation of cracked fiber-reinforced shotcrete panels. *ACI Materials Journal, 107*(5), 474–480.
5. Babafemi, A. J., & Boshoff, W. P. (2015). Tensile creep of macro-synthetic fibre reinforced concrete (MSFRC) under uni-axial tensile loading. *Cement and Concrete Composites, 55*(0), 62–69.
6. Zhao, G., Di Prisco, M., & Vandewalle, L. (2015). Experimental investigation on uniaxial tensile creep behavior of cracked steel fiber reinforced concrete. *Materials and Structures, 48*(10), 3173–3185.
7. British Standards Institution. (2006). BS EN 14889-2 fibres for concrete. Polymer fibres. Definitions, specifications and conformity. BSI. London, UK.
8. Vrijdaghs, R., di Prisco, M., & Vandewalle, L. (2017). Creep deformations of structural polymeric macrofibers. In P. Serna, A. Llano-Torre, & S. H. P. Cavalaro (Eds.), *Creep behaviour in cracked sections of fibre reinforced concrete: Proceedings of the international RILEM workshop FRC-CREEP 2016* (pp. 53–61). Dordrecht: Springer.
9. British Standards Institution. (2002). BS EN 12390-3 Testing hardened concrete – Part 3: Compressive strength of test specimens. BSI. London, UK.
10. British Standards Institution. (2005). BS EN 14651 Test method for metallic fibered concrete – Measuring the flexural tensile strength (limit of proportionality (LOP), residual). BSI. London, UK.
11. Vrijdaghs, R., di Prisco, M., & Vandewalle, L. (2016). Creep of cracked polymer fiber reinforced concrete under sustained tensile loading. In V. Saouma, J. Bolander, & E. Landis (Eds.), *9th international conference on fracture mechanics of concrete and concrete structures (FraMCoS-9)*. Berckeley.

# Chapter 49
# Dynamic Behavior of Textile Reinforced Polymer Concrete Using Split Hopkinson Pressure Bar

Mahmoud Abdel-Emam, Eslam Soliman, Amr Nassr, Wael Khair-Eldeen, and Aly Abd-Elshafy

Textile reinforced concrete (TRC) and textile reinforced mortar (TRM) have been introduced to the construction industry due to their relatively high tensile strength, high ductility, and ease of installation compared to other types of ordinary cementitious materials. A typical thin TRC plate consists of multidirectional fiber fabric reinforcement embedded in fine-grained cementitious concrete or mortar. One of the disadvantages of the TRC system is the potential early debonding that could occur between the fibers and the cementitious matrix. This paper discusses the possible use of textile fabric embedded in polymer matrix to form textile reinforced polymer concrete (TRPC) as an alternative material system to the conventional polymer concrete (PC) and TRC materials. The dynamic behavior of the new textile reinforced polymer concrete (TRPC) is quantified using modified split Hopkinson pressure bar (SHPB) system. Circular plates were prepared from TRPC with different number of fabric layers and centrally loaded using SHPB. In addition, the performance of TRPC specimens was compared to PC control specimen with no fiber fabric. The results show the ability of TRPC to withstand higher dynamic loads than the traditional PC. Such improvements in the dynamic behavior of the TRPC can benefit the design and construction of concrete panels against extreme loading scenarios.

M. Abdel-Emam · E. Soliman (✉) · A. Abd-Elshafy
Department of Civil Engineering, Assiut University, Assiut, Egypt
e-mail: esoliman@aun.edu.eg

A. Nassr · W. Khair-Eldeen
Department of Mechanical Engineering, Assiut University, Assiut, Egypt

## 1 Introduction

Over the last decades, rehabilitation and strengthening of existing reinforced concrete (RC) structures have become of great importance due to deterioration of these structures as a result of ageing, environmental conditions, and lack of maintenance or the need to meet the current design codes requirements. Common strengthening techniques include the use of concrete jackets or fiber reinforced polymer (FRP) composite materials. However, the behavior of such structural systems against extreme loading events (i.e., impact, blast, fire) remains questionable [1]. Textile reinforced concrete (TRC) materials started to gain popularity as strengthening systems over the last two decades due to their light weight, resistance to corrosion, resistance to high temperatures, high strength, and superior ductility [2, 3]. A typical TRC composite consists of high-strength fibers made of carbon, basalt, or glass in the form of textile fabrics embedded in fine-grained cementitious concrete or mortar. Applications for TRC in construction industry extend to thin concrete facades and claddings, parapet sheets, and noise/water protection walls [3]. Such structural applications are prone to extreme loading conditions.

The main disadvantage of using TRC is the potential early debonding between the fiber fabrics and the surrounding cementitious matrix. Several research articles have recognized and discussed the premature early debonding between fibers/cement matrix and its effect on the overall mechanical performance of TRC [4–6]. It was reported that the fiber pullout is the main load transfer mechanism in TRC due to fiber/matrix early debonding [6]. The poor bond between the fibers and cement matrix in TRC is attributed to the absence of chemical bonding mechanisms at the fiber/cement interface and the absence of contact between the inner filaments of the fiber threads and the cement matrix [7]. The problem of fiber/matrix debonding magnifies further when TRC is subjected to high loading rates. Materials subjected to high rate of loading exhibit a change in its mechanical response compared to static loading conditions [8], which may alter its structural capacity and failure mode [9]. In this paper, the use of polymer matrix as a replacement of cementitious matrix is suggested. The new textile reinforced polymer concrete (TRPC) is expected to have strong bond between the embedded fibers and polymer matrix due to the presence of chemical bond mechanism. As a result, TRPC would exhibit better dynamic response than that of the conventional TRC. In this study, the dynamic punching shear behavior of TRPC is investigated using Split Hopkinson Pressure Bar (SHPB). The effect of number of fiber fabrics of TRPC on the punch shear strength and absorbed energy is examined.

## 2 Experimental Program

### 2.1 Materials and Mixing Procedures

The polymer concrete mixture consisted of epoxy and filler. The epoxy was Kemapoxy 165 G, a two-component epoxy system obtained from Chemicals for Modern Building (CMP) Inc., Egypt. The epoxy system had resin-to-hardener ratio of 4:1. The filler was fine-grained aggregate with maximum nominal size of 4 mm. Approximately, 1700 kg of the fine-grained aggregates were added to 286 kg of epoxy to produce a cubic meter of polymer concrete. The average compressive strength of polymer concrete mixture after 7 days was 90 MPa. Bidirectional balanced glass fiber fabric is used to produce TRPC specimens. The glass fiber fabric had a mean yarn spacing of 5 mm, a Young's modulus of 72 GPa, and a tensile strength of 1700 MPa. The TRPC specimens were produced by first mixing the epoxy resin and hardener for 2 min using mechanical mixer rotating at speed of 300 r.p.m. The fine-grained filler was then added gradually and the mixing continued for an additional 5 min. Glass fabric layers were added during casting the fresh concrete at equal spacing through thickness to produce 27 mm thick, 100 mm diameter circular plates with four different numbers of embedded fabric layers. The produced specimens were the control TRPC specimens with no fabric layers (denoted as TRPC-0), and three other TRPC specimens with one (TRPC-1), two (TRPC-2), and three (TRPC-3) embedded fabric layers. The TRPC specimens were demolded after 24 h and were then cured at room temperature for 7 days before testing.

### 2.2 Punch Shear Test

The TRPC specimens were tested under dynamic punching shear. Split Hopkinson Pressure Bar (SHPB) setup was modified to enable applying dynamic punching shear load following Huang et al. 2011 test setup [10] as shown in Fig. 49.1. The SHPB consisted of 40 mm diameter striker, incident, and transmission bars. The TRPC specimens were inserted between the incident and transmission bars and were supported using a conical holder with inner diameter of 75 mm, as shown in Fig. 49.1. A 2.0 ± 0.25 bar pressure gas gun was used to launch the striker bar to impact the incident bar with impact velocity of 22 ± 3 m/s and to generate an elastic

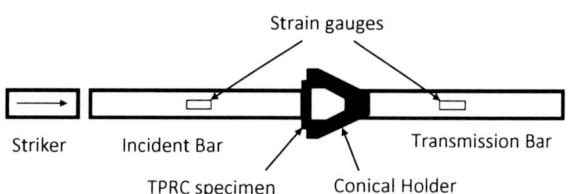

**Fig. 49.1** Schematic for SHPB modified test setup for punching shear load

compressive incident and reflected waves in the incident bar and transmitted wave in the transmission bar. To measure incident, reflected, and transmitted waves, two strain gauges were mounted on the incident and transmission bars. Digital oscilloscope with 1.0 MHz sampling rate was used to record the strain data. Based on strain and time measurements, the load–displacement response and the punch shear strength are determined. The punch shear force $P_1(t)$ is computed using Eq. 49.1 as follows:

$$P_1(t) = EA[\varepsilon_i(t) + \varepsilon_r(t)] \quad (49.1)$$

where $E$ is the Young's modulus of incident bar, $A$ is the cross-section area, $\varepsilon_i(t)$ is the incident strain, and $\varepsilon_r(t)$ is the reflected strain. The displacement $u(t)$ is calculated using Eq. 49.2:

$$u(t) = c \int_0^t (\varepsilon_i - \varepsilon_r) dt \quad (49.2)$$

where $c$ is the stress wave velocity of steel, which equals to 5100 m/s.

## 3  Results and Discussion

Figure 49.2a shows the load–displacement response of TRPC specimens with different numbers of fabric layers. The figure shows maximum dynamic punching shear forces of 231, 230, 211, and 239 kN for the TRPC specimens with zero, one, two, and three fabric layers, respectively. The changes in maximum dynamic forces with respect to the control specimen were 0, −8%, and +4% with the addition of one,

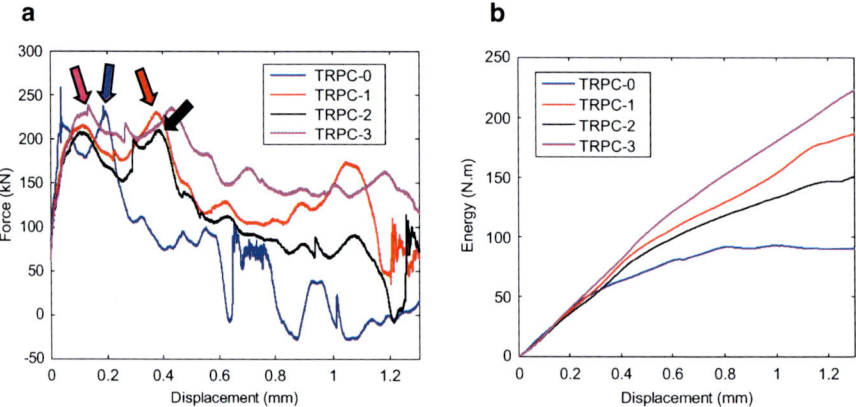

**Fig. 49.2** Dynamic behavior of TRPC specimens with various numbers of fabric layers: (**a**) load–displacement response and (**b**) absorbed energy

two, and three fabric layers, respectively. Such changes in maximum dynamic force can be considered insignificant given the nature of the test and the material statistical variability. It is therefore suggested that statistical analysis be conducted in many replicas of each specimens to confirm the significance of the difference in maximum force. It is also to be noted that for the control specimen, a sudden increase in a single data point for the force measurement is observed at a displacement of 0.03 mm. Such data point was excluded from the analysis as it does not represent a true mechanical behavior with a gradual increase in force to approach a maximum value.

On the other hand, Fig. 49.2a also shows that the TRPC specimens with fiber fabrics sustain high post-peak residual forces as opposed to the control TRPC specimen, which observed abrupt drop in the force after reaching the peak force. This observation indicates that the addition of fabric layers increases the ductility of the TRPC under dynamic shear loading. The observation is also in line with the results of energy absorption curves shown in Fig. 49.2b. The figure shows that the accumulative absorbed energy reached 96 J, 188 J, 155 J, and 237 J with adding zero, one, two, and three fabric layers, respectively. This corresponds to a significant increase in the absorbed energy reaching 96%, 62%, and 147% associated with the use of one, two, and three fabric layers, respectively. Such increase in absorbed energy demonstrates significant improvements in the ductility of TRPC materials, which in turn would yield similar improvements in the material response under dynamic loading. The results of the absorbed energy also show that the TRPC-1 observed higher absorbed energy than the TRPC-2. This could be attributed to the difference in the actual applied loading rate to the two specimens. To confirm this, the actual loading rates for the TRPC-1 and TRPC-2 specimens were obtained directly from the initial linear slope of the load–time curves and they were 10.1 kN/μs and 7.5 kN/μs, respectively. Clearly, the higher applied loading rate for the TRPC-1 yields higher maximum dynamic force and energy absorption than that of TRPC-2. More consistent loading rate is suggested for future testing.

The failure images for different TRPC specimens are shown in Fig. 49.3. The figure shows severe damage for the control TRPC specimen (TRPC-0) with the application of dynamic punching shear loading (Fig. 49.3a). The TRPC specimens with one and two layers observed lower level of damage compared to the control specimen as shown in Fig. 49.3b and c respectively. Furthermore, the TRPC specimen with three fabric layers observed little damage, with large pieces of the fractured specimens remaining intact after dynamic loading (Fig. 49.3d). The trend for the damage intensity of the four specimens is in agreement with the trend observed in the energy absorption between all the specimens.

## 4 Conclusions

In this study, dynamic punching shear test is conducted using modified SHPB on textile reinforced polymer concrete (TRPC) specimens incorporating various numbers of glass fiber fabric embedded. The load–displacement curves and the absorbed

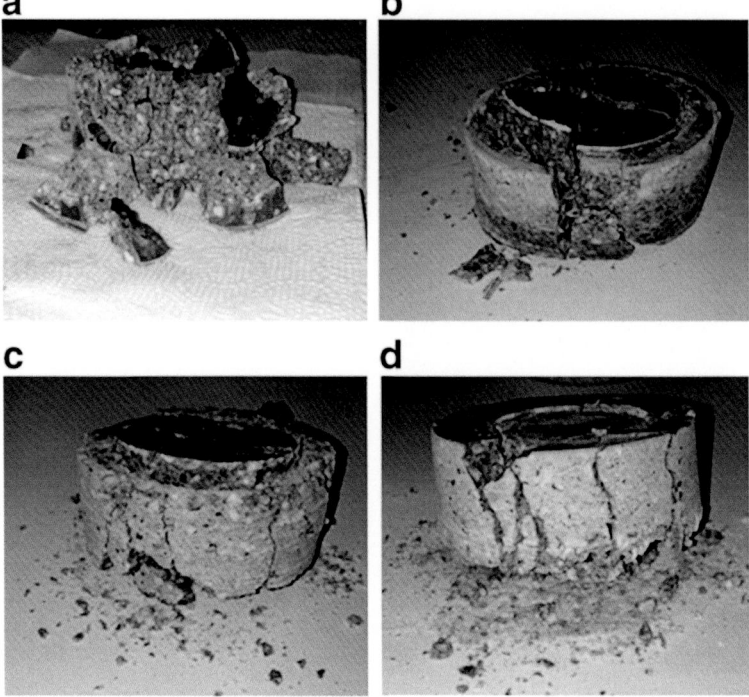

**Fig. 49.3** Failure images of TRPC specimens with various numbers of fiber fabrics: (**a**) control, (**b**) one layer, (**c**) two layers, and (**d**) three layers

energies are calculated and compared. It was found that by adding three layers of glass fiber fabric, the increase in absorbed energy reached 147%. Moderate improvements are observed with the use of two and one fabric layers. Such improvement demonstrates the benefit of using TRPC as a construction material subjected to extreme loading conditions.

# References

1. Kodur, V., Ahmed, A., & Dwaikat, M. (2009). *Modeling the fire performance of FRP-strengthened reinforced concrete beams. Composite & polycon*. Tampa: American Composites Manufacturers Association (ACMA).
2. Roye, A., Gries, T., & Peled, A. (2004). Spacer fabrics for thin walled concrete elements. In *6th international RILEM symposium on fibre reinforced concretes* (Vol. 139, pp. 1505–1514). Varenna: RILEM Publications SARL.
3. Brameshuber, W. (Ed.). (2006). *Report 36: Textile reinforced concrete-state-of-the-art report of RILEM TC 201-TRC* (Vol. 36). Bagneux: RILEM Publications.

4. Häußler-Combe, U., & Hartig, J. (2007). Bond and failure mechanisms of textile reinforced concrete (TRC) under uniaxial tensile loading. *Cement and Concrete Composites, 29*(4), 279–289.
5. Häußler-Combe, U., Jesse, F., & Curbach, M. (2004, April). Textile reinforced concrete-overview, experimental and theoretical investigations. In *Fracture mechanics of concrete structures. Proceedings of the fifth international conference on fracture mechanics of concrete and concrete structures, Ia-FraMCos, Vail* (Vol. 204, pp. 12–16).
6. Mobasher, B., Peled, A., & Pahilajani, J. (2006). Distributed cracking and stiffness degradation in fabric-cement composites. *Materials and Structures, 39*(3), 317–331.
7. Schorn, H., & Puterman, M. (2002). Polymer impregnated textile glass-fibre reinforcement for use in concrete. *Bautechnik, 79*(10), 671–675.
8. Henrych, J., & Major, R. (1979). *The dynamics of explosion and its use* (Vol. 569). Amsterdam: Elsevier.
9. Jones, N. (2011). Structural impact. Cambridge University Press, New York, NY, USA.
10. Huang, S., Feng, X. T., & Xia, K. (2011). A dynamic punch method to quantify the dynamic shear strength of brittle solids. *Review of Scientific Instruments, 82*(5), 053901.

# Chapter 50
# Steel-Fiber Self-Consolidating Rubberized Concrete Subjected to Impact Loading

**Mohamed K. Ismail, Assem A. A. Hassan, Katherine E. Ridgley, and Bruce Colbourne**

This investigation was carried out to evaluate the combined effect of crumb rubber (CR) and steel fibers (SFs) on improving the impact resistance of self-consolidating concrete (SCC) mixtures. Seven SCC mixtures were developed with varied percentages of CR (0–15% by volume of sand) and SF's volume of 0.35%. The performance of the developed mixtures was evaluated by testing the fresh properties, compressive strength, splitting tensile strength (STS), flexural strength (FS), and impact loading (drop weight on cylindrical and beam specimens). The results indicated that inclusion of CR decreased the compressive strength, STS, and FS of the tested mixtures, while the impact resistance obviously increased. Reinforcing CR mixtures with 0.35% SFs could compensate the reduction in the tensile strength resulting from adding rubber and further increase the resistance of mixtures to impact loading, achieving mixtures with promising properties for multiple structural applications.

## 1 Introduction

Combining crumb rubber (CR) and steel fibers (SFs) in the production of self-consolidating concrete (SCC) can be considered an innovative technique that combines the beneficial effects of CR and SFs and the desirable properties of SCC. Using aggregate with lower stiffness such as CR aggregates can contribute to decreasing the stiffness of the concrete composite and improving its energy absorption and impact resistance [1–4]. However, most research conducted on the use of rubber in concrete reported that the mechanical properties of concrete, including compressive strength, splitting tensile strength (STS), and flexural strength (FS), significantly

---

M. K. Ismail (✉) · A. A. A. Hassan · K. E. Ridgley · B. Colbourne
Faculty of Engineering and Applied Science, Memorial University of Newfoundland, St. John's, NL, Canada
e-mail: mohamed.ismail@mun.ca

decreased as the rubber content increased [1–6]. Using SFs in CR concrete can be an effective way to compensate for the reduction in STS and FS resulting from the addition of CR. Moreover, the inclusion of SFs can also improve the flexural toughness, impact strength, and ductility and limit the crack widths in concrete [7–10]. However, the addition of CR and SFs in SCC is considered a significant challenge due to their negative effect on the fresh properties of the mixture. Past studies reported that adding rubber in SCC caused an increase in $T_{50}$, V-funnel flow times, and viscosity of the mixtures [11]. Moreover, the low density of the rubber may easily encourage the rubber particles to float toward the concrete surface during mixing, thus increasing the risk of segregation [11, 12]. Similarly, the inclusion of SFs in SCC appeared to decrease the flowability and passing ability of the mixture due to the high blockage of SFs and coarse aggregates [10, 13]. Therefore, developing self-consolidating rubberized concrete (SCRC) containing SFs requires a balanced viscosity to improve the particle suspension and reduce the risk of segregation and thus achieve the acceptable fresh properties of SCC.

## 2 Materials

Type I Portland cement (ASTM C150), metakaolin (MK) (ASTM C618), and Type F fly ash (FA) (ASTM C618) were used to develop the designed mixtures. Crushed granite with a maximum size of 10 mm and crushed granite sand were used for the coarse and fine aggregates, respectively. The used crumb rubber in this investigation had a maximum size of 4.75 mm (see Fig. 50.1a), a specific gravity of 0.95, and negligible water absorption. Hooked-end SFs with a 35 mm length, a 65 aspect ratio, and a 0.55 mm diameter were used in this study (see Fig. 50.1b). Table 50.1 shows the designed mixtures.

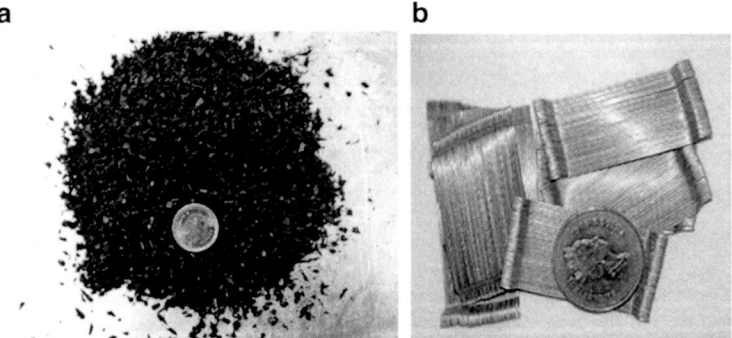

**Fig. 50.1** The used materials: (**a**) crumb rubber, (**b**) steel fibers

**Table 50.1** The developed mixtures

| Mix# | C | MK | FA | C.A. | F.A. | CR | HRWRA | SF |
|---|---|---|---|---|---|---|---|---|
| Mix1 | 275 | 110 | 165 | 620.3 | 886.1 | 0.0 | 3.43 | – |
| Mix2 | 275 | 110 | 165 | 620.3 | 841.8 | 16.2 | 3.43 | – |
| Mix3 | 275 | 110 | 165 | 620.3 | 797.5 | 32.4 | 3.75 | – |
| Mix4 | 275 | 110 | 165 | 620.3 | 753.2 | 48.6 | 3.75 | – |
| Mix5 | 275 | 110 | 165 | 616.5 | 836.7 | 16.1 | 4.63 | 27.48 |
| Mix6 | 275 | 110 | 165 | 616.5 | 792.7 | 32.2 | 4.63 | 27.48 |
| Mix7 | 275 | 110 | 165 | 616.5 | 748.6 | 48.3 | 4.63 | 27.48 |

Note: All mixtures have a 0.4 w/b ratio
*C.A.* coarse aggregates, *F.A.* fine aggregates, *CR* crumb rubber, *MK* metakaolin, *FA* fly ash, *SF* steel fiber

**Table 50.2** Fresh properties

| Mix# | Slump flow | | L-box H2/H1 | V-funnel | | Air % |
| | $D_s$ (mm) | $T_{50}$ (s) | | $T_0$ (s) | SR % | |
|---|---|---|---|---|---|---|
| Mix1 | 725 | 1.95 | 0.91 | 7.01 | 2.08 | 1.5 |
| Mix2 | 720 | 2.39 | 0.88 | 8.50 | 2.71 | 2.0 |
| Mix3 | 720 | 2.74 | 0.84 | 9.51 | 3.75 | 2.72 |
| Mix4 | 715 | 2.96 | 0.82 | 10.59 | 5.83 | 3.1 |
| Mix5 | 730 | 2.62 | 0.80 | 9.75 | 2.92 | 2.4 |
| Mix6 | 715 | 3.07 | 0.78 | 10.65 | 4.13 | 3.0 |
| Mix7 | 710 | 3.31 | 0.75 | 12.05 | 6.04 | 3.5 |

## 3 Fresh and Mechanical Properties Tests

The self-compactability of the developed mixtures in fresh state was evaluated as per tests recommended by the EFNARC [14]. The results of fresh properties tests including slump flow, L-box, and sieve segregation tests are shown in Table 50.2. The percentage of the entrained air in all tested mixtures was measured by following a procedure given in ASTM C231, as shown in Table 50.2. The mechanical properties tests included compressive strength, STS, and FS, according to ASTM C39, ASTM C496, and ASTM C78, respectively. The results obtained from mechanical properties tests are shown in Table 50.3.

## 4 Impact Resistance Tests

The impact resistance of the developed mixtures was evaluated using two different testing procedures. The first test was performed according to the drop-weight test procedures proposed by ACI 544 [15] (see Fig. 50.2a). The second test was conducted on beams using a 3-point loading setup to evaluate the energy absorption capacity of the developed mixtures under flexural impact loading (see Fig. 50.2b).

**Table 50.3** Results of mechanical properties and impact resistance

| Mix # | $f_c$ (MPa) | STS (MPa) | FS (MPa) | Drop-weight test | | | | | Flexural impact loading IE (J) |
|---|---|---|---|---|---|---|---|---|---|
| | | | | Number of blows | | | IE (J) | | |
| | | | | $N_1$ | $N_2$ | $N_2 - N_1$ | Initial | Failure | |
| Mix1 | 75.7 | 4.5 | 5.7 | 141 | 143 | 2 | 2813 | 2853 | 498 |
| Mix2 | 66.7 | 4.3 | 5.5 | 153 | 156 | 3 | 3052 | 3112 | 602 |
| Mix3 | 53.5 | 3.9 | 5.1 | 160 | 163 | 3 | 3192 | 3252 | 838 |
| Mix4 | 44.8 | 3.6 | 4.8 | 185 | 190 | 5 | 3691 | 3791 | 1002 |
| Mix5 | 67.6 | 5.3 | 6.7 | 337 | 391 | 54 | 6723 | 7800 | 1650 |
| Mix6 | 54.3 | 4.7 | 6.1 | 380 | 440 | 60 | 7581 | 8778 | 1853 |
| Mix7 | 44.9 | 4.3 | 5.6 | 470 | 534 | 64 | 9377 | 10,653 | 2056 |

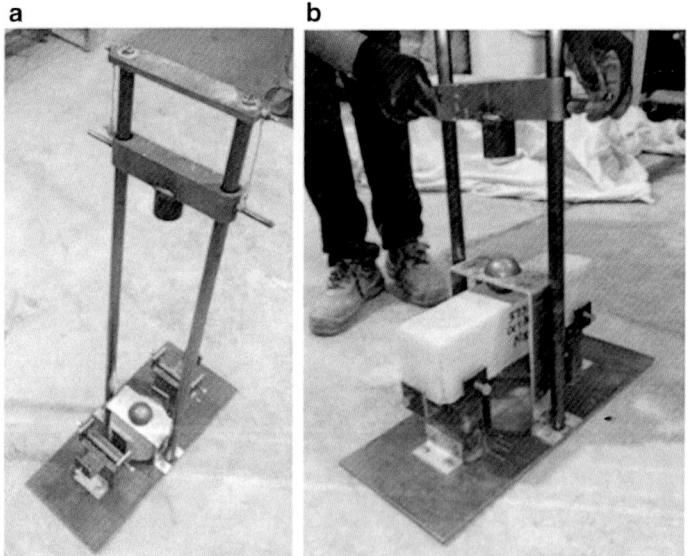

**Fig. 50.2** Impact tests: (**a**) ACI 544 drop-weight test, (**b**) flexural impact loading

For each mixture, three beams of 100 × 100 × 400 mm were tested with a loading span of 350 mm. The loading was applied by dropping a 4.45-kg drop hammer from a height of 150 mm onto the midspan of the tested beams. For the first test, the number of blows needed to produce the first visible crack (*N*1) was recorded to indicate the initial crack resistance. For the first and second test, the number of blows needed to cause failure (*N*2) was recorded to indicate the ultimate crack resistance. Table 50.3 shows the results of impact resistance for the developed mixtures.

## 5 Results and Discussion

### 5.1 Mechanical Properties

The 28-day compressive strengths of the tested mixtures are shown in Table 50.3. As shown in mixtures 1–4, the 28-day compressive strength, STS, and FS decreased as the percentage of CR increased. Increasing the CR from 0% to 15% showed a decrease in the 28-day compressive strength, STS, and FS reaching up to 40.8%, 20%, and 15.8%, respectively. The reduction of the compressive strength with inclusion of CR may be attributed to (i) the poor strength of the interfacial transition zone (ITZ) between the rubber particles and surrounding mortar, as observed by many researchers [16] and (ii) the significant difference between the modulus of elasticity of the rubber aggregate and hardened cement paste [17]. In addition, the measured air content appeared to increase as the percentage of CR increased (Table 50.2), which may also adversely affect the mechanical properties of the mixtures. It is worth noting that although the compressive strengths obviously decreased with an increase in the CR content, mixtures with up to 15% CR could be developed with a compressive strength of almost 45 MPa.

From Table 50.3, it can be also seen that using SFs with a volume of 0.35% (35 mm length) did not show a considerable effect on the 28-day compressive strength. On the other hand, using 0.35% SFs (35 mm length) noticeably increased the 28-day STS and FS of the developed mixtures by an average of 21% and 19.4%, respectively. These increases could be attributed to the fact that SFs act as crack arrestors, which in turn restricts the development of cracks and also helps to transfer the loads at the internal micro-cracks using the bridging effect.

### 5.2 Impact Resistance

The results of impact resistance under the drop-weight test for all tested mixtures are shown in Table 50.3. It can be seen that using CR generally improved the impact resistance of tested mixtures in terms of the number of blows required to cause the first visible crack ($N1$) and ultimate failure ($N2$) of the specimen. Increasing the percentage of CR from 0% to 15% increased $N1$ and $N2$ by 1.31 and 1.33 times. This increase could be attributed to the low stiffness of the rubber particles, which in turn increases the flexibility of the rubber-cement composite and considerably enhances its energy absorption compared to concrete with no CR. The results also showed that increasing the CR content contributed to increasing the post-cracking resistance; as shown, the difference between the number of blows for ultimate failure and first crack ($N2 - N1$) increased as the percentage of CR increased (see Table 50.3). This finding indicates a reduction in the brittleness of SCRC mixtures with the addition of CR.

Using 0.35% SFs (35 mm length) raised $N1$ and $N2$ by an average of 2.37 and 2.67 times, respectively. This increase resulted from the beneficial effect of SFs on arresting the cracks induced due to impact loading. The difference between $N2$ and $N1$ also showed higher increases with the inclusion of SFs, indicating a significant enhancement in the post-cracking resistance and ductility performance of concrete. It is worth noting that mixture with 15% CR and 0.35% SFs exhibited an increase of 3.33 and 3.73 times in the absorbed energy at the first and failure cracks, respectively, compared to the mixture with no CR and SFs (control mixture).

Table 50.3 also shows the results of all tested mixtures at failure under flexural impact loading. It can be seen that adding CR to concrete helped to improve its ultimate impact energy. Increasing the percentage of CR from 0% to 15% increased the ultimate impact energy at failure by two times. Using 0.35% SFs showed an increase in the ultimate impact energy of tested beams by an average of 2.33 times compared to mixtures with only CR. Combining 15% CR and 0.35% SFs showed a 4.13 times increase in the ultimate absorbed energy compared to the control mixture.

## 6 Conclusions

This study evaluated the combined effect of CR and SFs on the impact resistance of SCC mixtures. The impact resistance was investigated using the drop-weight test recommended by the ACI committee 544. The results showed that increasing the percentage of CR in SCC mixtures apparently decreased the compressive strength, STS, and FS, while the impact resistance in both the drop-weight test and flexural impact loading test showed a great enhancement. Using 0.35% SFs (35 mm length) in SCRC mixtures increased the STS and FS by an average of 20.5% and 19.5% (compared to SCRC with no SFs), but with no significant effect on the compressive strength. The addition of 0.35% SFs also increased the ultimate impact resistance of the drop-weight test and flexural impact loading test by an average of 2.68 and 2.33, respectively.

## References

1. Al-Tayeb, M. M., Abu Bakar, B. H., Ismail, H., & Akil, H. M. (2012). Impact resistance of concrete with partial replacements of sand and cement by waste rubber. *Polymer-Plastics Technology and Engineering, 51*(12), 1230–1236.
2. Gupta, T., Sharma, R. K., & Chaudhary, S. (2015). Impact resistance of concrete containing waste rubber fiber and silica fume. *International Journal of Impact Engineering, 83*, 76–87.
3. Reda Taha, M. M., El-Dieb, A. S., Abd El-Wahab, M. A., & Abdel-Hameed, M. E. (2008). Mechanical, fracture, and microstructural investigations of rubber concrete. *Journal of Materials in Civil Engineering, 20*(10), 640–649.
4. Ismail, M. K., & Hassan, A. A. A. (2017). Impact resistance and acoustic absorption capacity of self-consolidating rubberized concrete. *ACI Materials Journal, 113*(6), 725–736.

5. Ismail, M. K., & Hassan, A. A. A. (2016). Use of metakaolin on enhancing the mechanical properties of self-consolidating concrete containing high percentages of crumb rubber. *Journal of Cleaner Production, 125*, 282–295.
6. Abdelaleem, B. H., Ismail, M. K., & Hassan, A. A. A. (2017). Properties of self-consolidating rubberised concrete reinforced with synthetic fibres. *Magazine of Concrete Research, 69*(10), 526–540.
7. Song, P. S., & Hwang, S. (2004). Mechanical properties of high strength steel fiber reinforced concrete. *Construction and Building Materials, 18*(9), 669–673.
8. Olivito, R. S., & Zuccarello, F. A. (2010). An experimental study on the tensile strength of steel fiber reinforced concrete. *Composites Part B: Engineering, 41*(3), 246–255.
9. Nia, A., Hedayatian, M., Nili, M., & Sabet, V. F. (2012). An experimental and numerical study on how steel and polypropylene fibers affect the impact resistance in fiber-reinforced concrete. *International Journal of Impact Engineering, 46*, 62–73.
10. Khaloo, A., Raisi, E. M., Hosseini, P., & Tahsiri, H. (2014). Mechanical performance of self-compacting concrete reinforced with steel fibers. *Construction and Building Materials, 51*, 179–186.
11. Ismail, M. K., & Hassan, A. A. A. (2015). Influence of mixture composition and type of cementitious materials on enhancing the fresh properties and stability of self-consolidating rubberized concrete. *Journal of Materials in Civil Engineering, 28*(1), 1–12.
12. Topçu, I. B., & Bilir, T. (2009). Experimental investigation of some fresh and hardened properties of rubberized self-compacting concrete. *Materials and Design, 30*(8), 3056–3065.
13. Ismail, M. K., & Hassan, A. A. A. (2017). Use of steel fibers to optimize self-consolidating concrete mixtures containing crumb rubber. *ACI Materials Journal, 114*(4), 581–594.
14. EFNARC. (2005). *The European guidelines for self-compacting concrete specification, production and use* (English ed.). Norfolk: European Federation for Specialist Construction Chemicals and Concrete Systems.
15. American Concrete Institute (ACI) 544.2R-89. (1999). *Measurement of properties of fiber reinforced concrete*. West 481 Conshohocken.
16. Najim, K. B., & Hall, M. (2012). Mechanical and dynamic properties of self-compacting crumb rubber modified concrete. *Construction and Building Materials, 27*(1), 521–530.
17. Najim, K. B., & Hall, M. R. (2010). A review of the fresh/hardened properties and applications for plain-(prc) and self-compacting rubberised concrete (SCRC). *Construction and Building Materials, 24*, 2043–2051.

# Chapter 51
# Effect of Fiber Combinations on the Engineering Properties of High-Performance Fiber-Reinforced Cementitious Composites

Dongyeop Han, Min-Cheol Han, Jong-Tae Lee, and Cheon-Goo Han

Although concrete has many advantages including ease of handling, economic benefits, and high compressive strength, there are disadvantages such as low tensile strength, brittleness, and drying shrinkage. High-performance fiber-reinforced cementitious composites (HPFRCC) which has high volume of fiber reinforcement was suggested as a solution of the drawbacks of normal concrete. However, reinforcing fiber in HPFRCC may cause a decreasing workability; therefore, in this paper, optimized fiber combination with either or both metal and organic fibers is suggested to provide better performance of HPFRCC in tensile strength and ductility. Test results indicate that for a given combination of fiber, using long steel fiber, short steel fiber, and long organic fiber, decreased water-to-binder ratio (w/b) caused increased toughness and energy absorption with increased tensile strength. Additionally, increased fiber content also contributed to increased toughness; strain-hardening behaviors were observed especially, when ternary fiber content was raised to 2.0%.

## 1 Introduction

High-performance fiber-reinforced cementitious composite (HPFRCC) shows outstanding performance [1, 2]. However, most research on this material has concentrated on using single-fiber types, either metal or organic. However, according to previous research, using combined fiber which consisted of two or more types of

---

D. Han
Department of Architectural Engineering, and Engineering Research Institute, Gyeongsang National University, Jinju, Gyeongsangnam-do, South Korea

M. -C. Han (✉) · J. -T. Lee · C. -G. Han
Department of Architectural Engineering, Chcongju University, Cheongju, South Korea
e-mail: twhan@cju.ac.kr

fiber performed more favorably in fluidity, strength, and toughness than using single-type fiber [3, 4]. Specifically, among the various combinations of fibers with four different components including short and long steel fibers, and short and long organic fibers, the most efficient fiber combination was ternary fiber having long and short steel fiber and long organic fiber [5].

Therefore in this study, using the ternary fiber as the most efficiently combined fiber, the optimum mix proportion is suggested from the result of the analysis of the mechanical properties including fluidity and tensile strength depending on various mix proportions of water-to-cement ratio and fiber content.

## 2 Experiment

### 2.1 Experimental Plan

The experimental plan is summarized in Table 51.1. Based on this plan, proportions of the mixtures were prepared. As a test condition, five different water-to-binder ratios (w/b)s were prepared for 1% of fiber content, and five different fiber contents were prepared for 0.25 of w/b.

To evaluate the performance of the mixtures depending on different mix conditions, flow and air content were planned to assess fresh state properties. Compressive, tensile, and flexural strengths were designed to measure hardened state properties. Specifically, when the tensile strength was measured, the deformation of the specimen was planned to measure calculating strain.

**Table 51.1** Experimental plan

| Items | | | Variables |
|---|---|---|---|
| Mixture[a] | Fixed conditions | Binder combination | C:FA:SF = 7:2:1 |
| | | B:S (volume) | 1:0.6 |
| | | Fiber combination | SL + SS + OL |
| | | Target flow (mm) | Practically available range |
| | | Target air content (%) | 2.0 ± 1.0 |
| | Controlled factors | w/b | 0.20, 0.25[b], 0.30, 0.35, 0.40 |
| | | Fiber content (%) | 0, 0.5, 1.0[b], 1.5, 2.0 |
| Test | Fresh mortar | | Flow test |
| | Hardened mortar | | Compressive strength (7, 28 days) Tensile strength (28 day) with measuring strain Flexural strength (28 day) |

[a]*C* cement, *FA* fly ash, *SF* silica fume, *B:S* binder-to-sand ratio, *w/b* water-to-binder ratio, *SL* long steel fiber, *SS* short steel fiber, and *OL* long organic fiber
[b]Regarding the two tests on influence of fiber content, and influence of w/b, the w/b and fiber content were fixed to 0.25% and 1.0%, respectively

**Table 51.2** Physical properties of fiber

| Types[a] | Length (mm) | Diameter (mm) | Aspect ratio | Tensile strength (MPa) |
|---|---|---|---|---|
| SL | 35 | 0.53 | 66 | 1108 |
| SS | 13 | 0.12 | 108 | 2650 |
| OL | 30 | 0.47 | 63 | 650 |

[a]*SL* long steel fiber, *SS* short steel fiber, *OL* long organic fiber

## 2.2 Materials

Since the tested mixture was the mortar phase, powders and fine aggregate with fibers were used. The cement used was ordinary Portland cement from the South Korean market which has similar properties to Type I cement of ASTM C150 [6]. The fly ash and silica fume used were also generally available products from the South Korean market. The fine aggregate used was natural river sand obtained from South Korea. Superplasticizer as a chemical admixture was a generic product from the South Korean market. Additionally, the properties of the long steel fiber (SL), short steel fiber (SS), and long organic fiber (OL) are shown in Table 51.2.

## 2.3 Test Methods

To evaluate the fresh and hardened states properties, all tests were conducted following ASTM methods such as ASTM C230 [7] and ASTM C39 [8] for flow and compressive strength, respectively. On the other hand, the tensile strength test was conducted by following Japan Society of Civil Engineers standard (JSCE-E-53) [9].

## 3 Results and Discussion

### 3.1 Flow

During flow tests, two measurements were conducted: (1) initial flow without dropping and (2) standard flow after 25 drops. The initial and standard flow values depending on w/b and fiber content are shown in Fig. 51.1a, b, respectively. Additionally, the shape of the initial flow depending on w/b and fiber content are shown in Fig. 51.2a, b, respectively. From the flow test results, generally increasing flow was observed with increased w/b, while changing the fiber content did not change the flow much. Especially, at over 0.30 of w/b, water segregation was observed at initial flow. It was learned that the decreased volume fraction of the cement paste caused viscosity loss and segregation occurred.

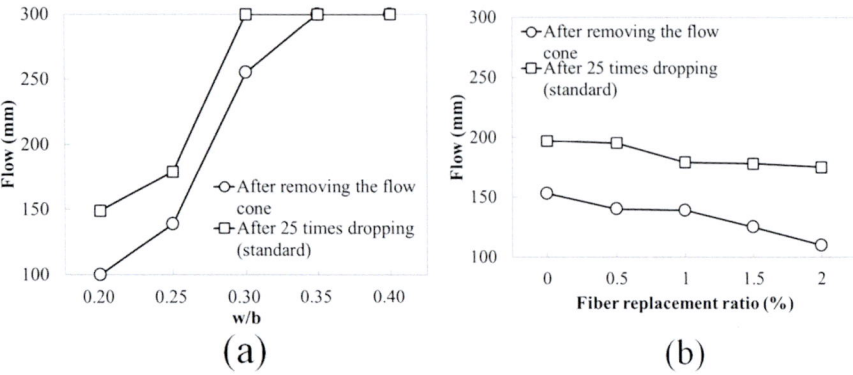

**Fig. 51.1** Influence of (**a**) w/b and (**b**) fiber content on flow

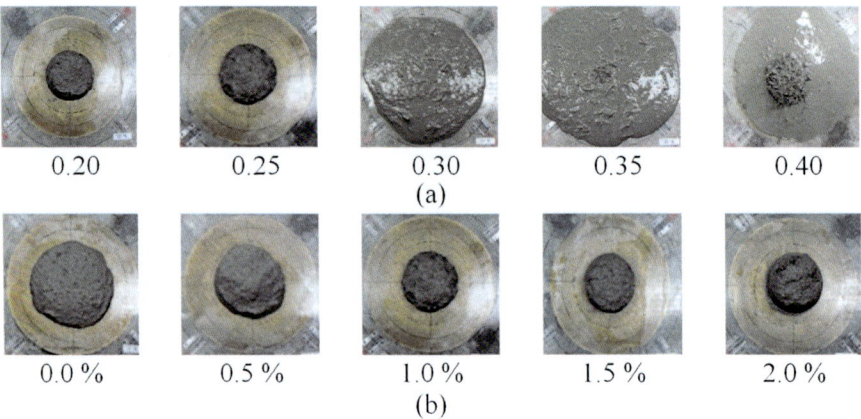

**Fig. 51.2** Shape of the initial flow depending on influence of (**a**) w/b and (**b**) fiber content

## 3.2 Compressive Strength

The compressive strength of the mixtures was measured at 7 and 28 days. The influence of different w/b and fiber content on compressive strength depending on ages are shown in Fig. 51.3a, b, respectively. First, as the w/b increased, compressive strength of the mixture decreased. Meanwhile, although the compressive strength increased with increased fiber content, there was no significant change with the fiber content. It was learned that the crack-controlling effect and confining effect of fiber reinforcement contributed to the increased compressive strength.

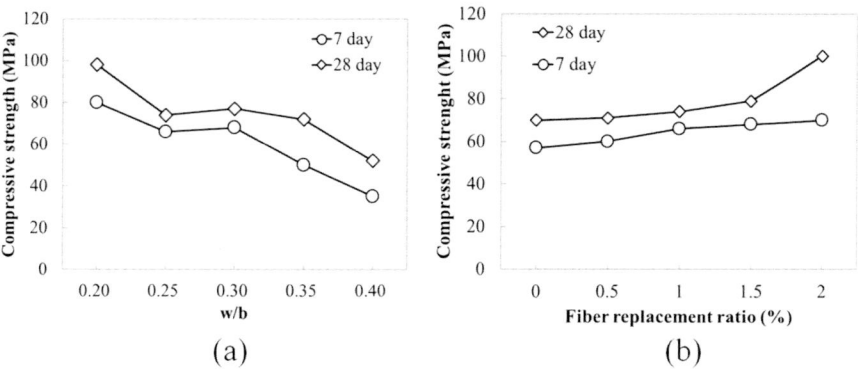

**Fig. 51.3** Influence of (**a**) w/b and (**b**) fiber content on compressive strength

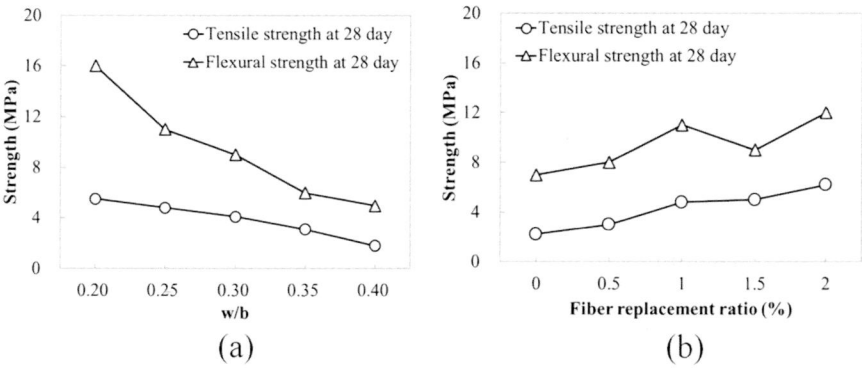

**Fig. 51.4** Influence of (**a**) w/b and (**b**) fiber content on tensile and flexural strength

## 3.3 Tensile and Flexural Strength

Both tensile and flexural strength of the mixtures were measured at 28 days. The tensile and flexural strength depending on w/b and fiber content are shown in Fig. 51.4a, b, respectively. Similar to the compressive strength results, depending on w/b, both tensile and flexural strengths decreased with increased w/b. However, unlike the compressive strength with fiber content, tensile and flexural strengths increased gradually with increased fiber content.

In addition, during the tensile strength test, strain was measured, and the strain-stress relation was obtained. Regarding the strain-stress curves, the influence of w/b and fiber content are shown in Fig. 51.5a, b, respectively. First as the w/b decreased, the toughness and energy absorption capacity increased. Interestingly, although the decreased w/b was expected to show brittle failure mode, because of the improved

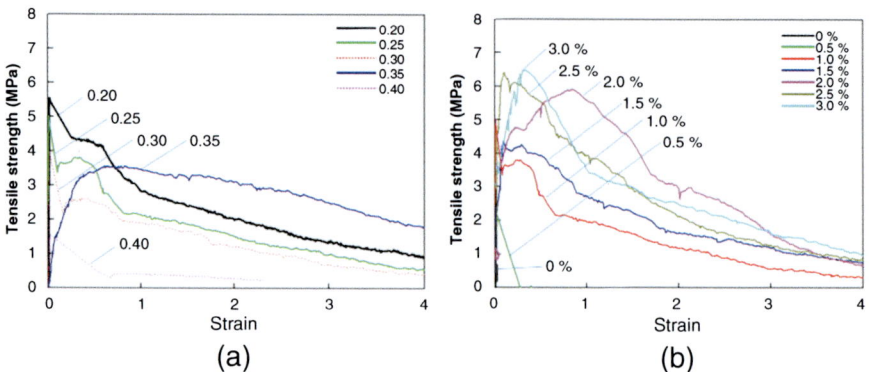

**Fig. 51.5** Influence of (**a**) w/b and (**b**) fiber content on strain-stress curves

adhesive performance between fiber and cement matrix, ductile failure mode was obtained. Furthermore, the fiber content changed the behaviors of the mixture more dramatically than w/b. In the case of 2% fiber content, strain-hardening behaviors were observed under applied tensile forces.

## 4 Conclusions

In this research, to discover the optimum mix design for practical usage of HPFRCC with ternary fiber, depending on various w/b and fiber content, fresh and hardened states properties were evaluated. A series of experiments was conducted, and results can be summarized as follows:

1. In fluidity aspect, higher than 0.30 of w/b caused segregation, while there was no significant problem with changing fiber content.
2. Compressive, tensile, and flexural strengths increased with decreased w/b and increased fiber content. Specifically, for compressive strength, the maximum value of 100 MPa was achieved with 0.20 w/b and 2% fiber content at 28 days. For tensile and flexural strengths, increased fiber content contributed with increasing the strengths. Thus, 100% and 67% of increased tensile and flexural strength values, respectively, were observed with increased fiber content from 0% to 2.0%.
3. According to the strain-stress relationship, decreased w/b increased toughness and energy absorption capacity with increased tensile strength. Additionally, increased fiber content also contributed to increased toughness; especially, when 2.0% of ternary fiber content, strain-hardening behaviors were observed.

Based on the results of the experiment, 0.25 of w/b and 2.0% of ternary fiber content can be suggested as optimal conditions in the aspects of fluidity, strength, and other mechanical properties.

## References

1. Kim, Y. D., Cho, B. S., kim, J. H., Kim, G. Y., Choi, K. Y., & Kim, M. H. (2003). An experimental study on the engineering properties of HPFRCC according to kinds, shapes and volume fraction of fibers. In *Proceeding of the Korea Institute of Building Construction Conference*, The Korea Institute of Building Construction (pp. 59–62).
2. Ali, H., Ali, K., Mohammad, S., Park, Y., & Ali, A. (2016). Ductile behavior of high performance fiber reinforced cementitious composite (HPFRCC) frames. *Construction and Building Materials, 115*, 681–689.
3. Kang, S. H., Ahn, T. H., & Kim, D. J. (2015). Effect of grain size on the mechanical properties and crack formation of HPFRCC containing deformed steel fibers. *Cement and Concrete Research, 42*, 710–720.
4. Naaman, A. E., & Reinhardt, H. W. (1996). Characterization of high performance fiber reinforced cement composites, High performance fiber reinforced cement composites: HPFRCC 2. In *Proceedings of 2nd International Workshop on HPFRCC, Chapter 41, RILEM*, No. 31, E. & FN Spon, London (pp. 1–20).
5. Han, D. Y., Han, M. G., Kang, B. H., & Park, Y. J. (2014). Effect of Hybrid Fibers on the Engineering Properties of HPFRCC. *Journal of the Korea Institute for Structural Maintenance and Inspection, 6*(6), 639–645.
6. ASTM International. (2012). *ASTM C150, standard specification for Portland cement*. West Conshohocken: ASTM International.
7. ASTM International. (2013). *ASTM C230, standard specification for flow table for use in tests of hydraulic cement*. West Conshohocken: ASTM International.
8. ASTM International. (2012). *ASTM C39, standard test method for compressive strength of cylindrical concrete specimens*. West Conshohocken: ASTM International.
9. Japan Society of Civil Engineers. (1995). *JSCE-E 53, test method for bond strength of continuous fiber reinforcing materials by pull-out testing*. Tokyo: Japan Society of Civil Engineers.

# Chapter 52
# Application of Fibre-Reinforced Polymer-Reinforced Concrete for Low-Level Radioactive Waste Disposal

Ricardo Lopes and Deon Kruger

This paper provides an overview of current research into the use of fibre and polymer reinforcement of concretes for the application in the field of low-level nuclear waste containment. This research investigates the effects of the use of polymers and fibre reinforcement in ordinary Portland cement binder concretes to create polymer-cement concrete or polymer-modified concrete, in comparison with geo-polymer concrete with polymer reinforcement and fibre reinforcement. This is to gain further knowledge into the effects polymer addition has on concrete durability under radiation conditions. In this research, various mixtures are designed and tested in order to determine the most feasible mixtures for producing durable, robust and sound containers for low-level nuclear waste storage. Such containers will be used for storage as well as for transportation of nuclear waste and therefore must also eliminate leaching and also resist microcracking. Other considerations for a suitable mix include ease of casting and long-term durability. Testing involved will include evaluating the mixtures through compressive and flexural tests, analysis of crack development, permeability and waterproof properties as well as resistance to gamma ray penetration. An opinion on which material is most suitable for the production of such low-level radioactive waste storage containers will be offered as a conclusion to the research paper.

R. Lopes · D. Kruger (✉)
Engineering and the Built Environment, University of Johannesburg, Johannesburg, South Africa
e-mail: dkruger@uj.ac.za

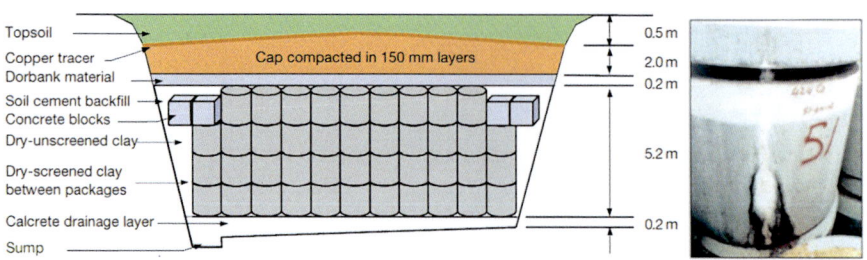

**Fig. 52.1** Vaalputs trench design and cracked container and leachate at Vaalputs [2]

## 1 Introduction and Background

South Africa (RSA) is one of the world's top producers of the Moly-99 isotope, and due to a constant growth in the industrial and medical usage of nuclear technology, adequate disposal is important. Indications are that some of the waste storage containers in use show signs of damage or microcracking when delivered by road from the supplier or that they develop such defects after being filled with waste and transported to the disposal site, probably because the current containers were designed mainly for storage and not to counter the forces caused by handling and transporting [1].

All hazardous nuclear waste in the RSA is stored at Vaalputs nuclear repository in the Northern Cape Province in near-surface trenches. Figure 52.1 shows a trench typically 100 m long, 7.7 m deep and 20 m wide at the bottom with walls having an 80° slope [2]. Figure 52.1 also shows a cracked container with visible leaching, the cause of which is uncertain. Several theories exist: excessive exposure to the elements causing premature container degradation, inadequate quality control of the containers, uncertainties and lack of understanding regarding the specifications, non-realistic lifetime expectancy and performance [2].

The current concrete containers contain no steel or polymer fibre, and the mixture is based on normal concrete. The containers are reinforced with a steel reinforcing mesh and are constructed to surpass minimum cover requirements [3].

## 2 Literature Review

### 2.1 Requirements of the RSA Nuclear Authority

In order to satisfy the specifications of the nuclear authority, the following tests are applicable: compressive strength, indirect tensile strength and durability indices including oxygen permeability, water sorptivity and chloride conductivity, water tightness, visual inspection and dimensions [4]:

- Minimum cementitious material content = 380 kg/m$^3$

- Compressive strength – 50 MPa, average of at least three test specimens (28 days)
- Tensile strength – 4.5 MPa, average of at least three test specimens (28 days)
- Relative density after 28 days curing and 3 days air drying above 2400 kg/m$^3$
- Water/binder ratio (binder includes all cementitious materials) = less than 0.40
- Chloride content by mass of cementitious content = less than 0.05%
- Oxygen permeability index – 10, variation of <5%/4 test specimens
- Water sorptivity – <6 mm/h., variation of <15%/4 test specimens
- Chloride conductivity – < 0.45 mS/cm, variation of <15%/4 test specimens
- Water tightness – no damp hair line cracks after holding water for 28 days
- Visual inspection – no cracks present (width > 0.1 mm), honeycombing, grout loss/voids

## 2.2 Considering Polymers as Co-binder or Micro-Reinforcement

There has been little input on the use of polymers in nuclear concrete containers. The information from the literature search permits a list of performance criteria to be created, and through consultation with industry professionals and performing research, the right combination of polymers and admixtures may be identified. Before lab work, a consideration of the theory was conducted along with some professionals in the industry. The following material considerations were selected to achieve the desired results:

- Cement 52.5 N: to meet compressive strength requirements
- Fly ash: to increase concrete mix density
- Superplasticiser: improving workability and effectiveness of compaction
- Latex pore blockers and others: to produce water tight concrete to overcome the water sorptivity and water tightness requirements
- Metallic fibres: to increase tensile strengths
- Polypropylene fibres: reduce microcracking ensuring smooth surface finish

## 3 Experimental Findings

Table 52.1 lists mixtures which acted as the baseline for analysis and were made using locally available aggregates with hand tamping, thus allowing for conservative benchmarks.

Table 52.2 lists the results observed that showed deviations from the hypothesised results; compressive and tensile strengths were lower than expected, while density was approximately the anticipated value. Figure 52.2 shows the strength development of each mixture.

**Table 52.1** Mixture designs

| Mix | Cement: coarse: fine[a] | Water/cement ratio | Cement type |
|---|---|---|---|
| A | 1:3:3 | 0.4 | LaFarge 52.5 N |
| B | 1:2:2 | 0.4 | LaFarge 52.5 N |
| C | 1:1.75:2.25 | 0.4 | LaFarge 52.5 N |

[a]Coarse: aggregate, 13 mm granite stone. Fine: aggregate, clean building sand

**Table 52.2** Summary of mixture design results

| Mix | Average compressive strength[a] | | | Average tensile strength[a] | Average density (kg/m$^3$) |
|---|---|---|---|---|---|
| | 1 day | 7 days | 28 days | 28 days | 28 days |
| A | 7.18 | 19.00 | 21.477 | 3.7 | 2447.33 |
| B | 16.70 | 26.30 | 37.64 | 4.52 | 2398.33 |
| C | 13.84 | 23.21 | 25.22 | 4.46 | 2486.33 |

[a]All readings for compressive and tensile strength are in MPa

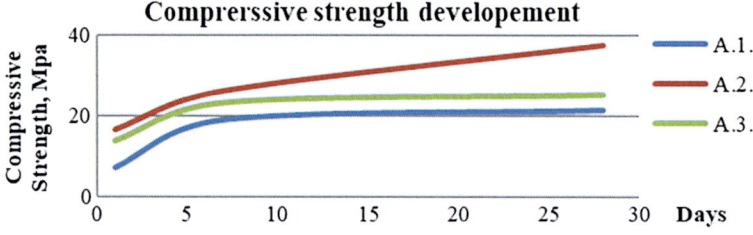

**Fig. 52.2** Compressive strength development graph

Using the information summarised in Fig. 52.2 a trend line revealed a formula which showed that a ratio of 1:1.87 (cement: aggregate) would achieve the desired 50 MPa compressive strength; however, this shows a rather ineffective and inefficient mix design, and due to the low water-cement ratio, it would still be difficult to ensure effective compaction by hand.

## 4  Discussion

The results show that acquiring a mix to meet all requirements using conventional materials is plausible; however, while the low water-cement ratio may theoretically increase strength and reduce cracking, practically, in observation, this resulted in poor workability and subsequently left voided concrete. We may also hypothesise that since the workability was low, compaction was inadequate; thus, internal voids produced lower densities. While cracks may not have been present, honeycombing and voids were noticed which decreased density and increased water sorptivity and permeability. Based on these results, it does not appear that concrete without polymers or admixtures is likely to produce containers that satisfy the requirements. The use of polymers in the concrete appears to be the most likely way forward.

## 4.1 Effects of the Addition of Polymers and Admixtures

The main focus or problems observed were workability and porosity. By increasing workability and thus allowing for improved compaction, the mixture could be designed to achieve the requirements effectively and efficiently.

Dynamon Easy 31 is an acrylic-based superplasticizer. Concrete in which it is used is easy to lay when fresh and has high performance characteristics when set. It is particularly suitable whenever the amount of water needs to be considerably reduced, making it ideal for use in waterproofing applications. It is a water solution with acrylic polymers, which separate the grains of cement and promote slow initial development of hydration [5].

Mapefluid N220 is suitable for applications which require high workability and low water-cement ratio. It has a slight retarding action on the hydration of cement and is therefore particularly suitable for precast steam-cured concrete elements. Ready-mixed concrete for watertight structures and ready-mixed quality concrete for structures with higher than 25 MPa [6].

Microcracking fibre reinforcement such as Mapefibre, which are 100% virgin monofilament polypropylene fibres, resistant to alkalis and do not absorb water could be considered. They reduce the formation of cracks induced by hygrometric shrinkage of the concrete at fresh and hardened stages. The special "monofilament" formation makes their distribution and position within the concrete mix easier, so that they are distributed uniformly, which resists the stresses induced by the setting/hardening process. These fibres are particularly suitable for precast-reinforced concrete panels and sections and come in different strand lengths, each with benefits to specific applications [7].

Industry professionals seemed to agree that while fibres may improve tensile strength of the material, they are not sufficient to eliminate reinforcing steel. Therefore, utilising fibres should be done in an attempt to reduce microcracking, and any additional tensile strength would be a bonus. It is also desirable to use a latex pore blocking admixture, 13 mm and small aggregates and about 30% cement replacement with fly ash to improve mechanical properties as well as density. In addition, PPC representatives recommended focusing on workability enhancement and compacting methods; streamlining this would be of utmost importance, as well as looking into developing self-compacting concretes [8–10].

Steel fibres could be considered to increase tensile strength properties in the container system. Steel fibres would be used to increase the tensile strength within the mixture. In general, steel fibres will have a tensile strength of 1244–1446 MPa. This will exponentially increase the tensile strength present in each mix but will naturally bring in a new set of issues and challenges to overcome [11].

To increase density and create watertight, waterproof and impermeable concrete, it is possible to consider Mapei's Idrocrete DM that is particularly suitable for applications where a high water repellent effect is required, but only in conditions in which little or no hydrostatic pressure is present. It can be used to prevent the efflorescence stains and is especially recommended as a primary hydrophobic waterproof for semidry concrete [12].

To increase density, Mapeplast FV improves the compacting properties of concrete when vibrated and, at the same time, creates a very even surface on the casting by eliminating air bubbles. It is a blend of special components which improves the cohesion of the cementitious matrix. Through improving cohesion, compaction and surface finish, density may also significantly improve [13].

Vibromix C2 is an active polymer-based admixture in an aqueous solution, which promotes the vibrocompression and extrusion action of moulds on fresh concrete. It also considerably increases mechanical strength after very short curing times; it limits the formation of surface efflorescence, to make components with higher performance levels and a more attractive appearance. A better compacting grade allows uniform components to be manufactured with fewer defects and better surface finish [14].

Furthermore in consultation with an expert from Mapei, a new product was identified, although it is not available in South Africa as of yet. The product, Planitop HPC, is a two-component fluid mortar suitable for casting into formwork without the risk of the mortar segregating. Areas up to 40 mm thick can be cast without using reinforcing steel. The following properties can be expected after a 28-day curing period: compressive strength of 130 MPa (40 MPa after 1 day), tensile strength of 8.5 MPA and a density of 2400 kg. Once hardened, Planitop HPC has the following characteristics: very high flexural and compressive strength, high ductility, high resistance to cyclical loads, impermeable to water and high resistance to wear due to abrasion or impact [15].

## 5 Future Work

The way forward will be to create a polymer-reinforced concrete from the benchmarks and to test the effects each of these products has on the concrete to determine if test results are in agreement with manufacturers' specifications. Once we have developed a polymer-reinforced concrete, it may be wise to compare it to products such as Planitop HPC in terms of mechanical properties and, most importantly, cost. Laboratory work, testing and analysis will be necessary; however, it seems that polymer-reinforced concretes are a very promising application for low-level radioactive waste disposal and could be expanded to work in toxic, hazardous or biohazardous waste containment structures. The authors look forward to expanding the topic and shedding light on the use of polymer-reinforced concrete and its applications.

## References

1. Carstens, D. A. (2017). *Personal communication, Meeting at NECSA,* Pelindaba, South Africa.
2. IAEA. (2005). Appendix: Examples of corrective actions implemented at repositories in member states. In IAEA (Ed.), *Upgrading of near surface repositories for radioactive waste* (pp. 35–128). Vienna: International Atomic Energy Agency.
3. Watson, J. (2017). *Phone Interview:* John Watson.
4. Shwarz, A. (2001). Manufacture of TES concrete waste disposal drums Rev 6. *South Africa: Eskom Koeberg nuclear power station design engineering.*
5. Mapei. (2014). *Dynamon Easy 31.* Great Britain: Mapei.
6. Mapei. (2014). *Mapefluid N220.* Great Britain: Mapei.
7. Mapei. (2015). *Mapefibre. Product Brochure,* United Kingdom.
8. Scheurwater, P. E. (2017). *Personal Communication, meeting PPC,* Johannesburg, South Africa.
9. Engelbrecht, S. L. (2017). *Personal communication, Meeting,* Sephaku Cement, Johannesburg, South Africa.
10. Lafarge. (2017). *Lafarge meeting.*
11. Grupo Fapricela Industria. (2016). *Technical data sheet DQ0809.* Freamunde, Portugal: Grupo Fapricela Industria.
12. Mapei. (2016). *Idrocrete DM.* Mapai product sheet, United Kingdom.
13. Mapei. (2014). *Mapeplast FV.* Mapai product sheet, United Kingdom.
14. Mapei. (2014). *Vibromix C2.* Mapai product sheet, United Kingdom.
15. Mapei. (2017). *Planitop HPC.* Mapai product sheet, United Kingdom.

# Chapter 53
# Efficiency of Polymer Fibers in Lightweight Plaster

**Jakob Sustersic, Andrej Zajc, and Gregor Narobe**

In the paper, results of the investigation of fiber-reinforced lightweight plaster (FRLP) with different types of polymer fibers are discussed. The following fibers were used: high-modulus aramid fibers (AR), mid-modulus and high-modulus polypropylene fibers (PP-M and PP-H), polyamide fibers (PA), polyacrylonitrile fibers (PAN), and polyvinyl alcohol fibers (PVA). The behavior of these composites under flexural and compressive load is observed at the age of 28 days and after accelerated aging. Therefore, toughness is evaluated to find out the efficiency of different types of the same volume percentage of fibers but with equal length. Mix proportions of investigated lightweight plasters were the same; only the fiber types were changed. Obtained results show that all used polymer fibers improve toughness of FRLP.

## 1 Introduction

Properties of composites (concrete, mortar, plaster) depend on the properties of their ingredients. During their use, a binder has the major influence on behavior of composites. Binder strength increases with aging, but brittleness increases as well. This means that crack propagates faster through weak points of a composite structure, which leads to momentary fracture of composite element. Therefore, fibers are added in composite, with the intention of strengthening those weak points and above all preventing crack propagation. Fibers are not connected to each other but are arbitrarily distributed throughout the mass. They are arranged along the aggregate

---

J. Sustersic (✉) · A. Zajc · G. Narobe
IRMA Institute for Research in Materials and Applications, Ljubljana, Slovenia
e-mail: jakob.sustersic@irma.si

grains and thus reinforce the interfaces, which represent the weakest part in the structure of the composite.

In practice, different fibers are used. For a good reinforcing effect, the fiber must have a high modulus of elasticity, high strength, suitable stretching, and good adhesion with a cement matrix and should be alkali-resistant (pH > 12). Favorable results give high molecular organic synthetic fibers with aromatic chains and crystals from balanced molecules, which are also too expensive. The reinforcement experiments are performed with synthetic fibers of high modulus and strength from conventional fibrous-forming polymers: polyacrylonitrile, polyamide, polyvinyl alcohol, and polyolefin (polyethylene and polypropylene).

Since the price of polypropylene fibers is lower compared to other fibers, their use in practice is more extensive than others. They are used for reinforcement of concrete, mortar, and plaster for new constructions and repair works. On the basis of experimental investigations carried out in shrinkage of high-performance polypropylene fiber-reinforced concrete, it can be concluded that their shrinkage is lower than that of the equivalent high-performance concrete without fibers. The reduction of shrinkage depends primarily on the volumetric share of applied fibers and their previous moistening [1]. After the Great Belt Tunnel fire in Denmark (1994), the Channel Tunnel fire (1996), and the Kaprun Tunnel fire in Austria (2000), designers become even more focused on the structural fire protection of concrete tunnel linings [2]. Results of many research projects show that addition of polypropylene fibers in concrete of secondary tunnel lining significantly improves fire resistance of the concrete [3–5].

This paper deals with the results of the investigations of the efficiency of various polymer fibers in lightweight plaster. Investigations were carried out as part of the development project of high-modulus polypropylene fibers for the reinforcement of cement products. The project was carried out by the Textile Department of the Faculty of Natural Sciences and Technology of the University of Ljubljana in cooperation with IRMA.

## 2 Experimental Dispositions

Mix proportions of fiber-reinforced lightweight plaster (FRLP) with different types of polymer fibers are shown in Table 53.1.

Six mixtures of FRLP were prepared. They were identified depending on the type of added fibers: FRLP-PA, FRLP-AR, FRLP-PP-M, FRLP-PP-H, FRLP-PAN, and FRLP-PVA. Lightweight plaster without fibers (LP) was prepared too, with the intention to draw comparisons with FRLP. Table 53.2 gives the results of the following properties for each type of fiber used: $\sigma_{sb}$, specific breaking stress; $\varepsilon_b$, breaking strain; $E_0$, elastic modulus; and $\varepsilon_0$, strain at which $E_0$ was read.

A beam with dimensions (10 × 10 × 50) cm was produced with each mixture. From each of these beams, six smaller beams with dimensions (4 × 4 × 16) cm were cut out. On three of these, the flexural behavior of FRLP and LP was investigated at

**Table 53.1** Mix proportion of FRLP with polymer fibers

| Parameters of mix proportion | | Values |
|---|---|---|
| (w/b)$_{tot.}$ (total water to binder ratio) | | 0.81 |
| Binder – cement/lime | | 1: 1 |
| Polymer fibers | | 0.2 vol.% |
| Aggregate (A) | 0–1 mm grinding polystyrene | 40 vol.% of A |
| | 1–3 mm calcite | 30 vol.% of A |
| | 2–4 mm calcite | 30 vol.% of A |
| Density of fresh FRLP | | 1940 kg/m$^3$ |

**Table 53.2** Properties of fibers

| | PA | AR | PP-M | PP-H | PAN | PVA |
|---|---|---|---|---|---|---|
| $\sigma_{sb}$ (cN/dtex) | 8.9 | 17.4 | 7.5 | 8.2 | 4.3 | 8.1 |
| $\varepsilon_b$ (%) | 20.3 | 2.4 | 17.2 | 9.7 | 22.0 | 8.3 |
| $E_0$ (GPa) | 3.9 | 112.2 | 7.2 | 14.4 | 5.9 | 17.9 |
| $\varepsilon_0$ (%) | 1.7 | 1.3 | 1.8 | 1.6 | 1.9 | 0.9 |

their age of 28 days. In this investigation, the beam was loaded with a concentrated force at the center of the distance between the supports, which was 10 cm. In this case, the load-flexural diagram was automatically recorded, from which the individual properties of FRLP and LP were determined. The investigation lasted until the beam had broken into two parts. On each of these, the behavior of the FRLP and LP at the compressive loading was investigated. Compressive load–vertical displacement diagrams were automatically recorded. Before the start of the investigation, the test specimens were stored in the air at room temperature of about +20 °C and relative humidity about 60%.

The remaining three beams from FRLP and LP were investigated in the same way after accelerated aging, which was carried out in the manner used in the measurement of the toughness durability of glass fiber-reinforced concrete [6]. At the age of 28 days, beams from FRLP and LP were kept for 14 days in water at a temperature of +80 °C.

The flexural behavior of the FRLP and LP to the deflection at the first crack $\delta_0$ was evaluated by the absorbed energy $G_f$ expressed in absolute value [Nm]. It was determined by measuring the area under the load–deflection ($P$–$\delta$) diagram up to the first crack. More typical and important for fiber-reinforced composites is the behavior after the first crack occurs. The form of dependence of $P$–$\delta$ after $\delta_0$ can be expressed by residual strength factors, which are determined according to ASTM C 1018-96. They are expressed as the proportions of the strength at the first crack. Two residual strength factors are defined, $R_{5,10}$ and $R_{10,20}$, which are calculated using the following equations:

$$R_{5,10} = 20 \times (I_{10} - I_5) \qquad (53.1)$$

$$R_{10,20} = 10 \times (I_{20} - I_{10}) \tag{53.2}$$

where $I_5$, $I_{10}$, and $I_{20}$ are toughness indices expressed as the ratio of the area of the diagram P–δ to $[(n + 1)/2] \times \delta_0$ and the area to the deflection at the first crack $\delta_0$:

$$I_n = \frac{\int_0^{(n+1/2)\times\delta_0} P(\delta)d\delta}{\int_0^{\delta_0} P(\delta)d\delta} \tag{53.3}$$

where $I$ is the toughness index; $n$ is equal to 5, 10, and 20; $\delta$ is the deflection [mm]; $\delta_0$ is the deflection at the first crack [mm]; and $P$ is the load [N].

The ultimate compressive strength $f_{cu}$ and the toughness index $TI_{10u}$ were determined from the compressive load–vertical displacement ($P_c$–$\delta_c$) diagrams. Toughness index $TI_{10u}$ is calculated using the following equation [7]:

$$TI_{10u} = \frac{G_{10u}(\text{FRLP})}{G_{10u}(\text{LP})} \tag{53.4}$$

where $G_{10u}$ (FRLP) is absorbed energy of FRLP and $G_{10u}$ (LP) is absorbed energy of LP. The absorbed energy $G_{10u}$ is determined by the area under the $P_c$–$\delta_c$ up to $5,5 \times \delta_{cu}$, where $\delta_{cu}$ is the vertical displacement at ultimate compressive strength.

## 3 Results and Their Discussion

$G_f$ of all FRLP is larger than LP. After accelerated aging, $G_f$ of all FRLP and LP increases, except $G_f$ of FRLP-PVA. The effect of fibers in the FRLP becomes much more apparent after the first crack, which is shown by the results of the residual strength factors $R_{5,10}$ and $R_{10,20}$ (Figs. 53.1 and 53.2).

Toughness of FRLP is much higher than that of LP, when flexural behavior of those composites is observed. Residual strength factors have been well suited to show post-crack behavior of FRLP. Residual strength factors of 28-day-old FRLP are close at higher flexures. FRLP-PVA shows the highest toughness. Residual strength factors of FRLP, which were exposed with accelerated aging, decrease more than those of 28-day-old FRLP, except FRLP-PVA. Toughness indices of FRLP-PVA increase after accelerated aging. After the test, the fibers were mechanically separated from the cement matrix. The surface of the fibers was visually assessed using the SEM (scanning electron microscope). Most cement particles (mostly calcium hydroxide crystals) were bound to PVA fibers and at least to AR fibers. The amount of particles bound to the rest of the fibers is between the two extreme cases.

The average results of $TI_{10u}$ are given in Fig. 53.3. Compressive strengths of all investigated plasters are more or less the same. Moderate increase of compressive

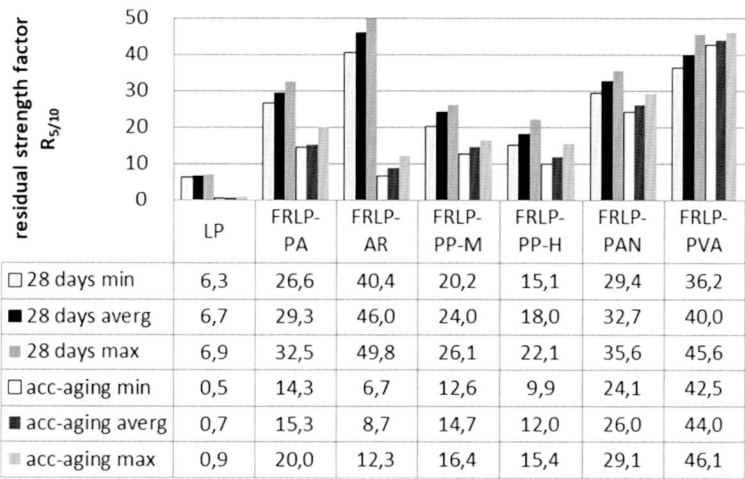

**Fig. 53.1** Average results of $R_{5,10}$ of FRLP and LP

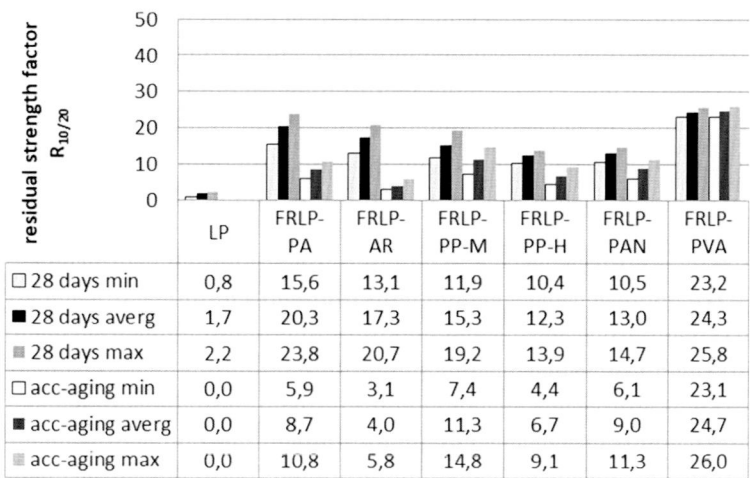

**Fig. 53.2** Average results of $R_{10,20}$ of FRLP and LP

strength of LP was obtained after accelerated aging. But toughness indices of FRLP are much higher than that of LP, and they increase after accelerated aging.

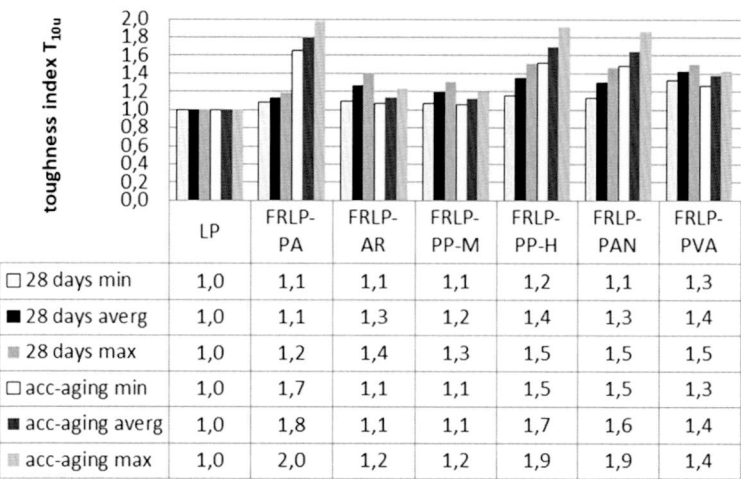

**Fig. 53.3** Average results of TI$_{10u}$ of FRLP and LP

## 4 Conclusions

Obtained results show that all used polymer fibers improve toughness of FRLP under flexural and compressive load at the age of 28 days and after accelerated aging. However, polyvinyl alcohol fibers show the highest influence on post-crack flexural behavior, among all used fibers. High-modulus polypropylene fibers, prepared in the laboratory in the frame of the project, show a pretty good influence on the increase in TI$_{10u}$.

## References

1. Saje, D., Bandelj, B., Sustersic, J., Lopatic, J., & Saje, F. (2011). Shrinkage of polypropylene fibre reinforced concrete. *Journal of Materials in Civil Engineering, 23*(7), 941–952.
2. Smith, K., & Atkinson, T. (2010) PP fibres to resist fire-induced concrete spalling. TunnelTalk. *Tummel-TECH*, Nov. 2010.
3. Kim, J.-H. J., Lim, Y. M., Won, J. P., & Park, H. G. (2010). Fire resistant behavior of newly developed bottom-ash-based cementitious coating applied concrete tunnel lining under RABT fire loading. *Construction and Building Materials, 24*, 1984–1994.
4. Behnood, A., & Ghandehn, M. (2009, November). Comparison of compressive and splitting tensile strength of high-strength concrete with and without polypropylene fibers heated to high temperatures. *Fire Safety Journal, 44*(8), 1015–1022.
5. Shuttleworth, P. (2001, April). Fire protection of precast concrete tunnel linings on the Channel Tunnel Rail Link. *Concrete, 35*, 38–39.
6. Shah, S. P., Ludirdja, D., Daniel, J. I., & Mobasher, B. (1988, September–October). Toughness-durability of glass fiber reinforced concrete systems. *ACI Materials Journal, 85*, 352–360.
7. Glavind, M., & Aarre, T. (1991) High-strength concrete with increased fracture-toughness. *Fiber-Reinforced Cementitious Materials Symposium, Boston* (pp. 39–46). Materials Research Society, Pittsburgh.

# Part V
# Polymer Concrete with Recycled Waste

# Chapter 54
# Properties of Ceramic Waste Powder-Based Geopolymer Concrete

Sama T. Aly, Dima M. Kanaan, Amr S. El-Dieb, and Samir I. Abu-Eishah

Geopolymer is an environmentally friendly emerging alternative to ordinary Portland cement concrete as its production requires less energy with lower carbon dioxide emissions. This is due to the implementation of industrial waste materials or by-products as binders replacing Portland cement. Geopolymer concrete is synthesized via the polycondensation reaction with alkali-activating materials rich in aluminosilicate. Several materials were utilized in producing geopolymer concrete such as fly ash, silica fume, metakaolin, and slag. Ceramic waste powder (CWP) produced from the final polishing process of ceramic tiles is produced in significant amounts and is composed mainly of silica and alumina and hence has a potential to be used as a geopolymer concrete ingredient. Therefore, in this study, the CWP's ability to produce geopolymer concrete is studied, and its compressive strength and durability characteristics were tested. Sodium hydroxide and potassium hydroxide were used as alkali activators with a pH concentration of 12 M. A curing temperature of 60 °C for 24 h was applied. The 7 and 28 days' results of the compressive strength, pores percentage, initial rate of water absorption, and bulk electrical resistivity showed the possibility of producing CWP-based geopolymer concrete.

---

S. T. Aly · D. M. Kanaan · A. S. El-Dieb (✉)
Civil and Environmental Engineering Department, United Arab Emirates University, Al Ain, UAE
e-mail: amr.eldieb@uaeu.ac.ae

S. I. Abu-Eishah
Chemical and Petroleum Engineering Department, United Arab Emirates University, Al Ain, UAE

# 1 Introduction

Concrete is being extensively used globally in the construction industry. Portland cement is responsible for approximately 17% of the total emissions related to construction and building industries [1]. To overcome this problem, alternative binders to Portland cement should be investigated and used to meet the growing demand of concrete. The geopolymer concrete offers a great opportunity for the reuse of industrial by-products and wastes such as fly ash and slag to fully replace cement [2]. Geopolymerization technology is based on the reaction of alkaline solutions such as sodium hydroxide (NaOH), potassium hydroxide (KOH), and sodium silicate ($Na_2SiO_3$) with the silica ($SiO_2$) and alumina ($Al_2O_3$) present in the by-product materials to produce cement-like binders. The cement replacement materials are rich in $SiO_2$ and $Al_2O_3$ but are low in calcium; thus, heat curing is needed for production of geopolymer concrete. These elevated temperatures would significantly speed up the geopolymerization reactions and hence produce the strength development rate required in the building industry. It has been reported that the optimal curing regime is around 24 to 48 h with a curing temperature between 60 °C and 80 °C [2]. Several studies have investigated the use of ceramic waste in the synthesis of geopolymer concrete [3, 4] using several alkaline solutions with different concentrations (i.e., NaOH, KOH, and $Na_2SiO_3$) and different curing periods and reported compressive strength ranging from 13 MPa up to 70 MPa.

The present paper focuses on understanding the behavior of ceramic waste powder (CWP)-based geopolymer, activated by a mixture of KOH and NaOH alkali solutions. Several preliminary mixtures were cast and tested to decide on the optimum ratios of the aggregate content, powder content, curing regime, and admixtures percentage. The final selected mixtures were tested at both 7 and 28 days for compressive strength and tested at 28 days for pores percentage, initial rate of water absorption, and bulk electrical resistivity.

# 2 Experimental Procedure

## 2.1 Materials

*Aggregate*: Crushed natural sand with fineness modulus of 3.9 was used.

*Solutions*: Sodium hydroxide (NaOH) and potassium hydroxide (KOH) were both used in the activation process of the synthesized geopolymer. The concentration of the solutions was 12 M according to the results reported in an earlier study [4].

*Ceramic Waste Powder (CWP)*: Mainly composed of silica (69.4%) and alumina (18.2%) with a total that exceeds 80% of the CWP weight. Other compounds include $Na_2O$, MgO, $K_2O$, CaO, $TiO_2$, and $Fe_2O_3$ with percentages 3.19%, 3.53%, 1.89%, 1.24%, 0.617%, and 0.83%, respectively. Specific surface by air permeability method was 555 $m^2/kg$.

**Table 54.1** Mixture designations and alkali solutions mixing ratios and regimes

| Mixtures designation | Alkali solutions% | | |
|---|---|---|---|
| | KOH | NaOH | Alkali solutions mixing regime |
| A | 0 | 100 | – |
| B | 100 | 0 | – |
| C | 20 | 80 | NaOH was added first and mixed with solids for 1 min; then KOH was added |
| D | 40 | 60 | NaOH and KOH solutions were mixed then added and mixed with solids |
| E | 60 | 40 | KOH was added first and mixed with solids for 1 min; then NaOH was added |

*Slag*: Ground granulated blast furnace slag (i.e., slag) was used as filler in some of the mixtures with a specific surface area 432 m$^2$/kg and a specific gravity of 2.93.

*Admixtures*: Polycarboxylic ether-based superplasticizer (Glenium® Sky 504) (ASTM C494 Type G and ASTM C1107 Type 2) was used.

## 2.2 Preparation of Specimens

Sodium hydroxide flakes and potassium hydroxide were dissolved separately in water to make a solution with the desired concentration (i.e., 12 M) at least 1 day prior to its use. The dry ingredients were first mixed together for about 1 min. Table 54.1 shows the regime followed in adding the alkali solutions to the mixture. The mixing ratio and mixing regime was based on the viscosity of the NaOH and KOH solutions. (i.e., NaOH solution is almost 12 times viscous than KOH solutions). Superplasticizer of the required dosage was added after the addition of all solutions. Concrete was cast into molds. The molds were covered with thin vinyl sheet to avoid loss of water due to evaporation during temperature curing.

## 2.3 Performance Evaluation

Flowability was measured according to ASTM C270. The compressive strength and bulk electrical resistivity were conducted on three 50 mm cube specimens for each mixture at 7 and 28 days. Pores percentage and initial rate of water absorption have been conducted on (100 mm diameter × 50 mm thickness) concrete discs cut from the middle third of concrete cylinders and measured at 28 days of age as per ASTM C642 and ASTM C1585, respectively.

## 3 Methodology

In this study, several parameters were investigated such as aggregate content, admixture dosage, inclusion of slag, and curing regime. First, the set of mixtures shown in Table 54.1 were performed with different CWP to sand ratios of 1:1.5, 1:2.0, and 1:2.5. At this stage, all mixtures were cured at 60 °C for 24 h. The mixtures were characterized by their flowability and 7 days' compressive strength. Second, all mixtures investigated in the first stage were performed with the use of superplasticizer admixture with 1.5% and 4.0% by weight of the CWP powder. Selection of mixture's aggregate and admixture content is based on the mixtures' flowability and 7 days' compressive strength. Third, the selected mixture was cured at 60 °C for 24 and 48 h to study the effect of curing time on 7 days' strength development. Fourth, the inclusion of slag as part of the binding material was investigated using the selected mixture. Slag was used to replace CWP with 10%, 20%, and 40% by weight. The mixtures with slag were exposed to different curing regimes: air curing, curing for 24 h at 60 °C and then air curing, and curing for 24 h at 60 °C and then water curing for 6 days. The slag inclusion level was selected based on the 7 days' compressive strength. Finally, the compressive strength, pores percentage, initial rate of water absorption, and bulk electrical resistivity were evaluated for the selected CWP mixture without and with slag.

## 4 Results and Discussions

### 4.1 Effect of Aggregate Content

The effect of aggregate content on flowability and 7 days' compressive strength is shown in Fig. 54.1a, b. It was noted that as the aggregate content increased, the flowability decreased. On the other hand, the inclusion of aggregate has improved the 7 days' compressive strength. The mixture with 1:2.5 CWP to aggregate ratio

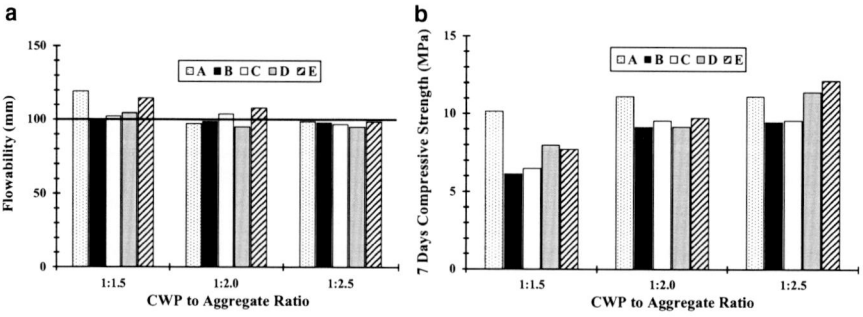

**Fig. 54.1** Effect of aggregate content (**a**) flowability and (**b**) 7 days' compressive strength

showed higher strength. Also, it was noted that the order of adding the alkali solutions affected the strength and flowability. The mixing regimes (D) and (E) (defined in Table 54.1) showed highest compressive strength. This could be attributed to the viscosity of the NaOH and KOH solutions and its effect on the flowability.

## 4.2 Effect of Admixtures

Upon using 1.5% by weight superplasticizer, variable improvements (10% up to 30%) were observed in flowability but slight improvement for the strength. The 1:2.5 CWP to sand ratio mixtures showed the highest strength (25%) improvement. Also, the same set showed the lowest improvement in flowability. Therefore, this set of mixtures were conducted using 4% by weight admixture. The flowability was improved with the increased dosage of the admixture. Among this set, mixing regime (D) and (E) showed the highest compressive strength compared to the other mixing regimes (A, B, and C).

## 4.3 Effect of Curing Time

The set of mixtures with 1:2.5 CWP to sand ratio were exposed to two curing times (24 h and 48 h) at 60 °C followed by air curing up to the test day (i.e., 7 days). Strength results were found to be higher for the 48-h curing period. However, although increasing curing time improved the strength, it was decided to limit the study on applying 24 h of curing for energy conservation purposes.

## 4.4 Effect of Slag Content and Curing Regime

In order to improve the mixture flowability, the same set of mixtures with CWP to aggregate ratio of 1:2.5 were cast with the incorporation of slag at three different inclusion levels (i.e., 10%, 20%, and 40% by weight). Also, three curing regimes were studied: air curing, curing for 24 h at 60 °C and then air curing, and lastly curing for 24 h at 60 °C and then water curing for 6 days. Figure 54.2 shows the effect of slag inclusion on flowability. It was noted in Fig. 54.2 that the flowability was significantly improved with the inclusion of slag and the highest improvement was for 40% slag inclusion level. The 7 days' compressive strength is presented in Fig. 54.3, which shows that the strength of the air cured mixtures was reduced with the inclusion of slag. The other two curing regimes improved the strength, and the improvement was proportional to the slag inclusion level and yielded comparable

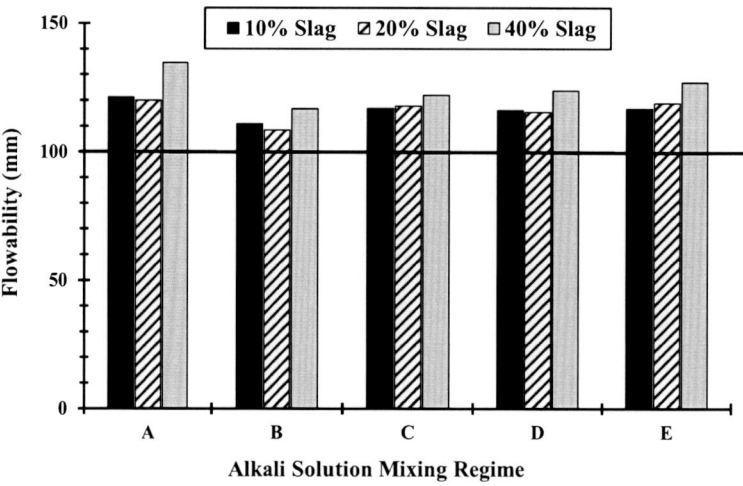

**Fig. 54.2** Effect of slag inclusion on flowability

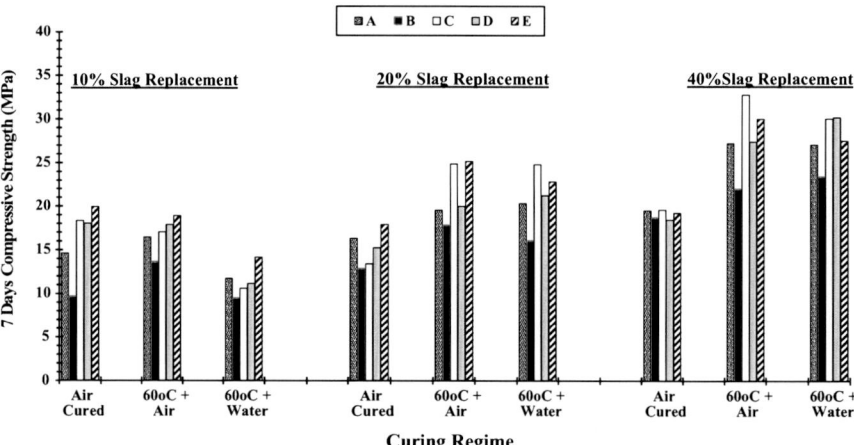

**Fig. 54.3** Effect of slag inclusion on 7 days' strength

strength values. Therefore, the curing regime which involves temperature curing for 24 h and then air curing was considered the optimum regime to be implemented.

## 4.5 Performance Evaluation

Based on the preliminary mixtures results, the binder to sand ratio of 1:2.5 and the 4% admixture dosage were selected. A mixture with 100% CWP (S-0) and a mixture with 40% slag (S-40) were selected for performance evaluation. Alkali solution

**Table 54.2** Test results of (S-0) and (S-40) mixtures

|  | S-0 | | S-40 | |
| --- | --- | --- | --- | --- |
|  | 7 Days | 28 Days | 7 Days | 28 Days |
| Compressive strength (MPa) | 17.21 | 17.91 | 39.30 | 40.73 |
| Pores (%) | – | 3.98 | – | 8.32 |
| Initial rate of absorption (mm/min$^{1/2}$) | – | 0.035 | – | 0.12 |
| Electrical bulk resistivity (k$\Omega$.cm) | – | 0.79 | – | 18.17 |

mixing regime (D) was chosen, and the curing for 24 h at 60 °C and then air curing regime was adopted. Table 54.2 summarizes the results for the studied mixtures. For both mixtures, the strength development from 7 to 28 days was not significant. This could be attributed to the fact that the main reaction is polymerization not hydration. Therefore, the polymerization is mainly achieved at early age. Also, the slag mixture showed significant increase in strength at both tested ages. Regarding the pores percentage and the initial rate of water absorption, it was noted that the results for S-0 mixture were significantly lower than those for S-40. Also, the inclusion of slag significantly increased the resistivity of the S-40 mixture compared to that of the S-0 mixture. This indicated high resistance to reinforcement corrosion.

## 5 Conclusion

CWP showed good potentials to be used in making geopolymer concrete. The increase of aggregate content improved the strength. The use of 4% superplasticizer is needed to improve the mixture flowability. The inclusion of 40% by weight slag as partial replacement of CWP significantly improved the strength and the bulk electrical resistivity. The application of curing for 24 h at 60 °C followed by air curing is the optimum curing regime to improve the performance of the produced CWP and slag geopolymer.

## References

1. Sustainable Buildings and Climate Initiative. (2009). *Buildings and climate change: Summary for decision-makers*. Paris: UNEP DTIE, Sustainable Consumption and Production Branch.
2. Provis, J. L., & Deventer, S. J. V. (2009). *Geopolymers-structure, processing, properties and industrial applications*. Great Abington/Cambridge, UK.: CRC Press/Woodhead Publishing.
3. Shehab, H. K., Eisa, A. S., & Wahba, A. M. (2016). Mechanical properties of fly ash based geopolymer concrete with full and partial cement replacement. *Construction and Building Materials, 126*, 560–565.
4. El-Dieb, A. S., & Shehab, I. E. (2014). Cementless concrete using ceramic waste powder. *International Conference on Construction Materials and Structures (ICCMATS)*, South Africa (pp. 487–494).

# Chapter 55
# Use of Recycled Polymers in Asphalt Concrete for Infrastructural Applications

Sook F. Wong

This paper reports the characterization of properties and performance of hot mix asphalt, focusing on stone mastic asphalt (SMA), with various types of recycled polymers (waste plastics). Investigations were conducted on the performance grade of bitumen (penetration value and softening point) as well as mixture strength (Marshall stability, Marshall flow, and Marshall quotient) for different waste plastics blends. The binders studied were conventional bitumen, target bitumen (polymer-modified bitumen), and waste plastics (LDPE, PP, PS, and HDPE). The target bitumen and polymer-modified bitumen with recycled polymers (10 wt% LDPE, 10 wt% PP, 25 wt% PS, and 2 wt% HDPE) were more superior binders than conventional bitumen due to decreased penetration values and increased softening points. The target mixture showed the greatest Marshall stability and Marshall quotient, whereas conventional mixture displayed the lowest values, suggesting the advantageous role of polymer-modified bitumen in target bitumen. SMA mixtures having optimum individual contents of 10 wt% LDPE, 10 wt% PP, 25 wt% PS, and 2 wt% HDPE were found to be feasible design formulations for mixture strength.

## 1 Introduction

Asphalt concrete consists of well-graded aggregate and bituminous binders which are heated and blended together in specified proportions in a hot mix plant. The conventional asphalt concrete comprises mineral aggregate bounded with asphaltic

---

S. F. Wong (✉)
Temasek Polytechnic, Singapore, Singapore

American Concrete Institute-Singapore Chapter, Singapore, Singapore
e-mail: drsfwong@tp.edu.sg

material. In recent years, there has been a strong move toward replacing some proportions of the mineral aggregate and asphaltic binder with recycled by-products.

Waste plastics are widely available polymers that can be used in asphalt concrete pavements that offer good engineering and economic advantages. Asphalt concrete incorporating recycled polymers can offer a very practical solution to waste plastics management and is of interest to industrialized nations throughout the world. In road construction, polymer-modified bitumen (PMB) (bitumen with polymeric material) has been used successfully as a paving material.

Wong et al. [1] studied the use of 5–10 wt% waste LDPE (low-density polyethylene) blends in stone mastic asphalt (SMA), as compared to conventional bitumen (control mixture with 0 wt% LDPE). They reported that the asphalt mixture with 10 wt% LDPE blend as polymer modifier could be a promising material for use of SMA in highways. Mahfouz et al. [2] found that ABS (acrylonitrile butadiene styrene) recycled polymer could be utilized as an effective modifier for bitumen to be used in flexible pavements. Köfteci [3] investigated the effect of recycled HDPE (high-density polyethylene) added to bitumen as modifier at 1, 2, 3, and 4 wt% of mixture and discovered that the best performance was attained at 4 wt% HDPE in relation to Marshall stability values.

Based on Marshall test results, Jassim et al. [4] concluded that waste plastics with fine particle size (passing sieve 1.18 mm), thin thickness (0.2 mm), and 15 wt% of aggregate could increase the Marshall stability and index of retained strength by 20% and 15%, respectively, more than the conventional mixture. Gawande et al. [5] noted that the addition of 8 wt% waste plastics significantly improved the Marshall stability, strength, and fatigue life of the modified bitumen and resulted in savings of 0.4% bitumen by weight of mixture or about 9.6% bitumen per $m^3$ of mixture.

Singapore is an island nation in Southeast Asia with a relatively dense population of 5,784,538 residents in 2017 with land area of 719.1 $km^2$. The solid waste generated in 2016 increased to 7.81 million tonnes, up by 140,700 tonnes from 7.67 million tonnes in 2015. The plastics waste generation in Singapore has increased rapidly over the past 15 years, from 546,537 tonnes in 2001 to 822,200 tonnes in 2016, which made up about 11% of total waste generated, i.e., 7,814,200 tonnes in 2016 [6]. With 762,700 tonnes of waste plastics discarded and only 59,200 tonnes recycled in 2016, the annual recycling rate hits a new low of 7% in 2015 and 2016 compared to the already low recycling rates of 9% and 11% in 2013 and 2014, respectively.

Despite the merits of waste plastics in asphaltic materials discussed above, they have not received the technical attention they warrant from the infrastructural sector. To date, related studies have been limited and fragmental, and none has proceeded to pilot-scale studies. So far, no integrated system has been established to convert as-received waste plastics that are contaminated and nonhomogeneous in nature into high value-added end products for infrastructural applications, especially in tropical (hot and humid) climates and under heavy traffic conditions as experienced in Singapore.

## 2 Objectives

In view of the above issues, the objectives of this study are:

- To examine the effect of waste plastics type on the performance grade of polymer-modified bitumen (PMB) at various waste plastics replacements
- To determine the optimum asphaltic mixtures incorporating various waste plastics type and content

## 3 Methodology

**Materials** The raw materials used were:

- *Aggregates* – The aggregates were clean, well-graded, angular stone in compliance with the requirements by the authority [7].
- *Binders* – The binders studied were (1) conventional bitumen, (2) target bitumen PMB (polymer-modified bitumen) with commercial polymer modifier known as PG-76, and (3) waste plastics (LDPE, low-density polyethylene; PP, polypropylene; PS, polystyrene; and HDPE, high-density polyethylene).

**Asphalt Premix** All asphalt mixtures were of 60/70 penetration grade [7]:

1. SMA with conventional bitumen
2. SMA with target bitumen (PMB) – *0 wt% waste plastics (control mixture)*
3. SMA with waste plastics – *10 wt% of optimum LDPE blend*
4. SMA with waste plastics – *10 wt% of optimum PP blend*
5. SMA with waste plastics – *25 wt% of optimum PS blend*
6. SMA with waste plastics – *2 wt% of optimum HDPE blend*

**Preparation Method** Aggregates were sieved, weighed, and mixed according to LTA specifications for SMA gradation [7]. The wet process was adopted to prepare all SMA mixtures. For mixtures 3–6, waste plastics (LDPE, PP, PS, or HDPE) were blended with bitumen at 160–170 °C, after which the aggregates were mixed with the modified bitumen. A total of three specimens were prepared for each test, to be conducted at average ambient temperature of 30 °C ± 3 °C.

**Tests for Bitumen** The following standard test methods were carried out:

- *ASTM D5* – to obtain penetration value (dmm or 0.1 mm), vertical penetration distance by a standard needle into sample under standard conditions of loading, time, and temperature. *Greater penetration value denotes softer consistency.*
- *ASTM D36* – using the ring and ball apparatus to measure the softening point (°C), being the temperature at which the sample attains a certain degree of softening under specified test conditions. *A higher softening point shows greater binding capability of bituminous material under specified heating conditions.*

**Table 55.1** Summary of test results for bitumen

| Mixture no. | Measured penetration (dmm) | LTA standards [7] (dmm) | Measured softening point (°C) | LTA standards [7] (°C) |
|---|---|---|---|---|
| 1 (Conventional) | 66.0 | 60–70 | 52.3 | 48–56 |
| 2 (Target 0 wt% plastics) | 57.2 | ≥50 | 82.8 | ≥80 |
| 3 (10 wt% LDPE) | 52.0 | ≥50 | 80.5 | ≥80 |
| 4 (10 wt% PP) | 53.5 | ≥50 | 86.7 | ≥80 |
| 5 (25 wt% PS) | 52.0 | ≥50 | 82.5 | ≥80 |
| 6 (2 wt% HDPE) | 54.9 | ≥50 | 82.7 | ≥80 |

**Tests for Asphalt Premix** The following standard tests were conducted:

- *ASTM D6927* – to measure the Marshall stability (kN), which is the maximum force recorded during compression of an asphalt specimen. *A greater Marshall stability value represents a stiffer asphalt mixture, thereby demonstrating more superior properties and performance.*
- *ASTM D6927* – to determine the Marshall flow (mm), which is the deformation recorded at the maximum force. *A smaller Marshall flow value indicates a stiffer asphalt mixture.*
- *ASTM D1559 and ASTM D6927* – the Marshall quotient MQ (kN/mm) is the ratio of Marshall stability (kN) to Marshall flow (mm). It is an indicator of the stiffness of asphalt mixture; *the higher the MQ, the stiffer the mixture.*

## 4 Results and Discussion

**Bitumen** The bitumen test results from this study are shown in Table 55.1.

A bitumen blend with decreased penetration value and increased softening point is suitable for tropical climates as experienced in Singapore because it can withstand high temperatures when used for road construction. Table 55.1 shows that the conventional bitumen had the highest penetration value (hence the softest consistency), followed by target bitumen (0 wt% waste plastics), 2 wt% HDPE and 10 wt% PP, whereas both 10 wt% LDPE and 25 wt% PS blends recorded the smallest penetration value of 52.0 dmm.

On the other hand, from Table 55.1, 10 wt% PP was found to display the highest softening point (thus the greatest binding capability), followed by the target bitumen (0 wt% waste plastics), 2 wt% HDPE, 25 wt% PS, 10 wt% LDPE, and the conventional bitumen.

**Table 55.2** Summary of Marshall test results for asphalt premix

| Mixture no. | Measured density (kg/m$^3$) | Marshall stability (kN) | Marshall flow (mm) | Marshall quotient (kN/mm) |
|---|---|---|---|---|
| 1 (Conventional) | 2240 | 8.11 | 6.97 | 1.16 |
| 2 (Target 0 wt% plastics) | 2290 | 15.93 | 3.59 | 4.44 |
| 3 (10 wt% LDPE) | 2260 | 11.34 | 3.66 | 3.10 |
| 4 (10 wt% PP) | 2260 | 13.22 | 3.58 | 3.69 |
| 5 (25 wt% PS) | 2250 | 12.04 | 3.56 | 3.38 |
| 6 (2 wt% HDPE) | 2260 | 12.63 | 3.25 | 3.89 |

The penetration value and softening point of conventional bitumen were within the allowable LTA requirements of 60–70 dmm and 48–56 °C, respectively, while the penetration values and softening points of target bitumen and mixtures containing waste plastics were in compliance with LTA standards of $\geq 50$ dmm and $\geq 80$ °C, respectively.

**Asphalt Premix** The Marshall test results are presented in Table 55.2.

A lower density implies asphalt mixture of lighter weight. This is favored in road construction due to smaller dead loads and hence smaller-sized sections required to support them. An asphalt premix with increased Marshall stability and quotient as well as reduced Marshall flow is ideal for infrastructural applications due to its strength, integrity, durability, and resistance against fatigue and rutting.

Table 55.2 revealed that conventional mixture had the lowest density (2240 kg/m$^3$) followed by 25 wt% PP; 10 wt% LDPE, 10 wt% PP, and 2 wt% HDPE had similar density (2260 kg/m$^3$), while target mixture had the highest density (2290 kg/m$^3$). The target SMA mixture exhibited the greatest Marshall stability and quotient followed by 10 wt% PP and 2 wt% HDPE which were comparable, 25 wt% PS, 10 wt% LDPE, and conventional mixture. On the other hand, the measured Marshall flow was the lowest for 2 wt% HDPE, followed by 25 wt% PS, 10 wt% PP, target mixture, 10 wt% LDPE, and conventional mixture.

## 5 Conclusions

- Based on a lower penetration value and a higher softening point, the target bitumen was found to be a better binder than conventional bitumen. The polymer-modified bitumen with recycled polymers (10 wt% LDPE, 10 wt% PP, 25 wt% PS, and 2 wt% HDPE) also provided more superior binders than conventional bitumen due to decreased penetration values and increased softening points.
- The optimum individual contents of 10 wt% LDPE, 10 wt% PP, 25 wt% PS, and 2 wt% HDPE gave satisfactory performance of polymer-modified bitumen, in compliance with standard requirements and specifications.

- The asphalt SMA mixtures investigated had densities in the range of 2240 to 2290 kg/m$^3$. The target mixture showed the greatest Marshall stability and Marshall quotient, whereas conventional mixture displayed the lowest values, suggesting the advantageous role of polymer-modified bitumen in target bitumen.
- The asphalt mixtures incorporating recycled polymers (10 wt% PP, 2 wt% HDPE, 25 wt% PS, 2 wt% HDPE) recorded higher Marshall stability and Marshall quotient as well as lower Marshall flow compared to conventional mixture, indicating the significant role of these waste plastics as polymer modifiers in SMA mixtures.
- SMA mixtures with optimum individual contents of 10 wt% LDPE, 10 wt% PP, 25 wt% PS, and 2 wt% HDPE were found to be feasible design formulations for mixture strength (e.g., Marshall stability, Marshall flow, and Marshall quotient).

**Acknowledgments** The author appreciates the research and development grant awarded by Singapore Ministry of Education Translational and Innovation Fund (MOE TIF) (Project code, MOE2014-TIF-1-G-012). The author is also thankful to the technical staff from the School of Applied Science, Temasek Polytechnic and from the collaborator Yun Onn Company (Pte) Ltd for their continuous support.

# References

1. Wong, S. F., Htwe, A. A., Oh, S. H., Leo, T. Y., Cheng, J., & Tay, B. K. (2017). Utilization of waste plastics in stone mastic asphalt for infrastructural applications. *Materials Science Forum (MSF) – Civil and Building Materials, 902,* 55–59.
2. Mahfouz, H., Tolba, I., El Sayed, M., Semeida, M., Mawsouf, N., El Laithy, N., Saudy, M., Khedr, S., & Breakah, T. (2016). Using recycled plastics as hot mix asphalt modifiers. In *Proceedings of Conference on Resilient Infrastructure* (pp. 1–10). London, UK, 1–4 June 2016.
3. Köfteci, S. (2016). Effect of HDPE based wastes on the performance of modified asphalt mixtures. *Procedia Engineering, 161,* 1268–1274.
4. Jassim, H. M., Mahmood, O. T., & Ahmed, S. A. (2014). Optimum use of plastic waste to enhance the Marshall properties and moisture resistance of hot mix asphalt. *International Journal of Engineering Trends and Technology (IJETT), 7,* 18–25.
5. Gawande, A., Zamre, G. S., Renge, V. C., Bharsakale, G. R., & Tayde, S. (2012). Utilization of waste plastic in asphalting of roads. *Scientific Reviews & Chemical Communications, 2,* 147–157.
6. NEA (National Environment Agency, Singapore). (2017). *Waste statistics and overall recycling: Waste statistics and recycling rate for 2016.* Retrieved June 04, 2017, from: http://app2.nea.gov.sg/topics_wastestats.aspx
7. LTA (Land Transport Authority, Singapore). (2010). *Materials and workmanship specification for civil and structural works – Document No. E/GD/09/104/A1* (LTA Engineering Group 2010).

# Chapter 56
# Influence of Method of Preparation of PC Mortar with Waste Perlite Powder on Its Rheological Properties

Grzegorz Adamczewski, Piotr Woyciechowski, Paweł Łukowski, Joanna Sokołowska, and Beata Jaworska

The aim of this research is to determine the influence of waste perlite powder quantity on rheological properties of polymer–cement mortar mixture. Influence of dosing sequence of waste perlite powder on mortar mixture properties was especially taken into consideration. In the research, eight mortar mixtures were prepared using different quantities of perlite powder in mortar and by using different dosing sequence of mixture components. During the research, the influence of waste perlite powder on consistency and workability of mortar mixture was evaluated. By analyzing the various dosage sequences of the ingredients, the problem of difficult homogenization of the mortar and excessive dusting of the waste perlite powder were investigated. The obtained results show that changing the dosing moment of the key component – waste perlite powder – can influence technological properties of polymer–cement mortars.

## 1 Introduction

One of the fundamental directions that almost all industries' development has taken is the pursuit of innovative solutions that will not only affect the production process and the quality of the product itself but will also protect our environment. As far as building construction is concerned, we are increasingly encountering regulations that require, for example, the use of secondary materials to improve the ecosystem and avoid producing enormous quantities of waste materials that are costly and require substantial storage space. For example, BREEAM and LEED certifications, which place a number of demands on newly designed and built constructions, place

---

G. Adamczewski (✉) · P. Woyciechowski · P. Łukowski · J. Sokołowska · B. Jaworska
Warsaw University of Technology, Faculty of Civil Engineering, Warsaw, Poland
e-mail: gad@il.pw.edu.pl

particular emphasis on the use of renewable energy sources, reuse of building materials, or the use of recycled materials to produce them. For years, research has been carried out on how to make more efficient use of raw materials, for example, as a result of old buildings' demolition. Meanwhile, the further development of science and the introduction of new technologies to meet market demands have led to the widespread use of materials that have been until recently considered useless. This paper presents one of the methods of using such waste – waste perlite powder – in the design and construction of polymer–cement mortar. Waste perlite powder is the by-product formed during the expanded perlite production process. The expanded perlite has found many applications and thus a large amount of by-product is produced. Besides its application in construction, one of the many extensive uses of perlite is in horticulture [1]. However, when it comes to building materials engineering, its ability to absorb water is a disadvantage, because it means increased water demand [2]. Waste perlite powder has remained unused for years, and its storage has become increasingly burdensome. In order to make the most efficient use of this waste and to not adversely affect workability, the waste quantity and dosing point during the preparation of mixture were analyzed.

In the technological process of the production of building composites, the order of dispensing ingredients and the way and time of mixing are very important. These variables affect both the workability and consistency of the mixture as well as the mechanical properties of the mature material. The optimum composite properties can be obtained by mixing the ingredients and uniformly distributing them in the mixture as quickly as possible. Additional importance have the physico-chemical properties of additives, such as superplasticizers or water demand of waste perlite.

The mechanically mixing technology of concrete allows three methods of dosage and mixing of ingredients:

- Single stage – all the ingredients are mixed at the same time.
- Two stage – cement slurry or mortar is mixed separately and aggregate is added in the second step.
- Multistage – cement slurry is separately prepared and then mixed with fine aggregate, and coarse aggregate is added in the last step.

It is equally important to plan the timing and method of water dispensing. An example is the production of lightweight concrete, where in order to achieve the best mixture and hardened composites properties, the aggregate should first be mixed with two third of the prepared volume of the mixing water. It is only in the next phases that cement is added and at the very end the rest of the water [3]. Proper dosing of the components of the liquid mortar controls its workability and affects the time of complete mixing. Inadequate selection of component dosing sequences, especially for components containing additional ingredients compared to standard mortar or concrete, may inhibit the achievement of correct consistency and consequently reduce the composite properties. An example is the use of superplasticizers. Studies show that the best effect of using this additive is when it is added at the very end to the previously mixed primary components [4]. In addition, by analyzing the order of component dosing, the effects of the individual components on the bundling rate or the gelation time, as in case of polymers, should be taken into consideration. Inadequate time of addition, for example, of the retarding or accelerating admixtures can make it difficult to stiffen the mixture and affect the mechanical properties.

## 2 Experimental Study

The aim of the present research was to analyze the influence of the waste perlite powder addition on the workability and consistency of the polymer–cement mortar in the variable order of dosing of the individual components. The study program consisted of preparing mixtures containing different waste content, followed by observations of changes occurring during mixing and dispensing of components.

The program of the research assumed the dosage of waste perlite powder at 10% and 20% of cement mass. For each of these waste contents three different dosage sequences of the different mortar components were prepared. In addition, to determine the effect of the superplasticizer on the workability of the mix, two mortars without the presence of perlite but with varying amounts of superplasticizer were prepared. The mass content of the remaining ingredients remained constant. The mortar components were mixed with an automatic blender, which allowed for two stirrer speeds of 140 and 285 revolutions per minute. Consistency of mortars was measured in two methods – the cone precipitating and the flow table [5].

Portland cement CEM I 42,5R in accordance with PN-EN 196-1 [6] and standard sand in accordance with PN-EN 196-1 were used. As a polymer modifier, epoxy resin has been used with a low viscosity, which is conductive to obtaining liquid consistency of the mortar mixture. It requires additional hardener. As recommended, the proportions are: One part of hardener for five parts of polymer. As an admixture, a superplasticizer was used. This allows for a strong liquefaction of the mortar mixture. This effect is necessary when using the waste perlite powder, which very strongly absorbs water and retains it in its grains. The waste perlite powder used was in the form of a very fine white powder (Fig. 56.1).

**Fig. 56.1** Microscopic image of waste perlite powder grains [7]

**Table 56.1** Mortar compositions

| Nr | Cement, g | Water, g | Fine aggregate, g | Epoxy resin + hardener, % (g) | Superplasticizer, % (g) | Waste perlite powder, % (g) |
|---|---|---|---|---|---|---|
| K0P0A | 450 | 225 | 1350 | 10 (45) | 1 (4,5) | 0 (0) |
| K0P0B | 450 | 225 | 1350 | 10 (45) | 2 (9) | 0 (0) |
| K1P10 | 450 | 225 | 1350 | 10 (45) | 2 (9) | 10 (45) |
| K1P20 | 450 | 225 | 1350 | 10 (45) | 2 (9) | 20 (90) |
| K2P10 | 450 | 225 | 1350 | 10 (45) | 2 (9) | 10 (45) |
| K2P20 | 450 | 225 | 1350 | 10 (45) | 2 (9) | 20 (90) |
| K3P10 | 450 | 225 | 1350 | 10 (45) | 2 (9) | 10 (45) |
| K3P20 | 450 | 225 | 1350 | 10 (45) | 2 (9) | 20 (90) |

Determined order of components dosing is marked with symbols K0, K1, K2, and K3. The diagrams for the dosing sequences are presented below:

- K0 – water + cement + fine aggregate + epoxy resin + superplasticizer
- K1 – water + cement + fine aggregate + epoxy resin + superplasticizer + waste perlite powder
- K2 – water + cement + fine aggregate + waste perlite powder + superplasticizer + epoxy resin
- K3 – (cement + fine aggregate + waste perlite powder) + (water + superplasticizer) + epoxy resin

The test was performed using standard mortar compositions. The markings of the tested PC mortar (Table 56.1) are as follows:

- K0P0A, K0P0B – without waste perlite powder, with different superplasticizer content, components dispensed in the same order
- K1P10, K2P10, K3P10 – 10% (by cement mass) of waste perlite powder addition, K1, K2, K3 correspond to the dosage sequence
- K1P20, K2P20, K3P20 – 20% (by cement mass) of waste perlite powder addition, K1, K2, K3 correspond to the dosage sequence

## 3 Results and Discussion

Waste perlite powder was a component that determines the obtaining of the homogenous mixtures. The greater the amount of waste, the longer is the mixing time needed to achieve homogeneous consistency. The influence of dosing sequence of waste perlite powder on consistency is presented in Table 56.2.

For mixtures containing 10% of waste perlite powder, time needed to achieve good consistency, was similar, regardless of the dosing sequence and was between 14–17 min. Due to the presence of waste, the time in which the mixture went into

**Table 56.2** Influence of dosing sequence of waste perlite powder on consistency of PCC mixtures

| Nr. | Time of waste perlite powder addition, min | Time of liquefying, min | Total mixing time, min | Precipitating the cone, cm | Flow table, cm |
|---|---|---|---|---|---|
| K0P0A | – | – | – | – | – |
| K0P0B | – | – | – | 10 | 27 |
| K1P10 | 5 | 11 | 17 | 9.7 | 20 |
| K1P20 | 10 | 22 | 29 | 6 | 16 |
| K2P10 | 3 | 11 | 14 | 11.2 | 26.5 |
| K2P20 | 3 | 17 | 20 | 7.7 | 17 |
| K3P10 | 0 | 12 | 14 | 8.3 | 20 |
| K3P20 | 0 | 25 | 28 | 8.5 | 16 |

plasticization phases increased considerably. This is due to the very high water demand of perlite powder, especially seen in the case of the K3P20 composition. It was only with the extension of the mixing time and the activation of the superplasticizer that the composition P20 went through plasticization phases, finally achieving a fluid state. In addition to the significance of the waste dosing sequence, the liquid phase dosing separation also influences the consistency of the mixture. Addition of water, admixture, and polymer, separated by waste addition, allowed to faster achieve liquidation state is clearly seen for K2. Addition of 10% of waste perlite powder, in case for K1 and K2, did not significantly reduce consistency of the mixture, compared to the reference composition – K0P0B. The K2 composition was characterized by the highest consistency measured on the flow table. For K1 and K3, the values were markedly lower than for K2, which in conjunction with the result of the precipitation in the cone test method shows that the K2 mixtures were characterized by the highest fluidity.

## 4 Conclusion

The results of the presented study suggested that waste perlite powder can be used as an addition to polymer–cement mortars. Maintaining proper mixing ratios and conditions allows obtaining a product with improved technical characteristics compared to unmodified standard mortar. Based on the results of the study, the following conclusion can be formulated:

- According to the state of the knowledge, irrespective of the dosing sequence, the increase in the waste perlite powder content results in a decrease in the liquidness of the mixtures and the time needed to homogenize the mixture is considerably longer.
- The optimal dosing sequence of components was obtained when the perlite was added to the wet mixture of solid components and the liquefiers (superplasticizer and resin) were added as the last components – K2 sequence of dosing.

- With a waste content of 10% of cement mass, the most preferred dosing sequence of ingredients is K2, which allows for the most liquid consistency of the mixture and also the shortest time needed to be liquefied.
- In the case of increased waste perlite powder content, regardless of the dosing order of ingredients, the same degree of liquidness of the mixture was obtained, but again the homogeneity of the mixture was obtained fastest in the case of K2 dispensing order.
- From the production point of view, K2 order is the most advantageous because the liquefaction is the fastest and finally the highest.

The obtained results show that changing the dosing moment of the key component – waste perlite powder – can influence technological properties of polymer–cement mortars. When analyzing the results it should be remembered that due to the previously mentioned properties of waste, especially low bulk density, even the use of low waste amount by mass allows to neutralize its large volume. This is extremely important not only when it comes to environment protection but also for the growing number of perlite manufacturers and their problem with the storage of this waste.

# References

1. Ciullo, P. A. (1996). Industrial minerals and their uses (Vol. 580). Noyes Publications, Westwood, New Jersey, USA.
2. Jaworska, B., Sokołowska, J. J., Łukowski, P., & Jaworski, J. (2015). Waste mineral powders as a component of polymer-cement composites. *Archives of Civil Engineering, LXI*(4), 199–210.
3. Sokołowska, J. J., Łukowski, P., Woyciechowski, P., & Kida, K. (2015). Effect of perlite waste powder on chemical resistance of polymer concrete composites. *Advanced Materials Research, 1129*, 516–522.
4. Łukowski, P. (2016). Polymer-cement composites containing waste perlite powder. *Materials, MDPIAG, 9*(10), 1–11.
5. PN-85/B-04500:1985. Zaprawy budowlane – badania cech fizycznych i wytrzymałościowych.
6. PN-EN 196–1:206. Metody badania cementu.
7. Łukowski, P., Sokołowska, J. J., Adamczewski, G., & Jaworska, B. (2013). Waste perlite powder as the potential microfiller of polymer composites. In S. Jemioło & M. Lutomirska (Eds.), Mechanics and materials (pp. 193–200), Politechnika Warszawska, Warsaw, Poland.

# Chapter 57
# Design and Manufacture of a Sustainable Lightweight Prefabricated Material Based on Gypsum Mortar with Semirigid Polyurethane Foam Waste

Sara Gutiérrez González, Carlos Junco, Veronica Calderon, Ángel Rodríguez Saiz, and Jesús Gadea

The characterization of a new gypsum mortar-based composite material that incorporates various combinations of polyurethane waste in its matrix is reported in this paper. The new material, a lightweight plate for use in internal ceilings, is characterized in a series of standardized tests: bulk density, mechanical behavior, and the reaction to fire test. Moreover, the study details the industrial manufacturing process linked to the integration of waste from the plastics industry. Increased quantities of polymer waste caused significant reductions in bulk density and mechanical strength, although with reasonable behavior of the plates up to substitution levels of 50%. The non-combustibility test demonstrated the potential of the new material for interior use in buildings. The technology of the new gypsum mortar-based material and polyurethane waste could potentially maximize the reuse and the lifespan of this type of waste that would otherwise be incinerated or dumped on landfill sites.

## 1 Introduction

In accordance with the last report published by PlasticsEurope—the Facts 2015 [1], demand for plastic in Europe was somewhere around 47.8 MTn in 2014, 3.1% higher than in 2013. In the case of polyurethane (PU) in Europe, demand for this type of plastic underwent an increase of 0.1% in 2014 with regard to the preceding year. Of the 47.8 MTn, 7.5% was PU, which implies a demand of 3.58 MTn. Approximately 70% of the 3.58 MTn is presented as foam, of which 30% consists of PU elastomers and other products. From the 2.50 MTn of PU foam, 27% of waste

---

S. G. González (✉) · C. Junco · V. Calderon · Á. R. Saiz · J. Gadea
Department of Architectural Construction, Higher Polytechnic School, University of Burgos, Burgos, Spain
e-mail: sggonzalez@ubu.es

© Springer International Publishing AG, part of Springer Nature 2018
M. M. Reda Taha (ed.), *International Congress on Polymers in Concrete*
*(ICPIC 2018)*, https://doi.org/10.1007/978-3-319-78175-4_57

(675,000 Tn) is generated, of which 459,000 Tn are dumped in landfill sites. The coolant industry, the construction sector (waste from the production process of insulation panels), and the automobile sector (excess from headliner production processes) all share very similar characteristics and properties, and together their total PU waste levels reach 102,500 Tn. Of this last figure, 21% (42,525 Tn) is recovered through recycling and 11% (22,275 Tn) incineration. So, 68% (137,700 Tn) of PU waste is dumped in landfill sites with no defined use. As regards the use of PU waste in construction materials, in 2008, Mounanga et al. completed an experimental study on the incorporation of (PU) waste from the destruction of insulation panels, in cement mixtures, with the purpose of producing lightweight concrete in cement mortars [2]. The Spanish patent ES 2386116A1 is exploited at present by the firm DIUREAN, S.L. [3]. There are previous studies on an acceptable interaction of PU waste in gypsum, which have demonstrated high lives of thermal insulation and an improvement in the permeability of the final material [4]. References also exist on the manufacture of lightweight gypsum board, with the inclusion of expanded polystyrene foam (ES 2277776 B1), for use as precast elements, as well as the use of PU waste in precast components for use in interior partitions [5].

This work is centered on the physical-mechanical characterization of a new lightweight plate for use in internal ceilings, and the demonstration of its viability on an industrial scale, which would permit its application in the future.

## 2 Raw Materials

Gypsum conglomerate is classified as A/14/3.5 in Standard EN 13279-1 [6], the specifications of which stipulate an initial setting period of over 14 min, with a compression resistance of $\geq 3.5$ N/mm2. According to the manufacturer's specifications, this gypsum presents a purity value of 92%.

Polyurethane foam waste (PFW) was taken from the waste generated in the manufacture of insulation panels in the automobile industry. Following shredding, it presented itself as a dust with a granulometry of between 0 and 0.5 mm and with real density and bulk density values of 1080 kg/m3 and of 72 kg/m3, respectively.

## 3 Experimental Procedure

The procedure to make the gypsum mortar plates (Y Board) consisted in the progressive substitution of gypsum with polyurethane foam waste (PFW) by volume. Both components were dried in an oven to a constant weight and then mixed until reaching homogenization. Subsequently, the mix water was added, and the mixture was prepared in accordance with standard EN 13279-2 [7]. Finally, after curing at a temperature of 24 °C and at a relative humidity of 50 ± 1% for 7 days, all

 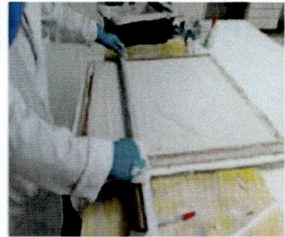 

**Fig. 57.1** Manufacturing and curing of gypsum plates in the laboratory

**Table 57.1** Bulk density

| Sample | Bulk density at 7 days(Kg/m$^3$) |
|---|---|
| Y0 Board | 1488 |
| Y0.5 Board | 1264 |
| Y1 Board | 1127 |
| Y2 Board | 882 |
| Y3 Board | 679 |
| Y4 Board | 541 |

the test specimens were then dried to a constant mass at 40 ± 2 °C and cooled to room temperature prior to testing (Fig. 57.1).

Plasterboard specimens were characterized by their bulk density, flexural strength, and total water absorption. The tests followed the instructions in Standards 14246:200622 and EN 520 [8], which establish the specifications and the test methods for gypsum plasterboard and plate for internal ceilings.

After establishing the characteristics of the new material, manufacturing of the precast components was performed on an industrial scale at a precast concrete factory.

## 4 Results and Discussion

### 4.1 Bulk Density

The results of the bulk density test are shown in Table 57.1 for the different types of plates. It may be seen that, as the volume of PU in substitution of gypsum increases, there is a drop in density, at 15% in examples of Y0.5 Board (1 part gypsum and 0.5 PU), even reaching reductions of up to 64% in examples of Y4 Board. This fact fundamentally occurs because of the replacement of the gypsum matrix (2960 kg/m$^3$) by PU (1080 Kg/m$^3$). In addition, PU requires high quantities of water to arrive at an acceptable workability according to EN-13279, which implies a high volume of pores and, in consequence, a reduction in the density of the material.

During the flexural strength test
UNE 14246:200622 Standard

After testing:
Optimal results were obtained:
Test piece without footprint

**Fig. 57.2** Flexural strength test on sample Y2 Board

**Table 57.2** Total absorption test

| Sample | Total water absorption (Avg. %) |
|---|---|
| Y0 Board | 26.36 |
| Y0.5 Board | 31.66 |
| Y1 Board | 39.81 |
| Y2 Board | 60.59 |
| Y3 Board | 73.21 |
| Y4 Board | 82.01 |

## *4.2 Flexural Strength*

The test of the mechanical behavior of the different plates was carried out following the methods defined in Standard EN 14246:2007 [9]. The test was optimal in the Y0.5 Board, Y1 Board, and Y2 Board type plates, in which there was no footprint in any of the boards after the test (Fig. 57.2).

## *4.3 Total Water Absorption Test*

As indicated in Table 57.2, an increased presence of waste in the mixture implies higher water absorption. This absorption capacity increased considerably in the case of the Y2 Board (60%), 50% higher than in commercial tiles. The absorption capacity of the polymer explains this behavior.

## *4.4 Fire Reaction Test*

The results of the non-combustibility test confirmed that the samples that incorporated PU in their composition, specifically the Y1 and Y2 Board samples, presented

**Table 57.3** Fire reaction test

| Parameter | Sample Y1 | Sample Y2 |
|---|---|---|
| Temperature rise of furnace (°C) | 0.6 | 2.4 |
| Duration of sustained flaming (s) | <5 | <5 |
| Loss of mass (%) | 26.1 | 27.1 |
| Euroclass classification | A2 | A2 |
| CTE-D-SI | (B-s2-d0) | (B-s2-d0) |

*Class A1* non-combustible; *Class A2* non-combustible as no flashover

flaming times of less than 20 s with a temperature increase of below 50 °C and a loss of mass of less than 50% (Table 57.3). This result indicates that, even if we only consider the contribution of the materials to fire development, its composition corresponded to Euroclass A2 (non-combustible), in accordance with the European fire reaction classification of building materials for homogeneous products [10].

## 4.5 Simulation of a Real Industrial Process

The process that is defined contemplates the introduction of the precast manufacturing system of a new polyurethane-gypsum mortar product, ceiling tiles, from the time of the PU waste collection to the packaging of the final product. The PU waste arrives at the factory in a compressed form to reduce its volume during transport. In the precast industry, the hemihydrate is received in a hopper, adapted to its density and volume, which feeds it into the doser and mixer mechanism. The waste processing is done in a grinding machine. The shredder has a cylinder connected to a pressurized system to prevent volatilization of the waste. The homogenized waste is fed into the volumetric doser with weight control to be mixed with the other components (gypsum, water, and eventual additives). The mixture is transported to the mold with a pump adapted to the density and the viscosity of the mixture, providing continuous pouring onto a conveyor belt of mold carriers (specific product molds). It was necessary to adjust the production line to accommodate the thickness of the new product. Traveling along the conveyor belt, it arrives at the grinding zone, where the tile surface is given the desired finish. Subsequently, it is taken to a forced air drying zone, or to go on, if necessary, to other (finishing and painting) processes (Fig. 57.3).

## 5 Conclusions

- Reductions in the bulk density were greater when larger amounts of waste were included in the final mixture.

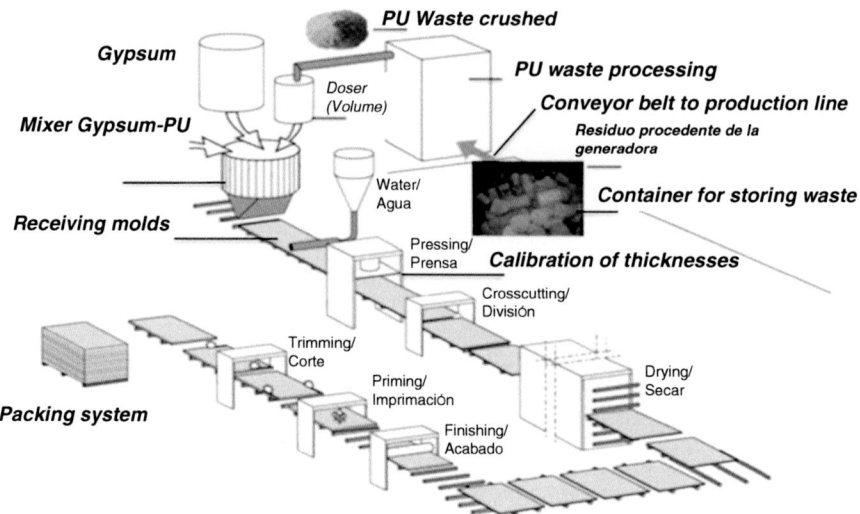

**Fig. 57.3** Real scale simulation of the industrial process

- Although the waste weakens the strength of the material, the results obtained in the flexural strength test comply with the standard on tiles with high percentages of waste (Y2).
- Plasterboards Y0.5, Y1, and Y2 have shown good behavior in their reaction to the fire test, which guarantees the safety of the material for use in ceilings.
- Real scale industrial simulation has shown the possibility of reproducing the process to potential manufacturers of precast gypsum and by doing so of promoting sustainability in the management of plastic PU waste in the industrial sector.

In comparison with current products, PU-gypsum board is more sustainable because it uses less natural resources and it has lower carbon footprint which is caused by the gypsum manufacturing as well as it reduces the amount of PU landfill sites in an economic , more profitable, and cost-efficient way.

**Acknowledgments** This study has been completed within the framework of the LIFE REPOLYUSE Project, LIFE 2016, Environment Life Programme by the European Commission.

# References

1. Plastics-the Facts 2015. An analysis of European plastics production, demand and waste data.
2. Mounanga, P., Gbongbon, W., Poullain, P., & Turcry, P. (2008). Proportioning and characterization of lightweight concrete mixtures made with rigid polyurethane foam wastes. *Cement and Concrete Composites, 30*(9), 806–814.

3. Gutiérrez-González, S., Gadea, J., Rodríguez, A., Junco, C., Calderón, V., Martín, A., & Campos, P. L. (2012). *Spanish patent ES 2 386 116 B2*.
4. Gutiérrez-González, S., Gadea, J., Rodríguez, A., Junco, C., & Calderón, V. (2012). Lightweight plaster materials with enhanced thermal properties made with polyurethane foam wastes. *Construction and Building Materials, 28*(1), 653–658.
5. Alameda, L., Calderón, V., Junco, C., Rodríguez, A., Gadea, J., & Gutiérrez-González, S. (2016). Characterization of gypsum plasterboard with polyurethane foam waste reinforced with polypropylene fibers. *Materiales de Construcción, 66*(324), e100.
6. EN 13279-1:2009 Gypsum binders and gypsum plasters. Definitions and requirements.
7. EN 13279-2:2004 Gypsum binders and gypsum plasters. Test methods
8. EN 520:2005 Gypsum plasterboards. Definitions, specifications and test methods.
9. EN 14246:2007 Gypsum elements for suspended ceilings – Definitions, requirements and test methods.
10. Official Journal of the European Communities No L, p. 50. Directive 89/106/EEC as regards the classification of the reaction to fire performance of construction products 23.2.2000.

# Chapter 58
# Cement Mortars Lightened with Rigid Polyurethane Foam Waste Applied On-Site: Suitability and Durability

Carlos Junco, Sara Gutiérrez, Jesús Gadea, Veronica Calderón, and Ángel Rodríguez

The durability of using cement mortars lightened with rigid polyurethane foam waste is analysed, adding to previous research on their characterization, preparation, and manufacture carried out by the Construction Materials Research Group, University of Burgos. For this work a storage hut was built. The mortars used two types of cement: CEM II and IV; the sand was partially substituted by shredded rigid polyurethane foam waste (PFW), with substitutions by volume of sand of 50% and 75%, in accordance with the use and application of the mortar. Suitability tests were performed, after 28 days had elapsed, in relation to adhesion and, after 3 months, in relation to hardness. Once the investigation was finished, it was concluded that the mortars made with aggregates of rigid polyurethane foam waste easily mixed and placed on-site. At 28 days from application of the mortars, a tile coating was executed on one wall and on the floor. Adherence tests and surface hardness test were performed on the "in situ" rendering. Four years later new tests sustain that the adherence and the surface hardness of the mortars applied on-site was similar to laboratory levels and is sufficient for the application of these commonly used mortars.

## 1 Introduction

The use of polymeric waste in total or partial substitution of sand in cement mortars has been profusely studied over recent years with very positive results [1, 2] because of its good resistant behaviour and durability [3]. Various percentages of sand were replaced by previously shredded rigid polyurethane foam waste, using different

---

C. Junco (✉) · S. Gutiérrez · J. Gadea · V. Calderón · Á. Rodríguez
Department of Construction, University of Burgos, Burgos, Spain
e-mail: cjunco@ubu.es

© Springer International Publishing AG, part of Springer Nature 2018
M. M. Reda Taha (ed.), *International Congress on Polymers in Concrete (ICPIC 2018)*, https://doi.org/10.1007/978-3-319-78175-4_58

types and quantities of cement in the mixes. The compressive strength of the laboratory tested samples was less than the strengths of the conventional mortars, although the laboratory samples were sufficiently strong for use as a building mortar, for plastering, and for rendering. However, their performance in terms of hardness and fatigue was similar [4].

Ease of workability for on-site placement of these materials is essential for their use in civil works and construction, using traditional or manual methods in principle and only then adapting the material for application with more advanced mechanical methods. In this research work, the aim is to contribute results on the viability of applying cement mortars with partial substitution of sand by previously shredded rigid polyurethane foam waste, so as to confirm the suitability of these mortars for use in building works and civil engineering projects.

## 2 Characterization of Materials

### 2.1 Raw Materials

**Cement** Two types of standardized commercial cement were used in the mixes: CEM II/B-L-32,5R and CEM IV/B (V) 42,5 N with an average strength of 32.5–42.5 MPa at 28 days, complying with standard EN 197–1 [5], with EC markings.

**Sand** The sand was extracted from an aggregate quarry and received no treatment after reception. It was previously dried for the laboratory tests and was used without drying for the work on-site. In all cases, the sand complied with the regulations on mortar aggregates – Standard EN 13139 [6] aggregate for mortars – and presented an acceptable granulometric line.

**Rigid Polyurethane Foam Waste** Offcuts from rigid polyurethane foam waste were collected from two factories in Burgos, which use the polymer in the manufacture of various components. As these two materials were different waste products, they were labelled PUR "A" and PUR "B". They had to be shredded to obtain maximum particle sizes of 4 mm for use in substitution of sand [7, 8].

No type of additive was employed as no substantial improvements were achieved with reference to its on-site application. Water from the municipal mains supply was used in the mixes.

### 2.2 Mortars

Cement mortars were manufactured from siliceous sand and water that served as reference specimens against which to contrast the characteristics of the mortars

**Table 58.1** Dosages of the mortars

| Label | CEM type | Cement/sand [by weight] Before | After | Subst volume of sand PUR [%] "A" | "B" | [By weight] W/C | P/C |
|---|---|---|---|---|---|---|---|
| MII4R | II | 1/4 | 1/4 | – | – | 0.89 | – |
| MIV4R | IV | 1/4 | 1/4 | – | – | 0.84 | – |
| MIV6R | IV | 1/6 | 1/6 | – | – | 1.27 | – |
| MII4PU50A | II | 1/4 | 1/2 | 50 | – | 0.84 | 0.065 |
| MII4PU50B | II | 1/4 | 1/2 | – | 50 | 0.89 | 0.085 |
| MIV4PU50B | IV | 1/4 | 1/2 | – | 50 | 0.83 | 0.086 |
| MIV6PU50A | IV | 1/6 | 1/3 | 50 | – | 1.22 | 0.097 |
| MIV6PU75A | IV | 1/6 | 1/1,5 | 75 | – | 1.43 | 0.146 |
| MIV6PU75B | IV | 1/6 | 1/1,5 | – | 75 | 2.14 | 0.192 |

**Table 58.2** Location, thickness, and finishing system of the mortars in the construction of a small hut

| Label | Application | Location | Thickness | Finishing |
|---|---|---|---|---|
| MII 4R | Ext. rend. | East wall | 0.6 + 0.6 cm | Trowel |
| MIV 4R | Int. rend | East wall | 0.6 + 0.6 cm | Trowel |
| MIV 6R | Bonding | Long walls | 1.3 cm | Trowel |
| MII 4PU 50A | Ext. rend. | Short north wall | 0.6 + 0.6 cm | Trowel polish |
| MII 4PU 50B | Int. rend | Short south wall | 0.6 + 0.6 cm | Trowel polish |
| MIV 4PU 50B | Bonding | Short north wall | 1.3 cm | Trowel |
| MIV 6PU 50A | Bonding | Long west wall | 1.3 cm | Trowel |
| MIV 6PU 75A | Floor | Hut floor | 6.0 cm | Screed |
| MIV 6PU 75B | Floor | Hut floor | 6.0 cm | Screed |

containing polyurethane foam. Then, the mortars were prepared with partial substitution of sand by shredded polyurethane foam, at percentage substitutions of 50% and 75%, by volume, with the following labels and dosages (Table 58.1). The amount of water doses in the mortars aimed to obtain an acceptable consistency for normal on-site application (plastic consistency) in accordance with norm EN 1015-3 [9].

## 3 Experimental Procedure: On-Site Application

The raw materials were transported from Burgos to Garray, where a small storage hut was built at a location close to the provincial capital (Soria), 1011 m above sea level, where a typically continental climate prevails. The storage hut was rectangular with exterior dimensions at its base of 4.25 m by 2.70 m and an average height of 2.25 m. The location of the mortar bonds and renderings used in the work, their thicknesses, and their finishing are listed in Table 58.2. The motive for using these mortars on the floor was their low density with no requirement for high strength.

**Fig. 58.1** Application of the different mixtures

The application of the different mixtures was done over 2 days under clear skies at temperatures of between 26 °C at 10 h in the morning and 33 °C at midday with a minimum temperature of 15 °C at dawn. Mixing was done as follows: previous mixing of both arid with approximately half of the water necessary for the mortar, for 30 s. The previous mixing of the sand with the foam was done to help with the homogenization of the sand. Finally, the cement was added along with the remaining amount of water. In this case, the duration of mixing was 3 min. The application of the different mortars was done in a traditional way. The finishing of the mortars on the floor was done with a screed, and compacting presented no difficulties other than dry conventional mortars (Fig. 58.1).

So as to compare the test results of both our laboratory mixes and our on-site mixes, strength, bending, and compression tests were conducted on the on-site mixtures, filling specimens of 160 mm × 40 mm × 40 mm EN 1015-11 [10]. At 28 days and 4 years after, adherence tests were also performed on the "in situ" rendering, by anchoring the device, similar to a template, on the wall. Furthermore Shore "C" surface hardness tests UNE 102039 [11] were performed on the wall renderings, which, although specifically for testing plaster work, might provide valid results to estimate the hardness of the rendering, by comparing them with the results for the reference mortar renderings. Finally at 28 days from application of the mortars, a tile coating was executed on one wall and on the floor, using cementitious adhesive type "C" EN 12004 [12] in thin layer. Adherence tests were performed on the "in situ" rendering.

## 4 Results and Discussion

The application of the mortars in their different forms presented a similar workability to the reference mortars and in some cases a better one. The mechanical strengths of the mortars prepared on-site were similar to the mortars prepared in the laboratory (Table 58.3).

The drop in strength of the mortars with substitution of 75% sand may be due to excess water that had to be added to the mixtures because of the high ambient temperature. Nevertheless, the compressive strengths of the rendering mortars yielded values higher than 7.5 MPa, values far above type CS IV, which is the most demanding class in standard EN 998-1 [13] for rendering and plastering

**Table 58.3** Bending, compression strength, and density at 28 days mortars applied on-site

| Label | Bending strength [MPa] | | Compressive strength [MPa] | | Density [g/cm$^3$] |
|---|---|---|---|---|---|
| | Lab. | On-site | Lab. | On-site | |
| MII 4PU 50A | 1.92 | 2.78 | 7.01 | 9.46 | 1.34 |
| MII 4PU 50B | 2.23 | 2.41 | 6.39 | 8.43 | 1.63 |
| MIV 6PU 75A | 1.15 | 1.10 | 4.03 | 2.84 | 1,10 |
| MIV 6PU 75B | 0.72 | 0.75 | 3.01 | 2.29 | 1.18 |

**Table 58.4** Resistance to adhesion at 28 days and at 4 years, mortars on-site

| Label | Adhesion resistance [MPa] | | |
|---|---|---|---|
| | Laboratory | On-site [28 days] | On-site [4 years] |
| MII 4PU 50A | 0.24 | 0.18 | 0.27 |
| MII 4PU 50B | 0.27 | 0.29 | 0.53 |

**Table 58.5** Shore "C" surface hardness units

| Label | Shore "C" Surface hardness | |
|---|---|---|
| | 28 days | 4 years |
| MII IV 4R | 85.0 | 86.0 |
| MII 4PU 50A | 84.9 | 82.9 |
| MII 4PU 50B | 87.0 | 85.2 |
| MIV 6PU 75A | 63.4 | 65.9 |
| MIV 6PU 75B | 46.2 | 57.7 |

mortars, widely used (GP) in standard UNE 41302 [14]. The adherence tests showed that the adhesion strength of the rendering mortars manufactured on-site was similar to the adhesion strength of those manufactured in the laboratory, with variations that in no case exceeded 25% (Table 58.4).

It should be indicated that the adhesion tests have a significant difficulty in their implementation, as the results obtained in a single test with eight determinations yielded very different values, so the average results were not very reliable, all the more so as an added torque effect occurred when the discs were detached with the devices, as indicated in the standard. Finally, "Shore C" surface hardness tests were performed in accordance with the standard. The hardness values of the mortars with a substitution of 50% of the volume of sand for polyurethane foam (Table 58.5), which were used in a vertical rendering, were similar to those obtained for the reference mortars placed in that same construction.

The surface hardness of the mortars used as a screed layer for the floor was much lower, in comparison with the reference mortars. This weakness may be due to the significant amount of polyurethane waste on the mortar surface. To test the cementitious adhesive behaviour, tiles were pulled out with a claw, being able to observe that the thin layer peeled off both in the cement mortar and on the back side of the tile (Fig. 58.2).

**Fig. 58.2** Adhesion test of cementitious adhesive

## 5 Conclusions

The results of these preliminary tests on placing mortars with partial substitution of sand by rigid polyurethane foam waste has produced mortars of good workability, at least equal to the reference mortars. The mortars mixed on-site, in a conventional concrete mixer, acquired similar strengths to those mixed in the laboratory and would be sufficient for use on factory floors and renderings in accordance with current regulations. The adherence of the mortars applied on-site, 4 years after, was similar to laboratory levels and is sufficient for the application of these commonly used mortars. The surface hardness of the mortar renderings with substitutions of 50% foam, sufficient for unprotected use, was similar to their reference specimen counterparts, meaning that those mortars are appropriate for application using traditional methods.

## References

1. Czarnecki, L. (2015). Use of polymers to enhance concrete performance. *Advanced Materials Research, 1129*, 49–58.
2. Garbacz, A., & Sokołowska, J. J. (2013). Concrete-like polymer composites with fly ashes. Comparative study. *Construction and Building Materials, 38*, 689–699.
3. Wong, S. F. (2015). Polymers in concrete towards innovation, productivity and sustainability in the built environment. *Proceeding of 15th International Congress on Polymers in Concrete (ICPIC 2015), Singapore*.
4. Hoover, C. G., & Ulm, F. (2015). Experimental chemo-mechanics of early-age fracture properties of cement paste. *Cement Concrete Research, 75*, 42–52.
5. EN 197-1. (2011). Cement composition, specification and conformity criteria. Part 1. Common cements. *European Committee for Standardization. Bruxelles*.
6. EN 13139. (2003). Aggregates for mortar. *European Committee for Standardization. Bruxelles*.
7. Junco, C., Rodríguez, A., Gadea, J., & Calderón, V. (2015). Deformability of mortars incorporating polyurethane foam waste under cyclic compression fatigue tests. *Advanced Materials Research, 1129*, 477–483.
8. Ruiz-Herrero, J. L., Velasco-Nieto, D., López-Gil, A., Arranz, A., Fernández, A., Lorenzana, A., Merino, J. A., & Rodríguez-Pérez, M. A. (2016). Mechanical and thermal performance of concrete and mortar cellular materials containing plastic waste. *Construction and Building Materials, 104*, 298–310.
9. EN 1015-3. (1999). Methods of test for mortar for masonry. Part 3. Determination of consistence of fresh mortar (by flow table). *European Committee for Standardization. Bruxelles*.

10. EN 1015-11. (2000). Methods of test for mortar for masonry. Part 11. Determination of flexural and compressive strength of hardened mortar. *European Committee for Standardization. Bruxelles.*
11. UNE 102-039-85. (1985). BUILDING plaster and building mulains plaster. Determination of shore C hardness and brinell harness. *Aenor. Asociacion Española de Normalizacion. Madrid.*
12. EN 12004. (2008). + A1:2012. Adhesives for tiles – Requirements, evaluation of conformity, classification and designation. *European Committee for Standardization. Bruxelles.*
13. EN 998-1. (2010). Specification for mortar for masonry. Rendering and plastering mortar. *European Committee for Standardization. Bruxelles.*
14. UNE 41302. (2013). IN Instructions for the application of rendering and plastering mortars. *Aenor. Asociacion Española de Normalizacion. Madrid.*

# Chapter 59
# Latex-Modified Concrete Overlays Using Recycled Waste Paint

Aly M. Said and Oscar Quiroz

Concrete overlays are exposed to aggressive environment and deicing salts combined with traffic loads. They are required to have high strength while exhibiting adequate durability. Furthermore, in the case of bridge deck overlays, they are required to have high abrasion resistance and low permeability to survive their service life and to avert salts leaching to the deck and the rest of the bridge superstructure. Poor performance of bridge overlays can result in bridge deck corrosion leading to replacement which is costly and can cause traffic interruption. Polymer-modified concrete is typically used in bridge deck overlays due to its superior tensile strength and low permeability. Polymer modification of concrete is performed by incorporating an appropriate dose of polymer during the mixing process, which adds to the cost of concrete. In this study, the use of waste latex paint to produce an economical polymer-modified concrete is performed. The resulting concrete is then compared to polymer-modified concrete produced using commercially available polymers in the market.

## 1 Introduction

Bridge decks are normally exposed to harsh deicing salts and antifreeze brine, which can cause degradation of bridge structural elements. The use of low-permeability overlays to minimize such problem can be achieved using latex-modified concrete. Typically, latex-modified concrete overlays are produced using styrene-butadiene

---

A. M. Said (✉)
The Pennsylvania State University, University Park, PA, USA
e-mail: aly.said@engr.psu.edu

O. Quiroz
Slater Hanifan Group, Las Vegas, NV, USA

© Springer International Publishing AG, part of Springer Nature 2018
M. M. Reda Taha (ed.), *International Congress on Polymers in Concrete (ICPIC 2018)*, https://doi.org/10.1007/978-3-319-78175-4_59

rubber polymer admixtures. The use of these commercially available admixtures is limited to the most critical area due to their cost. This includes the overlay, which separates the bridge superstructure elements from the corrosive deicing salts.

Waste latex paint stands as the largest, by volume, liquid hazardous material in the United States totaling over 16 million gallons on a yearly basis. Waste latex paint is hard to recycle and harmful to the environment since it contains volatile organic compounds. Various approaches exist for waste latex paint disposal including drying and discarding in a landfill, paint swaps, combustion, and by placing pigments in cement. Many communities are adopting waste latex paint disposal by drying and placement in a landfill as the most economical approach. Nonetheless, this is the least favorable technique for the environment since it generates a need for new resources. Other studies [1] proposed different approaches to recycle paint including one that is known as processed latex pigment (PLP). Furthermore, combustion or incineration chambers are also utilized to dispose of waste latex paint. PLP is a patented process for recycling leftover latex paint among other wastes.

The use of WLP in concrete was studied by Nehdi and Sumner [2]. They investigated the possibility of using WLP to substitute for commercial latex and to enhance normal concrete for use in sidewalks. The study indicated that increasing the WLP content in concrete caused a reduction in flexural strength and compressive strength in some mixtures. Additionally, Nehdi and Sumner found that the mixtures containing WLP displayed improved compressive strength, flexural strength, and chloride penetrability for the sidewalk mixtures. Furthermore, the investigation was followed by a field demonstration which highlighted the positive impact of WLP addition had on concrete workability and surface finishing.

Mohammed et al. [3] study the recycling of waste latex paint in concrete. Various percentages of waste latex paint were incorporated in concrete then compared to virgin latex. Results indicated that waste latex paint enhanced the properties of ordinary concrete and generated concrete comparable to commercial latex-modified concrete. In this investigation, waste latex paint concrete exhibited comparable rapid chloride penetrability, compressive strength, and air content to latex-modified concrete. Additionally, waste latex paint had improved properties over the control mixture in chloride penetrability, flexural strength, and compressive strength. The investigation emphasized the use of WLP in general purpose concrete for applications like sidewalks and nonstructural elements. Such applications usually do not require latex-modified concrete, they use normal concrete. Consequently, limited economic benefits are accomplished, generally via the decrease in admixture dosage. Nonetheless, in special applications that necessitate latex-modified concrete use, waste latex paint can present an economical alternative.

The current investigation explores the possibility of producing latex-modified concrete using waste latex paint for use in concrete overlays. The study evaluates the performance of concrete produced using waste latex paint to that produced using commercially available styrene-butadiene rubber additives for bridge deck overlays.

**Table 59.1** Proportions of the studied mixtures

|  | Control | SBR | WLP1 | WLP2 |
|---|---|---|---|---|
| Cement (lb/yd$^3$) | 658 | 658 | 658 | 658 |
| Water (lb/yd$^3$) | 248.15 | 90.35 | 177.22 | 209.55 |
| Styrene-butadiene rubber (lb/yd$^3$) | 0 | 205.63 | 0 | 0 |
| Waste latex paint 1 (lb/yd$^3$) | 0 | 0 | 170.17 | 0 |
| Waste latex paint 2 (lb/yd$^3$) | 0 | 0 | 0 | 205.74 |
| Total water (lb/yd$^3$) | 248.15 | 197.28 | 248.69 | 312.42 |
| Coarse aggregate (lb/yd$^3$) | 1149.22 | 1149.22 | 1149.22 | 1149.22 |
| Fine aggregate (lb/yd$^3$) | 1892.46 | 1892.46 | 1892.46 | 1892.46 |
| Superplasticizer oz/100 lb cement | 9.5 | 0.5 | 5 | 4 |
| Water-cement ratio (*w/c*) | 0.38 | 0.30 | 0.38 | 0.47 |
| Polymer-cement ratio (*p/c*) | – | 0.31 | 0.26 | 0.31 |
| Solid content (%) | – | 48 | 58 | 50 |

## 2 Experimental Program

Materials used in the study were provided by a regional concrete ready mix producer. It comprises type V Portland cement, coarse aggregate with size No. 8 grading, well-rounded fine aggregate sand (specific gravity of 2.78, water content of 0.13%, absorption of 0.80%, and fineness modulus of 3.00), and high-range water reducer. The No. 8 coarse aggregate has a standard gradation with aggregates between sieves with square openings of 3/8 in. to No. 8 (0.374–0.093 in.). The styrene-butadiene rubber is commercially available with a density of 8.4 lbs./gal (1 kg/L) and 48% solid content. The styrene-butadiene rubber's suggested uses encompass latex-modified overlays, toppings, bridge decks, stucco, parking decks, and highways. The paint used in this study came from two sources. The first, WLP1, had a tested solid content of 58% and a density of 77.8 lbs/ft$^3$, and it came in the eggshell white. The second, WLP2, had a tested solid content of 50% and came in interior eggshell enamel. The first paint was obtained in a sealed pail while the second paint was from a leftover of an unknown source in an open pail, with the possibility of some drying. Table 59.1 illustrates the mixes investigated and their composition. The investigated properties included abrasion, compressive strength, and chloride penetrability characteristics.

## 3 Results and Discussion

Table 59.2 shows that initially, the normal concrete control mixture had a slightly higher strength compared to WLP1 mixture. On day 3, the control, SBR, and WLP1 mixtures all had a compressive strength that is adequate for many applications that require 3000 psi. Nonetheless, WLP2 mixture did not reach an adequate strength until 28 days. The excess water in WLP2 lowered concrete compressive strength in

**Table 59.2** Average compressive strength data for the tested mixtures

| Mixture | Compressive strength | | |
|---|---|---|---|
| | 3 days (psi) | 7 days (psi) | 28 days (psi) |
| Control | 4863 | 5914 | 6792 |
| SBR | 3103 | 3912 | 6268 |
| WLP1 | 4580 | 5873 | 7613 |
| WLP2 | 1874 | 2815 | 4673 |

**Table 59.3** Results of the rapid chloride penetration tests for the studied mixtures

| Mixture | Passed charge, coulombs | Penetrability evaluation |
|---|---|---|
| Control | 1751 | Low |
| SBR | 703.5 | Very low |
| WLP1 | 738.5 | Very low |
| WLP2 | 745.5 | Very low |

**Table 59.4** Abrasion testing results for the tested mixtures

| Mixture | Volume abraded $in^3$ | Area of surface abraded $in^2$ | Abrasion coefficient $in^3/in^2$ |
|---|---|---|---|
| Control | 0.15 | 3.31 | 0.05 |
| SBR | 0.08 | 1.89 | 0.04 |
| WLP1 | 0.15 | 3.71 | 0.04 |
| WLP2 | 0.17 | 1.63 | 0.11 |

its mixture. At 28 days, WLP1 mixture surpassed the control mixture as the highest compressive concrete mixture.

The rapid chloride penetration test (RCPT) posted good results for SBR, WLP1, and WLP2 mixtures. Table 59.3 illustrates that the SBR mixture had the lowest permeability expressed in the current passing through the specimen with only 703.5 coulombs. The WLP1 mixture had the second lowest passing current with 738.5 coulombs followed by the WLP2 mixture having 745.5 coulombs passing. The three latex mixtures were evaluated as having very low penetrability. The control had 1751 coulombs passing which is a low penetrability rating. The use of waste latex paint lowered the chloride penetrability in a comparable level to the SBR.

Sand blasting technique was used to perform abrasion testing. WLP1 and SBR mixtures had the same performance expressed through the abrasion coefficient, as presented in Table 59.4. The mixture containing SBR had an abraded volume and an area of surface abraded nearly half the size of WLP1. The control mixture had an abraded volume 3% larger than WLP1. WLP2 had a largest abrasion coefficient among mixtures as shown in Table 59.3. WLP2 had a lower resistance to abrasion than WLP1 and the control mixture. WLP1 had a large abraded area with very shallow cuts into the specimen. The only difference between WLP1 and the control mixtures is the incorporation of the waste latex paint. Subsequently, the enhancement in abrasion resistance of WLP1 can be ascribed to the inclusion of waste latex paint.

The abrasion test results indicated that usually a decrease in $w/c$ is associated with an increase in the abrasion resistance in a concrete sample. Furthermore, an increase in the polymer-cement ratio $(p/c)$ is associated with an increase in the abrasion resistance of concrete. Consequently, to create a waste latex paint mixture with higher abrasion resistance, the mixture needs an increase $p/c$ coupled with a decrease in $w/c$.

## 4 Conclusions

The current study showed that the use of waste latex paint as a replacement for commercially available styrene-butadiene rubber is a feasible concept. The performance of one type of waste latex paint was on par with commercially available products and outperformed normal concrete mix overall. Mixture WLP1 had a higher strength gain rate and a higher ultimate strength than SBR1 and SBR2 which can put it in service rapidly. The abrasion coefficient of WLP1 was on par with that of SBR1 and SBR2, despite its higher abraded surface compared to SBR1 and SBR2. The investigators believe that, generally, the mixture WLP1 outperformed mixtures incorporating commercial SBR products. Nonetheless, the performance of mixture WLP2 was lower than that of commercial SBR product but overall satisfactory. Further research is needed to formulate a methodology for qualifying the use of waste latex paint in concrete applications.

## References

1. Segala, L. M. (2003). Recycling of nonhazardous industrial paint sludge, nonreusable leftover latex paint, and similar materials. *Metal Finishing, 101*(3), 38–40.
2. Nehdi, M., & Sumner, J. (2003). Recycling waste latex paint in concrete. *Cement and Concrete Research, 33*(6), 857–863.
3. Mohammed, A., Nehdi, M., & Adawi, A. (2008). Recycling waste latex paint in concrete with added value. *ACI Materials Journal, 105*(4), 367–374.

# Chapter 60
# Effect of Using Kaolin and Ground-Granulated Blast-Furnace Slag on Green Concrete Properties

Kamal G. Sharobim, Hassan A. Mohamadien, Omar M. Omar, and Mostafa M. Geriesh

Cement production causes pollution to the environment and the surrounding society especially at Helwan and Suez cities (Egypt) due to the emission of $CO_2$ gas resulting from fuel combustion in cement factories. The research studies the implementation of supplementary cementitious materials (SCM), namely, kaolin (KA) and ground-granulated blast-furnace slag (GGBS), into concrete as a partial replacement of cement and evaluates their effect on concrete mechanical behavior. The study consists of two parts; part I includes activating KA thermally at 750 °C for 3 h to produce metakaolin (MK) and activating GGBS by alkaline solution of 10 molar-concentrated KOH and $K_2SiO_3$ with a ratio of 1:2.5, respectively, then investigating their replacing effect on the compressive strength of cement mortars at 3, 7, and 28 days. In part II, the best replacement percentages in terms of compressive strength are selected, and their effects on concrete mechanical behavior at 56 days are investigated. The results showed that 10% MK replacement significantly improves concrete mechanical behavior at 28 days. Moreover, 30% GGBS replacement can be used in production of concrete of 25~35 MPa.

## 1 Introduction

Concrete is not environmentally friendly or compatible with the demands of sustainable development due to its large consumption of energy, raw materials, and emission of $CO_2$ gas. Green concrete can be defined as the concrete with material as a partial or complete replacement for cement or fine or coarse aggregates. A greener

---

K. G. Sharobim · H. A. Mohamadien (✉) · M. M. Geriesh
Faculty of Engineering, Suez Canal University, Ismailia, Egypt

O. M. Omar
Faculty of Industrial Education, Suez University, Suez, Egypt

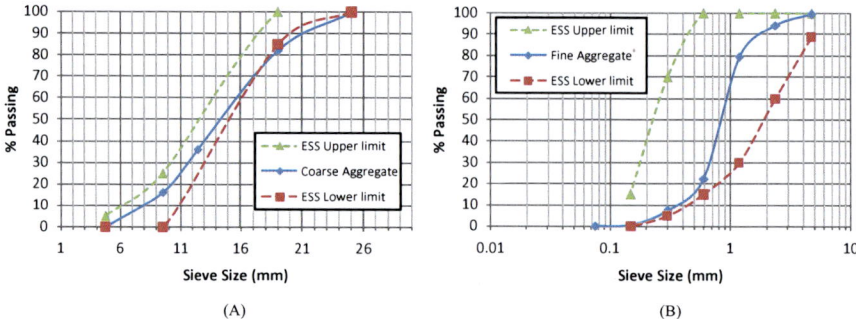

**Fig. 60.1** Particle size distribution of (**a**) coarse and (**b**) fine aggregate

**Table 60.1** Chemical composition of Kaolin and GGBS

| Chemical composition (%) | Kaolin | GGBS |
|---|---|---|
| $SiO_2$ | 47 | 38.83 |
| $Al_2O_3$ | 37 | 7.31 |
| $Fe_2O_3$ | 0.2 | 0.44 |
| CaO | 0.2 | 35.11 |
| MgO | 0.02 | 8.23 |

concrete involves less $CO_2$ production, less resource consumption, and usage of waste materials, e.g., fly ash, silica fume, marble dust, kaolin, ground-granulated blast-furnace slag [1], etc. Kaolin is traditionally used in the manufacturing of porcelain and is a fine-grained material. Research has shown that partial replacement of cement by Metakaolin improves concrete durability [2, 3]. Blast-furnace slag is a by-product that results from iron industry; it is formed when iron ore, cork, and a flux (e.g., dolomite) are melted together in a blast furnace; when the formed slag is rapidly cooled by large quantities of water, it produces a sand-like granule that is primarily ground into a product commonly known as ground-granulated blast-furnace slag (GGBS) [4]. These by-products were suggested to be used in concrete by partially replacing cement to produce green concrete.

## 2 Materials

The materials used are Type 1 Portland cement of grade 42.5 N and comply with ASTM C150, potable water, crushed dolomitic limestone of maximum nominal size (MNS) 19.5 mm with specific gravity of 2.6 as coarse aggregate, and natural siliceous sand of fineness modulus of 2.89 with specific gravity of 2.68 as fine aggregate. Figure 60.1a, b shows the grading curves for coarse aggregate and fine aggregate, respectively. Kaolin is provided by Middle East company for mineral industries at Qantra City of approximate particle size between 0.5 and 4 microns, and its chemical composition is presented in Table 60.1. Ground-granulated blast-

Table 60.2 Part I cement mortar mix designs

| Mortar mixes | Mix composition (as % of powder content) | | | | | | |
|---|---|---|---|---|---|---|---|
| | Cement | MK | GGBS | S[a]/P | W/P | AS[a]/P | HRWR |
| M8 | 100 | 0 | 0 | 3 | 0.5 | – | 1 |
| M9/MK5 | 95 | 5 | 0 | 3 | 0.5 | – | 1.5 |
| M10/MK10 | 90 | 10 | 0 | 3 | 0.5 | – | 1.8 |
| M11/MK15 | 85 | 15 | 0 | 3 | 0.5 | – | 2.5 |
| M12/AG30 | 70 | 0 | 30 | 3 | 0.5 | 0.15 | 3 |
| M13/AG45 | 55 | 0 | 45 | 3 | 0.5 | 0.23 | 3 |
| M14/AG60 | 40 | 0 | 60 | 3 | 0.5 | 0.3 | 3 |

[a]S, sand; AS, alkaline solution of 10 molar KOH and $K_2SiO_3$ with a ratio of 1:2.5

Table 60.3 Part II concrete mix designs

| Concrete mixes | Mix proportioning (kg/m$^3$) | | | | | | | |
|---|---|---|---|---|---|---|---|---|
| | C (kg) | W (kg) | CA[a] (kg) | FA[a] (kg) | KA (kg) | MK (kg) | GGBS (kg) | HRWR (Liters) |
| C1/CM | 350 | 158 | 1181 | 738 | – | – | – | 1.8 |
| C2/K10 | 315 | 158 | 1181 | 738 | 35 | – | – | 2.6 |
| C3/MK10 | 315 | 158 | 1181 | 738 | – | 35 | – | 2.6 |
| C4/G30 | 245 | 151 | 1181 | 738 | – | | 105 | 2.6 |
| C5/G45 | 192 | 144 | 1181 | 738 | – | – | 158 | 2.6 |

[a]CA, coarse aggregate; FA, fine aggregate

furnace slag is produced by Egyptian iron and steel company at Helwan City of particle size less than 70 microns and its chemical composition is presented in Table 60.1. Two superplasticizers produced by Sika Corporation which have commercial names of Sika ViscoCrete® 3425 and Sikament® R2004.

## 3 Mix Proportion

Mix designs for part I and part II are given in Tables 60.2 and 60.3, respectively.

## 4 Mechanical Tests

Compression test was conducted on 70 × 70 × 70 mm cement mortar specimens at ages 3, 7, and 28 days and on 150 × 150 × 150 mm concrete specimens at ages 7, 28, and 56 days. All specimens were cured in water at ambient temperature till testing date. Alkaline-activated GGBS mortar mixes were subjected to an additional thermal curing at 90 °C for 48 h. Compression results are shown in Fig. 60.2. From Fig. 60.2a, it is noticed that 10% MK replacement has the highest compressive

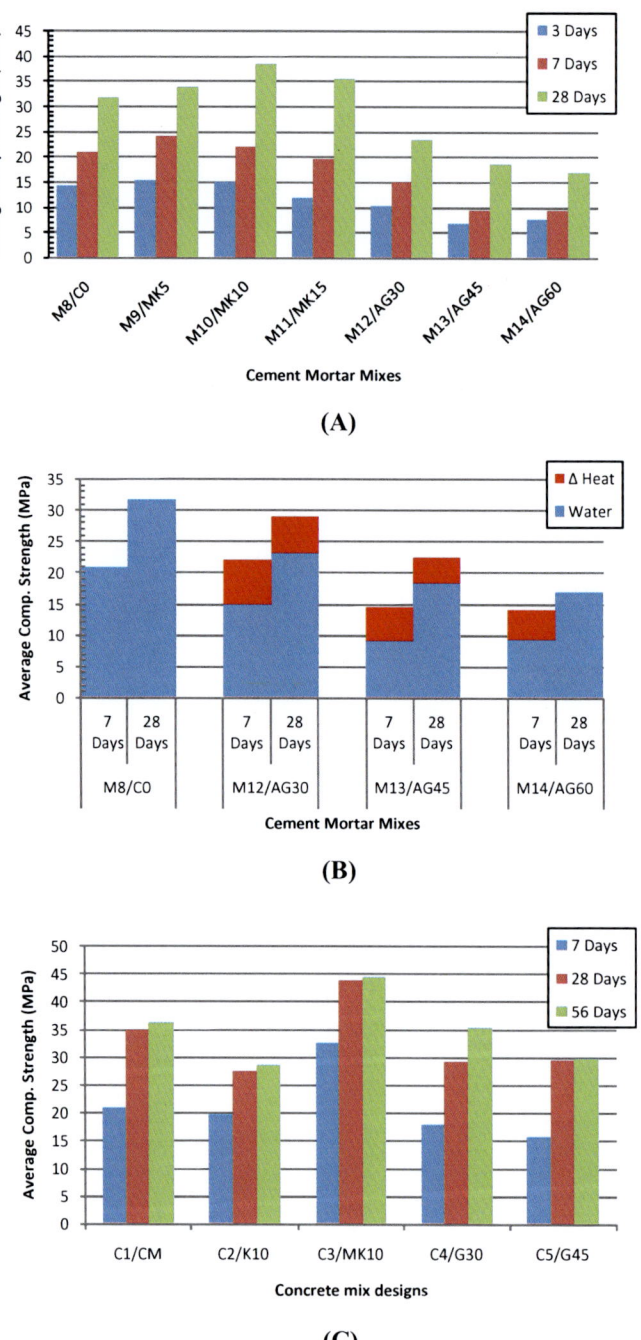

**Fig. 60.2** Compressive strength results of (**a**) mortar mixes subjected to water curing condition, (**b**) alkaline-activated GGBS mortar mixes and its increase after thermal curing at 90 °C for 48 h, and (**c**) concrete mixes

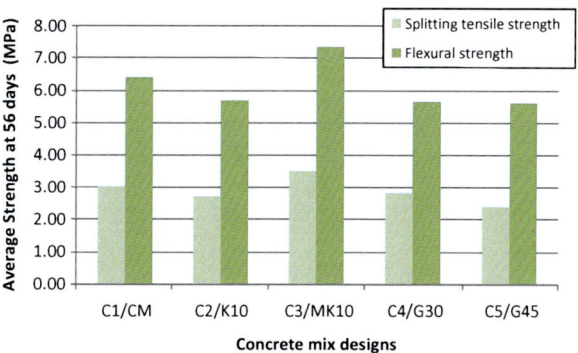

**Fig. 60.3** Splitting tensile and flexural strengths results of concrete mix

strength at 28 days [5]. However, alkaline-activated GGBS has shown lower strengths than the control mix at all ages with an inverse relation to the replacement percentage. From Fig. 60.2b, it is noticed that thermal curing has resulted in an increase in compressive strength; however, this increase is reduced by increasing the replacement percentage of GGBS until thermal curing becomes of no effect at 60% replacement at 28 days. The best result for GGBS is observed at 30% replacement where the compressive strength almost reached that of the control mix at 28 days. From Fig. 60.2c, it is noticed that replacing cement with KA reduced the compressive strength at all ages; however, the early strength gain in KA mix is higher than that of the control mix and even higher in MK mix [6]. The late strength gain of KA and MK is marginal at 56 days. Ten percent MK replacement has the highest compressive strength results at 28 days. On the other hand, GGBS has showed reduced compressive strength at ages of 7 and 28 days. However, at 56 days, the compressive strength of 30% GGBS replacement reached that of the control mix. The GGBS late strength gain is high.

Splitting tensile test was carried out on 150 × 300 mm cylinder specimens at the age of 56 days. Flexural strength test was carried out on 100 × 100 × 500 mm beam specimens at the age of 56 days. The flexural load structure consisted of a simply supported beam subjected to a three-point loading system of clear span 300 mm. Splitting tensile and flexural strengths results are shown in Fig. 60.3. From Fig. 60.3, it is noticed that using 10% MK replacement improved the splitting tensile strength by 16%. However, using 10% KA replacement reduced the splitting tensile by 11%. Furthermore, using 30% and 45% GGBS replacement reduced the splitting tensile by 7% and 22%. It is noticed that, using 10% MK replacement increased the flexural strength by 15%, this is concurrent with the splitting tensile result. However, using 10% KA replacement reduced the flexural strength by 11%. Using 30% and 45% GGBS replacement reduced the flexural strength by 11% and 12%, respectively.

## 5 SEM Analyses

Figure 60.4a, b shows the difference in microstructure between control and metakaolin mixes, respectively, analyzed by SEM at the age of 56 days.

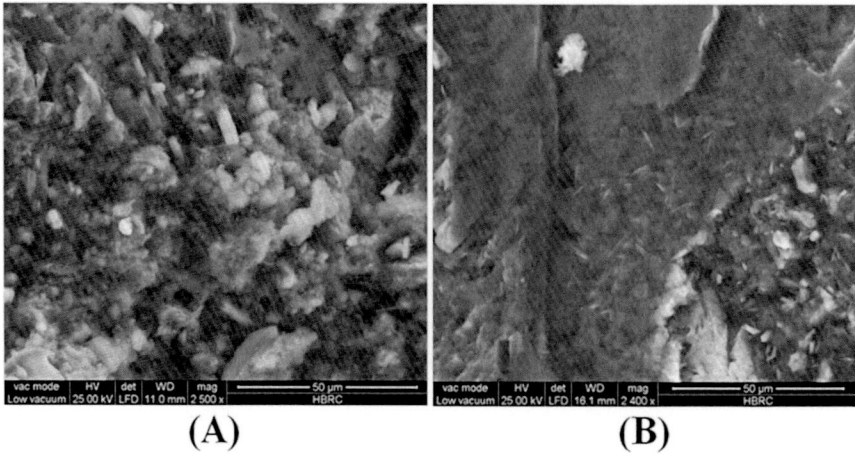

**Fig. 60.4** SEM morphology of (a) control and (b) MK mixes at 56 days

SEM of the control mix shows the presence of $Ca(OH)_2$ crystal plates in light color which indicates incomplete hydration reaction of dicalcium silicates, the high presence of ettringite in the form of needles which indicates fast setting at early ages, the high presence of pores in deep dark spots which indicates a less dense microstructure, and the nonuniform distribution of the hydration products especially calcium silicate hydrate in the form of gray gel which is responsible of strength gain in concrete. On the other hand, SEM of metakaolin mix shows the less presence of $Ca(OH)_2$ crystals indicating better hydration reaction, higher presence of ettringite indicating early strength gain, the less presence of pores indicating a more dense structure, and the uniform distribution of calcium silicate gel. Figure 60.5a, b shows the interfacial transition zone (ITZ) for control and metakaolin mixes, respectively, at the age of 56 days. It can be noticed the formation of $Ca(OH)_2$ crystals near the aggregate surface indicating weak bonding between aggregate surface and surrounding cement gel; however, the ITZ of metakaolin mix shows an extended connection of cement gel to the aggregate surface in a more homogenous pattern indicating a better bond between aggregates and cement gel.

## 6 Conclusions

From the previous experiments, the following conclusions can be drawn:

1. Using thermal activation of KA at 750 °C is very efficient in increasing its reactivity and enhancing the mechanical properties, while using alkaline activation of GGBS by 10 molar KOH and $K_2SiO_3$ with ratio 1:2.5 is inefficient due to its fast setting time effect and its negative effect on compressive strength unless it is cured thermally for 48 h at 90 °C.

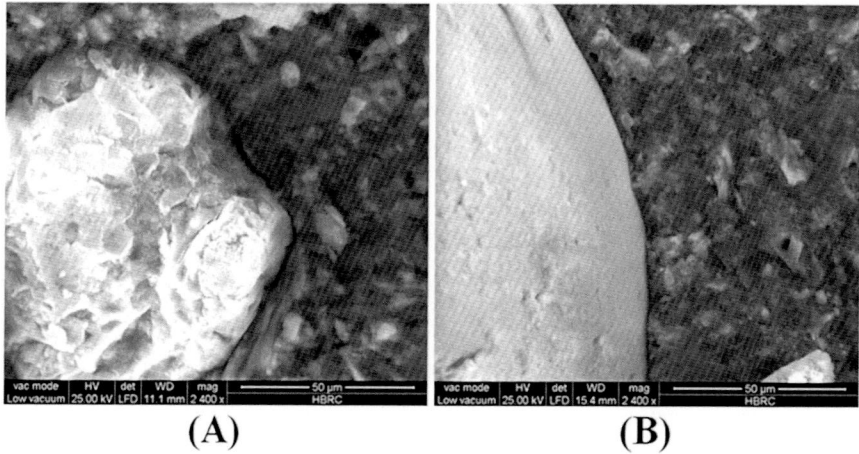

**Fig. 60.5** SEM showing ITZ of (**a**) control and (**b**) MK mixes at 56 days

2. For hardened concrete, MK has a faster interaction rate than cement and has a higher early strength gain. Replacing cement by 10% MK increases the compressive strength at 28 days by 21–25% and enhances the mechanical behavior of concrete. However, replacing cement by 10% KA results in a reduction in the compressive strength at 28 days by 21% and degrades the mechanical behavior of concrete.
3. For hardened concrete, GGBS has a slower interaction rate than kaolin. Thirty percent GGBS replacement mixes reached the design compressive strength of the control mix but at 56 days. However, 45% GGBS replacement mixes have shown a compressive strength lower by 16% than that of the control mix at 28 days with no further strength development at late ages.

## References

1. Siddique, R., & Bennacer, R. (2012). Use of iron and steel industry by-product (GGBS) in cement paste and mortar. *Resources, Conservation and Recycling, 69*, 29–34.
2. Oner, A., & Akyuz, S. (2007). An experimental study on optimum usage of GGBS for the compressive strength of concrete. *Cement and Concrete Composites, 29*(6), 505–514.
3. Dhinakaran, G., Thilgavathi, S., & Venkataramana, J. (2012). Compressive strength and chloride resistance of metakaolin concrete. *KSCE Journal of Civil Engineering, 16*(7), 1209–1217.
4. Siddique, R. (2014). Utilization (recycling) of iron and steel industry by-product (GGBS) in concrete: Strength and durability properties. *Journal of Material Cycles and Waste Management, 16*(3), 460–467.
5. Kadri, E. H., Kenai, S., Ezziane, K., Siddique, R., & De Schutter, G. (2011). Influence of metakaolin and silica fume on the heat of hydration and compressive strength development of mortar. *Applied Clay Science, 53*(4), 704–708.
6. Zeng, J., Shui, Z., & Wang, G. (2010). The early hydration and strength development of high-strength precast concrete with cement/metakaolin systems. *Journal of Wuhan University of Technology Materials Science Edition, 25*(4), 712–716.

# Chapter 61
# Lightweight Structural Recycled Mortars Fabricated with Polyurethane and Surfactants

Verónica Calderón, Raquel Arroyo, Matthieu Horgnies, Ángel Rodríguez, and Pablo Luis Campos

This paper reports on the properties of a new range of structural and lightweight mortars, including a low fraction of soluble nonionic surfactant, manufactured with different substitution rates of sand by polyurethane foam wastes. The fabrication is carried out by substituting sand by polyurethane wastes and adding low fractions of non-ionic surfactants with respect to the amount of cement. As characterized by the evolution of the mechanical strengths and porosity, the properties of the hardened mortars can vary significantly according to the chemical structure – hydrophilic–lipophilic balance value – of the nonionic surfactant. This new range of materials containing polymer wastes complies with the principle of sustainable development and contributes to a greener business model within the building sector.

## 1 Introduction

Following the policy toward zero waste plastics in landfill, the substitution of varying amounts of aggregates by recycled and reusable compounds, such as polymer foam wastes, is therefore of great interest in the production of new construction materials.

To further this research, the aim of this work was to reduce the use of traditional aggregates by recycling the polyurethane foam wastes in order to obtain structural construction materials with enhanced properties. More precisely, the addition of

---

V. Calderón (✉) · R. Arroyo · Á. Rodríguez · P. L. Campos
Departamento de Construcciones Arquitectónicas e Ingenierías de la Construcción y del Terreno, Universidad de Burgos, Burgos, Spain
e-mail: vcalderon@ubu.es

M. Horgnies
LafargeHolcim R&D Center, Quentin-Fallavier, France

polymer wastes is done in parallel with the use of nonionic surfactants, which are known to modify the microstructure after hydration [1] and to promote the adhesion of pigments at the surface of mortar [2]. By coupling several substitution rates of sand by the polymer wastes and the use of several nonionic surfactants (with different hydrophilic–lipophilic balance (HLB)), this research provides the compressive and flexural strengths, the bulk density, the porosity, and the microstructure that can be achieved for these new lightweight structural mortars.

## 2 Materials and Methods

1. The cement used was a CEM I 42.5 R with at least 95% clinker and 5% minority components, a density of 3065 kg/m$^3$, and Blaine specific surface of 3500 cm$^2$/g, according to EN 197-1.
2. The river sand used was commercial sand with particles size between 0 and 4 mm, and bulk volumic density of 2700 kg/m$^3$.
3. Polyurethane foam wastes were obtained from ground panels, used in the automotive industry with particle size between 0 and 4 mm before being used as an aggregate substitute. The bulk density was evaluated as 26 kg/m$^3$, according to EN 1097-6, and the real density as 121 kg/m$^3$, using the Pycnometer method (with isopropyl alcohol). The results deduced from elemental analysis and X-ray diffraction established a composition based on 65.5% carbon, 19.0% oxygen, 7.2% nitrogen, 6.2% hydrogen, 1.0% calcium, and 1.1% other minor components.
4. Four different types of nonionic surfactants (C13-oxo alcohol ethoxylate type) provided by BASF group were used in a liquid state. They are characterized by hydrophilic–lipophilic balance (HLB), using Griffin's method, as described in Table 61.1.
5. Mixtures and dosages. Mortars made with Portland cement, water, sand, PU foam wastes, and nonionic surfactants with different hydrolysis grades were mainly examined. A 1/3 cement/aggregate ratio was dosed, where the aggregate was the sum of sand plus polymer wastes. PU foam wastes were substituted for sand in rates that varied between 25% and 100% in volume. A specific dosage of 2% of nonionic surfactant by weight of cement was used.

**Table 61.1** HLB values for the different nonionic surfactants used

| Nonionic surfactant | HLB value | Comments |
| --- | --- | --- |
| SURF1 | 6.1 | Very "hydrophilic" |
| SURF2 | 12.1 | Less "hydrophilic" |
| SURF3 | 14.1 | Less "hydrophilic" |
| SURF4 | 10.5 | Less "hydrophilic" but bigger pendant chains |

## 3 Results

One of the most interesting properties of these materials is the improvement of the compressive strength with respect to the porosity. These properties are function of the presence of the different surfactants that provides a variation of the cement hydration behaviour [3]. According to Table 61.2 and by considering the Series 1, the progressive substitution of sand by the PU wastes logically induced a proportional reduction in density, yielding finally a lightweight mortar. This effect on the density had no consequence on mechanical strength. One explanation might be related to the positive effect of the mixture of both polymeric types, such as the hydrophilic characteristics of the surfactant used (SURF1) and the recycled polyurethane. Considering Series IV, using SURF4 surfactant, which although a less hydrophilic surfactant (with higher HLB values), has much more ramified chains.

The microporosity (covering the pore diameter range from about 200 μm to 3 nm) was measured by MIP and the macroporosity (with pore diameter range bigger than 200 μm) was obtained with the analysis of the 3D reconstruction images following computerized tomography tests.

As showed in Table 61.3, the microporosity in the mortar is bigger and proportional to the amount of PU, although the macroporosity decreased. One would

Table 61.2 Hardened bulk density and mechanical strength of the samples

| | Sample | Hardened bulk density (kg/m$^3$) | Flexural strength (MPa) | Compressive strength (MPa) |
|---|---|---|---|---|
| Series I (SURF1) | RTO3 | 2070 | 4.6 | 19.9 |
| | PU25TO3 | 2030 | 5.9 | 22.9 |
| | PU50TO3 | 2000 | 5.5 | 27.4 |
| | PU75TO3 | 1970 | 6.1 | 35.7 |
| | PU100TO3 | 1760 | 5.4 | 38.3 |
| Series II (SURF2) | RTO7 | 1980 | 5.9 | 18.9 |
| | PU25TO7 | 1740 | 3.5 | 8.7 |
| | PU50TO7 | 1630 | 3.3 | 7.1 |
| | PU75TO7 | 1280 | 2.3 | 8.4 |
| | PU100TO7 | 1250 | 1.3 | 2.4 |
| Series III (SURF3) | RTO11 | 650 | 2.2 | 5.8 |
| | PU25TO11 | 1550 | 2.0 | 4.6 |
| | PU50TO11 | 1160 | 0.3 | 2.4 |
| | PU75TO11 | 1140 | 0.1 | 2.2 |
| | PU100TO11 | 1060 | 0.1 | 2.3 |
| Series IV (SURF4) | RXL40 | 2110 | 5.2 | 21.7 |
| | PU25XL40 | 2095 | 6.1 | 28.6 |
| | PU50XL40 | 2000 | 5.3 | 37.1 |
| | PU75XL40 | 1860 | 5.0 | 30.2 |
| | PU100XL40 | 1800 | 3.8 | 34.8 |

**Table 61.3** Porosity characteristics

| Sample | Microporosity (MIP) (%) | Macroporosity (entrained air) (%) | Total porosity (%) |
|---|---|---|---|
| RTO3 | 19.4 | 3.4 | **22.8** |
| PU25TO3 | 17.4 | 3.0 | **21.9** |
| PU50TO3 | 18.9 | 2.6 | **20.0** |
| PU75TO3 | 20.3 | 2.4 | **22.7** |
| PU100TO3 | 24.0 | 2.0 | **26.0** |
| RXL40 | 18.6 | 12 | **30.6** |
| PU25XL40 | 18.1 | 7.5 | **25.6** |
| PU50XL40 | 19.5 | 6.4 | **25.9** |
| PU75XL40 | 27.6 | 5.8 | **33.4** |
| PU100XL40 | 29.6 | 3.3 | **32.9** |

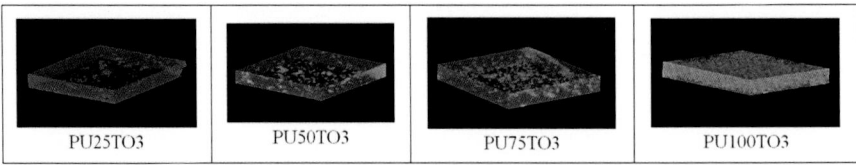

**Fig. 61.1** Polymer dispersion in yellow for samples of Series I (SURF1)

hypothesize that the nonionic surfactants showing the lowest HLB values contribute to a better distribution of the particles that can increase the cohesion [4].

The computerized tomography tests done of the same specimens confirmed the results in terms of dispersion (Fig. 61.1).

Consequently, taking into account the final properties, the combined effects of the mixture of both nonionic surfactants with the polymer wastes on the compressive strength of the mortar depend on the grade of hydrolysis of the polymer chains that induce significant variations in terms of air entraining [5].

# 4 Conclusions

The properties of the hardened mortars can vary significantly according to the chemical structure of the nonionic surfactant. The HLB value establishes the final properties of these materials according the replacement of sand by polymer.

The classification of the mortars by several methods established that the very hydrophilic surfactants (with the lowest HLB values of this study) induced a much more compact matrix than the hydrophobic surfactants (that tended to generate many more air voids by acting as air entrainers).

## References

1. Kong, X., Emmerling, S., Pakusch, J., Rueckel, M., & Nieberle, J. (2015). Retardation effect of styrene-acrylate copolymer latexes on cement hydration. *Cement and Concrete Research, 75*, 23–41.
2. Gueit, E., Darque-Ceretti, E., Tintillier, P., & Horgnies, M. (2012). Surfactant-induced growth of a calcium hydroxide coating at the concrete surface. *Journal of Coatings Technology and Research, 9*, 337–346.
3. Wang, R., Yao, L., & Wang, P. (2013). Mechanism analysis and effect of styrene–acrylate copolymer powder on cement hydrates. *Construction and Building Materials, 41*, 538–544.
4. Boutti, S., Urvoy, M., Dubois-Brugger, I., Graillat, C., Bourgeat-Lami, E., & Spitz, R. (2007). Influence of low fractions of styrene/butyl acrylate polymer latexes on some properties of ordinary Portland cement mortars. *Macromolecular Materials and Engineering, 292*, 33–45.
5. Hoover, C. G., & Ulm, F. (2015). Experimental chemo-mechanics of early-age fracture properties of cement paste. *Cement and Concrete Research, 75*, 42–52.

# Chapter 62
# Hydration in Mortars Manufactured with Ladle Furnace Slag (LFS) and the Latest Generation of Polymeric Emulsion Admixtures

Ángel Rodríguez, Sara Gutiérrez-González, Isabel Santamaría-Vicario, Veronica Calderón, Carlos Junco, and Jesús Gadea

Ladle furnace slags have successfully been used in the manufacture of masonry mortars. However, this type of steelmaking waste is not easily mixed with the other components, resulting in separation of the mix water and insufficient hydration, which prevents proper setting and hardening of the mixes. Two types of new-generation admixtures were used, in order to facilitate the interaction of the LFS with the cement paste: polyacrylic ester PAE polymers and a polymeric emulsion with a high molecular weight in different proportions. Their influence on the mortar properties in the fresh state, dosed with variable amount of LF slag, is studied. The results show the greater efficiency of the polyacrylic ester PAE admixtures at retaining the mix water, guaranteeing proper hydration of the mortar components and dispersion of the cement paste and the complete hydration of the LF slag.

## 1 Introduction

Steelmaking slags are by-products generated in the steelmaking manufacturing processes, both in the integral process (blast furnace slags or BFS and basic oxygen furnace slags or BOFS) and in electric arc furnaces (electric arc furnace slags or EAFS) and in refinement processes of secondary metallurgy in ladle furnaces (ladle furnace slags or LFS). Europe, one of the most prolific global steel producers, generates significant amounts of waste in the form of different types of slags, principally BFS. In contrast, the highest quantities of by-products produced in Spain, because of its steelmaking technology, are EAFS and LFS [1].

---

Á. Rodríguez (✉) · S. Gutiérrez-González · I. Santamaría-Vicario · V. Calderón · C. Junco · J. Gadea
Department of Construction, University of Burgos, Burgos, Spain
e-mail: arsaizmc@ubu.es

The use of steelmaking waste in the manufacture of construction materials for their use in civil works and building, as pavements, asphalt pavements, masonry mortars, and concretes, has been the subject of multiple investigations [2–4].

One problem observed with the use of LFS in the manufacture of mortars and concretes is the hydrophobic character of LFS and the difficulty of mixing it with the cement paste, which means that the hydraulic and pozzolanic properties of this waste are not exploited in the long term, as its total hydration is not guaranteed [5, 6].

The use of new-generation polymeric admixtures can improve the behavior of LFS, by modifying its surface tension, thereby facilitating its homogeneous mixture with the other mortar components, to increase its cohesion by modifying the rheology of the material and the viscosity of the cement paste [7, 8].

The investigation reported in this work studies the use of two types of new-generation polymeric admixtures to ensure the perfect combination and the integration of the LFS with the other mortar components, using polymeric emulsion admixtures. Cement mortars incorporating LFS were designed, adding different dosages of admixtures A and B, and their properties were analyzed in the fresh state, in order to explore the effect of the admixtures.

## 2 Raw Materials

Portland CEM I 42,5 R cement was used in the mortars, with a density of 3150 Kg/m$^3$ and composed of 95% clinker.

The aggregate used in the mixtures consisted of natural siliceous aggregate and LFS:

- Silica sand taken from sedimentary beds was used as a natural aggregate, previously washed and sorted by size. The real density of the sand, composed of $SiO_2 > 96\%$, was 2600 kg/m$^3$.
- The LFS was from secondary steelmaking processes, obtained from a ladle furnace in the refinement process of molten steel from the electric arc furnace. The slag, before its use, was stored in the laboratory to guarantee the total hydration of CaO and MgO, avoiding possible expansive effects when they transformed into portlandite and brucite, respectively. It presented a real density of 2673 Kg/m$^3$ and principally consisted of $SiO_2$, CaO, MgO, and $Al_2O_3$.

**Particle Size Distribution**

The natural aggregate and the LFS was sieved and separated by sizes, with the objective of designing a specific granulometry for the mortars in the study, in accordance with the granulometric line shown in Fig. 62.1. The aggregate used in the mortars under study, composed of silica sand and LF slag, was designed in accordance with the weight percentages shown in Table 62.1 for each sieve.

Taking into account the distribution of sizes and the quantities by weight of each one of the components, the proportion of LFS of the total aggregate was 35.5%.

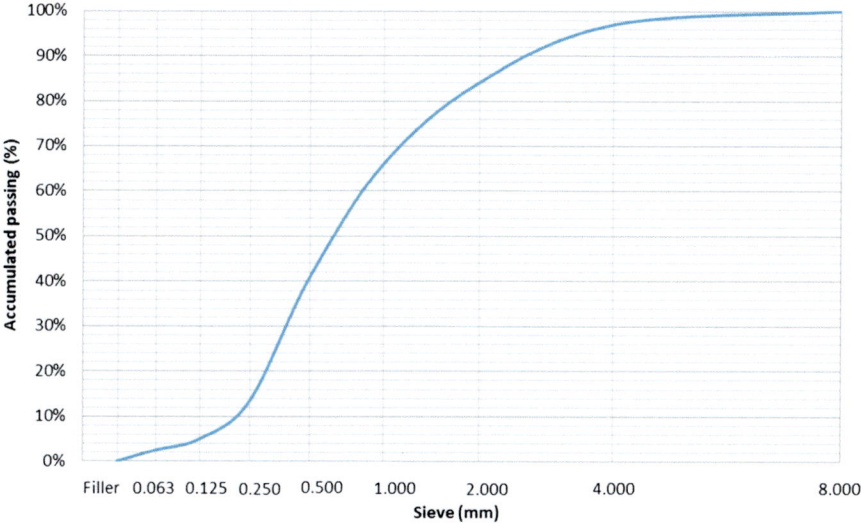

**Fig. 62.1** Particle size distribution

**Table 62.1** Distribution of aggregates

| Sieve (mm) | 8.00 | 4.00 | 2.00 | 1.00 | 0.50 | 0.25 | 0.125 | 0.063 | Filler |
|---|---|---|---|---|---|---|---|---|---|
| NA (%) | – | 100 | 100 | 100 | 50 | 50 | 50 | – | – |
| LFS (%) | – | – | – | – | 50 | 50 | 50 | 100 | 100 |

**Admixtures**

Two types of new generation polymeric emulsions were used:

- *Admixture A*: composed of polyacrylic ester (PAE), this admixture is characterized to facilitate the process of hydration of the cement, ensuring the dispersion of the cement paste over the aggregate surface.
- *Admixture B*: a polymeric emulsion with a high molecular weight, the effect of this admixture is to maintain the cohesion of the LFS with the other mortar components, reducing the risk of separation in the different phases of preparation of the mixtures.

## 3 Mortars Design

The mortars were dosed with proportions by weight (1:4:2) for the components of cement, aggregate, and water. The water added in each mixture was the necessary amount to achieve a slump on the shaking table of 175 ± 10 mm, according to EN 1015-3 [9].

**Table 62.2** Mortar design

| Sample | CEM I (gr) | NA (gr) | LF (gr) | Admixture A (gr) | Admixture B (gr) | w/c |
|---|---|---|---|---|---|---|
| MLF | 400 | 1032 | 568 | – | – | 1.0500 |
| MLF-A-0.5 | 400 | 1032 | 568 | 2 | – | 0.9375 |
| MLF-A-2.5 | 400 | 1032 | 568 | 10 | – | 0.8125 |
| MLF-B-0.5 | 400 | 1032 | 568 | – | 2 | 0.9625 |
| MLF-B-0.8 | 400 | 1032 | 568 | – | 3.2 | 0.9625 |

A reference mortar with no admixtures (MLF) was dosed, in order to compare the results obtained when adding the admixtures selected in the study. The mortars dosed for the study are shown Table 62.2.

## 4 Experimental Methods

Laboratory tests were performed to observe the mortar behavior in the fresh state, in accordance with the test procedures specified in the mortar standards:

- *Bulk density of fresh mortar*, according to the procedure of the standard EN 1015-6:1999 [10]
- *Air content of fresh mortar*, as per EN 1015-7:1999 [11]
- *Mortar workability*, according to the procedure of the standard EN 1015-9:2000 [12]
- *Water retention capacity*, as per UNE 83816:1993 EX [13]
- *Measuring the flow time of mortars* using a workability meter, according to the procedure of the NF P18-452 [14]

## 5 Results and Discussion

### 5.1 Bulk Density, Water/Cement Ratio, and Air Content of Fresh Mortar

In accordance with the results obtained in the tests, the admixtures in use modified the properties of the reference mortar (MLF), yielding in each case mortars with different characteristics (Tables 62.2 and 62.3).

The mortars dosed with Admixture A (MLF-A) presented higher quantity of occluded air than the mortars dosed with Admixture B (MLF-B), since Admixture A is an air entrainment agent. At high doses of this admixture (MLF-A-2.5), the density in the fresh state was the lowest (1755 Kg/m$^3$) of the whole series of mortars that were designed, while at a lower dosage, the density was slightly higher than that of reference MLF (1862 Kg/m$^3$). In contrast, mortars with Admixture B, a polymeric

**Table 62.3** Mortars characterization

| Sample | Bulk density of fresh mortar (kg/m$^3$) | Air content (%) | Water retentiveness (%) | Workable life (min) | Flow time (s) |
|---|---|---|---|---|---|
| MLF | 1827 | 15.0 | 87.1 | 273 | 7 |
| MLF-A-0.5 | 1862 | 14.5 | 90.3 | 197 | 5 |
| MLF-A-2.5 | 1755 | 21.0 | 96.0 | 224 | 4 |
| MLF-B-0.5 | 1931 | 11.7 | 89.8 | 190 | 13 |
| MLF-B-0.8 | 1951 | 10.5 | 87.8 | 166 | 12 |

emulsion of high molecular weight, were obtained with densities close to 2000 Kg/m$^3$, regardless of the quantity of admixture added.

If the quantity of water used in each mix to achieve the plastic design consistency is analyzed, the mortars dosed with Admixture A (MLF-A) recorded the lowest w/c ratios, even with regard to the reference mortar MLF. The water retention effect of Admixture A was observed but also good water retention behavior, a guarantee of good hydration of both the cement and the aggregates.

## 5.2 Workability, Water Retention Capacity, and Flow Time of Fresh Mortar

If the handling times of the design mortars are analyzed, mortar MFL-A-2.5 recorded the lengthiest time of use (224 min), only surpassed by the reference mortar MLF (273 min). This behavior permits a longer time for the cement hydration process and for the dispersion of the paste around the aggregates, which guarantees the hydration of the LFS.

MLF-A-2.5 is moreover a very workable mortar, because the workability meter flow time is the lowest of all the mixtures (4 s), with good fluidity but without losing the cohesion of its components. Mortars MLF-B presented shorter handling times and worse water retention in the mix. At higher doses of Admixture B (0.8%), the water retention of the mix was lower (87.8%). Both mortar MLF-B-0.5 (190 min) and mortar MLF-B-0.8 (166 min) recorded lower values than those of the reference mortar MLF. Likewise, both mortar MLF-B-0.5 (13 s) and mortar MPB-B-0.8 (12 s) also flowed more slowly in the workability meter.

The presence in the mortar MFL-A-2.5 (96,0%) of important proportions of occluded air and higher indices of water retained allows us to affirm that the air bubbles have the effect of retaining the water.

## 6 Conclusions

The results obtained in the tests with the design mortars in the fresh state allow us to establish the following conclusion:

Good behavior of the MLF-A dosed with polyacrylic ester (PAE) polymers was observed, because they allow the mix water to be retained over longer time spans, even with lower w/c ratios. Higher water retention permits the complete hydration of the cement, which guarantees the dispersion of the paste on the aggregate surface and, in consequence, the complete hydration of the LF.

As a continuation of the investigation reported in this paper, a detailed study of the microstructure of the design mortars would be advisable, as well as the properties in the hardened state and their durability in aggressive environments.

## References

1. EUROSLAG. (2013). Position paper on the status of ferrous slag Germany. The European Slag Association. Duisburg, Germany.
2. Fuente-Alonso, J. A., Ortega-López, V., Skaf, M., Aragón, Á., & San-José, J. T. (2017). Performance of fiber-reinforced EAF slag concrete for use in pavements. *Construction and Building Materials, 149,* 629–638.
3. Faleschini, F., Fernández-Ruíz, M. A., Zanini, M. A., Brunelli, K., Pellegrino, C., & Hernández-Montes, E. (2015). High performance concrete with electric arc furnace slag as aggregate: Mechanical and durability properties. *Construction and Building Materials, 101,* 113–121.
4. Rodríguez, Á., Manso, J. M., Aragón, Á., & González, J. J. (2009). Strength and workability of masonry mortars manufactured with ladle furnace slag. *Resources, Conservation and Recycling, 53*(11), 645–651.
5. Santamaría-Vicario, I., Rodríguez, A., Gutiérrez-González, S., & Calderón, V. (2015). Design of masonry mortars fabricated concurrently with different steel slag aggregates. *Construction and Building Materials, 95,* 197–206.
6. Ohama, Y. (1995). *Handbook of polymer-modified concrete and mortars: Properties and process technology.* William Andrew. Norwich, NY, United States.
7. Afridi, M. U. K., Ohama, Y., Iqbal, M. Z., & Demura, K. (1995). Water retention and adhesion of powdered and aqueous polymer-modified mortars. *Cement and Concrete Composites, 17*(2), 113–118.
8. Frías, M., San-José, J. T., & Vegas, I. (2010). Steel slag aggregate in concrete: The effect of ageing on potentially expansive compounds. *Materiales de Construcción, 60*(297), 33–46.
9. EN 1015-3. (2000). Methods of test for mortar for masonry. Part 3: Determination of consistence of fresh mortar (by flow table). *European Committee for Standardization.* Bruxelles.
10. EN 1015-6:1999/A1:2007. (2007). Methods of test for mortar for masonry - Part 6: Determination of bulk density of fresh mortar. *European Committee for Standardization.* Bruxelles.
11. EN 1015-7. (1999). Methods of test for mortar for masonry - Part 7: Determination of air content of fresh mortar. *European Committee for Standardization.* Bruxelles.
12. EN 1015-9. (2000). Methods of test for mortar for masonry - Part 9: Determination of workable life and correction time of fresh mortar. *European Committee for Standardization.* Bruxelles.
13. UNE 83816. (1993). Test methods. Mortars. Fresh mortars. Determination of water retentivity. *Asociación Española de Normalización y Certificación AENOR.* Madrid
14. NF P18-452 (1988) Mesure du temps d'écoulement des bétons et des mortiers aux maniabilimètres. In Norme Européenne-Association Française de Normalisation AFNOR. Paris.

# Chapter 63
# Chemical Resistance of Vinyl-Ester Concrete with Waste Mineral Dust Remaining After Preparation of Aggregate for Asphalt Mixture

Joanna J. Sokołowska and Piotr P. Woyciechowski

The aim of the research was to test the ability of vinyl-ester concrete containing waste powder to resist chemical attacks of sulfuric acid and sodium hydroxide. Tested composites contained mineral dust remaining after preparation of aggregates for mineral-asphalt composites. It was experimentally confirmed that from the point of view of chemical composition, granulometry, specific surface area, and density, the waste dust appeared to be a good substitute for commercial quartz powder used in polymer concrete. The evaluation of chemical resistance of polymer concrete with waste dust was done in terms of mass loss and change in compressive strength of composite after its exposition to aggressive media. The tests were performed according to the modified method described in guidelines for the precast elements production. Analysis of test results showed that even the high substitution of quartz powder with waste mineral dust (up to ~85%) in vinyl-ester concrete is possible and such composite presents high mechanical properties (though in some cases it requires using larger amounts of polymer) – not much worse than in case of non-modified composites.

## 1 Introduction and Research Scope

Polymer concretes (PC) are considered to be very chemically resistant due to their component properties, particularly the polymer binder that covers the aggregate with solid durable and resistant coating. The properties of the binder (usually liquid synthetic resin) make possible to experiment with the fillers (suspended and tightly sealed in the polymer matrix). Taking into account the expectations that PC should

---

J. J. Sokołowska (✉) · P. P. Woyciechowski
Warsaw University of Technology, Faculty of Civil Engineering, Warsaw, Poland
e-mail: j.sokolowska@il.pw.edu.pl

present high mechanical strength, it does not seem reasonable to interfere with the petrographic composition and the granulometry of the fine and coarse aggregate mix, as this could weaken the composite. On the other hand, there are no obvious contraindications against using unconventional types of the aggregate finest fraction, a microfiller. Based on classic PC design principles, microfiller is produced by grinding the rock used in coarse and fine aggregate. However, due to the need to reduce the consumption of nonrenewable raw materials, fine-grained waste materials (e.g., glass waste [1]) or industry by-products (e.g., GGBS, fly ashes [2–4], or, lately, waste perlite powder [5]) are increasingly used as microfillers. Some wastes can be used also in ordinary concrete (OC) or in polymer cement concrete (PCC) as they present hydraulic properties [6, 7]. Others, as their chemical composition excludes them as co-binders or active additives of type II, are considered to be inert fillers, better suited for use in PC rather than OC or PCC. Authors have already used conventional and fluidized fly ashes [8–10] and powdered waste of expanded perlite production [5–11] as PC microfillers. Their use had proven to produce good results; the authors have attempted to apply in PC another fine-graded waste, i.e., mineral dust remaining after the preparation of aggregates for mineral-asphalt composites. First results achieved in the framework of realized research were published in [12], where the influence of dust on vinyl-ester concrete mix properties (setting and consistence) was discussed. This paper discusses the impact of waste dust on vinyl-ester concrete chemical resistance in terms of basic characteristics: mass loss and change in compressive strength of composite after its exposition to aggressive media ($H_2SO_4$ and $NaOH$).

## 2 Statistical Design and Materials

Research was conducted according to statistical design. Authors have discussed and successfully used such designs to optimize the polymer composite compositions before [9, 10, 13]. The calculated models proved to be well suited to the experimental data, although statistical design concept assumed a reduction in the number of experimental points [14]. Design used in presented research required two material variables and nine experimental points with two-hold repetition of central point for good accuracy (for specific details, see [12]). Variables were relative mass ratios of components, polymer binder/microfiller (B/M) and dust powder/microfiller (P/M). The first variable ranged between 0.40 and 0.60, the second – in the total range (0.00–1.00; substitution by mass as values of quartz and waste dust density was not much different – respectively, 2650 kg/m$^3$ and 2621 kg/m$^3$).

As explained above, microfiller in tested PC consisted of commercial quartz powder and waste dust mixed in various proportions. Quartz powder was prepared by grinding of quartz sand. Waste dust was a residue of crushing and fractioning aggregates for mineral-asphalt composites. Chemical composition of the dust was therefore the same as for limestone (basic rock in the aggregate the dust was derived from). The XRD and EDS analysis (Fig. 63.1) showed that the dust main mineral

# 63 Chemical Resistance of Vinyl-Ester Concrete with Waste Mineral Dust...

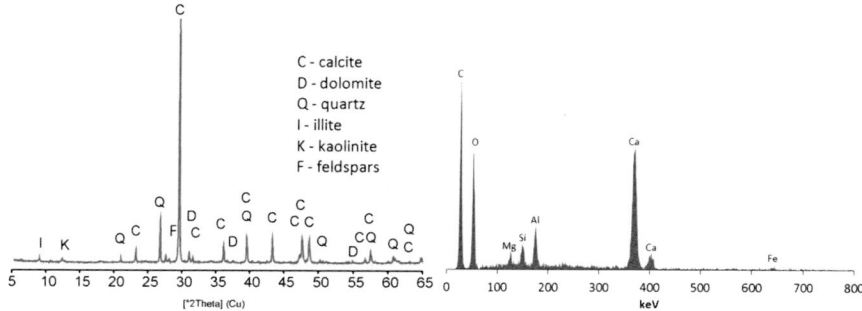

**Fig. 63.1** X-ray diffractogram (left) and energy dispersive spectrum (right) of waste mineral dust of limestone aggregate for asphalt mixture [12]

**Fig. 63.2** SEM observations of waste mineral dust: 1, aluminum silicate; 2, calcium carbonate; magnification 4000× (left) and 10,000× (right) [15]

**Table 63.1** Waste dust and quartz powder – size distribution characteristics

| Property | Quartz powder | Waste dust |
|---|---|---|
| Size range, μm | 0.58–152.45 | 0.15–394.24 |
| Mode; average size, μm | 31.96; 28.03 | 0.26 and 94.84; 60.00 |
| $D_{50}$, μm; $D_{90}$, μm | 22.80; 67.52 | 39.23; 174.62 |
| Specific surface area, $cm^2/cm^3$ | 6950 | 14,585 |

component (about 85%) was calcite (C), complemented by noticeable amounts of dolomite (D) and quartz (Q) and trace amounts of clay minerals – illite (I) and kaolinite (K). Extensive characteristics of the dust, including grain size distribution depending on variable parameters of aggregate drying process, are described by the authors in [12, 15]. The SEM observations showing the morphology of the dust particles are given on Fig. 63.2. The dust was thicker than quartz powder. Basic parameters of grain size distribution of dust and quartz powder are given in Table 63.1. Although the average and maximal size of mineral dust particles is greater than of quartz powder, the waste material contained particles smaller than

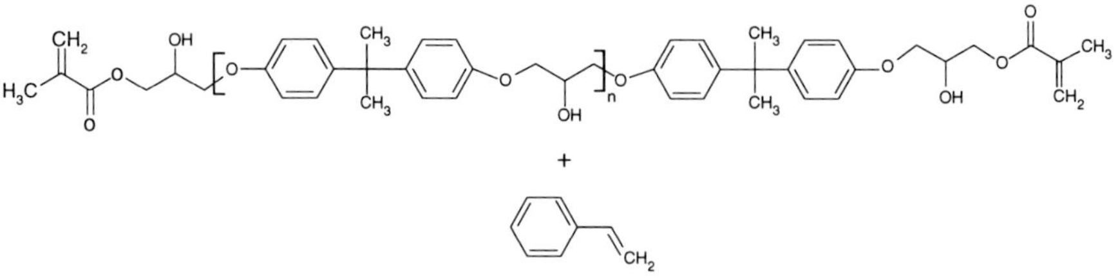

**Fig. 63.3** Vinyl-ester resin before cross-linking

0.58 μm. The presence of these particles resulted in the dust specific surface area being over twice as large as in case of quartz powder. This is also the reason for the need to add more polymer to the mix to ensure a good dust coverage and provide good workability of the mix with dust.

The basic aggregate used to prepare tested PC consisted of standard sand (in conformity with EN 196-1) and natural gravel (fraction 4–8 mm) in a ratio of 1:2 (total mass, 2000 kg/m$^3$). The gravel was washed and dried to remove any other dust.

Polymer binder used to prepare tested concretes was the commercial vinyl-ester. Composites made from this resin retain high mechanical strength in long-term contact with aggressive media, which makes the resin suitable for production of elements, that require high chemical resistance. The chemical formula of vinyl-ester resin is presented on Fig. 63.3.

## 3 Testing, Results, and Discussion

Testing was performed on halves of beams (40 × 40 × 160 mm) remaining after bending test. The applied test method included exposition to aggressive media (31 days in 1 M sulfuric acid or 4% sodium base) and measurements of change in mass ($\Delta m$) and compressive strength ($\Delta f_c$) of specimens not exposed and exposed to chemical attack. Table 63.2 contains test results (average values) obtained for modeled compositions described by material variables. As expected for PC, the base agent was more aggressive. However, using waste material did not significantly increase the vulnerability of the modified composite to the NaOH.

In order to quantify the results, a mass loss permitted by Polish ZUAT-15/X.05/2009 guidelines for testing the PC precast sewer elements, i.e., 0.25%, can be recalled as a satisfactory criterion. Only two of the tested composites did not fulfill this requirement. Both were rich in waste dust (composition No 6, pure dust; No 8, over 85% of dust), but at the same time less polymer binder was used (compare Table 63.2).

It seems that testing the compressive strength gives a better view of the effect of PC modifications on its chemical resistance. Analysis of compressive strength again showed that the base agent was more destructive. But these results do not indicate as

**Table 63.2** Change in mass and compressive strength of tested PC (described by variables $B/M$ and $P/M$) after exposition to sulfuric acid and sodium base

| No. | $B/M$, g/g | $P/M$, g/g | $\Delta m_{H2SO4}$, % | $\Delta f_{c,\ H2SO4}$, % | $\Delta m_{NaOH}$, % | $\Delta f_{c,\ NaOH}$, % |
|---|---|---|---|---|---|---|
| 1 | 0.43 | 0.15 | 0.01 | 3.9 | 0.18 | 18.4 |
| 2 | 0.57 | 0.85 | 0.02 | 18.3 | 0.14 | 24.0 |
| 3 | 0.40 | 0.50 | 0.09 | 7.9 | 0.22 | 16.0 |
| 4 | 0.60 | 0.50 | 0.06 | 3.1 | 0.00 | 19.1 |
| 5 | 0.50 | 0.00 | 0.05 | 3.1 | 0.05 | 21.6 |
| 6 | 0.50 | 1.00 | 0.00 | 14.2 | 0.43 | 14.7 |
| 7 | 0.50 | 0.50 | 0.06 | 15.0 | 0.01 | 13.8 |
| 8 | 0.43 | 0.85 | 0.07 | 16.9 | 0.32 | 11.5 |
| 9 | 0.57 | 0.15 | 0.01 | 15.3 | 0.10 | 21.2 |
| 10 | 0.50 | 0.50 | 0.12 | 9.4 | 0.10 | 17.9 |

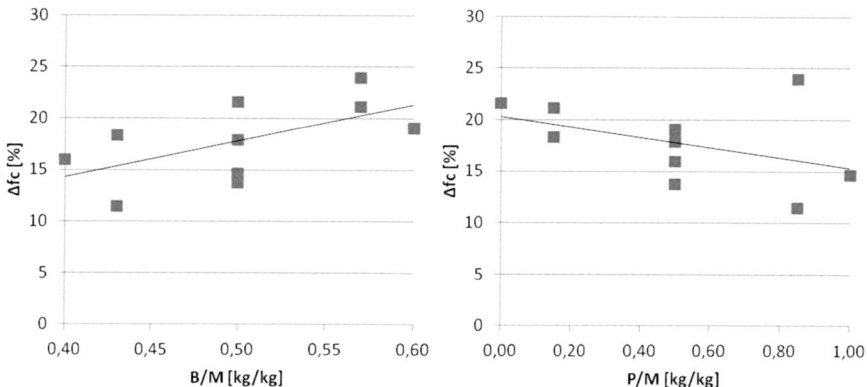

**Fig. 63.4** $B/M$ and $P/M$ vs. PC compressive strength loss after NaOH activity

optimal the same composition as in case of mass change study. It is difficult to conclusively state how the waste dust presence influenced the composite ability to maintain mechanical strength after $H_2SO_4$ attack. But trends can be observed in the case of NaOH activity (Fig. 63.4). Paradoxically, the more polymer was in the composition, the greater was the decrease in strength after chemical attack, and the more dust was in the microfiller, the smaller was that decrease. From the point of view of the reasonableness of presented study, these results seem very promising.

## 4 Conclusions

The presented results showed that using waste mineral dust remaining after preparation of aggregate for mineral-asphalt composites in vinyl-ester concretes is possible as waste dust did not significantly reduce the chemical resistance of modified composites. The research confirmed that PC, although rich in waste material, showed less sensitivity to sulfuric acid than the sodium base. Surprisingly, results obtained by two methods lead to slightly different conclusions about the optimal composition. In case of counteracting mass loss, the optimal compositions were rich in polymer binder. The good coverage of the microfiller grains with polymer layer, not the microfiller composition itself, seemed to be the key issue. This is understandable, but on the other hand there is a need to obtain high compressive strength. As polymer itself presents lower compressive strength than polymer filled with mineral fillers, introducing too much resin into the system is not favorable. Testing compressive strength of PC with waste dust showed clearly that the less polymer was in the composition, the smaller was the decrease in compressive strength after a NaOH attack. These tests also showed that the more dust was in the microfiller, the stronger was the composite. After analyzing the results, the question is raised whether mass loss testing in case of PC is meaningful. Such tests give some information about the chemical resistance (especially in the context of existing production guidelines), but often it is difficult to find any clear relation between mass loss and other properties as shown in this study.

# References

1. Bignozzi, M. C., Saccani, A., & Sandrolini, F. (2004). Glass waste valorization in advanced composite materials. In *Proceedings of the 11th int. congress on polymers in concrete – ICPIC '04* (pp. 587–595). Berlin.
2. Varughese, K. T., & Chaturvedi, B. K. (1996). Fly ash as fine aggregate in polyester based polymer concrete. *Cement & Concrete Composites, 18*, 105–108.
3. Gorninski, J. P., Dal Molin, D. C., & Kazmierczak, C. S. (2007). Strength degradation of polymer concrete in acidic environments. *Cement & Concrete Composites, 29*, 637–645.
4. Harja, M., Bărbuță, M., & Rusu, L. (2009). Obtaining and characterization of the polymer concrete with fly ash. *Journal of Applied Sciences, 9*, 88–96.
5. Łukowski, P., Sokołowska, J. J., Adamczewski, G., & Jaworska, B. (2013). Waste perlite powder as the potential microfiller of polymer composites. *Mechanics and Materials, 1*, 201–211.
6. Łukowski, P., & Salih, A. (2015). Durability of mortars containing ground granulated blast-furnace slag in acid and sulphate environment. *Procedia Engineering, 108*, 47–54.
7. Jaworska, B., Sokołowska, J. J., Łukowski, P., & Jaworski, J. (2015). Waste mineral powders as a components of polymer-cement composites. *Archives of Civil Engineering, 61*(4), 199–210.
8. Czarnecki, L., Garbacz, A., & Sokołowska, J. J. (2010). Fly ash polymer concretes. In *Proceedings of the 2nd international conference on sustainable construction materials and technologies* (pp. 28–30). Ancona.
9. Garbacz, A., & Sokołowska, J. (2010). Effect of calcium fly ash on PC properties. In *Proceedings of the 13th international congress on polymers in concrete – ICPIC 2010* (pp. 281–288). Funchal-Madeira.
10. Garbacz, A., & Sokołowska, J. J. (2013). Concrete-like polymer composites with fly ashes – Comparative study. *Construction and Building Materials, 38*, 689–699.
11. Sokołowska, J. J., Woyciechowski, P., Łukowski, P., & Kida, K. (2015). Effect of perlite waste powder on chemical resistance of polymer concrete composites. *Advanced Materials Research, 1129*, 516–522.

12. Sokołowska, J. J. (2016). Technological properties of polymer concrete containing vinyl-ester resin waste mineral powder. *Journal of Building Chemistry, 1*(1), 84–91.
13. Czarnecki, L., & Sokołowska, J. J. (2011). Optimization of polymer-cement coating composition using material model. *Key Engineering Materials, 466*, 191–199.
14. Czarnecki, L., & Sokołowska, J. J. (2015). Material model and revealing the truth. *Bulletin of the Polish Academy of Sciences Technical Sciences, 63*, 7–14.
15. Kępniak, M., Woyciechowski, P., & Franus, W. (2017). Chemical and physical properties of the waste limestone powder as a potential microfiller of polymer composites. *Archives of Civil Engineering, 2*(53), 67–78.

# Part VI
# Geopolymers

# Chapter 64
# Performance Studies on Self-Compacting Geopolymer Concrete at Ambient Curing Condition

Krishneswar Ramineni, Narendra Kumar Boppana, and Manikanteswar Ramineni

The present experimental work aims to study mechanical parameters of the self-compacting geopolymer concrete (SCGPC) developed and cured at ambient temperatures. Self-compacting geopolymer concrete is developed by partial replacement of class F fly ash with ground-granulated blast-furnace slag (GGBS) ranging from 0% to 60% with an interval of 10% of binder content. The rheological properties of developed self-compacting geopolymer concrete mixes are evaluated for various mixes, and mechanical properties like compressive strength test, split tensile strength test, and flexural strength test are carried out. Based on the results of mechanical and rheological properties, the dosage of GGBS is optimized. Results indicated that at 50% replacement of the GGBS as binder has satisfied the fresh properties of the self-compacting geopolymer concrete as per European Federation of National Associations Representing for Concrete (EFNARC) guidelines and has obtained better strength results compared to other SCGPC mixes. This paper also presents the initial and final setting time curves of geopolymer paste prepared using class F fly ash and GGBS with alkaline activators for various concentration of sodium hydroxide ranging from 10 molar to 14 molar with an interval of 2 molar.

---

K. Ramineni · N. K. Boppana (✉)
Department of Civil Engineering, VNRVJIET, Hyderabad, India
e-mail: narendrakumar_b@vnrvjiet.in

M. Ramineni
Department of Civil, Environmental and Sustainable Engineering, Arizona State University, Arizona, USA

# 1 Introduction

Self-compacting geopolymer concrete (SCGPC) is the special concrete developed by combining the principle of self-compatibility with geopolymer concrete for better durability and to reduce environmental pollution. Geopolymer concrete (GPC) is a relatively new sustainable and ecofriendly binding material which is produced from industrial by-products such as fly ash and slags by replacing 100% of cement in concrete. Geopolymers are gaining interest for its effective use of waste materials and also for low emission of greenhouse gases when compared to ordinary Portland cement (OPC). Geopolymer is a unique class of inorganic aluminosilicate polymer which is manufactured by the activation of various solid state aluminosilicate materials of geological origin or by-product materials like fly ash, metakaolin, and blast-furnace slag with a highly alkaline activating solution using elevated temperature conditions resulting in a three-dimensional polymeric chain [1]. The structural behaviors of the reinforced structural elements cast with heat-cured fly ash-based geopolymer concrete were found to be similar or superior to that of members made of OPC concrete [2]. Some researchers attempted to eliminate the requirement of curing for fly ash-based geopolymer at elevated temperature by enhancing the reactivity of fly ash in alkaline environment and by adding some calcium-containing materials. A previous study indicated that the addition of calcium oxide as a substitute of fly ash has deteriorated the mechanical properties of geopolymer concrete specimens cured at elevated temperatures but has showed improved performance of mechanical properties for ambient cured samples [2]. This gave the scope for the practical application of ambient cured geopolymer concrete to widen its application in the construction industry and eliminate the cost associated with heat curing. Thus, this study aimed to develop SCGPC mixtures at ambient curing conditions. The rheological properties, setting time parameters, and mechanical properties of geopolymer paste and SCGPC mixes were studied.

# 2 Experimental Investigation

**Materials Used**
Geopolymer is an inorganic polymeric concrete developed by reaction of alkaline solution with the source materials like clay, fly ash, microsilica, GGBS (ground-granulated blast-furnace slag), etc., which are rich in silicates and aluminates. In geopolymer concrete, the binding action is developed by the formation of alumino-silicate polymer [3]. In the present experimental work, low-calcium fly ash (or class F fly ash) and GGBS are used as the source materials in the development of SCGPC mixes. The class F fly ash is obtained from locally available dealer and GGBS from JSW cement dealer. The chemical compositions of both powder materials are shown in Table 64.1.

## 64 Performance Studies on Self-Compacting Geopolymer Concrete at Ambient...

**Table 64.1** Chemical composition of fly ash and GGBS

| Sample | $SiO_2$ | $Al_2O_3$ | $Fe_2O_3$ | CaO | MgO | $K_2O$ | $Na_2O$ | $SO_3$ | LOI |
|---|---|---|---|---|---|---|---|---|---|
| Fly ash | 60.54 | 26.20 | 5.87 | 1.91 | 0.38 | 1.02 | 0.62 | – | 2 |
| GGBS | 37.73 | 14.42 | 1.11 | 37.34 | 7.21 | – | – | 0.37 | 1.59 |

**Table 64.2** Physical properties of aggregates

| S. no | Property | Coarse aggregate | Fine aggregate |
|---|---|---|---|
| 1. | Specific gravity | 2.63 | 2.65 |
| 2. | Bulk density (kg/m$^3$) | 1549.57 | 1591 |
| 3. | Fineness modulus | 7.12 | 2.56 |

The locally available river sand is used as fine aggregate (F.A) and granite chips ranging from size 4.75 mm to 20 mm are used as coarse aggregate (C.A). The physical properties like dry density, water absorption, fineness modulus, and specific gravity of the aggregates are evaluated and are shown in Table 64.2. The mixture of sodium hydroxide (NaOH) and sodium silicate ($Na_2SiO_3$) is used as an alkaline liquid. The caustic soda flakes were used in case of sodium hydroxide, since caustic soda chemically contains 95–97% of sodium hydroxide and industrial water glass, which is also known as sodium silicate, is used. The alkaline solution is prepared by mixing water, sodium hydroxide, and sodium silicate which is calculated based on the sodium hydroxide molarity and the ratio of sodium silicate to sodium hydroxide. In the present work, the morality of the sodium hydroxide is considered as 12 M and the ratio of sodium silicate to sodium hydroxide as 2.5. To improve the workability, polycarboxylic ether-based superplasticizer (SP) is used, and the dosage of SP is 2% of the powder content. The mix proportions of the SCGPC mixes are calculated according to the Nan su mix design [4] and are shown in Table 64.3. The nomenclature of the SCGPC mix is designated by SCGPC 1 to SCGPC 7 representing replacement of GGBS as binder and dosages of 0–60% of the powder content with an interval of 10% of the powder content, respectively. For example, SCGPC 1 indicates 0% of GGBS content and 100% of class F fly ash.

### Concrete Batching, Mixing, and Placing

The mix proportions of the SCGPC mixes shown in Table 64.3 are measured using weight batching. The SCGPC mixes are produced in electrically operated pan mixer. Initially, the 12 molar sodium hydroxide solution is prepared by mixing caustic soda flakes in water. Then the solution is mixed with sodium silicate solution to make the alkaline solution. Then all the dry materials like coarse aggregate, fine aggregate, and binder materials, i.e., fly ash and GGBS, are mixed until the uniform distribution is observed. Secondly, water, chemical admixtures like SP, and premixed alkaline solution are added to the running mixer in about three to four intervals limiting the running time to 3 min. The mixer was kept still for 30 s and at last the extra water was sprinkled by mixing thoroughly for 2 min until uniform color and consistency was obtained. Once the mixing time was completed, the rheological tests starting with slump flow test, V-funnel test, and L-box test were performed in quick succession,

**Table 64.3** Mix proportions of SCGPC mixes

| Mix designation | Fly ash (kg/m$^3$) | GGBS (kg/m$^3$) | F.A (kg/m$^3$) | C.A (kg/m$^3$) | NaOH solution kg/m$^3$ | Na$_2$SiO$_3$ kg/m$^3$ | W/G ratio[a] | W/P ratio[b] | Extra water% |
|---|---|---|---|---|---|---|---|---|---|
| SCGPC 1 | 500 | – | 925 | 805 | 50 | 125 | 0.20 | 0.22 | 12 |
| SCGPC 2 | 450 | 50 | 925 | 805 | 50 | 125 | 0.20 | 0.22 | 12 |
| SCGPC 3 | 400 | 100 | 925 | 805 | 50 | 125 | 0.20 | 0.22 | 12 |
| SCGPC 4 | 350 | 150 | 925 | 805 | 50 | 125 | 0.20 | 0.22 | 12 |
| SCGPC 5 | 300 | 200 | 925 | 805 | 50 | 125 | 0.20 | 0.22 | 12 |
| SCGPC 6 | 250 | 250 | 925 | 805 | 50 | 125 | 0.20 | 0.22 | 12 |
| SCGPC 7 | 200 | 300 | 925 | 805 | 50 | 125 | 0.20 | 0.22 | 12 |

[a]Water to geopolymer solids ratio
[b]Water to powder ratio

Table 64.4 Acceptance criteria for SCC as per EFNARC guidelines [5]

| Method | Slump flow test | | V-funnel test | | L-box test |
|---|---|---|---|---|---|
| | Dia (mm) | $T_{50}$ (s) | $T_f$ (sec) | $T_{5\,min}$ (s) | $H_2/H_1$ |
| Range of values | 650–800 | 2–5 | 6–12 | $T_f + 3$ | 0.8–0.1 |

and mix performance is evaluated by comparing results with EFNARC guidelines as shown in Table 64.4. Then, the fresh concrete mixes were poured in to the molds of 150 mm side cubes for compressive strength, 150 mm diameter and 300 mm depth cylinder for split tensile strength, and 100 × 100 × 500 mm prisms for flexural strength tests for evaluating harden properties. The specimens are covered with polythene sheets and kept in molds for 24 h at room temperature of 25 ± 2 °C. After 24 h the specimens were removed from molds, and SCGPC specimens were cured at ambient conditions until the testing age under standard conditions.

## 3 Results and Discussions

**Fresh Properties and Compressive Strength**
The rheological properties of SCGPC mix proportions shown in Table 64.3 are evaluated as per EFNARC specifications, and the details of the fresh properties are presented in Table 64.5. From Table 64.5, it can be observed that the diameter of the flow decreases with an increase in GGBS dosage and similarly the L-box ratio also decreases. But in case of time, time increases with respect to increase in dosage of GGBS. This indicates that as the GGBS content increase as the substitute of fly ash, the workability of SCGPC mixes are decreased and the setting properties of the SCGPC at ambient temperature are enhanced when the other variables in the mixture are maintained the same. Similar observation is found in a previous study and also stated this is mainly due to the angular shape of GGBS particles as compared to spherical shape of fly ash particles and also the accelerated reaction of the calcium present in slag [2]. Only SCGPC1 – SCGPC6 have satisfied the EFNARC specifications, and mix SCGPC7 could not satisfy the EFNARC specifications.

From Table 64.6, it can be noticed that as the dosage of the GGBS content increases in SCGPC mixes the strength properties of SCGPC mix increases. The SCGPC 6, i.e., 50% replacement of GGBS with fly ash, has attained optimum strength, and later beyond 50% of GGBS, it found that there is a decrease in strength values. There is an increase of 40.3% and 33.97% in compressive strength, 37.6% and 36.7% in split tensile strength, and 26.5% and 33.15% in flexural strength of concrete samples at the end of 7 days and 28 days, respectively, for the SCGPC 6 mix when compared with SCGPC 1 mix. From Table 64.6, it is also observed that the gain of compressive strength from 7 to 28 days is decreased with increase in dosage of GGBS because as the GGBS content increases the calcium and silicates in mix tries to react to form C-S-H gel imparting more early strength to concrete and later strength due to polymeric reaction. Where in the case of SCGPC 1 mix, the

**Table 64.5** Fresh properties of SCGPC mixes

| Mix designation | Flow table | | V funnel | | L box |
| --- | --- | --- | --- | --- | --- |
| | Dia (mm) | $T_{50}$ (sec) | $T_f$ (sec) | $T_{5\ min}$(s) | |
| SCGPC 1 | 805 | 3 | 4 | 6 | 0.98 |
| SCGPC 2 | 792 | 3 | 6 | 9 | 0.98 |
| SCGPC 3 | 751 | 3 | 8 | 10 | 0.95 |
| SCGPC 4 | 708 | 4 | 9 | 12 | 0.87 |
| SCGPC 5 | 687 | 5 | 11 | 13 | 0.85 |
| SCGPC 6 | 654 | 5 | 11 | 14 | 0.82 |
| SCGPC 7 | 593 | 7 | 14 | 18 | 0.78 |

**Table 64.6** Strength properties of SCGPC mixes in MPa

| Mix designation | Compressive strength | | Split tensile strength | | Flexural strength | |
| --- | --- | --- | --- | --- | --- | --- |
| | 7 days | 28 days | 7 days | 28 days | 7 days | 28 days |
| SCGPC 1 | 18.95 | 27.93 | 2.12 | 2.76 | 3.01 | 3.81 |
| SCGPC 2 | 20.32 | 31.28 | 2.29 | 2.98 | 3.11 | 4.01 |
| SCGPC 3 | 23.73 | 34.91 | 2.46 | 3.12 | 3.38 | 4.25 |
| SCGPC 4 | 27.56 | 38.82 | 2.78 | 3.74 | 3.65 | 4.49 |
| SCGPC 5 | 28.93 | 40.18 | 3.04 | 4.03 | 3.81 | 5.01 |
| SCGPC 6 | 31.75 | 42.3 | 3.40 | 4.36 | 4.1 | 5.7 |
| SCGPC 7 | 32.51 | 39.01 | 3.15 | 3.89 | 3.88 | 4.26 |

silicates and aluminates tries to react, but it is a long-term process, so the early strength gain in case of SCGPC 1 is less when compared to SCGPC 6.

**Setting Time of Geopolymer Paste**

The initial and final setting time of geopolymer paste were evaluated for three various concentrations of NaOH solutions (10, 12, and 14 molar) by replacing the fly ash with GGBS ranging from 0% to 50% with 10% interval. From Fig. 64.1, it can be observed that as the concentration of the sodium hydroxide increases, the setting time decreases but only small durations. This is because the alkalinity of geopolymer is not affected by increase in concentration of sodium hydroxide solution [6]. From Fig. 64.1, it is observed that geopolymer pastes containing only fly ash as the binder takes a long duration of time to set since at ambient temperatures the rate of chemical reaction for polymerization is slow. But, when GGBS was incorporated in the mixture, both initial and final setting time of geopolymer pastes are decreased. This is mainly due to calcium present in GGBS which accelerates the setting time of geopolymer concrete in ambient conditions by quickly reacting with water molecules. Thus, the results match with the previous research that GGBS can be an effective binder to improve the setting parameters of geopolymer concrete in ambient condition.

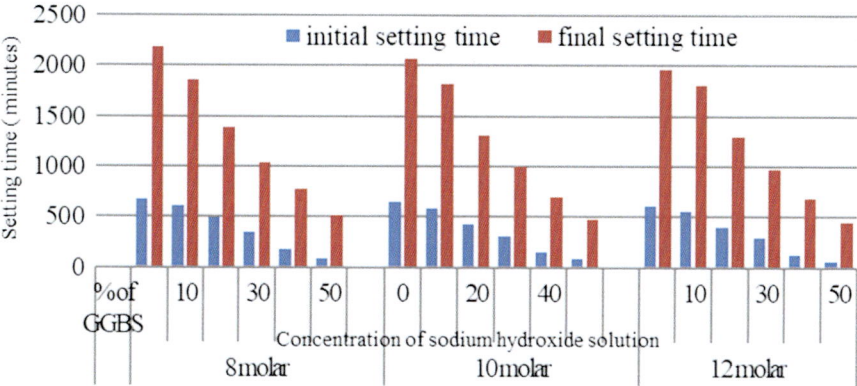

**Fig. 64.1** Graph showing initial and final setting times vs various concentration of sodium hydroxide solution

## 4 Conclusions

Based on the experimental investigations, it is found that 50% replacement of fly ash with GGBS improved the fresh and hardened properties confirming the use of these sustainable materials in concrete. There is an increase of 40.26% and 33.97% in compressive strength of concrete cubes at the end of 7 days and 28 days, respectively, for SCGPC 6 when compared to SCGPC 1. This combination has given solution for elimination of heat curing in case of geopolymer since the combination of fly ash and GGBS has drastically reduced the setting time and also enhanced the mechanical properties of the concrete. With this solution, the SCGPC developed with fly ash, and GGBS combination can be used in precast constructions and cast in situ constructions at ambient conditions. Since SCGPC also contains self-compatibility, it enables the use of concrete for high-rise structures and increases the durability parameters.

## References

1. Hadi, M. N. S., Farhan, N. A., & Neaz Sheikh, M. (2017). Design of geopolymer concrete with GGBFS at ambient curing condition using Taguchi method. *Construction and Building Materials, 140*, 424–431.
2. Nath, P., & Sarker, P. K. (2014). Effect of GGBFS on setting, workability and early strength properties of fly ash geopolymer concrete cured in ambient condition. *Construction and Building Materials, 66*, 163–171.
3. Narayanan, A., & Shanmugasundaram, P. (2016). An experimental investigation on flyash-based geopolymer mortarunder different curing regime for thermal analysis. *Energy and Buildings, 138*, 539–545.
4. Nan, S., Hsu, K.-C., & Chai, H. W. (2001). A simple mix design method for self-compacting concrete. *Cement and Concrete Research, 31*, 1799–1807.

5. EFNARC. (2002). *Specification and guidelines for self-compacting concrete*. www.efnarc.org/pdf/SandGforSCC.PDF (p. 8).
6. Patankar, S. V., Ghugal, Y. M., & Jamkar, S. S. (2014). Effect of concentration of sodium hydroxide and degree of heat curing on fly ash-based geopolymer mortar. *Indian Journal of Materials Science*, 938789, 1–6.

# Chapter 65
# Effect of 3D Printing on Mechanical Properties of Fly Ash-Based Inorganic Geopolymer

Biranchi Panda, Nisar Ahamed Noor Mohamed, and Ming Jen Tan

The inorganic polymer, e.g. geopolymer, is well known as a green construction material with its remarkable mechanical and durability properties compared to other ceramic and Portland cement-based materials. It is worth promising to explore such geopolymer properties processed by a recent innovative technology, i.e. *3D concrete printing*, which has established its ground in the construction industry. This study is focused on examining the directional effect of 3D printing process on compressive strengths of geopolymer mortar made from different industry by-products such as fly ash, slag, silica fume, etc. For a comparison, the same geopolymer mix was casted and cured in ambient temperature prior to its mechanical test with printed samples. Results from experimental tests revealed an anisotropy in the compressive strengths caused due to layered deposition of the geopolymer mortar.

## 1 Introduction

In 1972, French Prof. Joseph Davidovits developed a new kind of inorganic polymer which resulted from the polycondensation reactions of industrial aluminosilicate wastes such as slag, fly ash and user-friendly alkaline reagents [1]. This "inorganic polymer" material was firstly named "polysialate" at 1976 (Symposium of Macromolecules, IUPAC), and 9 years later, Prof. Davidovits coined another term "geopolymer" in his US patent [2] to stand for this family of inorganic polymers. Nowadays, the term "geopolymer" has been wildly accepted and getting lot of attention in both academia and industry as a sustainable cementitious material. Literature reveals that geopolymer materials were well studied by several pioneer

---

B. Panda (✉) · N. A. N. Mohamed · M. J. Tan
Singapore Centre for 3D Printing, School of Mechanical & Aerospace Engineering, Nanyang Technological University, Singapore, Singapore
e-mail: biranchi001@e.ntu.edu.sg

© Springer International Publishing AG, part of Springer Nature 2018
M. M. Reda Taha (ed.), *International Congress on Polymers in Concrete (ICPIC 2018)*, https://doi.org/10.1007/978-3-319-78175-4_65

**Fig. 65.1** Examples of 3D concrete printing: (**a**) counter crafting, USA; (**b**) Apis Cor, Russia; (**c**) Winsun, China; (**d**) Loughborough University, UK; (**e**) TU/e, Netherlands; (**f**) NTU, Singapore; (**g**) CyBe

researchers all over the world in terms of its application and fundamental properties; however, the correct mechanism of this inorganic material is still under several investigations [3]. Due to significant improved properties compared to ordinary Portland cement (OPC), geopolymers are being used in many core industries and especially in building and construction sites. A recent application of this material has been found in 3D printing process that is motivating many young researchers to develop thixotropic cementitious binders for large-scale concrete printing [4].

3D concrete printing started from an innovative idea of Prof. Khoshnevis [5] who initially put up a thought of layer-by-layer manufacturing of concrete habitat in Mars for ongoing astronauts, and then later it became a technology for additive manufacturing construction components. The main challenge of concrete printing is to develop a new material that can be extrudable and retain its shape after the extrusion. Also, the deposited layer should gain enough strength as soon as possible to avoid structural collapse which can be caused due to the weights of subsequent deposited layers [6]. Keeping these facts in mind, several research groups from different continents have come up with new cementitious materials and custom-made 3D printing setups that can be used for printing freeform digital model without any human intervention and tooling. Figure 65.1 shows an overall picture of different concrete printing processes developed by few leading industries and universities working in this area [7].

From the above examples, it is clear that 3D concrete printing can have significant impact on our current construction practices since it can save cost, time and human resources involved in the construction work and able to achieve freeform design manufacturing that are limited by our traditional approaches. In terms of material, it is noted that most of the mortars developed for concrete printing application are made from OPC binders, and this OPC has been criticized in the past by various environmentalists as one of the main causes of greenhouse effect. Therefore, this study aims to use and modify the fly ash-based geopolymer cement for its use in 3D printing process (thixotropic) with an aim of sustainable built environment.

Following highlighted sections will introduce the *material and methods* involved in designing printable geopolymer followed by their mechanical testing and few suggestions for further improvements in their properties.

## 2 Experimental Methods

Class F grade fly ash (FA) was collected from PR China as a main binder of the inorganic geopolymer. 10–20% FA was replaced by ground-granulated blast furnace slag (GGBS) and microsilica fume (SF) for ambient curing and improving the mortar rheology. Liquid potassium silicate (molar ratio = 2.0) supplied by Noble Alchem Pvt. Ltd. was used as received as alkaline reagent by mixing in a specified ratio with FA-Slag-SF dry powder. The mixing process was continued for 2 min to have uniform mixture of the binder. Then, river sand (max particle size <2 mm) was directly added into the binder as fine aggregates according to the mix design shown in Table 65.1. The whole mix was further mixed for another 1 min with slight addition of tap water and some thixotropic additives for the 3D printing application.

Prior to 3D printing, the developed geopolymer mix was characterized for rheology using a rotational rheometer (Viskomat XL) from Schleibinger Testing System, Germany. By mimicking the printing process, the final mortar was sheared under three different shear rates/rotational velocity, i.e. at first slowly ramping up to 60 rpm in 2 min followed by constant 60 rpm for 2 min and finally again ramping down to zero rpm in 2 min. The main motivation of performing this rheology test was to obtain a thixotropic behaviour which will be helpful for smooth extrusion and good shape retention properties.

In terms of 3D printing, a 4-axis gantry system was used in connection with a screw pump and numeric control system as shown in Fig. 65.2. The process starts from digital model of the sample in combination with material properties and printer parameters. The material (geopolymer mortar) was extruded through a 30 × 15 mm rectangle nozzle and simultaneously laid down on building platform layer by layer as defined in the digital file. We printed a geopolymer block having length 350 mm, width 200 mm and height 600 mm with 80 mm/s speed and later nine 50 × 50 × 50 mm$^3$ cube samples were cored out from it for the directional testing. The same mix was casted in 50 × 50 × 50 mm$^3$ mould for comparing the final strength with printed samples. All the samples were cured in room temperature and tested in 28 days at a loading rate of 0.6 MPa/s.

**Table 65.1** Mix design for formulating geopolymer mortar

| Materials | FA | Slag | SF | $K_2SiO_3$ | $H_2O$ | Sand | Additives |
|---|---|---|---|---|---|---|---|
| (%age) | 19.27 | 2.14 | 4.28 | 10.70 | 4.28 | 58.88 | 0.42 |

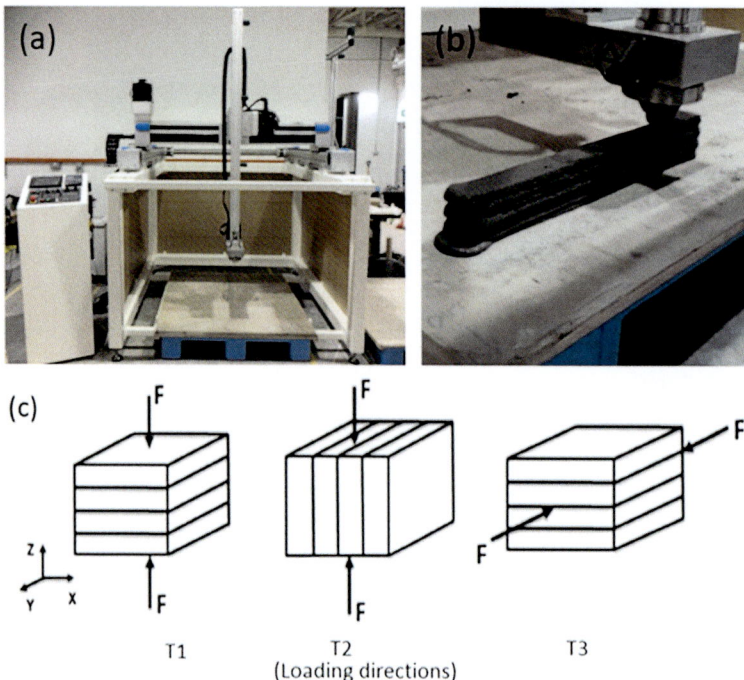

**Fig. 65.2** (**a**) 3D Concrete printer, (**b**) geopolymer printing, and (**c**) loading directions

## 3 Results and Discussion

Figure 65.3 shows the rheology of geopolymer mortar obtained from Viskomat XL, and it is clear that, like other Bingham fluid, geopolymer also possesses (static) yield stress in the beginning, and its apparent viscosity quickly went down to lower value when it was sheared at 60 rpm velocity.

This is an interesting behaviour of fluids, known as the time-dependent fluids, where apparent viscosity changes with time as it is continuously sheared. In principle, there is a need of such kind of material for 3D printing application that can hold high yield stress in the beginning and low viscosity when flowing under shear/pressure force. As a result, material can extrude out smoothly without any disruptions and, once it is deposited, also able to maintain its shape because of high yield stress property at rest. Unlike OPC, in geopolymer, there is lack of such behaviour due to absence of colloidal interactions, and in this regard, some thixotropic additives can be helpful to improve the printing performances without disturbing the core geopolymer mechanism.

Comparing the densities of printed and casted samples, it was found that printed geopolymers have lower density which can be assumed because of some voids inclusion during the printing process. The average density of the mould cast samples was 2100 kg/m$^3$ whilst that of printed specimens was a little lower at 1920 kg/m$^3$.

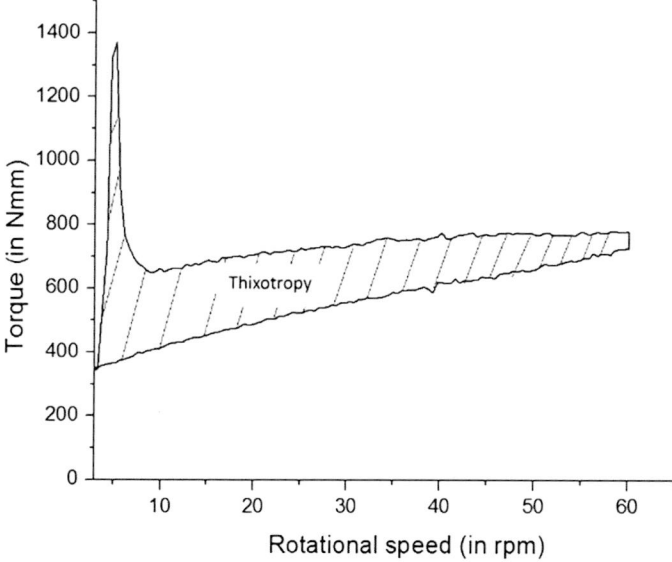

**Fig. 65.3** Plot of Torque versus rotational velocity of the geopolymer

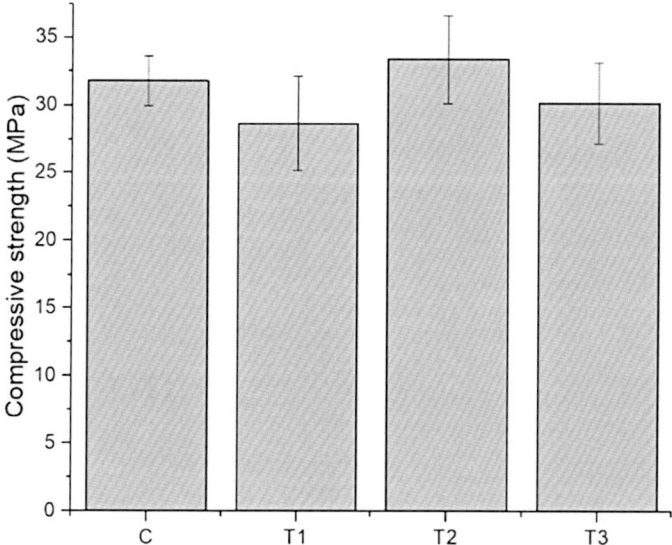

**Fig. 65.4** Compressive strength of printed (T1, T2, T3) and casted (C) geopolymer

From the results of compression test in Fig. 65.4, it was noticed that printed samples are weaker in T1 and T3 directions compared to the casting strength whilst T2 direction strength was highest amongst them. The average results were quite variable, and this was expected by the nature of layered extrusion process

[8–10]. The strength difference between T2 and T3 directions can be explained because of splitting of an interface layer (due to 30 × 15 nozzle size, 50 mm cube has an interface layer) during the T3 loading which can be further verified by considering micromechanics of composite material where it has been seen that for loading in a plane perpendicular to layers allow direct transfer and distribution of load uniformly throughout the cross section, but in case of loading in transverse direction, the beads start to separate from each other causing lower strength in sample.

## 4 Conclusion

A novel inorganic geopolymer was formulated and 3D printed in this research with an aim for sustainable built environment. The printed samples were characterized in their fresh, hardened stages, and the results were compared with traditional mould-casted samples. The performance of printed samples was found to be quite variable, and it depends on how good is the printing quality, number of layers, material open time and of course the direction of loadings. As an innovative digital construction technology, here, we have explored the directional properties of 3D-printed geopolymer samples, and in future, additional research can be carried out to minimize the anisotropic nature of the 3D printing process.

**Acknowledgement** The authors would like to acknowledge National Research Foundation Singapore (NRF) and Sembcorp Design and Construction Pte. Ltd. for funding this project.

## References

1. Davidovits, J. (2008). *Geopolymer chemistry and applications.* Geopolymer Institute, Saint-Quentin, France.
2. Davidovits, J., & Sawyer, J. L. (1985). *U.S. Patent No. 4,509,985.* Washington, DC: U.S. Patent and Trademark Office.
3. Liew, Y. M., Heah, C. Y., & Kamarudin, H. (2016). Structure and properties of clay-based geopolymer cements: A review. *Progress in Materials Science, 83*, 595–629.
4. Tay, Y. W. D., Panda, B., Paul, S. C., Mohamed, N. A. N., Tan, M. J., & Leong, K. F. (2017). 3D printing trends in building and construction industry: A review. *Virtual and Physical Prototyping, 12*(3), 1–16.
5. Khoshnevis, B., Bukkapatnam, S., Kwon, H., & Saito, J. (2001). Experimental investigation of contour crafting using ceramics materials. *Rapid Prototyping Journal, 7*(1), 32–42.
6. Panda, B., Paul, S. C., Mohamed, N. A. N., Tay, Y. W. D., & Tan, M. J. (2017). Measurement of tensile bond strength of 3D printed geopolymer mortar. *Measurement, 113*, 108–116.
7. Bos, F., Wolfs, R., Ahmed, Z., & Salet, T. (2016). Additive manufacturing of concrete in construction: Potentials and challenges of 3D concrete printing. *Virtual and Physical Prototyping, 11*(3), 209–225.
8. Panda, B., Paul, S. C., Hui, L. J., Tay, Y. W. D., & Tan, M. J. (2017). Additive manufacturing of geopolymer for sustainable built environment. *Journal of Cleaner Production, 167*, 281–288.

9. Le, T. T., Austin, S. A., Lim, S., Buswell, R. A., Law, R., Gibb, A. G., & Thorpe, T. (2012). Hardened properties of high-performance printing concrete. *Cement and Concrete Research, 42*(3), 558–566.
10. Panda, B., Paul, S. C., & Tan, M. J. (2017). Anisotropic mechanical performance of 3D printed fiber reinforced sustainable construction material. *Materials Letters, 209*, 146–149.

# Chapter 66
# Optimization of Fly Ash-Based Geopolymer Using a Dynamic Approach of the Taguchi Method

Takeomi Iwamoto, Kozo Onoue, Yasutaka Sagawa, and Ryosuke Tsutsumi

A geopolymer is an inorganic polymer devised by Davidovits. In the field of construction materials, hardened bodies resulting from polymer reactions among metallic ions, such as $Si^{4+}$ and $Al^{3+}$, and alkaline solutions, such as sodium hydroxide solution and sodium silicate solution, have been studied. Geopolymers have several superior characteristics, such as high strength and high resistance to sulfuric acid. Thus, geopolymers can be applied to precast concrete products, such as sewage pipes. There are several kinds of design parameters to be considered when manufacturing geopolymers, which makes it difficult to establish a reasonable design for this material. The present study presents an optimization of fly ash-based geopolymers using a dynamic approach of the Taguchi method. As an input value, the volume ratio of the active filler and the alkaline solution is adopted. Coal fly ash and ground-granulated blast-furnace slag are used as active fillers. Six design parameters are taken into account. Type of fly ash and the institutes where the experiments were conducted are considered as noise conditions. As a result, the applicability of this method to the optimization of fly ash-based geopolymer is confirmed. The optimized geopolymer mortar was also found to have a stable long-term compressive strength and a superior resistance to sulfuric acid.

---

T. Iwamoto · R. Tsutsumi
Graduate School of Science and Technology, Kumamoto University, Kumamoto, Japan

K. Onoue (✉)
Faculty of Advanced Science and Technology, Kumamoto University, Kumamoto, Japan
e-mail: onoue@kumamoto-u.ac.jp

Y. Sagawa
Department of Civil Engineering, Kyushu University, Fukuoka, Japan

# 1 Introduction

The geopolymer is an inorganic polymer proposed by Davidovits [1]. In the field of construction materials, research has been conducted on solidified bodies generated by the polymer reaction between alkali solutions, based on sodium silicate or sodium hydroxide, and metallic ions, such as $Si^{4+}$ and $Al^{3+}$, eluted from powders called active fillers (e.g., fly ash) [2]. Various parameters, such as the type of alkali solution and active filler, the mixing method, the curing method, and so on, exist in the production of geopolymers, making their design difficult. In the present study, focusing on the parameter design of the dynamic characteristics of the Taguchi method [3], an attempt was made to optimize the composition of fly ash-based geopolymers. Here, the dynamic characteristics refer to the output varying within a certain range with a varying input. Several studies (e.g., [4]) have shown the optimization of geopolymers by a static approach of the Taguchi method. In the static approach, a set of design parameters give one optimal solution for an output. However, in practical applications, it is more useful to vary output values, corresponding to the required performance, by changing the input. Note that Onoue and Bier [5] presented the applicability of the dynamic method for optimizing alkali-activated materials using natural pozzolan and ground-granulated blast-furnace slag.

# 2 Methodology

## 2.1 Materials

Two different types of fly ashes (JIS A 6201 [6] type II, produced from the same thermal power plant, whose physical properties are shown in Table 66.1), ground-granulated blast-furnace slag (specific surface area, 4160 $cm^2/g$; density, 2.91 $g/cm^3$), sodium silicate solution (specified in the JIS K 1408 [7] as No. 3, density, 1.41 $g/cm^3$), sodium hydroxide solution, and standard sand (density: 2.64 $g/cm^3$) were used. The procedure to prepare the sodium hydroxide solution is as follows: (1) A predetermined amount of pellet-shaped sodium hydroxide, corresponding to the

**Table 66.1** Physical properties of fly ashes

| Quality | | Type of fly ash | |
|---|---|---|---|
| | | a | b |
| $SO_2$ amount (%) | | 58.6 | 66.7 |
| Ignition loss (%) | | 2.2 | 1 7 |
| Density (g/cm³) | | 2.33 | 2.28 |
| Specific surface area (cm²/g) | | 3990 | 4050 |
| Flow value ratio (%) | | 108 | 108 |
| Activity index (%) | 28-day | 89 | 93 |
| | 91-day | 105 | 107 |

**Table 66.2** Levels of input value and design parameters

| Input value and design parameters | 1 | 2 | 3 | 4 | 5 | 6 |
|---|---|---|---|---|---|---|
| Input value ($F/L$) | 0.9 | 1.0 | 1.1 | | | |
| (A) SS/SH ratio | 0.5 | 1.0 | 1.5 | 2.0 | 2.5 | 3.0 |
| (B) Concentration of sodium hydroxide solution (M) | 3 | 6 | 9 | | | |
| (C) BFS replacement ratio | 0.05 | 0.15 | 0.25 | | | |
| (D) Mixing time (min) | 4 | 6 | 8 | | | |
| (E) Curing temperature (°C) | 60 | 75 | 90 | | | |
| (F) Cumulative temperature in heat curing (°C-h) | 640 | 1000 | 1350 | | | |

required concentration, was dissolved in distilled water in a plastic container, (2) the container was placed in a room at 20 °C and 60% relative humidity (RH) for about 24 h, and (3) distilled water was added to obtain a sodium hydroxide solution with the intended concentration. The sodium silicate solution and sodium hydroxide solution were mixed during the mixing process. All the materials were placed in a room at 20 °C and 60% RH at least 24 h before the experiments to reduce the influence of temperature.

## 2.2 Experimental Conditions

The following design parameters were considered: (A) the mass ratio of the sodium silicate solution to the sodium hydroxide solution (SS/SH ratio), (B) the concentration of the sodium hydroxide solution, (C) the mass replacement ratio of the ground-granulated blast-furnace slag (BFS) to the fly ash, (D) the mixing time, (E) the holding temperature for heat curing, and (F) the cumulative temperature of heat curing. The selected parameter magnitudes are shown in Table 66.2. Based on the results of past experiments, design parameter A was expected to have a large impact on the results. Thus, only the values of parameter A were set at six different values. Combinations of experiments were made with the L18 orthogonal table as shown in Table 66.3. Here, the numbers 1–3 in the table represent the values of the design parameters. The volume ratio of the active filler ($F$) and the alkaline solution ($L$), $F/L$, was taken as an input value, referring to a previous study [5]. Based on the results of the preliminary experiments on the flow value, the levels of the $F/L$ were set at 0.9, 1.0, and 1.1. The institutions where the experiments were conducted (Kumamoto University, Kyushu University) and the types of fly ash (a, b) were considered as noise conditions. The combinations were $N_1$ (Kumamoto University + fly ash a) and $N_2$ (Kyushu University + fly ash b).

**Table 66.3** L18 orthogonal table

| No. | A | B | C | D | E | F | G |
|---|---|---|---|---|---|---|---|
| 1 | 1 | 1 | 1 | 1 | 1 | 1 | 1 |
| 2 | 1 | 2 | 2 | 2 | 2 | 2 | 2 |
| 3 | 1 | 3 | 3 | 3 | 3 | 3 | 3 |
| 4 | 2 | 1 | 1 | 2 | 2 | 3 | 3 |
| 5 | 2 | 2 | 2 | 3 | 3 | 1 | 1 |
| 6 | 2 | 3 | 3 | 1 | 1 | 2 | 2 |
| 7 | 3 | 1 | 2 | 1 | 3 | 2 | 3 |
| 8 | 3 | 2 | 3 | 2 | 1 | 3 | 1 |
| 9 | 3 | 3 | 1 | 3 | 2 | 1 | 2 |
| 10 | 4 | 1 | 3 | 3 | 2 | 2 | 1 |
| 11 | 4 | 2 | 1 | 1 | 3 | 3 | 2 |
| 12 | 4 | 3 | 2 | 2 | 1 | 1 | 3 |
| 13 | 5 | 1 | 2 | 3 | 1 | 3 | 2 |
| 14 | 5 | 2 | 3 | 1 | 2 | 1 | 3 |
| 15 | 5 | 3 | 1 | 2 | 3 | 2 | 1 |
| 16 | 6 | 1 | 3 | 2 | 3 | 1 | 2 |
| 17 | 6 | 2 | 1 | 3 | 1 | 2 | 3 |
| 18 | 6 | 3 | 2 | 1 | 2 | 3 | 1 |

## 2.3 Experimental Procedure

For the mixing of the mortar, a Hobart mixer with a 2 L capacity was used. The amount of fine aggregate was kept constant at 1350 g, and the amounts of the other constituents were calculated based on the design parameters. The volume of fresh mortar was 0.88 L per batch. Sodium silicate solution, sodium hydroxide solution, fly ash, and blast-furnace slag were put into the bowl, in that order, and mixed at low speed for 30 s. Then, 1 bag of standard sand (1350 g) was added, followed by 30 s of mixing at low speed and 30 s of high-speed mixing. After a pause of 90 s, the fresh mortar was mixed at high speed for 60 s, 180 s, or 300 s, depending on design parameter D regarding mixing time (4, 6, and 8 min). Immediately after mixing, the flow value after 15 drops of fresh mortar was measured. Then, the fresh mortar was cast into a steel mold to produce 3 specimens with dimensions of 40 mm × 40 mm × 160 mm on the vibrating table, according to JIS R 5201 [8]. After scraping the surface of the specimen with a steel trowel, the mold, with the specimen inside, was wrapped three times with a wrapping film, placed in a constant temperature apparatus for approximately 3 h, and subjected to heat curing at a predetermined temperature/time corresponding to design parameters E and F. The rates of temperature rise and fall were set to 20 °C/h. After demolding, the specimens were placed for 24 h in a room at 20 °C and 60% RH. Afterward, flexural and compressive strengths were measured in accordance with JIS R 5201 [8].

## 2.4 Analysis of the Experimental Results

Based on the experimental results, the SN ratio (signal-to-noise ratio) and the sensitivity (gradient of input-output relationship) were calculated. In the present study, the optimization of the system was performed essentially based on the SN ratio, by combining the magnitudes of the design parameters having the highest SN ratios.

## 2.5 Long-Term Strength Test and Sulfuric Acid Resistance Test

For long-term tests on the strength and sulfuric acid resistance, specimens with dimensions of 40 mm × 40 mm × 160 mm were prepared under the optimized conditions while setting $F/L$ at 1.0. The compressive strength was measured when the material age reached 2, 7, 28, and 91 days. The curing conditions after demolding were atmospheric or underwater. For the sulfuric acid resistance test, 24 h after demolding, the specimens were immersed in a 10% sulfuric acid solution. The liquid/solid ratio was about 1.70. The mass change of the specimens after 7, 28, 63, and 91 days of immersion was measured. The 10% sulfuric acid solution was discarded and replaced with a new one after each measurement. The same test was performed on ordinary Portland cement mortar (OPC mortar, cured underwater for 28 days before the immersion) for comparison.

## 3 Results and Discussions

Figure 66.1 shows factor effect diagrams of the SN ratios and the sensitivities for the flow value and the flexural and compressive strengths. Focusing on the influence of the parameter A on the sensitivity of the flow value, Fig. 66.1 shows that it decreases with increasing SS/SH ratio. This indicates that the viscosity of fresh mortar is increased by increasing the SS/SH ratio. In addition, it is observed that the higher the concentration of sodium hydroxide solution and the GGBFS replacement ratio, the higher the sensitivities of the flexural and compressive strengths. Prioritizing the magnitude of the SN ratio regarding compressive strength, an optimized combination of the design parameters is obtained as $A_4B_2C_2D_2E_2F_2$.

Figure 66.2 shows the relationship between the input and output obtained under the optimized conditions. For only the flow value, $L/F$ is taken as an input value. Each plot in the figure is the average of the results obtained at the two different institutes. Using these expressions, it is possible to design geopolymers corresponding to the required performances.

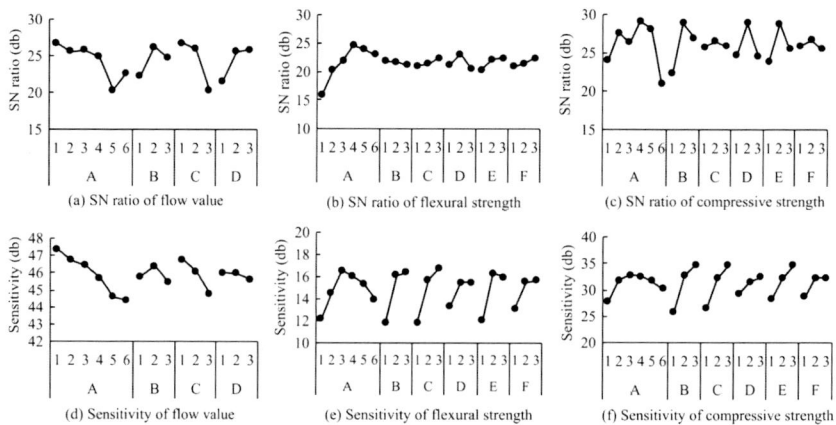

**Fig. 66.1** Factor effect diagrams of the SN ratios and sensitivities

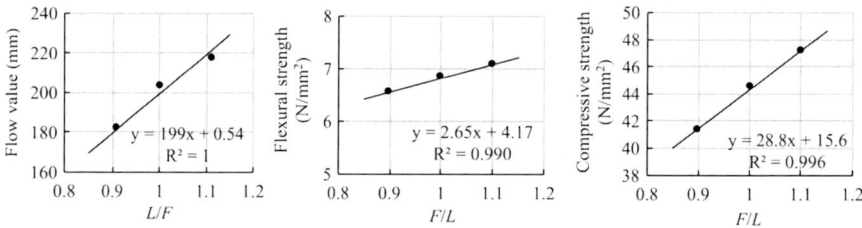

**Fig. 66.2** Relationship between input and output under optimized conditions

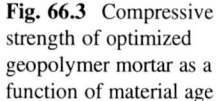

**Fig. 66.3** Compressive strength of optimized geopolymer mortar as a function of material age

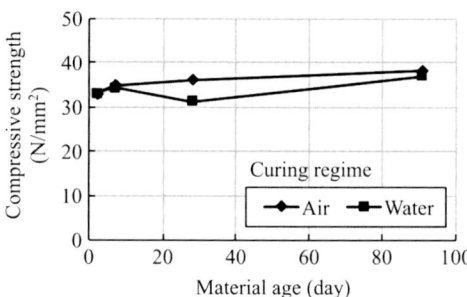

Figure 66.3 shows the results of the long-term strength test. It can be seen that the compressive strength of the fly ash-based geopolymer optimized in the present research does not change significantly over time. It is also observed that the curing regime has little effect.

Figures 66.4 and 66.5 show the results of the sulfuric acid resistance test. For the OPC mortar, it was observed that white corrosion product precipitated on the surface. After 63 days of immersion, the masses of the specimens started to decrease. On the other hand, the optimized fly ash-based geopolymer did not show significant

**Fig. 66.4** Mass change of specimens immersed in 10% sulfuric acid solution

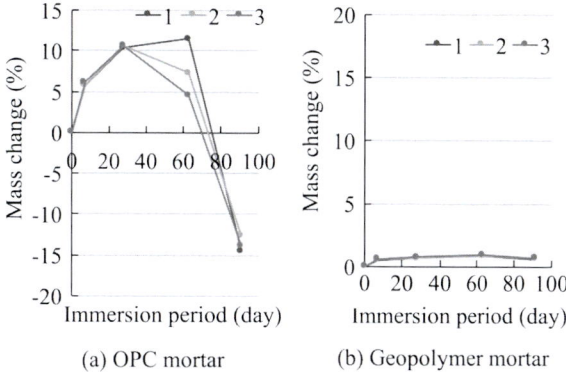

(a) OPC mortar  (b) Geopolymer mortar

**Fig. 66.5** Appearance change of specimens immersed in 10% sulfuric acid solution

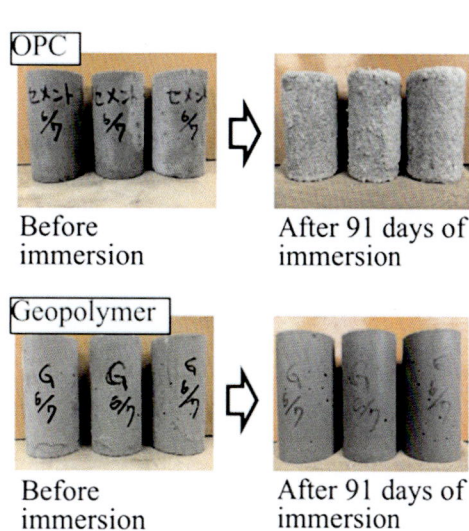

change in appearance or mass. A superior resistance of this material to sulfuric acid can be confirmed.

## 4 Conclusions

A fly ash-based geopolymer was optimized using a dynamic approach of the Taguchi method. The input-output relationships regarding flow value after 15 drops and the flexural and compressive strengths under optimized conditions were obtained. It was confirmed that the geopolymer manufactured with the optimized conditions exhibits stable long-term strength and superior resistance to a sulfuric acid solution.

**Acknowledgment** This work was supported by JSPS KAKENHI Grant Number JP16K06442.

# References

1. Davidovits J. (1982). US Patent No. 4349386. Mineral polymers and methods of making them.
2. Ichimiya, K., et al. (2011). Fundamental study on mixture proportion and manufacture method of geopolymer mortar. *Proceedings of the Japan Concrete Institute, 33*(1), 575–580. (in Japanese).
3. Tatebayashi, K. (2014). *Nyumon Taguchi method* (pp. 39–75). Tokyo: Union of Japanese Scientists and Engineers Publishing Company. (in Japanese).
4. Olivia, M., & Nikraz, H. (2012). Properties of fly ash geopolymer concrete designed by Taguchi method. *Materials & Design, 36*, 191–198.
5. Onoue, K., & Bier, T. A. (2017). Optimization of alkali-activated mortar utilizing ground granulated blast-furnace slag and natural pozzolan from Germany with the dynamic approach of the Taguchi method. *Construction & Building Materials, 144*, 357–372.
6. JIS A 6201. (2015). *Fly ash for use in concrete*. Tokyo: Japanese Industrial Standards Committee. (in Japanese).
7. JIS K 1408. (2016). *Sodium silicate*. Tokyo: Japanese Industrial Standards Committee. (in Japanese).
8. JIS R 5201. (2015). *Physical testing methods for cement*. Tokyo: Japanese Industrial Standards Committee. (in Japanese).

# Chapter 67
# Microstructural and Strength Investigation of Geopolymer Concrete with Natural Pozzolan and Micro Silica

Muhammed Kalimur Rahman, Mohammed Ibrahim, and Luai M. Al-Hems

In pursuit of finding sustainable building material, geopolymer concrete is developed utilizing aluminosilicate materials such as fly ash. Silica and alumina are the main precursors of alkali activation. The amount of silica and alumina in the source materials plays a significant role in the strength and microstructural development of such a concrete. If the source materials are supplemented by the addition of these precursors, the properties can be enhanced. Therefore, the reported study investigates the effect of incorporating micro silica at 5%, 7%, and 10% by weight, as partial replacement to natural pozzolan on the strength and microstructural properties of geopolymer concrete. Compressive strength was determined on the specimens cured in the oven maintained at 60 °C as well as at room conditions. Scanning electron microscopy (SEM) was utilized to determine the morphology of the developed alkali-activated paste (AAP). The results indicated that the natural pozzolan could be utilized, without any silica fume addition, to develop geopolymer concrete with reasonable strength that could be used for construction purposes if cured at elevated temperature. Further, concrete developed by replacing natural pozzolan with silica fume exhibited improved strength and microstructural characteristics. Seven percent micro silica replacement showed better compressive strength results and denser microstructure compared to the ones prepared with other replacement levels. The results of this study provided important information to synthesize natural pozzolan-based sustainable building material with enhanced properties.

M. K. Rahman (✉) · M. Ibrahim · L. M. Al-Hems
Center for Engineering Research, Research Institute, King Fahd University of Petroleum and Minerals, Dhahran, Saudi Arabia
e-mail: mkrahamn@kfupm.edu.sa

## 1 Introduction

In search of alternative building materials to ordinary Portland cement, geopolymer concrete is being developed utilizing source materials that predominantly consist of silica and alumina [1–3]. Such a concrete is being synthesized by alkali activation of source materials. These materials include natural and industrial by-products. In the past decade or so, fly ash has been extensively used in developing geopolymer concrete [4]. Amount of silica and alumina in the source material along with the concentration of alkaline activators plays a significant role in enhancing the properties of geopolymer concrete [5, 6]. Natural pozzolan is one of the other supplementary cementitious materials possessing potential of being used as precursor material in synthesizing geopolymer concrete. However, silica content in the locally available natural pozzolan was considerably lower as compared to fly ash and natural pozzolan from other parts of the world. Therefore, the reported study investigates the effect of incorporating micro silica at 5%, 7.5%, and 10% by weight, as partial replacement to natural pozzolan on the strength and microstructural properties of geopolymer concrete.

## 2 Materials and Methods

Natural pozzolan used in this study was powdered form of basaltic rock. Micro silica was acquired from the local supplier. Composition of natural pozzolan and micro silica is given in Table 67.1, determined by X-ray fluorescence (XRF) technique. The specific surface area and average particle size of NP used are 442 m$^2$/kg and 30 μm, respectively. A mixture of sodium silicate and 12 M sodium hydroxide solution was used as alkaline activator. Silica modulus of sodium silicate was 3.3, and its composition was $H_2O$, 62.50%; $SiO_2$, 28.75%; and $Na_2O$, 8.75%. Fine aggregate (FA) used was dune sand with a specific gravity of 2.62 in saturated surface dry condition. Lime stone aggregate was used as coarse aggregate (CA) having specific gravity of 2.56 in saturated surface dry condition.

**Table 67.1** Chemical composition of natural pozzolan

| Constituent | Natural pozzolan Weight (%) | Micro silica Weight (%) |
|---|---|---|
| Silica ($SiO_2$) | 40.48 | 90.2 |
| Alumina ($Al_2O_3$) | 12.90 | 0.84 |
| Ferric oxide ($Fe_2O_3$) | 17.62 | 0.29 |
| Lime (CaO) | 11.83 | 0 |
| Magnesia (MgO) | 8.33 | 0.97 |
| Potassium oxide ($K_2O$) | 1.67 | 1.32 |
| Sodium oxide($Na_2O$) | 3.60 | 0.65 |
| Phosphorus oxide ($P_2O_5$) | 1.37 | – |
| Loss on ignition | 1.6 | – |

**Table 67.2** Constituent materials of concrete with micro silica (in kg/m³)

| Mix | Natural pozzolan | Micro Silica | Sodium silicate | Sodium hydroxide | Fine aggregate | Coarse aggregate |
|---|---|---|---|---|---|---|
| M0 | 400 | 0 | 150 | 60 | 642 | 1193 |
| M1 | 380 | 20 | 150 | 60 | 640 | 1188 |
| M2 | 372 | 28 | 150 | 60 | 639 | 1186 |
| M3 | 360 | 40 | 150 | 60 | 638 | 1185 |

In total, four concrete mixes were prepared with 0%, 5%, 7%, and 10% replacement of natural pozzolan with micro silica having a total binder content of 400 kg/m³. Sodium silicate (SS) to sodium hydroxide (SH) weight ratio of 2.5 was maintained in all the mixes. Table 67.2 summarizes constituent materials for preparing geopolymer concrete specimens incorporating micro silica to determine compressive strength. For SEM analysis, alkali-activated paste (AAP) was prepared. Coarse aggregate to total aggregate and fine aggregate to total aggregate ratios were 0.65 and 0.35, respectively. Free water-to-pozzolanic material ratio of 0.05 was used in all the geopolymer mixtures. Alkaline activator-to-binder ratio was 0.525.

Compressive strength of concrete was determined on 50 mm concrete cube specimens. These specimens, after 24 h of casting, were de-molded and placed in plastic bags to avoid evaporation of moisture after which one set was kept in the oven maintained at 60 °C and the other stored in the laboratory maintained at 25 ± 2 °C. Compressive strength was determined according to ASTM C150 [7] after 0.5, 1, 3, 7, 14, and 28 days of curing for the samples cured at elevated temperature, while it was measured after 3, 7, 14, 28, 56, and 90 days in room curing regime. Alkali-activated paste (AAP) was cured for 14 days in the oven.

## 3 Results and Discussion

Evolution of compressive strength of geopolymer concrete prepared with and without micro silica cured at elevated temperature is given in Fig. 67.1. The data indicates that the strength development in the geopolymer concrete prepared with 100% natural pozzolan was rapid up to 3 days of curing as compared to the mixes incorporating micro silica. As the curing progressed, strength development was slowed down in this mix and maximum ultimate strength was achieved after 14 days of curing. Although the strength development in the concrete mixes containing micro silica was rather slow at the onset of curing, it remarkably increased as the curing progressed. For instance, 7-day compressive strength in the concrete mixes containing 0%, 5%, 7%, and 10% micro silica was 30.72, 27.16, 20.72, and 21.08 MPa, respectively, whereas it increased to 31.25, 36.54, 38.62, and 36.2 MPa after 14 days of curing. This increase was about 1.72%, 34.53%, 86.38%, and 71.72% in the concrete mixes prepared with 0%, 5%, 7%, and 10% micro silica,

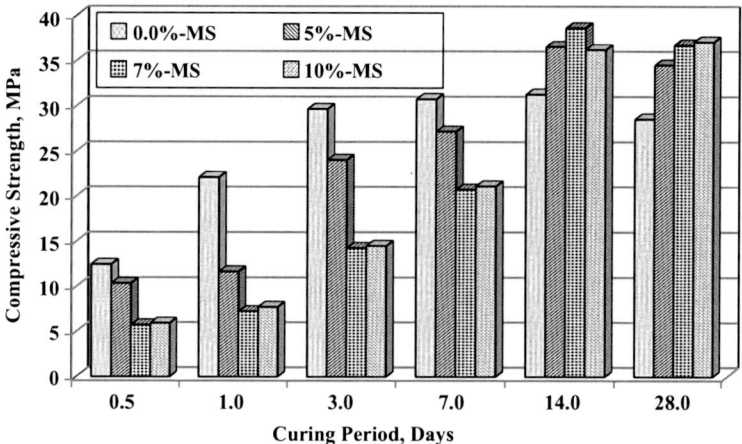

**Fig. 67.1** Compressive strength development in the AAC specimens

**Table 67.3** Strength development in the specimens cured at room conditions

| Mix | Micro silica, % | Compressive strength, MPa | | | | | |
|---|---|---|---|---|---|---|---|
| | | 3d | 7d | 14d | 28d | 56d | 90d |
| M0 | 0 | 4.04 | 7.22 | 11.22 | 12.26 | 12.54 | 12.86 |
| M1 | 5 | 5.26 | 6.38 | 8.04 | 6.9 | 10.36 | 14.56 |
| M2 | 7 | 3.24 | 3.54 | 4.66 | 12.32 | 18.08 | 20.68 |
| M3 | 10 | 4.16 | 5.22 | 6.78 | 10.48 | 15.78 | 18.74 |

respectively. Maximum strength was measured in the mix containing 7.5% micro silica after 14 days of curing.

Table 67.3 gives the compressive strength of concrete cured in the room condition. The highest strength was measured in the concrete specimens prepared with 7% replacement of natural pozzolan with micro silica after 90 days of curing. As the strength gain in these specimens was lower, room curing cannot be used to produce structural concrete with these constituent materials.

Figure 67.2 presents the SEM images of geopolymer paste incorporating micro silica cured at elevated temperature. Morphology of alkali-activated paste was not carried out for the specimens cured at room temperature. Microstructure of the specimen prepared without micro silica was very porous in nature with widespread cracks. As micro silica was added in the mixture, microstructure started to improve with reduction in pore volume. In the mixtures prepared by incorporating 7% and 10% micro silica, micrographs showed denser microstructure with homogenous gel-like matrix. Compared with other replacement levels of micro silica, the microstructure of 7% was uniform and denser with less unreacted particles and continuous gel matrix without clear particle boundaries. This more homogenous gel-like matrix without distinct voids most probably constituted of pure polymeric binder. However, 10% replacement level appeared to be excessive as unreacted particles with the clear

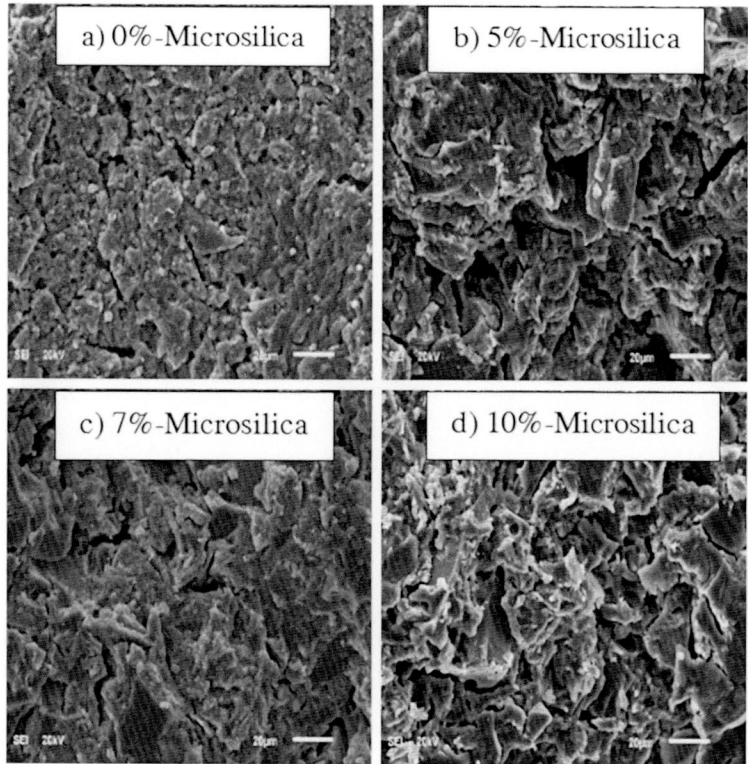

**Fig. 67.2** SEM of AAP specimens prepared with and without micro silica

particle boundaries were seen in the microstructure which did not beneficially help in improving the properties further. According to these findings, it can be stated that there was partial filling of voids in the specimens prepared with 5% micro silica, while the addition of 10% caused agglomeration of the particles which did not improve the microstructure and subsequently the strength of geopolymer matrix. These are the apparent observations of micrographs developed; however, pore size quantification is required for supporting these results. These results are consistent with the compressive strength data where 7% replacement of natural pozzolan with micro silica resulted in superior strength.

It can be inferred from these results that the initial sluggishness in the development of strength in the concrete mixes containing micro silica was due to the increase in the silica content for a given amount of alkaline solution which may have contributed in delayed polymerization, particularly in the specimens cured at elevated temperature [8]. However, as the curing continued due to the availability of additional soluble, silica accelerated the polymerization process which resulted in enhanced transformation of source materials to the polymeric gel [9]. This phenomenon was predominant in the mix containing 7% micro silica.

The findings of this study are consistent with the data presented elsewhere [10] in which 7% micro silica addition in the mixes prepared with source materials such as metakaolin and waste concrete produced from demolition showed enhancement in the properties such as strength and microstructure by creating more long-chain silicate oligomers. These results are indicative of the fact that in the process of polymerization due to the availability of soluble silica, particularly in the mixes incorporating micro silica, resulted in the formation of CSH or CASH gels along with the NASH products [11].

## 4 Conclusions

The objective of this work was to utilize locally available natural pozzolan and to enhance the properties of geopolymer concrete by incorporating micro silica. The result shows that:

- The strength of geopolymer concrete developed utilizing 100% natural pozzolan was rapid at the beginning of curing and slowed down as it continued.
- Incorporation of micro silica in the mixture of geopolymer concrete, particularly cured at elevated temperature, although resulted in sluggish strength development at the onset of curing, remarkably improved compressive strength as the curing continued.
- Compressive strength development in the geopolymer concrete was low at room curing even after 90 days.
- 7% addition of micro silica resulted in superior compressive strength in both the curing regimes as well as homogenous and denser microstructure with the enhanced transformation of source materials to polymeric gel.
- In summary, the NP-based geopolymer concrete developed is suitable for construction applications without incorporating any additives when cured at elevated temperature. However, partial replacement of NP with micro silica exhibited significant improvement in the properties of final product.

## References

1. Davidovits, J. (2008). *Geopolymer chemistry and applications*. Geopolymer Institute, France.
2. Kumar, S., Kumar, R., & Mehrotra, S. P. (2010). Influence of granulated blast furnace slag on the reaction, structure and properties of fly ash based geopolymer. *Journal of Materials Science, 45*(3), 607–615.
3. Li, C., Sun, H., & Li, L. (2010). A review: The comparison between alkali-activated slag (Si+ Ca) and metakaolin (Si+ Al) cements. *Cement and Concrete Research, 40*(9), 1341–1349.
4. Ankur, M., & Rafat, S. (2016). An overview of geopolymers derived from industrial by-products. *Construction and Building Materials, 127*, 183–198.
5. Villa, C., Pecina, E. T., Torres, R., & Gómez, L. (2010). Geopolymer synthesis using alkaline activation of natural zeolite. *Construction and Building Materials, 24*(11), 2084–2090.

6. Hardjito, D., Wallah, S. E., Sumajouw, D. M., & Rangan, B. V. (2004). On the development of fly ash-based geopolymer concrete. *Materials Journal, 101*(6), 467–472.
7. ASTM C150. (2005). Standard specification for Portland cement. In *Annual book of ASTM standards* (Vol. 4.01). Philadelphia: American Society for Testing and Materials.
8. Panias, D., Giannopoulou, I. P., & Perraki, T. (2007). Effect of synthesis parameters on the mechanical properties of fly ash-based geopolymers. *Colloids and Surfaces A: Physicochemical and Engineering Aspects, 301*(1), 246–254.
9. Oh, J. E., Monteiro, P. J., Jun, S. S., Choi, S., & Clark, S. M. (2010). The evolution of strength and crystalline phases for alkali-activated ground blast furnace slag and fly ash-based geopolymers. *Cement and Concrete Research, 40*(2), 189–196.
10. Khater, H. M. (2013). Effect of silica fume on the characterization of the geopolymer materials. *International Journal of Advanced Structural Engineering, 5*(1), 12.
11. Moon, J., Bae, S., Celik, K., Yoon, S., Kim, K. H., Kim, K. S., & Monteiro, P. J. (2014). Characterization of natural pozzolan-based geopolymeric binders. *Cement and Concrete Composites, 53*, 97–104.

# Chapter 68
# Performance of Steel Fiber-Reinforced High-Performance One-Part Geopolymer Concrete

Zahra Abdollahnejad, Tero Luukkonen, Paivo Kinnunen, and Mirja Illikainen

The high $CO_2$ emissions of ordinary Portland cement (OPC) production have led to increasing the efforts on developing eco-efficient alternative binders. Geopolymers are inorganic binders proposed as an alternative to OPC, which are mainly based on aluminosilicate by-products and alkali activators. Higher utilization of industrial waste materials, such as ceramic manufacturing waste, could be enabled by geopolymers. In ceramic industry, around 30% of raw materials end up in waste streams, and therefore, an attempt is made to recycle these materials. The ceramic wastes are rich in silicate and aluminate and have therefore high potential to be used in the geopolymeric concrete. In the present paper, the porcelain ceramic waste was used as 10% of total binder weight in substituting ground-granulated blast-furnace slag (GGBFS). The results showed that the resulting binders have comparatively high compressive strength ($\geq$60 MPa) and show brittle behavior, which is typical to inorganic binders with no fiber reinforcement. Microsteel fibers were used to improve the flexural performance of these binders at three different fibers by mass of binder (0.5%, 1%, and 1.5%). After curing, mechanical performances were investigated by measuring the compressive and flexural strength. The results showed that the addition of steel fibers significantly improved the flexural behavior. In addition, it was revealed that these fiber-reinforced binders had a deflection hardening behavior due to the bridging action of steel fibers.

Z. Abdollahnejad (✉) · T. Luukkonen · P. Kinnunen · M. Illikainen
University of Oulu, Fibre and Particle Engineering Research Unit, Oulu, Finland
e-mail: Zahra.Abdollahnejad@oulu.fi

# 1 Introduction

Ordinary Portland cement (OPC) contributes significantly to global anthropogenic $CO_2$ release so that the released carbon dioxide was approximately 5–7% of total global $CO_2$ in the 2000s [1, 2]. Consequently, the development of alternative low-carbon binders is recognized as an effective alternative to reduce the $CO_2$ emissions [3, 4]. Geopolymers are categorized as a subgroup of alkali-activated binders [5]. These binders are commonly activated by using the alkaline solutions. The impracticalities related to handling large amounts of viscous, corrosive, and hazardous alkali activator solutions have put pressure on developing one-part or "just add water" geopolymers that could be used similarly to OPC-based mixtures [6]. The dry ingredients are mixed with water in one-part geopolymers. Like concrete made from OPC, the plain alkali-activated binders presented a brittle behavior under the submitted loads. Introduction of the fiber improves mechanical properties and reduces the drying shrinkage rate in the reinforced mix compositions. One of the main drawbacks of slag compared to other supplementary cementitious materials is the high drying shrinkage rate, which can be significantly reduced using fibers. Fibers provide a bridging action in the fractured cross section, transferring tensile stresses across the crack. The crack-bridging efficiency of fibers depends on the various parameters such as fiber length, content, and type. Among the fibers, steel fiber was one of the earliest and most effective materials for improving the mechanical properties and impact resistance of cementitious composites [7]. Therefore, short-length steel fiber with superior mechanical properties was used to reinforce one-part slag/ceramic-based geopolymer mortars [8].

# 2 Materials and Methods

The designed geopolymer mortar composed of GGBFS, ground ceramic waste (porcelain), standard sand, and anhydrous sodium silicate ($Na_2SiO_3$) with a silica modulus of $SiO_2/Na_2O = 0.9$. Ceramic waste material was provided by IDO-Geberit, Finland, which produces sanitary porcelain. Slag was provided by Finnsementti (Finland). In this study, short-length steel fibers with three different contents of 0.5%, 1%, and 1.5% (mass of binder) was used to reinforce the mixtures. Table 68.1 lists the physical and mechanical properties of the microsteel fibers. Additionally, porcelain ceramic waste and short-length steel fiber used in this experimental study are shown in Fig. 68.1a, b, respectively. Moreover, the adopted flexural test setup and the fractured surface of the tested specimens are presented in

**Table 68.1** The physical and mechanical properties of short-length steel fiber

| Length to diameter | Elastic modulus (GPa) | Tensile strength (MPa) | Elongation at break (%) | Density (g/cm$^3$) |
|---|---|---|---|---|
| 333 | 200 | 2200 | 3 | 7.88 |

**Fig. 68.1** (a) Porcelain ceramic waste; (b) used short-length steel fiber

**Fig. 68.2** (a) Adopted flexural test setup; (b) fractured surface

**Table 68.2** Mix proportions of one-part slag/ceramic geopolymer

| Slag/binder | Sodium silicate/binder | Ceramic waste/binder | Water/binder | Sand/binder |
|---|---|---|---|---|
| 0.80 | 0.10 | 0.10 | 0.35 | 2.00 |

Fig. 68.2a, b, respectively. Table 68.2 mentions the material proportions of the geopolymeric mixtures.

In the batching process of mixtures, the dry ingredients (slag, ceramic waste, anhydrous sodium silicate, and fine aggregates) were mixed for 1 min. Then, water was added, and the obtained mixture was remixed for 3 min. Finally, the fiber-reinforced mixtures were prepared through gradually adding the fibers to the fresh mixture. Afterward, the mixtures were cast into the prismatic beams (40 × 40 × 160 mm), and then specimens were demolded after 24 h and then covered using the plastic bags for 28 days.

## 3 Test Procedures and Setups

Twelve prismatic beams with the dimension 40 × 40 × 160 mm were tested to assess the flexural performance of the reinforced and plain beams made with one-part slag/ceramic geopolymers under three-point bending (TPB) test in accordance with the ASTM C78 recommendation [9]. Prismatic beams evaluated under flexural load with a deflection rate of 0.6 mm/min.

Regarding the ASTM C349 recommendation, the compressive strength of mixtures obtained using the portions of prismatic beams broken in the flexure test [10]. The compressive load was measured with a load cell of 100 kN capacity and displacement rate of 1.8 mm/min. The compressive strengths were obtained from averaging six tested specimens.

## 4 Results and Discussion

Figure 68.3a indicates the effects of adding steel fiber on the 28 days compressive strength addition of fiber had no great impact on the compressive strength. The maximum reduction due to introduction of steel fibers was lower than 10%, as compared to the plain mixture. This reduction could be justified by increasing the porosity of the mixture. The minimum and the maximum compressive strengths were recorded around 60 and 65 MPa for the specimens reinforced with 1% and 0%

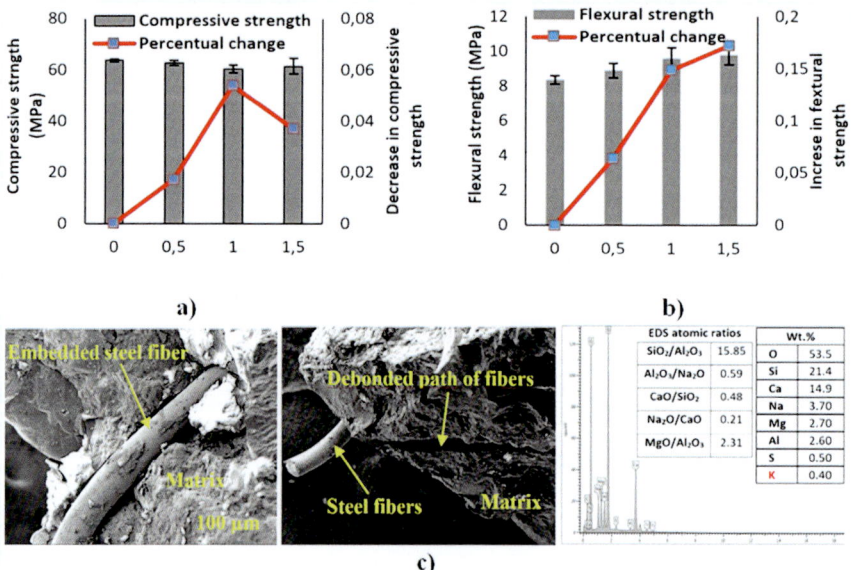

**Fig. 68.3** (a) Effect of steel fiber on the compressive strength; (b) effects of steel fiber on the flexural strength; (c) SEM image of the embedded steel fiber into the matrix and EDS analysis of the matrix

steel fiber, respectively. Figure 68.3b represents the influences of adding steel fibers on the flexural strength. The results revealed that the flexural strength was increased consistently by increasing fiber content. The minimum and the maximum flexural strengths were registered about 8.5 and 10 MPa for the specimens reinforced with 0% and 1.5% steel fiber, respectively. Based on the obtained results increasing the amount of steel fiber up to 1.5% resulted in recording about 20% increase of the flexural strength. This improvement was caused by forming hard bond between steel fibers and matrix; as shown in Fig. 68.3c, the hard bond could be derived from mechanical anchorage of steel fibers, as indicated in Fig. 68.3c. Moreover, the EDS atomic ratios also indicated that C-N-A-S-H gel formed in the matrix, which led to forming a dense matrix. Moreover, Fig. 68.2b indicates that the steel fibers were dominantly debonded from their surrounded matrix.

It is worth mentioning that previous findings demonstrated that slag/ceramic geopolymers indicated large drying shrinkage compared to other types of geopolymers, and this could form some cracks on the surface of specimen, which significantly affects mechanical and durability performances [11–13]. Visual monitoring confirmed that adding fibers controlled the drying shrinkage as no crack was observed on the specimens. Regarding the experimental results, the properties of the mixtures could be correlated through the empirical relations. In order to achieve this aim, the linear regression analysis was used, and the empirical relations were developed with high coefficient of determination ($R^2 = 0.83$). Figure 68.4a depicts the correlation among the compressive strength, flexural strength, and steel fiber content. With respect to the results, it was observed that the compressive strength was reduced by higher rate than the increase rate of the flexural strength. Moreover, it was revealed that there is a higher correlation between the flexural strength and steel fiber content, when compared to the compressive strength and fiber content. The results in Fig. 68.4b indicate that there is a linear relation between increasing the flexural strength and decreasing the compressive of the mixtures containing steel fiber, considering a high value of the coefficient of determination ($R^2 = 0.83$).

In Fig. 68.4, Fr is the flexural strength, c is fiber content, and Fc is the compressive strength.

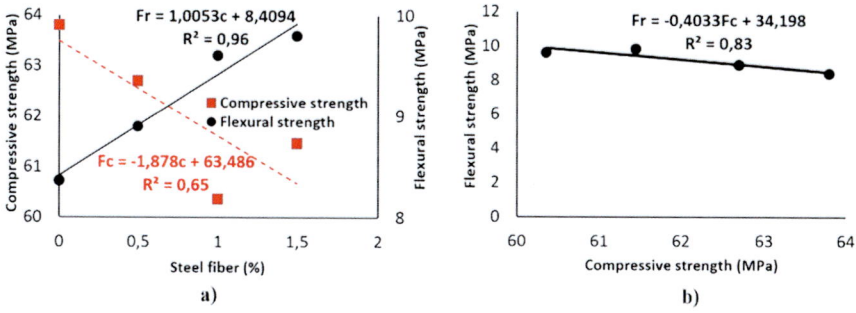

**Fig. 68.4** (a) Correlation among the compressive strength, flexural strength, and fiber content; (b) Correlation between the compressive and flexural strengths

## 5 Conclusions

This paper presents results from an experimental investigation on mechanical performances of the reinforced one-part slag/ceramic geopolymers with short-length steel fibers. Mechanical performances were addressed in terms of the compressive and flexural strength. Regarding the results, following remarks could be highlighted:

1. Forming C-N-A-S-H gel provided a dense matrix with reducing the porosity and strengthening the transition zone, which affects the bond properties at the fiber/matrix interface with increasing the friction pull-out behavior.
2. A linear correlation between mechanical characterizations and fiber was found.
3. Higher reduction rate was recorded for the compressive strength due to the addition of steel fibers, as compared to the increase rate of the flexural strength.
4. Regarding the results, the developed geopolymer is amenable to use in practical applications due to acceptable mechanical characterizations. However, further investigations on the durability performance are needed.

**Acknowledgments** This work was supported by the Finnish Funding Agency for Technology and Innovation (Tekes) [grant number 1105/31/2016] (project GEOBIZ).

## References

1. Ali, M. B., Saidur, R., & Hossain, M. S. (2011). A review on emission analysis in cement industries. *International Journal of Renewable and Sustainable Energy Reviews, 15*, 2252–2261.
2. Barcelo, M., Kline, J., Walenta, G., & Gartner, E. (2014). Cement and carbon emissions. *International Journal of Material and Structure, 47*, 1055–1065.
3. Flatt, R. J., Roussel, N., & Cheeseman, C. R. (2012). Concrete: An eco-material that needs to be improved. *International Journal of European Ceramic Society, 32*, 2787–2798.
4. Gartner, E., & Hirao, H. (2015). A review of alternative approaches to the reduction of $CO_2$ emissions associated with the manufacture of the binder phase in concrete. *International Journal of Cement and Concrete Research, 78*, 126–142.
5. Provis, J. L. (2014). Introduction and scope. In J. L. Provis & J. S. J. Van Deventer (Eds.), *Alkali activated materials, state-of-the-art report, RILEM TC 224-AAM* (pp. 1–9). Dordrecht: Springer.
6. Luukkonen, T., Abdollahnejad, Z., Yliniemi, J., Kinnunen, P., & Illikainen, M. (2018). One-part alkali-activated materials: A review. *Journal of Cement and Concrete Research, 103*, 21–34.
7. Havlikova, I., Merta, I., Schneemayer, A., Vesely, V., Simonova, H., Korycanska, B., & Kersner, Z. (2015). Effect of fiber type in concrete on crack initiation. *Applied Mechanics and Materials, 769*, 308–311.
8. Mastali, M., & Dalvand, A. (2016). Use of silica fume and recycled steel fibres in self-compacting concrete (SCC). *Construction and Building Materials, 125*, 196–209.
9. ASTM C78. (2016). *Standard test method for flexural strength of concrete (using simple beam with third-point loading)*. West Conshohocken: ASTM International.
10. ASTM C349-14. (2014). *Standard test method for compressive strength of hydraulic-cement mortars (using portions of prisms broken in flexure)*. West Conshohocken: ASTM International.

11. Abdollahnejad, Z., Mastali, M., Mastali, M., & Dalvand, A. (2017). A comparative study on the effects of recycled glass fiber on drying shrinkage rate and mechanical properties of the self-compacting concrete and fly ash/slag geopolymer concrete. *Journal of Materials in Civil Engineering*. https://doi.org/10.1061/(ASCE)MT.1943-5533.0001918.
12. Mastali, M., Dalvand, A., Sattarifard, A., & Abdollahnejad, Z. (2018). Effects of using recycled glass fibers with different lengths and dosages on fresh and hardened properties of self-compacting concrete (SCC). *Magazine of Concrete Research,* https://doi.org/10.1680/jmacr.17.00180
13. Askari, M. A., Mastali, M., Dalvand, A., & Abdollahnejad, Z. (2017). Development of deflection hardening cementitious composites using glass fibres for flexural repairing/strengthening concrete beams: Experimental and numerical studies. *European Journal of Environmental and Civil Engineering*. https://doi.org/10.1080/19648189.2017.1327888.

# Chapter 69
# Effect of Different Class C Fly Ash Compositions on the Properties of the Alkali-Activated Concrete

Eslam Gomaa, Simon Sargon, Cedric Kashosi, Ahmed Gheni, and Mohamed ElGawady

Class C fly ashes from two different coal-fired power plants were used to manufacture alkali-activated concrete. The workability and the compressive strength were studied in this paper. The workability was measured by the slump test. The compressive strengths at different ages of 1, 7, and 28 days were measured. Three different curing regimes including elevated heat curing at 70 °C, laboratory ambient curing at 23 ± 2 °C, and moist curing in the moisture room at 23 ± 2 °C were applied to identical mixtures to investigate the curing regime effects. Both types of fly ashes showed high slump of 212.5 and 225 mm. The results revealed that the compressive strength of the specimens that cured at the ambient or moist conditions increased with increasing the calcium content in the fly ash. However, the compressive strength of the specimens that were cured at 70 °C decreased when increasing the calcium content of the fly ash. The compressive strength of the concrete based on fly ash having higher calcium content at 28 days reached to 34.78, 36.62, and 51.46 MPa for oven-, ambient-, and moist-cured specimens, respectively. Furthermore, the compressive strength of the concrete based on fly ash having relatively lower calcium content at 28 days reached to 36.43, 30.79, and 47.45 MPa for oven-, ambient-, and moist-cured specimens, respectively.

## 1 Introduction

The production of one ton of the ordinary Portland cement released approximately one ton of $CO^2$ emissions into the atmosphere [1]. Therefore, finding an alternative binder system is an urgent need.

E. Gomaa · S. Sargon · C. Kashosi · A. Gheni · M. ElGawady (✉)
Missouri University of Science and Technology, Rolla, MO, USA
e-mail: elgawadym@mst.edu

In the last few decades, alkali-activated materials (AAMs) were developed as one of the potential alternatives to the ordinary Portland cement. AAMs are materials which are rich in silica ($SiO_2$) and ($Al_2O_3$) such as the fly ash (FA). These are dissolved in the existence of alkali activators (Alk) such as sodium hydroxide (SH) and sodium silicate (SS) in the presence of water (W) and elevated heat temperature to form networks of mineral molecules connected with covalent bonds in a geopolymerization process [2].

FA is used as the source of silica and alumina due to its availability around the world [3]. FA is classified per the ASTM C618-15 into class C and class F based on the chemical composition. Class F has higher aluminosilicate content. High calcium FAs harden at ambient temperature due to the formation of calcium silicate hydrate (CSH) and/or calcium aluminate silicate hydrate (CASH) [4]. The workability and compressive strength of alkali-activated concrete (AAC) are affected by several variables like water-to-fly ash ratio (W/FA), alkali solution-to-fly ash ratio (Alk/FA), sodium silicate-to-sodium hydroxide ratio (SS/SH), curing conditions, and calcium content in the FA [5]. In this research, class C FAs from two different sources were used to produce AAC. This research investigated the workability and compressive strength of AAC subjected to three different curing regimes: elevated heat curing at 70 °C (oven curing), laboratory ambient curing at 23 ± 2 °C (ambient curing), and moist curing in the moisture room at 23 ± 2 °C (moist curing).

## 2 Material and Methods

### 2.1 Fly Ash, Aggregate, and Alkali Activators

Class C FAs sourced from Labadie (LB) and Kansas City (KC) in Missouri State, USA, were used during this research. The major components of the two FAs (Table 69.1) are silicon dioxide, calcium oxide, and aluminum oxide.

Saturated surface dry condition (SSD) limestone and Missouri river sand were used in this study as the coarse aggregate and fine aggregate. The results of the sieve analysis of the coarse and fine aggregates showed that the grain sizes were in between the maximum and minimum limits of the ASTM C33-16.

Sodium silicate and sodium hydroxide were mixed in a 1:1 ratio by weight based to form the alkaline solution used in this study. A constant molarity of 10 M of SH solution was prepared prior to mixing. The commercial SS solution consisted of 55.9% water, 29.4% silicon dioxide ($SiO_2$), and 14.7% sodium oxide ($Na_2O$).

**Table 69.1** Chemical compositions of the fly ashes using X-ray fluorescence

|    | $SiO_2$ | $Al_2O_3$ | $Fe_2O_3$ | CaO   | MgO  | $Na_2O$ | $K_2O$ | $TiO_2$ | LOI[a] |
|----|---------|-----------|-----------|-------|------|---------|--------|---------|--------|
| LB | 36.89   | 13.99     | 3.52      | 36.96 | 4.80 | 1.62    | 0.62   | 0.87    | 0.50   |
| KC | 42.25   | 17.91     | 4.73      | 25.86 | 4.74 | 1.58    | 0.56   | 1.44    | 0.12   |

[a]*LOI* Loss on ignition

**Table 69.2** Mix design

| Material | CA | Sand | FA | SH | SS | W |
|---|---|---|---|---|---|---|
| Weight (kg/m$^3$) | 959 | 799 | 450 | 67.5 | 67.5 | 82.8 |

## 2.2 Mix Proportions and Mixing Procedure

Table 69.2 shows the mix design of the AAC. The alkali solution-to-water ratio was 0.3 by weight. The water-to-fly ash ratio was 0.35. The mixing procedure was as follows. The CA and sand were mixed together for 1 min before adding the FA and mixing for another 1 min followed by adding water gradually and mixing for 2 min. The alkali solution was prepared just prior to the mixing time. The mixed alkali solutions were added gradually for 5 min. Then the mixture was left to be mixed for another 5 min.

## 2.3 Curing Methods

The first regime consisted of an elevated heat curing in oven at 70 °C for 24 h after resting time of 2 h. Each cylinder was covered by an oven bag to prevent water evaporation from the samples. The second regime consisted of an ambient curing at laboratory temperature, i.e., 23 ± 2 °C. The cylinders were demolded 2 days after the casting and stored in oven bags until the testing day. The third regime consisted of moist curing in a moisture room which had a temperature of 23 ± 2 °C and relative humidity of 95 ± 5% after demolding the specimens.

## 2.4 Workability and Compressive Strength Tests

The slump test was performed according to ASTM C143-15 directly after the mixing. The compressive strength of the cylinders was conducted according to ASTM C39-17. The reported compressive strength results in the next section are the average of three concrete cylinders per each mixture. The compressive strength was measured at 1, 7, and 28 days of the mixing day.

## 3 Results

Both the concrete mixtures showed a high slump values. The slump values of the AAC based on LB and KC FAs were 187.5 and 200 mm, respectively. The higher slump value of the KC FA-based concrete is attributed to the lower calcium content compared to that of the LB FA. Additional nucleation sites for precipitation of the

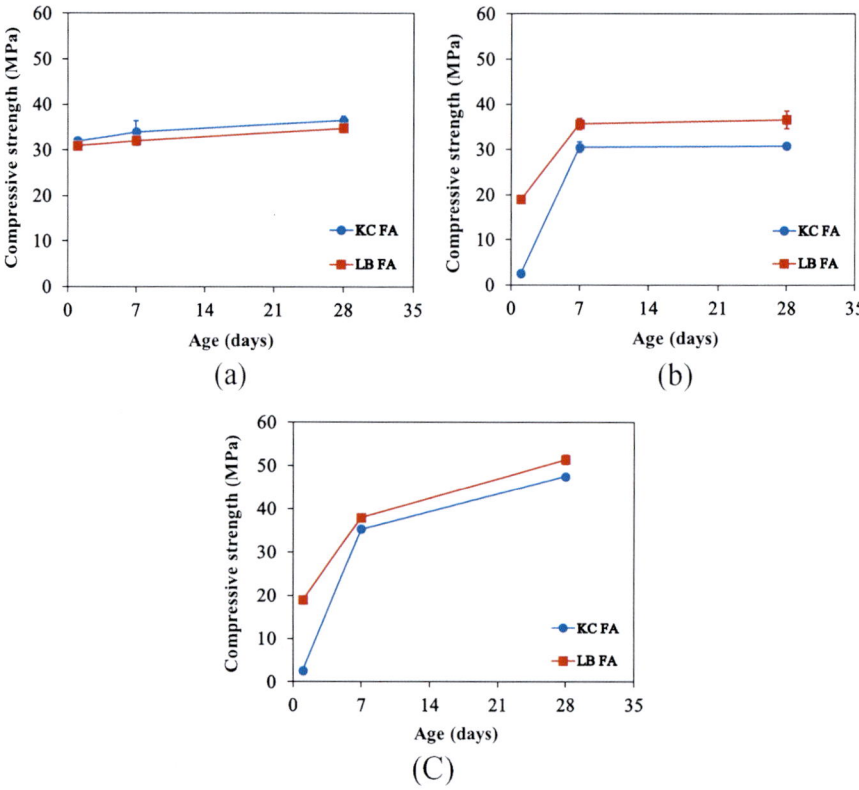

**Fig. 69.1** Compressive strength of (**a**) oven-, (**b**) ambient-, and (**c**) moist-cured alkali-activated concrete with time

FA dissolved species were created with higher calcium content which accelerated hardening process of AAC [6, 7].

The compressive strengths of the AAC mixtures are shown in Fig. 69.1. The compressive strengths of the AAC-based LB FA at 28 days were 34.78, 36.62, and 51.46 MPa for oven-, ambient-, and moist-cured specimens, respectively. The compressive strengths of the AAC-based KC FA at 28 days were 36.43, 30.79, and 47.45 MPa for oven-, ambient-, and moist-cured specimens, respectively. The compressive strength of the alkali-activated concrete based on LB FA showed higher compressive strength than the KC FA in case of ambient and moist curing regimes; however, it had lower compressive strength in the case of oven curing regime.

The difference in strengths was related to the reaction mechanism of the concrete. In the case of Portland cement-based concrete, the hydration reaction is responsible for gaining the strength over time, and the presence of the water during that reaction is necessary. However, in the alkali-activated systems based on FA class F or metakaolin which is rich in $SiO_2$ and $Al_2O_3$, the geopolymerization process is the main reaction in the system. The ambient- and moist-cured specimens have the

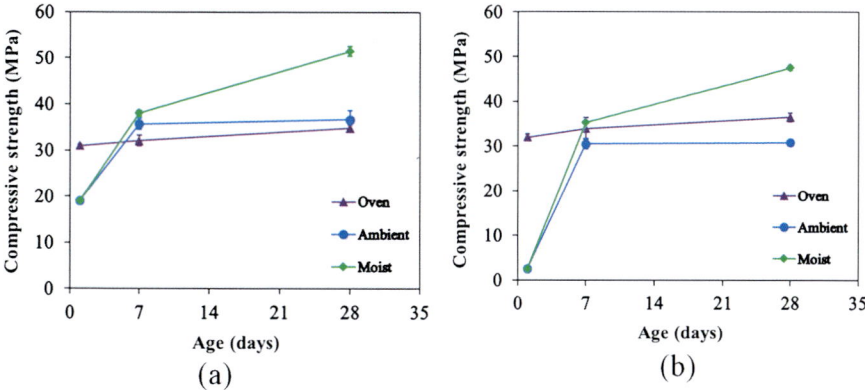

**Fig. 69.2** Compressive strength of alkali-activated concrete for different curing regimes with time of (**a**) LB and (**b**) KC FAs

ability to gain strength in the presence of high calcium content [4], where the calcium reacts with water to form CSH or CASH [8]. Therefore, the LB FA showed a higher compressive strength in the case of ambient and moist curing due to the presence of higher calcium content. However, the KC FA showed a higher compressive strength in the case of oven curing due to the presence of higher aluminosilicate content.

Figure 69.2 shows comparisons between the compressive strengths of the same mixtures subjected to different curing regimes. As shown in the figure, the compressive strength of the oven-cured specimens showed higher early (1-day) compressive strengths compared with the ambient- and moist-cured specimens for both types of FAs. However, the compressive strength of the moist-cured specimens was higher than the similar specimens cured in the oven and ambient regimes at 28 days. This is mainly due to predominant reaction mechanism. In the case of oven curing regime, the strength was gained due to the geopolymerization process which required a high temperature to take place [9]; however, the calcium was considered as a contaminant for the geopolymerization process and alters the microstructure [4].

The hydration reaction that forms CSH or CASH in the alkali-activated systems requires the presence of water and needs time to gain the full strength. So, the moist curing showed the highest compressive strength due to the presence of moisture during the curing period.

The ambient-cured specimens showed a lower compressive strength compared to those moist-cured specimens due to the lack of presence of moisture that is essential for the continuous hydration. Therefore, there was an improvement in the compressive strength from 1 to 7 days; then, the compressive strength remained constant beyond 7 days because the moisture content was consumed already at the early age and there was no free water available in the specimen to allow more hydration reaction to take place, so the reaction stopped.

## 4 Conclusion

The effect of different curing regimes on the compressive strength was investigated. Three different curing regimes were applied on AAC based on high-calcium FAs: an elevated heat (oven) curing at 70 °C, ambient curing at 23 ± 2 °C, and moist curing at 23 ± 2 °C and 95 ± 5% relative humidity regimes. Two FAs locally available obtained from two power plants located in Labadie and Kansas City, the state of Missouri, USA, were used in this study. The following conclusions can be drawn:

- The AAC can be cured at the ambient temperature in the presence of high calcium content (CaO).
- The oven curing regime is more beneficial if applied on a FA that has high aluminosilicate ($SiO_2 + Al_2O_3$) and low calcium contents.
- The ambient and moist curing regimes are more beneficial if applied on a FA that has low aluminosilicate and high calcium contents.
- The oven curing regime showed a higher early (1-day) compressive strength than the ambient and moist curing regimes.
- The moist curing regime showed the highest compressive strength among the other two curing regimes at 28 days due to the continuous reaction between the calcium in the presence of the silicon and aluminum and the moisture that produced more CSH or CASH which resulted in a higher compressive strength with time.
- The ambient curing regime showed an improvement in the compressive strength from 1 to 7 days. However, the strength remained constant due to the lack of free water that was consumed at the early age. Therefore, the hydration reaction stopped.

## References

1. Temuujin, J., Minjigmaa, A., Davaabal, B., Bayarzul, U., Ankhtuya, A., Jadambaa, T., & MacKenzie, K. J. D. (2014). Utilization of radioactive high-calcium Mongolian flyash for the preparation of alkali-activated geopolymers for safe use as construction materials. *Ceramics International, 40*(10), 16475–16483.
2. Davidovits, J., & Davidovics, M. (2008). Geopolymer: Room-Temperature Ceramic Matrix for Composites. In *Proceedings of the 12th Annual Conference on Composites and Advanced Ceramic Materials*: Ceramic Engineering and Science Proceedings, John Wiley & Sons, Inc. p. 835–841
3. Topark-Ngarm, P., Chindaprasirt, P., & Sata, V. (2014). Setting time, strength, and bond of high-calcium fly ash geopolymer concrete. *Journal of Materials in Civil Engineering, 27*(7), 04014198.
4. Temuujin, J. V., Van Riessen, A., & Williams, R. (2009). Influence of calcium compounds on the mechanical properties of fly ash geopolymer pastes. *Journal of Hazardous Materials, 167*(1), 82–88.

5. Gomaa, E., Sargon, S., Kashosi, C., & ElGawady, M. (2017). Fresh properties and compressive strength of high calcium alkali activated fly ash mortar. *Journal of King Saud University-Engineering Sciences, 29*(4), 356–364.
6. Pangdaeng, S., Phoo-ngernkham, T., Sata, V., & Chindaprasirt, P. (2014). Influence of curing conditions on properties of high calcium fly ash geopolymer containing Portland cement as additive. *Materials & Design, 53*, 269–274.
7. Lee, W. K. W., & Van Deventer, J. S. J. (2002). The effects of inorganic salt contamination on the strength and durability of geopolymers. *Colloids and Surfaces A: Physicochemical and Engineering Aspects, 211*(2), 115–126.
8. Guo, X., Shi, H., Chen, L., & Dick, W. A. (2010). Alkali-activated complex binders from class C fly ash and Ca-containing admixtures. *Journal of Hazardous Materials, 173*(1), 480–486.
9. Bakharev, T. (2005). Geopolymeric materials prepared using Class F fly ash and elevated temperature curing. *Cement and Concrete Research, 35*(6), 1224–1232.

# Chapter 70
# Effect of Curing Temperatures on Zero-Cement Alkali-Activated Mortars

Simon P. Sargon, Eslam Y. Gomaa, Cedric Kashosi, Ahmed A. Gheni, and Mohamed A. ElGawady

Three alkali-activated mortars (AAM), or what is called geopolymer, mix with different fly ashes sources were tested during the study. X-ray fluorescence was carried out on AAM samples to determine their chemical composition. Flowability and setting times of the AAMs were tested. Compressive strength was analyzed for five different temperatures of 30, 40, 55, 70, and 85 °C under five different time intervals of 4, 8, 16, 24, and 48 h. The compressive strength results indicate that the calcium content and ratio of silica to alumina played a pivotal role in the optimum curing conditions for each of AAMs.

## 1 Introduction

Fly ash is a toxic substance that is detrimental to the surrounding environment and the atmosphere around the world. The steady usage of coal over the last 100 years goes hand in hand with the steady increase in population. This has created a significant threat to a stabilized environment. To ease the pressure on the industry, an effective solution must be found to make use of this by-product.

Fly ash is currently used in part in many concrete mix designs due to its ability to improve durability [1]. However, this replacement is not enough to make an impact on its effect on the landfills. The 100% replacement of cement in mortar with fly ash and standard chemicals to achieve requisite strength is referred to as alkali-activated mortar. The chemicals are a combination of hydroxide and silicate compounds. Silicates act as a dissolution mechanism of $Si4+$ and $Al3+$ ions, and hydroxides

---

S. P. Sargon · E. Y. Gomaa · C. Kashosi · A. A. Gheni · M. A. ElGawady (✉)
Missouri University of Science and Technology, Rolla, MO, USA
e-mail: elgawadym@mst.edu

increase the solubility of these ions. They work together in various proportions to achieve strength in the alkali-activated mixes.

Alkali-activated mortars have been noted to have insufficient strength over a shorter period of time that is a requisite characteristic in the construction industry. To counteract this issue, elevated curing conditions can be investigated to be used in precast mortar and concrete. Further research is required to implement more confidence in the industry to use fly ash on larger scales up to a 100% replacement of ordinary Portland cement. The first step in finding a cost-effective solution is exploring the most efficient method of curing these 100% fly ash mixes [2].

In this study, mixes made of three different fly ashes were cured at five elevated temperatures and tested at five intervals of time. This helps us identify the most cost-effective curing conditions for each source.

## 2 Material Properties

### 2.1 Fly Ash

Fly ash was obtained from three coal power plants, Labadie, Jeffery, and Thomas hill in Missouri. These fly ashes will be referred to as FA1, FA2, and FA3. According to ASTM: C618 from X-ray fluorescence results, all three samples are class C fly ash.

### 2.2 Alkali Activators

Sodium hydroxide (NaOH) and sodium silicate type D ($Na_2SiO_3$) were used as the reagents.

### 2.3 Sand

Missouri river sand was used as fine aggregate and prepared according to ASTM gradation found in ASTM: C144.

## 3 Experimental Program

### 3.1 Mix Proportions

Three mixes were selected as in Table 70.1 to be made for each of the fly ashes. Based on the literature, the alkaline/fly ash ratio, water/fly ash ratio, and silicate/

**Table 70.1** Mix proportions

| Mix no. | 1 | 2 | 3 |
|---|---|---|---|
| Alkaline/fly ash ratio | 0.275 | 0.3 | 0.275 |
| Water/fly ash ratio | 0.38 | 0.38 | 0.4 |
| Silicate/hydroxide ratio | 1.0 | 1.0 | 2.0 |

hydroxide ratio were changed to study their impact on the mechanical properties of AAM.

## 3.2 Mixing Procedure

Sand and fly ash were mixed together to achieve a homogeneous condition initially. Water is added gradually over a period of 60 s at 136 rpm. Sodium silicate is mixed into the sodium hydroxide solution and added together gradually over a period of 5 min at 281 rpm. The mix is then allowed to continue at 281 rpm for an additional 5 min. The total mixing period upon adding chemicals for alkali activation is 10 min which, according to the previous literature, is optimum for higher strengths [3]. The fresh geopolymer mortar is placed into 50 × 50 × 50 mm brass cube molds according to ASTM C109.

## 3.3 Curing Procedure

Following tamping procedure according to ASTM C109, the mortar was allowed to rest at an ambient temperature of 22 °C for 2 h before placing it in the designated elevated temperature. The molds were covered in oven bags to reduce any loss of moisture from the mortar in the oven. All mixes were cured for time periods of 4, 8, 16, 24, and 48 h.

## 3.4 Setting Time and Flow

Setting time was tested according to ASTM: C191 using Vicat apparatus and flow was tested according to ASTM: C1437 using the flow table.

## 3.5 Compressive Strength

The compressive strength for each of the mixes was tested at periods of 4, 8, 16, 24, and 48 h at each curing temperature.

**Table 70.2** Chemical composition of fly ash

| ANID | $Na_2O$ (%) | MgO (%) | $Al_2O_3$ (%) | $SiO_2$ (%) | $P_2O_5$ (%) | $K_2O$ (%) | CaO (%) | $Fe_2O_3$ (%) |
|---|---|---|---|---|---|---|---|---|
| FA1 | 1.62 | 4.80 | 13.99 | 36.89 | 0.70 | 0.62 | 36.96 | 3.52 |
| FA2 | 1.85 | 8.00 | 17.44 | 37.94 | 0.71 | 0.39 | 28.79 | 3.67 |
| FA3 | 1.17 | 9.39 | 17.53 | 40.43 | 0.79 | 0.48 | 24.07 | 4.72 |

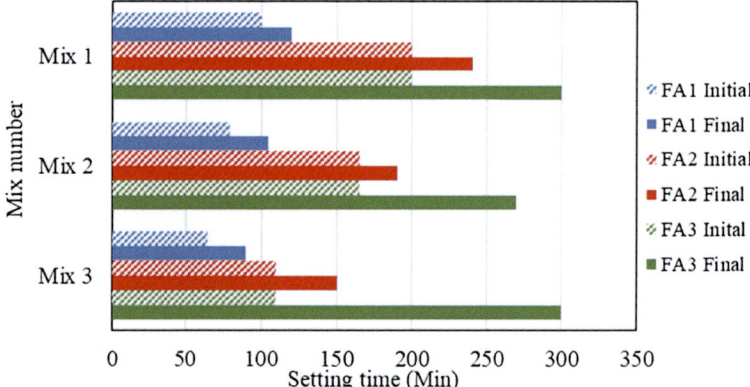

**Fig. 70.1** Final setting time of fly

## 4 Results and Discussion

### 4.1 XRF Analysis

X-ray fluorescence was carried out on all three fly ashes. This test gives us a holistic chemical composition of the fly ashes. It is considered to be a broad analysis of the sample. Loss on ignition was determined by firing samples at 700 °C for 2 h. From the acquired XRF results, displayed in Table 70.2, the three fly ashes were found to be class C, according to ASTM specifications for fly ash [4].

### 4.2 Setting Time and Flow

The initial and final setting time results are represented in Fig. 70.1. The calcium content was determined to be the decisive factor in the setting time. The lower calcium content had higher setting times as a result.

The flow results indicated in Fig. 70.2 suggest that increasing the ratio of alkali activators reduces the workability of the geopolymer mixes. Increasing the water content was found to improve workability.

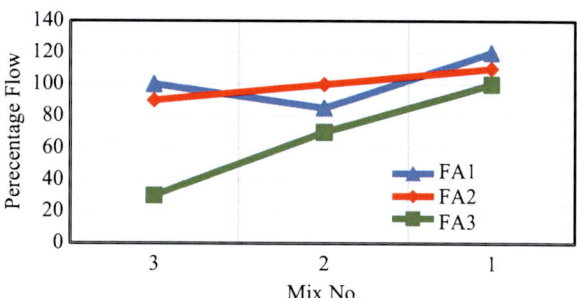

**Fig. 70.2** Flowability of fly ash

## 4.3 Compressive Strength

As shown in Fig. 70.3, the results for FA1, mix 1, indicate that curing at 40 °C for 48 h gave the best results. In the case of FA1 mix 2 and mix 3, the highest strengths were obtained from curing at 85 °C for 48 h. However, the difference in strength between the lower and higher temperatures was only 10%. This attributed to the higher calcium content of this particular fly ash to be cured at closer to ambient temperatures. The strength gain at lower temperatures indicates that they still had some potential for strength over a few more days. Increasing the amount of alkaline activator in the mix led to a shift of higher strengths for lower temperatures to a higher temperature of 85 °C. Increasing the ratio between silicate and hydroxide also performed similarly to increasing the alkaline activators. Fly ash 2 was found to gain preferable strength at 55 °C over a period of 48 h. However, 30 °C cured for 48 h had a less than 10% difference in strength at higher alkaline activator content. Altering between mixes 1, 2, and 3 had no effect on the behavior of the fly ash at different temperatures. Altering the silicate to hydroxide ratio improved its performance at higher temperatures.

Fly ash 3 suggests that the optimum temperature for curing was at 70 °C for 16 h. This would help reduce the cost of oven curing by more than half. The improved performance at elevated temperatures was caused due to a favorable ratio between $SiO_2/Al_2O_3$ and the reduced calcium content in the fly ash. Fly ash 3 did not perform well over a period of 48 h at lower temperatures. Increasing the silicate to hydroxide ratio reduced the overall strength gain of the mixes at higher temperatures.

Overall, fly ash 1 performed well at both higher and lower temperatures with around 10% difference in strength to stronger mixes, whereas fly ash 2 performed well at lower temperatures. Fly ash 3 was consistently weaker than the other two fly ashes due to its lower calcium content even though it had the highest ratio of silica to alumina. This contradicts some studies based on higher strengths at higher silica to alumina ratios giving higher strengths. Testing at lower temperatures over a very short interval was found to be ineffective for all the fly ashes as at least 16 h was needed to yield considerable results.

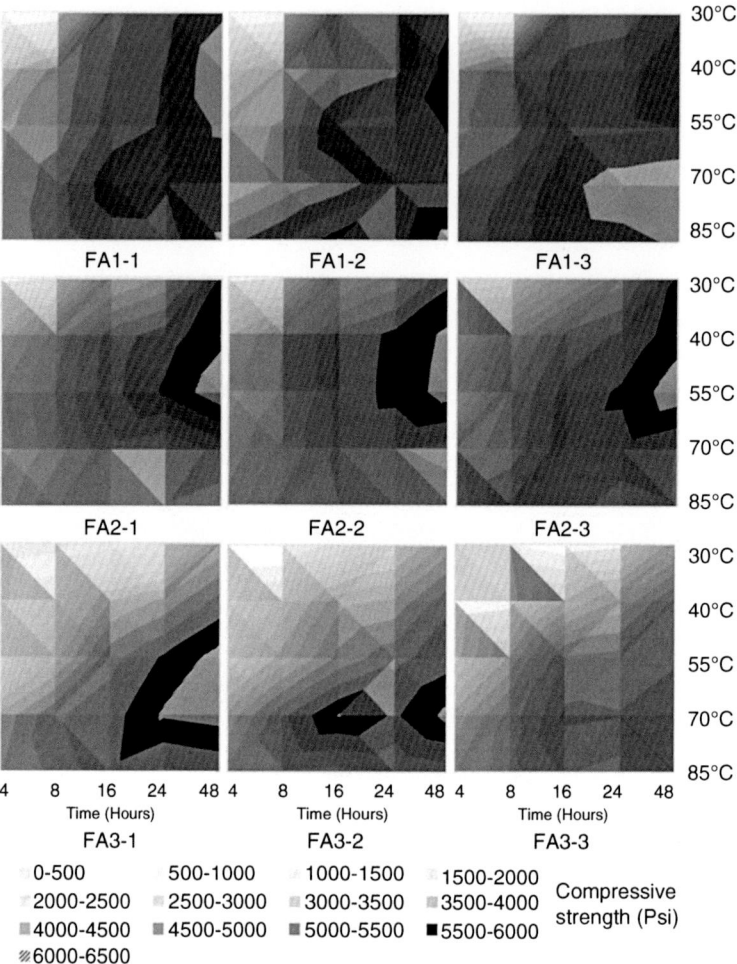

**Fig. 70.3** Compressive strength for six mixtures under varied curing temperatures and periods

## 5 Conclusion

Testing fly ashes from different sources at various time intervals and temperatures helped determine the optimum ratios of chemicals, curing times, and curing temperatures for particular chemical compositions. This will save time and money in the practical application of geopolymers in the field.

- Higher calcium content of the fly ash improved performances at lower temperatures.
- There was considerable strength gain over lower temperatures for all the mixes. The results indicate strength gain would continue even after 48 h.

- Fly ash with low calcium content had a favorable Si/Al ratio but was still consistently outperformed in strength contradictory to previous literature.
- Fly ash with high calcium content performed similarly well for all mix designs at different temperatures regardless of the water or chemicals added.

# References

1. Kong, D. L., & Sanjayan, J. G. (2010). Effect of elevated temperatures on geopolymer paste, mortar and concrete. *Cement and Concrete Research, 40*(2), 334–339.
2. Gomaa, E., Sargon, S., Kashosi, C., & ElGawady, M. (2017). Fresh properties and compressive strength of high calcium alkali activated fly ash mortar. *Journal of King Saud University-Engineering Sciences, 29*(4), 313–320.
3. Chindaprasirt, P., De Silva, P., & Hanjitsuwan, S. (2014). Effect of high-speed mixing on properties of high calcium fly ash geopolymer paste. *Arabian Journal for Science and Engineering, 39*(8), 6001–6007.
4. American Society for Testing and Materials. Committee C-9 on Concrete and Concrete Aggregates. (2005). *Standard specification for coal fly ash and raw or calcined natural pozzolan for use in concrete.* ASTM International, Pennsylvania, USA.

# Chapter 71
# Properties of PVA Fiber Reinforced Geopolymer Mortar

Wei Li and Hongjian Du

This study first investigates the mechanical and transport properties of PVA fiber reinforced geopolymer mortar, in which alkali-activated class F fly ash plays the role of binder instead of ordinary Portland cement (OPC). The fly ash was activated by 10 M NaOH solution and mortar cured in 80 °C for 3 days. The mix proportion was 0.4: 1: 2.75 for activator: fly ash: sand, by mass. PVA fiber was added at 2% by volume of the mortar mixture. Flexural toughness and compressive strength were determined at 7, 14, and 28 days. Transport properties of geopolymer mortar, including water sorptivity and resistance to chloride penetration, were also studied at 28 days.

## 1 Introduction

Concrete is the most widely used building material, in which cement is the most energy- and raw-material-intensive ingredient. For the manufacture of 1 t of cement, 5.31 GJ of energy is consumed, 1.25 t of $CO_2$ is emitted, and 1.63 t of raw materials are required. With the increasing concern with regard to climate change, environmental protection, and sustainable development of human society, there is an urgent need to develop a new generation of building materials that consume much less energy and raw materials and are more ductile and durable as well..

Geopolymer concrete in which alkali-activated binder is used as an alternative for cement has been receiving more interest recently [1–3]. The reaction of an alkali source with an alumina- and silica-containing solid precursor forms a solid material with performance comparable to hardened OPC, being able to bind other ingredients

---

W. Li (✉) · H. Du
Department of Civil and Environmental Engineering, National University of Singapore, Singapore, Singapore
e-mail: liwei@nus.edu.sg

in concrete. Compared with OPC, the alkali-activated binder can be entirely from waste-stream materials, such as fly ash, slag, and silica fume. At the same time, geopolymer also exhibits low ductility like ordinary cementitious composites. Geopolymer is promising not only because of its high greenness but also due to its lower cost. According to Duxson et al. [1], the cost of geopolymer is generally lower than OPC by a factor of 10–30%.

It is expected that fiber reinforced geopolymer (FRG) can combine the advantages of higher greenness and higher ductility. Previous research mainly focuses on the tensile strength, flexural toughness, and impact resistance of FRG. Durability of FRG is of great significance as well to increase its sustainability. However, the durability of FRG is seldom studied. Therefore, this study aims to investigate the mechanical and transport properties of FRG mortar. The results show the flexural toughness, water sorptivity, and resistance to chloride penetration of FRG, which is significant for better understanding the long-term performance of FRG.

## 2 Materials and Methods

Class F fly ash was alkali-activated by 10 M NaOH solution. The chemical composition of fly ash was shown in Table 71.1. Natural sand had a fineness modulus of 2.94 as per ASTM C136/C136M [4] and specific gravity of 2.65 g/cm$^3$ in accordance with ASTM C128 [5]. The PVA fiber used had a diameter of 39 μm and a length of 8 mm, which was provided by supplier. The mix proportion was 1: 0.4: 2.75 for fly ash: alkali-activator: sand, by mass. Mixing method is in accordance with BS EN 480-1 [6]. Fly ash and sand was dry mixed in Hobart mixer for 2 min, followed by the addition of alkali-activator. PVA fibers were added into the mortar mixture after wet mixing of 3 min. Superplasticizer was also used to increase the workability since fiber would decrease the flowability of mortar. The mortar mixture was poured into steel molds and cured in an oven at 80 °C (as shown in Fig. 71.1a) for 3 days and then continued to cure in ambient air until the test age.

**Table 71.1** Chemical composition of class F fly ash

| Chemical composition | Fly ash (%) | ASTM C618 [7] requirement (%) |
|---|---|---|
| $SiO_2$ | 38.9 | – |
| $Al_2O_3$ | 29.15 | – |
| $Fe_2O_3$ | 19.64 | – |
| $SiO_2 + Al_2O_3 + Fe_2O_3$ | 87.69 | 70.0 min |
| CaO | 2.5 | – |
| MgO | 2.1 | 5.0 max |
| $SO_3$ | 0.19 | 5.0 max |
| $Na_2O$ | 0.26 | 1.5 max |
| $K_2O$ | 0.48 | – |

(a) Specimens cured in 80 °C oven.  (b) Flexural test setup.

(c) Bridging effect of fibers at cracks. (d) Water sorptivity test.  (e) Specimens sprayed with 0.1 M AgNO$_3$ after RCPT.

**Fig. 71.1** Test photos of PVA fiber reinforced geopolymer mortar. (**a**) Specimens cured in 80 °C oven. (**b**) Flexural test setup. (**c**) Bridging effect of fibers at cracks. (**d**) Water sorptivity test. (**e**) Specimens sprayed with 0.1 M AgNO$_3$ after RCPT

Six 40 × 40 × 160 mm prisms were casted for flexural and compressive strength test, according to ASTM C348 [8] and C349 [9], respectively. Three 50 × 50 × 50 mm cubes were prepared for water sorptivity as per ASTM C1585 [10]. Three Φ100 × 50 mm cylindrical slides were used for rapid chloride penetration test in accordance with ASTM C1202 [11]. After 6 h of RCPT, the specimens were split and the chloride penetration depth was measured by spraying 0.1 M AgNO$_3$ solution. The chloride migration coefficient could be calculated from the penetration depth [12]. Pore size distribution of FRG at different ages was studied by using mercury intrusion porosimetry (MIP). MIP was conducted by using micrometitics AutoPore III, with a maximum pressure of 412.5 MPa. The minimum pore diameter reached under the maximum pressure was about 3.8 nm assuming a contact angle of 141.3° and a mercury surface tension of 0.485 N/m [13]. X-ray diffraction (XRD) patterns were also collected to examine the reaction at different ages; XRD spectra of the FRG was obtained by Shimadzu X-ray diffractometer (XRD-6000) with Cu Kα radiation at 40 kV and 30 mA with a scan speed of 0.5°/min between 2θ of 10° and 60°.

## 3 Results and Discussion

### 3.1 Mechanical Properties

The flexural and compressive strength was determined at 7, 14, and 28 days, as shown in Fig. 71.1b. The results are shown in Fig. 71.2. Both flexural and compressive strengths showed no increase with age of FRG, indicating that the strength

**Fig. 71.2** Flexural and compressive strength and toughness of FRG

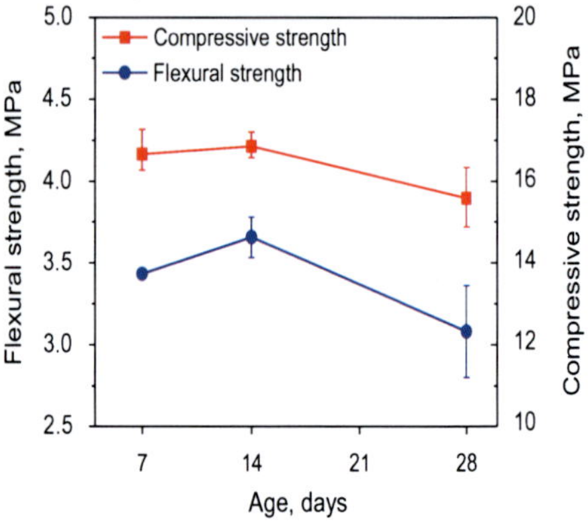

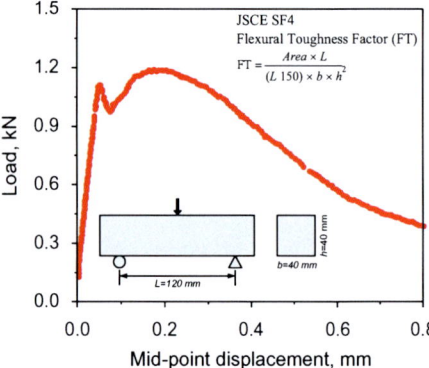

**Fig. 71.3** Representative flexural load–deflection curve for FRG

would stop gaining after 3 days of curing at elevated temperature. This might be due to the fast chemical reaction between alkali-activator and fly ash, that is, the polymerization process. The results are consistent with previous results by Hardjito et al. [14]. It should be noted that the compressive strength obtained in this study is lower than some results previously reported [1], in which more than 50 MPa was reported. The low strength may be caused by the lack of water glass in alkali-activator and the low activator to fly ash ratio [14].

Typical relationship between load and displacement at the middle point is shown in Fig. 71.3. The FRG prism did not break immediately after the first crack since the pull-out resistance of fibers at the crack can transfer the stress to the matrix and thus FRG can continue to carry the load, as shown in Fig. 71.1c. This has the effect of increasing the work of fracture, referred to as toughness and is represented by the area under the load–deflection curve [15]. The flexural toughness and flexural toughness factor (FT) according to JSCE SF4 method is shown in Fig. 71.4. Flexural toughness varied slightly with the age of FRG, similar as strength development.

**Fig. 71.4** Flexural toughness and flexural toughness factor of FRG

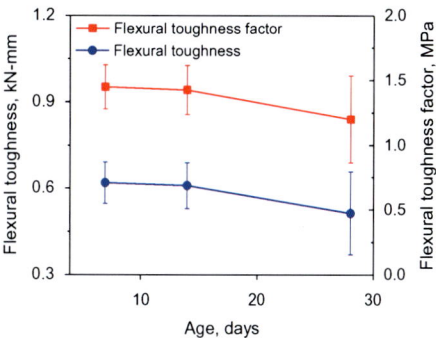

**Fig. 71.5** Water sorptivity of FRG

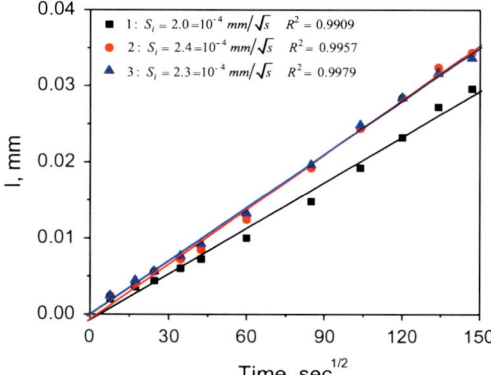

**Table 71.2** RCPT result for FRG

| RCPT, coulombs | x/d | Migration coefficient, $\times 10^{-12}$ mm$^2$/s |
|---|---|---|
| 6216 ± 752 | 0.18 ± 0.03 | 47.29 ± 8.81 |

x: chloride penetration depth; d: thickness of the specimen

## 3.2 Transport Properties

The water sorptivity test is illustrated in Fig. 71.1d and the results are shown in Fig. 71.5. The FRG exhibits an average initial sorptivity of $2.2 \times 10^{-4}$ mm/s$^{0.5}$. This result is lower than that obtained by Thokchom et al. [16], in which the fly ash based geopolymer showed a sorptivity between 3.0 and $6.89 \times 10^{-4}$ mm/s$^{0.5}$. The RCPT (Rapid Chloride Permeability Test) and chloride migration coefficient results are listed in Table 71.2. The white portion of the cross section in Fig. 71.1e represents the chloride penetration depth after RCPT. No available data from literature can be directly compared with the results obtained in this study. It is noted that although the migration coefficient was not high, the RCPT result indicated that FRG had a high permeability according to ASTM C 1202. This may be due to the high ionic concentration (i.e., Na$^+$, OH$^-$) in FRG.

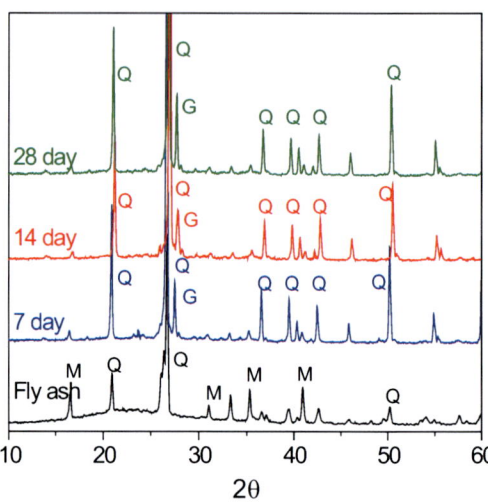

**Fig. 71.6** XRD patterns for FRG at different days

## 3.3 XRD

Patterns of FRG were collected at 7, 14, and 28 days, as shown in Fig. 71.6. The major crystalline phases in the class F fly ash were quartz ($\alpha$-$SiO_2$) and mullite ($Al_6Si_2O_{13}$). Regarding the XRD results for the alkali-activated sample, there was no significant change from 7 days to 28 days. The major crystalline phases in the alkali-activated sample were quartz ($\alpha$-$SiO_2$). Also a trace of calcium aluminum silicate (Gismondine) was detected while no zeolite Na-P1 was found.

## 3.4 Distribution of Pores

The pore structures of FRG were determined at 7, 14, and 28 days, as shown in Fig. 71.7. Consistent with the mechanical performance, there was no obvious change in the pore distribution with age of FRG, indicating that the polymerization process stopped after the oven curing. The critical pore diameter was determined to be 452 nm, larger than OPC mortar which normally has a critical pore diameter smaller than 100 nm [17]. The possible reason is the partial reaction between fly ash and alkali-activator that contains no water glass. This characteristic agrees with the compressive strength result.

## 4 Conclusions and Recommendations

Based on the experimental study of fiber reinforced geopolymer mortar, the following conclusions were reached:

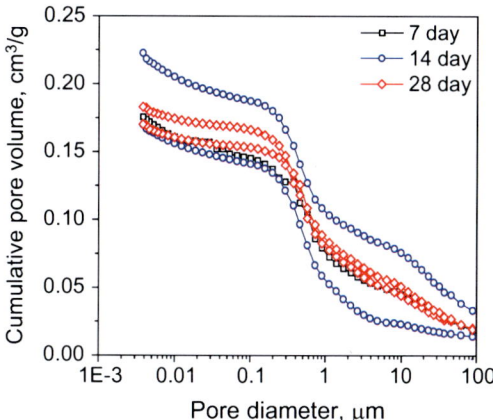

**Fig. 71.7** Pore distribution of FRG at different days

1. Mechanical properties only varied slightly with age, indicating that the polymerization process is quite fast. During the chemical reaction, Gismondine was found in the formed gel. PVA fiber can improve the flexural toughness due to the bridging effect at cracks.
2. Water sorptivity was determined to be lower than plain geopolymer. The critical pore diameter was 452 nm, larger than OPC mortar since there is lack of water glass in the alkali-activator.

To better understand and improve the performance of FRG, the following recommendations are proposed:

1. Water glass should be added into the alkali-activator, which is pure NaOH in this study. It is expected that the water glass can contribute to the polymerization reaction.
2. The effect of curing temperature and the alkali-activator to fly ash ratio should also be studied for higher performance of geopolymer.
3. Optimum fiber volume should be studied in future. The effect of fiber content on mechanical and transport properties should be considered to determine this optimum value.

# References

1. Duxson, P., Fermandez-Jimenez, A., Provis, J. L., Lukey, G. C., Palomo, A., & van Deventer, J. S. J. (2007). Geopolymer technology: The current state of the art. *Journal of Materials Science, 42*(9), 2917–2933.
2. Juenger, M. C. G., Winnefeld, F., Provis, J. L., & Ideker, J. H. (2011). Advances in alternative cementitious binders. *Cement and Concrete Research, 41*(12), 1232–1243.
3. Shi, C., Fernandez-Jimenez, A., & Palomo, A. (2011). New cements for the 21st century: The pursuit of an alternative to Portland cement. *Cement and Concrete Research, 41*(7), 750–763.

4. ASTM C136/C136M-14. (2014). Standard Test Method for Sieve Analysis of Fine and Coarse Aggregates, ASTM International, West Conshohocken, PA.
5. ASTM C128-15. (2015). Standard Test Method for Relative Density (Specific Gravity) and Absorption of Fine Aggregate, ASTM International, West Conshohocken, PA.
6. BS EN 480-1. (2014). Admixtures for concrete, mortar and grout. Testing methods. Reference 148 concrete and reference mortar for testing.
7. ASTM C618-17a. (2017a). Standard Specification for Coal Fly Ash and Raw or Calcined Natural Pozzolan for Use in Concrete, ASTM International, West Conshohocken, PA.
8. ASTM C348-14. (2014). Standard Test Method for Flexural Strength of Hydraulic-Cement Mortars, ASTM International, West Conshohocken, PA.
9. ASTM C349-14. (2014). Standard Test Method for Compressive Strength of Hydraulic-Cement Mortars (Using Portions of Prisms Broken in Flexure), ASTM International, West Conshohocken, PA.
10. ASTM C1585-13. (2013). Standard Test Method for Measurement of Rate of Absorption of Water by Hydraulic-Cement Concretes, ASTM International, West Conshohocken, PA.
11. ASTM C1202-17a. (2017a). Standard Test Method for Electrical Indication of Concrete's Ability to Resist Chloride Ion Penetration, ASTM International, West Conshohocken, PA.
12. Shi, C., & Wu, Y. (2005). Mixture proportioning and properties of self-consolidating lightweight concrete containing glass power. *ACI Materials Journal, 102*(5), 355–363.
13. Ramachandran, V. S., & Beaudoin, J. J. (1999). *Handbook of analytical techniques in concrete science and technology* (p. 964). Norwich: N.Y. Noyes Publications.
14. Hardjito, D., Wallah, S. E., Sumajouw, D. M. J., & Rangan, B. V. (2004). On the development of fly ash-based geopolymer concrete. *ACI Materials Journal, 101*(6), 467–472.
15. Mehta, P. K., & Monteiro, P. J. M. (2006). *Concrete: Microstructure, properties, and materials*. New York: McGraw-Hill.
16. Thokchom, S., Ghosh, P., & Ghosh, S. (2009). Effect of water absorption, porosity and sorptivity on durability of geopolymer mortars. *APRN Journal of Engineering and Applied Science, 4*(7), 28–32.
17. Halamickova, P., & Detwiler, R. J. (1995). Water permeability and chloride ion diffusion in Portland cement mortars: Relationship to sand content and critical pore diameter. *Cement and Concrete Research, 25*(4), 790–802.

# Chapter 72
# Thermal Performance of Fly Ash Geopolymeric Mortars Containing Phase Change Materials

M. Kheradmand, F. Pacheco Torgal, and M. Azenha

This paper reports experimental results on the thermal performance of fly ash-based geopolymeric mortars containing different percentages of phase change materials (PCMs). These materials have a twofold eco-efficient positive impact. On one hand, the geopolymeric mortar is based on industrial waste material. And on the other hand, the mortars with PCM have the capacity to enhance the thermal performance of the buildings. Several geopolymeric mortars with different PCM percentages (10%, 20%, 30%) were studied for thermal conductivity and thermal energy storage.

## 1 Introduction

Climate change-related effects are associated mainly to the emissions of energy sector [1]. This in turn is dependent on the population rise that will be responsible for a very high increase of electricity demand [2]. The energy needs of the building sector are expected to grow more than 70% [3]. The European Union adopted very ambitious plans in order to tackle this paramount problem. The European Energy Performance of Buildings Directive (EPBD) 2002/91/EC has [4] required that by the end of 2018, all new buildings must have a nearly zero-energy consumption. The use of innovative materials like PCMs will make it easier for this target to be met [5]. These materials use chemical bonds to store or release heat thus allowing for a reduction on the energy consumption. The capability to store or release thermal

---

M. Kheradmand · M. Azenha
University of Minho, Guimarães, Portugal

F. Pacheco Torgal (✉)
University of Minho, Guimarães, Portugal

University of Sungkyunkwan, Suwon, Republic of Korea
e-mail: torgal@civil.uminho.pt

energy in these materials depends strongly on the heat storage capacity, thermal conductivity, the melting temperature and the outdoor environment. Recently the use of PCMs on OPC-based materials has merit increased attention [6–8]. Also according to the Roadmap to a Resource Efficient Europe, all waste is to be managed as a resource [9]. This is a very important goal concerning the circular economy and zero-waste target [10]. Thus, materials that have the ability for the reuse of several types of wastes such as geopolymers must receive a special attention on this context [11]. This includes waste like fly ash because they are generated in high amount [12]. In this context this paper reports experimental results on the thermal performance of fly ash geopolymeric mortars containing PCMs because this is a research line that so far has received little attention.

## 2 Experimental Programme

The binder precursor was composed by 90% of fly ash and 10% of calcium hydroxide. Solid sodium hydroxide, which was obtained from commercially available product of Ercos, SA, Spain, was used to prepare the 12M NaOH solution. The chemical composition of the sodium hydroxide was 25%$Na_2O$ and 75%$H_2O$. The sodium silicate liquid was supplied by MARCANDE, Portugal. The chemical composition of the sodium silicate was 13.5%$Na_2O$, 58.7% $SiO_2$ and 45.2% $H_2O$. The fly ash was obtained from the PEGO Thermal Power Plant in Portugal, and it was classified as class F according to ASTM-C618 standard [13]. It was used as the base material for the production of the geopolymers. The chemical composition of the fly ash is presented in Table 72.1.

Calcium hydroxide was supplied by LUSICAL H100 and contains more than 99% CaO. The sand was used as inert filler provided from the MIBAL, Minas de Barqueiros, SA, Portugal. The superplasticizer was commercially available in polyacrylate from Acronal series, with a density of 1050 kg.m$^3$ from BASF. One type of organic microencapsulated PCM was considered: BSF26 with melting temperature of 26 °C. The properties of the selected PCM for this study are provided by the manufacturer and are presented in Table 72.2.

**Table 72.1** Major oxides in fly ash (%)

| $SiO_2$ | $Al_2O_3$ | $Fe_2O_3$ | CaO | MgO | $Na_2O$ | $K_2O$ | $TiO_2$ |
|---|---|---|---|---|---|---|---|
| 60.8 | 22.7 | 7.6 | 1.0 | 2.2 | 1.5 | 2.7 | 1.5 |

**Table 72.2** Properties of PCMs

| Operating temperature range (°C) | Latent heat of fusion (J/g) | Melting point (°C) | Apparent density at solid state (kg/m3) | Particle size distribution range (µm) |
|---|---|---|---|---|
| 10–30 | 110 | 26 | 350 | 5–90 |

The specimens were cured in laboratory conditions (25 °C and 65% relative humidity (RH)). The thermal conductivities of the mortars were determined in four representative measurements of each mortar formulation, using a steady-state heat flow metre apparatus (ALAMBETA, Model Sensora), following recommendation of ISO 8301:1991 [14]. Mortars were casted into cylinder moulds with diameter of 10 cm and length of 1 cm. Then thermal conductivity of the specimen is calculated based on heat conduction heat transfer theory according to [15]. Specific enthalpies of the mortars were determined by submitting the samples into the differential scanning calorimeter-DSC testing (Model NETZSCH 200 F3 Maia) and measure the corresponding heat fluxes at controlled environment. Based on this, the specific heat as a function of temperature can be obtained, and the specific enthalpy is determined. The DSC has an accuracy of ±0.2 °C for temperature measurements. All the specimens were tested within aluminium crucibles with volume of 40 μL under nitrogen ($N_2$) atmosphere with a flow of 50 mL/min. The specimens were weighted using analytical balance (model PerkinElmer AD-4) with accuracy of ±0.01 mg. Each specimen was sealed in the pan by using an encapsulating press. An empty aluminium crucible was considered a reference in all measurements. A heating/cooling rate of 5 °C/min was considered for all experiments.

## 3 Results and Discussion

The thermal conductivity results are presented in Table 72.3. The lowest thermal conductivity is noticed for the mixtures based on a sodium silicate/sodium hydroxide ratio of 2.5 and an activator/binder ratio of 0.7. For a similar activator/binder ratio, the reduction of the sodium silicate/sodium hydroxide ratio to 2.0 leads to highest results of thermal conductivity. Results show that the addition of PCM into the different mortars results in a consistent reduction of thermal conductivities. The highest reduction is noticed for the mixtures based on a sodium silicate/sodium hydroxide ratio of 2.0 and an activator/binder ratio of 0.7. The DSC curves for the testing of mortars at heating/cooling rate of 5 °C/min are shown in Fig. 72.1.

**Table 72.3** Thermal conductivity of mortars

| Formulations | Group name | Thermal conductivity (W/m. K) |
|---|---|---|
| 12M_2.5S/H_0.8A/B | A | 0.77 |
| 10PCM_12M_2.5S/H_0.8A/B | A | 0.70 |
| 20PCM_12M_2.5S/H_0.8A/B | A | 0.69 |
| 30PCM_12M_2.5S/H_0.8A/B | A | 0.44 |
| 12M_2.5S/H_0.7A/B_1.0SP | B | 0.52 |
| 10PCM_12M_2.5S/H_0.7A/B_1.5SP | B | 0.47 |
| 20PCM_12M_2.5S/H_0.7A/B_1.5SP | B | 0.44 |
| 30PCM_12M_2.5S/H_0.7A/B_1.5SP | B | 0.42 |
| 12M_2.0S/H_0.7A/B_1.0SP | C | 0.94 |
| 10PCM_12M_2.0S/H_0.7A/B_1.5SP | C | 0.90 |
| 20PCM_12M_2.0S/H_0.7A/B_1.5SP | C | 0.77 |
| 30PCM_12M_2.0S/H_0.7A/B_1.5SP_3.0 W | C | 0.35 |

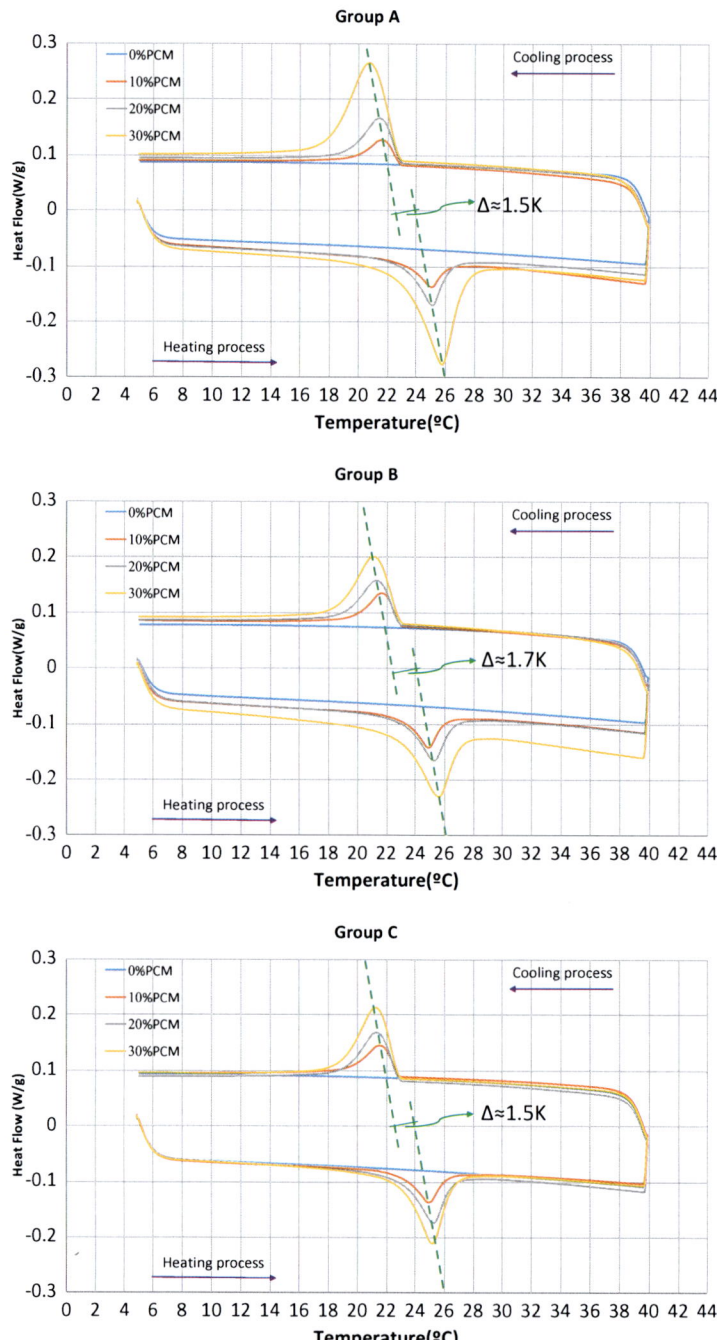

**Fig. 72.1** DSC curves of the alkali-activated mortars with and without PCM upon a cooling and a heating cyclic test with a rate of 5 °C/min: (**a**) group A based on 12M_2.5S/H_0.8A/B; (**b**) group B based on 12M_2.5S/H_0.7A/B_1.0SP; (**c**) group C based on 12M_2.0S/H_0.7A/B_1.0SP

Overall, the results suggested that the PCM peak temperature shifts in the direction of the imposed flux and further confirming higher peaks for mortars with higher mass fraction of PCM into the mix. The two dashed lines per graphic have been plotted by uniting the peak temperatures of all heating and all cooling thermograms: there is a clear linear relationship between the peak temperature of the thermogram and the percentage of PCM embedded. When the two dashed lines for a given group are compared, it can be noticed that they are approximately parallel and that the distance between them ranges from $\Delta \approx 1.5$ K to $\Delta \approx 2.5$ K. The difference observed in this hysteresis is known to depend on the internal thermal gradients upon the tested sample, which tend to lag or raise heat exchange from DSC. The average specific enthalpies for all the studied groups are $\approx$1.5 J/g, $\approx$2.5 J/g and $\approx$4 J/g for the mortar with 10%PCM, 20%PCM and 30%PCM, respectively.

# 4 Conclusions

The lowest thermal conductivity is noticed for the mixtures based on a sodium silicate/sodium hydroxide ratio of 2.5 and an activator/binder ratio of 0.7. For a similar activator/binder ratio, the reduction of the sodium silicate/sodium hydroxide ratio to 2.0 leads to highest results of thermal conductivity. Results show that the addition of PCMs results in a consistent reduction of thermal conductivities. The average specific enthalpies for all the studied groups are $\approx$1.5 J/g, $\approx$2.5 J/g and $\approx$4 J/g for the mortar with 10%PCM, 20%PCM and 30%PCM, respectively.

# References

1. King, D., Browne, J., Layard, R., O'Donnell, G., Rees, M., Stern, N., & Turner, A. (2015). A global Apollo programme to combat climate change. Centre for Economic Performance, London School of Economics and Political Science. http://cep.lse.ac.uk/pubs/download/special/Global_Apollo_Programme_Report. pdf. [Accessed 9 October 2015].
2. World Bank. (2014). World development indicators: Electric power consumption per capita in 2011. http://wdi.worldbank.org/table/5.11
3. Ürge-Vorsatz, D., Cabeza, L., Serrano, S., Barreneche, C., & Petrichenko, K. (2015). Heating and cooling energy trends and drivers in buildings. *Renewable and Sustainable Energy Reviews, 41*, 85–98.
4. European Union. (2010, June). Directive 2010/31/EU of the European Parliament and of the Council of May 19th, 2010 on the energy performance of buildings (recast). *Official Journal of the European Union.*
5. Pacheco-Torgal, F. (2014). Eco-efficient construction and building materials research under the EU Framework Programme Horizon 2020. *Construction and Building Materials, 51*, 151–162.
6. Jelle, B., & Kalnæs, S. (2017). Phase change materials for application in energy efficient buildings. In F. Pacheco-Torgal, C. G. Granqvist, B. P. Jelle, G. P. Vanoli, N. Bianco, & J. Kurnitski (Eds.), *Cost-effective energy efficient building retrofitting: Materials, technologies, optimization and case studies* (pp. 57–118). Cambridge: Woodhead Publishing.

7. Cunha, S., Aguiar, J., & Tadeu, A. (2016). Thermal performance and cost analysis of mortars made with PCM and different binders. *Construction and Building Materials, 122*, 637–648.
8. Cunha, S., Aguiar, J., & Pacheco-Torgal, F. (2015). Effect of temperature on mortars with incorporation of phase change materials. *Construction and Building Materials, 98*, 89–101.
9. European Commission. (2011). Roadmap to a resource efficient Europe. *COM(2011)* 571. EC, Brussels.
10. COM. (2014, July 2). 398 final. Towards a circular economy: A zero waste programme for Europe. Communication from the Commission to the European Parliament, the Council. *The European Economic and Social Committee and the Committee of the Regions.* Brussels.
11. Payá, J., Monzó, J., Borrachero, M., & Tashima, M. (2014). Reuse of aluminosilicate industrial waste materials in the production of alkali-activated concrete binders. In F. Pacheco-Torgal, J. Labrincha, A. Palomo, C. Leonelli, & P. Chindaprasirt (Eds.), *Handbook of alkali-activated cements, mortars and concretes* (pp. 487–518). Cambridge, UK: WoddHead Publishing.
12. American Coal Ash Association. (2016). https://www.acaa-usa.org/Publications/ Production-Use-Reports
13. ASTM C618 – 15. Standard specification for coal fly ash and raw or calcined natural pozzolan for use in concrete, *ASTM International*, West Conshohocken.
14. ISO:8301. (1991). Thermal insulation: determination of steady state thermal resistance and related properties, heat flow meter apparatus.
15. Lecompte, T., Le Bideau, P., Glouannec, P., Nortershauser, D., & Le Masson, S. (2015). Mechanical and thermo-physical behaviour of concretes and mortars containing phase change material. *Energy and Buildings, 94*, 52–60.

# Chapter 73
# Development of Fiber-Reinforced Slag-Based Geopolymer Concrete Containing Lightweight Aggregates Produced by Granulation of Petrit-T

Mohammad Mastali, Katri Piekkari, Paivo Kinnunen, and Mirja Illikainen

Using by-products as alternatives to ordinary Portland cement (OPC) is attracting growing attention in the sustainable construction material sectors. Alkali-activated binders have been proposed and emerged as an alternative to OPC binders, which seem to have acceptable mechanical and durability performances in addition to positive environmental impacts. These alternative binders, also named "geopolymer," use a wide range of aluminosilicate precursors, with differing availabilities, reactivates, costs, and $CO_2$ emissions. The usage of various materials results in obtaining the locally adaptable mix compositions, which establishes a broader toolkit. In this study, Petrit-T as a by-product from manufacturing sponge iron with fine particle-size distribution and rich in calcium was used to prepare the structural lightweight aggregates. Moreover, ground-granulated blast-furnace slag (GGBFS) as the binder was activated by a combination of sodium hydroxide and sodium silicate as the alkali activator. The effects of using different fiber types, including PVA, PP, and basalt, on mechanical properties were investigated. Mechanical properties were addressed in terms of the compressive and flexural strengths. The results showed that reinforcing the composition significantly affected the flexural performance. Moreover, it was revealed that using the granulated Petrit-T presented a lightweight concrete, with density $\rho \leq 1600$ kg/m³.

---

M. Mastali (✉) · K. Piekkari · P. Kinnunen · M. Illikainen
Fibre and Particle Engineering, Faculty of Technology, University of Oulu, Oulu, Finland
e-mail: Mohammad.mastali@oulu.fi

## 1 Introduction

Alkali-activated slag cements are one of the activated cementitious materials, which are produced by using waste product of steel companies and are rich in calcium content. These materials have shown great potential to reach high compressive strength at early age, great durability performance especially against chemical (acid and sulfate) attacks, rapid setting and hardening, low hydration heat, high-temperature resistance, and lower $CO_2$ emission compared to OPC [1–3]. These materials still have some drawbacks such as volumetric instability and the high drying shrinkage compared to OPC. Various solutions proposed to minimize the drying shrinkage in the alkali-activated slag cements such as using fibers as one of the most effective ways. Nevertheless, fibers could also have different effects on mechanical characteristics of the reinforced mix compositions, regarding the fiber type, length, and content. Therefore, an experimental investigation was executed in the present paper to clarify the effects of using different fibers on mechanical properties and forming cracks due to large drying shrinkage (visually). Moreover, lightweight aggregates were used in the mixtures to reduce the density. These aggregates were produced by alkali activation of Petrit-T by-product.

## 2 Experimental Program

The geopolymeric mortars were prepared through mixing slag, structural aggregates, and alkaline activator. The structural aggregates were prepared by mixing pre-wetted Petrit-T, sodium silicate with silica modulus of 2.5, and borax as retarders. To prepare the structural aggregates, the ingredients were granulated using an Eirich high-intensity mixer for 2–3 min, similarly to earlier alkali granulation. Afterward, the aggregates were covered with plastic bags for 24 h. Figure 73.1. depicts the particle-size distributions of the granulated Petrit-T. Moreover, the density of the aggregates was measured to be lower than 900 kg/m³. The lightweight structural aggregates should meet the requirements of the ASTM C 330 recommendation,

**Fig. 73.1** The granulated Petrit-T using Eirich high-intensity mixer

which bulk density should be less than 1120 kg/m³ for fine aggregates and less than 880 kg/m³ for coarse aggregates [4]. Thus, the granulated Petrit-T could be classified as the structural lightweight aggregates. The binder consisting of ground-granulated blast-furnace slag as the binder was activated by a combination of sodium hydroxide and sodium silicate as the alkali activator with a ratio of 2. The NaOH solution was also prepared through dissolving NaOH pellets in water with a molar concentration of 12 mol/L and cooled down to room temperature. Sodium silicate solution was used as a liquid sodium silicate with modulus of 2.5 (molar ratio $SiO_2/Na_2O = M$). The alkali activator to binder ratio was equal to 0.56.

Previous researches have demonstrated that the drying shrinkage of the alkali-activated slag binders could be much higher than cementitious compositions or other alkali-activated binders [5, 6]. Large drying shrinkage results in crack formation, which significantly affects mechanical and durability performances of the binders. Using fibers is one of the most effective solutions to control the drying shrinkage, since it not only reduces the drying shrinkage but also improves mechanical characteristics. Therefore, three different fibers (basalt, polypropylene (PP), polyvinyl alcohol (PVA)) were employed to reinforce the mixture in order to control shrinkage and enhance mechanical properties. The fiber content used to reinforce the mixtures was constant and equal to 2% of binder mass. The physical and mechanical properties of the fibers are listed in Table 73.1. Additionally, the used fibers are shown in Fig. 73.2a.

In the batching process of the mixtures, slag (530 g/L) was mixed with the lightweight aggregates (530 g/L) for 1 min. The alkali-activated solution (297 g/L)

**Table 73.1** The physical and mechanical properties of the fibers

| Fiber type | (L/D) | Elastic modulus (GPa) | Tensile strength (MPa) | Elongation at break (%) | Density (g/cm³) |
|---|---|---|---|---|---|
| PVA | 200 | 41 | 1600 | 6 | 1.3 |
| PP | 833 | 9.6 | 910 | <12 | 0.91 |
| Basalt | 333 | 100 | 4500 | 3.1 | 2.63 |

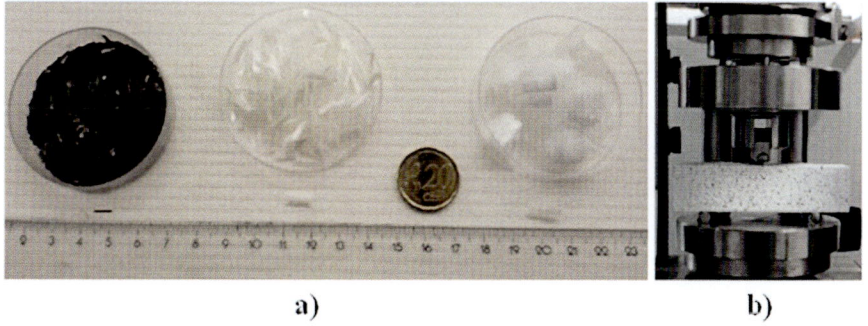

**Fig. 73.2** (a) The used fiber in reinforcing the mixtures; (b) device to execute the flexural strength test

was then added, and the obtained mixture was remixed for 2 min. Finally, the fiber-reinforced mixtures were prepared through gradually adding the fibers to the fresh mixture while mixing as long as desirable percentages of fiber are reached. During the preparation of the mixture, the fibers were gradually added to avoid balling. Afterward, the mixtures were cast into the prismatic beams (40 × 40 × 160 mm), demolded after 24 h and covered with plastic until the age of 28 days. Mechanical properties were characterized by measuring the compressive and flexural strength.

Prismatic beams were cast and tested under three-point bending (TPB) load conditions regarding the ASTM C78 recommendation [7]. The flexural load was submitted to the beams with a displacement rate of 0.6 mm/min, and the flexural load was measured using a load cell of 1000 kN capacity. For each mix composition, the flexural strengths represent the average of three replicated prismatic beams. The compressive test was carried out on the prisms broken in the flexural test according to the ASTM C349 recommendation [8]. The compressive load was imposed to the beams with a displacement rate of 1.8 mm/min. The compressive strengths of the compositions were obtained by averaging six prisms broken in the flexural test.

## 3 Results and Discussion

The influences of using fibers on mechanical properties are illustrated in Fig. 73.3. Regarding the results, the addition of fibers (regardless of the fiber type) increased the compressive strength. The addition of fiber increases the air voids in the mixture,

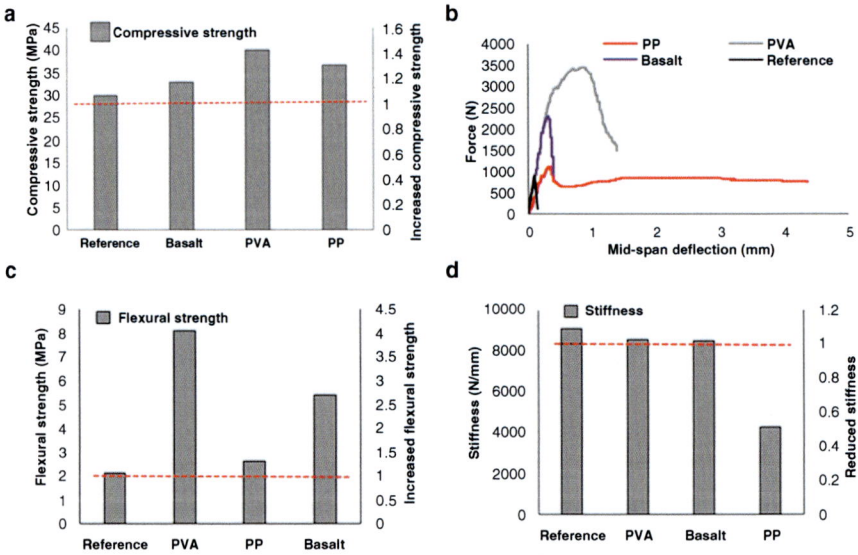

**Fig. 73.3** Effects of fiber on the (**a**) compressive strength, (**b**) force deflection, (**c**) flexural strength, (**d**) stiffness

and at the same time, fibers can limit the propagation of cracks. The interaction of these two effects could either increase or decrease the compressive strength. On the other hand, the addition of fibers could limit the further crack propagation, and subsequently, the compressive strength increased. As depicted in Fig. 73.3a, the minimum and maximum increase of the compressive strength compared to the plain mixture were recorded about 10% and 35% for the reinforced mixtures with basalt and PVA fibers, respectively. Fiber bridging action transfers the tensile stress across crack, and this phenomenon leads to increasing the flexural strength in the fiber-reinforced mixtures. According to the results obtained in Fig. 73.3b, c, it was revealed that the minimum and maximum improvements in the flexural strength compared to the plain mixture were measured around 25% and 4 times in the reinforced mixtures with PP and PVA fibers, respectively. After forming the initial crack, a deflection hardening behavior observed in the reinforced matrix with PVA, as showed in Fig. 73.3b.

The surface of the fibers is covered by hydrated particles due to the excellent bond of PVA fibers to matrix, as shown in Fig. 73.4a. Fewer hydration products covered the surfaces of PP and basalt fibers, which could be justified by the smoother surfaces of fibers compared to PVA fiber. The reinforced mixtures with basalt fibers indicated a brittle behavior after forming the initial crack, and they suddenly entered the softening phase, which may be caused by short length of basalt fibers. The reinforced mixtures with PP fibers presented an almost linear behavior to form the initial crack at the flexural stress of 2.5 MPa, followed by an almost perfectly plastic behavior up to a mid-span deflection of about 4.5 mm. This behavior could be explained by debonding of long PP fibers from its surrounded matrix. Regarding the result in Fig. 73.3d, except the mixtures containing PP fiber, introduction of PVA and basalt fibers had no great impact on reducing the stiffness. Using PP fibers decreased the stiffness more than 50%, which could be due to increase of porosity.

Furthermore, it is worth stating that no crack was visually monitored on the surface of the specimens reinforced with fibers, due to the drying shrinkage. This finding demonstrates that fibers were successful at mitigating the negative effects of drying shrinkage.

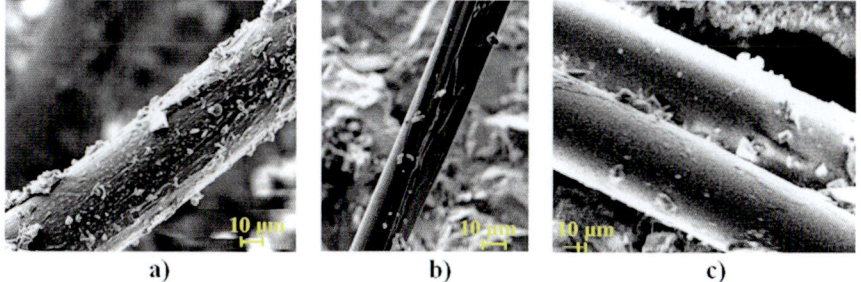

**Fig. 73.4** SEM images from the surfaces of (**a**) PVA fiber, (**b**) PP fiber, (**c**) basalt fiber

## 4 Conclusion

This paper presents results from an experimental investigation on mechanical properties of slag-based geopolymer concrete containing lightweight aggregates produced by alkali granulation of Petrit-T. The geopolymeric mixtures were reinforced by three different fibers, PVA, PP, and basalt. Mechanical properties were addressed in terms of their compressive and flexural strength. Regarding the results reported in this study, the following conclusions could be drawn:

1. Regardless of fiber type, the addition of fibers increased both compressive and flexural strengths.
2. PVA fibers indicated the greatest impact on enhancing mechanical properties compared to other fibers, which was caused by efficient bonding with the matrix.
3. After forming the initial crack, a hardening deflection behavior was observed for the reinforced mixtures with PVA fiber, while the reinforced mixtures with basalt and PP indicated softening and almost perfectly plastic behavior, respectively.
4. Using hybrid PVA/PP fibers could significantly reduce the material costs and increase the ductility of the fiber-reinforced mixtures. Therefore, further investigations on using hybrid fibers to obtain higher ductility with lower material costs are warranted using these materials.

**Acknowledgments** The study presented in this paper is a part of the research project "MINPET: mineral products from Petrit-T sidestream" that has received funding from the EIT RawMaterials, under grant agreement No. EIT/ EIT RAW MATERIALS/SGA2016/1.

## References

1. Ye, H., Cartwright, C., Rajabipour, F., & Radlinska, A. (2014). Effect of dying rate on shrinkage of alkali –activated slag cements. *4th international conference on the durability of concrete structures, 24–26 July 2014*, Purdue University, West Lafayette.
2. Collins, F. G., & Sanjayan, J. G. (1999). Workability and mechanical properties of alkali activated slag concrete. *Cement and Concrete Research, 29*, 455–458.
3. Atiş, C. D., Bilim, C., Çelik, Ö., & Karahan, O. (2009). Influence of activator on the strength and drying shrinkage of alkali-activated slag mortar. *Construction and Building Materials, 23*, 548–555.
4. ASTM C330/C330M-17a. (2017). *Standard specification for lightweight aggregates for structural concrete*. West Conshohocken: ASTM International.
5. Abdollahnejad, Z., Mastali, M., Mastali, M., & Dalvand, A. (2017). A comparative study on the effects of recycled glass fiber on drying shrinkage rate and mechanical properties of the self-compacting concrete and fly ash/slag geopolymer concrete. *Journal of Materials in Civil Engineering*. https://doi.org/10.1061/(ASCE)MT.1943-5533.0001918.
6. Askari, M. A., Mastali, M., Dalvand, A., & Abdollahnejad, Z. (2017). Development of deflection hardening cementitious composites using glass fibres for flexural repairing/strengthening concrete beams: Experimental and numerical studies. *European Journal of Environmental and Civil Engineering*. https://doi.org/10.1080/19648189.2017.1327888.
7. ASTM C78 / C78M-16. (2016). *Standard test method for flexural strength of concrete (using simple beam with third-point loading)*. West Conshohocken: ASTM International.
8. ASTM C349-14. (2014). *Standard test method for compressive strength of hydraulic-cement mortars (using portions of prisms broken in flexure)*. West Conshohocken: ASTM International.

# Chapter 74
# Applications of Geopolymers in Concrete for Low-Level Radioactive Waste Containers

Kyle D. Poolman and Deon Kruger

The purpose of this mini-thesis research paper is to investigate the applications of geopolymer concrete as a replacement for ordinary Portland cement (OPC) concrete currently used for the containment and transportation of low-level radioactive waste in South Africa. Geopolymers are materials which are high in aluminosilicates and use polymerisation reactions to bond together, as opposed to the conventional calcium-silicate-hydrate hydration reaction that occurs in traditional cement. The nuclear waste applications of geopolymers are investigated in the South African context – using locally available fine and coarse aggregates and class S fly ash originating from Kriel Power Station in Mpumalanga. Three geopolymer mix designs are tested – all of which use the same 1:2:2 mix proportion but with differing aggregate sizes, a common water/cement ratio of 0.5 and an 8 M or 12 M NaOH and $Na_2SiO_3$ activator. The aim of this research is to reach a density of greater than 2400 kg/m$^3$, compressive strength of 50 MPa and a tensile strength of 4.5 MPa using readily available industry materials. The mixes were cast and cured in an oven at 80° for 1 day and 3 days, respectively, after which the 8 M NaOH-only mix produced the best results at 11.9 MPa compressive strength, a 1.0 MPa tensile strength and a density of 2070 kg/m$^3$. These results have not met the South African Nuclear Energy Corporation (NECSA)'s standards as of yet, and the mixes will have to be further developed as a possible precast low-intermediate radioactive waste container solution.

---

K. D. Poolman · D. Kruger (✉)
University of Johannesburg, Gauteng, South Africa
e-mail: dkruger@uj.ac.za

## 1 Introduction

The revitalisation of South Africa's nuclear energy programme stems from an outcry for alternate energy production aside from coal – the government having already signed a deal in December 2016 for the construction of nuclear power stations that will generate around 10 GW of electricity, as well as the radioactive waste that accompanies it [1]. The nuclear plants currently operating in South Africa as well as those which are planned to be built will require the adequate handling of radioactive (or nuclear) waste. The development of improved materials, container designs and processes is thus vital to protect the public from the by-products of said *greener* energies.

Radioactive waste is a broad term referring to materials that either make contact with or radiate radioactivity [2]. Radioactive waste is classified into distinct categories, ranking from high-level waste (HLW) to low-level waste (LLW) as defined by the International Atomic Energy Agency (IAEA) [3]. Purposeful measures are taken when storing HLW due to the elevated temperatures and radioactivity of spent nuclear cores; however, LLW is put through a far less stringent cooling/observation period before being placed into reinforced concrete dry casks and transported to above or below ground storage facilities for monitoring and protection from the public. These dry containment casks are currently manufactured using ordinary Portland cement (OPC). Portland cement is commonly used as a binder in conjunction with various other substances to create blended cements; this is in an effort to reduce the output of carbon dioxide caused by OPC manufacturing. Cement extenders are a broad term used to classify these 'other substances' and usually comprise of a combination of *fly ash* and *ground-granulated blast-furnace slag* (GGBS) in South Africa.

Cement extenders may be used to create a polymerisation (hardening) reaction and thus take on the role of a primary binder. This reaction requires an alkali solution, in the place of pure water, coupled with materials that have a high proportion of silicates and aluminates [4]. The binders react with the solution and create polymeric bonds instead of the calcium-silicate-hydrate bonds formed in ordinary concrete. This type of reaction is referred to as a geopolymeric reaction, which further implies that the materials used are known as geopolymers. Geopolymeric molecular structures have been suggested as having surprisingly strong resistance to radioactive wave propagation due to their zeolitic microstructure and may have added benefits in use as a radioactive waste container material; however, this branch of research falls outside of the scope of this submission.

This research paper is part of what will be an ongoing partnership between the University of Johannesburg's Faculty of Engineering and the Built Environment and NECSA. The aim of this paper is to test a hypothesised geopolymer mix design against certain requirements for low-level radioactive waste containment and storage. This is done by limiting certain variables (such as activator type, type of extender and w/c ratio) and focusing on varying the geopolymer concrete mix design using only materials available in South Africa.

**Table 74.1** Relevant NECSA dry cask concrete container specifications [5]

| | |
|---|---|
| Binders | |
| *Aggregates* | |
| Coarse | Greywacke, quartzite and any quartzitic derivatives may *not be used* |
| | The preferred coarse aggregate is crushed granite |
| | Maximum ALD is 19 mm |
| Fine | Preferred to be from 'natural disintegration of rock', such as Klipheuwel sand |
| *Strength and durability* | |
| 28 days compressive | 50 MPa |
| 28 days tensile | 4.5 MPa |
| Density | >2400 kg/m$^3$ |
| Water/cement ratio | <0.4 |

## 2 Experimental Procedure

The procedure followed focuses on testing the mechanical properties of the fly ash-based geopolymer concrete against the requirements set out by NECSA – it must be noted that the requirements below may not suite geopolymer concretes and must be observed with a certain degree of experimental scepticism. The full list of OPC concrete requirements for dry cask containers is summarised in Table 74.1.

It must be noted that the experimental procedure is specifically designed in such a way so as to limit the variability involved with creating geopolymer concrete mixes and thus only test the effects of one aspect change at a time.

### 2.1 Experimental Tests

A compressive strength test is performed in accordance with SANS 5860:2006 and SANS 5863:2006 on each mix. This test will produce multiple 100 mm cube specimens per mix, which are tested for compressive strength gains over various curing periods [6, 7].

A tensile strength test is performed in accordance with SANS 5860:2006 and SANS 6253:2006 [8].

A density test is performed in accordance with SANS 6250:2006 on each mix [9].

### 2.2 Materials

The aggregates selected for this research have been selected for their commonality regarding use in the concrete industry; the fine aggregate is thus selected to be river sand, and the two sets of coarse aggregates are selected as 19 mm and 6.7 mm crushed granite. The binder material chosen is a *class S fly ash*, which originates from the Kriel coal power station in Mpumalanga, South Africa. Kriel power station is one of the few power stations in the country which has a dedicated coal mine supplying coal via conveyor belt to the power station. This should provide a material which is consistent should experiments continue for years to come. The fly ash used for experiments is notably fine as it has a 10%–11.5% (by mass) retention on a 45-micron sieve [10]. The chosen alkaline activator is a sodium hydroxide (NaOH) and sodium silicate ($Na_2SiO_3$) solution with molar weights of 39.987 g/mol and 122.06 g/mol, respectively, the solvent being distilled water and the solute being a combination of sodium hydroxide and sodium silicate pellets.

## *2.3 Mix Proportions*

Mix A is cast and cured in accordance with SANS 5861:2006 to reach an appropriate strength [11]. Mix B and Mix C are cast in accordance with SANS 5861:2006; however, the mixes will be cured in an oven at 80 °C in a sealed mould for the allotted time required. This is done whilst being sealed in plastic (to maintain the presence of water) [12]. A nominal mix proportion is adopted from the Concrete Institute of Southern Africa's recommendations, the binder/sand/stone proportion being 1:2:2. The mix designs are represented in Table 74.2.

## 3 Results and Discussion

The geopolymer mix results are represented in Table 74.3 below; the results seem to conclude that the mix design using only the 8 M NaOH produced the best strength development. However, the use of sodium hydroxide in conjunction with sodium

Table 74.2 Mix designs

| Mix A (OPC comparable) | 52.5 N ordinary Portland cement (5 kg)<br>19 mm crushed granite agg. (10 kg)<br>River sand (10 kg)<br>w/c = 0.5 |
|---|---|
| Mix B (geopolymer) | SFA binder (5 kg)<br>19 mm crushed granite agg. (10 kg)<br>River sand (10 kg)<br>w/c = 0.5 |
| Mix C (geopolymer) | SFA binder (5 kg)<br>6.7 mm crushed granite agg.(10 kg)<br>River sand (10 kg)<br>w/c = 0.5 |

**Table 74.3** Geopolymer results

| Mix design | Mean density ( kg/m$^3$) | Mean comp strength (MPa) | Mean tensile strength (MPa) |
|---|---|---|---|
| 12 M NaOH, Na$_2$SiO$_3$; 6.7 mm | 2151.5 | 4.2 | 0.2 |
| 12 M NaOH, Na$_2$SiO$_3$; 19 mm | 2051.8 | 7.4 | 0.2 |
| 8 M NaOH; 19 mm | 2070 | 11.9 | 1.0 |

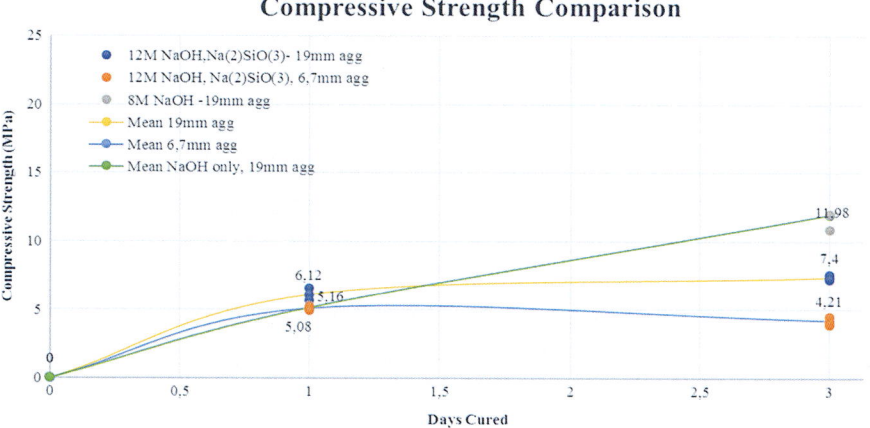

**Fig. 74.1** Result comparison

silicate may require a longer curing period to reach full strength and will thus be tested at 7 days and 14 days curing stages to ascertain the period required to reach the same strength as the 8 M sodium hydroxide mix.

Obtained compressive strength experimental results are shown in Fig. 74.1.

The creation of geopolymer mixes includes more than five variables (them being activator molarity, silicate/sodium ratios, mix design, water/cement ratio and NaOH/Na$_2$SiO$_3$ ratio) which present a difficulty in producing accurate results; developing a precast geopolymer which meets the NECSA mechanical strength requirements may thus take further developments and research.

## 4 Conclusion

The water/cement ratio used for the mix exceeds NECSA's requirements, as fly ash requires a large amount of liquid (due to its fineness) to be workable, especially as the sole binder constituent. The mixtures are slightly workable at this w/c ratio and thus should be increased in future tests. The geopolymers have significant strength gain within the first 3 days; however, the best mix thus far is the 8 M NaOH and

19 mm aggregate mixture which outperforms the others in tensile and compressive strength. The density of all three mixes is, however, below the NECSA density requirement (2400 kg/m$^3$) and thus requires a relook at the nominal mix design (1:2:2) that has been used. The use of geopolymers as a low-level concrete waste container material requires further testing and experimentation; however, it must also be noted that the specialised curing scheme, care and materials required to create this unique concrete are not a hindrance – as it is quite possible to reproduce this process in a precast LLW container factory once a sufficient mix design has been concluded. Producing a consistent geopolymer material for the benefit of handling the by-products of nuclear energy should therefore be pursued in the future.

## References

1. Kemm, K. (2017). *Engineering News*, [Online]. Available: http://www.engineeringnews.co.za/article/nuclear-build-initiative-now-rolling. [Accessed March 2017].
2. Horiuchi, C. (2007). Managing nuclear waste. In *Handbook of globalizaton and the environment* (pp. 381–399). San Francisco: University of San Francisco.
3. International Atomic Energy Agency (IAEA). (2005). Upgrading of near surface repositories for radioactive waste, *Technical Reports* (Vol. I, no. 433).
4. Adam, A. A., Molyneaux, T. C. K., Patnaikuni, I., & Law, D. W. (2010). Strength, sorptivity and carbonation in blended OPC-GGBS, alkali activated slag, and fly ash based geopolymer concrete. In *ISEC-5 2009* (pp. 563–568). CRC Press, London, UK.
5. Koeberg Nuclear Power Station. (2005). *Manufacturing of TES concrete waste disposal drums*. Cape Town.
6. SABS Standards Division. (2006). *SANS5860:2006 – Concrete tests*. Pretoria: SABS.
7. SABS Standards Division. (2006). *SANS 5863:2006 strength tests* (1.1 ed.). Pretoria: SABS.
8. SABS Standards division. (2006). *SANS 6253:2006* (1.1 ed.). Pretoria: SABS.
9. SABS Standards Division. (2006). *SABS 6250:2006 concrete density test* (1.1 ed.). Pretoria: SABS.
10. Ulula Ash (Pty)Ltd. (2016). *Test certificate for class S fly ash (SFA)*. Pretoria: PPC Group Laboratory Services.
11. SABS Standards Division. (2006). *SANS 5861. 2006. (Part 1–3)* (1.1 ed.). Pretoria: SABS.
12. Lee, S., Van Riessen, A., & Chon, C. M. (2016). Benefits of sealed-curing on compressive strength of fly ash-based geopolymers. *Materials, 9*(7), 598.

# Part VII
# Fiber Reinforced Polymers (FRP)

# Chapter 75
# Microstructure and Mechanical Property Behavior of FRP Reinforcement Autopsied from Bridge Structures Subjected to In Situ Exposure

Wei Wang, Omid Gooranorimi, John J. Myers, and Antonio Nanni

This paper examines the microstructure and mechanical property behavior of FRP reinforcement extracted from several FRP-reinforced concrete bridges built more than a decade ago. These include bridges constructed in Missouri and Texas. In particular, the following tests were performed including SEM, FTIR, EDX, glass transition temperature ($T_g$), and short shear test (SBS). Results from the evaluation demonstrated that the FRP bars are performing at a level with no observed degradation. The paper collects data to study the implications of the current ACI code knockdown factors relative to field-related data.

## 1 Introduction

An overwhelming number of studies have focused on FRP durability with exposure to simulated concrete pore water solution at elevated temperature environmental conditions. These tests are usually conducted in an alkaline environment because the pH value of a concrete environment is roughly 12 to 13. However, this accelerated aging alkaline environment is different with that presented in field concrete members [1, 2]. Therefore, monitoring the characteristics of existing projects is considered as a real demonstration of FRP reinforcement durability. Several core samples with encapsulated GFRP bars were extracted from the Southview Bridge and Walker Bridge in Missouri and the Sierrita de la Cruz creek bridge in Texas after 15 or more

---

W. Wang · J. J. Myers (✉)
Department of Civil, Architectural, and Environmental Engineering, Missouri University of Science and Technology, Rolla, MO, USA
e-mail: jmyers@mst.edu

O. Gooranorimi · A. Nanni
Department of Civil, Architectural, and Environmental Engineering, University of Miami, Coral Gables, FL, USA

years of in-service exposure. All of the applied GFRP bars were produced by Hughes Brothers Inc. of Seward, Nebraska, prior to field construction. SEM, EDX, FTIR, $T_g$ measurement, and SBS of GFRP samples were tested.

## 2 Preparation of GFRP Samples

Two #6 (0.75 in. diameter) GFRP bars were extracted from Southview Bridge in Rolla, Missouri, USA. One #5 (0.625 in. diameter) and two #6 GFRP bars were collected from Sierrita de la Cruz Bridge near Amarillo, Texas, USA. Four #2 GFRP bars were obtained from Walker Bridge in Rolla, Missouri, USA. Figures 75.1 and 75.2 illustrate the extracted GFRP bars from different concrete cylinders.

Small slices were cut to prepare the samples of SEM, EDS, and FTIR. They were ground carefully using sandpaper and micro cloth.

**Fig. 75.1** The extracted GFRP samples from Southview Bridges (left) and Sierrita de la Cruz Bridge (right)

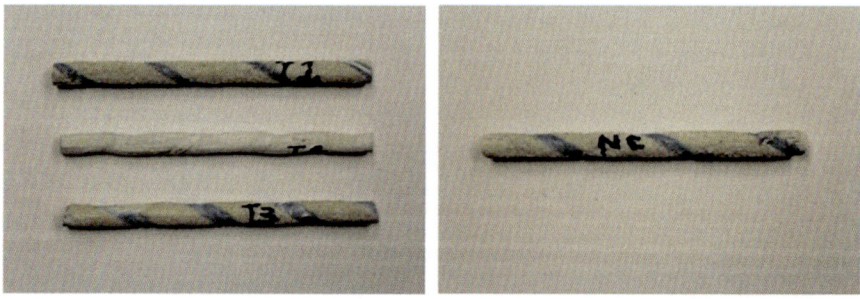

**Fig. 75.2** The extracted GFRP samples from core1 (left) and core2 (right) of Walker Bridge

## 3 Experimental Results and Discussions

### 3.1 SEM and EDS Analyses

The samples were scanned at different levels of magnification, and images were taken at random locations. More attention was focused on the areas in the vicinity of the analyzed GFRP bars and individual glass fibers in the epoxy because of possible deterioration. The representative SEM micrographs of the GFRP samples from these three bridges are exhibited in Figs. 75.3 and 75.4.

No evidence of deterioration on the glass fibers was observed from these micrographs. Individual fiber still maintained integrity. Figures 75.5, 75.6, and 75.7 show the results of EDS analysis. The $y$-axis presents the counts (number of X-rays

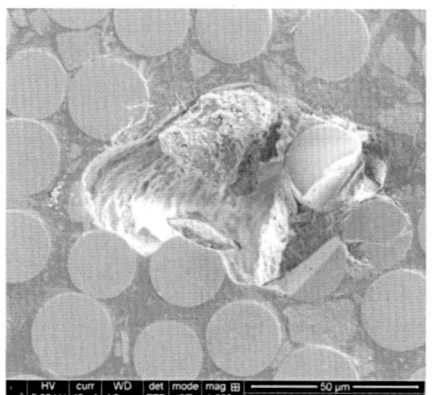 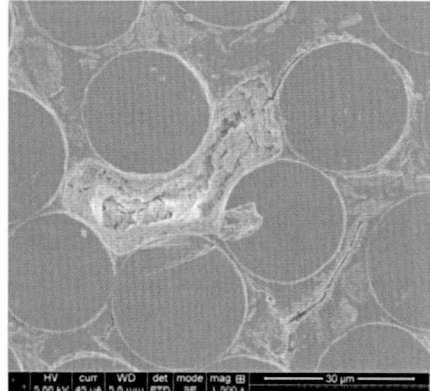

**Fig. 75.3** Glass fibers from Southview Bridge (left) and Walker Bridge (right) at magnification levels of 1000 × and 1500 ×, respectively

**Fig. 75.4** Glass fibers at magnification levels of 1500 × (Sierrita de la Cruz Bridge)

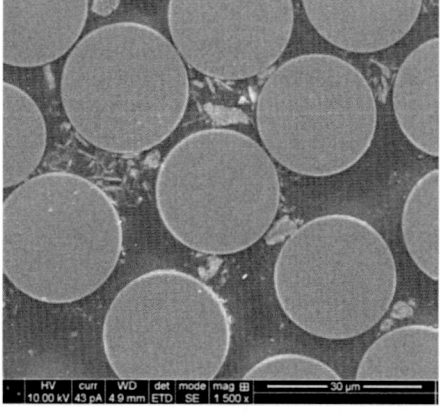

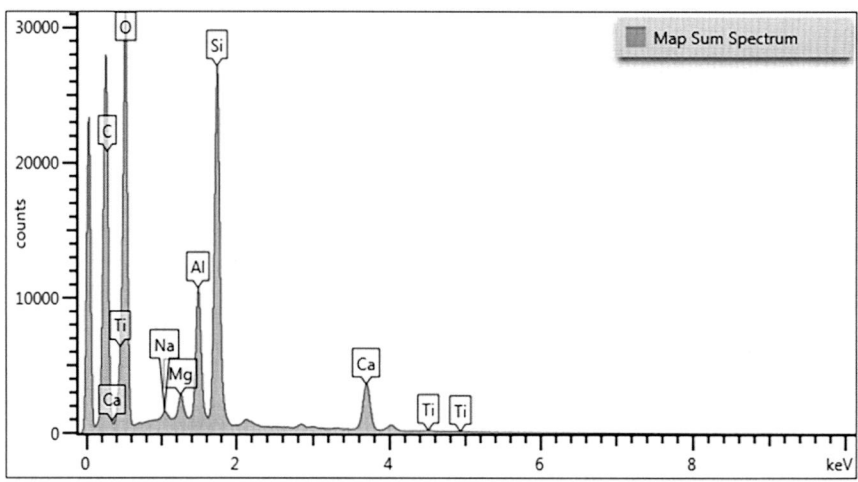

**Fig. 75.5** EDS analysis (Southview Bridge)

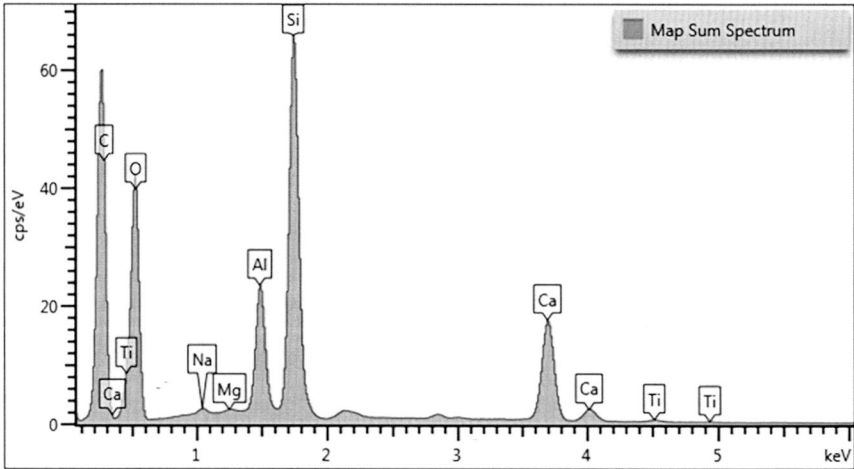

**Fig. 75.6** EDS analysis (Walker Bridge)

received and processed by the detector) and the *x*-axis shows the energy level of those counts.

There was no change in chemical composition of fiber and resin matrix when comparing the results of the extracted samples in this study and the findings of the in-service and control specimens from the report of UM [3]. The silica from glass fibers was not dissolved in the alkaline environment of concrete after several years of service.

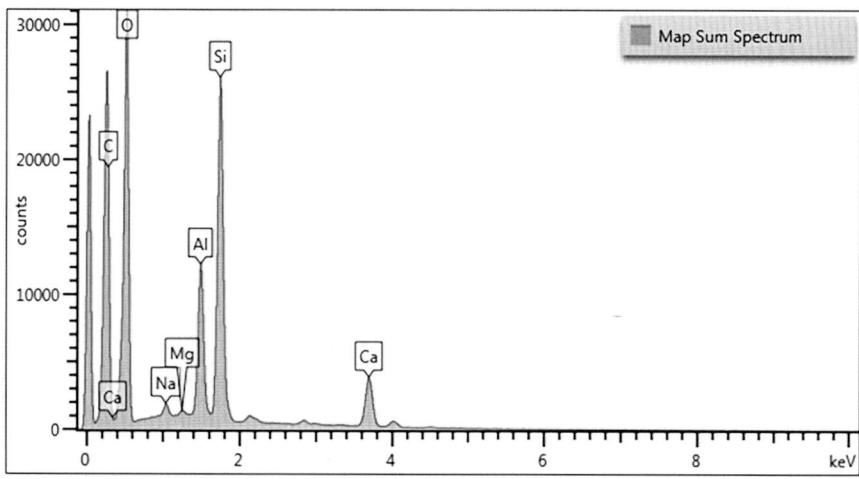

**Fig. 75.7** EDS analysis (Sierrita de la Cruz Bridge)

## 3.2 Fourier Transform Infrared (FTIR) Spectroscopy

All resins contain ester bonds that are susceptive to various processes because they are the weakest link of the polymer. Deterioration mechanism of the resin may be the alkaline hydrolysis of the ester linkages. It is well-known that concrete is an alkaline environment. When the hydrolysis reaction occurs, free hydroxyl ions (OH-) induce ester linkage attack, and the resin chain is destroyed. Finally, if the resin deteriorates, it will not transfer stresses to the glass fibers or protect the fibers against alkaline attack. Representative results of the FTIR analysis for the in-service GFRP samples from Southview Bridge are illustrated in Fig. 75.8.

It shows that there was no difference in the spectra of the two samples. Similar conclusions were drawn for the GFRP samples from other bridges. However, due to no control specimens tested in this time, it cannot be concluded that $OH^{-1}$ was from the hydrolysis reaction. Therefore, FTIR analysis of control samples will be recommended as future work.

## 3.3 $T_g$ Measurements

According to ASTM E1640–13 [4], differential scanning calorimetry (DSC) was used to evaluate the $T_g$ of the resins of the GFRP bars. Table 75.1 shows the results of $T_g$ measurements.

Due to lack of $T_g$ test results on GFRP samples prior to construction of these three bridges, there are no identical production lot references to compare to the results of Walker Bridge and Southview Bridge. However, the result of current control

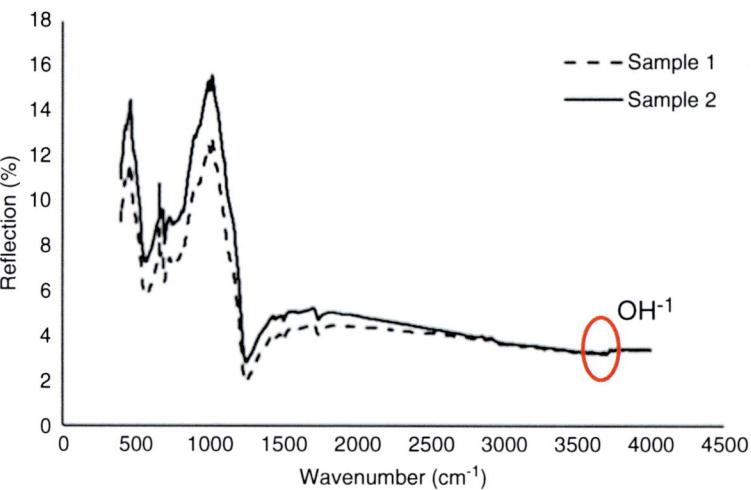

**Fig. 75.8** FTIR spectra for GFRP samples in Southview Bridge

**Table 75.1** Sample table for demonstration

| Diameter (in.) | Bridge | No. of samples | Ave. (°F) MST | Ave. (°F) control | Ave. (°F) UM |
|---|---|---|---|---|---|
| 0.25 | Walker Bridge | 3 | 186.0 | N/A | 177.8 |
| 0.75 | Southview Bridge | 6 | 176.6 | N/A | N/A |
| 0.625 | Sierrita de la Cruz Bridge | 3 | 187.4 | N/A | N/A |
| 0.75 | | 6 | 187.2 | 177.9 | 238.8 |

Note: Conversion units: 1 in. = 25.4 mm, °F = 1.8 °C + 32

specimens may be considered a reference. There is no significant decrease between the results of this study and the value from the control GFRP bars. In fact, the results of this work show an increase in the $T_g$ of the bars indicating added post-curing effects of the resin.

### 3.4 SBS Tests

An SBS test was performed based on ASTM D4475-02 (Reapproved 2016) [4]. The purpose of this test was to measure the interlaminar shear properties of GFRP bars. The results of these three bridges are summarized in Table 75.2. Table 75.3 illustrates the results of control from Sierrita de la Cruz Bridge.

The maximum loads of #5 and #6 GFRP samples are lower than the results of control samples tested at the time of construction, but it must be cautioned that the number of extracted specimens which were tested does not yield a significantly

Table 75.2 The results of SBS tests of the three bridges

| Bridge | Walker Bridge (1999) | | Sierrita de la Cruz Bridge (2000) | | Southview Bridge (2004) |
|---|---|---|---|---|---|
| Specimen | S1 | S3 | S1 | S2 | S1 |
| Diameter (in.) | 0.250 | 0.250 | 0.625 | 0.750 | 0.750 |
| Span length (in.) | 0.75 | 0.75 | 1.875 | 2.25 | 2.25 |
| Max. load (lb) | 399 | 354 | 2882 | 3281 | 2913 |

Note: Conversion units: 1 in. = 25.4 mm, 1 lb. = 0.0044 kN

Table 75.3 The results of the control from Sierrita de la Cruz Bridge

| Bridge | Sierrita de la Cruz Bridge | |
|---|---|---|
| Resource of data | Hughes Brothers (control, 2000) | |
| Specimen | 10 samples | 10 samples |
| Diameter (in.) | 0.625 | 0.750 |
| Max. load (lb) | 3009 | 4664 |

Note: Conversion units: 1 in. = 25.4 mm, 1 lb. = 0.0044 kN.

viable sample population. More SBS tests of extracted samples will be performed as future work from other projects.

## 4  Conclusions

No degradation was observed at the fibers and resin level of the GFRP samples extracted from these three bridges. There was no difference in the spectra between two GFRP samples that were extracted from the same bridge. However, due to the fact that no control specimens were concurrently tested at this time, it cannot be concluded that $OH^{-1}$ was from the hydrolysis reaction. For $T_g$ measurement and SBS test, more GFRP samples are needed to be evaluated to develop a larger more significant database of results. All three bridges appear to be performing very well after 15–17 years of field exposure in service.

## References

1. Mufti, A. (2005). Durability of GFRP reinforced concrete in field structures. *Proceeding of 7th international symposium on fibre reinforced polymer for reinforced concrete structures-FRPRCS-7* (pp. 889–895).
2. Uomoto, T. (1996). Durability of FRP as reinforcement for concrete structures. *Composite Materials Building Structures 3rd International Conference* (pp. 423–437).
3. Gooranorimi, O., Dauer, E., Myers, J. J., & Nanni, A. (2016). Long-term durability of GFRP reinforcement in concrete: A case study after 15 years of service. *Interim report, University of Miami.*
4. ASTM E1640-13. (2013). Standard test method for assignment of the glass transition temperature by dynamic mechanical analysis. *American Society for Testing and Materials.*

# Chapter 76
# The Influences of Mechanical Load on Concrete-Filled FRP Tube Cylinders Subjected to Environmental Corrosion

Song Wang and Mohamed A. ElGawady

Fiber-reinforced polymer (FRP) has been introduced into civil engineering since last century and finds tremendous applications in retrofitting and constructing infrastructures. Numerous studies have been done on its durability performance; however, most of the test specimens were not applied with mechanical loads while subjecting to environmental conditions, which didn't reflect the actual service load in a real application. This paper aims to investigate the influence of mechanical loads on concrete-filled FRP tube (CFFT) cylinders while they are under harsh environmental conditions. Both loaded and unloaded specimens were put into an environmental chamber and exposed to freeze/thaw cycles, wet/dry cycles, and heating/cooling cycles for 72 days. Compression tests and split-disk tensile tests were conducted on the CFFT cylinders and the outer FRP tubes, respectively, after the conditioning was completed. Experimental results showed slight reduction on the stress, but considerable decrease on the strain for loaded CFFT when comparing to unloaded specimens.

## 1 Introduction

Numerous researchers have experimentally studied the durability of FRP subjected to various environmental conditions; however, very limited studies have the sustained mechanical load applied to the specimens during conditioning. The load was shown to further degrade the mechanical properties of the FRP specimens [1, 2].

For the durability study of CFFT, to the best knowledge of the authors, only one researcher has applied the sustained axial load to CFFT cylinders during

---

S. Wang · M. A. ElGawady (✉)
Civil, Architectural and Environmental Engineering Department, Missouri University of Science and Technology, Rolla, MO, USA
e-mail: elgawadym@mst.edu

**Table 76.1** Dimensions and mechanical properties of the GFRP tube

| Parameter | $OD^a$ (mm) | $t^a$ (mm) | $f_L^a$ (MPa) | $E_L^a$ (MPa) | $\varepsilon_L^a$ (%) | $f_H^a$ (MPa) | $E_H^a$ (MPa) | $\varepsilon_H^a$ (%) |
|---|---|---|---|---|---|---|---|---|
| Lab | 167.6 | 3.2 | 58.1 | 10,501 | 0.71 | 151.7 | 13,348 | 1.73 |

[a]OD outer diameter, $t$ wall thickness, $f_L$ and $f_H$ ultimate tensile strength in longitudinal and hoop directions, respectively, $E_L$ and $E_H$ elastic modulus in longitudinal and hoop directions, respectively, $\varepsilon_L$ and $\varepsilon_H$ failure strain in longitudinal and hoop directions, respectively

conditioning. Fam et al. [3] exposed CFFT cylinders to 300 freeze/thaw cycles within 68 days while under sustained load corresponding to 30% of their ultimate confined strength. Test results showed the sustained load slightly improved the strength of the CFFT cylinders with low-strength concrete having 28-day compressive strength of 21 MPa due to creep effect. However, this effect did not happen on cylinders with medium-strength concrete having 28-day compressive strength of 41 MPa.

This study presents the effect of sustained mechanical load on the durability performance of CFFT cylinders subjected to the combined freeze/thaw, wet/dry, and heating/cooling cycles. Compression tests were conducted on cylinders, while hoop tensile tests were carried out on FRP rings, for both loaded and unloaded specimens.

## 2 Materials

Glass-fiber-reinforced polymer (GFRP) tubes used in this research were fabricated with polyester resin and glass fiber with winding angle being equal to ±53 °. The averaged mechanical properties along longitudinal and hoop directions of the tube tested in laboratory are listed in Table 76.1.

SCC was used to cast all of the CFFT cylinders and 102 mm × 204 mm unconfined concrete cylinders. The average 28th day compressive strength of three replicate concrete cylinders per ASTM C39 was 46.5 MPa.

## 3 Specimen and Test Setup Preparation

All CFFT cylinders were divided into three different sets: unconditioned unloaded (control), conditioned unloaded, and conditioned loaded (Fig. 76.1a) with symbols of UC (unconditioned), FT (freeze/thaw), and FT+S (freeze/thaw with stress), respectively. Each group had four specimens: three 305 mm high for compression tests and one 229 mm high for split-disk tensile tests. At least three 102 mm × 204 mm unconfined concrete cylinders went through the same environmental conditions as the CFFT, aiming to obtain the unconfined concrete strength of concrete at the day of CFFT test.

# 76 The Influences of Mechanical Load on Concrete-Filled FRP Tube Cylinders...

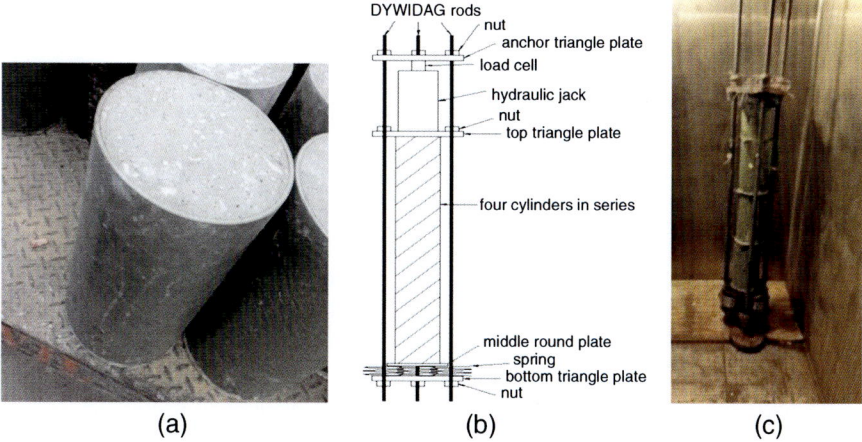

**Fig. 76.1** (a) CFFT cylinders, (b) test setup drawing, and (c) inside the environmental chamber

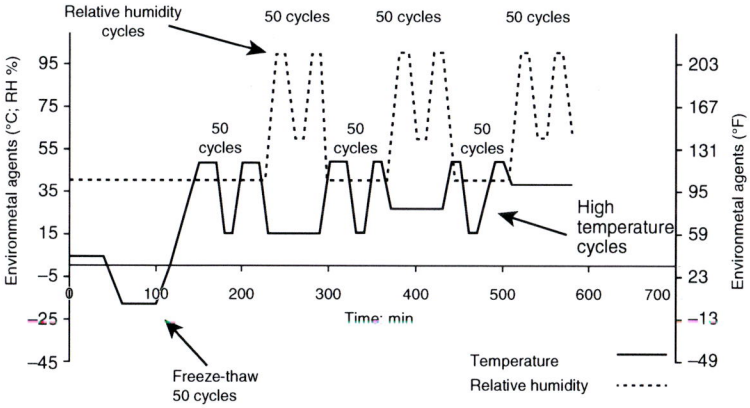

**Fig. 76.2** Exposure regime used in the environmental chamber

Four cylinders in series were axially loaded between triangular steel plates with three posttensioned Dywidag bars (Fig. 76.1b, c). A hydraulic jack was used to apply the load, and the load was monitored by a load cell. The target sustained stress was 6.9 MPa corresponding to approximately 10% of the CFFT's axial compressive strength, which is a common stress ratio for bridge columns [4]. Three sets of washer springs were placed between the bottom two plates to sustain the load. Strain gauges were instrumented on each Dywidag bar to monitor the load relaxation, and the setup was reloaded if necessary.

Specimens that were designed to sustain environmental conditions were placed inside the environmental chamber for 72 days and sustained a total of 350 cycles consisting of one set of 50 freeze/thaw cycles, three sets of 50 heating/cooling cycles, and three sets of 50 wet/dry cycles (Fig. 76.2). This regime was first proposed by Micelli and Nanni [5] to represent severe weather conditions in the Midwest of the USA for an exposure period of approximately 20 years.

## 4  Setup for Compression and Split-Disk Tensile Tests

The cylinders were removed from the environmental chamber after the end of conditioning and were kept under room conditions for at least 1 week before testing. Compression test was conducted on a MTS 250 load frame. Two linear variable displacement transformers (LVDTs) were instrumented to each tested specimen to monitor its global vertical deformations. Two strain gauges were bonded to the GFRP tube surface symmetrically at mid height along the hoop direction to measure the hoop strains of each GFRP tube (Fig. 76.3a). The compression load was applied monotonically at a 0.5 mm/min displacement rate until failure occurred. For the ring test, two semicircle steel plates were placed inside each GFRP ring while the loading head pulled apart the steel plates at a loading rate of 0.25 mm/min until GFRP ruptures. Two strain gauges were bonded to each ring specimen on each side (Fig. 76.3b).

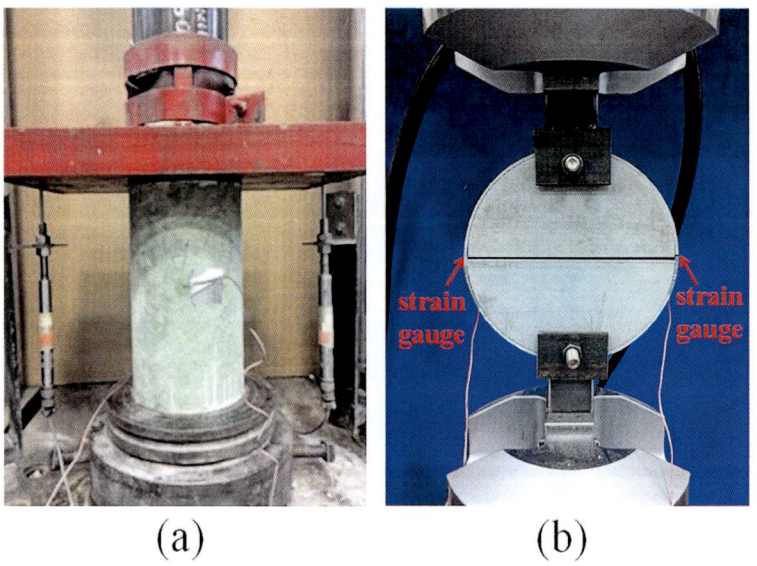

**Fig. 76.3** Test setup for (**a**) compression and (**b**) split-disk tensile tests

## 5 Results and Discussions

All the CFFT cylinders failed due to GFRP tubes rupture followed by diagonal shear fracture of the concrete core with only one or two major internal cracks developed. More catastrophic failures were observed for the conditioned specimens compared to the control specimens as the freeze/thaw cycles embrittled the resin of the GFRP [5–7]. The averaged maximum normalized stress, maximum vertical strain, and hoop strain for each CFFT cylinder are shown in Fig. 76.4, where the normalized stress is defined as the CFFT confined strength $f'_{cc}$ divided by $f'_c$ of this CFFT. The environmental exposures had negligible effect on the normalized maximum strength of CFFT but cause pronounced reduction on the maximum strain values.

The degraded strain could be attributed to matrix cracking and/or fiber-matrix debonding due to wet/dry cycling and temperature cycling [8]. The maximum strains of CFFT_FT+S cylinders were smaller than those of CFFT_FT specimens, reflecting that the sustained load could further deteriorate the properties of the GFRP, especially for the strain. The sustained axial load on the CFFT cylinders exerted radial pressure on the GFRP tube, resulting in sustained hoop stress on the GFRP tube in addition to the residual stresses caused by environmental conditioning. Microcracks and cracks were generated and propagated due to such stress state, which would attract extra moisture and eventually further degraded the strain of the material. However, the sustained axial load barely affected the maximum normalized stress.

The trends of mechanical properties' changes on GFRP rings are similar to that of CFFT cylinders under compression (Fig. 76.5), revealing that the mechanical performance of CFFT cylinder under compression is dominated by the properties of the outer GFRP tube, which corresponds to the failure mode where the CFFT crushed immediately after the outer GFRP tube ruptured. The sustained axial load on FT+S specimens did not affect the maximum stress of the GFRP tube significantly but did cause pronounced deterioration on the maximum strain compared to FT rings.

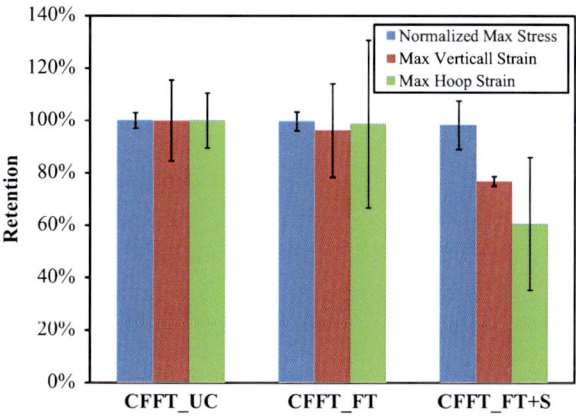

**Fig. 76.4** Averaged normalized max stress and max strains for CFFT cylinders with ± one standard deviation

**Fig. 76.5** Averaged tensile properties of GFRP rings with ± one standard deviation

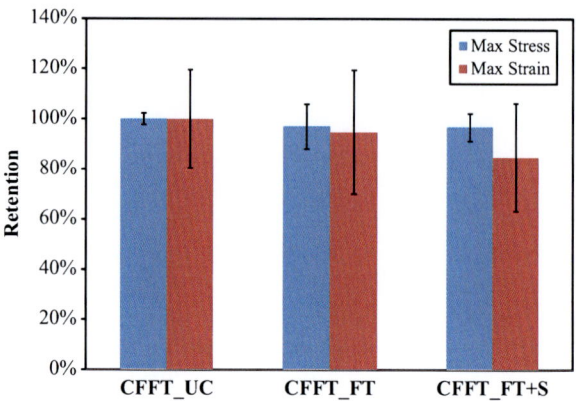

## 6 Conclusions

Experimental work has been done to investigate the effect of mechanical load on the CFFT cylinders subjected to combined freezing-thawing cycles, heating/cooling cycles, and wet/dry cycles. Several conclusions can be drawn and shown as below:

1. The combined environmental exposures with sustained load affected the CFFT cylinders by embrittling the FRP during freeze/thaw cycles, causing microcracks on matrix and fiber/matrix debonding due to wet/dry cycles, temperature cycles, and external load. The compressive strength was barely affected, while the maximum vertical strain and hoop strain were decreased by up to 23% and 39%, respectively.
2. The sustained load further deteriorated the strain capacity of the FRP by inducing more microcracks. However, it barely affected the strength of the FPR.

## References

1. Hollaway, L. C. (2010). A review of the present and future utilisation of FRP composites in the civil infrastructure with reference to their important in-service properties. *Construction and Building Materials, 24*(12), 2419–2445.
2. Wu, H., Fu, G., Gibson, R. F., Yan, A., Warnemuende, K., & Anumandla, V. (2006). Durability of FRP composite bridge deck materials under freeze-thaw and low temperature conditions. *Journal of Bridge Engineering, 11*(4), 443–551.
3. Fam, A., Kong, A., & Green, M. F. (2008). Effects of freezing and thawing cycles and sustained loading on compressive strength of precast concrete composite piles. *PCI Journal, 53*(1), 109–120.
4. Anumolu, S., Abdelkarim, O. I., & ElGawady, M. A. (2016). Behavior of hollow-core steel-concrete-steel columns subjected to torsion loading. *Journal of Bridge Engineering, 21*(10), 77–88.
5. Micelli, F., & Nanni, A. (2004). Durability of FRP rods for concrete structures. *Construction and Building Materials, 18*, 491–503.

6. American Concrete Institute (ACI). (2007). *Report on fiber-reinforced polymer (FRP) reinforcement for concrete structures*. ACI 440R-07. Farmington Hills.
7. Karbhari, V. M., Rivera, J., & Dutta, P. K. (2000). Effect of short-term freeze-thaw cycling on composite confined concrete. *Journal of Composites for Construction, 4*(4), 191–197.
8. Zaman, A., Gutub, S. A., & Wafa, M. A. (2013). A review on FRP composites applications and durability concerns in the construction sector. *Journal of Reinforced Plastics and Composites, 32*(24), 1966–1988.

# Chapter 77
# Finite Element Analysis of RC Beams Strengthened in Shear with NSM FRP Rods

Akram R. Jawdhari and Ali Hadi Adheem

This study presents three-dimensional FE models on RC T-beams strengthened in shear with near-surface mounted (NSM) fiber-reinforced polymer (FRP) rods, with various rod spacing, strengthening pattern, and internal shear reinforcement ratios. The FE models utilized different constitutive material relations for concrete, steel, and FRP and implemented a mixed-mode bond-slip law at the NSM FRP rod-concrete interface. Comparisons, such as ultimate loads and load-deflection history, validated the accuracy of the developed models. More importantly, the FE models were able to predict the brittle shear failure of the beams and debonding of the NSM rods. For an accurate simulation of shear failure, the shear retention factor ($\beta$) should be less than 0.2. The correlated models will be used in further research to examine the effects of various parameters expected to influence the behavior of the retrofitted beams, in an effort to strengthen the knowledge on this topic.

## 1 Introduction

One of the successful applications of fiber-reinforced polymer (FRP) is the shear strengthening/retrofit of reinforced concrete (RC) beams and girders, using externally bonded (EB) FRP sheets or prefabricated plates. Concrete members need an enhancement to their shear strength if they are deficient in shear or if the shear capacity falls below the flexural strength due to flexural retrofit [1].

---

A. R. Jawdhari (✉)
Department of Civil Engineering, Queens University, Kingston, ON, Canada
e-mail: akram.jawdhari@queensu.ca

A. H. Adheem
Kerbala Technical Institute, Al-Furat Al-Awsat Technical University, Kerbala, Iraq

© Springer International Publishing AG, part of Springer Nature 2018
M. M. Reda Taha (ed.), *International Congress on Polymers in Concrete*
*(ICPIC 2018)*, https://doi.org/10.1007/978-3-319-78175-4_77

Many experimental tests, numerical and analytical studies, and field applications have been carried out to examine the effectiveness of EB FRP shear strengthening [1, 2]. Significant increase of shear capacity, ranging between 29% and 135% [3–5], for concrete beams strengthened with EB FRP has been reported in the literature. Various parameters governing the performance of the strengthened members, such as spacing between strips, number of strips (or axial stiffness), inclination angle, the use of anchorage, and adhesive type, have been examined by a large number of studies [6].

In addition to EB, FRP composites have also been mobilized within the near-surface mounted (NSM) technique, referred to as NSM FRP. The technique has gained wide attraction and proved to be viable option for shear and flexural retrofit of concrete structures [6–9]. NSM FRP technique consists of cutting grooves or slits within the concrete cover and inserting FRP rods or strips, using an epoxy adhesive or cementitious mortar [8]. Available tests have confirmed the effectiveness of the method in increasing the shear capacity of concrete beams, [8–11]. For example, Al-Mahmoud et al. [8] reported an increase in shear strength in the range of 34.7–43.6%, for RC beams tested under three- and four-point loads.

Although there has been a large number of experimental tests and analytical studies on the shear strengthening of RC members with NSM reinforcement, the number of finite element (FE) studies is still very limited. The current study aims at expanding the current database on numerical simulations of NSM FRP used in shear strengthening of concrete. The FE models are developed for the experimental beams tested by De Lorenzis and Nanni [12]. The FE models present various nonlinear models for the concrete (in tension and compression), steel reinforcement, FRP rods, and the bond interface between the rod and concrete.

## 2 Summary of Experiments

The specimens in [12] were T-shaped RC beams with a total length of 10 m, tested under four-point bending. Flexural reinforcement consisted of two No. 9 [$\phi = 29$ mm] as tensile rebars for all specimens and two No. 4 [$\phi = 13$ mm] compressive rebars for specimens where internal steel stirrups are included. The FRP rods were No. 3 [$\phi = 10$ mm] carbon FRP (CFRP) deformed bars. Figure 77.1 shows the beam's dimensions and cross-section details. The text matrix included two control beams, one without shear stirrups and one with shear stirrups, and beams strengthened in shear with NSM CFRP rods according to the following patterns: (a) beam with rods spaced at 125 mm, (b) beam with rods spaced at 125 mm and extended to the flange of the beam (i.e., anchored), and (c) beam with rods spaced at 175 mm.

The material properties are as follows: average concrete strength ($f_c'$) = 31 MPa; yield strength of flexural and shear steel ($f_y$) = 414 and 345 MPa, respectively; and tensile strength and modulus of elasticity of CFRP rods ($f_u$) = 1875 MPa and ($E_f$) = 104.8 GPa, respectively. The tensile strength and modulus of the two-part

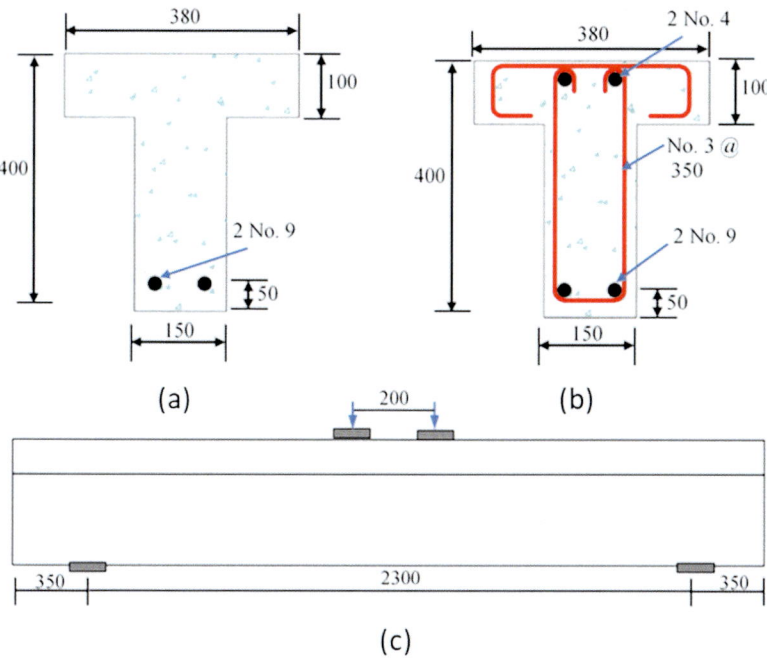

**Fig. 77.1** Geometry of T-beams, (**a**) cross-section [no stirrups], (**b**) cross-section [with stirrups], (**c**) elevation, dimensions in mm

epoxy used to fill the grooves and bond the NSM FRP rod to concrete are 13.8 MPa, and 2800 MPa, respectively.

## 3 Finite Element Models

The three-dimensional finite element (FE) models were created using the ANSYS software [13]. In the experiments, one half of the beam length [control side] was over-reinforced with shear stirrups to enforce the shear failure to occur on the half [test side]. Therefore, symmetry was not available in the beam's length direction. Taking advantages of symmetry in the width direction, the FE models were for the entire beam's length and half of its width. Concrete volume was modeled using special concrete elements available in ANSYS [Solid 65]. Steel plates, added at the support and loading locations, and adhesive were modeled using solid elements [Solid 185]. The steel reinforcement and CFRP rods were modeled with truss elements [Link 188]. Figure 77.2 shows several parts of the FE models. A contact-target pair (with CONTA 173 and TARGE 170 elements) is used to simulate the slipping and debonding of the NSM rod from concrete, along the adhesive-concrete interface. A bond-slip ($\tau$-$\delta$) law is required to define the interfacial behavior of the

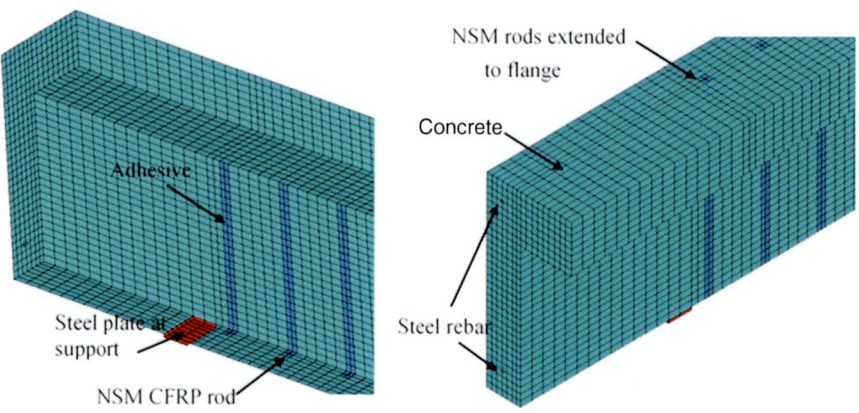

**Fig. 77.2** FE models, showing different components

**Table. 77.1** Ultimate load comparisons between the experiment and FE models

| Beam code[a] | Strengthening scheme | Maximum load, kN | | % diff.[b] |
|---|---|---|---|---|
| | | Experiment | FE | |
| BV | Control, no internal stirrups | 180 | 177 | 2.43 |
| BSV | Control, with internal stirrups | 307 | 308 | 0.37 |
| B90-5 | NSM rods at 125 mm | 257 | 237 | 8.00 |
| B90-5A | NSM rods at 125 mm, extended to flange | 373 | 350 | 6.50 |
| B90-7 | NSM rods at 175 mm | 231 | 236 | 2.44 |

[a]Beam code is from De Lorenzis and Nanni [12]
[b]% difference $= \frac{P_{exp}}{P_{FE}}$

interface. A mixed-mode [comprising both peeling and shear stresses] ($\tau$-$\delta$) law, used by Omran and El-Hacha [14] for simulating the debonding between NSM FRP rods and concrete, is utilized in this study.

## 4  Results and Discussions

Table 77.1 lists the experimentally reported ultimate loads vs. the ones extracted from the FE analysis, along with the percentage difference between the two datasets. As can be seen from the table, the percentage difference varied between 0.37% and 8% with an average (for all five specimens) of 3.95%. Figure 77.3 plots the load vs. mid-span deflection (P-$\Delta$) variation obtained from the tests and FE models. The FE models seem to give a good prediction of (P-$\Delta$) curve, although some fluctuations in stiffness exist between the experimental and FE predictions. Those fluctuations, provided they are negligible, can be expected in a FE analysis as a result of

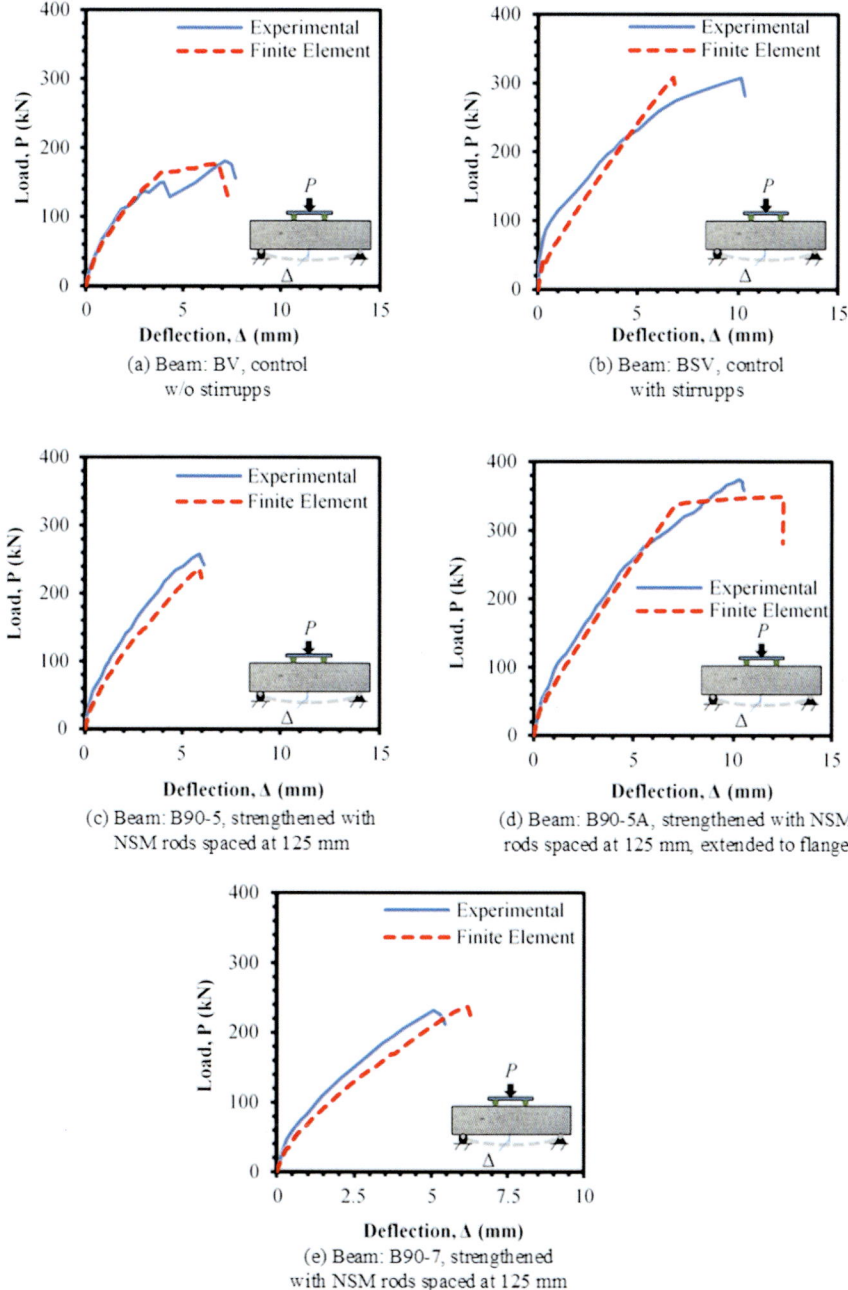

Fig. 77.3 Load mid-span deflection comparisons

the modeling assumptions considered such as the mesh size, material models, boundary conditions imposed, etc.

It was noticed that for the control beam without shear stirrups [Beam, BV], which failed in brittle shear, a shear retention factor ($\beta$) for open and closed crack has to be reduced in order to capture the failure mode. For this beam, a value of $\beta = 0.1$ was used for the open and closed crack. For other beams that didn't experience shear failure, a value of $\beta = 0.3$ was used for open crack and $\beta = 1.0$ for closed crack.

## 5 Conclusions

Three-dimensional finite element models have been developed in this article to examine the behavior of RC beams strengthened in shear with NSM CFRP rods, having various rod spacing, strengthening schemes, and internal shear reinforcement ratios. The FE models were compared to the data of several beams tested by other researchers. Comparisons, such as ultimate loads and load-deflection history, validated the accuracy of the developed models. More importantly, the FE models were able to predict the brittle shear failure of the beams and debonding of the NSM rods. For an accurate simulation of shear failure, the shear retention factor ($\beta$) should be less than 0.2.

## References

1. Teng, J. G., Lam, L., & Chen, J. F. (2004). Shear strengthening of RC beams with FRP composites. *Progress in Structural Engineering and Materials, 6*, 173–184. https://doi.org/10.1002/pse.179.
2. Chen, J. F., & Teng, J. G. (2003). Shear capacity of FRP strengthened RC beams: Fibre reinforced polymer rupture. *ASCE Journal of Structural Engineering, 129*(5), 615–625. [14 February 2001].
3. Lee, T. K., & Al-Mahaidi, R. S. H. (2003). Strength and failure mechanism of RC T-beams strengthened with CFRP plates. *Proceedings of the 6th international symposium on FRP reinforcement for concrete structures, Singapore* (pp. 247–256).
4. Khalifa, A., & Nanni, N. (2000). Improving shear capacity of existing RC T-section beams using CFRP composites. *Cement and Concrete Composites, 22*(3), 165–174. ISSN 0958-9465.
5. Triantafillou, T. C. (1998). Shear strengthening of reinforced concrete beams using epoxy-bonded FRP composites. *ACI Structural Journal, 95*(2), 107–115.
6. Adhikary, B. B., & Mutsuyoshi, H. (2004). Behavior of concrete beams strengthened in shear with carbon-fiber sheets. *ASCE Journal of Composites for Construction, 8*(3), 2004.
7. El-Hacha, R., & Rizkalla, S. (2004). Near-surface-mounted fiber reinforced polymer reinforcements for flexural strengthening of concrete. *ACI Structural Journal, 101*(5), 717–726.
8. Al-Mahmoud, F., Castel, A., François, R., & Tourneur, C. (2010). RC beams strengthened with NSM CFRP rods and modeling of peeling-off failure. *Composite Structures, 92*, 1920–1930.
9. Al-Mahmoud, F., Castel, A., Minh, T. Q., & François, R. (2015). Reinforced concrete beams strengthened with NSM CFRP rods in shear. *Advances in Structural Engineering, 18*(10), 1563–1574.

10. Barros, J., & Dias, S. J. E. (2003). Shear strengthening of reinforced concrete beams with laminate strips of CFRP. *Proceedings of the international conference composites in constructions CCC2003*, Cosenza (pp. 289–294).
11. Jalali, M., Sharbatdar, M. K., Chen, J. F., & Alaee, F. J. (2012). Shear strengthening of RC beams using innovative manually made NSM FRP bars. *Construction and Building Materials, 36*, 990–1000. ISSN 0950-0618.
12. De Lorenzis, L., & Nanni, A. (2001). Strengthening of reinforced concrete beams with NSM fiber-reinforced polymer rods. *ACI Structural Journal, 98*(1), 60–68.
13. ANSYS. (2016). *Release 17.2 documentation for ANSYS. Version 17.2*. Canonsburg: ANSYS Inc..
14. Omran, H., & El-Hacha, R. (2012). Nonlinear 3D finite element modeling of RC beams strengthened with prestressed NSM-CFRP strips. *Construction and Building Materials, 31*, 74–85.

# Chapter 78
# Effect of Sustained Load Level on Long-Term Deflections in GFRP and Steel-Reinforced Concrete Beams

Stephanie L. Walkup, Eric S. Musselman, and Shawn P. Gross

ACI 440.1R [1] applies an adjustment factor of 0.6 to the long-term deflection multiplier for steel-reinforced beams to reflect the experimentally observed differences between FRP and steel-reinforced members. The objective of this study is to evaluate the effect of the level of sustained load on the long-term multiplier in GFRP-reinforced beams. Five beams, including four GFRP-reinforced beams and one steel-reinforced control beam, were tested in four-point bending on a simply supported span with different temporary service loads ($M_a = 1.80\ M_{cr}$–$2.36\ M_{cr}$) and sustained load ($M_{sus} = 0.58\ M_a$–$0.85\ M_a$) levels. Mid-span deflections were recorded twice weekly over a period of 100 days, and the long-term deflection multiplier, $\lambda_\Delta$, was plotted vs. time. The results indicate that the temporary service load level, rather than the sustained load level, affects the long-term deflection multiplier with larger temporary service loads causing a smaller long-term multiplier. The current 0.6 multiplier on the time factor for FRP-reinforced beams also appears to underestimate long-term deflections. In addition, the multiplier to capture the effects of GFRP vs. steel-reinforcement may not be a constant value.

## 1 Introduction

Concrete beams reinforced with glass fiber-reinforced polymer (GFRP) bars are designed for both strength and serviceability criteria; however, serviceability criteria in the form of immediate and particularly long-term deflections generally govern their design. A significant amount of work, including both experimental and analytical, has been performed to determine the effect of long-term deflections on FRP

---

S. L. Walkup (✉) · E. S. Musselman · S. P. Gross
Villanova University, Villanova, PA, USA
e-mail: swalku01@villanova.edu

**Table 78.1** Long-term deflection results for Hall and Ghali [6] data

| Specimen | $M_a/M_{cr,exp}$ | $\Delta_i$, (mm) | $\Delta_{250\ days}$, (mm) | $\lambda_\Delta = (\Delta_{250\ days}/\Delta_i) - 1$ |
|---|---|---|---|---|
| Lg-S-3-1.5 | 1.49 | 4.6 | 9.9 | 1.2 |
| Lg-G-3-1.5 | 1.43 | 14.7 | 22.6 | 0.5 |
| Lg-S-3-3.0 | 2.47 | 9.4 | 16.5 | 0.8 |
| Lg-G-3-3.0 | 2.82 | 34.3 | 46.0 | 0.3 |

beams. However, to date, the *Guide for the Design and Construction of Structural Concrete Reinforced with Fiber-Reinforced Polymer (FRP) Bars*, ACI 440.1R [1], applies only a single multiplier of 0.6 to the long-term deflection multiplier, $\lambda_\Delta$, for steel-reinforced beams to reflect experimentally observed differences between members with FRP and steel reinforcement, respectively. The Canadian Standards Association's *Design and Construction of Building Components with Fiber-Reinforced Polymers,* CSA-S806 [2], prescribes a single 1.0 multiplier. The long-term multiplier for steel is directly related to the time factor, $\xi$, developed by Yu and Winter [3].

The 0.6 multiplier for FRP-reinforced members is close to the 0.55 value first introduced by Brown [4] who suggested that the multiplier should be within the range of 0.36–0.73 and that the multiplier is directly proportional to the concrete compression strain caused by the sustained load. Other researchers including Vijay and GangaRao [5], Hall and Ghali [6], and Mias et al. [7, 8] have studied the effects of different load levels on the long-term multiplier. Vijay and GangaRao [5] found that larger long-term multipliers occurred in specimens with lower sustained loads. Data from Hall and Ghali [6] demonstrated that steel- and GFRP-reinforced beams with larger sustained loads had larger initial deflections and larger additional deflections. Further analysis of their data (Table 78.1) shows that while the additional deflection was larger, the long-term multiplier for the beams with the smaller applied loads was higher.

While the research by both Vijay and GangaRao [5] and Hall and Ghali [6] found that larger long-term multipliers occurred in specimens with lower sustained loads, in both of these studies, the service load moment, $M_a$, and sustained load moment, $M_{sus}$, were equal, and the load that caused the variability in long-term multipliers could not be isolated.

Mias et al. [7, 8] tested a series of normal- and high-strength concrete GFRP-reinforced beams with different reinforcement ratios, $\rho$, and subject to long-term loading. Each beam was cyclically loaded to a temporary service load moment, $M_a$, to achieve 2000 με in the reinforcement, and then a sustained load moment to cause an extreme fiber compressive stress of either $0.30f'_c$ or $0.45f'_c$ was sustained for at least 250 days. This research demonstrated that the reinforcing ratio, $\rho$, rather than the sustained load level had an effect on the long-term multiplier for both normal and high strength concrete GFRP-reinforced concrete beams.

## 2 Objective

The objective of this study is to evaluate the effect of sustained load level on the long-term multiplier in GFRP-reinforced beams.

## 3 Experimental Method

Five 102-mm-wide by 203-mm-deep beams were tested in four-point bending on a 1.83 m simply supported span (Figs. 78.1 and 78.2).

The two applied loads produced a 152-mm-long constant moment region at mid-span. Four of the beams were reinforced with two No. 16 GFRP bars, and the remaining beam was a control specimen reinforced with two No. 10 steel bars. All bars had 19 mm clear cover. A simple concrete mix: 9.64% water, 18.54% cement, 33.27% fine aggregate, and 38.55% course aggregate, having a water-to-cement ratio of 0.52 with a targeted compressive strength of 31.0 MPa was used for all specimens. The measured modulus of elasticity averaged 37.8 GPa for all specimens. Steel reinforcement had a yield strength of 414 MPa and modulus of elasticity of 200 GPa. The manufacturer's reported guaranteed tensile strength and modulus of elasticity of the GFRP bars were 724 MPa and 46.2 GPa, respectively.

Testing was conducted in an environmentally controlled testing room with a temperature of 23.0 ± 2 °C and relative humidity of 50 ± 4%. All beams were subjected to an initial temporary load (often referred to in the literature as a "pre-cracking" load) producing a service moment, $M_a$, and then were held at various sustained loads, producing a sustained moment, $M_{sus}$, for a period of 100 days. Load was applied using a spring-loaded frame that was mounted to a strong floor. Loss of

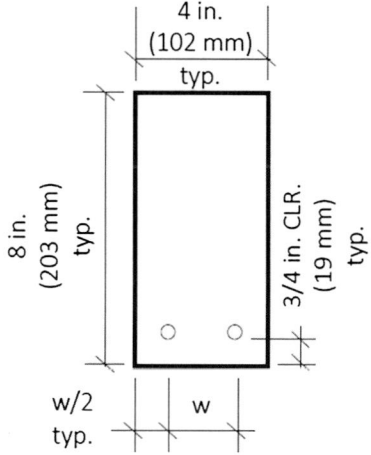

**Fig. 78.1** Specimen geometry

**Fig. 78.2** Load application

**Table 78.2** Service and sustained load moments

| Specimen | G5-1.8-58 | G5-1.8-71 | G5-1.8-85 | G5-2.4-65 | S3-2.2-65 |
|---|---|---|---|---|---|
| $M_a$ (kN-m) | 5.34 | 5.34 | 5.34 | 6.87 | 6.39 |
| $M_{sus}$ (kN-m) | 3.12 | 3.85 | 4.58 | 4.54 | 4.23 |

load in the spring was measured by monitoring spring elongation using DEMEC mechanical strain gage points applied at each end of the spring, and the load was reshored as necessary to ensure the applied moment was $M_{sus} \pm 2\%$.

Three GFRP-reinforced beams were loaded to 1.8 times their cracking moment and then sustained at either 58, 71, or 85 percent of that temporary service load. The 1.8 $M_{cr}$ service load and maximum of 0.85 $M_a$ sustained load level were selected to assure that the concrete strain remained within the linear-elastic limit. The fourth GFRP-reinforced beam and the steel-reinforced beam were initially loaded to 2.4 and 2.2 times their cracking moment, respectively, which corresponds to the permissible design dead and live loads based on limiting strength and serviceability design criteria. Sustained loads for these beams were 0.65 $M_a$, again remaining within the linear-elastic range. Table 78.2 shows the service and sustained load moments. Mid-span deflections measured with a 0.0254 mm precision dial gage were recorded twice weekly. At 100 days, the beams were reloaded to the full service load and then unloaded completely.

Specimen designations are Rb-P-S where R is the type of reinforcement (GFRP (G) or steel (S)), $b$ is the reinforcing bar size, $P$ is the ratio of the temporary service load moment, $M_a$, to the cracking moment, $M_{cr}$ ($M_a/M_{cr}$), and $S$ is the percent of the service load moment, $M_a$, that is sustained.

## 4 Results

Table 78.3 shows for each specimen the initial service load deflection, $\Delta_{i,\text{serv}}$; the initial sustained load deflection, $\Delta_{i,\text{sus}}$; the 100-day long-term deflection under sustained load, $\Delta_{\text{tot 100 days}}$; and the deflection observed upon reloading to the full service load at the conclusion of the 100-day test, $\Delta_{\text{serv 100 days}}$. As expected, initial service load deflections were larger for specimens with higher service loads, and the sustained load initial deflection was larger for specimens with higher sustained loads.

The initial unsustained live load deflections, $\Delta_{\text{LL,unsus}}$, and 100-day unsustained live load deflections, $\Delta_{\text{LL,unsus 100 days}}$, are similar for each specimen. For specimens G5-1.8-71, G5-1.8-85, and G5-2.4-65, these deflections are within 0.02 mm, and for specimens G5-1.8-58 and S3-2.2-65, these deflections are within 0.13 mm. The similar unsustained live load deflections confirm that creep, shrinkage, and other long-term effects do not affect unsustained live load deflections and that the overall stiffness of the specimen represented by the product of the modulus of elasticity of the concrete and the effective moment of inertia of the specimen, $E_c I_e$, remains unchanged over time when evaluating short-term loading.

A graph of the long-term multiplier, $\lambda_\Delta$, vs. time (Fig. 78.3) shows that the sustained load level does not affect the long-term deflection multiplier of GFRP-reinforced beams. The long-term deflection multiplier, $\lambda_\Delta$ (Table 78.3), is similar for all of the GFRP-reinforced beams with a temporary service load of 1.8 $M_{\text{cr}}$, regardless of the sustained load level ($\lambda_\Delta$ = 0.85, 0.92, 0.85 for specimens G5-1.8-58, G5-1.8-71, and G5-1.8-85, respectively), while $\lambda_\Delta$ = 0.76 for specimen G5-2.4-65 which had a higher temporary service load but similar sustained load level.

To evaluate the current 0.6 adjustment value used in ACI 440.1R, the ratio of the long-term multipliers for the GFRP-reinforced beams to that of the steel-reinforced beam ($\lambda_\Delta$ S3-2.2-65 = 0.87) was calculated. For specimens G5-1.8-58, G5-1.8-71, and G5-1.8-85, these ratios are 0.98, 1.06, and 0.98, respectively, within 6% of the 1.0

**Table 78.3** Beam specimen deflections and long-term multiplier, $\lambda_\Delta$

| Specimen | A<br>$\Delta_{i,\text{serv}}$, mm | B<br>$\Delta_{i,\text{sus}}$, mm | C<br>$\Delta_{\text{tot}}$<br>100 days, mm | D<br>$\Delta_{\text{serv}}$<br>100 days, mm | =A − B<br>$\Delta_{\text{LL}}$, unsus, mm | =D − C<br>$\Delta_{\text{LL, unsus}}$<br>100 days, mm | =C/<br>B − 1<br>$\lambda_\Delta$ |
|---|---|---|---|---|---|---|---|
| G5-1.8-58 | 2.26 | 1.52 | 2.82 | 3.68 | 0.74 | 0.86 | 0.85 |
| G5-1.8-71 | 2.34 | 1.83 | 3.51 | 4.04 | 0.51 | 0.53 | 0.92 |
| G5-1.8-85 | 2.64 | 2.41 | 4.47 | 4.70 | 0.23 | 0.23 | 0.85 |
| G5-2.4-65 | 3.84 | 2.79 | 4.93 | 5.97 | 1.04 | 1.04 | 0.76 |
| S3-2.2-65 | 2.57 | 1.96 | 3.66 | 4.39 | 0.61 | 0.74 | 0.87 |

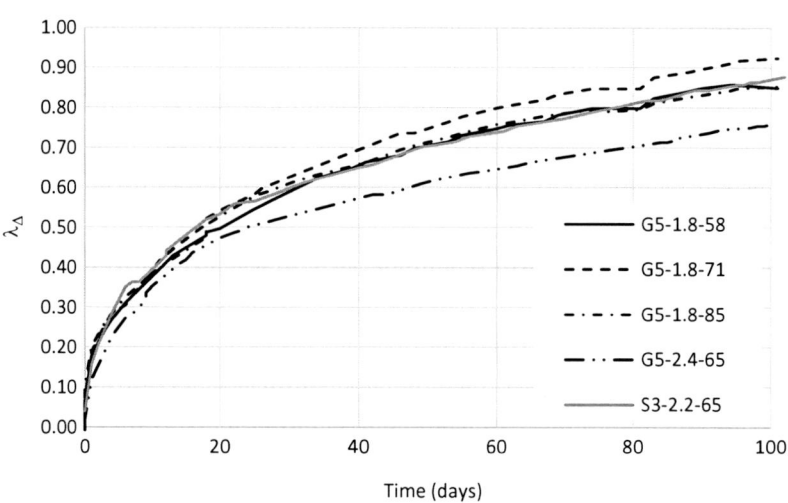

**Fig. 78.3** Effect of service and sustained load moments on the long-term deflection multiplier, $\lambda_\Delta$

value prescribed by CSA-S806 and much higher than the 0.6 ACI 440.1R value. However, for specimen G5-2.4-65, which was loaded to anticipated design loads, this ratio is 0.87, in between the 0.6 and 1.0 values prescribed by ACI 440.1R and CSA-S806, respectively. This smaller multiplier for G5-2.4-65 shows that the temporary service load level rather than the sustained load level may affect the long-term multiplier.

The present study as well as the past research conducted by Mias et al. [7, 8] included both temporary service load moments, $M_a$, as well as variable smaller sustained load moments, $M_{sus}$, and both found that the sustained load level does not affect the multiplier. Using this information, it can be concluded that the lower long-term multipliers observed for beams subject to higher loads in both the Hall and Ghali [6] and Vijay and GangaRao [5] studies are related to the higher service load moments, $M_a$, that were applied to the specimens and not the sustained load levels which were equivalent to $M_a$.

## 5 Conclusions

Sustained load level does not significantly affect the long-term deflection multiplier. However, the temporary service (pre-cracking) load does affect the long-term multiplier with larger pre-cracking loads causing a smaller long-term multiplier. The current 0.6 multiplier on the time factor for FRP-reinforced beams appears to underestimate long-term deflections for specimens with a service-to-cracking moment ratio of 1.80 and 2.35. Multipliers of 1.0 and 0.87 would be more appropriate for GFRP-reinforced beams subjected to $M_a/M_{cr}$ ratios of 1.80 and 2.35, respectively. This multiplier to capture the effects of GFRP vs. steel reinforcement may not be a constant value but rather may be a function of other factors such as the temporary service (pre-cracking) load level.

## References

1. ACI Committee 440. (2015). *Guide for the design and construction of structural concrete reinforced with Fiber-Reinforced Polymer (FRP) bars.* ACI 440.1R-15. Farmington Hills: American Concrete Institute.
2. CSA Standard CAN/CSA-S806-12. (2017). *Design and construction of building components with fibre-reinforced polymers.* Mississauga: Canadian Standards Association.
3. Yu, W. W., & Winter, G. (1960). Instantaneous and long-time deflections of reinforced concrete beams under working loads. *ACI Journal, 57*(2), 29–50.
4. Brown, V. (1997). Sustained load deflections in GFRP-reinforced concrete beams. In *Non-metallic (FRP) reinforcement for concrete structures (FRPRCS-3), proceedings of the 3rd international symposium on nonmetallic (FRP) reinforcement for concrete structures* (Vol. 2, pp. 495–502). Sapporo.
5. Vijay, P., & GangaRao, H. (1998). Creep behavior of concrete beams reinforced with GFRP bars. In *Proceedings of the first international conference on durability of composites for construction (CDCC'98)* (pp. 661–667). Sherbrooke.
6. Hall, T., & Ghali, A. (2000). Long-term deflection prediction of concrete members reinforced with glass fiber reinforced polymer bars. *Canadian Journal of Civil Engineering, Ottawa, Ontario, Canada, 27*, 890–898.
7. Mias, C., Torres, L., Turon, A., & Barris, C. (2013a). Experimental study of immediate and time-dependent deflections of GFRP reinforced concrete beams. *Composite Structures, 96*, 279–285.
8. Mias, C., Torres, L., Turon, A., & Sharaky, I. (2013b). Effect of material properties on long-term deflection of GFRP reinforced concrete beams. *Construction and Building Material, 41*, 99–108.

# Chapter 79
# Flexural Behavior and Cracks in Concrete Beams Reinforced with GFRP Bars

Naser Kabashi, Cene Krasniqi, Jakob Sustersic, Arton Dautaj, Enes Krasniqi, and Hysni Morina

Concrete beams reinforced with glass fiber-reinforced polymer (GFRP) bars exhibit large deflections and crack widths compared with concrete members reinforced with conventional steel. Current design methods for predicting deflections under the loading and crack widths developed for concrete structures reinforced with steel may not be used for concrete structures reinforced with GFRP, but some of parameters will be used. Based on this paper research work and past studies, a theoretical correlation for predicting crack width was proposed. Experimental results are focused in examining the two sets of concrete beams with different percent of reinforcement using the GFRP bars. In this case the compare parameters are deflections, cracks, and bearing capacity using the analytical and experimental results. The beams were tested under a static load to examine the effects of the reinforcement ratio and compressive strength of concrete on cracking, deflection, ultimate capacity, and modes of failure.

## 1 Introduction

The last years, use of the fiber-reinforced polymer (FRP) bars as internal reinforcement for concrete structures has been sustained and particularly encouraged in the international technical and research community and specific technical instructions

---

N. Kabashi (✉) · C. Krasniqi · A. Dautaj · E. Krasniqi
Faculty of Civil engineering and Architecture, Prishtine, Kosovo
e-mail: naser.kabashi@uni-pr.edu

J. Sustersic
Institut IRMA, Ljubljana, Slovenia

H. Morina
Institute IBMS, Prishtine, Kosovo

© Springer International Publishing AG, part of Springer Nature 2018
M. M. Reda Taha (ed.), *International Congress on Polymers in Concrete (ICPIC 2018)*, https://doi.org/10.1007/978-3-319-78175-4_79

and guidelines used in practice. Some of guidelines and rules will be used in this research paper in analyzing process and try to compare with experimental behavior of examined concrete beams. Fiber-reinforced polymer (FRP) reinforcement for concrete has been developed as a solution to replace steel and to avoid damage in RC structures due to corrosion, especially in aggressive environments. The proposed model for analysis, which developed the formula and methods using in steel bars, can't be directly applied to the FRP-reinforced concrete beams. The initial difference is on the mechanical properties: behavior of the steel bars and FRP bars which have the linear elasticity behavior; low modulus of elasticity, different stiffness, and different tensile resistance are the main indicator factors to develop and predict the formula and methods for calculations, predicted by models reference [1].

This work aims to give a contribution to the design procedures of FRP reinforced beams by a comparison between a large experimental database and theoretical predictions, both at service and ultimate conditions. Six concretes or two sets of beams reinforced with different GFRP reinforcement ratios were tested. The experimental results compared favorably with those predicted by the models [1, 12].

## 2 Design of FRP-RC Beams

The structural engineers are relatively unfamiliar with the design using the GFRP, because many factors are different. The behavior of the FRP reinforced bars is linear elastic to failure during the loading process, but brittle failure is different comparing with conventional steel reinforcement [1–3, 11].

### 2.1 Flexural Strength

The flexural strength of a reinforced beam section with FRP bars according to the ACI-Code requires:

$$M_u \leq \phi M_n \cdot M_n = 0.85 f'_c ab \left( d - \frac{a}{2} \right) \quad (79.1)$$

where $M_u$ is the required ultimate moment, $M_n$ is the nominal moment capacity of the beam section, and $\phi$ is the strength reduction factor ($\phi = 0.90$ for flexure).

During the design procedure in flexural for all design codes, it's important to calculate the three stages of reinforced section: (1) under-reinforced, (2) balanced, and (3) over-reinforced [1, 3, 6, 12].

## 2.2 Ductility Requirements

Ductility is a desirable structural property because it allows stress redistribution and provides warning of impending failure. In steel RC structures, ductility is defined as the ratio of post-yield deformation to yield deformation which usually comes from steel. Due to the linear-strain-stress relationship of FRP bars, the traditional definition of ductility cannot be applied to structures reinforced with FRP reinforcement [12].

The ductility requirements of the ACI-Code provisions limit the steel ratio

$$\rho_s = \frac{A_s}{b\,d} \qquad (79.2)$$

to lower and upper limits of $\rho_{smin} = \dfrac{1.4}{f_y}$ (where $f_y$ is in MPa) and $\rho_{smax} = 0.75\rho_{sbal}$;

$\rho_{sbal} = \dfrac{0.85\,\beta_1 f'_c}{0.67 f_{pu}}$; $f_{pu}$ – ultimate tensile strength of FRP bar.

## 2.3 Serviceability Requirements

A properly designed reinforced concrete element is to satisfy two requirements: strength and serviceability. At a cross-sectional level, two requirements limit the serviceability limit state (SLS): stresses in materials and cracking in structure. The stress in the FRP reinforcement should also be limited to avoid creep rupture or stress corrosion, which consists in the creep of the material under a constant load after a certain time. In general, cracking is controlled to ensure adequate structural performance as well as sufficient durability of the structure [1, 4, 8, 12].

The research paper present the developed of behavior the FRP-RC cracked rectangular sections. Assuming elastic behavior and that the Bernoulli hypothesis is satisfied, the curvature of a cracked section is:

$$K_{cr} = \frac{\varepsilon_c}{x} \qquad (79.3)$$

where $\varepsilon_c$ is the maximum concrete strain and $x$ is the distance from top surface to neutral axis.

Alternatively, the curvature can be:

$$K_{cr} = \frac{M}{E_c I_{cr}} \qquad (79.4)$$

where $M$ is the applied moment and $I_{cr}$ is the moment of inertia of the cracked section.

**Table 79.1** Properties of examined samples

| Sample | Type | Nominal diam. (mm) | Reinf. perc. (%) | Modulus of elasticity (GPA) | Tensile (N/mm$^2$) |
|---|---|---|---|---|---|
| "1" 2Ø6 mm | GFRP | 6.05 | 0.15 | 47.55 | 1022.1 |
| "2"2Ø10mm | GFRP | 10.05 | 0.42 | 38.45 | 1194.3 |

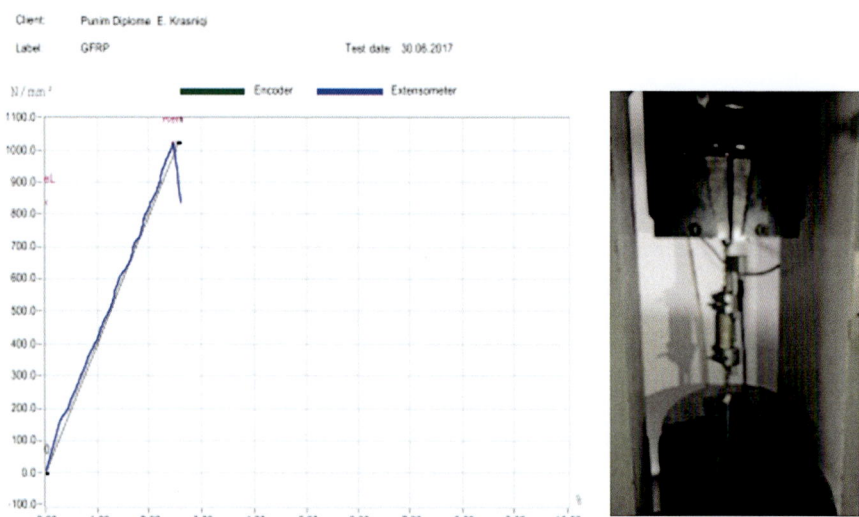

**Fig. 79.1** Examinations of mechanical properties of GFRP bars

## 3 Experimental Program

### 3.1 Materials

#### 3.1.1 Reinforced Bars

Two types of FRP reinforced bars are used in the experimental works: GFRP Ø6 mm and GFRP Ø10 mm with a helically wound fiber on the outside of the bar to produce surface deformations and are coated with a coarse silica sand to improve its bond to concrete. The mechanical properties are presented in Table 79.1 based on the experimental examinations presented in Fig. 79.1.

#### 3.1.2 Concrete

Common strength concrete, class C 30/37, was used with request to be used for aggressive environment.

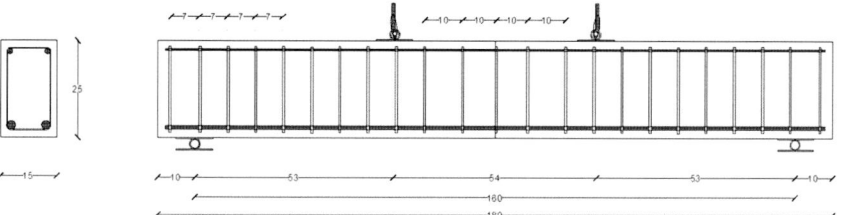

**Fig. 79.2** Geometric and reinforcement details of the test beams

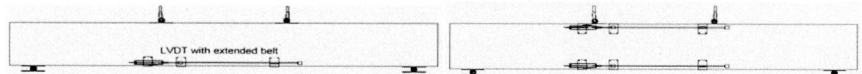

**Fig. 79.3** Typical beam test setup with deflections and cracks – experimental and analytical analysis

## 3.2 Concrete Beams for Testing

The reinforced concrete beam specimens were rectangular in cross section, with a width of 150 mm and a height 250 mm. Each reinforced beam specimen contained two reinforcing bars placed in a single layer presented in Fig. 79.2 [5, 7].

## 3.3 Testing Setup

The load was applied centrally by a 150 kN hydraulic jack, Controls MCC8, and a spreader beam was used to distribute the load to the two third-span points. The displacement transducers (LVDT) were used to measure deflections and cracks in critical positions presented in Fig. 79.3 [11, 12].

## 4 Experimental Results and Discussion

The beam testing results with regard to failure modes, load-deflection response, and moment-curvature response are provisory presented in this paper, as the primary focus is on the cracking behavior [1, 3, 6].

## 4.1 Crack Width and Spread of Cracks Using the GFRP Bars

The concrete beams reinforced with GFRP bars with different diameters, is observe the cracking behavior of the beams with focused to the size of crack widths and spread of the cracks. Crack width is calculated using different methods with objective tendency to compare with experimental results which are shown in Table 79.2 [9].

The comparison objective will be oriented on beams reinforced with GFRP-2Ø 6 mm, Series "1," and beams reinforced with GFRP-2Ø 10 mm, Series "2," with the beam behavior presented in Fig. 79.4. When the cross section exceeds about 40% of their bearing capacity, crack formation process begins rapidly, and "first crack phase" is emphasized in the sense of crack width and depth [9, 11, 12].

Comparing the behavior for different diameters of GFRP will be presented in Fig. 79.5 based on calculations methods presented in Table 79.2 [10, 11].

Different strength reduction factors are proposed in existing design documents to determine the appropriate limits depending on the different types of FRP reinforcement (different codes). Significant differences are found in design codes to limit the stress in the FRP reinforcement in order to avoid creep rupture or stress corrosion.

## 5 Conclusion

The conclusions are based on the comparison between experimental and analytical calculations, and based on that, we can conclude:

- The use of smaller-diameter GFRP bars offers the better deflection enhancement than larger-diameter bars for the same reinforcement ratio, in scope of behavior of the structural elements.
- The stiffness of beams was observed when increasing the reinforcement ratio or comparing the use of smaller bars or larger diameter bars.
- The GFRP in RC beams is designed for failure crushing in concrete, and capacity in flexure can be estimated using the method of over-reinforced sections.
- The crack spread prediction, analytical nonlinear analysis, FEM analysis, using the ATENA software, showed closed results compared with experimental results in our case study.
- The comparison analytical and experimental cracks analysis in all methods presented in research work gave the better prediction on the higher ratio-loading –crack width, compare with previous research works were are not specify.
- The usage of GFRP bars is limited only in few structures, due to its limitation of serviceability criteria.

**Table 79.2** Different methods on the crack analysis

| Series | Reinf. percent (%) | ULS Mr [kN m] | EC2 function [cracks-M/Mu] | Gergely-Lutz-SLS function[cracks-M/Mu] | Modified Gergely-Lutz Function [cracks-M/Mu] | ATENA relations [cracks-M/Mu] | Experimental results function[cracks-M/Mu] |
|---|---|---|---|---|---|---|---|
| "1" | 0.19 | 11.09 | 2.951 mm-72% | 1.598 mm-72% | 0.752 mm-72% | 1.722 mm-72% | 2.288 mm-72% |
| "2" | 0.46 | 20.56 | 1.967 mm-70% | 1.412 mm-70% | 1.014 mm-70% | 1.531 mm-70% | 1.739 mm-70% |

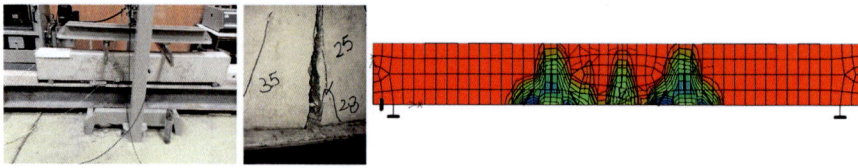

**Fig. 79.4** Crack mode failure under apply loads and a typical rupture pattern – ATENA software

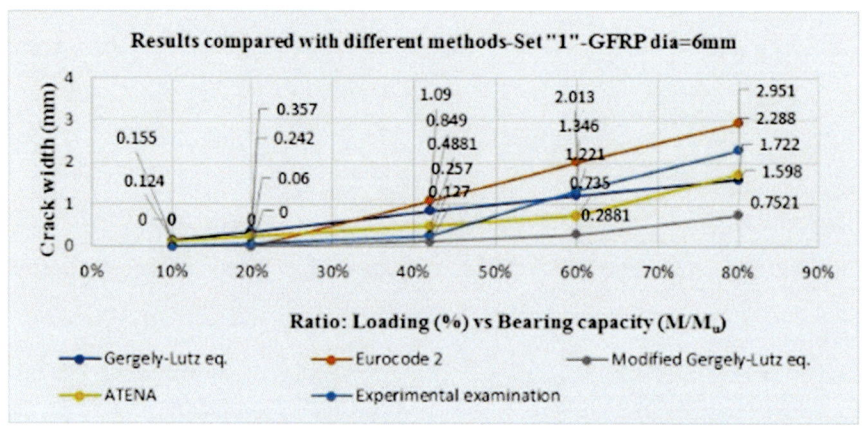

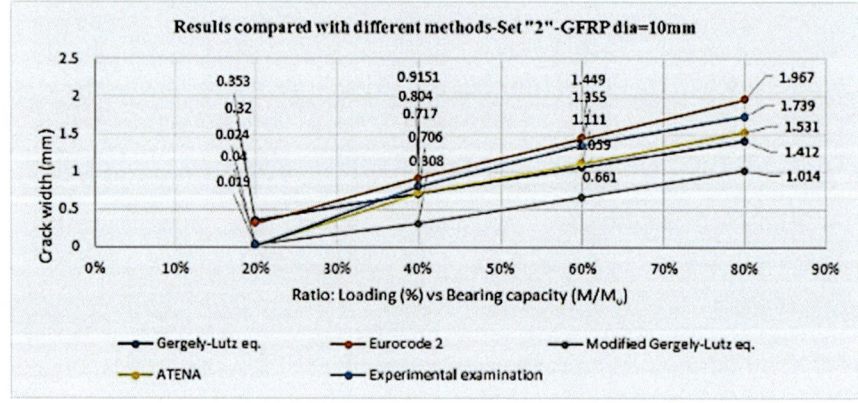

**Fig. 79.5** Behavior of beams – Sets "1" and "2" in crack development

# References

1. Bank, L. C. (2006). *Composites for construction: Structural design with FRP materials*. Hoboken: Wiley.
2. Teng, J. G., Chen, J. F., Smith, S. T., & Lam, L. (2002). *FRP strengthened RC structures*. Chichester: Wiley.
3. Arduini, M., & Nanni, A. (1997). Parametric study of beams with externally bonded FRP reinforcement. *ACI Structural Journal, 94*(5), 493–501.

4. Kabashi, N., Krasniqi, C., & Muriqi, A. (2013). Flexure behaviour the concrete beams reinforcement with polymer materials. In *Advanced materials research* (Vol. 687, pp. 472–479). Trans Tech Publications, Switzerland.
5. ACI Committee 440. (2007). *ACI 440R -07, Report on fiber-reinforced polymer (FRP) reinforcement for concrete structures*. American Concrete Institute, Farmington Hills, MI.
6. Bank, L. C. (1993). Properties of FRP reinforcement for concrete- in fiber-reinforced-plastic (FRP) reinforcement for concrete structures: Properties and applications. In A. Nanni (Ed.), *Developments in civil engineering* (Vol. 42, pp. 59–86). Amsterdam: Elsevier.
7. Toutanji, H. A., & Saafi, M. (2000). Flexural behavior of concrete beams reinforced with glass fiber-reinforced polymer (GFRP) bars. *ACI Structural Journal, 97*(5), 712–719.
8. Qu, W., Zhang, X., & Huang, H. (2009). Flexural behavior of concrete beams reinforced with hybrid (GFRP and steel) bars. *Journal of Composites for Construction, 13*(5), 350–359.
9. El-Nemr, A., ElSafty, A., & Benmokrane, B. (2016). *Flexural behavior of concrete beams reinforced with GFRP bars of different grades*. International Conference on Structural and Geotechnical Engineering.
10. Michaluk, C. R., Rizkalla, S. H., Tadros, G., & Benmokrane, B. (1998). Flexural behavior of one-way concrete slabs reinforced by fiber reinforced plastic reinforcements. *ACI Structural Journal, 95*, 353–365.
11. Yamini, Roja, S., Gandhi, P., Pukazhendhi, D. M., & Elangovan, R. (2014). Studies on flexural behaviour of concrete beams reinforced with GFRP bars. *International Journal of Scientific & Engineering Research, 5*(6), 82–89.
12. Alsayed, S. H. (1998). Flexural behaviour of concrete beams reinforced with GFRP bars. *Cement and Concrete Composites, 20*(1), 1–11.

# Chapter 80
# Flexural Rigidity Evaluation of Seismic Performance of Hollow-Core Composite Bridge Columns

Mohanad M. Abdulazeez and Mohamed A. ElGawady

This paper investigates experimentally the seismic behavior of two hollow-core fiber-reinforced polymer-concrete-steel (HC-FCS) columns under cyclic loading as a cantilever. The typical precast HC-FCS member consists of a concrete wall sandwiched between an outer fiber-reinforced polymer (FRP) tube and an inner steel tube. The FRP tube provides continuous confinement for the concrete wall along the height of the column. Five large-scale HC-FCS columns were investigated during this study to estimate the effective flexural (which is an important factor to define the buckling capacity and deflection of such columns) and the effective structural stiffness of the composite columns. These columns have the same geometric properties; the only difference was in the thickness of the inner circular steel tubes and the steel tube embedded length into the footing. A three-dimensional numerical model has been developed using LS-DYNA software for modeling these large-scale HC-FCS columns. The nonlinear FE models were designed and validated against experimental results gathered from HC-FCS columns tested under cyclic lateral loading and used to evaluate the effective stiffness's results. The estimated effective stiffness results that were obtained from the experimental work were compared with the FE results. This study revealed that the effective flexural and the effective structural stiffness for the HC-FCS columns need more investigation to be addressed in the standard codes, since the embedded hollow core steel tube socket connections cannot reach the fully fixed end condition to act as a cantilever member subjected to a lateral load with a fully fixed end condition. Moreover, the effective stiffness results were found to be highly sensitive to the steel tube embedded length and slightly to the unconfined concrete strength.

---

M. M. Abdulazeez · M. A. ElGawady (✉)
Department of Civil, Architectural, and Environmental Engineering, Missouri University of Science and Technology, Rolla, MO, USA
e-mail: elgawadym@mst.edu

© Springer International Publishing AG, part of Springer Nature 2018
M. M. Reda Taha (ed.), *International Congress on Polymers in Concrete (ICPIC 2018)*, https://doi.org/10.1007/978-3-319-78175-4_80

## 1 Introduction

It's estimated that Americans spend 14.5 million hours per day on traffic. Ten to fifteen percent of that congestion is caused by work zones even when it occurs in the off-peak, they increase traffic congestion [1]. The need for a new solution is highly requested to reduce the amount of time it takes to build our roads and bridges from months to several hours. Accelerated bridge construction (ABC) includes such elements as prefabricated modular units that are built off-site in a controlled environment and then transferred to the construction area for rapid installation. ABC reduces traffic disruptions and life-cycle costs and improves construction quality and safety, resulting in more sustainable development [2]. One technique to accelerate bridge construction is to use precast bridge columns with excellent seismic performance.

An excellent candidate for precast columns is the concrete-filled tube, which consists of a hollow tube made out of steel or fiber-reinforced polymer filled with concrete. Another candidate for precast columns is the hollow-core steel-concrete-steel (HC-SCS) columns consisting of two generally concentric tubes with concrete shell between them [3–8]. The concrete infill is confined by both tubes, resulting in high concrete confinement and column ductility [9]. All these mentioned research showed the superior seismic and axial capacity of HC-SCS columns.

Several regions around the world are susceptible to earthquake where large ductility demands are imposed on bridge columns. The design and the construction of a column-footing connection are crucial for precast columns to provide the ductility demands. Thus, under an earthquake event, all damage in the precast assembly is confined to the column, and the adjacent element experiences no damage, emulating cast-in-place construction performance. However, there is neither CFST nor HC-FCS columns that have satisfactory codes in standards such as *AASHTO LRFD Bridge Design Specifications* [10], the American Institute of Steel Construction (AISC) *Steel Construction Manual*, and the American Concrete Institute (ACI) [11]. Based on the available literature, limited studies have been focused on the design for CFST column-to-cap beam connections using different types that were proposed in the literature for CFST columns including welded and bolted steel plate and embedded base and rebars, and embedded structural steel connections were used for precast column-footing connections [12–14]. Likewise, limited studies have been conducted on understanding the performance of HC-FCS column [15] or on the column-footing connections [16]. These previous studies have concentrated mainly on determining the critical embedded length of the steel tube into the cap beam or the footing, but effect of the connection rigidity had not addressed comprehensively.

In this paper, a finite element model using LS-DYNA software was simulated to estimate the effective stiffness and flexural stiffness of HC-FCS column-footing connections in addition to the standard codes and compared to the experimental test results.

**Table 80.1** Summary of the investigated HC-FCS columns

| # | Column ID | Steel tube thickness $t_s$ (mm) | Steel tube embedded length $l_e$ (mm) |
|---|---|---|---|
| 1 [15] | F4-24-E344 | 12.7 | 635 |
| 2 [18] | F4-24-E324 | 6.35 | 635 |
| 3 | F4-24-E3(1.5)4 | 4.8 | 635 |
| 4 | F4-24-P1(0.8)4 | 2.8 | 508 |
| 5 | F4-24-E3(0.5)4 | 1.6 | 508 |

## 2 Experimental Work

Five 0.4-scale HC-FCS columns with different steel tube thicknesses and an embedment length were investigated in this study (Table 80.1). These columns were tested under constant axial load and lateral cyclic load. The tested HC-FCS columns have a circular cross-section with an outer diameter of 610 mm and a clear height of 2032 mm. The lateral load was applied at a height of 2413 mm with shear span-to-depth ratio of approximately 4.0. The column consisted of an outer filament-wound GFRP tube having a constant thickness of 9.5 mm along the height of the column. The inner steel tube had an outer diameter of 406 mm. A concrete wall having a thickness of 102 mm was sandwiched between the steel and FRP tubes (Fig. 80.1).

The columns' label used in the current experimental work consisted of four segments. The first segment is a letter F referring to flexural testing followed by the column's height-to-outer diameter ratio ($H/D_o$). The second segment refers to the column's outer diameter ($D_o$) in inch. The third segment refers to the GFRP matrix using E for epoxy and P for iso-polyester base matrices; this is followed by the GFRP thickness in 1/8 inch (3.2 mm), steel thickness in 1/8 inch (3.2 mm), and concrete wall thickness in inch (25.4 mm).

The HC-FCS column construction sequences and details have been illustrated in the literature [15, 17]. The mechanical properties of the steel tube, FRP tube, concrete mix design, and rebar are mentioned in details in the literature [15, 17].

### 2.1 Parametric Study

#### 2.1.1 FE Models Verification

Five-dimensional numerical models have been developed using LS-DYNA software for modeling of large-scale HC-FCS columns. The footing, concrete wall, and loading stub were modeled using solid elements with an average length of 25.4 mm and constant-stress one-point quadrature integration to reduce the computational time and increase the model stability. The outer FRP and inner steel tubes were simulated using shell elements with an average height of 25.4 mm. The hourglass type and coefficient used during this study were 5 and 0.03, respectively. The models have been described in details by Abdulazeez et al. [19].

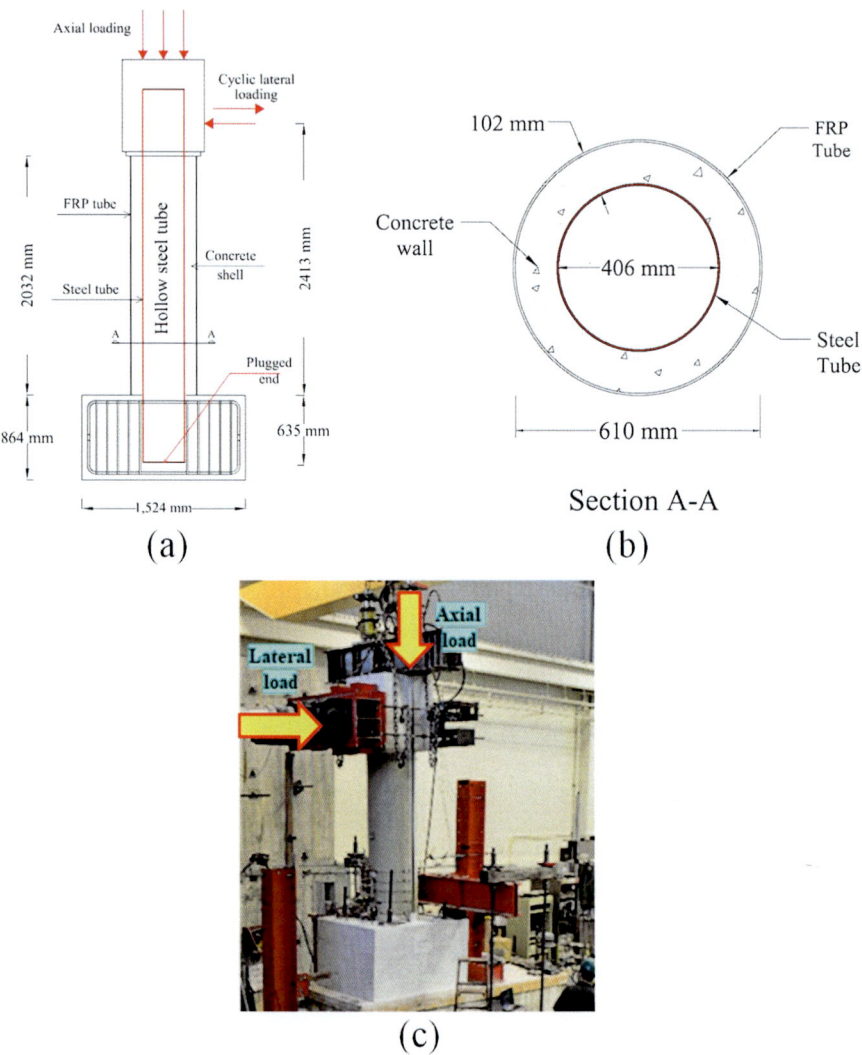

**Fig. 80.1** Construction layout of the column (**a**) elevation, (**b**) column cross-section, (**c**) HC-FCS column prior to the test

## 2.2 Loading Protocol and Test Setup

Constant axial load, $P$, of 489.3 kN (110 kips) corresponding to 5% of the calculated $P_o$ (of an equivalent RC-column with the same diameter 610 mm (24 inches) and 1% longitudinal reinforcement ratio was calculated as in [18]) was applied to the column using three external prestressing strands on each of the west and east sides of the column. After applying the axial load, static cyclic lateral load was applied in a

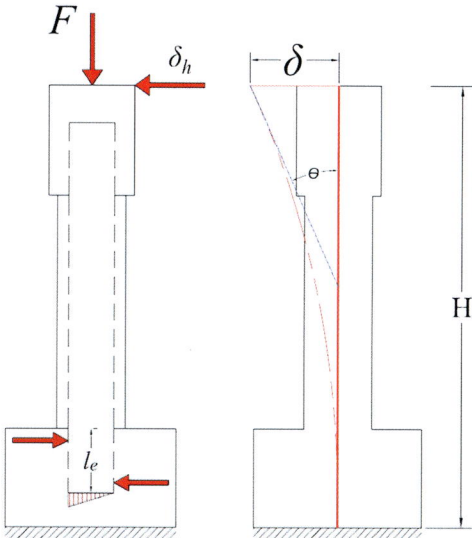

**Fig. 80.2** HC-FCS column load resisting mechanism

displacement control using two hydraulic actuators connected to the column loading stub. The loading regime is based on the recommendations of FEMA 2007. Two cycles were performed for each displacement amplitude [17].

## 3 Results and Discussion

### 3.1 Rigidity Evaluation

The combined load mechanism applied on the HC-FCS columns is described in Fig. 80.2. It is simply represented by a cantilever member subjected to a lateral load at the top with a fully fixed end condition at the bottom. The structural stiffness ($K$) for this assembly can be estimated using as Eq. (80.1). However, the embedded hollow core steel tube socket connections cannot reach the fully fixed end condition; the flexural stiffness of the embedded steel tube socket connection should be reduced from the fully fixed cantilever. Thus, more comprehensive study should be conducted using parametric analysis.

$$F = K\,\delta; \quad K = \frac{3EI_{\text{eff}}}{H^3} \tag{80.1}$$

The maximum elastic deflection is affected by the main parameters, including the embedded length, footing and column concrete strength, and thickness of the steel tube. The effective flexural stiffness has been estimated using the following standard provision expressions:

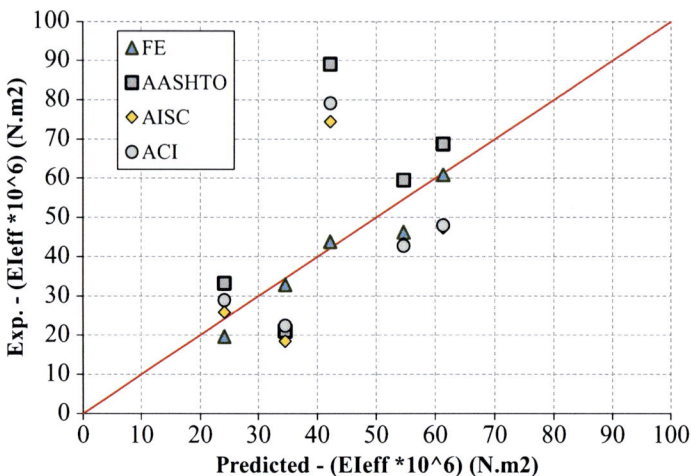

**Fig. 80.3** Effective flexural stiffness (experimental vs. predicted from FE and standard codes)

**Table 80.2** Summary of the predicted flexural stiffness results

| Column ID | Exp. | | FE | | ACI Eq. (80.2) | AASHTO Eq. (80.4) | AISC Eq. (80.3) |
|---|---|---|---|---|---|---|---|
| | $K$ | $EI_{eff}$ (N.m²) | $K$ | $EI_{eff}$ (N.m²) | $EI_{eff}$ (N.m²) | $EI_{eff}$ (N.m²) | $EI_{eff}$ (N.m²) |
| F4-24-E3 (0.5)4 | 2.2 | 34.5 | 2 | 32.8 | 23 | 21 | 18.4 |
| F4-24-P1 (0.8)4 | 1.5 | 24.2 | 1.4 | 19.6 | 29 | 33 | 26 |
| F4-24-E3 (1.5)4 | 3.3 | 54.3 | 2.8 | 47.2 | 43 | 59 | 43.3 |
| F4-24-E324 | 3.8 | 61.3 | 3.7 | 60 | 47.6 | 68 | 47 |
| F4-24-E344 | 2.7 | 42.3 | 2.5 | 41 | 78 | 89 | 74.5 |

$$EI_{eff} = E_s I_s + \frac{0.2 E_c I_g}{1 + \beta_d} \quad \text{ACI codes} \tag{80.2}$$

$$EI_{eff} = E_s I_s + C_3 E_c I_c; \quad C_3 = 0.6 + 2\left(\frac{A_s}{A_s + A_c}\right) \leq 0.9 \quad \text{AISC codes} \tag{80.3}$$

$$EI_{eff} = E_s I_s + 0.4 I_c \left(\frac{E_c A_c}{A_s}\right) \quad \text{AASHTO codes} \tag{80.4}$$

Figure 80.3 and Table 80.2 illustrate the evaluated results of $EI_{eff}$ and $K$ values for the investigated columns from the FE and the available codes compared to the experimental test results. As shown in Fig. 80.3, the AASHTO and AISC codes predicted larger $EI_{eff}$ than the ACI code and FE results compared to the obtained experimental test results especially for F4-24-E344 HC-FCS column. While, the predicted flexural and structural stiffness results from the FE models as well as the ACI code were close to the experimental test results.

The reason may be due to the insufficient steel tube embedded length that led to severe footing damage and steel tube pullout, and thereby not achieving full flexural behavior for the column [16]. The lowest values of $EI_{eff}$ were notice for columns F4-24-P1(0.8)4 and F4-24-E3(0.5)4 due to the short steel embedded length (508 mm). Consequently, the embedded length of the steel tube into the footing is a significant parameter that affects the stiffness of the entire socket connection and need to be addressed in the available codes for better HC-FCS column design.

## 4 Conclusion

The embedded length of the steel tube into the footing and the unconfined compressive strength for the column and the footing are significant parameters that affect the stiffness of the entire socket connection and need to be addressed in the available codes for better HC-FCS column socket connection design.

## References

1. Schrank, D. L., & Lomax, T. J. (2009). *2009 urban mobility report*. Texas Transportation Institute, Texas A & M University.
2. Dawood, H., Elgawady, M., & Hewes, J. (2014). Factors affecting the seismic behavior of segmental precast bridge columns. *Frontiers of Structural and Civil Engineering, 8*(4), 388–398.
3. Ozbakkaloglu, T., & Fanggi, B. L. (2013). Axial compressive behavior of FRP-concrete-steel double-skin tubular columns made of normal-and high-strength concrete. *Journal of Composites for Construction, 18*(1), 04013027.
4. Teng, J. G., Yu, T., Wong, Y. L., & Dong, S. L. (2007). Hybrid FRP–concrete–steel tubular columns: Concept and behavior. *Construction and Building Materials, 21*(4), 846–854.
5. Shakir-Khalil, H. (1991). Composite columns of double-skinned shells. *Journal of Constructional Steel Research, 19*(2), 133–152.
6. Idris, Y., & Ozbakkaloglu, T. (2013). Seismic behavior of high-strength concrete-filled FRP tube columns. *Journal of Composites for Construction, 17*(6), 04013013.
7. Teng, J. G., & Lam, L. (2004). Behavior and modeling of fiber reinforced polymer-confined concrete. *Journal of Structural Engineering, 130*(11), 1713–1723.
8. Abdelkarim, O. I., & ElGawady, M. A. (2016). Behavior of hollow FRP–concrete–steel columns under static cyclic axial compressive loading. *Engineering Structures, 123*, 77–88.
9. Anumolu, S., Abdelkarim, O. I., & ElGawady, M. A. (2016). Behavior of hollow-core steel-concrete-steel columns subjected to torsion loading. *Journal of Bridge Engineering, 21*(10), 04016070.
10. AASHTO, LRFD. (2008). "Bridge design specifications, customary US units, with 2008 interim revisions." American Association of State Highway and Transportation Officials, Washington, DC.

11. American Concrete Institute. (2005). *Building code requirements for structural concrete (ACI 318-05) and commentary (ACI 318R-05)*. Farmington Hills: American Concrete Institute.
12. Moon, J., Roeder, C. W., Lehman, D. E., & Lee, H. E. (2012). Analytical modeling of bending of circular concrete-filled steel tubes. *Engineering Structures, 42*, 349–361.
13. Lehman, D. E., & Roeder, C. W. (2012). Foundation connections for circular concrete-filled tubes. *Journal of Constructional Steel Research, 78*, 212–225.
14. Kingsley, A. M. (2005). *Experimental and analytical investigation of embedded column base connections for concrete filled high strength steel tubes* (Doctoral dissertation). University of Washington.
15. Abdelkarim, O. I., ElGawady, M. A., Gheni, A., Anumolu, S., & Abdulazeez, M. (2016). Seismic performance of innovative hollow-core FRP–concrete–steel bridge columns. *Journal of Bridge Engineering, 22*(2), 04016120
16. Abdulazeez, M. M., Abdelkarim, O. I., Gheni, A., ElGawady, M. A., & Sanders, G. (2017). Effects of Footing Connections of Precast Hollow-Core Composite Columns. Proceedings, Transportation Research Board (TRB) 96th annual meeting. No. 17-01256.
17. Abdulazeez, M. M., & ElGawady, M. A. (2017). *Seismic behavior of precast hollow-core FRP-concrete-steel column having socket connection*. Proceedings, SMAR 2017, ETH Zurich.
18. Abdelkarim, O. I., Gheni, A., Anumolu, S., & ElGawady, M. A. (2015). Seismic behavior of hollow-core FRP-concrete-steel bridge columns. In Structures Congress 2015, pp. 585–596.
19. Abdulazeez, M. M., & ElGawady, M. A. (2017). *Nonlinear analysis of hollow-core composite building columns*. Proceedings, SMAR 2017, ETH Zurich.

# Chapter 81
# Three-Dimensional Numerical Analysis of Hollow-Core Composite Building Columns

Mohanad M. Abdulazeez and Mohamed A. ElGawady

This paper presents a numerical study on the behavior of hollow-core fiber-reinforced polymer-concrete-steel (HC-FCS) columns with square steel tubes under combined axial compression and flexural loadings. The investigated HC-FCS column consisted of an outer circular fiber-reinforced polymer (FRP) tube, an inner square steel tube, and a concrete wall between them. Three-dimensional numerical models were developed using LS-DYNA software for modeling large-scale HC-FCS columns. The finite element (FE) models were designed and validated against experimental results gathered from HC-FCS columns tested under cyclic lateral loading. The FE results were in decent agreement with the experimental backbone curves. These models subsequently were used to conduct a parametric FE study investigating the effects of the confinement ratio and buckling instabilities on the behavior of the HC-FCS columns. In general, the HC-FCS columns with square steel tube failed by steel tube local buckling followed by FRP rupture. The obtained local buckling stresses that result from the FE models were compared with the values calculated from the empirical equations of the design codes. Finally, based on the FE model results, an expression was proposed to predict the square steel tubes local buckling stresses of HC-FCS columns.

## 1 Introduction

More recently, an advanced design version of hollow-core columns was proposed, where the outer steel tube was replaced with fiber-reinforced polymer (FRP) tube [1], later known as hollow-core FRP-concrete-steel (HC-FCS). The FRP tube fibers are

---

M. M. Abdulazeez · M. A. ElGawady (✉)
Department of Civil, Architectural, and Environmental Engineering, Missouri University of Science and Technology, Rolla, MO, USA
e-mail: elgawadym@mst.edu

mostly oriented in the hoop direction to increase the confinement of the concrete wall. Many studies have been conducted to investigate the effect of the new hybrid innovative columns on flexural, axial compression, and combined axial-flexural behavior. These studies demonstrated that HC-FCS columns could display superior performance under extreme loads as the concrete wall is confined by both FRP and steel tubes, resulting in a triaxial state of compression that increases the strength and strain capacity of the concrete. Furthermore, local and global buckling of the steel tube is restrained by the confined concrete wall and thereby increases the deformation and strength capacity of an HC-FCS member.

Experimental and theoretical studies have been conducted on bare steel tubes as well as concrete-filled box steel tube columns (CFBST) to investigate the behavior and accuracy of the local buckling instabilities [2–10]. The theoretical solutions in these studies have been developed to ascertain the initial local and post-local buckling and have relied mostly on the use of two methods: finite strip method (FSM) and effective width method (EWM). The FSM was introduced by Cheung (1976) [11] and then developed by invoking the semi-analytical finite strip method (SAFSM) by Uy and Bradford [8] and Uy (1998). This method is used in the initial local buckling capacity calculations by incorporating the effect of the residual stresses, which can be important in the elastic range of structural response [12]. The concept of EWM was first proposed by Von Karman et al. [2] for perfect plates, which accounts for post-buckling of stiffened plate elements by suggesting that the distribution of the design stresses was concentrated at the supporting edges. This method was then modified by Winter (1970) [7] to account for the reduction in real plate strength due to the effect of the imperfection. Formulas are widely available for calculating the buckling in this linear domain, but they are not available for nonlinear stage which requires advanced computer technology using FE analysis. Nonlinear buckling analysis provides greater accuracy than the linear elastic stage [13–16].

The presented paper introduces extensive parametric FE analysis study on large-scale HC-FCS columns with square steel tubes to better understand their behavior under combined lateral and axial loads. The parametric study includes the effects of the confinement ratio and local buckling instability on the behavior of HC-FCS columns. The FE analyses were developed using LS-DYNA software based on the nonlinear large displacement theory. The study proposed a simple expression to calculate the local buckling stresses (bifurcation point) of HC-FCS columns with square steel tubes.

## 2 FE Modeling

Solid elements with constant-stress one-point quadrature integration have been used for modeling the footing, concrete wall, and loading stub. The outer FRP and inner square steel tubes were simulated using Belytschko-Tsay four-node shell elements with six degrees of freedom per node. A sensitivity analysis was conducted to determine the different element sizes. The final model had 4144 elements and 5544 nodes. The hourglass type and coefficient used during this study were 5 and 0.03, respectively. The model materials, contact, and boundary conditions have been explained in details in the literature [17].

The column was loaded using half of the axial compressive load applied to the top of the loading stub due to the symmetry. Then, the column was loaded laterally with a linear ramp-up displacement at the middle nodes of the common surface between the column top and the bottom surface of the loading stub at a height of 1000 mm. Using displacement-controlled loading allows tracking of the post-buckled performance of the investigated column.

## 2.1 FE Parametric Study

The parametric study was carried out using the validated model to provide in-depth understanding of the performance of the full-scale HC-FCS columns having square inner steel tubes. The effects of the FRP confinement ratio and the steel tube local buckling instability parameters were examined, and a simple expression to calculate the local buckling stresses (bifurcation point) of HC-FCS columns with square steel tubes was proposed using the FE results.

Table 81.1 shows the variables of the investigated columns. All of the columns had an outer diameter ($D_o$) of 1524 mm and a height of 10,160 mm (Fig. 81.1). The lateral load was applied at a height ($H$) of 7620 mm from the footing top level resulting in an aspect shear span-to-diameter ratio ($H/D_o$) of 5. The outer FRP tube had a thickness $t_f$ ranging from 4.7 mm to 23.7 mm, resulting in a confinement ratio that ranged from 0.05 to 0.25. The FRP confinement ratio was calculated using Eq. 81.1:

$$\text{Confinement ratio} = \frac{f_l}{f'_c} = \frac{2 E_f t_f \varepsilon_f}{D_f f'_c} \tag{81.1}$$

where $f_l$ is the confining pressure; $f'_c$ is the concrete unconfined compressive strength; $E_f$ is the hoop modulus of elasticity of the FRP tube; $t_f$ is the total nominal thickness of the FRP tube; $\varepsilon_f$ is the hoop ultimate tensile strain of the FRP tube; and $D_f$ is the internal diameter of the FRP tube.

**Table 81.1** Summary of the parametric study results

| Column ID | Details | | FE moment capacity (kN-m) | FE lateral drift (%) |
|---|---|---|---|---|
| C9 | Confinement ratio (CR) | 0.05 | 8346 | 1.15 |
| C0 | | 0.10 | 9992 | 6.32 |
| C10 | | 0.15 | 11,395 | 6.88 |
| C11 | | 0.20 | 12,458 | 7.38 |
| C12 | | 0.25 | 12,298 | 7.08 |

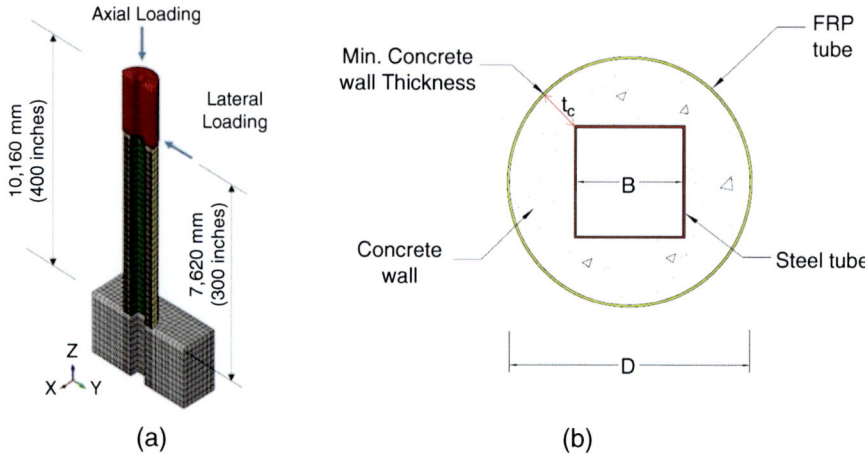

**Fig. 81.1** FE model of the square HC-FCS columns (**a**) large-scale model; (**b**) column's cross section

The axial load ($P$) was applied on each column with a percent of its axial capacity ($P_o$), which was calculated according to ACI-318 (2014) [18] using Eq. 81.2. The $P$ value ranged from 5% to 45% of the $P_o$.

$$P_o = A_s f_y + 0.85 A_c f'_c \tag{81.2}$$

where $f_y$ is the yield strength of the steel tube; $A_s$ is the cross-sectional area of the steel tube; and $A_c$ is the cross-sectional area of the concrete column.

## 2.2 FE Validation

The validated FE model using LS-DYNA software was presented by Abdulazeez and ElGawady for the HC-FCS column with square steel tube [17]. The model was validated with the column experimentally tested by Ozbakkaloglu and Idris (2014) [19]. The FE results had a good agreement with the experimental results with high accuracy [20].

## 3 Results and Discussion

### 3.1 Effect of Confinement Ratio $(f_l/f'_c)$

Five different columns with confinement ratios of 0.05, 0.1, 0.15, 0.2, and 0.25 were modeled to investigate the effects of the confining pressure on both the buckling

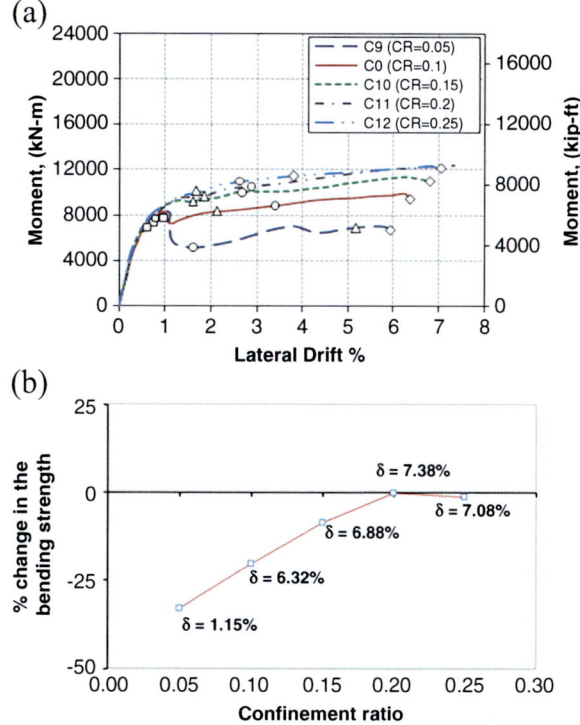

**Fig. 81.2** Nominal confinement ratio (CR) (**a**) moment versus lateral displacement capacity; (**b**) percentage change in the bending strength and maximum lateral displacement capacity □ Steel tube buckling in compression ○ Steel tube yielded in compression △ Steel tube yielded in tension ◇ FRP tube rupture

**Table 81.2** Summary of the used expressions

| References | Method | Expression |
|---|---|---|
| Uy and Bradford (1996) [8] | SAFSM | $F_{cr} = \frac{\pi^2 E_s}{12(1-v^2)} k \left(^t/_b\right)^2$ |
| AISC (2010) [21, 22] | Effective width | $F_{cr} = \frac{9 E_s}{\left(^b/_t\right)^2}$ |
| Ge and Usami (1994) [3] | Effective area | $\frac{f_b}{f_y} = \frac{1.2}{R} - \frac{0.3}{R^2} \leq 1.0, R = \frac{b}{t} \sqrt{\frac{12(1-v^2)}{4\pi^2}} \sqrt{\frac{f_y}{E_s}}$ |

instabilities and HC-FCS columns' bending capacity. The columns' concrete wall thickness $t_c$ is (254 mm) with steel tube thickness $t_s$ of (6 mm).

Figure 81.2a, b and Table 81.2 show that by increasing the confinement ratio by 300 % (from 0.05 to 0.2), the moment capacity increased by 49.3 % from 8346 kN-m to 12,458 kN-m. This moment strength increase was due to the increase of the compressive strength of the confined concrete as well as stiffness, which reduced the contribution of the steel tube in the axial compressive stresses. This behavior led to delaying the steel tube local buckling. However, it was found that columns C11 and C12 with respective confinement ratios of 0.2 and 0.25 had almost the same moment strength. This behavior indicated that the higher confinement ratio of 0.25 has no remarkable effect on increasing the confined concrete strength. The HC-FCS column with small CR weakened the concrete wall stiffness and thus its ability to sustain the compression loads. As shown in Fig. 81.3, at the same lateral drift, column C9 with CR of 0.05 had a higher concrete lateral pressure than column C11 with CR of 0.2, which led to form steel tube local buckling instabilities.

## 3.2 Buckling Strength Evaluation for HC-FCS

The instability of the HC-FCS members due to the local and global buckling is considered as a limit design state. Local buckling developed in the inner steel tube as compressive stress-initiated complex phenomena causes a case of redistribution of the generated stresses. Table 81.2 summarizes three known expressions by Uy and Bradford [8], American Institute of Steel Construction (AISC) manual, and Ge and Usami [3] used to determine the local buckling critical compressive stresses of steel plates. These expressions were used to calculate the local buckling stresses for the investigated HC-FCS columns with different $B/t_s$ of 30, 60, 90, 120, and 180. The obtained results were compared with the FE buckling stress outcomes [17].

The value of the first buckling stress has been identified on the equilibrium path by a decrease in the stress-strain curve.

where $F_{cr}$ is the buckling stress (MPa), $k$ is the buckling coefficient = 10.30 for sidewalls in concrete filled tubes, $E_s$ is the elastic modulus (GPa), $v$ is the Poisson's ratio, $b/t$ is the steel slenderness ratio, $f_b$ is the local buckling strength (MPa), and $R$ is the width-to-thickness ratio.

All the columns failed by buckling stresses developed at early loading stage except the HC-FCS column with $B/t_s = 30$, where the steel tube reached early to yielding state in the compression (loading) side due to the steel tube large thickness of 24 mm.

A nonlinear regression analysis was then performed on the calculated and FE critical stress, and the best fit for predicting the buckling stresses is presented in Fig. 81.4 and the proposed Eq. 81.3:

$$F_{cr} = 1.82 \, E_s \times \left(b/t\right)^{-1.65} \qquad (81.3)$$

where $F_{cr}$ is the buckling stress (bifurcation point) (MPa) of HC-FCS columns and $b/t$ is the steel slenderness ratio.

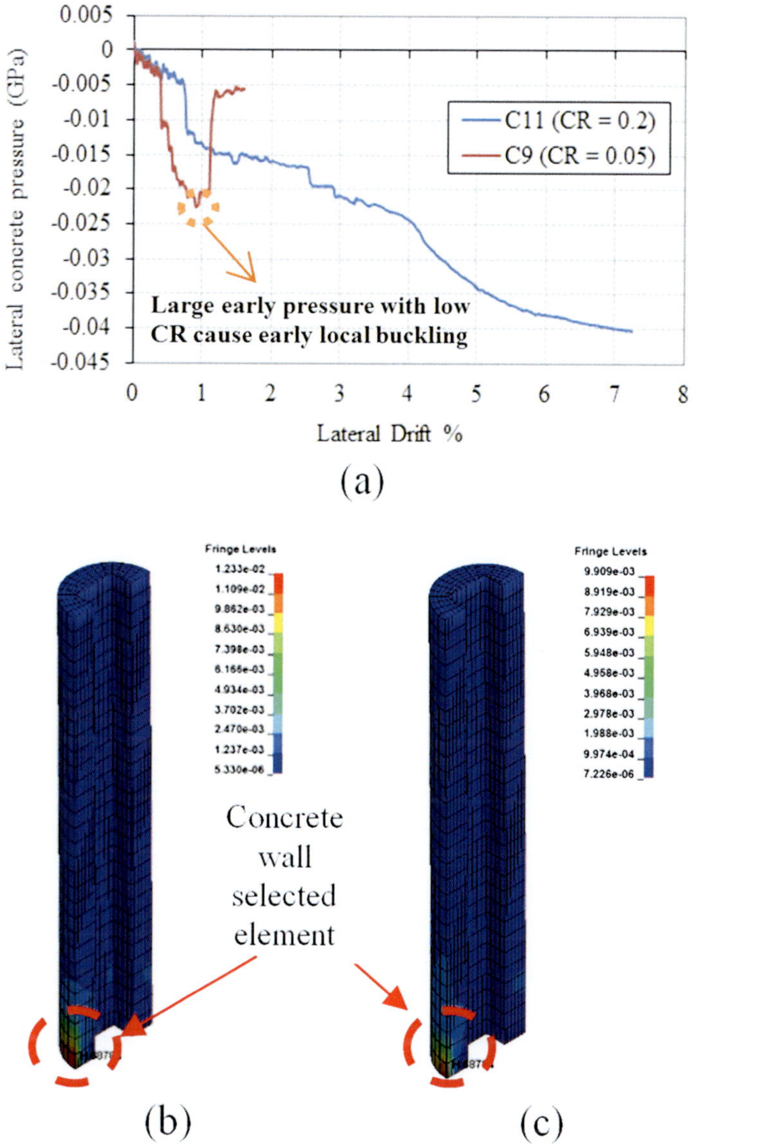

**Fig. 81.3** (a) Concrete wall-FRP tube lateral pressure for an element in the south direction at 254 mm above the footing top level; (b) C9; (c) C11

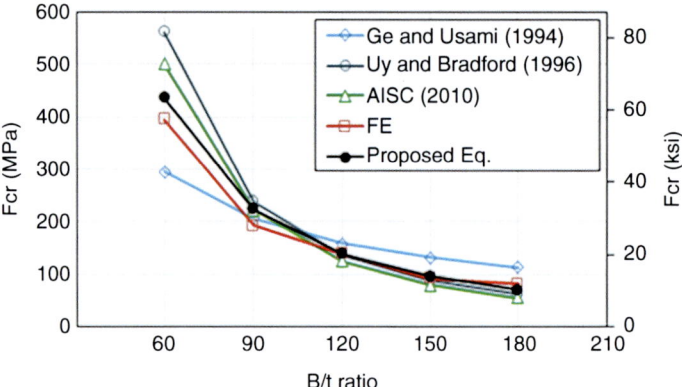

**Fig. 81.4** Buckling strength versus $B/t_s$ ratios

## 4 Conclusions

Based on the parametric study, observations, comparison, and the results demonstrated in this study, the following conclusions can be formed:

1. Increasing the confinement ratio $(f_l/f'_c)$ by 300% (from 0.05 to 0.2) resulted in an increase in the bending capacity by 49.3%. This moment strength increasing was due to the increase of the compressive strength of the confined concrete as well as stiffness, which reduced the contribution of the steel tube in the axial compressive stresses and delaying local buckling occurring.
2. The buckling strength increased with increasing the steel tube thickness. An expression was presented to predict the steel tube local buckling stress of the HC-FCS columns using nonlinear FE analysis.

## References

1. Teng, J. G., & Lam, L. (2004). Behavior and modeling of fiber reinforced polymer-confined concrete. *Journal of Structural Engineering, 130*(11), 1713–1723.
2. Karman, T. V. (1932). The strength of thin plates in compression. *Transactions of the ASME, 54*(2), 53.
3. Ge, H. B., & Usami, T. (1994). Strength analysis of concrete-filled thin-walled steel box columns. *Journal of Constructional Steel Research, 30*(3), 259–281.
4. Guo, L., Zhang, S., Kim, W. J., & Ranzi, G. (2007). Behavior of square hollow steel tubes and steel tubes filled with concrete. *Thin-Walled Structures, 45*(12), 961–973.
5. Teng, J. G., & Rotter, J. M. (Eds.). (2006). *Buckling of thin metal shells*. New York: CRC Press.
6. Winter, G. (1947). Strength of thin steel compression flanges. *Transactions of the ASCE, 112*, 527.
7. Winter, G. (1970). Commentary on the 1968 edition of the specification for the design of cold-formed steel structural members.

8. Uy, B., & Bradford, M. A. (1996). Elastic local buckling of steel plates in composite steel-concrete members. *Engineering Structures, 18*(3), 193–200.
9. Wright, H. D. (1995). Local stability of filled and encased steel sections. *Journal of Structural Engineering, 121*(10), 1382–1388.
10. AS 4100. (1998). Steel structures, Sydney: Standards Australia.
11. Cheung, Y. (1976). Finite strip method in structural mechanics. Pergoman, New York, USA.
12. Uy, B. (2001). Local and postlocal buckling of fabricated steel and composite cross sections. *Journal of Structural Engineering, 127*(6), 666–677.
13. Rust, W., & Franz, U. *Quasi-static limit load analysis by ls-dyna in combination with ANSYS*. CAD-FEM GmbH, Schmiedestraβe 31, D-31303 Burgdorf.
14. Pavlovčič, L., Froschmeier, B., Kuhlmann, U., & Beg, D. (2012). Finite element simulation of slender thin-walled box columns by implementing real initial conditions. *Advances in Engineering Software, 44*(1), 63–74.
15. Sussman, T., & Bathe, K. J. (1987). A finite element formulation for nonlinear incompressible elastic and inelastic analysis. *Computers & Structures, 26*(1–2), 357–409.
16. Byklum, E., & Amdahl, J. (2002). A simplified method for elastic large deflection analysis of plates and stiffened panels due to local buckling. *Thin-Walled Structures, 40*(11), 925–953.
17. Abdulazeez, M. M., & ElGawady, M. A. (2017). *Nonlinear analysis of hollow-core composite building columns*. Proceedings, SMAR 2017, ETH Zurich.
18. American Concrete Institute (ACI). (2014). Building code requirements for structural concrete and commentary. ACI 318-14, Farmington Hills, MI.
19. Ozbakkaloglu, T., & Idris, Y. (2014). Seismic behavior of FRP-high-strength concrete–steel double-skin tubular columns. *Journal of Structural Engineering*.
20. Abdulazeez, M. M., & ElGawady, M. A. (2017). *Seismic behavior of precast hollow-core FRP-concrete-steel column having socket connection*. Proceedings, SMAR 2017, ETH Zurich.
21. AISC Committee. (2010). *Specification for structural steel buildings (ANSI/AISC 360-10)*. Chicago: American Institute of Steel Construction.
22. Ziemian, R. D. (Ed.). (2010). *Guide to stability design criteria for metal structures*. Hoboken: Wiley.

# Chapter 82
# Pultruded GFRP Reinforcing Bars with Carbon Nanotubes

Rahulreddy Chennareddy, Amr Riad, and Mahmoud M. Reda Taha

Pultrusion is a renowned method in industry to produce glass fiber reinforced polymer (GFRP) reinforcing bars. Pristine multiwalled carbon nanotubes (P-MWCNTs) and multiwalled carbon nanotubes (MWCNTs) with carboxyl functional group (COOH-MWCNTs) were dispersed into the vinyl ester resin to produce GFRP bars. The GFRP bars were produced using a pultrusion prototype facility recently developed at the University of New Mexico. Direct tension and short beam shear tests were conducted to determine the mechanical properties of the nano-modified GFRP reinforcing bars. The experimental program shows the ability of MWCNTs to improve the mechanical behavior of GFRP reinforcing bars by 20% and 111% for the tensile and shear strength, respectively. Of particular interest is the absence of the typical broom failure observed in neat GFRP when functionalized MWCNTs were used. The proposed nano-modification of GFRP using MWCNTs might enable overcoming many of the current limitations of GFRP reinforcing bars.

## 1 Introduction

Corrosion caused by the use of deicing salts and severe climate conditions is responsible for numerous structurally deficient bridge decks [1]. GFRP reinforcing bars have become an acceptable alternative for typical steel bars when corrosion is a major problem [1]. Presently, GFRP is commercially available at a relatively low price in different configurations such as uni- and bidirectional laminates, reinforcing bars, and pultruded sections. GFRP reinforcing bars are used for both new

---

R. Chennareddy · M. M. Reda Taha
Department of Civil Engineering, University of New Mexico, Albuquerque, NM, USA

A. Riad (✉)
Department of Civil Engineering, Faculty of Engineering, Al-Azhar University, Cairo, Egypt

construction and for strengthening of existing structures. However, literature shows that GFRP exhibits premature tension failure due to weak interfacial bond between the glass fibers and the polymer matrix [2]. This weak interfacial bond results in a number of other potential limitations of GFRP, including limited fatigue strength and relatively low creep rupture stress [3]. Such mechanical limitations result in design code provisions limiting the maximum stress in GFRP bars in structural design [4]. More importantly, shear strength of GFRP is relatively low compared with carbon fiber reinforced polymer (CFRP) and steel bars. This limits the possible use of GFRP as dowels for bridge decks or slabs on grades and in shear critical regions [5]. Finally, all GFRP frame structures utilize pultruded GFRP sections for their light weight, easy construction, and corrosion resistance [6]. Structural design using these sections is typically governed by the limited shear strength of GFRP profiles at structural joints. Limited shear strength of GFRP thus represents a major limitation for its practical use in concrete and other structures.

Carbon nanotubes (CNTs) are the strongest materials available today [7]. With appreciable strength, low cost, and easy industrial availability, MWCNTs in small quantities are used to improve the strength and stiffness of the polymer composite materials [8]. When MWCNTs are dispersed in a polymer matrix, they act as reinforcement fibers at the microscale. However, the diameter of MWCNTs in nano scale allows them to interfere with the polymerization of the polymers, altering the polymer matrix. Furthermore, MWCNTs can be engineered by surface functionalization using active chemical groups to form covalent bonds with the matrix [9]. In the current study, we suggest using vinyl ester polymer nanocomposite by incorporating pristine (P-MWCNTs) at 2.0 wt.% of the vinyl ester resin and MWCNTs functionalized with carboxylic group (COOH-MWCNTs) at 0.5 wt.% of the vinyl ester resin. Our hypothesis is that incorporating MWCNTs into the polymer resin will improve the bond between the polymer matrix and the silane sizing on the surface of glass fibers. This will lead to improving the mechanical properties, specifically shear strength of GFRP.

## 2 Experimental Methods

Glass fiber spools used for pultrusion were Hybon® 2732, supplied by PPG Industries, Inc. Vinyl ester (700) with methyl ethyl ketone peroxide as the curing agent has been used as the polymeric matrix in fabricating the GFRP pultruded composite bars. P-MWCNTs, and COOH-MWCNTs were supplied by Cheap Tubed, Inc., Grafton, VT, USA. These MWCNTs have inner diameter of 5–10 nm and outer diameter of 20–30 nm with bulk density of 0.21 gm/cm$^3$ and 110 m$^2$/g specific surface area. For dispersing MWCNTs in the ester resin, ultrasonication at 40 °C for 60 min followed by mechanical stirring at 800 rpm for 120 min at 80 °C were used. Later the MWCNTs–vinyl ester nanocomposite was allowed to cool down to room temperature and then used in the pultrusion of GFRP bars. For the pultrusion process, a die with 9.5 mm diameter hole and 600 mm long with heating plates has been used to

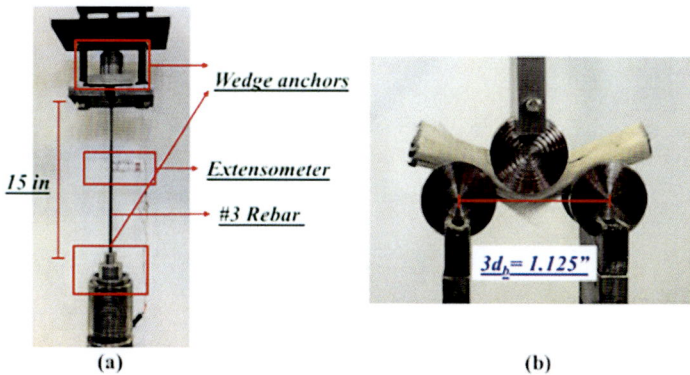

**Fig. 82.1** Test setup: (**a**) direct tension; (**b**) longitudinal shear test of GFRP bars incorporating MWCNTs

**Table 82.1** Test results

| Sample description | Tensile strength MPa | Tensile modulus GPa | Shear strength MPa |
|---|---|---|---|
| Neat GFRP | 694 ± 71 | 45.4 ± 0.29 | 24.6 ± 1.0 |
| 0.50 wt.% COOH-MWCNTs | 832 ± 42 | 45.5 ± 1.66 | 49.6 ± 2.4 |
| 2.0 wt.% P-MWCNTs | 708 ± 18 | 46.8 ± 0.28 | 37.8 ± 2.1 |

maintain a constant temperature of 150 ± 1 °C inside the die to cure the bar. A constant pull speed of 3.0 mm/s was used with a speed controlled gear motor. Post fabrication, the GFRP bars were cured at 130 °C for 2 h to ensure complete polymerization of the polymer matrix. GFRP bars with constant fiber volume fraction in three MWCNTs concentrations, 0.0 wt.% MWCNTs (neat), 0.5 wt.% COOH-MWCNTs, and 2.0 wt.% P-MWCNTS were fabricated. Three bars were mechanically tested for each type under uniaxial tension following ASTM D7205/D7205M and five bars for each type under longitudinal shear test using short beam bend test following ASTM D4475 [10, 11]. Figure 82.1a, b presents the experimental protocol for tensile and short beam shear test. The data for the two tests were acquired at 10 Hz interval. Fiber volume fraction of the GFRP bars with and without MWCNTs was determined using ASTM-D3171 [12].

## 3 Results and Discussion

The fiber volume fractions of the GFRP for neat, 0.5 wt.% COOH-MWCNTs, and 2.0 wt.% P-MWCNTs GFRP bars were 61.2%, 59.3%, and 60.4%, respectively. The results of the direct tension tests are presented in Table 82.1. The stress–strain behavior of GFRP with and without MWCNTs is shown in Fig. 82.2. Tension test results indicate that an improvement in tensile strength by 20% was achieved

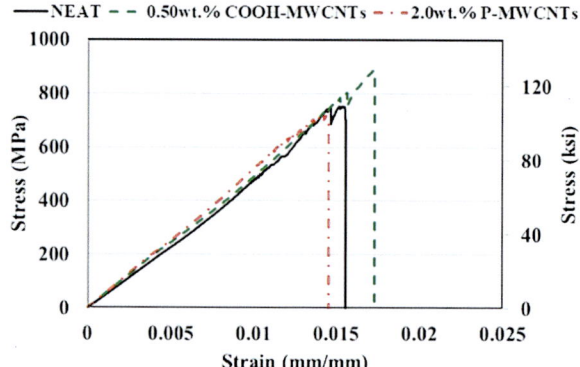

**Fig. 82.2** Stress–strain behavior of GFRP bars neat and with MWCNTs under uniaxial tension

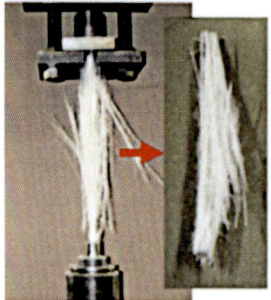
NEAT
Significant broom

0.50wt.% COOH-MWCNTs
No broom

2.0wt.% P-MWCNTs
Significant broom

**Fig. 82.3** Tension failure modes for GFRP bars with MWCNTs

compared with neat GFRP bars when functionalized COOH-MWCNTs were used. This improvement was proven to be statically significant with 95% confidence level using student t-test.

The stress–strain behavior of GFRP with MWCNTs showed a linear elastic behavior to failure with similar slopes for all the GFRP samples with and without MWCNTs. The strain at failure was higher for the samples with 0.5 wt.% COOH-MWCNTs, as shown in Fig. 82.2. This increase in the strain at failure can be attributed to the improved interfacial bond between the silane sizing on the glass fibers and the COOH functionalization on the MWCNTs. However, GFRP incorporating 2.0 wt.% P-MWCNTs showed a negligible improvement in tensile strength and strain compared with neat GFRP. This negligible improvement might be attributed to the absence of functional groups in P-MWCNTs to interfere with the polymerization and to improve the bond with glass fibers. Moreover, GFRP bars with 2.0 wt.% P-MWCNTs showed a similar stress–strain behavior to that of neat GFRP. More interestingly, the modes of failure in tension of GFRP bars incorporating MWCNTs are presented in Fig. 82.3. It is apparent that neat GFRP and GFRP

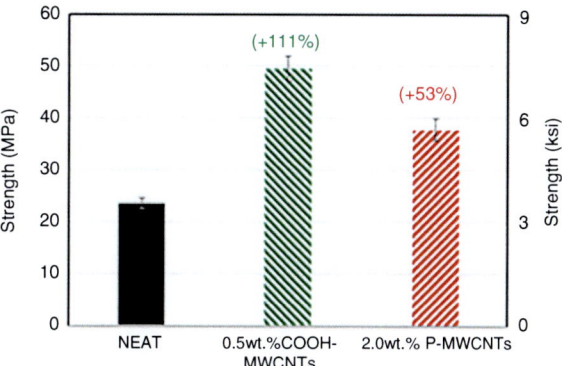

**Fig. 82.4** Short beam shear strength for GFRP bars incorporating MWCNTs

incorporating 2.0 wt.% P-MWCNTs showed the typical broom failure exhibited in GFRP as reported in the literature [2]. However, GFRP with 0.5 wt.% COOH-MWCNTs showed almost no broom failure. This might be explained by the apparent ability of COOH-MWCNTs to improve the interfacial bond between glass fibers and vinyl ester matrix. This, in its turn, resulted in an increased tensile strength and prevented the typical broom effect that follows fibers debonding from the matrix.

Furthermore, the short beam shear strength results are presented in Table 82.1. A significant improvement in shear strength by 111% and 53% was observed for GFRP incorporating 0.5 wt.% COOH-MWCNTs and 2.0 wt.% P-MWCNT, respectively, compared with neat GFRP. The results are summarized in a bar chart shown in Fig. 82.4. The shear strength improvements of GFRP bars with MWCNTs compared with neat GFRP were proved to be statistically significant with 95% confidence level using student t-test. As the longitudinal shear strength of the GFRP is matrix dominant behavior, it is obvious that MWCNTs can significantly improve the shear strength of GFRP bars. While the improvement using COOH-MWCNTs can be explained by the possible chemical reaction of COOH-MWCNTs and the vinyl ester matrix, the ability of P-MWCNTs to improve the shear strength of GFRP bars needs further discussion. It is apparent that the relatively high content of P-MWCNTs (2.0 wt.%) used in producing GFRP bars, enabled the P-MWCNTs to act as microscale reinforcement in the vinyl ester matrix and thus enabled improved transfer of shear stresses within GFRP composite bar. Further research is warranted to explain this observation. The above results indicate that using as low 0.5 wt.% COOH-MWCNTs well-dispersed in the vinyl ester matrix prior to fabrication of GFRP bar can significantly improve the tensile strength by 20% and shear strength by 111%. This high improvement in shear strength of GFRP can have significant economic benefits in design of GFRP dowels widely used in bridge decks and slabs on grades. Economic analysis of the above addition showed the use of MWCNTs can result in increasing GFRP cost by 10–15%. This is a very limited cost increase compared to the significant improvement in shear strength above 100% of neat GFRP bars. Further investigations are ongoing to examine other improvements and benefits in using those new GFRP bars [13].

Further investigations are warranted to examine the possible use of different MWCNTs concentrations both for COOH-MWCNTs and P-MWCNTs. Finally, microstructural analysis using scanning electron microscope (SEM) and Fourier transformation infrared spectroscopy (FTIR) are also warranted to prove the existence of chemical reaction between MWCNTs and the vinyl ester matrix.

## 4 Conclusions

Incorporating functionalized COOH-MWCNTs improved the tensile strength of pultruded GFRP reinforcing bars by up to 20% and improved the shear strength by 111% with an evident change in GFRP failure mode. Incorporating P-MWCNTs improved shear strength of GFRP reinforcing bars by 53% and made no effect on the tensile strength and the failure mode. Improvement in shear strength of GFRP bars is attributed to the potential chemical reaction of COOH-MWCNTs with the vinyl ester matrix producing improved bond with the silane sizing on glass fibers. Shear strength improvements with P-MWCNTs is attributed to the ability of P-MWCNTs to work as microscale fiber reinforcement, improving shear transfer within the GFRP bars. The significant improvement in shear strength of GFRP bars using MWCNTs is specifically useful for GFRP dowels used in bridge deck applications.

**Acknowledgments** This work is funded by the TranSET University Transportation Center (UTC). The authors greatly acknowledge this support.

## References

1. Nkurunziza, G., Debaiky, A., Cousin, P., & Benmokrane, B. (2005). Durability of GFRP bars: A critical review of the literature. *Progress in Structural Engineering and Materials, 7*(4), 194–209.
2. Kumar, M. S., Raghavendra, K., Venkataswamy, M. A., & Ramachandra, H. V. (2012). Fractographic analysis of tensile failures of aerospace grade composites. *Materials Research, 15*(6), 990–997.
3. Brown, D. L., & Berman, J. W. (2010). Fatigue and strength evaluation of two glass fiber-reinforced polymer bridge decks. *Journal of Bridge Engineering, 15*(3), 290–301.
4. CSA S6. (2010). *Canadian highway bridge design code (CHBCD)*. Mississauga: Canadian Standards Association International.
5. Yost, J. R., Gross, S. P., & Dinehart, D. W. (2001). Shear strength of normal strength concrete beams reinforced with deformed GFRP bars. *Journal of Composites for Construction, 5*(4), 268–275.
6. Sobrino, J. A., & Pulido, M. (2002). Towards advanced composite material footbridges. *Structural Engineering International, 12*(2), 84–86.
7. Sydlik, S. A., Lee, J. H., Walish, J. J., Thomas, E. L., & Swager, T. M. (2013). Epoxy functionalized multi-walled carbon nanotubes for improved adhesives. *Carbon, 59*, 109–120.

8. Eklund, P., Ajayan, P., Blackmon, R., Hart, A. J., Kibng, J., Pradhan, B., et al. (2007). *International assessment of research and development of carbon nanotube manufacturing and applications*. Baltimore: World Technology Evaluation Center.
9. Mylvaganam, K., & Zhang, L. C. (2007). Fabrication and application of polymer composites comprising carbon nanotubes. *Recent Patents on Nanotechnology, 1*(1), 59–65.
10. ASTM D7205/D7205M-06. (2016). *Standard test method for tensile properties of fiber reinforced polymer matrix composite bars*. West Conshohocken: ASTM International.
11. ASTM D4475-02. (2016). *Standard test method for apparent horizontal shear strength of pultruded reinforced plastic rods*. West Conshohocken: ASTM International.
12. ASTM D3171-15. (2015). *Standard test methods for constituent content of composite materials*. West Conshohocken: ASTM International.
13. Reda Taha, M., Chennareddy, R., & Riad, A. (2017) Smart ester-based high performance pultruded GFRP with self-sensing capabilities and methods of making, Provisional US Patent filed, October 2017.

# Chapter 83
# On Mechanical Characteristics of HFRP Bars with Various Types of Hybridization

Andrzej Garbacz, Elzbieta Szmigiera, Kostiantyn Protchenko, and Marek Urbanski

The principal objective of this study is to elaborate the reliable properties for hybrid fiber-reinforced polymer (HFRP) bars, which will have the potential to be considered as a competitive alternative to conventional reinforcement for concrete structures. Understanding mechanical performance of HFRP bars with various types of hybridization will allow for more precise design estimations that will balance safety and cost. Numerical modeling of tensile strength test was performed for hybrid carbon/glass fiber-reinforced polymer (HC/GFRP) bars and hybrid carbon/basalt fiber-reinforced polymer (HC/BFRP) bars with the use of finite element analysis (FEA) simulations. The variable parameters were two factors: the bar configuration and volume fraction of fibers. Results indicate that for both HC/GFRP and HC/BFRP bars, the location of carbon fibers in the near-surface region is more appropriate when the volume fraction of carbon fibers is less than volume fraction of glass (or basalt) fibers; otherwise, it is better to locate carbon fibers in the core region. Results of numerical modeling were considered for producing HFRP bars for further experimental investigations.

## 1 Introduction

The application of fiber-reinforced polymer (FRP) bars has progressed beyond the experimental stage, and already FRP bars are used in many concrete structures, such as prestressed concrete structures, foundations, road surfaces, parking, bridge designs, etc. [1]. An increasing use of FRP bars encourages introducing new types of FRP bars in terms of change of constituents and modification of their

A. Garbacz (✉) · E. Szmigiera · K. Protchenko · M. Urbanski
Faculty of Civil Engineering, Warsaw University of Technology, Warsaw, Poland
e-mail: a.garbacz@il.pw.edu.pl

characteristics. An investigation of mechanical and physical characteristics of new types of FRP bars will allow designers to consider more options and select an appropriate type of FRP bars.

The optimal solution that meets cost and mechanical requirements is the use of hybrid fiber-reinforced polymer (HFRP) bars composed of less expensive basalt or glass fibers with their partial substitution by carbon fibers. The HC/BFRP and HC/GFRP bars are characterized by better mechanical properties than BFRP and GFRP bars and more rentable cost than CFRP bars. The process of producing CFRP bars and their cost are relatively high [2, 3].

An extensive research on hybrid fiber-reinforced polymer (HFRP) bars is being conducted at the Warsaw University of Technology in conjunction with FRP manufacturing company upon the program of the National Centre for Research and Development in Poland. This research aims at characterizing newly developed types of HFRP bars, their mechanical and physical performance, as well as their performance in structural systems [4].

The numerical investigation described in this work includes two different HFRP bar types: hybrid carbon/glass fiber-reinforced polymer (HC/GFRP) bars and hybrid carbon/basalt fiber-reinforced polymer (HC/BFRP bars). The variable parameters were two factors: the bar configuration and volume fraction of fibers. Numerical simulation of the tensile test was performed in finite element analysis (FEA) software, and Young's modulus was determined for all types.

## 2 Materials Selection and the Reasons for the Investigation

The aim of this study is to investigate the influence of fiber substitution and their location on the mechanical properties of HFRP bars. The Young's modulus was a required output due to its importance in the design of concrete structures reinforced with FRP bars.

The carbon fibers are characterized by strong anisotropy and were selected due to its high properties in the longitudinal direction. Low-strength (LS) carbon fibers were chosen because apart from lower cost, its strain is approximately the same as the strain of selected basalt or glass fibers. The glass fibers are characterized by weak anisotropy that is not as strong or rigid as carbon fibers; however it is much cheaper and significantly less brittle when used in composites. Basalt fibers similarly to glass fibers are characterized by weak anisotropy, and the selection of basalt fibers is mainly based on its environmentally friendly manufacturing process. Table 83.1 describes properties of materials utilized in this work.

The properties of composite materials including Young's moduli can be calculated using the rule of mixtures (ROM) (axial loading – Voigt model) as it comes from the literature [5–7]. Based on this equation, the Young's moduli were determined for the HC/BFRP and HC/GFRP for different combinations depending on carbon to basalt and carbon to glass fiber volume fractions, respectively. Different levels of fibers substitution were proposed: 1:1, 1:2, 1:3, 1:4, and 1:9 (Fig. 83.1).

**Table 83.1** Properties of constituents utilized for preparing HFRP bars [2]

| Material property | Epoxy resin | Carbon fibers LS | Basalt fibers | Glass fibers E-type |
|---|---|---|---|---|
| Density (g/cm$^3$) | 1.16 | 1.90–2.10 | 2.60–2.80 | 2.40–2.50 |
| Diameter (μm) | – | 7.00–11.00 | 11.2–13.4 | 6.00–21.00 |
| $E_{11}$ (GPa) | 3.45 | 232.00 | 89.00 | 73.10 |
| $E_{22}$ (GPa) | 3.45 | 15.00 | 89.00 | 73.10 |
| $\nu_{12}$ | 0.35 | 0.279 | 0.26 | 0.22 |
| $\nu_{23}$ | 0.35 | 0.49 | 0.26 | 0.22 |
| $G_{12}$ (MPa) | 1.28 | 24.00 | 21.70 | 29.95 |
| $G_{23}$ (MPa) | 1.28 | 5.03 | 21.70 | 29.95 |
| $\sigma_{11}$ (MPa) | 55–130 | 2500–3500 | 1153–2100 | 600–1437 |

$E_{ii}$ is the Young's modulus along axis $i$, $\nu_{ij}$ is the Poisson ratio that corresponds to a contraction in direction $j$ when an extension is applied in direction $i$, and $G_{ij}$ is the shear modulus in direction $j$ on the plane whose normal is in direction $i$ and $\sigma_{ii}$ is the tensile strength in the direction $i$

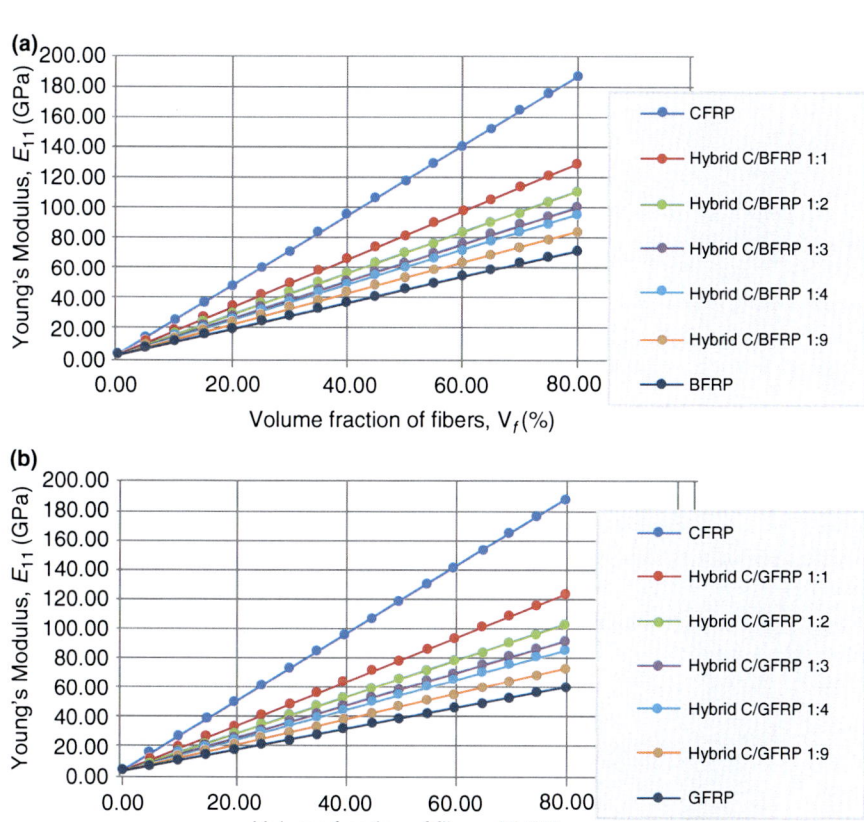

**Fig. 83.1** Fiber volume fraction and Young's modulus (**a**) HC/BFRP, (**b**) HC/GFRP

The assumption was made that volume fraction of epoxy resin should be not less than 20% to ensure bonding.

The results of calculations indicate that partial substitution of basalt or glass fibers by carbon fibers can increase the Young's modulus of FRP bars. A little higher values of Young's modulus were obtained in the case of modification of BFRP bars.

## 3 Numerical Simulation

The proposed equation does not consider the location of fibers. Experimental studies mentioned in [8] suggest that for HFRP bars the location of fibers can influence on the mechanical properties of final products. Therefore, authors consider two different combinations of bar configuration (Fig. 83.2). The main reason for the numerical simulation was to estimate the influence of fiber location on the mechanical properties of HFRP bars.

The numerical simulation of tensile strength test for HFRP bars was performed in finite element analysis (FEA) software ANSYS. The bars were modeled as cylindrical elements with the diameter of 8 mm and the length of 850 mm. The constant pressure of 500 MPa was applied on the side edges. One central point was fixed in $y$ and $z$ directions. The structure of the HFRP bars consisted of a core and surface regions, which were perfectly interconnected. As an example Fig. 83.3 demonstrates stress and strain distributions for the sample of HC/BFRP bars with different bar configuration, where the ratio of carbon fibers to basalt fibers was 1:3.

The obtained results were used for defining the Young's moduli for various HFRP bars depending on the constituents, their location, and the level of substitution of fibers by carbon fibers. Results obtained numerically (FEA) are in good convergence with Voigt model (ROM) (Table 83.2).

The obtained results of FEM simulation show that location of carbon fibers has not a big influence on a final value of Young's modulus. However, some technological problems were observed during production of the bars with carbon fibers located in a surface layer, which results in increased heterogeneity in fiber distribution (Fig. 83.4). In addition, the temperature rise during production caused local scorching. Consequently, it was decided that carbon fibers would be placed in the center of the bar.

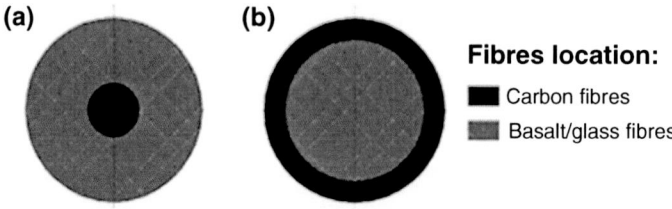

**Fig. 83.2** Regions of carbon fiber location (**a**) in the core and (**b**) in the near surface

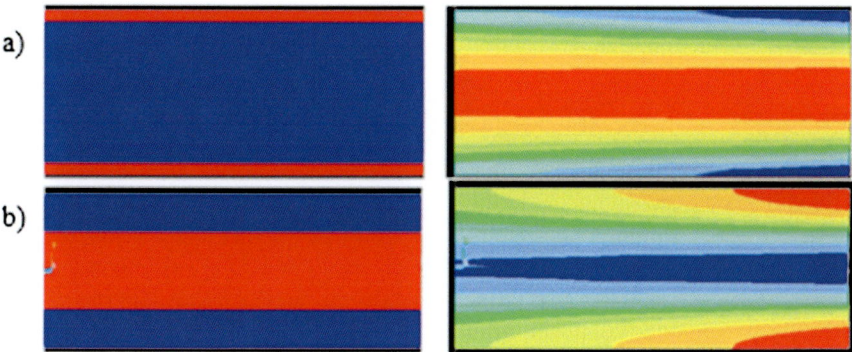

**Fig. 83.3** The stress and strain distributions using FEM model for HC/BFRP 1:3 (**a**) carbon fibers in the near-surface region (**b**) carbon fibers in core region

**Table 83.2** The Young's modulus obtained by ROM and FEA modeling (GPa)

| | HC/BFRP | | | HC/GFRP | | |
|---|---|---|---|---|---|---|
| Fibers substitution C:B or C:G fibers | | FEA C fiber location | | | FEA C fiber location | |
| | ROM | ○ | ● | ROM | ○ | ● |
| 1:9 | 83.3 | 83.0 | 82.9 | 71.9 | 71.2 | 71.2 |
| 1:4 | 94.8 | 93.6 | 93.6 | 84.6 | 83.7 | 83.7 |
| 1:3 | 100.5 | 101.1 | 100.1 | 90.9 | 91.9 | 90.0 |
| 1:2 | 110.0 | 110.5 | 109.1 | 101.5 | 102.6 | 100.5 |
| 1:1 | 129.1 | 127.8 | 129.2 | 122.7 | 121.5 | 124.0 |

*C* Carbon, *B* basalt, *G* glass

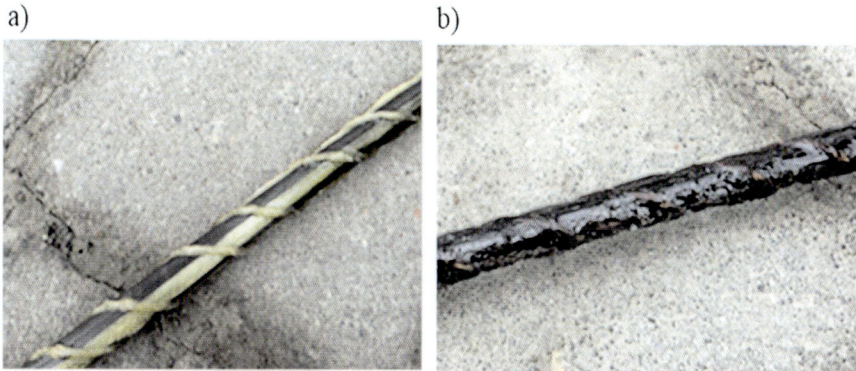

**Fig. 83.4** Imperfections of HFRP rods resulting from production technology: (**a**) nonhomogeneous distribution of carbon fibers, (**b**) scorching

## 4 Results and Discussion

On the basis of the analytical calculation and the computer simulations for hybrid FRP bars, the following main conclusions can be drawn:

1. It is possible to predict the final characteristics of HFRP bars through the modification of constituents and their volume fractions before it will be produced.
2. Results obtained numerically are in good convergence with Voigt model results.
3. Results of numerical modeling show that bar configuration is less important than volume fraction of fibers. The difference between various bar configuration can be 2% at maximum; meanwhile, the volume fraction of all analyzed combinations can influence on the final stiffness by 74.6%.

Although the results of numerical modeling suggest that location of carbon fibers in the near-surface region is better, however, it was decided that carbon fibers in HFRP bars will be located in the center of the bar due to the technological issues.

The authors wish to acknowledge the financial support of the project – "Innovative Hybrid – FRP composites for infrastructure design with high durability." NCBR: PBS3/A2/20/2015.

## References

1. ACI 440.4R-04. (2011). *Prestressing Concrete Structures with FRP Tendons (Reapproved)*. Farmington Hills, MI: American Concrete Institute.
2. Protchenko, K., Młodzik, K., Urbański, M., Szmigiera, E., & Garbacz, A. (2016). Numerical estimation of concrete beams reinforced with FRP bars. In *MATEC web of conferences* (Vol. 86, p. 02011). EDP Sciences. Moscow, Russia.
3. Garbacz, A., Urbański, M., & Łapko, A. (2015). BFRP bars as an alternative reinforcement of concrete structures – Compatibility and adhesion issues. *Advanced Materials Research, 1129*, 233–241.
4. Garbacz, A., Szmigiera, E., Urbański, M., Protchenko, K., & Kubas, M. (2017). Research on hybrid FRP reinforcement for concrete infrastructure. In *Inzynieria I Budownictwo* (Vol. 73/8, pp. 428–432). BazTech [in Polish]. Warsaw, Poland.
5. Voigt, W. (1889). Uber die beziehung zwischen den beiden elasticitatsconstanten isotroper korper. *Annalen der Physik, 38*, 573–587. [in German].
6. Black, T., & Kosher, R. (2008). Non metallic materials: Plastic, elastomers, ceramics and composites. In *Materials and processing in manufacturing* (10th ed., pp. 162–194). Wiley. ISBN 978-0470-05512-0. Indianapolis, USA.
7. Barbero, E. J. (2011). *Introduction to composite materials design* (2nd ed.). Boca Raton: CRC Press/Taylor & Francis Group.
8. Bakis, C. E., Nanni, A., Terosky, J. A., & Koehler, S. W. (2001). Self-monitoring, pseudo-ductile, hybrid FRP reinforcement rods for concrete applications. *Composites Science and Technology, 61*, 815–823.

# Chapter 84
# Fatigue Behavior Characterization of Superelastic Shape Memory Alloy Fiber-Reinforced Polymer Composites

Sherif M. Daghash and Osman E. Ozbulut

Fiber-reinforced polymer (FRP) composites have been frequently used for strengthening concrete structures. However, conventional FRPs exhibit brittle behavior with relatively low ultimate tensile strains and limited energy dissipation capacity, and possess limited fatigue life. Superelastic shape memory alloys (SMAs) are a class of metallic alloys that can recover strains between 6% and 8% upon load removal. SMAs possess excellent corrosion resistance, enhanced energy dissipation abilities, and high fatigue properties. Small-diameter superelastic SMA strands are new form of SMA elements. SMA strands can replace conventional fibers to produce resilient composites with enhanced ductility and deformability. These composites can be used for concrete and steel infrastructure strengthening and energy absorption applications.

This study investigates the fatigue behavior of composite material that consists of a thermoset polymer matrix reinforced with superelastic 0.35 mm-diameter SMA strands, which is composed of seven 0.117 mm-diameter wires. SMA-FRP coupons were fabricated and tested under cyclic loading protocols at various strain levels to characterize the fatigue behavior of the composite. Tests results were analyzed in terms of dissipated energy, equivalent viscous damping, and secant modulus. Results revealed that even under continuous cyclic loading, the SMA–FRP composite was still able to recover relatively large strains, dissipate energy, and sustain large portion of its initial mechanical properties.

---

S. M. Daghash (✉) · O. E. Ozbulut
Department of Civil and Environmental Engineering, University of Virginia, Charlottesville, VA, USA
e-mail: daghash@virginia.edu

# 1 Introduction

Fiber-reinforced polymers (FRPs) display numerous advantages such as high strength-to-weight ratio, high corrosion resistance, and good durability [1]. However, conventional FRPs including carbon and glass FRPs exhibit linear elastic stress–strain behavior up to rupture in a brittle manner with a limited ultimate tensile strain. They possess limited creep and fatigue strengths, and no ability to dissipate strain energy. Superelastic SMAs are metallic alloys that offer a unique ability of recovering large deformations between 6% and 8% upon removal of the load [2]. Superelastic SMAs also display high corrosion and fatigue properties. As a result, researchers embedded superelastic SMA elements in CFRP and GFRP composites to improve the composite impact and damage resistance and enhance the overall structural performance [3, 4]. The superelastic SMAs prevented complete perforation under impact loading and suppressed damage propagating in the composite [3]. They also increased the composite strain capacity, ductility, and dissipated elastic energy [4]. However, the composite tensile strength was considerably decreased as material defects were generated in the matrix by embedding the SMAs [5]. Additional researches have been conducted to explore utilizing superelastic SMA wires as the sole reinforcing fiber to fabricate SMA-FRP composites [6, 7]. In previous studies, the SMA fiber volume fraction ranged between 3% and 20.3%. Although the fabricated composites were able to recover maximum strains up to 7% with minimal residual deformations and obtained significantly high ultimate tensile strain between 10% and 12%, they displayed tensile strength below 150 MPa.

Recently, superelastic SMA strands were introduced as a new structural form of SMA elements [8]. SMA strands consist of thin SMA wires wrapped helically. The strands' formation provides damage-resistant redundancy if any of the composing wires break. Over a monolithic wire of comparable size, SMA strands have cost advantages and provide better surface roughness. To the authors' knowledge, no research has been performed on using small-diameter SMA strands in thermoset polymer composites. In structural applications, FRPs are sensitive to the repetitive stress cycles that create internal damage and failure of the composite system [9]. In order to enable more application areas of SMA–FRP composites, it is important to study their behavior under fatigue loading. Only a few studies in the literature have investigated the fatigue behavior of SMA–FRP composites. These studies mainly focused on composites reinforced with pre-strained heat-activated SMA wires [10, 11].

This study utilizes 0.35 mm-diameter superelastic SMA strands as the reinforcing fibers to fabricate SMA–FRP composite and explores the fatigue behavior of the developed composites. SMA–FRP coupons are tested under cyclic loading protocols at various strain levels. Results are evaluated in terms of dissipated energy, equivalent viscous damping, and secant modulus, in addition to modes of failure.

## 2 Methods

In this study, EPOTUF® epoxy resin system, supplied by Reichhold Chemicals, Inc., was used to form the body of the specimens and bind the reinforcing fibers together. The epoxy has a 100:70 resin-to-hardener mixing ratio by weight, and can reach its full strength after 7 days of curing at room temperature according to the manufacturer. Neat epoxy samples with no reinforcement were made and tested under monotonic tensile loading. The neat epoxy obtained average ultimate tensile stress of 49.3 MPa and average failure strain of 6%. The reinforcing fibers in this study were superelastic SMA strands supplied by Fort Wayne Metals, Inc. Each strand is composed of seven wires with a diameter of 0.117 mm for each wire and has a total diameter of 0.35 mm. According to the manufacturer, the strands have transformation stress of 774 MPa, tensile ultimate stress of 1316 MPa, and ultimate elongation of 14%. SMA-FRP coupons were fabricated using the vacuum-assisted hand layup technique as guided by the ASTM D5687 specifications [12]. Each coupon was 12.7 mm × 1.0 mm, and utilized 84 SMA strands to reach a fiber volume fraction of 50%. Before the cyclic tests, CFRP tabs were glued on the gripping area of each specimen to prevent crushing of this area.

The fabricated coupons were cycled until failure under three different levels of target strains. The specimens were cycled between target strains of approximately 3.0%, 6.0%, and 9.0% and back to zero strain using strain rate of 0.005/sec in a strain-controlled mode tests. As shown in Fig. 84.1, the tests were performed using MTS servo hydraulic machine. Test loads were recorded using the MTS data acquisition system, and displacements were captured by a laser extensometer attached to the system over a 50 mm gauge length at the middle portion of each specimen.

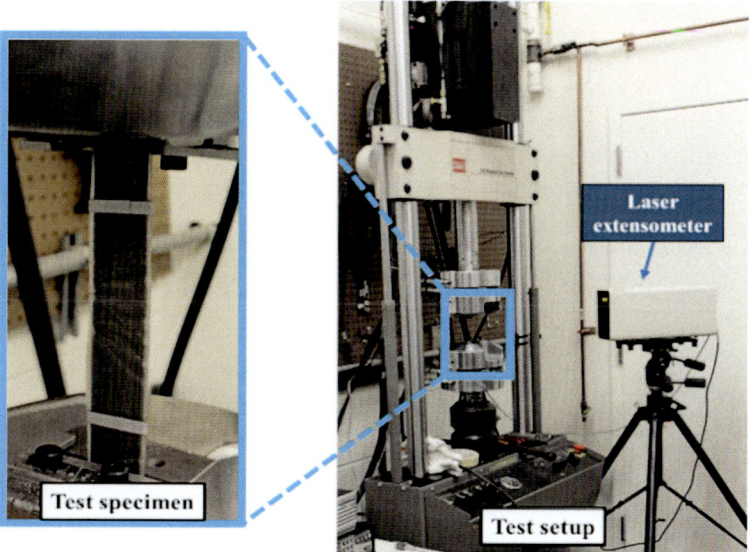

**Fig. 84.1** Test setup with SMA–FRP composite specimen

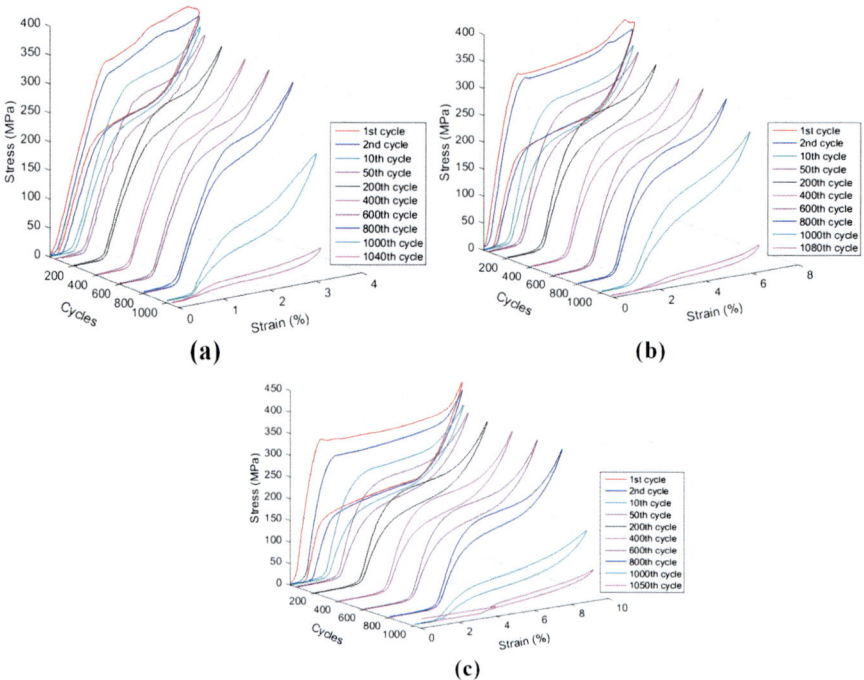

**Fig. 84.2** Stress–strain curves of composite specimens cycled under maximum strains of: (**a**) 3.0%, (**b**) 6.0%, and (**c**) 9.0%

## 3 Results

Figure 84.2 shows the stress–strain curves of SMA–FRP specimens cycled under three different target maximum strains of 3.0%, 6.0%, and 9.0%. It can be observed that the loading envelopes of the composite followed the typical loading trajectory of superelastic SMA elements. The composite response under cyclic loading starts with linear elastic portion, followed by constant plateau of the phase transformations stage. In the case of 9.0% specimens, the curves were extended to the strain-hardening (post-phase transformation) region. While unloading, the composite recovers large portion of the maximum strains, but with permanent residual deformations attributed mostly to the plastic deformations of the epoxy matrix. The curves also shows that under fatigue cyclic loading, the composites experienced degradation in the superelastic behavior, displayed as decrease in the maximum stress at each cycle, decrease in the area enclosed by the loading and unloading trajectories, and increase in the residual deformations. The degradation in the behavior mostly occurred through the first ten cycles and stabilized afterward until failure of the composite. The 3.0%, 6.0%, and 9.0% specimens obtained transformation stress of

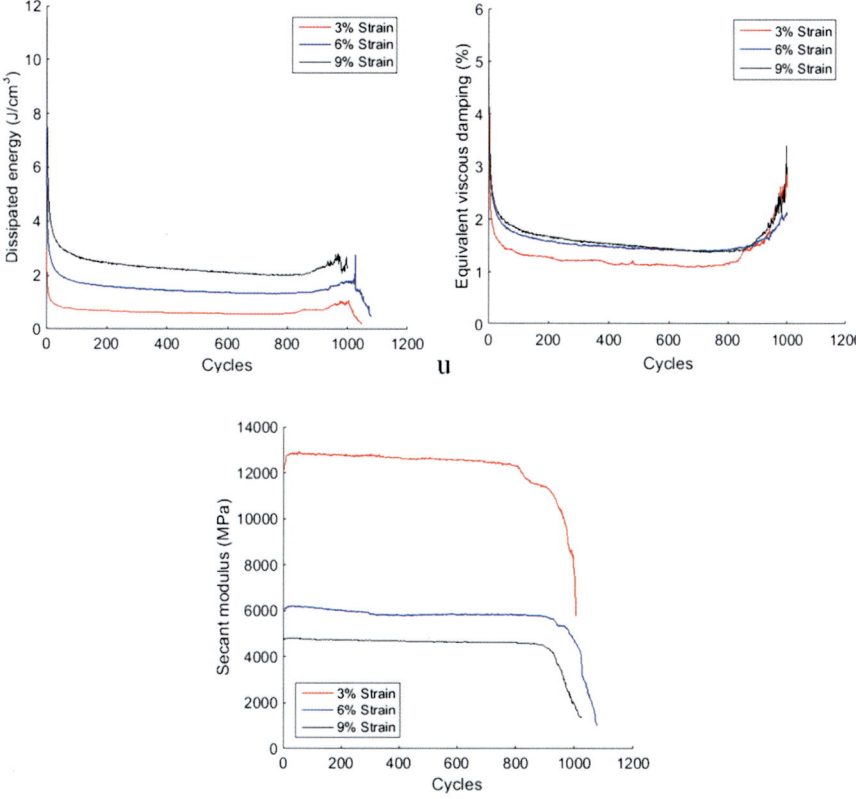

**Fig. 84.3** Dissipated energy, equivalent viscous damping, and secant modulus of tested specimens

320 MPa, and maximum strains of 3.26%, 6.51%, and 9.2%, respectively, with corresponding maximum residual deformations of 0.7%, 1.3%, and 2.1%.

To compare results of the tested specimens, the dissipated energy ($\Delta W$), equivalent viscous damping ($\zeta_{eq}$), and the secant modulus ($E_{sec}$) were calculated per cycle and plotted as shown in Fig. 84.3. The dissipated energy in any cycle is defined as the area enclosed by the loading and unloading branches of that cycle. The equivalent viscous damping and secant modulus were calculated using the following equations:

$$\zeta_{eq} = \frac{\Delta W}{4\Pi W} \qquad (84.1)$$

$$E_{sec} = \frac{\sigma_{max} - \sigma_{min}}{\varepsilon_{max} - \varepsilon_{res}} \qquad (84.2)$$

where $W$ is defined as the energy dissipated in equivalent linear system with the same maximum and minimum stresses $\sigma_{max}$ and $\sigma_{min}$ [MPa] and maximum and residual strains $\varepsilon_{max}$ and $\varepsilon_{res}$ of the load cycle that $\Delta W$ was calculated for. While the 9.0%

specimen obtained the highest dissipated energy of 11.53 J/cm$^3$ at the first cycle and decreased to 2.0 J/cm$^3$ at failure, the 6.0% specimen obtained dissipated energy of 7.4 J/cm$^3$ at the first cycle and decreased to 1.37 J/cm$^3$ at failure, and the 3.0% specimen obtained dissipated energy of 2.88 J/cm$^3$ at the first cycle and failed at 0.6 J/cm$^3$. The equivalent viscous damping of both 6.0% and 9.0% specimens stabilized at approximately 1.5%, and higher than the equivalent viscous damping of 3.0% specimen that stabilized at 1.2%. On the other hand, the 9.0% specimen exhibited the lowest secant modulus of 4808 MPa, while the 6.0% and 3.0% specimens exhibited secant moduli of 6190 MPa and 12,930 MPa, respectively.

The tested composite specimens failed at approximately the same cycle number, confirming that the fatigue failures of FRP composites are primarily dominated by the matrix behavior. During the early cycles of loading, debonded areas at the interface between the strands and epoxy matrix were initially created due to the transformation of the strands with the decrease in their diameters. Under continuous fatigue loading cycles, the debonded areas increased significantly, accompanied with horizontal and longitudinal matrix cracks. This resulted in failure of the composite system and significant loss of the specimen force capacity.

## 4 Conclusions

SMA–FRP composites with 50% fiber volume fraction were fabricated. The composites utilized 0.35 mm-diameter superelastic SMA strands as the reinforcing fibers. Each strand consists of seven 0.117 mm-diameter wires. The composite specimens were cycled under maximum strains of 3.26%, 6.51%, and 9.2% until failure, and the dissipated energy, equivalent viscous damping, and secant modulus were calculated for each specimen. The results proved the enhanced strain capacity of the composite at relatively high stresses and the superelastic behavior with the ability to recover approximately 80% of the applied strains. The composite also obtained high energy dissipation capacity. Results of this research proved the ability of the fabricated composite to be utilized in infrastructure strengthening and energy absorption applications.

## References

1. Bank, L. C. (2012). Progressive failure and ductility of FRP composites for construction. *Journal of Composites for Construction, 17*(3), 406–419.
2. Hurlebaus, S., & Gaul, L. (2006). Smart structure dynamics. *Mechanical Systems and Signal Processing, 20*(2), 255–281.
3. Kim, E. H., Lee, I., Roh, J. H., Bae, J. S., Choi, I. H., & Koo, K. N. (2011). Effects of shape memory alloys on low velocity impact characteristics of composite plate. *Composite Structures, 93*(11), 2903–2909.

4. Wierschem, N., & Andrawes, B. (2010). Superelastic SMA–FRP composite reinforcement for concrete structures. *Smart Materials and Structures, 19*(2), 025011.
5. Jang, B. K., Koo, J. H., Toyama, N., Akimune, Y., & Kishi, T. (2001). Influence of lamination direction on fracture behavior and mechanical properties of TiNi SMA wire-embedded CFRP smart composites. *SPIE's 8th annual international symposium on smart structures and materials* (pp. 188–197). International Society for Optics and Photonics.
6. Zafar, A., & Andrawes, B. (2013). Fabrication and cyclic behavior of highly ductile superelastic shape memory composites. *Journal of Materials in Civil Engineering, 26*(4), 622–632.
7. Daghash, S. M., & Ozbulut, O. E. (2016). Characterization of superelastic shape memory alloy fiber-reinforced polymer composites under tensile cyclic loading. *Materials & Design, 111*, 504–512.
8. Reedlunn, B., Daly, S., & Shaw, J. (2013). Superelastic shape memory alloy cables: Part I–isothermal tension experiments. *International Journal of Solids and Structures, 50*(20), 3009–3026.
9. Wang, Z., Xu, L., Sun, X., Shi, M., & Liu, J. (2017). Fatigue behavior of glass-fiber-reinforced epoxy composites embedded with shape memory alloy wires. *Composite Structures, 178*, 311–319.
10. Shimamoto, A., Zhao, H. Y., & Abe, H. (2004). Fatigue crack propagation and local crack-tip strain behavior in TiNi shape memory fiber reinforced composite. *International Journal of Fatigue, 26*(5), 533–542.
11. El-Tahan, M., & Dawood, M. (2015). Fatigue behavior of a thermally-activated NiTiNb SMA-FRP patch. *Smart Materials and Structures, 25*(1), 015030.
12. Standard, A. S. T. M. (2007). D5687/D5687M-07. Standard guide for preparation of flat composite panels with processing guidelines for specimen preparation.

# Chapter 85
# Strength Performance of Concrete Beams Reinforced with BFRP Bars

Elzbieta Szmigiera, Marek Urbanski, and Kostiantyn Protchenko

This paper presents an experimental and numerical investigation of the strength performance of concrete beams reinforced with Fiber-Reinforced Polymer (FRP) bars. The numerical model prepared in Finite Element Analysis (FEA) software was validated against the experimental data and found to be in good convergence. The experimental program consisted of the flexural tests of beams reinforced with Basalt Fiber-Reinforced Polymer (BFRP) bars and beams reinforced with steel bars for comparison. In both, numerical and experimental investigations, the deflections of beams reinforced with BFRP bars were significantly greater than deflections of beams reinforced by steel bars. Additionally, the destruction mechanism of beams was the crushing of concrete region in compression zone (apart from one case during the experimental program). The maximum deviation for the strength capacity of beams reinforced with BFRP bars does not exceed 9%. The implementation of the obtained FEA model in further analysis will allow better predicting of the strength performance of beams reinforced with FRP bars before being tested experimentally.

## 1 Introduction

One of the most common mechanisms of concrete structures' deterioration is the corrosion of reinforcement. Carbonization of concrete, penetration of chloride ions and sulfuric acids are the most common reasons for steel corrosion. To avoid adverse effects caused by this phenomenon, materials with improved corrosion resistance can be used instead of conventional anticorrosion protection [1].

---

E. Szmigiera · M. Urbanski (✉) · K. Protchenko
Institute of Building Engineering, Faculty of Civil Engineering, Warsaw University of Technology, Warsaw, Poland
e-mail: m.urbanski@il.pw.edu.pl

The use of Fiber-Reinforced Polymer (FRP) bars as internal reinforcement will be considered as an alternative to steel reinforcement when implemented in concrete structures, its unique characteristics being the decisive factor [2].

The available North American and European code guides provide recommendations for design of structural elements with composite materials such as Carbon–FRP (CFRP), Glass–FRP (GFRP), and Aramid–FRP (AFRP) bars. Some of the specifications for Basalt–FRP (BFRP) bars are provided in Russian norms [3]. BFRP has been assured to have advantages in achieving the goal by enhancing safety and reliability of structural systems compared with the conventional FRP composites [4, 5].

BFRP bars used in concrete structures are expected to provide more benefits that are comparable or superior to other types of FRP, while being significantly cost-effective [6, 7].

BFRP bars are produced from basalt roving, consisting of fibers fabricated from basalt rocks through melting process. Basalt fibers are inorganic fibers that are characterized by high tensile strength, better chemical resistance, extended operating temperature range, and better environmental friendliness when compared to E-glass FRP. The manufacturing process of BFRP bars is also considered to be environmentally friendly.

A significant number of investigations have been conducted on flexural performance of beams reinforced internally with FRP bars of various types. However, most of them have been experimentally based and studies on the use of BFRP bars are still limited. Therefore, numerical modeling of beams reinforced with BFRP bars using the Finite Element Analysis (FEA) was proposed for this work.

## 2 Methods and Materials

### 2.1 Experimental Program

The aim of the experimental testing was to verify suitability of BFRP bars as internal reinforcement for the concrete beams subjected to bending.

The experimental program consisted of four-point flexural tests for six simply supported beams made of concrete class *C30/37*. Three of the beams were reinforced at the bottom with BFRP bars of diameter 8 mm. Other beams were reinforced at the bottom with steel bars of the same diameter (reference beams). Two steel bars of the same diameter were used as the top reinforcement and steel stirrups were used for all of the samples. Assumed clear cover was 20 mm from all sides. Some material characteristics for reinforcement that were obtained experimentally are shown in Table 85.1. The scheme of tested samples is described in Fig. 85.1.

In the beams, which were reinforced with BFRP bars, the central bottom basalt bar was lengthened on both ends to enable the measurement of the slip in the process of loading. The middle part of tested samples was without stirrups and top reinforcement. The seven pairs of bench marks, with spacing 20 mm, were arranged on

## 85 Strength Performance of Concrete Beams Reinforced with BFRP Bars

**Table 85.1** Material characteristics for reinforcement

| Parameter | BFRP bars Ø8 mm | Steel bars Ø8 mm |
|---|---|---|
| Average modulus of elasticity, (GPa) | 39.05 | 200.00 |
| Average rupture strain (%) | 2.81 | 0.17 |
| Average tensile/yield strength (MPa) | 1051.79 | 348.00 |

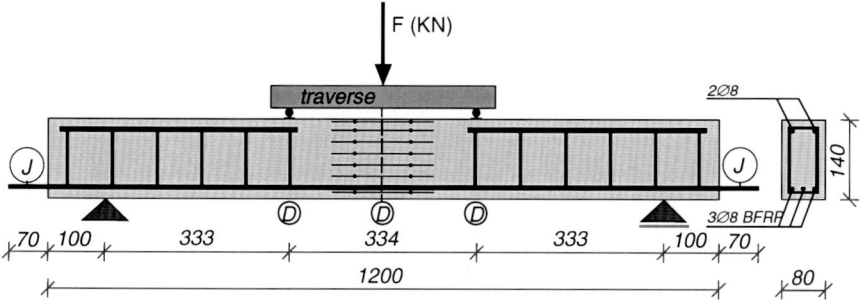

**Fig. 85.1** Scheme of tested samples reinforced with BFRP bars, *J* slip measurement sensor, *D* deflection gauge, dimensions in mm

**Fig. 85.2** Destruction of one of the samples reinforced with BFRP bars

the side surface of the beam. Registration of concrete strains was made with a mechanical extensometer with a measuring length of 100 mm.

Figure 85.2 shows the destruction of one of the samples reinforced with BFRP bars. The maximal loading was equal to 45 kN. It is noteworthy that there was no rupture of the flexural basalt bars and, therefore, they have not reached their tensile strength. The destruction of beam took place due to shear in support zone and it had brittle nature.

The strength capacity of beams reinforced with BFRP was higher than the strength capacity of reference beams. However, it is worth mentioning that beams reinforced with BFRP bars had higher deflection than reference beams. The axial stiffness is the reason for higher deflection. BFRP has an axial stiffness much lower than steel and hence a higher deflection.

Sensitivity analyses were not performed for the test. Experimental setup is explained more extensively in the literature [2, 9].

## 2.2 Numerical Modeling

The objective of the FEA simulation was to demonstrate numerical model that will be able to simulate overall structural performance of structures reinforced with FRP bars that will be validated against the experimental data.

The geometry, properties of materials, and application of loads in FEA are set according to experimental program. The Abaqus FEA software was used to prepare numerical analysis in a three-dimensional model. The elements type C3D8R was used to simulate the nonlinear behavior of concrete and steel supports. Uniaxial compression behavior of concrete was defined with the implementation of Hongestad's model [7]. The tensile behavior of concrete was assumed as linearly elastic up to the tensile strength. To simulate behavior of FRP and steel reinforcements, the element type T3D2 was used. The post-failure behavior of cracked concrete was defined with the help of tension-stiffening model.

The stress–strain relationship of FRP reinforcement can be obtained by Hooke's law. The tensile stress of FRP reinforcement is linearly proportional to its fractional extension in length or strain by the modulus of elasticity. The stress–strain relationship of steel reinforcement is assumed as linearly elastic up to yielding strength, beyond which it behaves as completely plastic [8].

The application of loads was done with the gradual accumulation by value 0.002, and as the first step it was loaded with the force $F_0 = 20$ MN/m$^2$. Assuming the load transfer from the strip with dimensions of 0.02 m × 0.08 m, the final applied loading from one strip was defined according to Eq. 85.1:

$$F = F_0 \times a \times b \qquad (85.1)$$

where $F_0$ is the initial load [MN/m$^2$] and $a$ and $b$ are the dimensions of the strip [m]. The total force applied to beam is equal to the load applied through two strips, which corresponds to the value of 64 kN.

Figure 85.3a and b show the distribution of stresses in the concrete element and BFRP bars, respectively, just before the destruction of the element. Figure 85.3c shows the distribution of strains for the entire element.

Table 85.2 describes the comparison between experimental and numerical results.

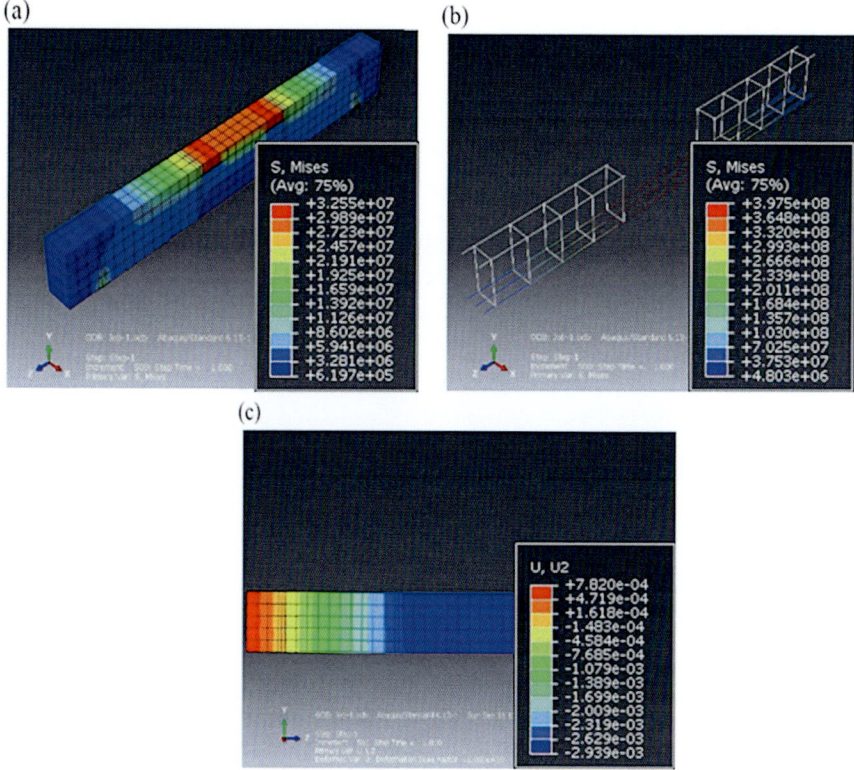

**Fig. 85.3** Distribution of stresses: (**a**) in concrete element, (**b**) in reinforcement, dimensions in *Pa*, (**c**) distribution of strains in entire element

**Table 85.2** Comparison between experimental and numerical results

| Parameter | Experimental results | | Numerical results | |
|---|---|---|---|---|
| | BFRP bars | Steel bars | BFRP bars | Steel bars |
| Maximal deflections (mm) | 19.54 | 4.48 | 24.50 | 6.20 |
| Strength capacity (kN) | 46.77 | 37.60 | 49.28 | 44.15 |

## 3 Conclusions

The numerical simulation for the beams reinforced with BFRP and steel bars was prepared and validated based on experimental data. The numerical estimation conforms to the results obtained from experimental testing that deflections of beams reinforced with BFRP bars were significantly higher than for beams reinforced with steel bars. However, the higher bearing capacity was achieved for the beams reinforced with BFRP bars (Table 85.2). This phenomenon can be explained by the following factors:

- The better bond characteristics between FRP bars and concrete than between steel bars and concrete.
- The equilibrium path of FRP reinforcement is more similar to concrete behavior. There is a lack of plasticity stage, unlike in steel bars.

The analyses were performed for short-span beams. The estimation of strength capacity for long-span beams was not considered.

During numerical simulation, it was also observed that with an increase of tensile strength of concrete, the carrying capacity of the beam increased significantly in comparison to strength of beams reinforced with steel bars.

The current analyses suggest that the BFRP reinforcement can be considered as a promising alternative to steel bars; however, comprehensive investigation is still required.

The authors wish to acknowledge the financial support for the project – "Innovative Hybrid – FRP composites for infrastructure design with high durability" NCBR: PBS3/A2/20/2015.

# References

1. Czarnecki, L., Łukowski, P., & Garbacz, A. (2017). *Naprawa i ochrona konstrukcji z betonu [English: Protection and repair of concrete structures]*. Comment to the norm PN-EN 1504. Wydawnictwo Naukowe PWN. Warsaw, Poland.
2. Garbacz, A., Urbański, M., & Łapko, A. (2015). BFRP bars as an alternative reinforcement of concrete structures – Compatibility and adhesion issues. In *Advanced material research* (Vol. 1129, pp. 233–241). Trans Tech Publications, Switzerland.
3. СП 63.13330. (2012). Бетонные и железобетонные конструкции. Основные положения. (English: СП 63.13330 Concrete and won concrete construction. Design requirements). Moscow, Russia: Research center of Construction.
4. Wu, Z., Wang, X., & Wu, G. (2012). Advancement of structural safety and sustainability with basalt fiber reinforced polymers. *Proceedings of 6th international conference on FRP composites in civil engineering (CICE)* (p 29). Rome.
5. Tomlinson, D., & Fam, A. (2014). Performance of concrete beams reinforced with basalt FRP for flexure and shear. *Journal of Composites for Construction, 19*(2), 14014036.
6. Elgabbas, F., Ahmed, E., & Benmokrane, B. (2015). Physical and mechanical characteristics of new basalt-FRP bars for reinforcing concrete structures. *Journal of Construction and Building Materials, 95*, 623–635.
7. Wei, B., Cao, H., & Song, S. (2010). Environmental resistance and mechanical performance of basalt and glass fibers. *Journal of Materials Science and Engineering: Part A, 527*, 4708–4715.
8. Hognestad, E., Hanson, N. W., & McHenry, D. (1955). Concrete stress distribution in ultimate strength design. *Journal in American Concrete Institute, 52*, 455–479.
9. Łapko, A., & Urbański, M. (2015). Experimental and theoretical analysis of deflections of concrete beams reinforced with basalt rebar. *Archives of Civil and Mechanical Engineering, 15* (1), 223–230. https://doi.org/10.1016/j.acme.2014.03.008.

# Part VIII
# Polymer Concrete with Nanomaterial

# Chapter 86
# A Comparative Study on Colloidal Nanosilica Incorporation in Polymer-Modified Cement Mortars

Niloufar Zabihi and M. Hulusi Özkul

## 1 Introduction

Nanosilica particles are known for improving mechanical properties and durability of the mixtures, through different mechanisms. In fresh state, accelerated hydration, remarkable increase in water demand, and yield strength of mixtures are common observations. The acting mechanisms through which nanosilica particles tend to change the properties are the filling effect, additional pozzolanic activity, and remarkable increase in total surface area [1–7].

In polymer-modified mixtures, based on the majority of the observations, hydration kinetics is slowed down, workability is increased, and there is a decrease in compressive strength. Rapid reaction between polymer molecules and cement particles and the additional polymer film formed in the internal structure of the mix are responsible for the induced properties [8–15].

In this research paper, mortar samples incorporating both nanosilica and polymer latexes were produced and tested mainly to investigate their mechanical properties. In the next sections, the employed materials, experimental design, and the plan of the research will be explained, followed by a more detailed experimental procedure. Afterward, the results of the experiments will be discussed, and finally, the main outcomes of this research will be summarized in the conclusion section.

---

N. Zabihi · M. H. Özkul (✉)
Building Materials Laboratory, Faculty of Civil Engineering, Istanbul Technical University, Istanbul, Turkey
e-mail: Hozkul@itu.edu.tr

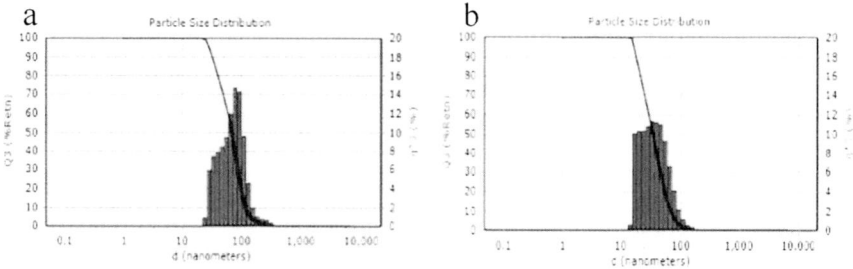

**Fig. 86.1** Particle size distribution of nanosilica and cement: (**a**) C8, peak diameter (70.5 nm), solid content (50%); (**b**) C17, peak diameter (34.3 nm), solid content (40%)

## 2 Experimental Studies

### 2.1 Materials

Portland cement CEM I PC 42.5 with specific weight of 3.14 gr/cm$^3$ was used to produce all the samples. Styrene butadiene rubber (SBR) with specific weight of 1.1 gr/cm$^3$ and PH of 10 was employed, with two different types of colloidal nanosilica (Cembinder 17 and Cembinder 8), having different average particle sizes. The particle size distribution of nanosilica colloids tested in Malvern Zetasizer is shown in Fig. 86.1. To design the composition of the mixtures, central composite design method was employed in order to numerically evaluate the significance of each component and their interaction in cement media.

Water-to-binder ratio of the mixtures was kept constant (0.5), and the samples were produced in such a way to have similar flowability of 80 ± 3% throughout the study. Therefore, in high percentages of nanosilica, a polycarboxylate-based superplasticizer (Sika ViscoCrete Hi-Tech 3051) was added to the fresh mix. Crushed sand, with the specific gravity of 2.78 g/cm$^3$ was also utilized, and to determine the flowability and consistency of the mixtures, flow table test device described in ASTM Standard of C 230/C 230 M was employed [16].

### 2.2 Methodology

Having two different admixtures in the mix, to investigate each one's effect, and the interaction between them, the systematic statistical method of central composite design was employed to design the experiments.

The collected data were analyzed through analyses of variance and then through response surface method, to evaluate the significance of each factor and optimize them. Response surface is a well-established method of designing the experiments, aiming to have an optimal experimental process, revealing the response to the changes in the independent factors, with relatively minimal charges and maximal

**Table 86.1** Central composite mix design

| Cement (g) | Water (g) | NS % | Polymer % | Sand (g) |
|---|---|---|---|---|
| 443.3 | 192.9 | 1.5 | 5 | 1350 |
| 450.0 | 199.6 | 0 | 5 | 1350 |
| 448.0 | 179.7 | 0.44 | 8.54 | 1350 |
| 448.0 | 215.6 | 0.44 | 1.46 | 1350 |
| 443.3 | 167.5 | 1.5 | 10 | 1350 |
| 443.3 | 218.3 | 1.5 | 0 | 1350 |
| 438.5 | 170.1 | 2.56 | 8.54 | 1350 |
| 438.5 | 206.1 | 2.56 | 1.46 | 1350 |
| 436.5 | 186.1 | 3 | 5 | 1350 |

results. Objectives of this method are to obtain a relationship between the independent variables (in this study, amounts of nanosilica and polymer) and response (tested variables) and to find how influential each factor is on the response. Furthermore, in optimizing problems, employing this method can give the optimal set of different independent variables resulting in extremum response [17]. Based on this design, levels of employing each independent factor are labeled (coded) according to the design with $-\alpha$, $-1$, 0, 1, $+\alpha$ (i.e., five levels of employing each factor).

The generated mix design has been shown in Table 86.1.

## 2.3 Experimental Procedures

To produce the samples, first cement and sand were mixed together, and then water, the polymer latex, and, if required, the superplasticizer were added to the fresh mix. Finally, colloidal nanosilica was added, and the mixing process was continued for about 3 min, until the mix was consistent and homogeneous. To determine the flowability, flow table test device described in ASTM standard of C 230/C 230 M was employed. Casted samples were demolded after 1 day and put in the water for the first 3 days and for the next 25 days in the air, to have the polymer film formed.

Flowability and the mechanical properties tests of compressive and flexural strengths were performed on the samples. The standard procedures of ASTM C109 [18] and ASTM C348 [19] were followed for compressive and flexural strength tests, respectively.

## 3 Results and Discussions

### 3.1 Flowability Test

Flow table test was employed as the indication of the consistency of the mortar samples. The flowability level of all the samples was kept equal; hence, the

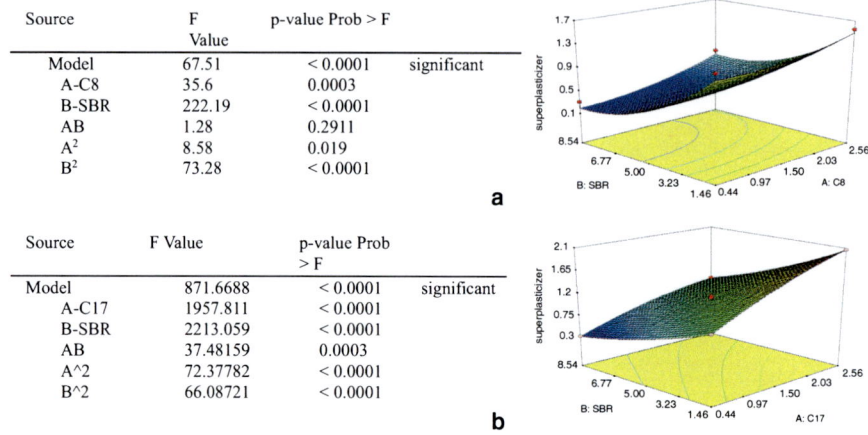

Fig. 86.2 Superplasticizer consumption results analyses of samples with (a) C8 and SBR (b) C17 and SBR

superplasticizer demand of each mixture was the clue to the flowability of its original state; the higher the superplasticizer demand, the lower the flowability. The following diagrams show (Fig. 86.2) the superplasticizer demand of the mixtures, in percentage, by weight of cement.

Quadratic pattern of superplasticizer demand is confirmed by ANOVA analyses. It is remarkable that when the amount of polymer is negligible, the superplasticizer demand is significant; by increasing the amount of polymer, this demand is moderated. This observation is indicated by the significant interaction term in the fitted models. Comparison between the results shows that samples with C17 consume higher superplasticizer than those with C8. As the average particle size of C17 is smaller, the higher surface area results in higher water (or superplasticizer) demand.

## 3.2 Compressive Strength Test

Compressive strength test was performed on the mortar samples of $4 \times 4 \times 4$ cm$^3$. The results are shown in Fig. 86.3.

For both types of nanosilica, a quadratic model was fitted, and the interaction term was remarkable. In the samples with both types of nanosilica, although it is obvious that by increasing the amount of SBR, their compressive strength decrease, increasing the amount of nanosilica moderates this reduction remarkably, which is an improving effect of nanosilica. For example, in samples with SBR and C8 (Fig. 2a), at low amounts of C8 (~0.44%), increasing the amount of SBR from 1.46 to 8.54% decreases the compressive strength by about 20%, while in high amounts of C8 (2.56%), the compressive strength of samples with 1.46% and 8.54% SBR are almost equal. The same discussion can be made about the samples with SBR and

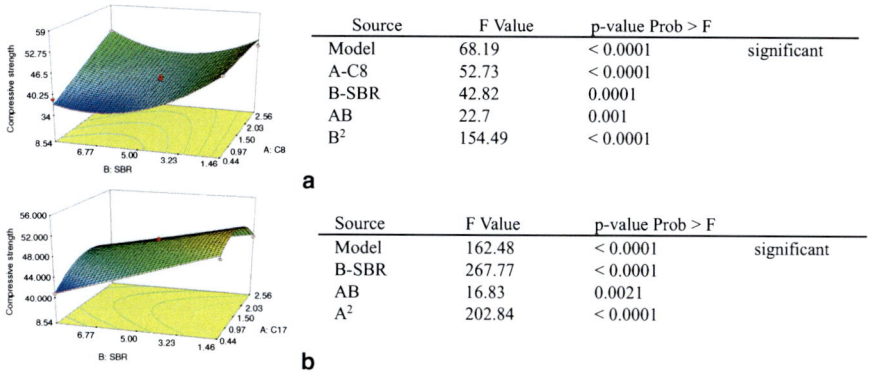

**Fig. 86.3** Compressive strength of mortar samples with SBR, (**a**) C8, (**b**) C17

C17. Increasing the amount of SBR at 0.44% C17 decreases the compressive strength by almost 20%, while in 2.56% C17, this increase in the amount of SBR reduces the compressive strength by 8%, which is significantly lower, which can be attributed to the presence of nanosilica particles.

## 3.3 Flexural Strength Test

This test was performed on $4 \times 4 \times 16$ cm$^3$ prismatic samples. The ANOVA and response surface analysis results are shown in Fig. 86.3, for both types of employed nanosilica.

From the diagrams of Fig. 86.4, both types of nanosilica have improving effects on the flexural strength of polymer-modified mixtures. Sharp decrease in flexural strength at low amounts of nanosilica is moderated in samples with C8. In samples with C17, this improving effect is even more significant with a remarkable improve in their flexural strength when almost 3% SBR and 1.5% C17 are added.

## 4 Conclusions

In this paper, the effect of having both nanosilica and polymer latexes in a cement mixture was investigated by having two types of nanosilica, with different average particle size, and one type of polymer latex (SBR). In the fresh state, flowability and, in the hardened state, compressive and flexural strength were the tested properties. Significant influence of both materials of nanosilica (both types) and polymer were observed on all the tested properties. In the fresh state, high superplasticizer consumption of nanosilica particles was moderated by incorporating the SBR latex. In

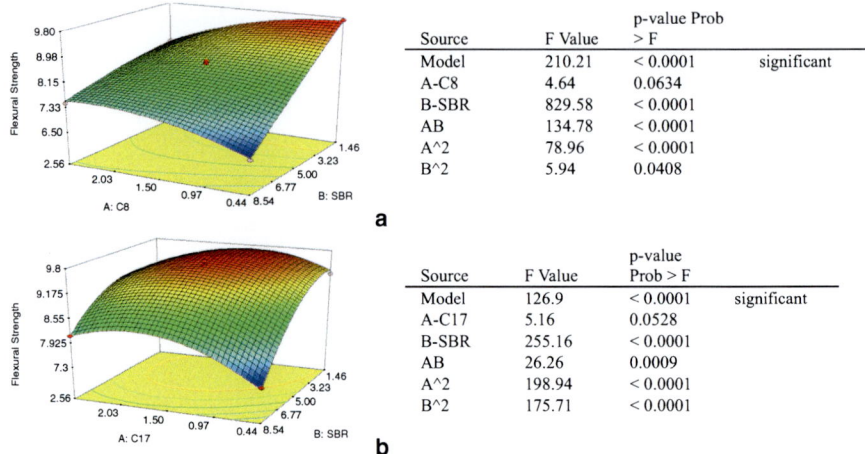

**Fig. 86.4** Flexural strength of mortar samples incorporating SBR, (**a**) C8, (**b**) C17

hardened state also, the decline in the tested mechanical properties were moderated and in one case (flexural strength of samples with C17) were improved, at the presence of nanosilica.

## References

1. Belkowitz, J. S., & Armentrout, D. (2010). An investigation of nano silica in the cement hydration process. In *2010 concrete sustainability conference*. Dubai.
2. Berra, M., Carassiti, F., Mangialardi, T., Paolini, A. E., & Sebastiani, M. (2012). Effects of nanosilica addition on workability and compressive strength of Portland cement pastes. *Construction and Building Materials, 35*, 666–675.
3. Quercia, G., & Brouwers, H. J. H. (2010). *Application of nano-silica (nS) in concrete mixtures* (pp. 431–436). Lyngby: 8th fib international Ph. D. symposium in Civil Engineering.
4. Quercia, G., Hüsken, G., & Brouwers, H. J. H. (2012). Water demand of amorphous nano silica and its impact on the workability of cement paste. *Cement and Concrete Research, 42*(2), 344–357.
5. Said, A. M., Zeidan, M. S., Bassuoni, M. T., & Tian, Y. (2012). Properties of concrete incorporating nano-silica. *Construction and Building Materials, 36*, 838–844.
6. Senff, L., Labrincha, J. A., Ferreira, V. M., Hotza, D., & Repette, W. L. (2009). Effect of nano-silica on rheology and fresh properties of cement pastes and mortars. *Construction and Building Materials, 23*(7), 2487–2491.
7. Zabihi, N., Ozkul, M. H., Parlak, N., & Özer, B. (2016). The hydration properties of nanosilica-polymer- Portland cement composites. In *12th international congress on advances in Civil Engineering (ACE2016)*. Istanbul.
8. ACI Committee 548. (2003). *Polymer-modified concrete*. Farmington Hills: American Concrete Institute.
9. Kuhlmann, L. A., & Walters, D. G. (1993). *Polymer-modified hydraulic-cement mixtures*. Philadelphia: ASTM Special Technical Publication.

10. Miller, M. (2008). *Polymers in cementitious materials*. Smithers Rapra Press, Shawbury, UK.
11. Morin, V., Moevus, M., Dubois-Brugger, I., & Gartner, E. (2011). Effect of polymer modification of the paste–aggregate interface on the mechanical properties of concretes. *Cement and Concrete Research, 41*(5), 459–466.
12. Ohama, Y. (1995). *Handbook of polymer-modified concrete and mortars: Properties and process technology*. Norwich: William Andrew.
13. Ramli, M., & Tabassi, A. A. (2012). Effects of polymer modification on the permeability of cement mortars under different curing conditions: A correlational study that includes pore distributions, water absorption and compressive strength. *Construction and Building Materials, 28*(1), 561–570.
14. Van Gemert, D., & Beeldens, A. (2013). Evolution in modeling cement hydration and polymer hardening in polymer-cement concrete. *Advanced Materials Research, 687*, 291–297.
15. Zeng, S., Short, N. R., & Page, C. L. (1996). Early-age hydration kinetics of polymer-modified cement. *Advances in Cement Research, 8*(29), 1–9.
16. ASTM, C. (2008). *230, Standard specification for flow table for use in tests of hydraulic cement*. West Conshohocken: ASTM International.
17. Hinkelmann, K., & Kempthorne, O. (2008). Design and analysis of experiments. In *Introduction to experimental design* (Vol. 1). New York: Wiley-Interscience.
18. ASTM, C. (2012). 109, StandardTest Method for Compressive Strength of Hydraulic Cement Mortars (Using 2-in. or [50-mm] Cube Specimens). *West Conshohocken, PA: ASTM International*.
19. ASTM, C. (2014). 348, Standard Test Method for Flexural Strength of Hydraulic-Cement Mortars. *West Conshohocken, PA: ASTM International*.

# Chapter 87
# Parametric Study on the Performance of UHPC and Nano-modified Polymer Concrete (NMPC) Composite Wall Panels for Protective Structures

Olaniyi Arowojolu, Ahmed Ibrahim, and Mahmoud Reda Taha

This study presents the performance of hybrid concrete walls incorporating ultrahigh-performance concrete (UHPC) and nano-modified polymer concrete (NMPC) under blast loading. UHPC has compressive strength with failure strain very close to that of conventional concrete. Such strain is insufficient to sustain large plastic deformations required for structures subjected to very high strain rates. However, NMPC made from cement replacement by epoxy polymer provides failure strain ten times higher than conventional concrete, tensile, and excellent bond strengths. The nonlinear finite element code LS-DYNA has been used to conduct the numerical analyses. The hybrid composite concrete walls were produced by interconnecting elements made of UHPC and NMPC and were assessed for post-blast damage. The key parameters considered were charge weights, standoff distances, and the composite action of the composite UHPC and NMPC wall panels. The proposed hybrid concrete walls showed outstanding resistance to blast scenarios.

## 1 Introduction

The behavior of reinforced concrete (RC) structures subjected to blast loading differs from when the structures are subjected to static loading because of the high-strain rate effect and the complex 3D states of stress. The absence of a library of material models describing material's behavior under high-strain rates made it further

O. Arowojolu · A. Ibrahim (✉)
Department of Civil and Environmental Engineering, University of Idaho, Moscow, ID, USA
e-mail: aibrahim@uidaho.edu

M. Reda Taha
Department of Civil Engineering, University of New Mexico, Albuquerque, NM, USA

© Springer International Publishing AG, part of Springer Nature 2018
M. M. Reda Taha (ed.), *International Congress on Polymers in Concrete (ICPIC 2018)*, https://doi.org/10.1007/978-3-319-78175-4_87

difficult to accurately simulate RC structure behavior under blast. In addition, the nonavailability of sufficient literature for explosive loaded structure has further challenged research in this area of study. Traditionally, conventional concrete and high-strength concrete are highly brittle materials which made them unsuitable for resisting high-strain loading [1, 2]. However, research has shown that the brittleness of concrete could be improved by the additional of short dispersed fibers in addition to other mixture proportioning to improve the concrete density and microstructure. This improvement gave rise to ultrahigh-performance fiber-reinforced concrete (UHPFRC). The UHPFRC has been reported to possess high ductility, high fracture energy of about 20,000–40,000 $J/m^2$, high compressive strength in the range of 100–200 MPa, and high tensile strength ranging from 20 to 40 MPa [3], in addition to being able to provide high performance in harsh environmental conditions such as severe freeze-thaw cycles. UHPFRC strain at failure is less than that of conventional concrete. Moreover, UHPFRC lacks uniformity in fiber distribution and is relatively expensive due to its high cement content [3]. Xu et al. [4] studied the behavior of UHPFRC columns subjected to blast loading in comparison. It was reported that UHPFRC specimens effectively resisted the overburden axial loads associated with blast and showed low permanent displacements. It was shown that UHPFRC performed better than the NSC in resisting blast loads [2, 5, 6]. The addition of steel fibers in making UHPFRC creates a homogenous concrete matrix which helps in controlling crack initiation and propagation. Furthermore, steel fibers proved to limit scabbing and spalling of concrete under blast loads compared with NSC [7].

## 2 Nano-modified Polymer Concrete

The replacement of cement binder by polymer materials gives rise to polymer concrete (PC). PC has a lower curing time, higher bond strength with steel and concrete substrates, and higher weather resistance and durability compared with NSC [8]. Recent research has shown the ability to significantly improve PC using nanomaterials. Successful efforts have shown the ability to produce nano-modified polymer concrete (NMPC) with superior mechanical properties with the addition of relatively small amount of alumina nanoparticles (ANPs) and multiwalled carbon nanotubes (MWCNTs) at relatively very low contents below 1 wt.% as ratio of the weight of the polymer binder [9, 10]. ANPs have a maximum nominal size of 50 nm, while MWCNTs have a diameter of 30 nm and a length ranging from 5 to 10 µm. The tensile strength and failure strain of concrete are two crucial criteria for blast resistance. UHPFRC has a tensile strength of about 6–8 MPa and tensile failure strain of 0.0225%, while NMPC has tensile strength of 9–12 MPa and a tensile failure strain of 5.0% compared with 0.1% tension failure strain for NSC [10].

## 3 Numerical Simulation

In most finite-element simulations where UHPFRC and NSC were used for protective structures, like those implemented in Autodyn, ABAQUS, and LS-DYNA, the stress-strain curve of UHPFRC, specifically the strain-softening phase, has not been fully captured with the simulation models. Results of such simulations displayed poor response predictions especially after the first cracking. The concrete damage model available in LS-DYNA is an example of such concrete models. Concrete damage is then defined by the characteristics strength of concrete determined by the unconfined concrete compressive strength at 28 days of age. On the other hand, continuous surface cap model (CSCM) is an alternative material model that has been used in LS-DYNA for dynamic simulation. CSCM showed satisfactory results for simulating normal and high-strength concrete behavior during blast [9].

In this study, MAT_Elastc_Plastic_Hydrodynamic material is adopted for simulating NMPC because the elastic behavior of the concrete is captured by the concrete Young's modulus, while the inelastic behavior is captured by the effective plastic strain and effective plastic stress measured by testing concrete cylinders. The MAT_Elastc_Plastic_Hydrodynamic material model has been reported to successfully capture the nonlinear softening of UHPFRC after yielding [10, 11]. In order to capture the volumetric stress and strain associated with pressure generated during blast, Gruneisen equation of state was used [12]. The equation of state adopted was for compressed material under cubic shock velocity-particle velocity.

## 4 Case Study

A structural concrete wall was simulated under blast loading. The wall is made of interlocking blocks of solid UHPC and NMPC blocks each of 200 mm by 200 mm by 100 mm thickness. The whole wall is 1200 mm in height, is 1200 mm wide, and has 100 mm thickness. A schematic of the wall is shown in Fig. 87.1a. An air burst with ground surface reflection type was used in this simulation. Different scaled distances ranging from 0.773 to 0.387 m/kg$^{1/3}$ were used in this simulation. The wall was simulated in LS-DYNA. A termination time of 60 ms, with a time step of 0.2 ms, was selected for this simulation. Concrete was modeled as 8-node brick element with a uniform mesh size of 10 mm; this time step and mesh size were selected after conducting mesh sensitivity analysis. The failure mode and displacement-time histories were observed and then analyzed. The stress-strain curve of NMPC as used in the simulation is shown in Fig. 87.1b. UHPC behavior was linear elastic to failure.

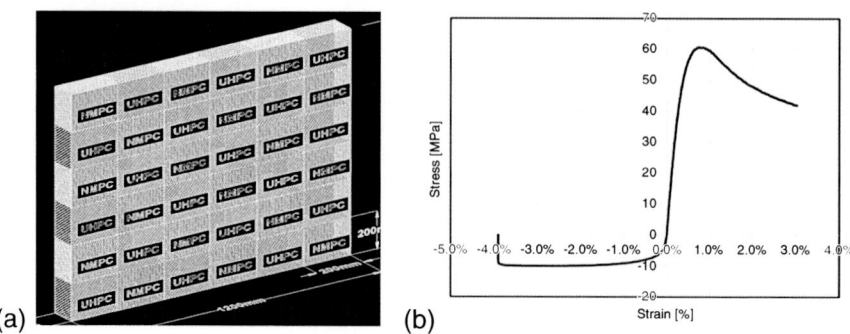

**Fig. 87.1** Hybrid concrete wall under blast (**a**) wall configuration (**b**) stress-strain behavior of NMPC used in simulation of wall behavior

**Table 87.1** Summary of hybrid wall test results

| Scaled distances (m/kg$^{1/3}$) | Maximum deflection (mm) | Residual deflection (mm) | Post-blast failure mode remarks |
|---|---|---|---|
| 0.773 | 5.6 | 3.0 | Minor horizontal shear cracks at supports |
| 0.613 | 9.0 | 5.8 | Horizontal shear cracks at the supports |
| 0.487 | 16.0 | 12.5 | Diagonal tension-shear cracks, at mid-span |
| 0.387 | 32.0 | 13.2 | Flexure crack, extending to shear cracks in two orthogonal directions at mid-span, extending to the supports |

## 5 Results and Discussion

The behavior of the hybrid wall subjected to blast loading includes the displacement-time histories and pressure-time histories for the scaled distances at the mid-span of the wall. The maximum mid-span displacement and residual displacement for each scaled distance are presented in Table 87.1. Different deformation modes were observed, ranging from shear cracks to flexural crack. A typical cracking of the hybrid wall is shown in Fig. 87.2a. The deformed shape of the wall under blast load is shown in Fig. 87.2b. No spalling, scabbing, nor crushing was observed in concrete in the hybrid wall. This shows that the combination of UHPFRC with NMPC can result in significant improvement in blast resistance.

The maximum displacement for all the simulated specimens occurred in the first cycle at a very short time period. The residual displacements were of relatively low magnitude and much lower than the maximum displacement. The residual displacement shall be considered in the design of the wall to avoid lateral buckling when subjected to blast load.

It is important to note that the effect of strain rate was not considered in the simulations since the material characteristics of UHPFRC and NMPC were extracted from static testing. Further research is warranted to examine the strain rate effect and

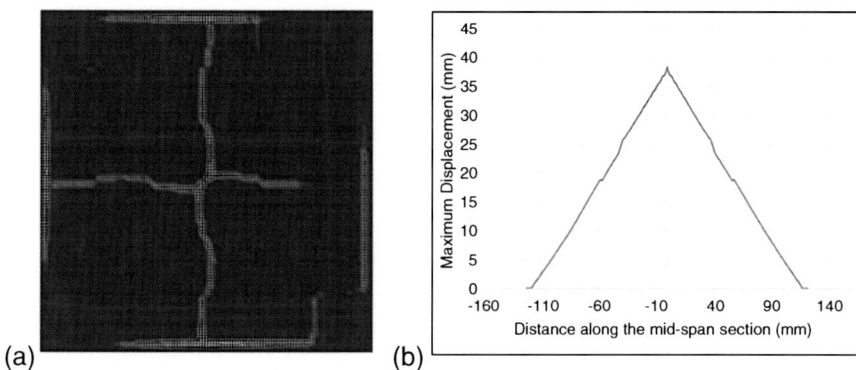

**Fig. 87.2** Behavior under 0.387 m/kg$^{1/3}$ charge: (**a**) cracking pattern (**b**) wall deformation

to examine the significance of geometrical distribution of the two materials on blast resistance of the wall.

## 6 Conclusions

In this study, a series of numerical simulations were conducted on hybrid concrete wall made from interlocking blocks made of UHPFRC and NMPC, to determine its performance against blast events in protective structures. Scaled distances ranging from 0.387 to 0.773 m/kg$^{1/3}$ using the same wall configuration were examined. No crushing, scabbing, or spalling in walls was observed during the simulation. As the scaled distance decreased, the displacement increased, but the walls remained intact. Full collapse of the wall was not observed. The effect of strain rate on the simulation results was not examined. The hydrodynamic material model parameters provided good prediction of the behavior of the proposed hybrid wall made of UHPFRC and the NMPC. Further studies are warranted to determine the optimal geometrical configuration of the wall to maximize its blast resistance.

**Acknowledgment** The authors greatly appreciate the support provided by the department of civil and environmental engineering at the University of Idaho, Moscow. Special thanks to Transpo Industries for providing funding research for NMPC.

## References

1. Bischoff, P. H., & Perry, S. H. (1991). Compressive behaviour of concrete at high strain rates. *Materials and Structures, 24*(6), 425–450.

2. Schenker, A., Anteby, I., Gal, E., Kivity, Y., Nizri, E., Sadot, O., Michaelis, R., Levintant, O., & Ben-Dor, G. (2008). Full-scale field tests of concrete slabs subjected to blast loads. *International Journal of Impact Engineering, 35*(3), 184–198.
3. Barnett, S. J., Lataste, J. F., Parry, T., Millard, S. G., & Soutsos, M. N. (2010). Assessment of fibre orientation in ultra high performance fibre reinforced concrete and its effect on flexural strength. *Materials and Structures, 43*(7), 1009–1023.
4. Xu, J., Wu, C., Xiang, H., Su, Y., Li, Z. X., Fang, Q., Hao, H., Liu, Z. Z., & Li, J. (2016). Behaviour of ultra high performance fibre reinforced concrete columns subjected to blast loading. *Engineering Structures, 118*, 97–107.
5. Li, J., Wu, C., Hao, H., Su, Y., & Liu, Z. (2016). Blast resistance of concrete slab reinforced with high performance fibre material. *Journal of Structural Integrity and Maintenance, 1*(2), 51–59.
6. Ngo, T., Mendis, P., & Krauthammer, T. (2007). Behavior of ultrahigh-strength prestressed concrete panels subjected to blast loading. *Journal of Structural Engineering, 133*(11), 1582–1590.
7. Brandt, A. M. (2008). Fibre reinforced cement-based (FRC) composites after over 40 years of development in building and civil engineering. *Composite Structures, 86*(1), 3–9.
8. Daghash, S. M., Tarefder, R., & Taha, M. M. R. (2015). A new class of carbon nanotube: Polymer concrete with improved fatigue strength. In *Nanotechnology in construction (pp. 285–290)*. Cham: Springer.
9. Arowojolu, O., Rahman, M. K., & Hussain, B. M. (2017). Dynamic response of reinforced concrete bridge piers subjected to combined axial and blast loading. In *Structures congress*, Colorado, USA.
10. Teng, T. L., Chu, Y. A., Chang, F. A., Shen, B. C., & Cheng, D. S. (2008). Development and validation of numerical model of steel fiber reinforced concrete for high-velocity impact. *Computational Materials Science, 42*(1), 90–99.
11. Wang, Z. L., Wu, J., & Wang, J. G. (2010). Experimental and numerical analysis on effect of fibre aspect ratio on mechanical properties of SRFC. *Construction and Building Materials, 24*(4), 559–565.
12. Grüneisen, E. (1912). Theory of solid state of monatomic elements. *Annalen der Physik, 39*, 257–306.

# Chapter 88
# Mechanical Characterization of Polymer Nanocomposites Reinforced with Graphene Nanoplatelets

Ugur Kilic, Sherif M. Daghash, and Osman E. Ozbulut

Fiber reinforced polymer (FRP) composites have been extensively used for strengthening concrete structures. To manufacture FRPs or bond them to concrete structures, usually thermoset polymers are used. The mechanical properties and integrity of these adhesives significantly affect the performance of FRP-strengthened structures. Graphene nanoplatelets (GNPs) are carbon-based functional fillers that possess large surface area and high aspect ratio. They are easy to be processed in the host matrix and have excellent material properties at a relatively low cost. This study investigates the tensile behavior of GNP-reinforced nanocomposites. Two different epoxy matrices, one ductile and another brittle, are considered. First, the effect of ultrasonication duration in dispersion of GNPs is studied. Then, specimens with different GNP concentration levels are prepared to assess the effect of GNP content on the developed nanocomposites. Monotonic uniaxial tensile tests are conducted to study the effect of GNP addition to tensile strength and tensile modulus of two different epoxy resins. Morphology of GNPs and the fracture surface of the developed nanocomposites are also observed using SEM to assess the dispersion of GNPs. Results shows that both tensile strength and tensile modulus of ductile epoxy increase with increasing GNP content up to 1 wt. %, while for brittle epoxy a significant increase in tensile modulus is observed with 2 wt. % GNP concentration together with a slight decrease in tensile strength.

---

U. Kilic (✉) · S. M. Daghash · O. E. Ozbulut
Department of Civil and Environmental Engineering, University of Virginia, Charlottesville, VA, USA
e-mail: uk7ud@virginia.edu

© Springer International Publishing AG, part of Springer Nature 2018
M. M. Reda Taha (ed.), *International Congress on Polymers in Concrete (ICPIC 2018)*, https://doi.org/10.1007/978-3-319-78175-4_88

## 1 Introduction

There is a growing demand for advanced composite materials with improved mechanical properties to meet various performance requirements in structural applications. Fiber reinforced polymer (FRP), a composite material composed of a polymer matrix and reinforcing fibers, has been widely used for reinforcing or strengthening concrete structures. Although various thermosetting polymers such as phenolic, polyester, and vinylester resins have been considered as a matrix in FRPs, epoxy resin has been one of the most commonly used matrices in fabrication of FRPs due to its good and versatile properties [1]. Adding nanofillers to epoxy resins can provide a polymer matrix with improved mechanical properties and additional functional properties such as self-sensing for FRPs.

Carbon nanomaterials exhibit superior mechanical and electrical properties [2, 3], which make them ideal fillers for polymer nanocomposites [4]. The most commonly studied carbon-based nanomaterials have been carbon nanotubes (CNTs). Due to their unique structures, CNTs possess excellent electrical, thermal, and electronic transport properties [5]. However, the poor dispersion and the high cost of CNTs are the two critical issues for their use in epoxy nanocomposites [6]. Graphene is one of the stiffest and strongest materials available today with ~1 TPa in Young's modulus and ~130 GPa in strength [7]. The unique size and platelet morphology of graphene make these particles especially effective at providing barrier properties, while their pure graphitic composition makes them excellent electrical and thermal conductors. Graphene has an outstanding thermal conductivity of around 5000 $Wm^{-1} K^{-1}$ [8] and electrical conductivity of around $10^8$ $S\ m^{-1}$. More recently, graphene nanoplatelets (GNPs), which consist of few layers of graphene, have emerged as one of the most attractive nanofillers for polymer matrices with an excellent balance between structural properties and cost. GNPs can be found in a variety of geometric features as a function of thickness, diameter, and number of atomic layers. GNPs have high aspect ratios and specific surface areas [9] and significantly lower costs compared to CNTs. If a good dispersion is achieved, efficient enhancement in mechanical properties can be obtained with a low amount of nanofiller.

The objective of this study is to investigate the relationship between the GNP content and mechanical properties of nanocomposites that are made from two different types of epoxy matrix. For this purpose, a suitable dispersion strategy and curing process have been developed. The structure of the fabricated nanocomposite was determined with optical microscope and scanning electron microscope (SEM). Analysis was presented on the effects of different GNP contents and dispersion on the structure and mechanical properties of the GNP/epoxy nanocomposite.

## 2 Experimental Procedure

### 2.1 Materials

In this study, GNPs were obtained from XG Sciences, Inc. with the commercial name of Grade M25, which indicates the particle diameters of 25 microns. Grade M25 GNPs have a typical surface area of 120–150 m$^2$/g and an average thickness of 6–8 nm. Two different types of epoxy matrices (one brittle, one ductile) were studied. Both matrices were based on Diglycidyl Ether of Bisphenol A (DGEBA). 635 Thin Epoxy System with 2:1 hardener ratio (by weight) was used as ductile epoxy matrix that has a low viscosity and was supplied by US Composites, Inc. For the brittle one, EPOTUF® 37–140 resin and EPOTUF® 37–650 hardener were used with a ratio of 3:1 by volume and provided by Reichhold.

### 2.2 GNPs Dispersion

An ultrasonic processor device (Cole-Parmer 750-Watt Ultrasonic Homogenizer) was used for dispersion of GNPs into epoxy matrices. In order to evaluate the effect of ultrasonication time in dispersion of GNPs, four different ultrasonication time periods (30 min, 1 h, 2 h, and 3 h) were applied for epoxy nanocomposites with 0.5 wt. % of GNPs. The optimal ultrasonication period was determined and used to prepare all the test samples for characterization of epoxy nanocomposites with different GNP ratios.

### 2.3 Nanocomposite Fabrication

Five sets of nanocomposite samples were manufactured with different concentrations of GNPs (0, 0.25, 0.5, 1, and 2 wt. %) for both epoxy matrices. The fabrication processes for ductile epoxy were carried out using the following steps. The appropriate amounts of GNPs were added to 635 Thin Epoxy Resin (ductile epoxy) and mixed by using the ultrasonic processor at an amplitude of 40% for 2 h. The resulting mixture was degassed inside a vacuum oven (29" Hg pressure) at 90 °C for 30 min. Then, the mixture was mechanically mixed with the hardener for 3–5 min using the ratio of 2:1. The mixture was cast into molds and rested at ambient temperature for 1 day and subsequently cured inside an oven at 121 °C (250 °F) for 2 h and post cured at ambient temperature for 6 more days.

To fabricate the nanocomposites with EPOTUF® 37–140 matrix (brittle epoxy), the same procedure described above was used except that the samples were not degassed in the vacuum oven since degassing caused considerable foam in that resin. The mixture then was mechanically mixed with the hardener for 3–5 min in the ratio of 3:1 and the molding and curing processes described above were applied. At least five samples were made and tested for each specimen type.

## 3 Results and Discussions

The effect of ultrasonication time on the dispersion of GNPs in epoxy resins was studied first. Figure 88.1 shows the optical microscope images of GNP–epoxy nanocomposites before the addition of hardener for ductile epoxy. It can be seen that the increase in the ultrasonication time generally resulted in better dispersion as agglomeration of GNPs was reduced. However, processing time beyond 2 h had almost no additional effect on the dispersion.

In addition, tensile properties of epoxy nanocomposites with 0.25 wt. % of GNPs were studied for the specimens that were fabricated using 1 h and 2 h of ultrasonication. The tensile properties were measured according to ASTM D638 [10] (Type I sample geometry; 165 mm long, 3.2 mm thick, 13 mm width) at a crosshead speed of 5 mm/min at an MTS load frame, and the strains were recorded using a laser extensometer. Figure 88.2a compares the mean tensile strength of GNP–epoxy nanocomposites for two different epoxy resins sonicated for 1 h and

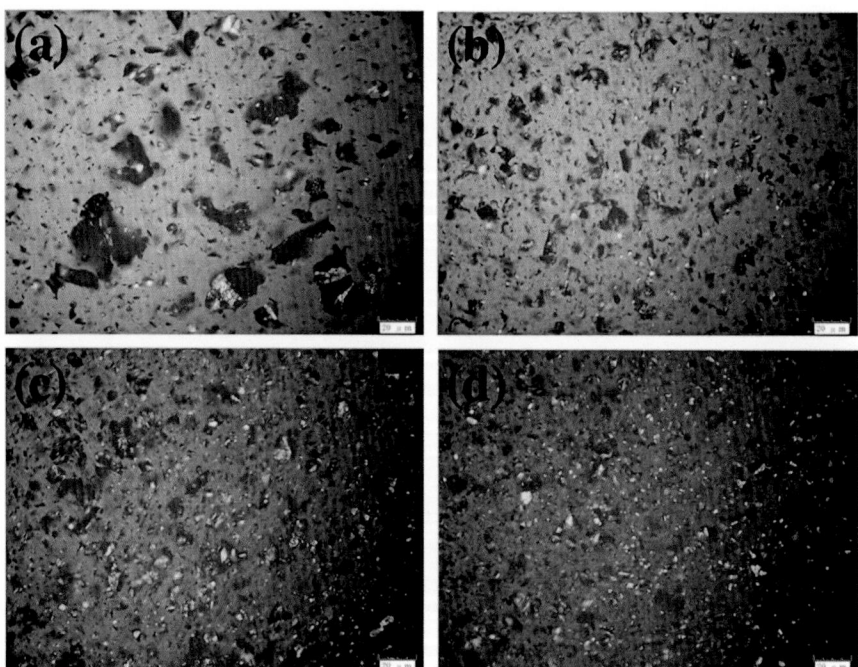

**Fig. 88.1** Optical microscope images of GNP–epoxy nanocomposites for ultrasonication time of (**a**) 30 min, (**b**) 1 h, (**c**) 2 h, and (**d**) 3 h

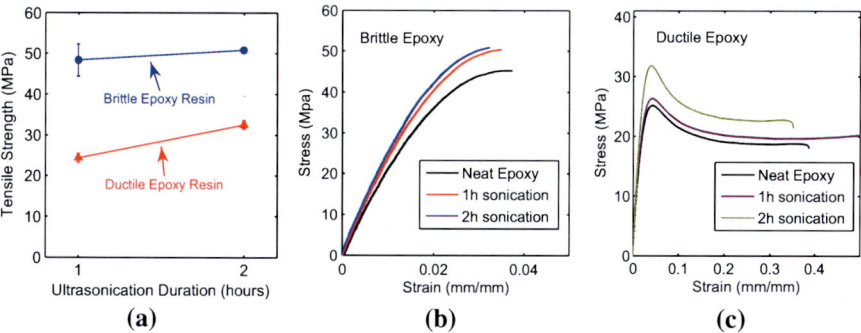

**Fig. 88.2** (**a**) Mean tensile strength of GNP–epoxy nanocomposites for 1 h and 2 h of ultrasonication; tensile stress–strain curves for (**b**) brittle and (**c**) ductile epoxy resins with and without GNPs

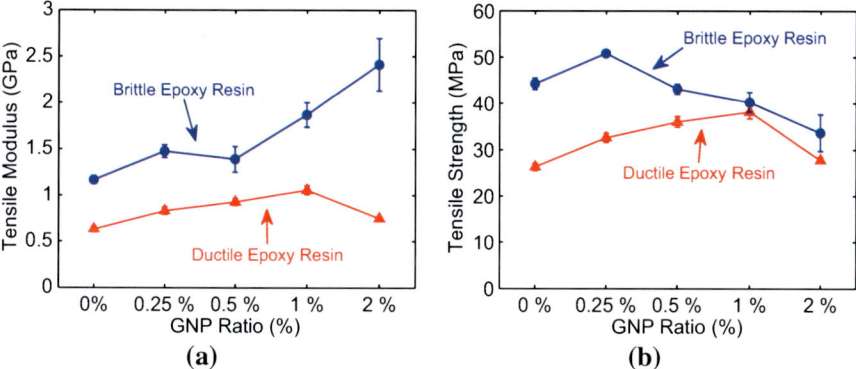

**Fig. 88.3** (**a**) Tensile modulus and (**b**) tensile strength for epoxy nanocomposites with different GNP concentrations

for 2 h. An increase in the tensile strength for both epoxy resins was observed when the ultrasonication time was increased from 1 h to 2 h. Moreover, Fig. 88.2b and c illustrates typical stress–strain curves for both epoxy resins without GNPs (neat epoxy) and with 0.25 wt. % GNPs (sonicated 1 h or 2 h). Since both optical images and tensile test results suggest a better dispersion with 2 h ultrasonication, an ultrasonication time of 2 h was used to prepare GNP–epoxy nanocomposites with different nanofiller ratios.

Figure 88.3 shows the change in the tensile modulus and tensile strength with various amounts of GNPs. Both the modulus and strength increase with the addition of GNPs for ductile epoxy up to 1 wt. % GNP concentration. The modulus for the nanocomposite fabricated from ductile epoxy increased from 0.63 GPa for neat epoxy to 1.06 GPa for the sample containing 1 wt. % GNP. The ultimate tensile strength also increased from 26.3 MPa (neat epoxy) to 38.2 MPa for the formulation containing 1 wt. % GNP.

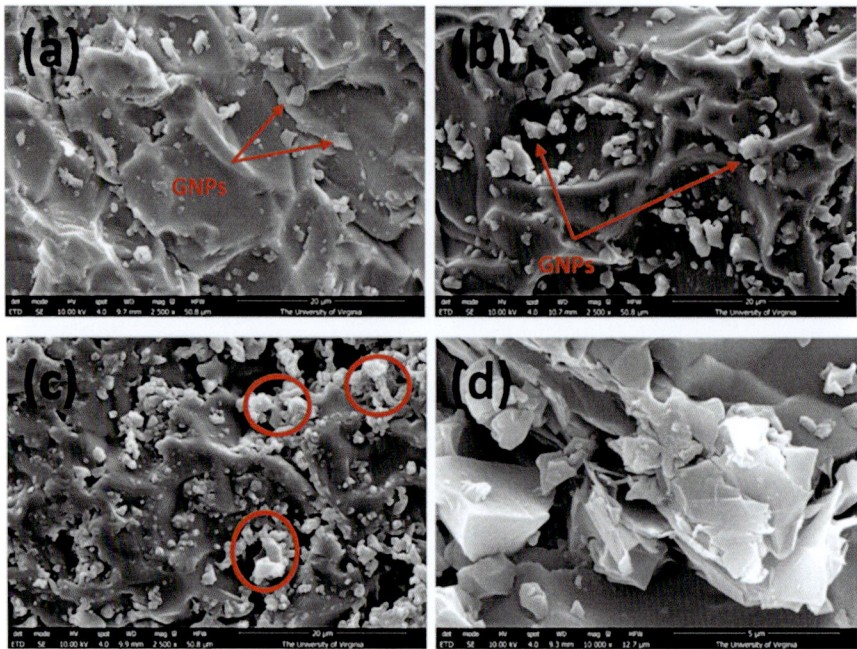

**Fig. 88.4** SEM images of (**a**) 0.25 wt.%, (**b**) 0.5 wt.%, (**c**) 1 wt.% of GNP, (**d**) GNP agglomerates

For the brittle epoxy, there was no significant change in the tensile modulus for 0.25 wt. % to 0.5 wt. % GNP concentrations. However, it increased up to 2.4 GPa for the samples containing 2 wt. % of GNP compared to tensile modulus of 1.2 GPa for neat epoxy. On the other hand, tensile strength for this epoxy increased from 44.2 MPa (neat epoxy) to 50.8 MPa for epoxy with 0.25 wt. % of GNP but an increased GNP ratio caused a decrease.

Figure 88.4 indicates the Scanning Electron Microscopy (SEM) images of GNP–epoxy nanocomposites with ductile epoxy resin and 0.25 wt. %, 0.5 wt. %, and 1 wt. % of GNPs. It can be seen that GNPs are generally well-dispersed at each concentration level but some agglomerates were observed in the structure (marked with a circle in Fig. 88.4c and zoomed in Fig. 88.4d) when the GNP content is 1 wt. %. Further increase in GNP content (i.e., 2 wt. %) caused additional agglomerates, which can explain the decrease in the tensile strength and modulus.

## 4 Conclusions

This study explored the tensile characteristics of one brittle and one ductile epoxy resin reinforced with graphene nanoplatelets. The effect of sonication duration in the dispersion of GNPs was also studied. For ductile epoxy, both the mean tensile modulus and strength increased by 68% and 45%, respectively, compared to neat epoxy when 1 wt. % GNP was added and a further increase in GNP content did not result in any additional improvement. For brittle epoxy, tensile modulus increased by 108% for epoxy nanocomposites with 2 wt. % GNP ratio. There was also an initial slight increase (15%) in tensile strength when 0.25 wt. % GNP was added but increasing GNP content to 1 wt. % and 2 wt. % decreased the tensile strength by 9% and 24%, respectively. Further studies will be conducted to study the effect of GNP concentration on flexural modulus and fracture toughness of the developed GNP–epoxy nanocomposites.

# References

1. Shokrieh, M., Esmkhani, M., Shahverdi, H. R., & Vahedi, F. (2013). Effect of graphene nanosheets (GNS) and graphite nanoplatelets (GNP) on the mechanical properties of epoxy nanocomposites. *Science of Advanced Materials, 5*(3), 260–266.
2. Xu, M., Futaba, D. N., Yamada, T., Yumura, M., & Hata, K. (2010). Carbon nanotubes with temperature-invariant viscoelasticity from −196 °C to 1000 °C. *Science, 330*(6009), 1364.
3. Geim, A. K. (2009). Graphene: Status and prospects. *Science, 324*(5934), 1530.
4. Sengupta, R., Bhattacharya, M., Bandyopadhyay, S., & Bhowmick, A. K. (2011). A review on the mechanical and electrical properties of graphite and modified graphite reinforced polymer composites. *Progress in Polymer Science, 36*(5), 638.
5. Chatterjee, S., Nafezarefi, F., Tai, N. H., Schlagenhauf, L., Nuesch, F. A., Chu, B. T. T., Thostenson, E. T., Ren, Z., & Chou, T.-W. (2012). Size and synergy effects of nanofiller hybrids including graphene nanoplatelets and carbon nanotubes in mechanical properties of epoxy composites (2001) Advances in the science and technology of carbon nanotubes and their composites: a review. *Composites Science and Technology, 61*(13), 1899–1912.
6. Coleman, J., Khan, U., & Gun'ko, Y. (2006). Mechanical reinforcement of polymers using carbon nanotubes. *Advanced Materials, 18*, 689–706.
7. Lee, C., Wei, X., Kysar, J. W., & Hone, J. (2008). Measurement of the elastic properties and intrinsic strength of monolayer graphene. *Science, 321*, 385–388.
8. Balandin, A. A., Ghosh, S., Bao, W., Calizo, I., Teweldebrhan, D., Miao, F., & Lau, C. N. (2008). Superior thermal conductivity of single-layer graphene. *Nano Letters, 8*(3), 902–907.
9. ASTM D638 – Standard test method for tensile properties of plastics.

# Chapter 89
# Oil Well Cement Modified with Bacterial Nanocellulose

Christian M. Martín, Ignacio Zapata Ferrero, Patricia Cerrutti, Analía Vázquez, Diego Manzanal, and Teresa M. Pique

During oil well completion, cement slurries are pumped into the annular area between the casing and the borehole. These are used to contain and prevent corrosion and erosion of the casing and to hydraulically isolate the various formations through the borehole, avoiding the filtration of pollutants in aquifers, loss of productivity, etc. Oil well cementing is a complex task. Incorrect isolation or loss of circulation causes damages, both economically and environmentally.

Cement slurry is a mixture of cement, water, and additives. In particular, high-performance materials are used for oil wells. It is further characterized by having a high w/c ratio and a relatively low viscosity. The aim of this work is to optimize the performance of the slurries with the addition of bacterial nanocellulose (BNC). In order to study the effect of BNC in cement slurries, its properties were studied in fresh and hardened state. In particular, density, fluidity, hydration degree, and mechanical properties were evaluated. The results confirm the positive effect of the addition of BNC to the slurry, presenting low percentage of free liquid and an increase of the compressive strength.

---

C. M. Martín · D. Manzanal
Facultad de Ingeniería, Universidad de Buenos Aires, Ciudad de Buenos Aires, Argentina

I. Z. Ferrero
ITPN, UBA-CONICET, Ciudad de Buenos Aires, Argentina

P. Cerrutti · A. Vázquez · T. M. Pique (✉)
Facultad de Ingeniería, Universidad de Buenos Aires, Ciudad de Buenos Aires, Argentina

ITPN, UBA-CONICET, Ciudad de Buenos Aires, Argentina
e-mail: tpique@fi.uba.ar

© Springer International Publishing AG, part of Springer Nature 2018
M. M. Reda Taha (ed.), *International Congress on Polymers in Concrete (ICPIC 2018)*, https://doi.org/10.1007/978-3-319-78175-4_89

## 1 Introduction

Primary cementing is the activity in which cement is pumped into the annulus between the casing and the borehole. It is a necessary labor that must hold and restrain the casing, protect the casing against erosion and corrosion, hydraulically isolate the various formations crossed by the well, avoid pollutant filtration into aquifers and productivity losses, prevent gas migration, and guarantee the borehole stability [1].

To adapt the cement slurry to the different environments where the wells are located, many different additives are used. Bacterial nanocellulose (BNC) is a nontraditional material in the cement industry which is gaining relevancy every year [2]. It can be synthesized as a primary metabolite by some bacteria as *Gluconacetobacter xylinus*. Initially, it is formed in the interior of the bacterial cells whereupon it is ejected forming fibers with a rectangular cross-section around 3–10 nm in thickness, 30–100 nm in width, and 1–9 mm in length. This material has outstanding physical-chemical and mechanical properties as its natural origin, its biodegradability, its nanometer size, its high elastic modulus, and its remarkable tensile strength [3].

In this work, properties of the cement slurry with the addition of BNC have been evaluated through tests of significant relevancy in this matter, such as fluidity, free fluid, isothermal calorimetry, and compressive strength.

## 2 Materials

Class G cement from company "Cemento Loma Negra SA," was used as received. Its composition is summarized in Table 89.1.

BNC was provided by the company "Nanocellu-ar" in a mixture with 98% water. This material was added to the mixing water after being subjected to a mechanical dispersion process. A superplasticizer (SP) ADVA 175 LN from "GCP SRL" was used to maintain constant fluidity for all mixtures. This work describes the properties of three mixtures. Their compositions are described in Table 89.2.

**Table 89.1** Class G cement composition

| $C_3S$ | $C_2S$ | $C_3A$ | $C_4AF$ | Free CaO | MgO |
|---|---|---|---|---|---|
| 50.0% | 30.0% | 5.0% | 12.0% | 1.1% | 1.9% |

**Table 89.2** Mixtures relative proportions in mass

|  | CP | CP + BNC 0.1% | CP + BNC 0.2% |
|---|---|---|---|
| Cement | 1 | 1 | 1 |
| Water | 0.440 | 0.440 | 0.440 |
| BNC | 0.000 | 0.001 | 0.002 |

## 3 Methods

Flowability was kept constant to all mixtures with the addition of the SP. It was measured with the mini slump tests, and the plastic yield was calculated using Roussel et al. [4] equation (Eq. 89.1).

$$\zeta_0 = \frac{225 \cdot \rho \cdot g \cdot V^2}{128 \cdot \pi^2 \cdot R^5} \tag{89.1}$$

where $\zeta_0$ is the plastic yield [Pa], $\rho$ is the specific gravity of the slurry experimentally measured in the fresh state [kg/m³], $g$ is the gravity acceleration [9.81 m/s²], $V$ is the volume of the truncated cone [m³], and $R$ is the average measured diameter [m].

Free fluid was determined according to the procedure described in API 10B [5]. It was calculated using Eq. 89.2.

$$\varphi = \frac{V_{FF} \cdot \rho}{m_S} \cdot 100 \tag{89.2}$$

where $\varphi$ is the free fluid percentage [%], $V_{FF}$ is the free fluid volume [cm³], $\rho$ is the specific gravity of the slurry [g/cm³] and $m_S$ is the initial mass of the slurry [g].

The maximum free fluid value given by API 10B is 5.9% [5].

Compressive strength was measured after 28 days of humid curing of 5 cm blocks on an INSTRON 5980. Isothermal calorimetry was conducted in a TAM AIR, water and water + BNC was added to the sample holder inside the equipment and automatically mixed.

## 4 Results and Discussions

Slump and plastic yield values ($\zeta_0$) (necessary shear stress to make the cement slurry flow) for each mixture, with and without SP, are reported in Table 89.3. The addition of BNC notoriously decreased the flowability. An extremely high amount of SP had to be added to reach the same flowability as the unmodified cement slurry (0.1% of SP for CP + BNC 0.1% and 0.5% of SP for CP + BNC 0.2%). With these amounts of SP, CP + BNC 0.1% and CP + BNC 0.2% presented the same plastic yield as CP.

Furthermore, the increase in $\zeta_0$ between the different cement slurries was remarkable. With an addition of 0.1% of BNC, this value is increased to 2000% compared

**Table 89.3** Mixtures' slump and plastic yield values

|  | Slump [cm] | $\zeta_0$ [Pa] |
|---|---|---|
| CP | 23 | 0.13 |
| CP + BNC 0.1% without SP | 13 | 2.46 |
| CP + BNC 0.2% without SP | 8 | 25.29 |
| CP + BNC 0.1% with SP | 23 | 0.13 |
| CP + BNC 0.2% with SP | 20 | 0.13 |

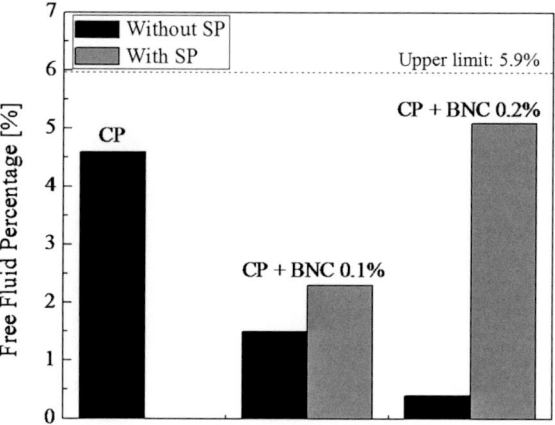

**Fig. 89.1** Free fluid results

to the reference cement slurry, and, with an addition of 0.2%, it increased to 1000% with respect to the CP + BNC 0.1% slurry. The influence of the percentage of BNC in the slurries' plastic yield can be reproduced with an exponential equation (Eq. 89.3) with $R^2 = 0.995$.

$$\tau_0 = 0.14 e^{26.3(\%BNC)} \qquad (89.3)$$

Figure 89.1 presents the free fluid percentage obtained for each of the cement slurries, differentiating those that have SP from those that have not. As can be seen, for all cases, the condition established by the API [5] is verified; no value exceeded 5.9% even with high dosages of SP.

These results show the important water retention imparted to the cement slurries by the addition of BNC, as an increase in its addition means a decrease in the free fluid percentage. Furthermore, it can be noticed that the addition of SP is related with an increase in this value, which has to be taken into account when dosing oil well cement slurries.

Finally, Fig. 89.2 shows the results of compressive strength after 28 days of humid curing of the studied cement slurries. It can be appreciated that the compressive strength of slurries with BNC increased in comparison with the reference one. When only 0.1% of BNC was added, it increased 16% and 19% for 0.2%.

Lu et al. [6] established that BNC accelerates CSH production during hydration reaction. This could be the cause of the noteworthy increase in compressive strength results. Nevertheless, no changes that would confirm this hypothesis were found in the hydration reaction of CP and CP + BNC measured by means of isothermal calorimetry, as it can be seen in Fig. 89.3.

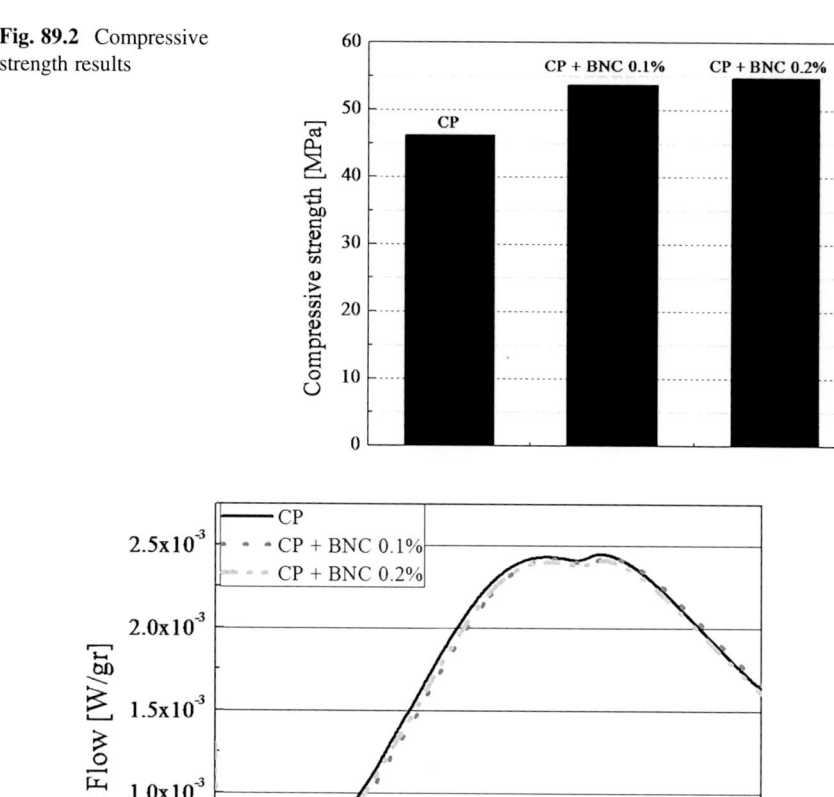

**Fig. 89.2** Compressive strength results

**Fig. 89.3** Isothermal calorimetry

## 5 Conclusions

In this work, the properties of cement slurries modified with BNC, in both fresh and hardened state, were studied in order to analyze its possible application as an oil well cement additive. Initially, the exponential increase of the plastic yield demonstrated the excellent ability of the BNC as a viscosity modifier admixture. Even the lowered percentages of free fluid of cement slurries with BNC demonstrate this. The free fluid percentage of the cement slurries modified with BNC and SP showed that caution should be taken when incorporating a high amount of SP. Nevertheless, values were within the acceptable range because of BNC addition.

Furthermore, an improvement in the compressive strength was appreciated as a result of the addition of BNC to the cement slurries. No information about this increase was found by means of isothermal calorimetry. More research should be done to explain this behavior.

# References

1. Abbas, R., Jarouj, H., Dole, S., Junaidi, E. H., El-Hassan, H., Francis, L., & Messier, E. (2003). A safety net for controlling lost circulation. *Oilfield Review, 15*, 20–27.
2. Cerrutti, P., Roldán, P., García, R. M., Galvagno, M. A., Vázquez, A., & Foresti, M. L. (2016). Production of bacterial nanocellulose from wine industry residues: Importance of fermentation time on pellicle characteristics. *Journal of Applied Polymer Science, 133*, 43109.
3. Charreau, H., Foresti, M. L., & Vázquez, A. (2013). Nanocellulose patents trends: A comprehensive review on patents on cellulose nanocrystals, microfibrillated and bacterial cellulose. *Recent Patent on Nanotechnology, 7*, 56–80.
4. Roussel, N., Stefani, C., & Leroy, R. (2005). From mini-cone test to Abrams cone test: Measurement of cement-based materials yield stress using slump tests. *Cement and Concrete Research, 35*, 817–822.
5. American Petroleum Institute. (2010). *API 10A: Specification for cements and materials for well cementing* (24th ed.). API Publishing Services, Washington D.C.
6. Lu, S., Chen, N., Pei, Z., Peng, Y., & Huang, T. (2011). Preparation and properties of bacterial cellulose reinforced cement composites. *China Powder Science and Technology, 17*, 57–60.

# Chapter 90
# Effect of Incorporating Nano-silica on the Strength of Natural Pozzolan-Based Alkali-Activated Concrete

Mohammed Ibrahim, Muhammed K. Rahman, Megat Azmi M. Johari, and Mohammed Maslehuddin

A new class of concrete with precursors such as fly ash and pozzolan, totally replacing ordinary Portland cement (OPC), together with alkaline activators is being extensively researched in producing alkali-activated concrete (AAC). The strength development of AAC depends on curing temperature, composition, and fineness of the source materials. Reactivity of the binders with the alkaline activators increases with the fineness, leading to the improved mechanical and microstructural properties. This study focuses on the development of AAC utilizing natural pozzolan (NP) as source material. In order to enhance the properties, NP is partially replaced with nano-silica up to 7.5% in the AAC mixes. Compressive strength was measured on the specimens cured for 0.5, 1, 3, 7, 14, and 28 days in the oven maintained at 60 °C. Scanning electron microscopy (SEM) was used to determine the morphology of the developed alkali-activated paste (AAP). The results indicated that AAC with NP as a binder gained reasonable strength after 3 days of curing at elevated temperature. Further, NP replaced with nano-silica (NS) exhibited improved strength and microstructural characteristics. AAC with 5% nano-silica showed better compressive strength results and denser microstructure compared to the ones prepared with other replacement levels.

---

M. Ibrahim
School of Civil Engineering, Universiti Sains Malaysia, Engineering Campus, Nibong Tebal, Pulau Pinang, Malaysia

Center for Engineering Research, Research Institute, King Fahd University of Petroleum and Minerals, Dhahran, Saudi Arabia

M. K. Rahman (✉) · M. Maslehuddin
Center for Engineering Research, Research Institute, King Fahd University of Petroleum and Minerals, Dhahran, Saudi Arabia
e-mail: mkrahamn@kfupm.edu.sa

M. A. M. Johari
School of Civil Engineering, Universiti Sains Malaysia, Engineering Campus, Nibong Tebal, Pulau Pinang, Malaysia

## 1 Introduction

Concrete research is currently focused on developing sustainable alternative binders to OPC to enable mitigation of greenhouse gas emissions associated with the production of OPC. Making headway in this research is the development of alkali-activated binder in which OPC is replaced totally with supplementary cementitious materials (SCMs) that are rich in silica and alumina [1]. The majority of research studies conducted utilized fly ash as precursor in developing AAC [2]. The key engineering properties of AAC depend on the chemical composition and fineness of the source materials as well as concentration of alkaline activators [3]. As the fineness of source material increases, so does the reactivity with the alkaline activators, resulting in improved properties [4]. In this respect nanomaterials can play a significant role as they are highly reactive due to their enormous specific surface area. Natural pozzolan is another class of naturally available SCM, which has a strong potential for use as a source material in synthesizing AAC. Therefore, this study focusses on developing NP-based alkali-activated concrete, and subsequently the properties of this concrete were improved by partially replacing it with nano-silica.

## 2 Materials and Methods

Natural pozzolan used in this study is a powder form of volcanic rock available locally in the Western Saudi Arabia. The nano-silica used is an aqueous dispersion of colloidal silica approximately 50% solids by mass supplied by AkzoNobel Germany. The chemical composition of NP as determined by X-ray fluorescence (XRF) technique is shown in Table 90.1. The specific surface area and average particle size of NP used are 442 $m^2$/kg and 30 µm, respectively. The alkaline activators used are a combination of aqueous sodium silicate (SS) and 14 M sodium hydroxide (SH) solution. Silica modulus of sodium silicate was 3.3, and its composition was $H_2O$, 62.50%; $SiO_2$, 28.75%; and $Na_2O$, 8.75%. Dune sand with a specific gravity of 2.62 in saturated surface dry (SSD) condition was used as fine aggregate. Limestone

**Table 90.1** Chemical composition of natural pozzolan

| Constituent | Weight (%) |
|---|---|
| Silica ($SiO_2$) | 40.48 |
| Alumina ($Al_2O_3$) | 12.90 |
| Ferric oxide ($Fe_2O_3$) | 17.62 |
| Lime (CaO) | 11.83 |
| Magnesia (MgO) | 8.33 |
| Potassium oxide ($K_2O$) | 1.67 |
| Sodium oxide ($Na_2O$) | 3.60 |
| Phosphorus oxide ($P_2O_5$) | 1.37 |
| Loss on ignition | 1.6 |

**Table 90.2** Constituent materials for AAC mixtures with nano-silica (in kg/m$^3$)

| Mix | Natural pozzolan | Nano-silica | Sodium silicate | Sodium hydroxide | Fine aggregate | Coarse aggregate |
|---|---|---|---|---|---|---|
| M0 | 400 | 0 | 150 | 60 | 650 | 1206 |
| M1 | 396 | 8 | 150 | 60 | 646 | 1200 |
| M2 | 390 | 20 | 150 | 60 | 640 | 1188 |
| M3 | 380 | 40 | 150 | 60 | 630 | 1170 |
| M4 | 370 | 60 | 150 | 60 | 620 | 1152 |

aggregate having a specific gravity of 2.56 in SSD condition was used as coarse aggregate.

All concrete mixes were prepared with a constant binder content of 400 kg/m$^3$ having SS/SH by weight ratio of 2.5. Table 90.2 summarizes constituent materials for preparing AAC specimens incorporating nano-silica for compressive strength tests. For SEM analysis AAP specimens were prepared. Coarse aggregate to total aggregate and fine aggregate to total aggregate ratios were 0.65 and 0.35, respectively. Free water to pozzolanic material ratio of 0.25 was used in all the AAC mixtures. Alkaline activator to binder ratio was 0.525.

After 24 h of casting, specimens were de-molded, placed in plastic bags to avoid evaporation of moisture, and kept in the oven maintained at 60 °C. AAP specimens were cured for 7 days in the oven after which morphological studies were carried out. Compressive strength of concrete was determined on 50 mm cube specimens according to ASTM C 150 after 0.5, 1, 3, 7, 14, and 28 days of curing at 60 °C.

## 3 Results and Discussion

Figure 90.1 shows the compressive strength development in the nano-silica-modified AAC. The data indicate that the strength development in the AAC specimens prepared with higher dosage of nano-silica was slow as compared to that of the lower ones in the initial stages of curing. For instance, 1-day compressive strength of AAC prepared with 0%, 1%, 2.5%, 5%, and 7.5% nano-silica was 22.92 MPa, 22.32 MPa, 14.24 MPa, 12.04 MPa, and 7.32 MPa, respectively. About 90% of strength development was recorded after 3 days of curing in the concrete without nano-silica when compared with that of 7 days strength. However, further curing of AAC specimens, particularly containing nano-silica, remarkably improved its strength. Seven-day compressive strength of specimens prepared with varying dosage of nano-silica was in the range of 37.52–44.52 MPa, maximum value being for the mixture prepared with 5% nano-silica replacement, whereas the lowest was in the control mix. Compared to the strength of concrete, prepared without nano-silica, there was about 18% higher strength recorded in the specimens containing 5% nano-silica. It is interesting to note that the 7-day strength was 12.6%, 34.4%, 43.3%,

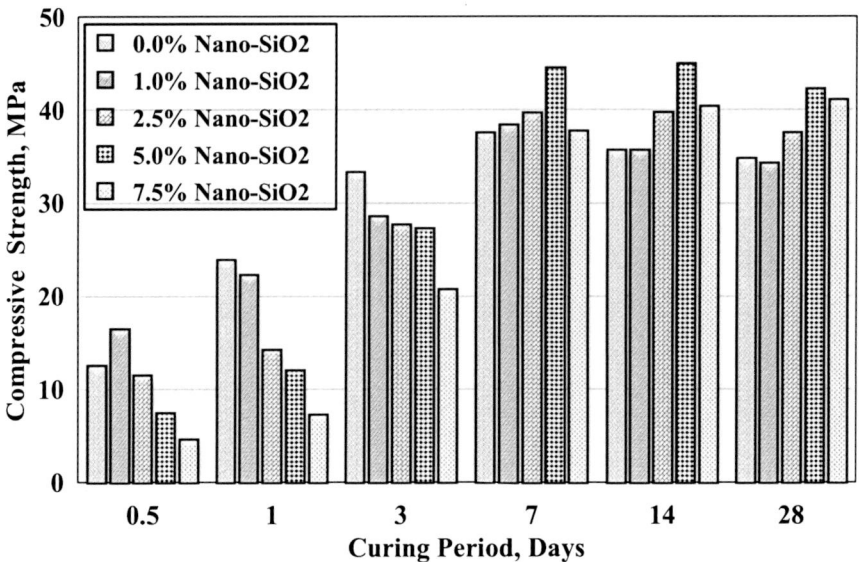

**Fig. 90.1** Compressive strength development in the AAC specimens

63.0%, and 81.5% higher as compared to 3 days, respectively, for the specimens prepared with 0%, 1%, 2.5%, 5%, and 7.5% nano-silica.

What appears from the findings is that as the dosage of nano-silica increased, compressive strength development was delayed. This phenomenon is attributed to the increase in the amount of silica in the system for a given alkaline activator content resulting in delayed polymerization of these particles. When the pozzolanic material containing nanoparticles encounters the alkaline solution, dissolution of the base material takes place initially, and the resulting products are deposited on the highly reactive nanoparticles leading to the formation of nucleation sites. Nucleation of hydration products on the nanoparticles further nurtures and accelerates the dissolution of pozzolanic material, due to which properties of the materials improve significantly [5].

Figure 90.2 illustrates SEM images of NP-based AAP prepared with varying dosages of nano-silica. The NP-based AAP without nano-silica has a porous microstructure, and the matrix was coarser in nature with widespread micro-cracks and voids compared to that containing nano-silica. As the NP was partially replaced with nano-silica from 1.0% to 7.5% by weight, the microstructure started densifying with a reduction in the pore volume. Compared with other replacement levels of nano-silica, the microstructure of 5% was more homogenous and denser having less unreacted particles with continuous gel matrix without clear particle boundaries. This more homogenous gel-like matrix without distinct voids is constituted of pure polymeric binder. On the other hand, the SEM micrograph of AAP containing 7.5% nano-silica, shown in Fig. 90.2, appeared to be also dense with unreacted particles embedded in the structure indicating that the quantity of nano-silica was excessive,

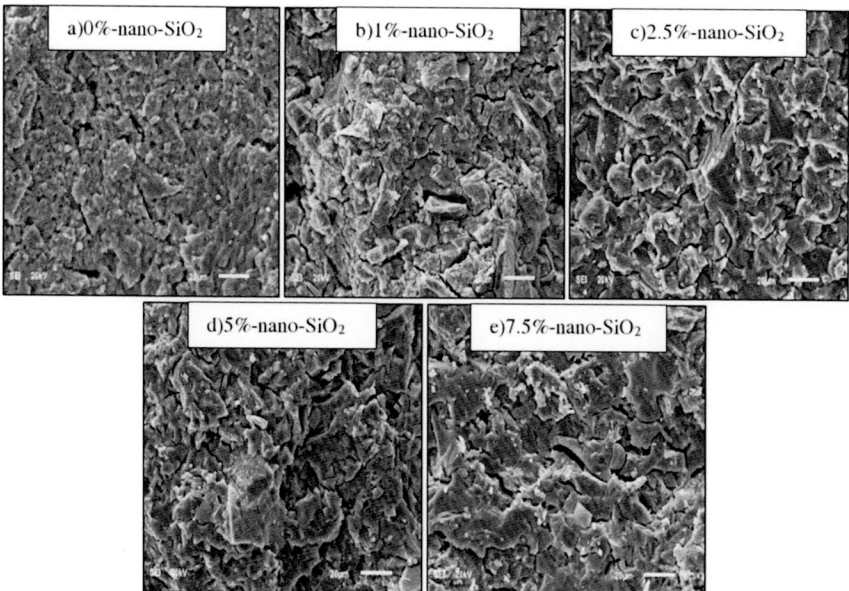

**Fig. 90.2** SEM of AAP specimens prepared with and without nano-silica

and it remained unreacted together with the particles of NP. Unreacted particles are identified clearly in the matrix with the boundaries. Therefore, it appears from the microstructure that there was only partial filling of voids in the specimen prepared with nano-silica up to 2.5% replacement, while the addition of 7.5% of it caused agglomeration of the nanomaterial, which in turn did not beneficially assist the microstructural development and strength.

The microstructure of AAP specimens prepared by partial replacement of NP with nano-silica became homogenous and denser due to increase in the Si species, which enhanced polymerization in the system generated in the process of reaction between Si, Al, and highly alkaline materials [6]. In case of AAP prepared with 5% nano-silica, all the aluminosilicate materials are consumed in the process of alkali activation leaving minimal unreacted particles in the mixture resulting in the enhancement of microstructure. These findings are complemented by the compressive strength results.

These results are in good agreement with the data presented by Adak et al. [7] in which 6% of nano-silica replacement of fly ash resulted in better compressive strength and compact microstructure in comparison to the other replacement levels due to enhanced transformation of amorphous compounds to the crystalline. The significant improvement in the compressive strength and microstructure of specimens containing 5% and 7.5% nano-silica is attributed firstly, to the enhancement in the reaction products in the process of polymerization due to the presence of highly reactive nano-silica yielding additional CSH or CASH gels along with NASH and secondly due to the filler effect [8].

AAC using NP as a main binder has strong potential for a concrete without OPC, with nano-silica contributing to significant strength enhancement. The advantages of the developed concrete include reduction in greenhouse gas emissions, utilization of natural and industrial byproducts, superior strength, and durability characteristics. The concrete, however, has disadvantages, as it requires heat curing, and polymerization process is sensitive to the concentration of alkaline materials. The cost impact of nano-silica could be another minus point.

## 4 Conclusions

The aim of this study was to improve the properties of NP-based AAC by incorporating nano-silica. The investigation conducted shows that:

- Strength development was slow in the mixes containing higher amounts of NS as compared to the lower levels at the onset of curing. As the curing progressed, concrete specimens prepared with NS exhibited a significant enhancement in the compressive strength and microstructural characteristics as compared to the specimens prepared without NS.
- Replacement of NP by 5% NS in the AAC was an optimal level, achieving superior strength as well as homogenous and denser microstructure.
- NP-based AAC is suitable for construction applications without incorporating NS. However, NP partially replaced with NS exhibited significant improvement in the properties of the final product.

## References

1. Ryu, G. S., Lee, Y. B., Koh, K. T., & Chung, Y. S. (2013). The mechanical properties of fly ash-based geopolymer concrete with alkaline activators. *Construction and Building Materials, 47*, 409–418.
2. Ankur, M., & Rafat, S. (2016). An overview of geopolymers derived from industrial by-products. *Construction and Building Materials, 127*, 183–198.
3. Paloma, A., Fernandez-Jimenez, A., Kovalchuck, G., Ordonez, L. M., & Naranjo, M. C. (2007). OPC fly ash cementitious systems: Study of gel binders produced during alkaline hydration. *Journal of Materials Science, 42*, 2958–2966.
4. Wang, P. Z., Trettin, R., & Rudert, V. (2005). Effect of fineness and particle size distribution of granulated blast furnace-slag on the hydraulic reactivity in cement systems advanced cement research. *Advanced Cement Research, 17*, 160–166.
5. Bjornstrom, J., Martinelli, A., Matic, A., Borjesson, L., & Panas, I. (2004). Accelerating effects of colloidal nanosilica for beneficial calcium-silicate-hydrate formation in cement. *Chemical Physics Letters, 392*(1–3), 242–248.
6. Chindaprasirt, P., De Silva, P., Sagoe-Crenstil, K., & Hanjitsuwan, S. (2012). Effect of $SiO_2$ and $Al_2O_3$ on the setting and hardening of high calcium fly ash-based geopolymer systems. *Materials Science, 47*, 4876–4883.

7. Adak, D., Sarkar, S., & Mandal, S. (2014). Effect of nanosilica on strength and durability of fly ash based geopolymer mortar. *Construction and Building Materials, 70*, 453–459.
8. Yip, C. K., Lukey, G. C., & Van Deventer, J. S. J. (2012). Coexistence of geopolymeric gel and calcium silicate hydrate at the early stage of alkali activation. *Cement and Concrete Research, 35*, 1688–1697.

# Part IX
# Strengthening and Restoration Using Polymers

# Chapter 91
# Review of Polymer Coatings Used for Blast Strengthening of Reinforced Concrete and Masonry Structures

Girum S. Urgessa and Mohammadjavad Esfandiari

This paper presents a literature review on the use of polymer coatings in strengthening reinforced concrete and masonry structures against the effects of blast. The use of glass and carbon fiber-reinforced polymers (GFRP and CFRP) in blast strengthening applications has been studied very well in the past two decades. However, experimental and numerical studies of polymer coatings used in blast strengthening applications are scarce in literature. This paper compiles the available experimental and numerical studies that utilized polymer coatings as protective layers, including, but not limited to, polyurea and polyurethane spray-on polymers.

## 1 Introduction

In the past two decades, there have been considerable interest and research in developing cost-effective strengthening methods for resisting or mitigating the effect of structural and secondary damages caused by man-made or accidental explosions. These blast hardening methods provide increased strength for resisting out-of-plane loadings with improved ductility [1, 2]. The use of fiber-reinforced polymers (FRP) has been one of the popular and cost-effective strengthening methods [3, 4]. However, the majority of FRP retrofits have some drawbacks including premature failure due to debonding and delamination as witnessed in full-scale blast tests [5]. As an alternative to FRP retrofits, researchers have been investigating the use of polymer coatings for smaller size blast loads and for use in non-load-bearing elements. This paper compiles available experimental and numerical studies that utilized polymer coatings as protective layers including, but not limited to, polyurea and polyurethane

---

G. S. Urgessa (✉) · M. Esfandiari
George Mason University, Fairfax, VA, USA
e-mail: gurgessa@gmu.edu

spray-on polymers. First, a short description is provided about the types of polymer coatings used in blast strengthening applications. Second, a summary is presented on experimental and numerical studies that used polymer coatings for retrofitting masonry and reinforced concrete structures against blast. Note that this review paper does not include polymer coatings used for strengthening of steel structures.

## 2 Elastomeric Polymers

Elastomeric polymers, including polyurea and polyurethane, are composed of long polymer chains that are cross-linked by chemical bonds. These chains allow elastomeric polymers to be reversibly stretchable within a significant range of deformation. The resulting ductile property makes elastomeric polymers attractive in blast retrofit applications [6].

Polyurea polymers are derived from the reaction of an isocyanate (–NCO) component and a polyamine having two or more primary amino groups (–NH2). The presence of strong hydrogen bonding between chains of the polyurea results in a nano-composite microstructure consisting of discrete domains distributed within a matrix. As such, polyurea exhibits highly ductile property and significant rate dependency due to its viscoelastic nature. It can be categorized as a typical large strain elastic-plastic material. Polyurea coatings have been used widely as truck bed liners and for coatings of pipelines due to their high durability and water tightness [7].

Polyurethane polymers are derived from the reaction of a diisocyanate containing at least two isocyanate (–NCO) components and a diol containing at least two alcohol (hydroxyl or –OH) components in the presence of a catalyst. Polyurethane is an attractive material due to the possibility of modifying its microstructure, and it leads to a wide range of mechanical behavior. Particularly, thermoplastic polyurethanes are highly elastomeric, exhibiting resistance to impact, abrasion, and weather. Description of chemistry and molecular level structure of polymer coatings are presented in [8].

## 3 Blast Testing and Analysis of Elastomeric Polymer Retrofits

Elastomeric polymers can be applied on the surfaces of masonry or reinforced concrete structures by spraying or brushing. Earlier full-scale applications of elastomeric polymers as an exclusive retrofitting option were carried out by the US Air Force Research Laboratory (AFRL) at the turn of this century [9, 10]. The genesis of this early testing program was the need for developing expedient retrofit methods for unreinforced infill concrete masonry walls of existing buildings. The test program initially evaluated 21 different polymers, including 7 extruded from thermoplastic sheet materials, 13 spray-on polymers, and 1 brush-on polymer. A polyurea-based polymer was selected for the final use in full-scale blast tests after careful consideration of cost, ease of installation, strength, stiffness, and toxicity of the polymers. The full-scale blast tests showed that polymer coatings can be effective in reducing the vulnerability of unreinforced non-load-bearing concrete masonry unit (CMU) walls. Furthermore, the tests showed that a tenfold increase in peak pressures can be resisted without catastrophic collapse when using unreinforced CMU walls retrofitted with polymer coatings. In order to predict wall deflections at various explosive yields and standoff distances, an expedient single-degree-of-freedom (SDOF) model was developed. The SDOF model idealizes the dynamic response of the wall by calculating the time history motion of the center point of the wall. The SDOF model results were then compared to a maximum deflection criterion associated with an adequate level of protection. These experimental and numerical studies laid a foundation for researchers investigating polymer coating for blast retrofits.

Hoo Fatt et al. [11] developed an equivalent SDOF model to predict the dynamic response of polymer-retrofitted CMU walls when subjected to blast effects. The coupling of bending and membrane resistance of CMU walls was captured in the model. The SDOF predictions were then compared with finite element analysis results obtained from ABAQUS. The walls under consideration were coated with 2.1 mm polyurea. A maximum deflection of approximately 178 mm was recorded during the blast trials which imparted a peak pressure of 5.8 kPa associated to a pulse duration of 20 ms. The prediction of deflection from the SDOF model and the FE code was found to be in close agreement.

Baylot et al. [12] investigated the use of a two-part sprayed-on polyurea as part of a large experimental program investigating three different types of CMU blast retrofits. The polyurea coating resulted in a 3.2 mm thick retrofit at the back of a CMU wall and extended to the top and bottom of the test reaction structure by 50.8 mm. The walls were subjected to C4 explosives targeting desired peak pressures and impulses. The polyurea retrofits were considered a success because they prevented debris from entering the structures after the blast experiments. Complementary numerical studies were conducted using the Wall Analysis Code, a SDOF-based code that can be used to predict the response of structural elements to blast loads.

Davidson et al. [13] presented their findings from three full-scale explosive tests conducted to determine if the use of polymer coating is an effective method for blast hardening of unreinforced masonry (URM) walls. The three types of polymers studied range from 12,700 to 14,100 kPa in tensile strength. It was reported that the use of polymers improved the performance of URM walls on the order 12 times. The research results were considered to be significant and spurred a series of follow-up tests with the aim of developing a retrofitting method.

Hrynyk and Myers [14] conducted eight tests on unreinforced masonry walls retrofitted in two different schemes. The first scheme involved the exclusive use of 3 mm thick spray-on polyurea retrofit, which was allowed to overlap 51 mm onto the surrounding reinforcing concrete framing members. The second scheme involved

10 mm thick GFRP grids embedded within a polyurea material. The exclusive polyurea retrofit was shown to increase the deflection capacity of the walls with significant improvement in energy dissipation. The GFRP-polyurea retrofit failed prematurely due to lack of anchorage. However, both schemes were shown to reduce or prevent masonry debris scatters. In addition to the experimental work, a simplified analytical model was presented for estimating the capacity of the retrofitted walls at ultimate limit state.

Ciornei [15] tested four quarter-scale masonry walls under various shock tube induced blast pressures. Two walls were built as non-load-bearing infill walls, and the other two walls were built as load-bearing walls. A polyurea-based retrofit was used for each category of walls. One of the two infill walls was sprayed with a layer of polyurea, and one of the two load-bearing walls had a retrofit system comprised of smooth steel wires sprayed with a layer of polyurea. The test results indicated that the load capacity and the stiffness of the masonry walls significantly increased when the retrofits were incorporated. Polyurea proved to be an excellent retrofit material for dissipating blast-induced energy while providing ductility to the system and changing the failure mode from brittle to ductile. The use of polyurea was also shown to efficiently control fragmentation. In addition to the testing program, numerical analysis was conducted by developing displacement-resistance curves and incorporating it into a SDOF model. It was shown that SDOF predicted mid-height displacements were in close agreement with experimentally recorded displacements.

Irshidat et al. [16] presented experimental and analytical studies on the use of nanoparticle-reinforced polymeric materials for blast hardening of URMs. Two types of nanoparticles, exfoliated graphene nano-platelets (XGnP) and polyhedral oligomeric silsesquioxane (POSS), were used in the study with a polyurea coating. Based on a reduced size physical testing, it was reported that the POSS-reinforced polyurea played a major role in improving the response of the URM to blast loads. However, the XGnP-reinforced polyurea did not present a significant change. The authors validated the test results using the AUTODYN finite element analysis program. Furthermore, they developed a simplified single-degree-of-freedom model that incorporated resistance functions and load-deflection graphs for the reinforced masonry wall. The single-degree-of-freedom model was determined to be helpful in analyzing URM walls.

Rivera [17] conducted the second phase of the research completed by Irshidat et al. [16] to evaluate whether the use of polymer coatings can increase the risk of fire during a blast event. Polyurea, polyurea with POSS, and polyurea with exfoliated graphene platelets were investigated. Testing was initially performed using a cone calorimeter heat release rate (HRR) measurements. Flammability characterization and heat flux generation for the structural components and system were then determined using the Fire Dynamic Simulator model, which exposed concrete columns and masonry walls to an existing fire. The polyurea appeared to perform better than a typical epoxy that is known to be flammable and producing burning fragments. The presence of graphene was shown to mitigate and delay ignition times.

The use of elastomeric polymer coatings for blast strengthening of reinforced concrete structures is limited when compared to masonry structures. In reinforced concrete structures, the polymer coatings can have the added benefit of serving as a protection layer for the reinforcing steel. Lim and Peijun [18] conducted blast tests using ½ kg TNT explosives on 100 mm thick concrete panels to study whether concrete spalling during a blast event can be arrested using sprayed-on polymer coatings. They found that extensive spalling occurred for a control slab without the sprayed-on polymeric coating. In comparison, a 3–4 mm thick layer of sprayed-on polymer was able to arrest spalling. Numerical models of the reinforced concrete slab were developed using AUTODYN, and results were shown to be correlated very well with the tests, including crater formation in the front and back of the slabs.

Raman et al. [19] presented a numerical study evaluating the application of polyurea coating applied to reinforced concrete panels using LS-DYNA. Their selection of panel configurations and explosive loads was guided by an earlier experiment conducted on unretrofitted reinforced concrete slab subjected to blast. They optimized the thickness and location of polyurea coatings informed by the numerical study. They showed that increasing the thickness of the polyurea coating after a certain extent does not seem to contribute significantly in reducing the displacement response. They also noted that the application of a polyurea coating on the non-blast-facing (tension) face of the panel tends to be more effective in terms of displacement control. Nonetheless, the application of the coating on the blast-facing face or on both surfaces of a structure was shown to contribute to the energy absorption of the system. From the numerical study, it was evident that the use of polyurea coating on the reinforced concrete panels enhances the blast resistance of the slabs.

Motivated by their numerical findings, Raman et al. [20] performed experiments to determine the behavior of polyurea coated reinforced concrete panels subjected to blast effects. The tested panels had dimensions of 1700 × 1000 × 60 mm. One of the panels had 4 mm polyurea coating on the face opposite to the blast loading, and two panels were covered with 4 mm polyurea coating on both sides. All the panels were subjected to blast loads resulting from the detonation of 1 kg ammonite charge placed at a 1 m standoff distance. The experiments showed that the polymer coating increased the resistance of the panels to the applied blast load compared to a control panel with no polymer coating. In addition, the two panels with polymer coatings on both sides of the panel had shown better resistance when compared to the one panel where the polymer coating was applied only on one side.

## 4 Conclusion

This review compiles experimental and numerical studies on the use of polymer coatings for protecting structures against blast loading. The application of polymer coatings as an exclusive retrofit method has been continuing to grow in the past decade. The majority of the applications were found to be on retrofitting masonry walls with a few studies reported on retrofitting reinforced concrete structures. Overall, polymer coatings have been shown to be an expedient and cheap option for retrofitting lightly loaded masonry and reinforced concrete structures.

# References

1. Ward, S. P. (2004). Retrofitting existing masonry buildings to resist explosion. *ASCE Journal of Performance of Constructed Facilities, 18*(2), 95–99.
2. Urgessa, G. S. (2009). Finite element analysis of composite hardened walls subjected to blast loads. *Journal of Engineering and Applied Sciences, 2*(4), 804–811.
3. Buchan, P. A., & Chen, J. F. (2007). Blast resistance of FRP composites and polymer strengthened concrete and masonry structures. *Composites: Part B, 38*, 509–522.
4. Urgessa, G. S., & Maji, A. K. (2009). Dynamic response of retrofitted masonry walls for blast loading. *ASCE Journal of Engineering Mechanics, 136*(7), 858–864.
5. Maji, A. K., Brown, J. P., & Urgessa, G. S. (2008). Full-scale testing and analysis for blast-resistant design. *ASCE Journal of Aerospace Engineering, 21*(4), 217–225.
6. Connell, J. D. (2002). *Evaluation of elastomeric polymers for retrofit of unreinforced masonry walls subjected to blast* (MS thesis). The University of Alabama at Birmingham, Birmingham.
7. Somarathna, H. M. C. C., Raman, S. N., Badri, K. H., Mutalib, A. A., Mohotti, D., & Ravana, S. D. (2016). Quasi-static behavior of palm-based elastomeric polyurethane: For strengthening application of structures under impulsive loadings. *Polymer, 8*(5), 202.
8. Grujicic, M., Dentremont, B. P., Runt, J., Tarter, J., & Dillon, G. (2011). Concept-level analysis and design of polyurea for enhanced blast-mitigation performance. *Journal of Materials Engineering and Performance, 21*(10), 2024–2037.
9. Knox, K. J., Hammons, M. I., Lewis, T. T., & Porter, J. R. (2000). *Polymer materials for structural retrofit*. Tyndall AFB: Force Protection Branch, Air Expeditionary Forces Technology Division, Air Force Research Laboratory.
10. Porter, J. R., Dinan, R. J., Hammons, M. I., & Knox, K. J. (2002). Polymer coatings increase blast resistance of existing and temporary structures. *AMPTIAC Quarterly, 6*(4), 47–52.
11. Hoo Fatt, M. S., Ouyang, X., & Dinan, R. J. (2004). Blast response of elastomer coatings. In *Proceedings of the eighth international conference on structures under shock and impact*, Crete (pp. 129–138).
12. Baylot, J. T., Bullock, B., Slawson, T. R., & Woodson, S. C. (2005). Blast response of lightly attached concrete masonry unit walls. *ASCE Journal of Structural Engineering, 131*(8), 1186–1193.
13. Davidson, J. S., Porter, J. R., Dinan, R. J., Hammons, M. I., & Connell, J. D. (2004). Explosive testing of polymer retrofit masonry walls. *ASCE Journal of Performance of Constructed Facilities, 18*(2), 100–106.
14. Hrynyk, T. D., & Myers, J. (2008). Out-of-plane behavior of URM arching walls with modern blast retrofits: Experimental results and analytical model. *ASCE Journal of Structural Engineering, 134*(10), 1589–1597.
15. Ciornei, L. (2012). *Performance of polyurea retrofitted unreinforced concrete masonry walls under blast loading* (MS thesis). University of Ottawa, Ottawa.
16. Irshidat, M., Al-Ostaz, A., Cheng, H. D., & Mullen, C. (2011). Nanoparticle reinforced polymer for blast protection of unreinforced masonry wall: Laboratory blast load simulation and design models. *Journal of Structural Engineering, 137*(10), 1193–1204.
17. Rivera, H. K. D. (2013). *Nanoenhanced polyurea as a blast resistant coating for concrete masonry walls* (MS thesis). The University of Mississippi, Oxford, MS.
18. Lim, B., & Peijun, H. (2008). Sprayed-on polymer as concrete spall shield. *Solid State Phenomena, 136*, 145–152.

19. Raman, S. N., Ngo, T., Mendis, P., & Pham, T. (2012). Elastomeric polymers for retrofitting of reinforced concrete structures against the explosive effects of blast. *Advances in Materials Science and Engineering*, 8, Article ID 754142. https://doi.org/10.1155/2012/754142.
20. Raman, S. N., Jamil, M., Ngo, T., Mendis, P., & Pham, T. (2014). *Retrofitting of RC panels subjected to blast effects using elastomeric polymer coatings*. Proceedings of concrete solutions, 5th international conference on concrete repair (pp. 353–360). Belfast, United Kingdom.

# Chapter 92
# Evaluation of Polymer-Modified Restoration Mortars for Maintenance of Deteriorated Sewage Treatment Structures

Wanki Kim and Sunhee Hong

The purpose of this study is to find appropriate polymer-modified repair mortars for polymer lining process in deteriorated sewage treatment structure. The polymer-modified repair mortars are prepared with polymer-binder ratio of 0%, 10%, and 20%; ground-granulated blast-furnace slag (BFS) contents of 0%, 25%, and 50%; fly ash (FA) contents of 0%, 10%, and 20%; and an antifoamer content of 3% and tested for strengths, adhesion, drying shrinkage, and sulfate resistance. It is concluded from the test results that the redispersible polymer powder (RPP)-modified repair mortars with BFA and FA cause improvements in its sulfate resistance, depending on the polymer-binder ratio. The properties of the RPP-modified mortars are satisfied with the KS (Korean Standard) F 4042 quality requirements. However, the RPP-modified repair mortars using BFS and FA require its improvement in strength.

## 1 Introduction

Deterioration in sewer treatment RC structures is caused by acid excretion which etches the surface of concrete, penetrating the mortar surface. Millions of dollars are being spent worldwide on the repair and maintenance of sewer systems and wastewater treatment plants. Korea also needs an enormous cost for repairing sewage treatment structures. And there are more than 500 old wastewater and sewage treatment structures, which are mostly constructed from 1976 to 2010 in Korea. This study is to evaluate the performance of dry-mixed cross-sectional restoration

W. Kim (✉)
Department of Architectural Engineering, Hyupsung University, Bongdam, Hwaseong, Gyeonggi-Do, South Korea

S. Hong
Sampyo R&D Center, Gyeongchundae-ro, Gwangju-si, Gyeonggi-do, South Korea

© Springer International Publishing AG, part of Springer Nature 2018
M. M. Reda Taha (ed.), *International Congress on Polymers in Concrete (ICPIC 2018)*, https://doi.org/10.1007/978-3-319-78175-4_92

mortar with acid resistance. Redispersible polymer powders (RPP) are mostly added to the dry-mixed binder and fine aggregate, and then water is mixed with the mixture [1]. And mineral admixtures such as BFS and FA improve its sulfate resistance of deteriorated sewage treatment structures [2]. The dry-mixed RPP-modified restoration mortars using mineral admixtures are prepared with various polymer-binder ratios and BFS and FA contents and tested for strengths, adhesion, drying shrinkage, and sulfate resistance. From the results, the qualities of polymer-modified restoration mortars on the deterioration of concrete structure for sewage treatment facilities are evaluated according to KS F 4042 [3].

## 2 Materials

### 2.1 Cement and Fine Aggregate

Ordinary Portland cement and siliceous sand (size: 0.075~1.2 mm) were used for all the mortar mixes.

### 2.2 Cement Modifiers

Redispersible polymer powder used as a cement modifier was a vinyl acetate-ethylene-methyl methacrylate (VE/E/MMA) terpolymer powder. The properties of the redispersible VA/E/MMA terpolymer powder are given in Table 92.1. Before mixing, to control entrained air, hydrocarbon-based powdered antifoamer was added to the VA/E/MMA terpolymer powder in a ratio of 3% to the total solids of polymer powder.

### 2.3 Mineral Admixtures

Ground granulated blast-furnace slag (BFS) and fly ash (FA) were used as mineral admixtures. The properties of the BFS and FA are given in Table 92.2.

**Table 92.1** Properties of VA/E/MMA terpolymer powder

| Appearance | Particle size (μm) | Glass transition point (°C) | pH (20 °C) |
|---|---|---|---|
| White powder | 400 | 12 | 7.0 |

**Table 92.2** Properties of ground granulated blast-furnace slag and fly ash

|     | Density (g/cm$^3$) | Blaine-specific surface (cm$^2$/g) | Percent flow (%) | Activity index (%, 28d) |
| --- | --- | --- | --- | --- |
| BFS | 2.93 | 4250 | 96 | 105 |
| FA | 2.19 | 3860 | 102 | 85 |

**Table 92.3** Mix proportions of RPP-modified restoration mortars

| Binder: sand (by weight) | Polymer-binder ratio (%) | BFS content (%) | FA content (%) | Flow (mm) |
| --- | --- | --- | --- | --- |
| 1:2 | 0, 10, 20 | 0, 10, 20, 30 | 0, 10, 20 | 170 ± 5 |

## 3 Testing Procedures

### 3.1 Preparation of Specimens

According to KS (Korean Standard) F 2476 (Test Method for Polymer-Modified Mortar), mortars were mixed with the mix proportions given in Table 92.3, and their flow was adjusted to be constant at 170 ± 5 mm. Beam specimens 40 × 40 × 160 mm and cylindrical specimens 75 × 150 mm were molded and then subjected to a 1-day, 20 °C, 90% R.H. moist plus 27-day, 20 °C, 60% R.H. air-dry cure. Concrete substrates and specimens for adhesion test were prepared as follows: According to KS F 4001 (Precast Concrete Paving Flags), concrete substrates 300 × 300 × 60 mm were molded and then subjected to a 1 day, 20 °C, 0% R.H. moist plus 6-day water plus 7-day, 20 °C, 60% R.H. dry cure. The cured concrete substrates were bonded in a size of 300 × 300 × 10 mm by the mortars mixed with the mix proportions and then given a 3-, 7-, and 28-day, 20 °C, 60% R.H. dry cure.

### 3.2 Strength Test

The cured beam specimens were tested for flexural and compressive strengths in accordance with KS F 2476.

### 3.3 Adhesion Test

The cured bonded specimens were tested for adhesion in tension in accordance with KS F 4042 by using a manually direct pull-gage machine.

## 3.4 Length Change Test

Immediately after beam specimens were given a 1-day, 20 °C, 90% R.H. moist cure, their original length was measured. Then the mortar specimens were cured at 20 °C, 60% R.H. for 28 days, and their drying shrinkage was measured by the comparator method specified in KS F 2424 (Standard Test Method for Length Change of Mortar and Concrete).

## 3.5 Acid Resistance

The cured cylindrical specimens were immersed in 5% sulfuric acid ($H_2SO_4$) at 20 °C for 28 days. The test solutions were changed every 7 days up to an immersion period of 28 days. After 28-day immersion, the attacked portions of the specimens were cleaned out by tap water. And then, the cylindrical specimens were cut, and the cross sections were sprayed with 1% phenolphthalein alcoholic solution. The depth of the rim of each cross section with color change was measured with slide calipers. Then the penetration depth was calculated as follows:

$$\text{Penetration depth} = (A - B)/2 \tag{92.1}$$

where $A$ is the original diameter (mm) of the specimens and $B$ is the non-attacked average depth (mm) with color change.

# 4 Test Results and Discussion

Figure 92.1 shows the flexural and compressive strengths of RPP-modified mortars with various BFS and FA contents at a polymer-binder ratio of 10%. The flexural and compressive strengths of the RPP-modified mortars with BFS and FA contents decrease with an increase in the BFS content. Also, the flexural and compressive strengths of the RPP-modified mortars tend to decrease with an increase in the FA content. It is considered that BFS and FA did not exhibit the micro-filler effect for improving the strength with the specific surface area similar to ordinary Portland cement. However, regardless of the BFS and FA content, the flexural strength of the RPP-modified repair mortar is higher than the KS F 4042-specified value (6 MPa) at the curing period of 3 days, and the compressive strength is higher than the KS-specified value (20 MPa) at the curing period of 7 days. In general, the flexural and compressive strengths of the RPP-modified mortars are improved over unmodified mortar, and the extent of the strength improvement is marked in the flexural strength rather than the compressive strength [1].

Figure 92.2 shows the adhesion in tension of RPP-modified mortars various polymer-binder ratios and BFS contents at a FA content of 20%. The adhesion in

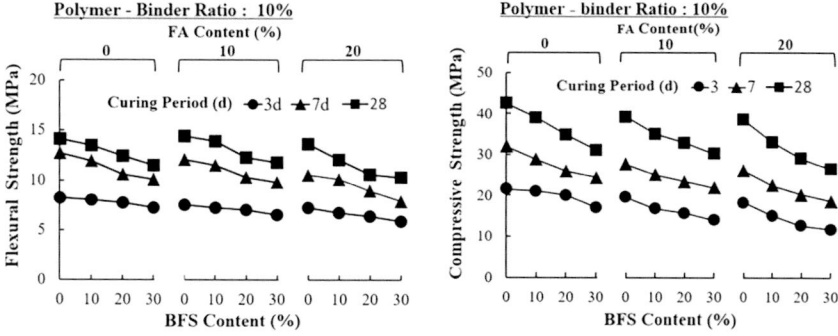

**Fig. 92.1** BFS content vs. flexural and compressive strengths of RPP-modified restoration mortars

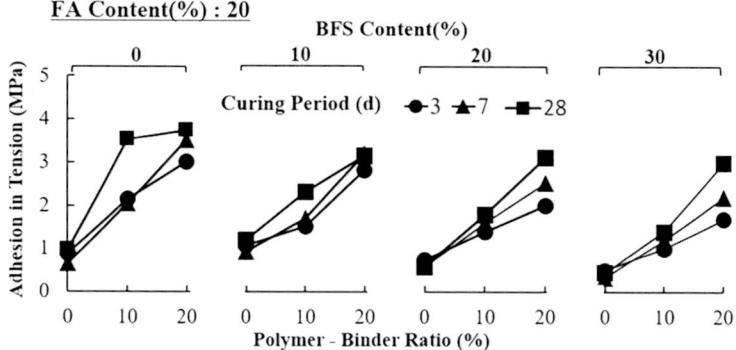

**Fig. 92.2** BFS content vs. adhesion in tension of RPP-modified mortars

tension of the RPP-modified mortars with FA content of 20% increases sharply at polymer-binder ratios up to 20%. By contrast, the adhesion in tension of the RPP-modified mortars with BFS decreases fairly with an increase in the BFS content. However, the adhesion of RPP-modified mortars with BFS is much higher than the quality requirement (1 MPa) at polymer-binder ratios of 10% or more. In general, the failure modes in the adhesion test in tension of the bonded RPP-modified mortars to cement concrete were adhesive failure at BFS and FA content of 20% or more and cohesive failure in the RPP-modified mortars at polymer-binder ratio of 10% or higher. In particular, the failure modes of the RPP-modified mortar with polymer-binder ratio of 10% and 20% at curing period of 28 days were completely cohesive failure in the cement concrete.

Figure 92.3 shows the drying shrinkage of RPP-modified mortars with various BFS and FA contents at a polymer-binder ratio of 10%. In general, most RPP modified mortars show a considerably large drying shrinkage compared to polymer dispersion-modified mortars [1]. The drying shrinkage of the

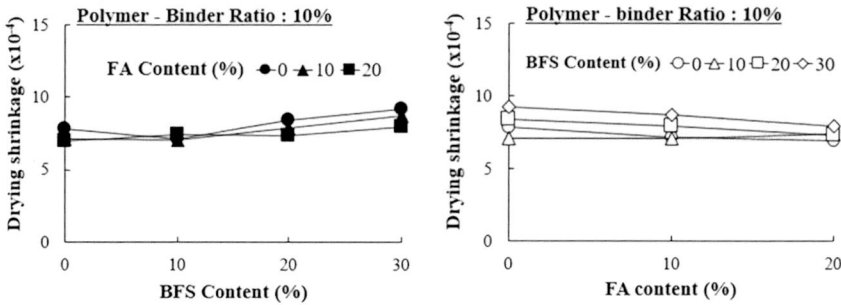

**Fig. 92.3** BFS content vs. 28-day drying shrinkage of RPP-modified mortars

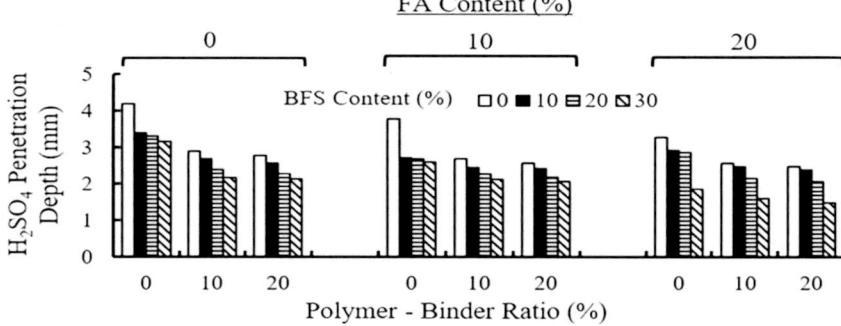

**Fig. 92.4** $H_2SO_4$ penetration depth of RPP-modified mortars immersed in 5% $H_2SO_4$ for 28 days

RPP-modified mortars with a polymer-binder ratio of 10% tends to increase with an increase in the BFS content. However, it tends to decrease with an increase in the FA content except for BFS content of 10%. In this study, the range of drying shrinkage of the RPP-modified mortars with BFS and FA content at a polymer-binder ratio of 10% are 6.9–9.2 × $10^{-4}$. All of the RPP-modified mortars meet quality requirement of 15 × $10^{-4}$.

Figure 92.4 shows the $H_2SO_4$ penetration depth of RPP-modified mortars immersed in 5% $H_2SO_4$ for 28 days. The $H_2SO_4$ penetration depth of the RPP-modified mortars with various BFS and FA contents decreases considerably with an increase in the polymer-binder ratio. The acid resistance of the RPP-modified mortars with BFS and FA content immersed in 5% $H_2SO_4$ solution for 28 days is improved with increase in the BFS and FA content and is more governed by polymer-binder ratio. The $H_2SO_4$ penetration depth of the RPP-modified mortars with polymer-binder ratio of 10% and 20% is less than 3 mm irrespective of the BFS and FA content. All of the RPP-modified mortars meet the quality requirement of 3 mm.

## 5 Conclusions

The following conclusions can be obtained from the test results:

1. The flexural and compressive strengths and adhesion of RPP-modified mortars with BFS and FA content are improved with an increase in the polymer-binder ratio and are fairly decreased with an increase in the BFS content.
2. The drying shrinkage of RPP-modified mortars with a polymer-binder ratio of 10% is decreased with increasing FA content. However, it tends to increase with increasing BFS content.
3. The acid resistance of RPP-modified mortars with BFS and FA content immersed in 5% $H_2SO_4$ solution for 28 days is improved with an increase in the BFS and FA content and is more governed by polymer-binder ratio.
4. In conclusion, the properties of RPP-modified mortars using BFS and FA are satisfied with the KS F 4042 quality requirements. However, the strengths and adhesion of RPP-modified mortars tend to decrease with increasing BFS and FA content. Accordingly, the RPP-modified restoration mortars with BFS and FA should be improved in strength and adhesion.

This work was supported by the National Research Foundation of Korea Grant funded by the Korean Government (NRF-2016935934).

## References

1. Ohama, Y., Demura, K., & Kim, W. K. (1994). Properties of polymer-modified mortars using redispersible polymer powders. In *Proceedings of the first East Asia symposium on polymers in concrete* (pp. 81–90). Chuncheon.
2. Bondar, D., Lynsdale, C. J., Milestone, N. B., & Hassani, N. (2015). Sulfate resistance of alkali activated pozzolans. *International Journal of Concrete Structures and Materials, 9*(2), 145–158.
3. Korean Standard Association (2017). *Korean Standard F 4042(Polymer-modified cement mortar for maintenance in concrete structure)*. Seoul, Korea.

# Chapter 93
# Finite Element Modeling of CFRP-Strengthened Low-Strength Concrete Short Columns

Khaled A. Alawi Al-Sodani, Muhammed K. Rahman,
Mohammed A. Al-Osta, and Ali A. H. Al-Gadhib

Carbon fiber-reinforced polymer (CFRP) sheets and plates are now being extensively used as retrofitting/strengthening system, due to high strength, low weight, corrosion resistance, and ease and speed of application. Structures with very low-strength reinforced concrete columns, constructed during the early 1970s and 1980s, are rampant in many countries. These structures are prone to collapse during a seismic event. It is important to investigate if these low-strength concrete columns can be made safe by CFRP sheet strengthening. An experimental and numerical investigation conducted on short circular low-strength concrete column is presented. The results of this study showed that the confinement equations for concrete columns as per ACI and CSA codes give erroneous results for these columns. The CSA code equations require satisfaction of certain constrains, which are not applicable to such columns. Control specimen (unconfined specimens) results showed that both CSA and ACI equations overestimate the ultimate loads of low-strength RC columns by an order in magnitude. For CFRP-confined specimens, the ACI and CSA code equations overestimate the strength by about 13% and 24%.

K. A. Alawi Al-Sodani · M. A. Al-Osta
Department of Civil and Environmental Engineering, King Fahd University of Petroleum and Minerals, Dhahran, Saudi Arabia

M. K. Rahman (✉)
Center for Engineering Research, Research Institute, King Fahd University of Petroleum and Minerals, Dhahran, Saudi Arabia
e-mail: mkrahamn@kfupm.edu.sa

A. A. H. Al-Gadhib
Department of Civil and Environmental Engineering, King Fahd University of Petroleum and Minerals, Dhahran, Saudi Arabia

Center for Engineering Research, Research Institute, King Fahd University of Petroleum and Minerals, Dhahran, Saudi Arabia

## 1 Introduction

Steel plates have been traditionally utilized for strengthening of seismically deficient concrete structures. However, lack of corrosion resistance and difficulty in handling and application led to the development of lightweight high-strength CFRP sheets for retrofitting of concrete columns and structural elements [1]. CFRP strengthening technique is not devoid of disadvantages which include high costs, vulnerability at high temperature, lack of vapor permeability, and difficulty in post-earthquake assessment behind FRP jackets [2–6]. Strengthening and repair of earthquake-damaged reinforced concrete columns by confinement using CFRP sheets are currently the predominantly adopted technique. Several studies have been conducted on the efficiency of CFRP sheets as confining reinforcement for columns having rectangular or circular sections [7, 8].

Large stocks of older concrete structures, constructed in the early 1970s and 1980s in Turkey, Saudi Arabia, and several other countries, have very low-strength concrete columns. Limited studies exist on the behavior of very low-strength columns strengthened by wrapping CFRP sheets. This paper presents the experimental and numerical investigations conducted on low concrete strength circular columns strengthened by CFRP sheets and applicability of the current ACI and CSA code equations for predicting the CFRP-confined concrete columns.

## 2 Materials and Methods

Two circular columns 15 cm in dia. and 100 cm in height, one without CFRP (control specimen) and the other strengthened with CFRP, were tested under axial load. The columns have 3ɸ12 (steel ratio $a_s = 2\%$) bars as longitudinal reinforcement and 8 mm dia. Ties at 150 mm centers with a clear cover of 35 mm. ordinary Portland cement, sand, and aggregate of the maximum size of ¾ inch were used to make the mix. The concrete compressive strength of concrete-based tests conducted on 7.5 × 15 cm cylinders was $f'c = 10.3$ MPa. Steel bars ($f_y = 420$ MPa) were used as main steel reinforcements and ties. CFRP sheets SikaWrap Hex 330C and Sikadur 330 resin were used for strengthening of columns. The properties of SikaWrap Hex 330C and its cured laminate are given in Table 93.1. The columns were tested under

**Table 93.1** Properties of SikaWrap Hex 330C and its cured laminate

|  | SikaWrap Hex 330C | SikaWrap Hex 330C cured laminate |
|---|---|---|
| Sheet width [mm] | 600 | 600 |
| Thickness [mm] | 0.33 | 0.33 |
| Tensile strength [MPa] | 3450 | 960 |
| Tensile modulus elasticity [MPa] | 230,500 | 73,100 |
| Ultimate elongation % | 1.5 | 1.33 |

axial load in a loading frame up to failure, and the ultimate load capacity of the columns and failure modes was measured.

## 3  Finite Element Simulation

Numerical modeling of the columns tested in the experimental program was carried out using FE software (ABAQUS). For verification, a circular column test from literature was analyzed. The effect of confinement was used in finite element modeling (confined concrete Model, Mander et al. 1988) [9]. This model considers separate stress-strain curves for unconfined cover and confined concrete core. Using this strategy in FE modeling, FE model in this paper considers the following effects: tension stiffening effects in tension and confinement effects in compression, stress-strain relation of the steel, and variation in the geometry of column section due to cumulative concrete cover spalling under high strain. Figure 93.1 displays the confined and unconfined points in a circular column (used in defining confined and confined regions in ABAQUS).

The nonlinear behavior of the columns (C1 and C2) was simulated using ABAQUS/Standard software. In this study, the elastic modulus, $E$, steel stress, $f$, and Poisson's ratio were taken from the experimental work. A solid eight-node element with linear-reduced Gauss integration points was used to model the concrete, whereas 3D truss elements were utilized to model both longitudinal and transverse reinforcements. The axial force in the steel plate is applied with zero eccentricity (i.e. at the center of the column) and increased incrementally until the failure occurs. Fixed and free boundary conditions were adopted at the bottom and top end of the columns, respectively. Concrete damage plasticity (CDP) material model was adopted for concrete. The material parameters needed for this CDP model were based on experimental tests and with some default values recommended in ABAQUS as shown in Table 93.2. The properties of CFRP sheets used in finite element simulation are shown in Table 93.3. Dynamic explicit analysis was utilized, and a workable mesh density with 11 mm element size was selected after several trials. To avoid stress concentration and to get equal stress distributions, steel plates

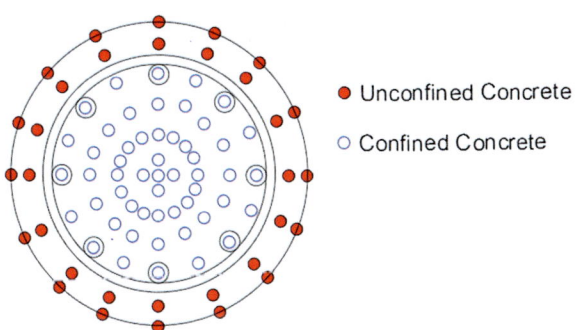

**Fig. 93.1** Confined and unconfined points in a circular column cross section

**Table 93.2** Parameters utilized in plastic damage model

| Mass density [Tone/mm$^3$] | Young's modulus [MPa] | Poisson's ratio | Dilation angle $\psi$ [Degree] | Eccentricity $\varepsilon$ | $f_{b0}/f_{c0}$ | K |
|---|---|---|---|---|---|---|
| 2.1E-009 | 15,083 | 0.2 | 36 | 0.1 | 1.16 | 0.67 |

**Table 93.3** Proprieties of CFRP Lamina

| $t_{ply}$ [mm] | No. of layers | $E_1$ [MPa] | $E_2$ [MPa] | $\nu_{12}$ | $G_{12}$ [MPa] | $G_{13}$ [MPa] | $G_{23}$ [MPa] | $\sigma_u$ [MPa] |
|---|---|---|---|---|---|---|---|---|
| 0.33 | 1 | 73,100 | 7310 | 0.3 | 5200 | 5200 | 3200 | 960 |

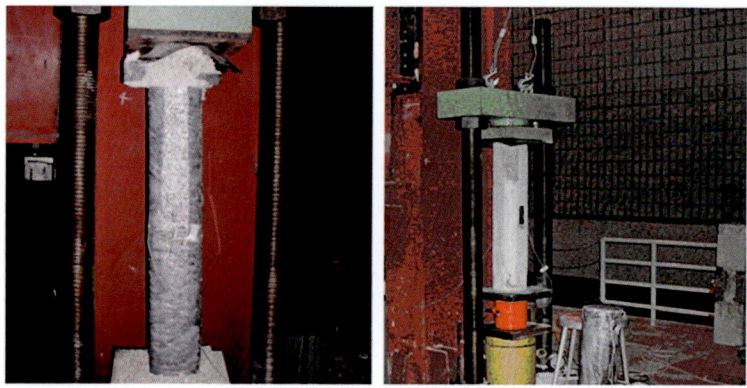

**Fig. 93.2** Failure modes of confined and unconfined columns

with 15 cm diameter were used at the ends. Linear hexahedral elements of type C3D8R were utilized to model the steel plates.

## 4 Results and Discussion

Two circular columns were tested under concentric axial load to failure (Fig. 93.2 shows the failure mode of confined and unconfined columns tested in the laboratory). Each column was tested under uniaxial compression up to failure. The load was applied in small increments.

Table 93.4 shows a comparison between experimental and calculated values of (using CSA and ACI 440.2R) ultimate loads for unconfined (control) and CFRP-confined RC columns. Control specimen results show that both CSA and ACI code equations overestimate the ultimate loads of low-strength RC columns by about two times higher than the experimental results. This can be attributed to the fact that these equations do not take into account the effect of confinement in the stress-strain curve and defining unconfined cover and confined core by separate stress-strain curves. For CFRP-confined column specimen, both ACI 440.2R and CSA codes overestimate

# 93 Finite Element Modeling of CFRP-Strengthened Low-Strength Concrete Short Columns

**Table 93.4** Ultimate loads from experimental and ACI/CSA codes for unconfined and CFRP confined columns

| No. | ACI 440.2R $P_n$ [kN] | CSA | Experiment [kN] | Change with respect to ACI [%] | Change with respect to CSA [%] |
|---|---|---|---|---|---|
| C1 | 294.25 | 291.49 | 131.2 | −55.41 | −54.99 |
| C2 | 359.51 | 409.6 | 312.2 | −13.16 | −23.78 |

**Table 93.5** Ultimate loads from experimental and FE simulation for unconfined and CFRP-confined columns

| No. | Type | Experiment [kN] | FEM (ABAQUS) [kN] | Difference FE and Expt. [%] |
|---|---|---|---|---|
| C1 | Unconfined | 131.2 | 134 | 2.13 |
| C2 | Confined | 312.2 | 307.25 | −1.59 |

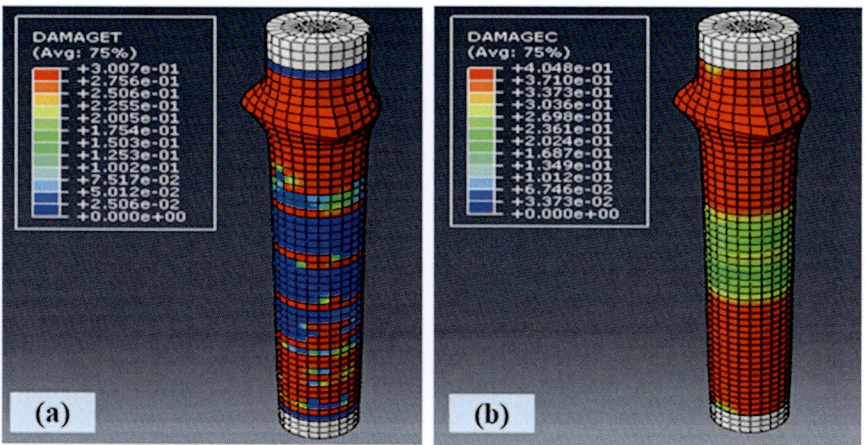

**Fig. 93.3** Concrete principal stresses on column (C1) (**a**), compression damage (**b**), and tension damage

the axial load capacity, but the difference is much smaller compared to the control column. The axial load capacity from ACI 440.2R and CSA codes was −13.2% and 23.8% higher compared to the experimental results. Consideration of confinement provided by CFRP in the model reduces the error. The difference from experimental results may be attributed to neglecting the effect of transverse steel (stirrups) and violation of some constraints associated with ACI 440.2R and CSA equations.

Numerical modeling using FE software (ABAQUS) was carried out for the two columns. Table 93.5 shows a comparison of ultimate loads obtained by experimental and FE simulation for the unconfined/confined low-strength RC columns. The ultimate load values from FE simulations are in excellent agreement with the experimental results. The percentage variation is very low being +2.1% and −1.6% for the control and CFRP-confined columns. Figure 93.3 shows the mode of failure (compression and tension damage) of unconfined column. The failure mode from FE models captures the failure mode of the experimental tests.

## 5 Conclusions

Based on the results presented in this investigation, the following conclusions could be drawn:

1. Control specimen results show that both CSA and ACI equations overestimate the ultimate loads of low-strength RC columns by an order in magnitude.
2. For confined low-strength specimens, also both ACI 440.2R and CSA equations overestimate the axial load capacity by $-13.2\%$ and $-23.8\%$, respectively. The confinement provided by CFRP is considered, but the confinement by steel stirrups is neglected.
3. Finite element simulation of the two columns considering appropriate confinement effects captures the experimental axial load capacity with high accuracy. The error being less than 2%.

## References

1. Teng, J. G., Chen, J. F., Smith, S. T., & Lam, L. (2003). Behaviour and strength of FRP-strengthened RC structures: A state-of-the-art review. *Proceedings of the institution of civil engineers Structures and Buildings, 156*(1), 51–62.
2. Bailey, C. G., & Yaqub, M. (2012). Seismic strengthening of shear critical post-heated circular concrete columns wrapped with FRP composite jackets. *Composite Structures, 94*(3), 851–864.
3. Yaqub, M., Bailey, C. G., & Nedwell, P. (2011). Axial capacity of post-heated square columns wrapped with FRP composites. *Cement and Concrete Composites, 33*(6), 694–701.
4. Bisby, L. A., Chen, J. F., Li, S. Q., Stratford, T. J., Cueva, N., & Crossling, K. (2011). Strengthening fire-damaged concrete by confinement with fibre-reinforced polymer wraps. *Engineering Structures, 33*(12), 3381–3391.
5. Pellegrino, C., & Modena, C. (2010). Analytical model for FRP confinement of concrete columns with and without internal steel reinforcement. *Journal of Composites for Construction, 14*(6), 693–705.
6. Rocca, S., Galati, N., & Nanni, A. (2008). Review of design guidelines for FRP confinement of reinforced concrete columns of noncircular cross sections. *Journal of Composites for Construction, 12*(1), 80–92.
7. Campione, G., & Miraglia, N. (2003). Strength and strain capacities of concrete compression members reinforced with FRP. *Cement and Concrete Composites, 25*(1), 31–41.
8. Rousakis, T. C. (2005). *Mechanical behaviour of concrete confined by composite materials* (Doctoral dissertation, PhD Thesis).
9. Mander, J. B., Priestley, M. J., & Park, R. (1988). Theoretical stress-strain model for confined concrete. *Journal of Structural Engineering, 114*(8), 1804–1826.

# Chapter 94
# Improvement Works to Existing Column Stumps by Fiber-Reinforced Polymer Strengthening System

Jin Ping Lu and Sook Fun Wong

How to improve the durability of concrete structures has long been studied in many countries, especially in the island country of Singapore, where marine concrete structures are very common along coastal areas. The service life of marine concrete structure can be enhanced by different ways, such as the addition of admixtures and using high-grade concrete. Another effective method is by using the FRP system to strengthen the concrete structure. This study deals with improvement works carried out on the column stumps of existing concrete structures by adopting the fiber-reinforced polymer (FRP) strengthening system. Various standard test methods and computer modeling were conducted to examine the effectiveness of the FRP strengthening system using glass fiber composite. The results showed that the FRP wrapping for column stumps of 200 × 200 mm can increase the ultimate strength by an average of 124%.

## 1 Introduction

Currently, the durability of concrete under marine conditions can be enhanced by the addition of admixtures and using high-grade concrete. The fiber-reinforced polymer (FRP) jackets have been successfully used as an external mean to strengthen existing RC columns in recent years with very promising results, among others [1–3]. Several studies [4–6] on the performance of FRP-wrapped columns have been conducted, using both experimental and analytical approaches. However, the majority of such

---

J. P. Lu (✉)
Admaterials Technologies Pte Ltd, Singapore, Singapore
e-mail: jinping@admaterials.com.sg

S. F. Wong
Temasek Polytechnic, Singapore, Singapore

studies are limited to the improvement of FRP to columns of circular cross section. In the recent years, the studies on the effectiveness of FRP on the columns of square or rectangular cross sections [7] have increased, but the available data are still limited. This study is a continuous work from our previous paper [7] with a series of tests on different rectangular shapes and sizes for plain concrete (PC) and reinforced concrete (RC) columns. Glass fiber-reinforced polymer (CFRP) sheets were used to strengthen the columns. Three horizontal wrapping were prepared for concrete columns with the size of 200 mm × 200 mm × 700 mm, and the columns were tested under compression to verify the effectiveness of FPR wrapping. Computer model was developed for the theoretical calculation of the improvement of loading capacity. The calculated results were compared with the testing results. The results by predictions of the proposed computer model are very close to test data.

## 2 Test Specimens

All the specimens were cast in G20 concrete. Eight specimens were with dimensions of nominal size 200 mm (W) × 200 mm (L) × 700 mm (H). Four (4) columns with only two end bands were wrapped and dental stone capped were prepared as control specimens (Fig. 94.1). Four (4) specimens were wrapped with three layers of horizontal FRP of MapeiWrap G Uni-AX900 COMPOSITE strengthening system (Fig. 94.2).

To prevent damage at both ends during the test, the two end bands were wrapped for control specimens as shown in Fig. 94.1. For the wrapped specimens, three layers of MapeiWrap G Uni-AX900 COMPOSITE Strengthening system were used for wrapping. The thickness of each layer is about 2–3 mm. The two end bands of width 75 mm were also wrapped as control specimens to prevent the damage at both ends. The specimens wrapped with FRP are shown in Fig. 94.2. If the original column ends are not even, it was capped with dental stone as shown in the photos.

**Fig. 94.1** Control specimens, where two end bands were wrapped and dental stone capped

**Fig. 94.2** Wrapped specimen with three layers of horizontal FRP and two end bands wrapped

**Fig. 94.3** Cross section with rebar arrangement

## 3 Calculation by Computer Model

**Calculation of Loading Capacity of RC Column Without Wrapping**

The below basic parameters of concrete and steel bars were used for calculation:

Concrete compressive strength $f_{cu}$ = 30 MPa, $\gamma$ = 1.5, maximum aggregate size = 20 mm.

Steel tensile strength fy = 460 MPa, $\gamma$ = 1.15, diameter of link bar size is 6 mm. The location of rebar is shown in Fig. 94.3.

Concrete cover thickness is 30 mm.

Column section size: high = 200 mm, width = 200 mm (Fig. 94.4).

The loading capacity of the column calculated from the model is shown in Table 94.1. It can be seen that the loading capacity is in the range from 614 KN (for T8 bar) to 847 kN (for T16 bar). Figure 94.5 shows the interaction chart.

**RC Column Concrete Confined by FRP Wrapping**

For the input of computer model, the basic parameters shown in Table 94.3 are used. As shown in Fig. 94.4 [4], confining stress will be concentrated around the corners when rectangular concrete column was confined with FRP wrapping. The

**Fig. 94.4** Confinement in square section

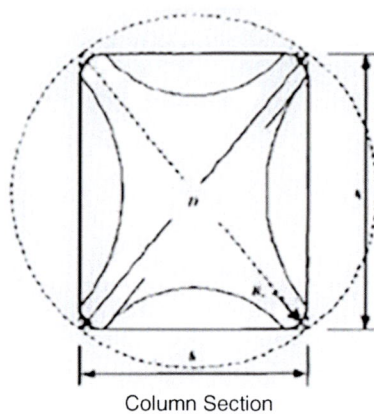

Column Section

**Table 94.1** Calculation results for control RC columns

| Type | Bar Ø | Asc % | Link Ø | Bar c/c | Nuz (kN) |
|---|---|---|---|---|---|
| T | 16 | 2.01 | 6 | 112 | 847 |
| T | 14 | 1.54 | 6 | 114 | 774 |
| T | 12 | 1.13 | 6 | 116 | 711 |
| T | 10 | 0.79 | 6 | 118 | 657 |
| T | 8 | 0.50 | 6 | 120 | 614 |
| T | 8 | 0.50 | 6 | 120 | 614 |

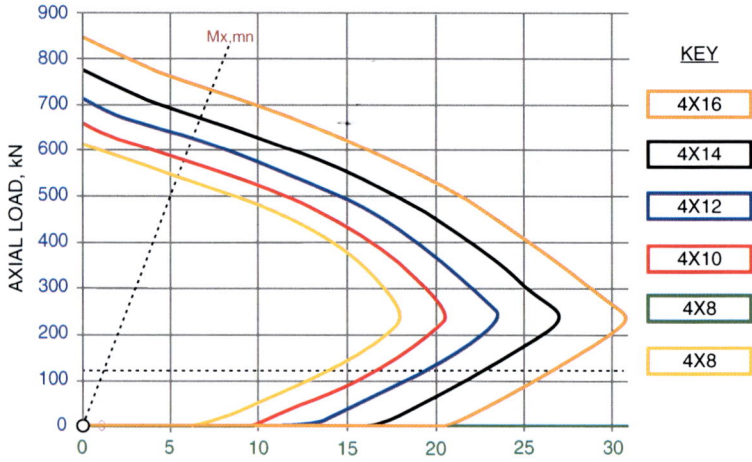

**Fig. 94.5** Interaction chart

experimental results [5] have confirmed that the effectiveness of an FRP wrapping will be reduced for a square and rectangular section compared to a circular section. For square concrete column wrapped by FRP sheets, the failure is generally by FRP rupture, despite this reduced effectiveness [5].

The result of confinement stress calculated from Eq. (94.1) is 47 N/mm$^2$:

$$f_i = \frac{2f_{frp}t_{frp}}{D} \qquad (94.1)$$

The total area of 4 T16 longitudinal reinforcement bars is $A_s = 804$ mm$^2$. The gross area of the cross section is

$$A_g = bh - (4 - \pi)R_c^2 = 39914 \text{ mm}^2 \qquad (94.2)$$

To calculate the ratio of area confined effectively, Eq. (94.3) can be calculated from, and the result is 0.44:

$$\frac{A_e}{A_c} = \frac{1 - \left(\frac{\frac{b}{h}(h-2R_c)^2 + \frac{h}{b}(b-2R_c)^2}{3A_g}\right) - \rho_{sc}}{1 - \rho_{sc}} \qquad (94.3)$$

where

The steel content ratio $\rho_{sc} = 0.0201$
Coefficient of confinement for circular column $k_1 = 2$.

Shape modification factor is

$$k_s = \frac{bA_e}{hA_c} = 0.4477 \qquad (94.4)$$

Equation (94.5) based on the model proposed by Lam and Teng [6] can be used to calculate the cylinder strength with FRP confinement:

$$\frac{f'_{cc}}{f'_{co}} = 1 + k_1 k_s \frac{f_1}{f'_{co}} \qquad (94.5)$$

where:

| | | | |
|---|---|---|---|
| Cylinder strength with FRP confinement | = | $f'_{cc}$ | 58 | N/mm$^2$ |
| Equivalent concrete cube strength | = | $f_{cc}$ | 72 | N/mm$^2$ |

The design ultimate axial capacity can be calculated from Eq. (94.6), and the result is 2324 KN:

$$N_u = \left(\frac{0.67}{\gamma_c}f_{cu} + k_1 k_g k_s \frac{f_1}{\gamma_{frp}}\right)A_c + \frac{f_y}{\gamma_s}A_{sc} \qquad (94.6)$$

Table 94.2 summarizes the effectiveness of FRP on the rectangular concrete column, and the ultimate loading capacity can be increased by 27%.

**Table 94.2** Material properties for calculation

| Width of column section | $b$ | 200 | mm |
|---|---|---|---|
| Height of column section | $h$ | 200 | mm |
| Equivalent diameter | $D$ | 200 | mm |
| Tensile strength of MapeiWrap | $f_{EG900}$ | 2560 | N/mm² |
| Ultimate tensile force of MapeiWrap, three layers | $F_{EG900}$ | 2711 | N/mm |
| Tensile strength of MapeiWrap System | $f_{FRP}$ | 460 | N/mm² |
| Total thickness of FRP applied | $t_{FRP}$ | 3.42 | mm |
| Ultimate tensile force of MapeiWrap System | $F_{FRP}$ | 1573.2 | N/mm |
| Concrete cube strength without FRP confinement | $f_{co}$ | 30 | N/mm² |
| Equivalent cylinder strength | $f'_{co}$ | 24 | N/mm³ |

**Table 94.3** Calculation results by computer model

| S/N | Steel reinforcement | Description | Design ultimate axial load capacity (KN) | Strength gain % for design |
|---|---|---|---|---|
| 1 | 4 T16 | With no FRP wrapping | 847 | Not applicable |
| 2 | 4 T16 | With three horizontal FRP wrapping | 2324 | 147% |

## 4 Experiments

**Specimen Preparation** The column specimen must be cast and cured for at least 28 days and must be placed vertically, and if the top of the column is uneven, dental stone should be used for capping. Then the specimen will be placed on the trolley of RBU5000 machine, a compression loading frame in four-column design with electric crosshead adjustment. The testing height of this RBU5000 can be adjusted, and it has sufficient axial stiffness even when the maximum height is utilized. The trolley will be pushed into RBU5000 machine, and the piston of the machine will be engaged. The crosshead will be lowered down to touch the column and locked.

**Pre-Loading** The test specimen will be pre-loaded with not more than 25 kN. Load cell and displacement device will be connected from RBU machine to a data logger.

**Loading** The test will be started with a loading rate of 1 mm/min. Loading will be continued until the specimen fails.

**Testing Results** The testing results are shown in Table 94.3. It can be seen that the average ultimate load is 1021 kN for the control specimens and 2288 kN for the strengthened specimens with FRP wrapping. The ultimate load of the FRP strengthened specimens has been increased 124% compared to control specimens, slightly lower than the value calculated by computer model which is 147. The load (KN) versus displacement (mm) relationship curves for each concrete specimens are shown in Fig. 94.6.

**Fig. 94.6** Load-displacement curves of columns of 200 × 200 cross section

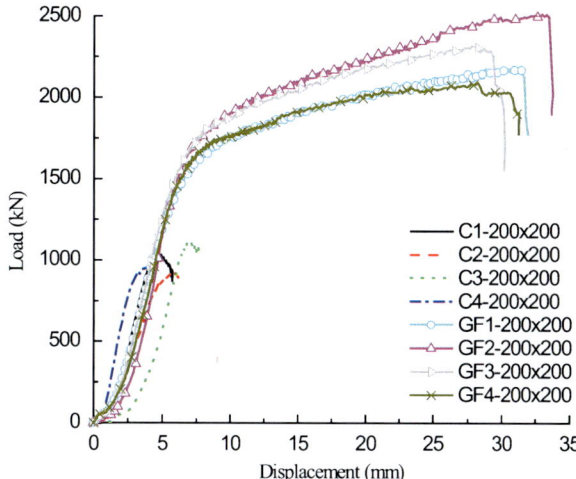

## 5 Conclusion

An average of 124% improvement of ultimate compressive strength has been achieved for the square column with 200 × 200 mm cross section by glass fiber FRP wrapping. The theoretical calculation results by the proposed computer model are very close to the actual testing results. For the strengthened specimens, the failure occurred in the middle of the specimen due to delamination of FRP.

## References

1. Saadatmanesh, H., Ehsani, M. R., & Li, M. W. (1994). Strength and ductility of concrete columns externally reinforced with composites straps. *ACI Structural Journal, 91*(4), 434–447.
2. Mirmiran, A., et al. (1998). Effect of column parameters on FRP-confined concrete. *ASCE Journal of Composites for Construction., 2*(4), 175–185.
3. Benzaid, T. C., Chikh, N. E., & Mesbah, H. (2008). Behaviour of square concrete columns confined with GFRP composite wrap. *Journal of Civil Engineering and Management, 14*(2), 115–120.
4. Youssef, M. N., Feng, M. Q., & Mosallam, A. S. (2007). Stress-strain model for concrete confined by FRP composites. *Composites Part B: Engineering, 38*, 614–628.
5. Rochette, P., & Labossière, P. (2000). Axial testing of rectangular column models confined with composites. *ASCE Journal of Composites for Construction, 4*(3), 129–136.
6. Lam, L., & Teng, J. G. (2002). Strength models for fiber-reinforced plastic-confined concrete. *Journal of Structure Engineer, 128*(5), 612–623.
7. Lu, J. P., Tan, K. H., & Zheng, D. (2015). Strengthening of column stump using glass fiber composite strengthening system. *Advanced Materials Research: Polymers in Concrete Towards Innovation, Productivity and Sustainability in the Build Environment, 1129*, 353–360.

# Chapter 95
# Silicone Resin Enclosing Method Applied for the Maintenance of Steel Bearings

Makoto Kawakami, Fujio Omata, Atsushi Toyoda, and Shingo Kato

Recently, the corrosion of steel bearings of bridges due to leakage of water from expansion joints has been a serious issue in Japan, affecting the lifetime of newly constructed and existing bridges. Particularly in snowy and cold regions, leaking water-containing chloride ion that remains on the pavement surface from spraying of deicing agents is the main corrosion factor of steel bearings. In this study, a transparent elastic silicone resin was applied for enclosing the steel bearing in order to control and monitor the corrosion and prevent deterioration. This newly developed maintenance method using silicone resin is effective due to the following characteristics: (1) by enclosing with silicone resin, corrosion factors such as intrusion of water and chloride ion are perfectly prevented, (2) the high transparency of silicone resin enables monitoring and thus allowing preventive action against any possible corrosion of steel bearings, and (3) due to high flexibility and deformability of the resin, the function and movement of the steel bearings are not restricted.

## 1 Introduction

Bearings are the members at the contact point between the bridges' superstructure such as decks and girders and substructure such as piers and abutments. In addition to transferring vertical dead and live loads from the superstructure, they should be

---

M. Kawakami (✉)
Akita University, Akita-shi, Japan
e-mail: kawakami@gipc.akita-u.ac.jp

F. Omata
Maintenance Technology Co. Ltd., Tokyo, Japan

A. Toyoda · S. Kato
Suncoh Consultant Co. Ltd., Tokyo, Japan

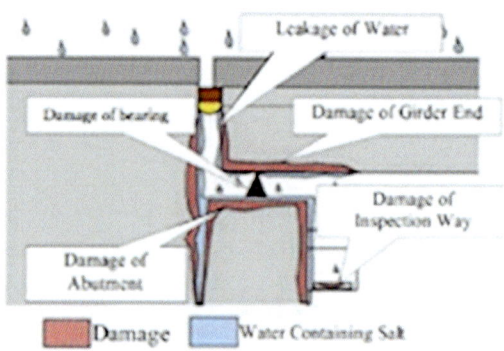

**Fig. 95.1** Process of water leak

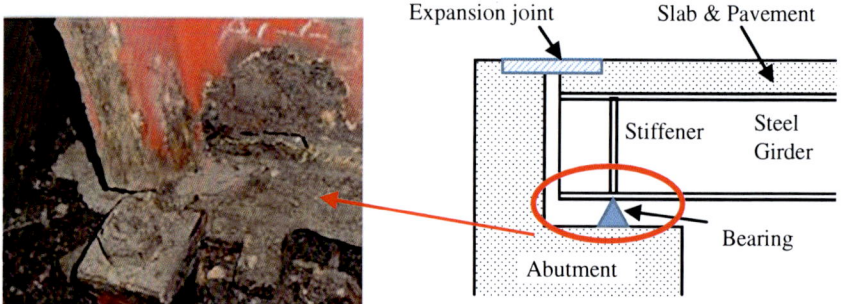

**Fig. 95.2** Typical corrosion example of steel bearing

able to sustain horizontal loads of wind and earthquake and transfer them to the substructure. Furthermore, bearings should allow the movement and rotation of the superstructure and absorb the mutual differential displacement between the super and substructures [1]. Bearings used for highway bridges are made of either steel or rubber. Most bearings of existing bridges in Japan are steel, with the usage of the rubber bearings increasing gradually for considerations of earthquake resistance. From the viewpoint of durability of steel bearings, it is desirable that they maintain a corrosion-free state during their service period. However, many steel bearings are corroded due to leakage of water from expansion joints as illustrated in Fig. 95.1. Especially in snowy and cold regions, leaking water contains chloride ion from the pavement, which causes severe corrosion as shown in Fig. 95.2.

The present study proposes a method to prevent steel bearings corrosion by enclosing the bearing with transparent and flexible silicone resin, which has been normally applied for joint sealing and painting, and reports on the physical characteristics and performance of this material.

**Fig. 95.3** Water leakage from expansion joint

## 2 Current Status of Deterioration of Steel Bearings Located at Girder End

In Japan, the corrosion of steel road bridges is evaluated according to the following four stages [2]. If the corroded area is below 0.05%, it is considered that "There is no corrosion, and the protective coating paint film is sound", and evaluated as Stage 1. If the corroded area is above 0.05% and below 0.5%, it is considered that "There is slight corrosion, but the coating film maintains its corrosion protection function", and evaluated as Stage 2. If the corroded area is above 0.5% and below 8.0%, it is considered that "Corrosion is apparent, and the coating film lost part of its corrosion protection function", and evaluated as Stage 3. If the corroded area is above 8.0%, it is considered that "Corrosion is progressing, and the coating film lost its corrosion protection function", and evaluated as Stage 4.

An example of water leakage from the expansion joint installed at the end of the girder is shown in Fig. 95.3. Water leakage to the underside of the bridge from the expansion joint happens frequently and causes corrosion to the steel bearings. This is a serious problem and the bearing corrosion is usually dealt with by repainting or spray zinc galvanization. However, these methods cannot be considered as thorough solutions.

## 3 Physical Properties of Transparent Silicone Resin

Any measure to address steel bearing corrosion is required to have the following functionalities: (1) can be applied in confined and narrow locations, and allow visual investigation after installation; (2) allow the bearing to maintain the required design displacement and movement; and (3) prevent further corrosion of the bearing.

In order to satisfy the above required functionalities, a method of entirely enclosing the bearing itself with a transparent silicone resin was developed, and the applicability of such method was researched. The standard and tested values of the physical properties of the silicone resin used are shown in Table 95.1.

**Table 95.1** Physical properties of silicone resin [3–5]

| Test item | Test method | Unit | Standard value | Measured value | Curing condition |
|---|---|---|---|---|---|
| Viscosity | JIS K 6833 | mPa/s | <1000 | 800 | 23 °C |
| Pot life | – | hours | 4 or more | 20 | 20 °C |
| Specific gravity | JIS A 6024 | – | 0.97 ± 0.1 | 0.97 | – |
| Elongation | JIS K 6251 | % | 400% or more | 600 | 20 °C -7 days |

From Table 95.1, it is clear that the proposed transparent resin has 600% elongation property, which allows it to sufficiently follow the displacement of the bridge. When enclosing the bearing in actual execution, the amount of resin used is much more than the specimen tested above for physical properties and in order to avoid breaking of the resin, it is recommended to use this resin enclosing method at line bearings without so much elongation movement.

Furthermore, because the silicone resin uses platinum catalyst, coming in contact with water, amin or sulfur compounds may inhibit its hardening. For that reason, the adhering properties of the resin were investigated by the tensile strength tests [6]. The results are shown in Table 95.2.

From Table 95.2, it was confirmed that adherence characteristics may change at the interface between the bearing and the mortar of the bearing seat, leading to failure at the interface. Then the use of single component urethane resin primer on the surface of the steel bearing and the mortar of the bearing seat (No. 3 & 6) was appropriate. By using this primer, the transparent silicone resin undergoes a cohesive failure, which means that enclosing the steel bearing with this resin while using the primer leads to good cohesion, thus restraining the corrosion.

## 4 Performance of Transparent Silicone Resin Applied to Steel Bearing

One of the main concerns regarding the enclosing of steel bearing with transparent silicone resin is the possibility of losing transparency due to coloration from ultraviolet rays of the sun. Therefore, a durability test was performed by accelerated irradiation using xenon arc lamp for 3000 h. According to JSCE Guidelines [7], it is proposed that 3000 h of xenon arch lamp is equivalent to 2000 h of sunshine carbon arc lamp. Thus, the 3000 h of xenon arc lamp are representative of roughly 13 years of outdoor exposure.

The appearance of the transparent silicone resin after exposure to the xenon arc lamp irradiation is shown in Fig. 95.4, and elongation test results after 3000 h are given in Table 95.3. From both Fig. 95.4 and Table 95.3, it is clear that the transparency of the resin has not changed and its elongation performance has not deteriorated.

After that, the test specimens as mockups for real steel bearings and bearing seat mortar were manufactured and the repeated cyclic push–pull test was done in order to confirm if the transparent silicone resin can follow the bearing displacement and

**Table 95.2** Tensile adhesion test for various adherents [6]

| No. | Adherent | | Treatment | $Ds^a$ at max. Load | $Ds^a$ at failure | Failure mode | Evaluation |
| --- | --- | --- | --- | --- | --- | --- | --- |
| | Part | Surface condition | | | | | |
| 1 | Steel | Painting | Urethane resin | No | 39.4 | 53.3 | Cohesive failure | Good |
| 2 | Steel | Blasting | No Corrosion | No | 46.8 | 55.7 | Interfacial failure | Not good |
| 3 | Steel | Blasting | No Corrosion | Primer[b] | 43.4 | 50.9 | Cohesive failure | Good |
| 4 | Steel | Blasting | Corrosion | Wire brush | 48.6 | 52.2 | Cohesive failure | Good |
| 5 | Cement mortar | Sanding | | No | 23.9 | 46.6 | Interfacial failure | Not good |
| 6 | Cement mortar | Sanding | | Primer[b] | 40.1 | 49.1 | Cohesive failure | Good |

[a]Ds.: Displacement (mm)
[b]Primer: single component urethane resin primer

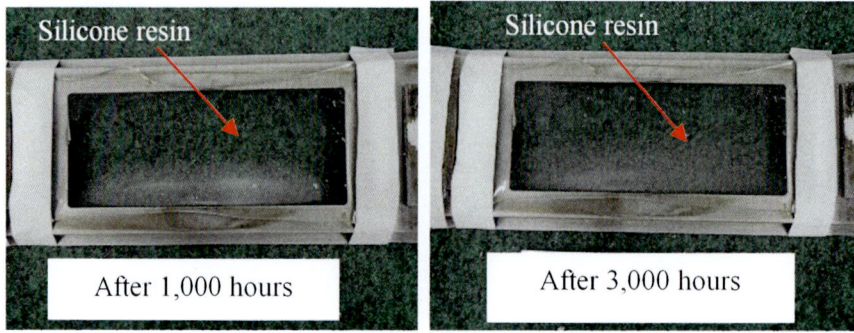

**Fig. 95.4** Accelerated weathering and exposure to artificial irradiation

**Table 95.3** Elongation performance of transparent silicone resin after accelerated weathering and exposure to artificial irradiation [5]

| Material | Exposure time | Elongation performance |
|---|---|---|
| Transparent silicone resin | 0 h | 600% |
|  | 3000 h | 600% |

**Fig. 95.5** Outline of cyclic load test

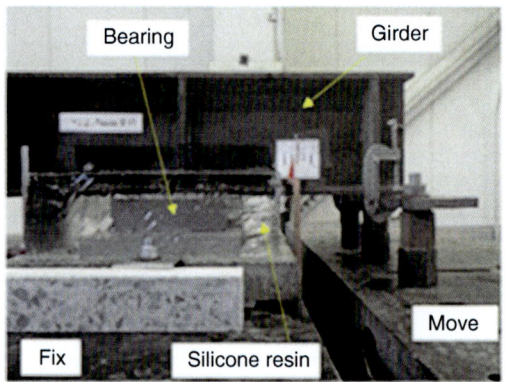

movement. Outline of the test is given in Fig. 95.5, and test method and apparatus are illustrated in Fig. 95.6. Note from Fig. 95.5 as to how the silicone resin has enough transparency to sufficiently allow visual observation of the enclosed bearing. The silicone resin was simply poured around the bearing.

The speed of the push–pull cyclic test was 100 mm/min. At first, the push–pull amount was 5 mm (±2.5 mm) for the initial 1000 cycles, after which breakage and separation of the transparent silicone resin were checked. After that the amount was gradually increased from 10 mm (±5 mm) to 40 mm (±20 mm) in 5 mm increments with 1000 cycle each, and the breakage or separation was checked after each increment. One cycle, for 5 mm (±2.5 mm), for example, consists of the following repetitions: $0 \rightarrow +2.5 \rightarrow 0 \rightarrow -2.5 \rightarrow 0$. After all above cycles, the final push–pull10 mm (±5 mm) for 11,000 cycle, equivalent to 30 years (at 1 cycle per day

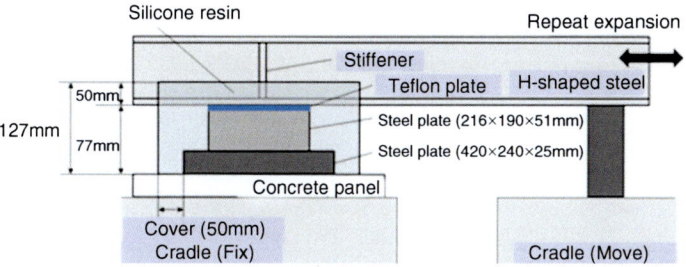

**Fig. 95.6** Cyclic load test method

rate), was done, and breakage and separation were checked. These tests confirmed that the transparent silicone resin does not break or separate from the bearing and follow the bearing in its movement and displacement.

## 5 Conclusion

To enhance road bridges' durability, and as a measure against the steel bearings losing their original performance due to corrosion, a method of enclosing the bearing with transparent silicone resin is newly developed. The following were confirmed by various tests:

1. The elongation of the transparent silicone resin is 600% and is confirmed through cyclic tests that it does not break or separate from the bearing but rather allows and follows the bearings' movement and displacement.
2. The silicone resin maintains its transparency as tested using an accelerated durability test by xenon arc lamp for 3000 h, equivalent to 13 years of outdoor exposure, confirming that visual checking of the bearing is possible after enclosing.
3. By using one component urethane resin primer, the transparent silicone resin maintains strong adherence to the steel bearing and the mortar of bearing seat. Consequently, it prevents the corrosion of the steel bearing due to complete sealing.
4. The transparent silicone resin is applied by simply pouring around the bearing, which allows application in the confined and narrow areas around actual bearings.

## References

1. Japan Road Association. (2004). Road Bridges Bearing Manual, Chapter 2. *Bearing Design Fundamentals*, 6-8.
2. Japan Road Association (2014), Steel Bridges Corrosion Prevention Manual, Chapter 6.3.4 Evaluation Method, II105.

3. JIS K 6833: Adhesives – General testing methods.
4. JIS A 6024: Epoxy adhesives for repairing and reinforcement in buildings.
5. JIS K 6251: Rubber, vulcanized or thermoplastic – Determination.
6. JIS A 1439: Testing methods of sealants for sealing and glazing in buildings.
7. Japan Society of Civil Engineers, Recommendation for Concrete Repair and Surface Protection of Concrete Structures, Guidelines for design and construction of surface protection methods.

# Chapter 96
# Bio-Based Polyurethane Elastomer for Strengthening Application of Concrete Structures Under Dynamic Loadings

Sudharshan N. Raman, H. M. Chandima C. Somarathna, Azrul A. Mutalib, Khairiah H. Badri, and Mohd. Raihan Taha

Feasibility of application of a bio-based elastomeric polyurethane (PU) coating to improve the dynamic resistance of concrete specimens by enhancing their energy absorption capability was investigated. A series of experimental investigation were conducted using scaled concrete specimens with dimensions of 160 × 40 × 40 mm, which were coated with eight different coating configurations by varying the coating thickness and location. Three-point bending test was conducted under quasi-static and dynamic conditions, by varying the strain rates (0.00033 $s^{-1}$ and 0.067 $s^{-1}$). The maximum flexural stress, failure strain, and strain energy density characteristics were used to assess the effectiveness of the proposed retrofitting technique. Polymer layers of 1–4 mm thick provided 2.9–8.9 times enhancement in failure strain, 3.0–11.3 times enhancement in strain energy density, and a marginal enhancement in the maximum flexural stress under dynamic conditions compared to the dynamic response of uncoated concrete specimens. In addition, the dynamic response of concrete specimens was improved when the thickness of the PU coating was increased and when the coating was applied on both faces.

S. N. Raman (✉)
Department of Architecture, Universiti Kebangsaan Malaysia, Bangi, Malaysia
e-mail: snraman@ukm.edu.my

H. M. C. C. Somarathna
Department of Civil Engineering, University of Jaffna, Jaffna, Sri Lanka

Department of Civil and Structural Engineering, Universiti Kebangsaan Malaysia, Bangi, Malaysia

A. A. Mutalib · M. R. Taha
Department of Civil and Structural Engineering, Universiti Kebangsaan Malaysia, Bangi, Malaysia

K. H. Badri
School of Chemical Sciences and Food Technology, Universiti Kebangsaan Malaysia, Bangi, Malaysia

# 1 Introduction

Interruption and failure of civilian buildings and critical infrastructures due to dynamic loadings have been widely reported, and these have created negative impact to other sectors due to the interdependent nature of the services and industries. These incidents have highlighted the need for feasible and resilient strengthening solutions for existing concrete structures which have been constructed using normal-strength concrete [1, 2]. A number of advanced engineering solutions have been investigated and introduced as active protective feature for structural elements including concrete, and they have demonstrated both pro and con characteristics [3, 4]. Application of materials which exhibit high stiffness together with high strain capacity shows a sound approach to enhance energy absorption capacity of structural elements [1]. Researches that have been conducted to investigate the feasibility of elastomeric coatings for structural retrofitting applications have indicated the potential in enhancing the dynamic resistance of structural elements [1, 5]. Furthermore, this technique offers the solution for some limitations of conventional and other advanced techniques, by making it sustainable, cost-effective, and practical to existing structures [6–8]. Polyurethane (PU) polymers have been considered for such applications due to their reliable and customizable physical and mechanical properties and their ease of application [5, 9–11]. The present experimental study investigates the feasibility of bio-based (palm-based) elastomeric PU coating to enhance dynamic response of concrete structures, through dynamic flexural test on elastomeric PU-coated scaled concrete specimens.

# 2 Experimental Program

## 2.1 *Materials*

Scaled concrete specimens of 32 MPa (cylinder) strength at 28 days were produced using locally sourced ordinary Portland cement (CEM-I), 5 mm downgraded river sand fine aggregate and crushed stones (maximum size of 9.5 mm) coarse aggregate. The PU resins were prepared from the rapid reaction between palm-based polyol (PKO-p) and 4,4-diphenylmethane diisocyanate (MDI) in the presence of polyethylene glycol (PEG: Mw 200 Da) as the chain extender via a solution casting technique; acetone (industrial grade) was used as a solvent in the pre-polymerization technique [5].

**Table 96.1** Details of the test specimens and their dynamic flexural properties

| Specimen designation | Thickness of PU coating (mm) | | Dynamic flexural properties | |
|---|---|---|---|---|
| | Top | Bottom | Maximum stress (MPa) | Failure strain |
| CON | – | – | 8.6 | 0.004 |
| T1 | 1 | – | 8.5 | 0.013 |
| T2 | 2 | – | 8.7 | 0.019 |
| T4 | 4 | – | 9.5 | 0.027 |
| B1 | – | 1 | 8.4 | 0.012 |
| B2 | – | 2 | 9.0 | 0.013 |
| B4 | – | 4 | 9.4 | 0.021 |
| T1B1 | 1 | 1 | 9.2 | 0.032 |
| T2B2 | 2 | 2 | 9.9 | 0.036 |

**Fig. 96.1** (a) PU-coated concrete specimen (T2). (b) Dynamic flexural test setup

## 2.2 PU-Coated Concrete Test Specimens and Dynamic Flexural Test

Concrete specimens with dimensions of 160 (L) × 40 (W) × 40 (H) mm were prepared. PU coating was applied with nine different coating configurations as described in Table 96.1. The effect of the PU coating thickness on the dynamic resistance of concrete samples was analyzed by varying the coating thickness with 1, 2, and 4 mm coatings, corresponding to 2.5, 5, and 10% thickness compared to the depth of the concrete specimen (Fig. 96.1a). The position of the coating also plays a major role in the effectiveness of this technique, because the mechanism of dynamic resistance is considerably different depending on the direction of the loading and material state (compression or tension) of the PU layer [12]. Positions of PU coatings were varied as top face, bottom face, and both faces with equal total coating thicknesses (volume). Three-point bending test was conducted using Zwick 100 kN ProLine materials testing machine, in accordance to ASTM C 78 (Fig. 96.1b). The specimens were tested under two strain rates, 0.067 $s^{-1}$ and 0.00033 $s^{-1}$, which correspond to dynamic and quasi-static conditions, respectively,

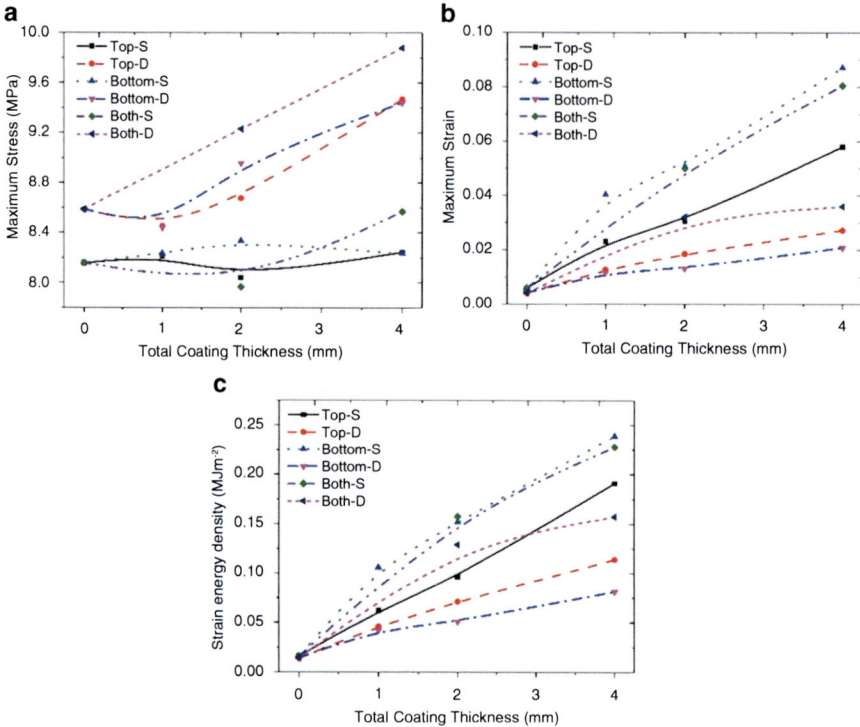

**Fig. 96.2** The comparison of flexural properties of test specimens: (**a**) maximum flexural stress, (**b**) failure strain, and (**c**) strain energy density

by varying crosshead speed. The two strain rates were denoted with S and D that corresponded to quasi-static and dynamic conditions, in the specimens' designation. The test was continued until the ultimate failure of the test specimens. The stress-strain response of the specimens was analyzed, and the findings are discussed. Three specimens were tested for each case.

## 3 Experimental Results and Discussion

### 3.1 Maximum Flexural Stress

Figure 96.2a plots the average maximum flexural stress of the tested specimens. It shows that the maximum flexural stress under dynamic condition is higher than the quasi-static condition for all cases. This behavior is referred as strain rate sensitivity, and the enhancement of the strength values is due to two phenomena: viscous (free water) effect, cross-aggregate cracks, and structural (inertia forces and confinement) effect [13]. Under quasi-static condition, a maximum of 4% enhancement in flexural stress was observed when 2-mm-thick coatings were applied on both faces while showing minor deviations (<2%) in other coating configurations. However under dynamic conditions, a clear enhancement of the maximum flexural stress was shown, and a maximum of 15% stress enhancement was observed when 2-mm-thick PU coatings were applied on both faces. Coating on both faces exhibited a higher value compared to coating on individual faces under each (total) coating thickness.

## 3.2 Failure Strain

PU-coated specimens exhibited higher failure strain, which can be attributed to the elastomeric PU layer which possesses high strain capacity, and enhanced the strain capacity of the concrete as a result of the additional confinement effect provided by the PU matrix within the concrete. Figure 96.2b plots the comparison of the average failure strains, and it shows that the failure strain is increasing remarkably with the thickness of the PU layer when the coating(s) are applied on either side of the concrete specimen. Each case clearly showed that the failure strain was increased with coating thickness. Under quasi-static conditions, higher strain enhancement was observed when coating was applied on the bottom face than coating on the top face or both faces. The highest enhancement was recorded when 4 mm coating was applied on the bottom face, which is 14.8 times than the uncoated specimen. However, the strain enhancement under dynamic condition was higher when the coating was applied on both faces compared to coating only on an individual face. The strain enhancement varied between 2.9 and 8.9 times under different coating configurations, compared to the control under dynamic condition.

## 3.3 Strain Energy Density

The strain energy was calculated by integrating the area underneath the stress-strain curve. Figure 96.2c plots the variation of ultimate strain energy density, and it indicated clear enhancements in the strain energy density of the PU-coated specimens under both loading conditions. This demonstrates that the energy dissipation is shared between concrete and PU coating; and the specimens deflected more, which verifies the beneficial contribution of the PU layer in terms of energy absorption. The strain energy density was increased gradually with the increase in PU layer thickness, caused by the enhanced strain work. This strain energy enhancement was observed under both loading conditions when coating was applied on either side. This evidences that the presence of a PU layer on either face provided higher structural capacity and dynamic resistance to the concrete element. Under quasi-strain condition, higher strain energy density was observed when PU was coated on the bottom face, while under dynamic condition, the strain energy density was higher when coating was applied on both faces. Therefore, the application of the PU coating on both faces contributed positively in enhancing the strain energy under dynamic conditions. It can be deduced that the ductility of the concrete specimens was improved with the presence of the PU coating and subsequently increases with the thickness of the PU layer. The dynamic resistance of the concrete elements was more dominant when both faces were coated.

## 4 Conclusions

The dynamic resistance of concrete elements was enhanced with the application of PU coating on either side of the face with respect to the direction of loading. Failure strain and strain energy density were enhanced by factors of 2.9–8.9 and 3.0–11.3, respectively, with a marginal enhancement in the maximum flexural stress under dynamic conditions, compared to the dynamic response of uncoated specimens. The response of specimens was improved when the thickness of the PU coating was increased and when the coating was applied on both faces with equal mass, compared to the application on an individual face.

**Acknowledgment** The authors would like to express their gratitude to the Ministry of Higher Education, Malaysia, and Universiti Kebangsaan Malaysia, for providing the necessary funding for this research through the FRGS Grant (FRGS/1/2015/TK01/UKM/02/1) and the ERGS Grant (ERGS/1/2013/TK03/UKM/02/6).

## References

1. Somarathna, H. M. C. C., Raman, S. N., Mutalib, A. A., & Badri, K. H. (2015). Elastomeric polymers for blast and ballistic retrofitting of structures. *Jurnal Teknologi (Sciences & Engineering), 76*, 1–13.
2. Ngo, T., Mendis, P., Gupta, A., & Ramsay, J. (2007). Blast loading and blast effects on structures – An overview. *Electronic Journal on Structural Engineering. Special Issue: Loading on Structures, 7*, 76–91.
3. Malvar, L. J., Crawford, J. E., & Morrill, K. B. (2007). The use of composites to resist blast. *ASCE Journal of Composites for Construction, 11*(6), 601–610.
4. Buchan, P. A., & Chen, J. F. (2007). Blast resistance of FRP composites and polymer strengthened concrete and masonry structures – A state-of-the-art review. *Composites Part B: Engineering, 38*(5–6), 509–522.
5. Somarathna, H. M. C. C., Raman, S. N. Badri, K. H., Mutalib, A. A., Mohotti, D., & Ravana, S. D. (2016). Quasi-static behavior of palm-based elastomeric polyurethane: For strengthening application of structures under impulsive loadings. *Polymers, 8*(5), 202, 23p.
6. Bahei-El-Din, Y. A., & Dvorak, G. J. (2007). Behavior of sandwich plates reinforced with polyurethane/polyurea interlayers under blast loads. *Journal of Sandwich Structures and Materials, 9*(3), 261–281.
7. Amini, M. R., Isaacs, J., & Nemat-Nasser, S. (2010). Investigation of effect of polyurea on response of steel plates to impulsive loads in direct pressure-pulse experiments. *Mechanics of Materials, 42*(6), 628–639.

8. Mohotti, D., Ngo, T., Mendis, P., & Raman, S. N. (2013). Polyurea coated composite aluminium plates subjected to high velocity projectile impact. *Materials and Design, 52*, 1–16.
9. Somarathna, H. M. C. C., Raman, S. N., Badri, K. H., & Mutalib, A. A. (2015). Analysis of strain rate dependent tensile behaviour of polyurethanes. In *Proceedings of the 6th International Conference on Structural Engineering and Construction Management 2015*. Kandy, Sri Lanka (7p).
10. Pztrovic, Z. S., & Ferguson, J. (1991). Polyurethane elastomers. *Progress in Polymer Science, 16*(5), 695–836.
11. Badri, K. H. (2012). Chapter 20, Biobased polyurethane from palm kernel oil-based polyol. In F. Zafar & E. Sharmin (Eds.), *Polyurethane* (pp. 447–470). Rijeka: INTECH.
12. Raman, S. N. (2011). *Polymeric coatings for enhanced protection of reinforced concrete structures from the effects of blast* (PhD Thesis). 308p. Department of Infrastructure Engineering, The University of Melbourne, Australia.
13. Eibl, J., & Schmidt-Hurtienne, B. (1999). Strain-rate-sensitive constitutive law for concrete. *ASCE Journal of Engineering Mechanics, 125*(12), 1411–1420.

CPI Antony Rowe
Eastbourne, UK
May 22, 2019